THE HISTORICAL AND PHYSICAL FOU[illegible] MECHANICS

The Historical and Physical Foundations of Quantum Mechanics

Robert Golub
North Carolina State University, Raleigh, NC, USA

Steven K. Lamoreaux
Yale University, New Haven, CT, USA

OXFORD
UNIVERSITY PRESS

Great Clarendon Street, Oxford, OX2 6DP,
United Kingdom

Oxford University Press is a department of the University of Oxford.
It furthers the University's objective of excellence in research, scholarship,
and education by publishing worldwide. Oxford is a registered trade mark of
Oxford University Press in the UK and in certain other countries

Published in the United States of America by Oxford University Press
198 Madison Avenue, New York, NY 10016, United States of America

British Library Cataloguing in Publication Data

Data available

Library of Congress Control Number: 2022951663

ISBN 978–0–19–882218–9
ISBN 978–0–19–882219–6 (pbk.)

DOI: 10.1093/oso/9780198822189.001.0001

Printed and bound by
CPI Group (UK) Ltd, Croydon, CR0 4YY

We dedicate this book to the memories of Prof. J.M. Pendlebury,
Dr. V.K. Ignatovich, Prof. A. Steyerl, and Prof. H.G. Dehmelt.

Preface

There is an enormous number of books and other writings concerned with explaining and interpreting quantum mechanics. Standard texts tend to concentrate on methodology and applications to specific problems, while discussions of interpretation and the historical development tend to contain a minimum of mathematics. The idea behind this book is that to gain a real understanding of the subject, some acquaintance with the historical development is essential; after all, that history is the narrative of how humanity learned quantum mechanics. The ideas were not found written on tablets on a farm in e.g., New York State, but were slowly and painstakingly developed by people just like us. We also provide accompanying discussions for the various interpretations that have been suggested, providing sufficient mathematical illustrations that highlight the respective features and differences.

Acknowledgements

We would like to thank our friends and family for their understanding and support as our attention was drawn away to the writing of this book.

R. G. would like to thank his wife Ekaterina Korobkina, E. David Davis for contributions to Chapter 21 and help with the book, and Roland Gähler. In addition, he thanks Prof. Chueng Ji for discussions at the early stages of the book.

S. K. L. thanks his wife, Melissa, and daughter, Zoe, his Friday Owl Shop cigar friends Carl J. Frano and James Surprenant, and Dr. Sidney B. Cahn for their unfailing and unflappable moral support. He also thanks Mr. Edward S. McCatty (B.A. (Amherst), M.Div. (Yale), M.A. Lit. (UCL)) for editorial comments on Chapter 1 and for providing inspiration throughout the project.

We especially thank Dr. Yulia Gurevich, whose expert editing, together with her vast knowledge of physics, clarified and strengthened many parts of this book. Her artistic talent is evident in many of the figures.

Yale University provided support for the preparation and editing of the manuscript. NCSU also provided support with a sabbatical for R. G.

Contents

APPENDICES

Part I

Basis of the Theory

The first part of this book provides a historical background and brings us to the modern theory.

Part I

Basis of the Theory

The first part of this book provides a historical background and framework for the modern theory.

1 Introduction

"You have nothing to do but mention the quantum theory and people will take your voice for the voice of science and believe anything you say." George Bernard Shaw, 1938[1]

1.1 Overview

A search on Amazon.com for books on "quantum theory" returns over 10,000 hits while searching for "quantum physics" returns over 20,000. This corresponds to one book a day for 30 years. These books range from advanced mathematical treatises to books without a single equation, from deep philosophical debates between authors with different understandings of the subject to textbooks teaching the methodology and various applications. In addition, there are vast numbers of papers in historical and philosophical journals concerned with the development and philosophical implications of the theory. For those interested, there are also many volumes of collected correspondence and many online archives of oral and written material.[2]

While there is little dispute over the mathematical apparatus of the theory and its application to physical problems there is a wide spectrum of divergent opinions about what the theory is trying to tell us concerning the nature of reality. For a long time following WWII, there was little interest among physicists for such questions as attention was turned to the frenetic development of different technologies. However, recent decades have seen, in addition to an amazing range of applications of the theory, an ever-increasing attention to what is called the "interpretation" of quantum mechanics. There is now a bewildering forest of these interpretations each of which has a group of supporters as well as opponents. As, to this date, none of the interpretations has been able to convince a majority of working physicists (who, it should be said, mostly ignore these discussions, an attitude that has been summed up as "shut up and calculate") of its correctness or necessity. It is almost as if physics is splitting into a number of cults uniting supporters and critics in a never-ending embrace.

It is striking that all of the proposed interpretations are concerned with the original form of the theory, the Schrödinger theory supplemented by the Dirac transformation theory, seemingly ignoring the most advanced form of the theory, i.e., that involving the

[1]Quoted by Simon, D.R., On the Power of Quantum Computation, 35th Annual Symposium on the Foundations of Computer Science, (1994) Santa Fe, NM and at www.greatest-quotations.com.

[2]See e.g., American Philosophical Society Library: Sources for the History of Quantum Physics, 1898-1950, https://search.amphilsoc.org/collections/view?docId=ead/Mss.530.1.Ar2-ead.xml.

The Historical and Physical Foundations of Quantum Mechanics. Robert Golub and Steven K. Lamoreaux, Oxford University Press.
 DOI: 10.1093/oso/9780198822189.003.0001

quantization of the non-relativistic Schrödinger equation, introduced by Jordan, with the support of Pauli and Wigner among others, that he called "second quantization." As we will see, this formulation solves several problems associated with the original form of the theory and almost supplies its own interpretation, as does relativistic quantum field theory where questions of interpretation, essentially whether particles or waves are prior, are much less prominent.[3] This concentration on a not fully matured version of the theory might be considered by some as an indication that the interpretation discussion is caught in a time-warp, devoting its attention to a theory that could be viewed as already superseded.

The purpose of this book is to take a step back and attempt to retrace the development of the theory by investigating original sources, the original published papers and letters, of the participants. This is the path by which humanity learned quantum mechanics and following it might hopefully lead to an improved understanding. Of course, the attempt by physicists themselves to approach the history of their subject is an exercise fraught with difficulties, as has been recognized by several practitioners. For example, Silvan Schweber, a theoretical physicist turned historian of science, recognizes[4] that "the history of science cannot escape some form of whiggism. The data is so rich that some selection must be made." A whig history of science is the view of the scientific winners who write as if their triumph was an inevitable result of the correctness of their ideas. Whig history of science displays the historical development as proceeding from a past ruled by ignorance to a glorious present without taking account of the actual state of knowledge in the past.

We can see the result of trying to overcome the limitation mentioned by Schweber and include all relevant publications along with biographical information on the many actors and excerpts from correspondence, in the heroic work made by Mehra and Rechenberg, who have completed a nine volume treatise, "The Historical Development of Quantum Mechanics," published between 1982 and 2001.[5] This has been an enormous help in writing the present volume

S.A. Goudsmit, the codiscoverer of electron spin,[6] was skeptical as to the utility of the history of science:[7] "Many historians have written very pretty stories about how a discovery should have been made, but it is unfortunately very improbable that the development was as logical as these fabricated stories would indicate. Luck and random events play a much larger role than people are ready to admit." In addition he complains that: "They (the historians of physics) present things as if the whole of physics was created by a handful of geniuses. This is completely unfair to the many physicists whose work enables the great discoveries of the geniuses."

[3]Weinberg, S. *The Quantum Theory of Fields, Vol. 1,* Foundations, CUP, (1995)

[4]Schweber, S.S., *QED and the Men Who Made It: Dyson, Feynman, Schwinger and Tomonaga,* Princeton, 1994

[5]Mehra, J. and Rechenberg, H., *The Historical Development of Quantum Mechanics Volumes 1-6,* comprising nine volumes in total as some of the volumes are printed in two parts, Springer - Verlag 1982–2001.

[6]Uhlenbeck, G.F. and Goudsmit, S., Naturwissenschaften, 13, 953, (1925) and *Spinning Electrons and the structure of Spectra,* Nature, 117, 264 (1926)

[7]Goudsmit, S.A., *The Discovery of the Electron Spin* (in German), Phys.Blaetter, 10, 4345 (1965)

He then goes on to state "Historians are often unjust with respect to the experimental physicists. Even though the evolution of ideas is very important for history, we should not neglect the geniuses among the experimental physicists whose discoveries and results are absolutely necessary for new ideas and their verification," and further makes the point that "Published articles are not very reliable as historical sources. In a good article, the author tries to convince the reader so he often chooses a different train of thought as that by which he came upon the idea." This is something that can be attested to by any experienced researcher.

Steven Weinberg (op. cit.) explicitly disdains the historical approach to teaching physical theories, preferring a logical development of the theory as it is presently understood. This, of course, vitiates the importance of direct observation of natural phenomena, and the fact that current physical theories were at one time tenuous hypotheses that required testing via the scientific method. As such, abandoning the historical approach appears as a throwback toward Scholasticism with its basis in dogmatism.

Albert Einstein was also skeptical of a historical approach:

> Only those who have successfully wrestled with problematic situations of their own age can have a deep insight into those situations, unlike later historians who find it difficult to make abstractions from those concepts and views which appear to his generation as established or even self evident.[8]

While there is certainly a large degree of truth in all of this the fact is that the original published papers are closer to the original ideas than a third-generation textbook and can be expected to reflect something of the then-contemporary zeitgeist as the result of the author's stated wish to persuade his readers. We also make use of letters and contemporary accounts when appropriate.

Thus, in this book, while being aware of these issues, we will attempt to trace the main lines of the development with the hope that this return to the roots will cast some light on what are today considered the difficulties of the theory.

1.2 The Prehistory of Quantum Mechanics: atomism

Quantum mechanics has its fundamental basis in the atomic theory of matter, which has its roots in atomism. Atomism was originally a *philosophical* theory that material objects are discontinuous, being constructed of indivisible distinct types of atoms—equivalently, quantized units of matter, that serve as building blocks. Atoms are now understood to be of limited variability (chemical elements, isotopes, periodic table), but each type of atom has unique and fixed properties, and all atoms of a given type are now understood to be identical and indistinguishable.

The concept of atomism has a long checkered if not tortuous history, one that is rarely expounded upon in physics books. We will present a very abbreviated overview of the development of modern ideas, and these are from a very Western perspective. There was likely widespread communication in the ancient world that allowed ideas to be spread, and it is not impossible that Greek atomism had its origin with the Indian

[8]Einstein, A., Reply to criticisms in Schilpp, P.A., ed., *Albert Einstein Philosopher-Scientist, Vol.II*, Harper, 1949, 1951.

sage and philosopher Acharya Kanad (Kashyap) who around 600 BCE speculated on the limit of divisibility of matter and proposed particles that could not be divided further, *anu* or *atoms*. Perhaps it stands to reason that any society with a merchant class has speculated on the degree of divisibility (hence minimum marketable unit) of material bodies; a vast body of history is never recorded, or lost—in the words of Roy Batty—like tears in the rain.

In the teaching of physics, in those rare instances where history is mentioned, Democritus (ca. 450 BCE) is often credited with the original formulation of the atomic hypothesis, and that is it, nothing more. The story is almost infinitely more complicated, and we will attempt to provide some highlights. Democritus was a student of Lucretius (ca. 475 BCE) with whom the atomic idea has its roots, which he formulated in response to Parmenides' deduction that reality is an illusion.[9]

According to Parmenides, for an object to move from one location to another, it would need to be destroyed at the first location and recreated at the new location. As this appears to be an impossibility, Parmenides made the logical leap that reality is an illusion. The notion that there is no reality, that all that exists is illusion, has come up many times since the ancient Greek philosophers—Shakespeare's "All the world is a stage," and more recently, the notion that we are living in a computer simulation is being taken seriously[10] and is an essential form of Idealism. The various Zeno paradoxes were put forward in support of Parmenides' assertion, to show that the physical universe as we believe we are observing it is indeed an impossibility. Of course, it is easy to believe everything is an illusion until a severe toothache starts on a Saturday night; reality is often in conflict with our beliefs, expectations, and prejudices, that are formed in the echo chambers of our minds, colleagues, and nowadays, social media (Facebook and Twitter).

Lucretius, followed by, and embellished by, Democritus, answered Parmenides' claim by inventing atoms, and equally important, the *void*, in which atoms move. The void is nothingness, and the argument against its existence continues today because we are faced with the problem of inventing a description for something that does not exist, which is an apparent self-contradiction. Nonetheless, the complete atomic picture was laid out by Democritus, in which objects are constructed of atoms of varying characteristics, and these atoms, collected together as objects, move together freely in the void. These are the basic tenets of the modern picture of the universe and matter, perhaps coincidentally, as this was a philosophical theory.

Jumping ahead some 100 years, Aristotle took a step backward in his adoption of Empedocles' notion (450 BCE) that the material world comprises four elements, earth, wind, fire, water, and further surmised that the natural state of matter was at rest. (The Greek notion of elements might have also been derived from the Hindu *Veda* which existed in oral form from 2 millennia BCE and in written form from 1 millennia BCE, in which the same four elements plus a fifth, the all-important void, are postulated.)

[9]Although it is tempting to ascribe the discontinuity of matter as assumed by Democritus as resulting from a lack of understanding of mathematical continuity, however, the development of the early philosophical theories follows a more complicated path. See, for example, Bernard Pullman, *The Atom in the History of Human Thought* (Oxford University Press, 2001).

[10]Bostrom, Nick (2003). *Are You Living in a Computer Simulation?*. Philosophical Quarterly **53**, (211): 243-255. doi:10.1111/1467-9213.00309.

A fifth element, quintessence, was introduced as the element from which heavenly bodies are constructed. By the medieval ages, a new notion (a form of Monism) was introduced that everything was a quintessence-like element, scraped together into a particular form, at which point the quintessence assumed the properties of the form, e.g., a pencil sharpener, the keyboard on which I am typing, etc. This notion was taken as a central principle or tenet by the Catholic Church, and provides a mechanism for transubstantiation. This tenet was important enough that atomism was specifically addressed by the Council of Trent (1545 to 1563) as anathema (heretical).

Galileo is of course known for the trials he endured concerning his promoting the heliocentric model of the solar system. The most puzzling aspect of the entire Galileo affair is that he had been well received by Pope Urban VIII, who was fully aware of and studied Galileo's writings. The results of Galileo's first trial in 1616 were limited to orders to cease holding, teaching, or defending heliocentric ideas. Up to this time, Galileo had a good relationship with the Jesuits, even the faction in charge of imposing church doctrine, which included the canons of the Council of Trent; this faction was also in charge of general education. Galileo's teachings were at odds with Aristotle and Scholasticism, so that a faction of Jesuits (for whom Aristotelian teachings were educational canon) became increasingly hostile toward Galileo; this hostility only increased with the minimal results of Galileo's first trial, especially when he did not cease promoting his scientific ideas and continued to write books. Pope Urban VIII, to appease these Jesuits in his efforts to consolidate power, acceded to their demands that Galileo be again brought to trial, before the Inquisition, for heresy. Recent discoveries in the Vatican records show that more charges, in addition to those associated with heliocentric theories, were being prepared to bring up Galileo's embracing atomism as an additional heresy.[11] To further inflame the situation, Simplicio in *Dialogue on the Two World Systems* was suggested as modeled on Urban VIII. In 1632, the Pope ordered another investigation against Galileo. This time he was prosecuted following the normal methods of the Inquisition, however, Galileo was then of advanced age and was therefore not subject to torture and death for being found guilty of heresy, but consequently was placed under house arrest for the rest of his life. A younger man, Giordano Bruno, who 30 years earlier (during the tenure of Pope Clement VIII) embraced heliocentricity, atomism, and many other heretical scientific and sociological notions, and taught them with abandon, was decreed guilty of heresy and on 17 February 1600 was hung upside down naked before being burned at the stake.

One interesting and important aside is that Galileo's and Bruno's writings were preserved in the Vatican Archives; this is one particularly astonishing aspect of the Catholic Church in that the writings of enemies were very often preserved, unlike most human institutions where the memories of adversaries are erased as a pathetic panacea against future threats. The Church did not initiate a campaign to collect up Galileo's books and ritualistically destroy them, in contradistinction to, for example, the Nazis' handling the works of enemies of the state by burning books in well-publicized bonfires, or the Memory Hole of Orwell's 1984 . This is not to say that it *never* happened, but it appears the preservation was a general matter of course.

[11] Pietro Redondi (Raymond Rosenthal, Translator), *Galileo- Heretic* (Princeton University Press, 1989).

The above, of course, is presented with the caveat that the understanding of historical events is fraught with difficulties; do we have the complete picture? What were all participants in the event thinking? What were the fundamental motivations, e.g., consolidation of power, controlling the masses, etc.? Regarding Galileo, Arthur Koestler comments:

But there existed a powerful body of men whose hostility to Galileo never abated: the Aristotelians at the universities. The inertia of the human mind and its resistance to innovation are most clearly demonstrated not, as one might expect, by the ignorant mass which is easily swayed once its imagination is caught—but by professionals with a vested interest in tradition and in the monopoly of learning. Innovation is a twofold threat to academic mediocrities: it endangers their oracular authority, and it evokes the deeper fear that their whole, laboriously constructed intellectual edifice might collapse. The academic backwoodsmen have been the curse of genius from Aristarchus to Darwin and Freud; they stretch, a solid and hostile phalanx of pedantic mediocrities, across the centuries. It was this threat, not Bishop Dantiscus or Pope Paul III which had cowed Canon Koppernigk into lifelong silence. In Galileo's case, the phalanx resembled more a rearguard—but a rearguard still firmly entrenched in academic chairs and preachers' pulpits.

as quoted in Pullman, op. cit., p 128., from Arthur Koestler.[12]

On a kinder note, as a philosophical theory, Aristotelianism is perfectly internally consistent. However, this does not mean it represents reality; as Bertrand Russell quipped,[13]

A philosophy [of nature] that is not self-consistent cannot be entirely correct, but one that is self-consistent may well be completely false.

According to Jeroen van Dongen, "Kuhn himself mentioned a kind of epiphany he had experienced when assisting Conant in teaching the history of science: Reading Aristotle, he shockingly discovered that his own Newtonian expectations were blocking him from seeing the consistency and integrity of Aristotle's physics. This experience put him on the path" to his famous book introducing the concept of paradigms.[14] Koestler, again, was less kind,

Aristotelian physics is really a pseudoscience, out of which not a single discovery, invention or new insight has come in two thousand years; nor could it ever come and that was its second profound attraction. It was a static system, describing a static world, in which the natural state of things was to be at rest, or to come to rest at the place where by nature they belonged, unless pushed or dragged; and this scheme of things was the ideal furnishing for the walled-in universe, with its immutably fixed Scale of Being.

However, this was not Aristotle's fault; to again quote Bertrand Russell,

In reading any important philosopher, but most of all in reading Aristotle, it is necessary to study him in two ways: with reference to his predecessors, and with reference to his successors. In the former aspect, Aristotle's merits are enormous; in the latter, his demerits are equally enormous. For his demerits, however, his successors are more responsible than he is. He came at the end of the creative period in Greek thought, and after his death it was two thousand years before the world produced any philosopher who could be regarded as approximately his equal. Toward the end of this long period his authority had become almost as unquestioned as that of the Church, and in science, as well as in philosophy, had become a serious obstacle to progress. Ever since the beginning of the seventeenth century, almost every serious intellectual advance has had to begin with an attack on some Aristotelian doctrine;

[12]Koestler, A., *The Sleepwalkers: A History of Man's Changing View of the Universe* (London, Arkana, 1959); the above quote is apparently a back-translation from French.

[13]Bertrand Russell, *History of Western Philosophy* (First published by George Allen and Unwin Ltd, London. 1946).

[14]van Dongen, J. *In Europe*, Physics in Perspective 22, 3-25 (2020).

in logic, this is still true at the present day. But it would have been at least as disastrous if any of his predecessors (except perhaps Democritus) had acquired equal authority. To do him justice, we must, to begin with, forget his excessive posthumous fame, and the equally excessive posthumous condemnation to which it led.

And finally, Crescenzo (as quoted by Pullman [15]) states,

For both Plato and Aristotle, who were constantly in search of the prime cause and ultimate purpose, it is as though Democritus had told them the plot of a comedy while skipping the first and last scenes.

What Galileo brought forward, as begun by Copernicus and Kepler, is the possibility of the use of mathematics to describe physical systems, and motion or dynamics, in particular. This applicability and effectiveness of mathematics in this endeavor is the basis of modern physical sciences and engineering, and it is not obvious that this should be possible.[16] Newton's work (circa 1700) carried the application of mathematics to a revolutionary new level, and was a harbinger of the end of Aristotelian dominance in Western thinking.

Even some alchemists at this time had moved on from the notion that everything is composed of the four primordial elements, earth, air, water, heat. For example, instead of combining elements to form gold, the German alchemist Henning Brand attempted to extract existing gold from urine; he reasoned that because urine is normally golden, it must contain gold, or might hold the key to finding the Philosopher's Stone. Some time around 1669, he embarked on the Augean task of boiling down 5,700 liters of putrefied human urine (there is no record of how he obtained this quantity), and then subjecting the residue to heat in the presence of carbon, which reduced phosphates to elemental phosphorus, a task that brings to mind Marie and Pierre Curie's later Augean task of extracting a fraction of a gram of radium from tons of pitchblende.

Phosphorus was the first element to be discovered that was not already known in ancient times, and the appearance of a continuous glow must have been astounding and awe-inspiring to the alchemist. Brand, of course in typical alchemist fashion, kept his discovery secret but ended up selling the recipe–and also tipped off Robert Boyle (who soon figured out his own extraction method) as to the source of phosphorus.

The final major blow to Aristotelianism came with the discovery that water could be created by combining hydrogen and oxygen, most notably as described by Lavoisier in 1789; he surmised that water is 85% oxygen and 15% hydrogen by weight–and is therefore hardly an element. Lavoisier's oxygen theory of combustion also brought down the phlogiston theory of combustion, and led to the law of conservation of mass.

One of the reasons for our delving into this history is that Scholasticism (education based on Aristotelian precepts) dominated much of Western Europe from before the twelfth century through the eighteenth and well into the nineteenth century in some regions, and had a profound effect on the development of atomic theory in physics, but less so on chemistry. One of the last times that someone of note made mention of the primordial elements was Napoleon, who quipped, "God created a fifth element especially for Poland–mud," after his army was mired during the 1806 campaign into

[15]Pullman, *loc. cit.*, p. 56

[16]Eugene Wigner, *The Unreasonable Effectiveness of Mathematics in the Natural Sciences*, Communications in Pure and Applied Mathematics, **13**, 1 (1960).

Fig. 1.1 *The Alchemist Discovering Phosphorus* by Joseph Wright of Derby (1771, reworked 1795)(Public Domain).

Poland.[17] By the mid-nineteenth century, Aristotelianism was relegated to a joke, in particular, Melville in *Moby Dick* describes the state of the *Pequod*, when her upper decks were overloaded during the search for a leaking whale-oil cask stored deep below deck, as, "Top-heavy was the ship as a dinnerless student with all Aristotle in his head."

It is noteworthy that teaching atomism was controversial and associated with a progressive outlook for more than a millennium. As late as 1624 the court of King Louis VIII of France threatened the teaching of atomism with the death penalty. General

[17] F. Loraine Petre, *Napoleon's Campaign in Poland* (Sampson Low, Marston and Company: London, 1901). p. 51.

questions about the infinitesimal were frowned upon well into the nineteenth century (such notions were at odds with Aristotelianism and Scholasticism), for example, Bernard Bolzano (1781–1848) was viewed as a progressive radical, thus unacceptable to the Austrian rulers (House of Hapsburg-Lorraine) of Bohemia and was ejected from his university position; it was 50 years after his death that Karl Weierstrauss found in his writings the foundation of the Bolzano-Weirstrauss theorem, which is essential to the notion of continuity and one of the theoretical underpinnings of calculus.

As we have already stated, this book is not a history of science, but an attempt to place the development of quantum mechanics in a historical framework. As a society, we like nice stories of how ideas were developed and introduced, but the truth is almost always more complicated. A silly example is the supposed invention of the sandwich by John Montague, the 4th Earl of Sandwich. It is hard to believe that in the 30 millennia that bread existed in one form or another, nobody ever placed a slice of meat between two pieces of bread. What the good earl accomplished was to make the consumption of such acceptable in polite company.

Even the relatively recent work of Planck has been hotly argued among science historians. In particular, Martin Klein and Thomas Kuhn really could not agree on Planck's personality; was he carefully conservative or a reckless revolutionary?[18] In fact, both are correct to some degree. Humans are complex, and often express different and seemingly incompatible views depending on the situation.

1.3 Religion and science

The separation between science and religion that developed during this period in Europe, that is, from the eleventh to the eighteenth century, and even into the nineteenth century, is quite remarkable and perhaps unique in the development of human societies and cultures. By the time of Newton, the need for reconciliation between scientific observation with the Bible largely disappeared from scientific literature; what is especially remarkable is that Newton had a literalist interpretation of the Bible and wrote extensively on the subject, however, he held his notions in close secret as his embracing monotheism was at odds with the doctrine of Trinity College, where he held his faculty position, as there were potential serious consequences for holding such views.

How this separation came about remains a mystery, although some credit Francis Bacon with developing the concept of empiricism and with the development of the scientific method, however, he was a contemporary of Galileo, and they were certainly aware of each other's work, so it appears difficult to assign credit to either. Bacon was especially against Aristotle's syllogism and rules of inductive enquiry; for Galileo, Aristotelian physics was simply incompatible with reality.[19] Others paved the way; recall Giordano Bruno, who perished for his science teaching some years before Galileo's predicament. Often the young are the ones leading the way toward revolution. Compare the situation of Bruno and Galileo with the anti-Vietnam war movement in the USA, originally fomented principally by college students until the fundamental hypocrisy

[18] Jeroen van Dongen, *In Europe, loc. cit.*

[19] W. Mays, *Scientific Method in Galileo and Bacon,* Indian Philosophical Quarterly **1**, vol. 3, 217 (1974).

and pathology of the war were finally revealed through the Kent State Massacre and the Pentagon Papers.

It is also wrong to say that before e.g., Francis Bacon there was no such thing as the scientific method. Testing by trial and error is part of the human psyche and has existed since the beginning of conscious thought. The Egyptians could not have constructed the pyramids and other structures if they did not have a system to study nature, record observations and methods, and transmit knowledge between generations. Records suggest that the Egyptian culture stagnated in that new ideas were not allowed to develop, and thus the society could not keep up with a changing environment, or threats from external political forces.

The separation probably was the result of Scholasticism dominating monastic medieval teaching, a critical method of philosophical analysis predicated upon Aristotle, with a Latin Catholic theism being separate and not subjected to logical argument and analysis but which was to be accepted as infallible and invariant doctrine. Such curricula dominated teaching in the European medieval universities from about 1100 to 1700. Those interested in scientific observation had to skirt around Aristotle, and in doing so, bypassed religious scrutiny and debate. The problems scientists faced are best illustrated by Galileo's interaction with the Church; as long as Galileo called his observations and conclusions "theoretical" there was no conflict with the Church or with its doctrine. In this sense, the conflict with Galileo was a battle to decide who gets to interpret scripture, or alternatively, who has political control.

In the history of Western science, all of this led to science being done outside of religious considerations, originally clearly to avoid conflict with the Church, and also with politics, which is almost the same thing. Later this separation became a matter of course and part of our now accepted scientific culture.

In nations governed by Sharia Law, science and doctrine are expected to be, and to remain, mutually compatible. During the Soviet era, Russian scientists had to at least obliquely acknowledge dialectical materialism. A well-known anecdote tells of Beria approaching Kurchatov, the head scientist of the Soviet atomic bomb project, regarding the fallacies of Einstein's theory of relativity as it is incompatible with the fundamental notions of dialectical materialism. Kurchatov replied that without relativity, there cannot be an atomic bomb. Apparently, the Soviet philosophers were able to unify away the incompatibilities.

The scientist-as-atheist is a modern Western notion that had its beginnings from dancing around the Church and Aristotle, but was later amplified by Darwin's theory of evolution. Darwin did not set a timeline for evolution, because he realized that estimates for the age of the Sun (30 million of years), due to the energy released from gravitational interactions, were not sufficiently long. Biologists argued that there needed to be another energy source for the Sun, as did geologists who needed more time for their sedimentary rocks to form, and they were correct.[20] The age of the earth due to Old Testament genealogy and the Jewish calendar is about 6,000 years, and this is viewed by many as a conflict between science and religion that arises from taking the ancient scriptures literally, instead of seriously. It is worth noting

[20]Bethe, H.A., *Energy Production in the Stars*, Nobel Lecture, Dec. 11, 1967. https://www.nobelprize.org/uploads/2018/06/bethe-lecture.pdf.

that some Kabbalists from Spain in the twelfth to thirteenth centuries calculated the Earth's age as in the range of 1 million to 2.5 billion years.[21]

The fundamental incompatibility between science, religion, and politics is that the basis of science is falsification. As academics, we tend to view conflicts between doctrine and scientific observation as being due to semantic issues, e.g., the six days of creation refers not to days, but perhaps to vast eons; the modern reader has no idea of what the original writer had in mind. This is about the best that can be done under the notion of the infallibility of ancient scripture. And what do we mean by falsification? In science, the breadth of this notion goes from two measurements of the same quantity being inconsistent due to experimental errors, to an entire theoretical construct being incorrect.

In response to Einstein's famous remark, "God does not play dice with the Universe," Bohr said "Einstein, stop telling God what to do." Perhaps this is a lesson regarding the strict interpretation of historical and philosophical documents.

1.4 Birth of the modern atomic theory of matter

The first modern kinetic theory of gases is due to James Hermann, who in 1716 deduced that the pressure exerted by a gas is proportional to the average squared velocity of the gas particles times the number density.

In 1729 Euler attempted to mathematically explain the behavior of gases with a kinetic theory based on Robert Boyle's gas data from 1662. He assumed that the gas particles would all move at the same speed. Daniel Bernoulli formulated a kinetic theory of gases, with the notion that the velocities would be statistically distributed, but did not specify the distribution, however, he anticipated the work of James Clerk Maxwell a century later. Bernoulli's work was not widely accepted, in part because

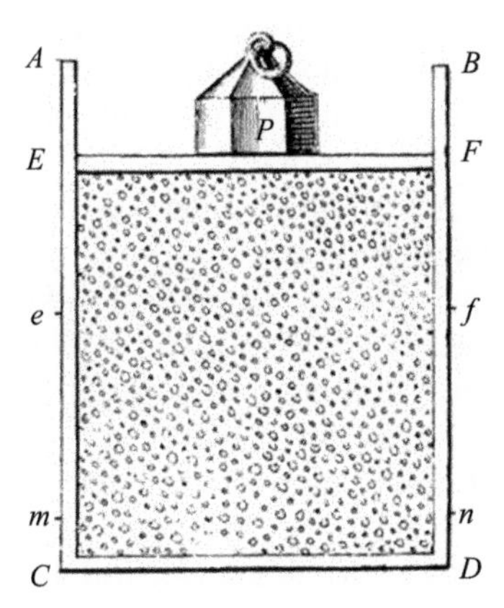

Fig. 1.2 Bernoulli's sketch of gas molecules holding up a weight via the force exerted on a piston, as still seen today in elementary thermodynamics books. (Public Domain)

[21]Dov Ginsberg, *The Age of the Earth From Judaic Traditional Literature*, Earth Sciences History **3**, vol. (2), 169, 173 (1984).

conservation of energy had not yet been established and it was not obvious that collisions between particles could be perfectly elastic. Roger Boskovich, a Croatian Jesuit, carried the scientific atomic theory further by surmising that atoms are influenced by interatomic potentials modeled on Newtonian gravity (1758) and provided the first insight that collisions might be elastic.

It is of interest to note that Benjamin Franklin was very interested in oil films on water, mostly because of their both anecdotal and actual effects to reduce the amplitude of wind-driven waves. Franklin would carry vials of oil with him which he would sometimes pour onto ponds or lakes to study the effects of films. He noticed that a teaspoon of olive oil would spread to an area of about one-half an acre. He did not estimate the size of a molecule based on this, but this result implies a molecular size of about a nanometer.[22] Later, Agnes Pockels (circa 1885) was the first to estimate the size of oil molecules based on her measurements of films.

Experiments with gases that first became possible at the turn of the nineteenth century led John Dalton (1766–1844) in 1803 to propose the basis of modern atomic theory based on the following assumptions:

1. Matter is made up of atoms that are indivisible and indestructible.
2. All atoms of an element are identical.
3. Atoms of different elements have different weights and different chemical properties.
4. Atoms of different elements combine in simple whole numbers to form compounds.
5. Atoms cannot be created or destroyed. When a compound decomposes, the atoms are recovered unchanged.

This is supplemented with Avogadro's hypothesis, named after Amedeo Avogadro, who, in 1812, stated that two given samples of a perfect gas, with the same volume and at the same temperature and pressure, contain the same number of molecules.

At this point, the development of atomic theory in physics deviates from its development in chemistry. With few notable exceptions, for chemists at this time atoms were becoming very real and brought new understanding to chemical reactions and compounds. Physicists were of less uniform opinion as to the reality of atoms, a state that persisted until well into the twentieth century.

In 1828 the chemist Frederich Wöhler synthesized urea from inorganic compounds, and disproved the vitalist hypothesis that "organic" compounds could be made only by living things. In 1855, August Kekulè formulated the ring structure of Benzene. When isomers of dibromobenzene were not discovered, he proposed that the double bonds in the rings oscillate between carbon atom pairs—does this mark the invention of quantum mechanics?

In 1869, the Russian chemist Dmitri Mendeleev developed a framework that would become the modern periodic table. While arranging the elements according to their atomic weight, he found that they tended to fall into columnar groups with similar

[22]See, e.g., Joost Mertens, *Oil on troubled waters: Benjamin Franklin and the honor of Dutch Seamen*, Physics Today **59**, 1, 36 (2006); https://doi.org/10.1063/1.2180175. Also, W. M. Klipstein, J. S. Radnich and S. K. Lamoreaux, *Thermally excited liquid surface waves and their study through the quasielastic scattering of light*, American Journal of Physics 64, 758 (1996); online: https://doi.org/10.1119/1.18174.

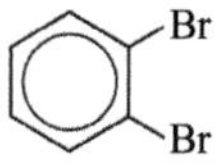

Fig. 1.3 a. 1,2-Dibromobenzene with fixed bonds in the ring, compared to b. where the bonds locations are delocalized, and considered to be the correct formula because the non-existence of two isomers of 1,2-Dibromobenzene.

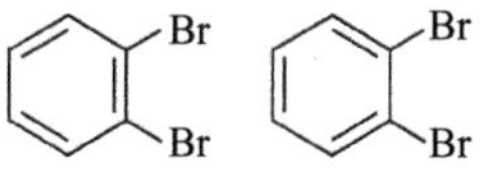

properties, and he inserted gaps for elements that he suspected were not yet discovered. Based on the properties of a group, Mendeleev predicted the properties of some undiscovered elements and gave them names such as "eka-aluminum" for an anticipated element with properties similar to aluminum. Eka-aluminum was later discovered as gallium. However, discrepancies remained; the position of certain elements, such as iodine and tellurium, could not be explained until the discovery of isotopes.

By this time, chemical industries were burgeoning, particularly in England and the United States, and especially in Germany. The acceptance of atoms and atomism by chemists was profitably productive; the wealth condensed in the flasks of organic chemists helped inspire and secure the funding of science in general. As such, the rejection of atoms by many physicists appears as particularly intellectually schizophrenic.

1.5 Atomism and physics

In 1860, James Clerk Maxwell, after reading a paper by Clausius[23] that introduced the notion of the mean free path, began his studies of kinetic theory and determined the velocity spectrum of speeds in an idealized gas by use of heuristic methods that were later fully developed by Boltzmann. At the time, the notion of a velocity distribution went against the conventional theory, which was that a range of velocities would be equalized by molecular collisions. Maxwell also investigated kinetic theory in general, and discovered "the curious result" that viscosity is independent of pressure, which was unexpected. He published an estimate of the mean free path based on air viscosity measurement that had been done by Stokes.[24] Maxwell and Katherine Clerk Maxwell (his wife) made the first reliable measurements of the dependence of gas viscosity on temperature and pressure. These measurements were performed in the attic of their house, with the temperature controlled by selective stoking of the fireplace. Their results, reported in 1866, supported the kinetic theory of gas viscosity and provided the first accurate measurement of the effective diameter of the atoms or molecules the gas comprises, based on Loschmidt's work, cited below.

[23]Clausius, R. (1857), "*Ueber die Art der Bewegung, welche wir Wärme nennen,*" Annalen der Physik, 100 (3): 353-379. English translation *The Nature of the Motion which we call Heat*, Philosophical Magazine, Vol. 14, pp. 108-27 (1857).

[24]Maxwell, J.C. (1860) *Illustrations of the dynamical theory of gases. Part I. On the motions and collisions of perfectly elastic spheres*, Philosophical Magazine, 4th series, 19: 19-32. Maxwell, J.C. (1860) *Illustrations of the dynamical theory of gases. Part II. On the process of diffusion of two or more kinds of moving particles among one another*, Philosophical Magazine, 4th series, 20: 21-37.

In a paper he wrote in 1866, Maxwell states that "Loschmidt had deduced from the dynamical theory the following remarkable proportion: –As the volume of a gas is to the combined volume of all the molecules contained in it, so is the mean path of a molecule to one-eighth of the diameter of a molecule,"[25] which relates the mean free path to the diameter as

$$8\epsilon\lambda = d \tag{1.1}$$

where ϵ is the ratio of a condensed volume of gas to the volume of its vapor, λ is the mean free path in the gas, and d is the molecular diameter. That is, to say, by measuring the volume of a condensed gas, assuming the atoms are tightly packed, and by use of the expansion on evaporation, the diameter of the molecules the gas comprises can be determined. Although Loschmidt did not determine the number of molecules in a unit volume (Loschmidt's number) this was a mathematical step that Maxwell provided, resulting in a number differing by a factor of two from the modern value.

Shortly thereafter (1867), George Johnstone Stoney published an estimate of the number of molecules in a volume of gas that he had determined in 1860, which can be used to determine the number of molecules in a mole. He subsequently (1874) invented the electron (he had various names, electrolion, electrine, and settled on electron (Lorentz's preference)) as the charged valency particle of electrolysis, and determined its electric charge by dividing Faraday's constant by the number of molecules in a mole. Although his estimation of the electron charge e is 1/16 its present value, this error can be traced to an error in his independent determination of Loschmidt's number. Stoney also invented natural dimensionless physical units (what we now call Planck units), and are in essence the same as Planck units up to factors of $\sqrt{\alpha}$.[26]

During this period, black body radiation was being studied with experimental techniques of increasing precision, and measurement by Tyndall led Joseph Stefan to, in 1869, conclude the power radiated from a black body scales as its temperature to the 4th power. Boltzmann was able to derive this relation from thermodynamic principles by considering an ideal heat engine with electromagnetic radiation as the working gas.

1.5.1 Atomism and anti-atomism: the emergence of atomic physics

Much has been written about the rejection of atomic theory in the late nineteenth century which continued well into the twentieth century.[27] The arguments against atomism were not put forward by a unified front, but by some valid and invalid concerns

[25] Loschmidt, J. (1865). "*Zur Grösse der Luftmoleüle*". Sitzungsberichte der Kaiserlichen Akademie der Wissenschaften Wien. 52 (2): 395-413. English translation: J. Loschmidt with William Porterfield and Walter Kruse, trans. (October 1995) *On the size of the air molecules*, Journal of Chemical Education, 72 (10) : 870-875.

[26] O'Hara, J. G. (1975). *George Johnstone Stoney, F.R.S., and the Concept of the Electron.* Notes and Records of the Royal Society of London. 29 (2): 265-276.

[27] This section is based on an assemblage of historical books, articles, and original publications, including David Lindley, *Boltzmann's Atom* (The Free Press, New York, 2001); John T. Blackmore, *Ernst Mach, His Work, Life, and Influence* (Univ. of California Press, Berkeley, 1972); E. Broda, *The Intellectual Triangle: Mach-Boltzmann-Planck-Einstein* CERN 81-10, July 1981; and *The Interaction of Boltzmann with Mach, Ostwald, and Planck, and His Influence on Nernst and Einstein,* 16th International Congress on the History of Science, Bucharest, 1981; and Pullman, loc. cit.

regarding the theory, which complicates the discussion. For example, Ernst Mach—known for the Mach number for specifying supersonic speeds, and Mach's principle that the origin of inertia was due to the mass distribution in the universe, and was among the most respected experimental physicists in Europe during this period—had a strongly held view that atoms are untestable theoretical constructs that cannot be observed and therefore have no place in science. His positivist views appear to have been formed in reaction to the dogmatic educational system of the Austro-Hungarian empire, then under the rule of the Hapsburgs (but that rule came to a dramatic conclusion with the assassination of Duke Ferdinand followed by its subsequent dissolution at the end of WWI). The educational dogma that was instilled in the Great Unwashed (Mach's family was of meager existence) of the empire was based on Scholasticism combined with Catholic doctrine and was spooned out as an intellectual pablum to keep the masses content with what Mach realized was a false and arbitrary worldview.

Mach's insistence that science be grounded in direct observation had a profoundly positive effect on science, and in particular Mach's views influenced Einstein in his development of relativity. When atoms were brought up, Mach was known to ask if anyone has ever seen one. Mach was aligned in his views in many ways with the great Nobel-prize-winning chemist Wilhelm Ostwald, who also rejected the notion of atoms, and put forward an alternative theory, energeticism, that energy was the ultimate entity and made no particular assumptions about the nature of matter. The goal of energeticism was to understand all physical processes through the concept of pure energy. Ostwald's view, being an expert on catalysis and equilibrium phenomena, was that atomism was too simple to be of use for such complicated problems, and he was mostly correct on this point. One has the impression that Ostwald was searching for the concept of the Gibbs free energy which, of course, was being developed by J.W. Gibbs in the US and was then largely unknown in Europe. As an aside, it was only later, after developing his classical notions of thermodynamics, including his phase change laws, that Gibbs embarked on studies of statistical mechanics using atoms and molecules–but Boltzmann suspected that he might have had atomic models in his mind earlier.

Planck was anti-atomism, apparently because he did not think atoms were needed or useful for the development of thermodynamics. Planck was also deeply religious, and although a Protestant, he was aware of the suffering of Catholics under Bismarck's Kulturkampf (1871–87), which subjected the Roman Catholic Church to state controls, and had its climax in 1875, when civil marriage was made obligatory throughout Germany. Atomism was still in conflict with Catholic teachings, and Planck as a leading scientist and an associate editor (from 1895–1907, and then one of two co-editors until 1943) of the leading physics journal of the time, *Annalen der Physik*, perhaps did not want to promote an agenda that could be viewed as anti-Catholic in the light of the recent suffering. As we have mentioned, Planck has been viewed as both a revolutionary (Klein) and a conservative (Kuhn). It should be noted that Planck published Einstein's papers, particularly on relativity, and was willing to publish Boltzmann's works, but subject to a proviso.

The atomists and anti-atomists of this period were, on a personal level, very friendly with each other despite their heated and sometimes public debate over atoms and kinetic theory, with the exception of the chilly relationship between Boltzmann and Planck. *Annalen der Physik* would not publish Boltzmann's papers on kinetic theory unless they contained a disclaimer that atoms were only convenient theoretical constructs. On the other hand, Planck and Boltzmann teamed up and provided a unified front against energeticism.

A climactic event in the relationship between Planck and Boltzmann occurred in 1896, regarding Boltzmann's H-theorem. Boltzmann formulated the H-theorem around 1871. Briefly, H is determined from the energy distribution function $f(E,t)dE$ of molecules at time t. The value $f(E,t)dE$ is the number of molecules that have kinetic energy between E and $E + dE$. H is defined as

$$H(t) = \int_0^{\infty} f(E,t)\left(\ln\frac{f(E,t)}{\sqrt{E}} - 1\right) dE. \tag{1.2}$$

For an isolated ideal gas (with fixed total energy and a fixed total number of particles), H is minimized when the particles have a Maxwell-Boltzmann distribution. If the particles are distributed some other way e.g., all having the same kinetic energy, the value of H will be larger. Boltzmann showed that when collisions between particles occur, other distributions are unstable and move irreversibly toward the minimum value of H, that is, toward the Maxwell-Boltzmann distribution.

Very soon after the H-theorem was published, Loschmidt, who was Boltzmann's colleague at Vienna, pointed out that it should not be possible to produce an irreversible process when the underlying dynamics are time-symmetric. All one needs to do is reverse time for an entropy-increasing process to show that there are states where H increases (equivalently, entropy decreases) over time—this is Loschmidt's paradox. The H-theorem is based on the assumption of "molecular chaos," that in kinetic theory particle motion is be considered independent and uncorrelated—so it would not generally be possible to reverse the motion. Boltzmann conceded to Loschmidt that such states were possible, while noting that they are so rare and unusual that in practice have no significant contribution. Because Boltzmann introduced H as a proxy for entropy, the H-theorem was the first attempt to use statistical mechanics (although the name statistical mechanics was only later invented by Gibbs) to derive the second law of thermodynamics from classical reversible dynamics, however, because of the assumption of molecular chaos, the reasoning was circular. Boltzmann worked to address this objection, leading to his entropy formula of 1877,

$$S = k \log W, \tag{1.3}$$

which relates entropy S to the number of possible configurations W of the molecules of an ideal gas.

A second criticism of the H-theorem was raised in 1896 by Ernst Zermelo,[28] a student/assistant of Planck. Briefly, Zermolo argued that according to an accepted theorem of Henri Poincaré, an ensemble of particles must access all the accessible

[28]Zermelo, E., "*On a law of dynamics and the mechanical theory of heat,*" Ann. d. Phys. 57, 485, 1896, (dated Dec. 1895) English translation in Brush, S.G., *Kinetic Theory: Selected Readings in Physics, Vol. 2*, Pergamon, 1966

configurations (that is, sample all regions of phase space as allowed by energy and momentum conservation), and must therefore return to the starting position infinitely many times, to any degree of precision. This means that an isolated system of particles must at some time return to a state of decreased entropy. This came to be known as the recurrence objection (as named by Paul and Tatyana Ehrenfest). Zermolo took the position that the law of entropy increase is absolutely correct (the classical thermodynamicist Planck would agree with this) and therefore the kinetic theory of gases must be invalid. After Zermolo's publication in *Annalen der Physik*, Boltzmann wrote a brilliant and astonishingly sarcastic response, stating that Zermolo would likely suspect that the dice were loaded because he failed to roll a six one thousand times in succession, because the likelihood of such an event is not exactly zero.[29] Boltzmann insisted that his response be published in *Annalen der Physik* without change, and it was accepted.[30] (Zermelo left physics and made seminal contributions to mathematics, in particular to set theory.)

In more recent subsequent studies of Poincaré's works, it has been discovered that in fact, he did not support the use of his theorem in this manner; in essence, it is impossible to maintain a real physical system to the degree of mathematical precision required for the theorem's strict validity.[31] For example, it is easy to imagine that the walls of the container holding a gas can permanently absorb that gas, or release an impurity gas to the sample; electric charges on the wall of the container can move around, disappear, or be generated by natural radioactivity or cosmic rays. Secondly, and perhaps more importantly, Boltzmann had already acceded the point that entropy can decrease for a short time, but the probability is so small that in the long term, the second law is valid. We now understand such processes through the generalized fluctuation-dissipation theorem.

As we will discuss at length in an upcoming chapter, in 1900 Planck used Boltzmann's statistical methods, together with the introduction of quantization through his constant h, to derive the black body spectrum.

Einstein's 1905 *annus mirabilis* papers includes one on Brownian motion and describes the consequences of Boltzmann's fluctuations.[32] This was also the year that Einstein completed his Ph.D. dissertation at the University of Zurich entitled "A New Determination of Molecular Dimensions." His dissertation, a scant 24 pages, outlines the theory and measurements of diffusion rates of sugar solutions; he determined a value of Avogadro's number to within a factor of three of the currently accepted value. This led to his most highly cited publication. Experimental work by Perrin, published in 1909, confirmed all of Einstein's Brownian motion predictions, and provided a determination of Avogadro's number in conflict with Einstein's (recall that Avogadro's number had already been obtained by Loschmidt and, with less accuracy, by Stoney 40 years earlier, and the sizes of molecules had been determined by Agnes Pockels in

[29]K. Mendelssohn, *The World of Walther Nernst: The Rise and Fall of German Science 1864-1941* (London: The Macmillan Press Ltd., 1973) p. 115.

[30]Boltzmann, L., *On energetics*, Ann. d. Phys., 58, 595, 1896

[31]Brush, S.G., *Poincaré and Cosmic Evolution*, Physics Today **33**, no. 3, 42 (1980).

[32]Einstein, A., *On the movement of small particles suspended in a stationary liquid required by the molecular kinetic theory of heat*, Ann. d. Phys, **17**, 549, 1905 (received 11 May 1905).

1892[33]). Einstein subsequently corrected several theoretical errors (the hydrodynamics of the system is complicated) and, together with improved diffusion measurements, obtained a value within 5% of the presently accepted value (his studies were completed around 1911).[34]

Atomism together with kinetic theory was by then accepted by most physicists, but with a notable exceptions which include Mach. We must recall that in chemistry, especially by organic chemists, the existence of atoms had already been largely accepted as a necessary fact, if not a matter of course, a century earlier.

One of the last feats in statistical mechanics before the invention of quantum mechanics was the calculation of the entropy of an ideal gas in 1912, independently by Otto Sakur (who died in a chemical explosion in Fritz Haber's laboratory in 1914) and Hugo Tetrode. They used the Planck constant cubed as the unit volume of a phase space element ($(dpdx)^3$) for an ideal gas (Planck had used this as the phase space volume for electromagnetic radiation), and provided a means to determine the number of states W in the Boltzmann entropy equation. The equation was experimentally verified using mercury vapor. Using h^3 as the unit volume of phase space anticipated de Broglie's wave hypothesis by 12 years.[35]

Thus the era of anti-atomism waned at an accelerating rate, with the exception of Mach, whose positivism was amplified into a dogmatism that rejected the use of hypothesis in science—this is of course untenable; as it has been said, no appeal to natural science that avoids hypothesis completely could ever succeed. Mach's earlier notions were not so extreme and were primarily a warning that models, and hypotheses, should be grounded in reality, and this should be taken as his lasting legacy. This lesson remains relevant today, and will so for as long as anything resembling science is allowed and practiced by human society.

As a closing of this historical interlude, it should be noted that there were no villains, and perhaps a few heroes, in the final development of the atomic theory of matter. For example, there has been a modern tendency to blame Mach for Boltzmann's depressive spiral that resulted in his death; certainly, Boltzmann had been distraught over the lack of general acceptance of atomic theory and the continued attacks on its validity. On the other hand, as Lise Meitner (who was a student of Boltzmann) noted, "He may have been wounded by many things a more robust person would have hardly noticed.... I believe he was a powerful teacher just because of his uncommon humanity." Boltzmann's demise was most likely due to his painfully deteriorating health and failing eyesight.[36] Planck later considered Mach to be a "false prophet," however, it's easy to accept a prophet's teachings when they correspond to what one wants to hear; on the other hand Planck and other atomic detractors eventually changed their minds as was ultimately necessary to embrace any sort of quantum theory of matter, and

[33]Agnes Pockels, *On the relative contamination of the water-surface by equal quantities of different substances*, Nature **47**, 418 (1892).

[34]A. Einstein, Justification for my work, *A new determination of molecular dimensions Annalen der Physik (ser. 4) 34, 591-592, 1911.*

[35]Walter Grimus, *100th anniversary of the Sackur-Tetrode equation*, Ann. Phys. (Berlin) **525**, No. 3, A32-A35 (2013). DOI 10.1002/andp.201300720.

[36]Blackmore, loc. cit., p. 208

radiation. Planck's attack resulted in Mach's leaving "The Church of Physics" to spend the rest of his life studying anthropology and psychology.[37] During a lecture in Paris, some months after the posthumous publication of Mach's *The Principles of Physical Optics* that, in its introduction, presents a rejection of relativity, "Which I [Mach] find to be growing more and more dogmatical," Einstein made a widely reported comment that Mach was "un bon mechanicien," but "un deplorable philosophe."[38]

With this, we arrive at the time when quantum mechanics begins in earnest, and marks the start of our formal discourse on the subject. We will first describe some observed "quantum" properties of the physical world.

[37] John Blackmore, *Ernst Mach Leaves "The Church of Physics"*, The British Journal for the Philosophy of Science Vol. 40, No. 4 (Dec., 1989), pp. 519-540.

[38] Holton, Gerald,*Mach, Einstein, and the Search for Reality.* Daedalus **97** (2), Historical Population Studies (Spring, 1968): 636-673.

2

Properties of the quantum world: indeterminacy, interference, superposition, entanglement

In this chapter we describe several phenomena that contradict the predictions of classical physics, sometimes in a most dramatic way. Most of these examples were discovered after the formulation of quantum mechanics as a complete theory. Several of the related experiments were proposed as challenges to quantum theory because the quantum mechanical predictions appeared so absurd that they seemed likely to indicate places where the theory breaks down. To date, quantum mechanics has been verified, rather than contradicted, by the experimental results. As many of these tests of quantum mechanics go right to the heart of "quantum wierdness" the following discussion will give some physical intuition for the main features of the quantum world.

2.1 Indeterminacy—random behavior

As Laplace was the first to popularize, if Newton's equations and Maxwell's equations accurately described the physical world, we would expect the future to be entirely determined by the past. If we imagine a super-intelligent demon who could know the position and velocity of every particle in the universe he could, in principle, calculate the positions of all particles in the future.[1] Boltzmann and others have shown that random behavior is consistent with classical physics for large many-body systems, but single particles also show random behavior in many situations, in contrast to the predictions of classical theories, and quantum mechanics provides an understanding of these phenomena. Quantum mechanics introduces indeterminacy as a fundamental (and, hence, unavoidable) feature of our view of nature, whereas a classical physicist could cling to the hope that nature is, in principle if not in practice, deterministic (as many still do).

As an example, some atomic nuclei are observed to decay by various processes. A nucleus consisting of Z protons and N neutrons with $Z+N=A$ can undergo β–decay such as

$$\begin{aligned} (Z, A) &\rightarrow (Z+1, A) + e^{-} \\ (Z, A) &\rightarrow (Z-1, A) + e^{+}, \end{aligned} \tag{2.1}$$

[1] Chaos theory has shown that even such a classical universe would not be predictable.

The Historical and Physical Foundations of Quantum Mechanics. Robert Golub and Steven K. Lamoreaux, Oxford University Press.
 DOI: 10.1093/oso/9780198822189.003.0002

where $e^{\mp}$ represents an electron or a positron. Some such nuclei can be produced in nuclear reactions, which allows us to measure the time between their creation and decay. As is well known, the atoms do not all decay at the same time after creation, and the decay times of individual atoms vary over a wide range. The decay process is found to be independent of external parameters such as temperature, physical state (solid, liquid, or gas), and any chemical compound that the atoms may be part of: "The decay rate depends only on the number of atoms." The observations are consistent with the idea that the same fraction of atoms decays in each time interval dt of a given length, so that if $n(t)$ is the average number of atoms remaining at time t after their creation, we have

$$\frac{dn(t)}{dt} = -\frac{n(t)}{\tau}, \tag{2.2}$$

where $1/\tau$ is a proportionality constant, $dn(t)$ is the mean number of nucleii decaying in the time interval between t and $t+dt$, and the minus sign indicates that the number of atoms is decreasing. The solution is

$$n(t) = N_0 e^{-\frac{t}{\tau}}, \tag{2.3}$$

where N_0 is the number of atoms at $t = 0$. The fraction of particles remaining at t or the probability of a particle remaining until t is

$$\frac{n(t)}{N_0} = P(t) = e^{-\frac{t}{\tau}}. \tag{2.4}$$

The number of particles, on average, decaying between $t = 0$ and t is given by

$$N_{dec} = \int_0^t dn(t') = \int_0^t \frac{dn(t')}{dt'} dt' = N_0 \int_0^t e^{-\frac{t'}{\tau}} \frac{dt'}{\tau} = N_0 \left(1 - e^{-\frac{t}{\tau}}\right). \tag{2.5}$$

The probability of a nucleus surviving until t and then decaying in the interval between t and $t + dt$ is given by the product of the probabilities of the two events

$$P(t) \frac{dt}{\tau} = e^{-\frac{t}{\tau}} \frac{dt}{\tau}, \tag{2.6}$$

and the average survival time is then

$$\langle t \rangle = \int_0^\infty t P(t) \frac{dt}{\tau} = \int_0^\infty t e^{-\frac{t}{\tau}} \frac{dt}{\tau} = \tau. \tag{2.7}$$

Similarly, $\langle t^2 \rangle = 2\tau^2$. The parameter τ called lifetime or decay time thus completely characterizes the decay.

Could it be that each nucleus is created with some property that determines its decay time? Such parameters, called hidden variables, will be discussed in detail later in the book. To date, no hidden variables determining the decay times of individual particles have been found and we have no idea how to look for such parameters. We are thus forced to conclude that nuclear decay is a completely random process and all the information we can have about this process is contained in the lifetime τ. Quantum mechanics provides ways to calculate τ, and so far no discrepancies have been found between the quantum mechanical predictions and experimental measurements. The decay of, say, the muon behaves in the same way.

2.2 The wave nature of light and matter and its connection with random behavior

2.2.1 Photons

In 1905, Einstein showed that the strange properties of the photoelectric effect[2] could be understood if we considered the energy of a light beam as residing in small packets (quanta), each containing an energy $h\nu$, where h is Planck's constant[3] and ν is the frequency of the light. This brings up the question of the relation between the quanta and the classical electromagnetic field theory, which gives an excellent description of optical phenomena such as interference and diffraction. In classical optics, we imagine that the electric and magnetic fields making up the light can be represented as plane waves given by the real part of

$$f\left(\vec{r},t\right) = Ae^{i\left(\vec{k}\cdot\vec{r}-\omega t\right)}, \tag{2.8}$$

where A is an amplitude, $\vec{k}$ is a vector (the wave vector) with magnitude $k = 2\pi/\lambda$ whose direction determines the direction of propagation of the wave, and $\omega = 2\pi\nu$. Here λ is the wavelength of the light and ν is its frequency. Note that the phase $\left(\vec{k}\cdot\vec{r} - \omega t\right)$ is constant for

$$\frac{x}{t} = \frac{\omega}{\left|\vec{k}\right|} = c, \tag{2.9}$$

where x is the distance traveled by a surface of constant phase in the direction of $\vec{k}$ in time, t. This means that the wave moves with a velocity c.

We will consider the classical treatment of such a light wave passing through a thin slit of width a and hitting a screen a distance L away, as shown in Fig. 2.1.

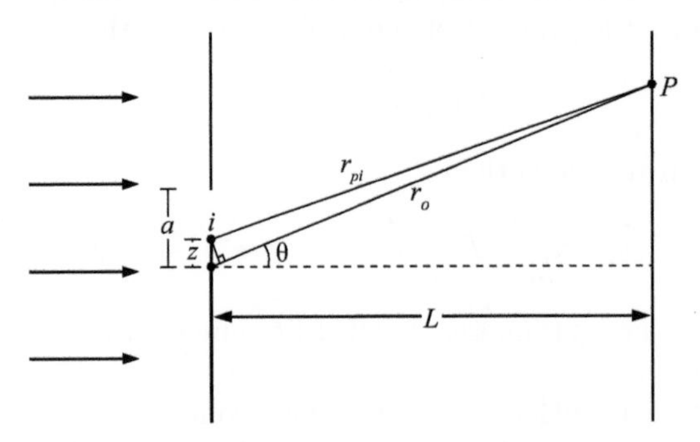

Fig. 2.1 Geometry for single-slit diffraction t.

[2]The photoelectric effect, in which metals in a vacuum emit electrons when exposed to light, will be discussed in the next chapter.

[3]This constant was introduced by Planck to get the correct form of the spectrum of electromagnetic radiation in thermal equilibrium (black body spectrum). See Chapter 3.

The wave arriving at a point P on the screen will be, according to Huygens' principle, a superposition of waves leaving each point in the slit. We write the distance from a point i in the slit to the point P on the screen as

$$r_{pi} = r_o - z\sin\theta, \tag{2.10}$$

where r_o is the distance to a reference point taken as the bottom of the slit. Then the wave amplitude at P at a fixed time $t = 0$ will be

$$\begin{aligned} E(P) = E_o \int_0^a e^{ikr_{pi}} dz &= E_o e^{ikr_o} \frac{\left(e^{-ika\sin\theta} - 1\right)}{-ik\sin\theta} \\ &= E_o e^{ikr_o} e^{-i\frac{ka\sin\theta}{2}} \frac{\left(e^{-i\frac{ka\sin\theta}{2}} - e^{i\frac{ka\sin\theta}{2}}\right)}{-ik\sin\theta} \\ &\approx aE_o e^{ikr_o} e^{-i\frac{ka\sin\theta}{2}} \frac{\sin\left(\frac{ka\theta}{2}\right)}{ka\theta/2}, \end{aligned} \tag{2.11}$$

where in the last line we have made the small angle approximation $\sin\theta \sim \theta$. The intensity of light striking point P, $I(P) = |E(P)|^2$, is plotted in Fig. 2.2.

It is much more interesting to consider two slits separated by a distance D, as shown in Fig. 2.3, which yields an interference pattern superimposed on the single slit diffraction pattern.

Assuming that $L \gg D$ so that the rays from both slits to P are nearly parallel, we can apply the same approach as for the single slit. Making the small angle approximation $\sin\theta \sim \theta$, we have

$$E(P) = E_o e^{ikr_o}\left[\int_0^a e^{-ikz\theta} dz + \int_D^{D+a} e^{-ikz\theta} dz\right] = \frac{E_o e^{ikr_o}}{-ik\theta}\left(e^{-ika\theta} - 1\right)\left(1 + e^{-ikD\theta}\right). \tag{2.12}$$

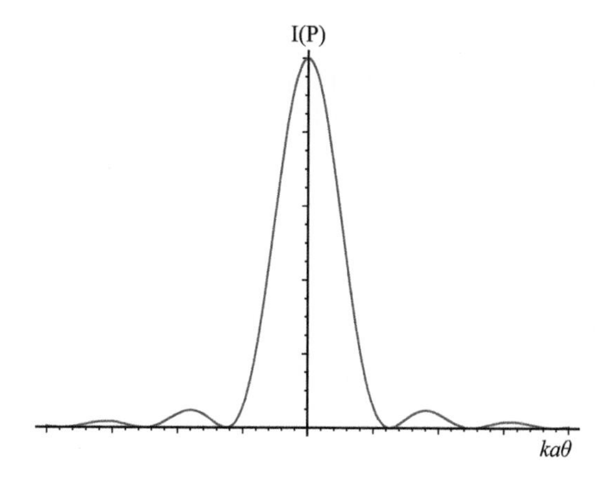

Fig. 2.2 Intensity on the screen for single-slit diffraction.

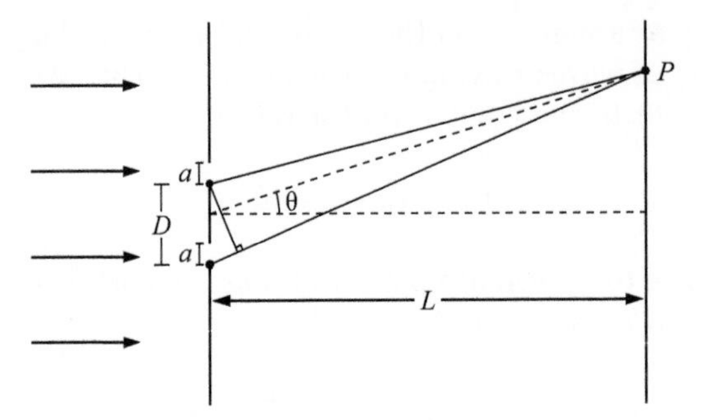

Fig. 2.3 Geometry for two-slit interference.

The intensity is then

$$I(P) = |E(P)|^2 \propto \left[\frac{\sin\left(\frac{ka\theta}{2}\right)}{ka\theta/2}\right]^2 \left(\cos\frac{kD\theta}{2}\right)^2. \tag{2.13}$$

This is plotted in Fig. 2.4 for the case $D = 6a$. The single slit diffraction pattern (square brackets in Eq. (2.13)) serves as an envelope for the much sharper two-slit interference fringes (round brackets in Eq. (2.13)).

When adding contributions from the different paths, the crucial element is the phase difference between the paths going through each of the two slits, which is given by the last term in Eq. (2.12).

We can perform the two-slit experiment with relatively intense light and use a photodiode to measure the light intensity at the screen. Since the photodiode current is proportional to the light intensity striking the photodiode's surface, the current will trace out the pattern shown in Fig. 2.4 as we move the detector along the screen.

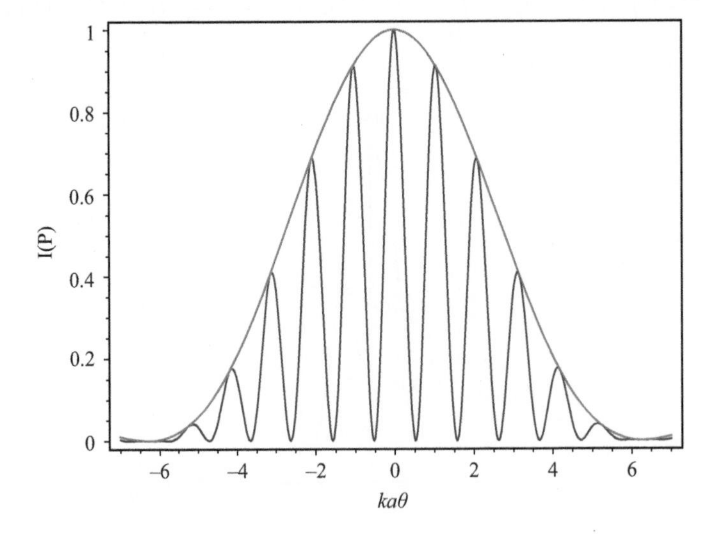

Fig. 2.4 Intensity on the screen for two-slit interference with $D = 6a$.

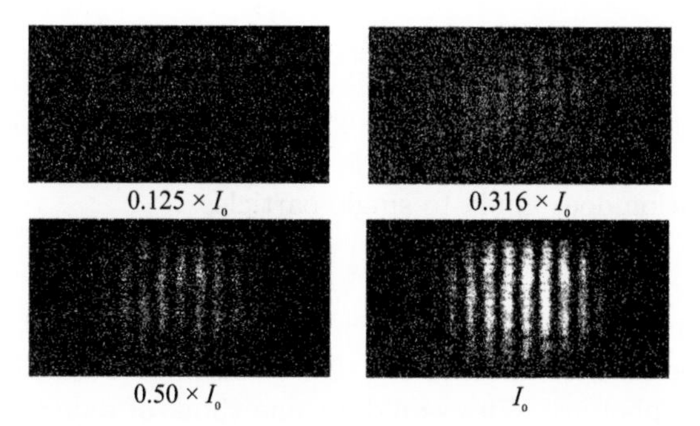

Fig. 2.5 A photon two-slit interference pattern, obtained with less than one photon in the system at any time, as a function of intensity (as set by neutral density filters). The integration time is the same for all images. (Lamoreaux Lab, Yale University)

But what if we perform the experiment with light so weak that only one of Einstein's photons can be in the apparatus (i.e., between the source and the detector) at a time? The results are depicted in Fig. 2.5.

The point where each individual photon hits the screen is random, but after many photons are observed, the distribution of photons on the screen agrees with the classically calculated intensity distribution. What is going on here? The positions where the photons are detected are randomly distributed, just like the decay times of the decaying atoms. However, the classical light intensity at a point gives the probability of a photon arriving at that point. Again, we might ask if the photons could be carrying some hidden variable that determines their trajectories. We will have much to say about this later in the book. In any case, because the photons arrive randomly, the hidden variables must be randomly distributed, and the observations force us to associate the classical intensity with the probability of arrival at a given point.

Can we ask which slit a particular photon went through on its way to the screen? The answer is no! Any attempt to determine which slit a given photon went through will destroy the interference pattern. We will see an example of how this works later in this chapter.

Now let's consider a single photon. Before it hits the screen, all we can say is that the probability of its arriving at a given point on the screen is given by Eq. (2.13), but after it hits the screen we know where it is, which means that the probability of its being at its point of detection is one and the probability of its being anywhere else is zero. In this sense, the wave representing the probability distribution of that particle in all of space has "collapsed" to a point. The instantaneous collapse of the probability distribution upon measurement was one of Einstein's early objections to quantum mechanics and has been the subject of intense discussion for nearly a hundred years now. We return to this later in the book. Although the probability distribution for each individual photon collapses upon measurement, the classical intensity distribution still provides the probability distribution for the following photons. This leads to another question that has been debated almost from the beginnings of quantum theory: whether the classical field as a probability distribution can be applied to single particles or only to ensembles of particles. We have seen that the interference pattern does not exist for a single photon; it only emerges when one observes an ensemble of many photons. If the probability distribution applied only to ensembles of particles, the collapse anomaly would not arise. However, in recent years it has become possible to perform experiments with single quantum objects, which indicate that the probability distribution does apply to single particles.

2.2.2 Electrons

If we perform the two-slit experiment with massive fundamental particles, say electrons, we obtain the results shown in Fig. 2.6[4], which are basically indistinguishable from the results for photons. Why would anyone think of doing such an experiment with electrons, which were thought to be point particles with mass and electric charge that behaved like tiny marbles? This experiment was only performed after quantum mechanics was a well-established theory. The first electron diffraction experiments were performed in 1927 by C. Davisson and L. Germer, but the *tour de force* of the two-slit diffraction experiment with electrons only became possible many years later.

As we will see in Chapter 5, de Broglie suggested that if light waves were accompanied by particle-like entities, particles might be accompanied by a wave-like motion, an idea crucial to the development of quantum mechanics. The wave-like property of electrons was confirmed by measurements of electron diffraction from a nickel crystal, which produced a diffraction pattern very similar to that previously observed with X-rays. Fig. 2.7 shows a diffraction pattern obtained with an electron beam diffracted from polycrystalline aluminum.[5] Since the lattice constant of aluminum is known to

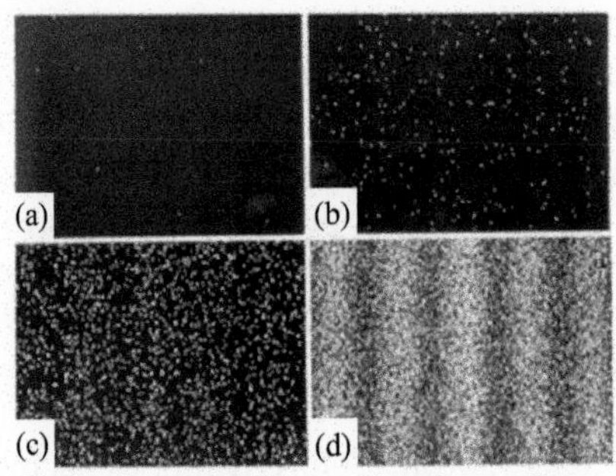

Fig. 2.6 Single electron events build up to form an interference pattern in a double-slit experiment ("electron biprism," which consists of two parallel plates and a fine filament at the center). The number of electron accumulated on the screen are: (a) 8 electrons; (b) 270 electrons; (c) 2,000 electrons; (d) 160,000. The total exposure time from the beginning to stage (d) is 20 min. (Reprinted courtesy of the Research & Development Group, Hitachi, Ltd., Japan, https://www.hitachi.com/rd/research/materials/quantum/doubleslit/)

[4]Tonomora, A., *The double slit experiment*, Physics World, 1 Sept., 2007 https://physicsworld.com/a/the-double-slit-experiment/.

[5]Laue, M., von, *Materiewellen ind Ihre Interferenzen*, Akademishe Verlages (Leipzig, 1944) p. 124, Fig. 36. Defunct publisher.

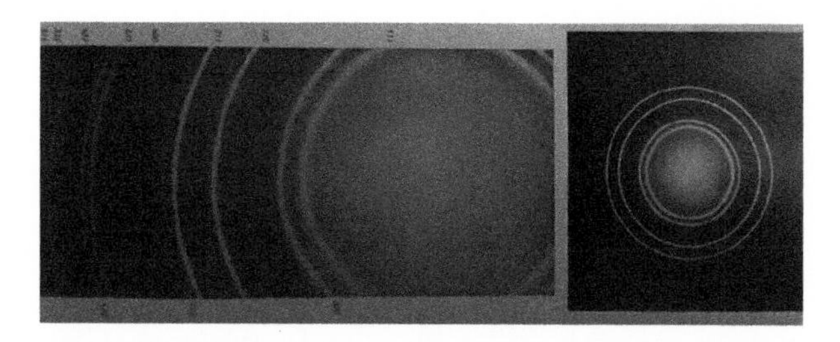

Fig. 2.7 Diffraction pattern observed with electrons on polycrystalline aluminum. (Public Domain)

be $a = 4.04$ Å from X-ray diffraction, the wavelength associated with the electrons can be calculated from the size of the diffraction rings. The results were found to agree with de Broglie's prediction

$$\lambda_{deB} = h/p, \tag{2.14}$$

where h is Planck's constant and p is the momentum of the electrons.

In the case of photons, the probability of finding a photon at a given position on the screen was given by the intensity of the light beam, which is proportional to the square of the electric field $\left|\vec{E}(\vec{r},t)\right|^2$. The electric field at a point on the screen was the sum of two terms, each term being associated with one of the slits. Thus, the probability was the square of a sum of terms. For electrons, quantum mechanics introduces a function of $(\vec{r},t)$, usually denoted by $\psi(\vec{r},t)$ using the notation introduced by Schrödinger, which is again a sum of terms, with each term corresponding to a possible path of the electron. This complex-valued function is called the probability amplitude, and the probability is given by the square of the magnitude of the amplitude. If we write $\psi(\vec{r},t) = \psi_1(\vec{r},t) + \psi_2(\vec{r},t)$, where $\psi_1(\vec{r},t)$ and $\psi_2(\vec{r},t)$ correspond to passage through the first or second slit, respectively. The probability of the electron being at a point $(\vec{r},t)$ on the screen is then given by

$$|\psi(\vec{r},t)|^2 = |\psi_1(\vec{r},t)|^2 + |\psi_2(\vec{r},t)|^2 + 2\mathrm{Re}\left[\psi_1^*(\vec{r},t)\,\psi_2(\vec{r},t)\right]. \tag{2.15}$$

It is the last term—the interference term—that is responsible for the observed interference pattern.

As with photons, the position of detection of an individual electron in the two-slit experiment appears random. It is only when we collect the results for many electrons that the interference pattern, and hence the wave nature of electrons, emerges. This wave-particle duality has been the subject of an enormous literature and continues to be the subject of much dispute.

Feynman called the two-slit interference pattern with electrons the "the only mystery" of quantum mechanics and stated that "it is impossible to explain in any classical way." Since the electron cannot go through both slits, the problem of reconciling definite trajectories with the observed interference pattern seems intractable. However, David Bohm has produced a model that does just that, as we shall see. Another approach, so-called "second quantization," denies that trajectories exist and considers the electrons as excitations of a wave field. This is discussed in Chapter 12.

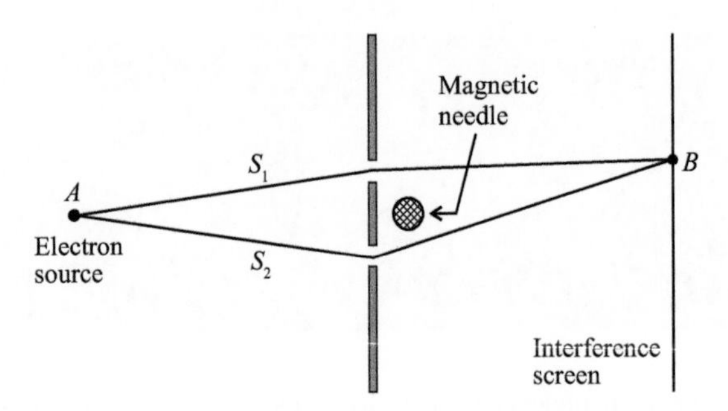

Fig. 2.8 Geometry for the two-slit interference experiment with electrons using a ferromagnetic needle to detect which slit each electron went through. The needle is a ferromagnetic rod perpendicular to the plane of the figure; it is shown in cross section.

Since electrons carry electric charge, we can attempt to determine which slit an electron goes through by placing a ferromagnetic needle between the two slits, as shown in Fig. 2.8. If the electron goes through the top slit, it will produce a magnetic field at the needle directed out of the paper; if it goes through the bottom slit, the field will be directed into the paper. By detecting the flux change in the needle, we can, in principle at least, determine which slit individual electrons pass through. If we assume the needle is a perfect ferromagnet so that the magnetic field is completely confined inside it, there can apparently be no back action on the electron, and we could determine the slit that any individual electron passed through without affecting the interference pattern. Although the magnetic field is zero in the regions traversed by the electrons, quantum mechanics predicts that any measurement capable of determining which slit an electron passed through does produce a back action, and on passing through the system shown in Fig. 2.8, the electron waves will acquire random phase shifts that destroy the interference pattern. The probability of finding an electron at a given point on the screen, previously calculated according to Eq. (2.13), will then be given by

$$|\psi(P)|^2 \propto \left[\frac{\sin\left(\frac{ka\theta}{2}\right)}{ka\theta/2}\cos\left(\frac{kD\theta}{2}+\delta\right)\right]^2 \tag{2.16}$$

where δ is a random phase shift between the electron waves going through each slit. The interference pattern will thus be washed out by averaging over δ. The consistency of quantum mechanics is thus preserved as determining which slit the electron passed through destroys the interference pattern.

The mechanism of this back action is discussed in Section 2.5.

2.3 Superposition and projection

2.3.1 Linearly polarized light

In this section we show how some classical behavior of a light wave leads to uncanny (quantum) results when applied to particles (photons).

Classically, light waves are pictured as consisting of electric and magnetic fields oscillating in space and time and undergoing a wave-like motion, as shown in Fig. 2.9.

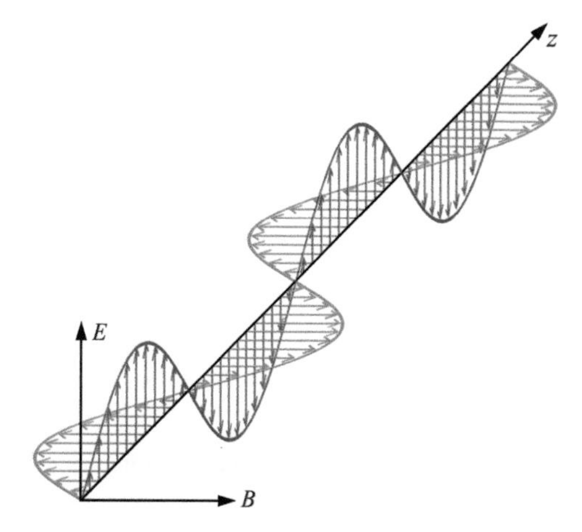

Fig. 2.9 A light wave propagating in the z direction with $\vec{E}$ and $\vec{B}$ perpendicular to the direction of propagation and to each other.

The figure shows the fields at a single instance of time. As time progresses the whole pattern moves along the z axis with velocity c. We are discussing what is called coherent light when the phase of the wave remains constant in time and space. Many light fields consist of incoherent light, that is the phase of the wave varies randomly in time and space over some range The magnetic field $\vec{B}$ is always perpendicular to the electric field $\vec{E}$. For a monochromatic plane wave, the space-time dependence is $\propto e^{i\left(\vec{k}\cdot\vec{r}-\omega t\right)}$. This represents a wave traveling in the direction of $\vec{k}$ with frequency ν and wavelength λ, where $\omega = 2\pi\nu$ and $\left|\vec{k}\right| = 2\pi/\lambda$. The electric field vector can point in any direction perpendicular to the direction of motion of the wave, $\hat{k} = \vec{k}/\left|\vec{k}\right|$. The direction of $\vec{E}$ is called the direction of polarization of the wave. Looking at a plane perpendicular to $\hat{k}$, we can decompose $\vec{E}$ into a sum of vectors along any two perpendicular directions, for example x' and y', as shown in Fig. 2.10. Then, at a fixed time ($t = 0$)

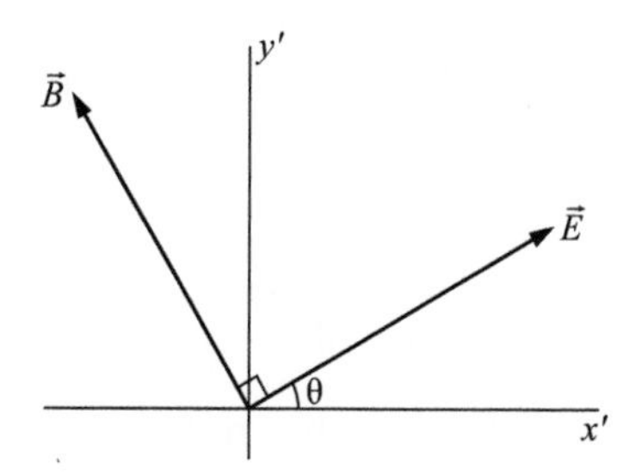

Fig. 2.10 Direction of polarization for a linearly polarized light wave propagating out of the paper.

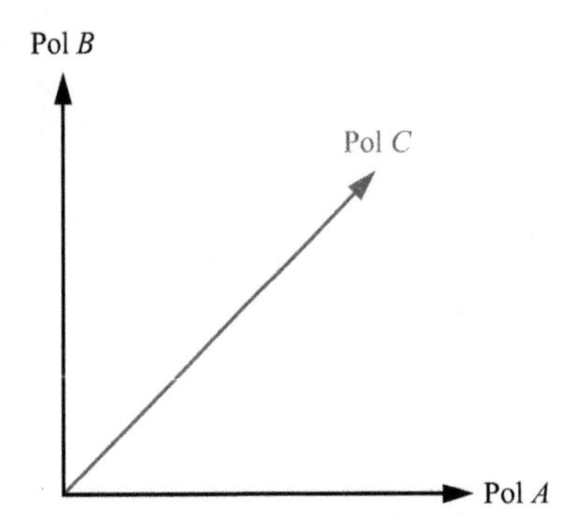

Fig. 2.11 Directions of the polarization axes of the polarizing foils A, B and C.

$$\begin{aligned} E'_x &= E\cos\theta \\ E'_y &= E\sin\theta, \end{aligned} \tag{2.17}$$

as shown in the figure, so we can consider light polarized along the x-axis as being a superposition of light polarized along the x' and y' axes. Some materials, such as Polaroid filters, only transmit light that is polarized along a particular direction in the material. If such a polarizer A with its transmission axis along x' is placed in the light beam, the transmitted light will have an electric field with a magnitude of $E\cos\theta$ and an intensity of $E^2\cos^2\theta$. The remaining light, polarized along y', will be absorbed.

If we follow the first polarizer with a second polarizer B oriented along the y' axis, no light will pass through B since the light transmitted through the first polarizer is polarized along x' and thus has no component along y'. If a third polarizer C (see Fig. 2.11) is placed between A and B with its axis at 45° to the other two axes, light of amplitude $E_C = E_A\cos(\pi/4)$ will be transmitted through C. This light will be polarized in the direction of the C axis. This light will then have a component $E_B = E_C\cos(\pi/4) = E_A\cos^2(\pi/4)$ along the B axis, so inserting the film C allowed some of the light previously blocked to travel through B. If we had inserted N polarizers, each rotated by an angle $\pi/(4N)$ with respect to its neighbor, the transmitted amplitude would be proportional to

$$\begin{aligned} E_A\left(\cos\left(\frac{\pi}{4N}\right)\right)^N &\to E_A\left(1-\frac{1}{2}\left(\frac{\pi}{4N}\right)^2\right)^N \sim E_A\left(1-\frac{1}{2N}\left(\frac{\pi}{4}\right)^2\right) \\ &\to E_A \end{aligned} \tag{2.18}$$

as $N\to\infty$. Thus, by making a very large number of successive measurements, we can rotate the plane of polarization with negligible loss of photons.

2.3.2 Circularly polarized light: an alternative description

We can also have a superposition of two waves polarized in perpendicular directions but out of phase by an angle ϕ. The first wave might be described by $E_x^{(1)} = E\cos\omega t$ while the second wave would have

$$E_y^{(2)} = E\cos(\omega t+\phi) = E(\cos\omega t\cos\phi - \sin\omega t\sin\phi). \tag{2.19}$$

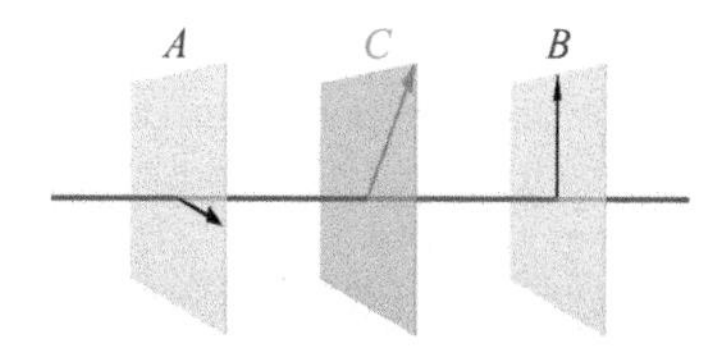

Fig. 2.12 Set of three polarizers A, B, and C that allows the transmission of light through two polarizers with perpendicular axes.

For $\phi = 0$, we have $E_x = E_y = E \cos \omega t$, so the light is linearly polarized along a direction making an angle of $\pi/4$ with respect to the x-axis:

$$\vec{E} = E\left(\hat{i} + \hat{j}\right) \tag{2.20}$$

where $\hat{i}, \hat{j}$ are unit vectors in the x and y directions. For $\phi = \pi/2$, we find $E_x^{(1)} = E \cos \omega t$ and $E_y^{(2)} = -E \sin \omega t$, so the vector sum of the two fields moves in a circle $\left(E_x^2 + E_y^2 = E^2\right)$ in the clockwise direction. We call this positive circular polarization, for which

$$E_x^{(+)} = E \cos \omega t \quad \text{and} \quad E_y^{(+)} = -E \sin \omega t. \tag{2.21}$$

If $\phi = -\pi/2$, the $\vec{E}$ field vector rotates in the opposite direction ($E_y^{(2)} = E \sin \omega t$); this corresponds to negative circular polarization. Adding the two oppositely circularly polarized waves together produces a linearly polarized wave, since

$$\begin{aligned} E_x &= E_x^{(+)} + E_x^{(-)} \\ &= 2E \cos \omega t \\ E_y &= E_y^{(+)} + E_y^{(-)} \\ &= 0. \end{aligned} \tag{2.22}$$

Adding two linearly polarized waves that are polarized along perpendicular directions and have a phase difference of $\phi = \pi/2$ produces a circularly polarized wave, and adding two waves with opposite circular polarization produces a linearly polarized wave. Sending a circularly polarized wave through a linear polarizer will transmit one of the two linearly polarized components, while sending a linearly polarized wave through a circular polarizer selects one of the circular polarization components, (+) or (−), that make up the linearly polarized wave. In this way we can switch back and forth between linear and circular polarization.

2.3.3 Photons

Everything we have done so far is just addition and decomposition of vectors applied to a classical field. But what about the photons? Since we need to think of light as a flow of individual photons, how can we describe these properties of the polarization?

We start with a linearly polarized wave incident on a polarizer with its axis at 45° to the direction of polarization of the beam. We know that a fraction $\cos^2(\pi/4) = 1/2$ of the incident light intensity will pass through the film. If we study the transmitted light, we will find that it consists of photons of the same energy ($E = h\nu$) as the incident light, but only half of the photons are transmitted. The photons are not split. We know from the classical discussion that the transmitted photons will be polarized parallel to the axis of the polarizer, which is rotated by 45° with respect to the original polarization direction. Thus, we are forced to say that there is a probability of 1/2 that a photon will be transmitted by the polarizer. Furthermore, since we know that the polarization of the classical wave can be decomposed into polarization components and superposed to create new types of polarization, the photons must have some property that can be superposed and decomposed in the same way. We call this property the *polarization state* of the photon.

We will designate the state of a photon linearly polarized along the x direction by $|x\rangle$. Then the state of a photon linearly polarized at an angle θ to the x-axis must be written as

$$\cos\theta\,|x\rangle + \sin\theta\,|y\rangle \tag{2.23}$$

if large numbers of single photons are to reproduce the known classical behavior. The fraction of photons passing through a polarizer oriented along the $\hat{x}$ ($\hat{y}$) direction must be $\cos^2\theta$ ($\sin^2\theta$) to agree with the classical result. Each photon that passes through a polarizer oriented along the $\hat{x}$ ($\hat{y}$) direction will be found to be polarized in that direction, so we are led to the conclusion that on passing through the polarizer, a photon initially in a superposition state described by Eq. (2.23) makes a sudden jump into one of the orthogonal states $|x\rangle$ or $|y\rangle$. Exactly which state cannot be predicted for any individual photon. We can only talk about the probability of each possible outcome.

Similarly, we can designate the state of a circularly polarized photon as $|+\rangle$ or $|-\rangle$. We must have

$$|x\rangle = \frac{1}{\sqrt{2}}\left[|+\rangle + |-\rangle\right] \tag{2.24}$$

$$|y\rangle = \frac{1}{i\sqrt{2}}\left[|+\rangle - |-\rangle\right] \tag{2.25}$$

for the photon description to agree with the classical behavior. The $i = e^{i\pi/2}$ indicates the $\phi = \pi/2$ phase shift as in Eq. (2.21). For a light beam made of photons to exhibit the same superposition properties as a classical wave, we must allow some property of the photons to be subject to the same superposition rules that apply to the classical wave. We have called that property the polarization state.

Eqs. (2.24) and (2.25) can be rewritten as

$$|+\rangle = \frac{1}{\sqrt{2}}\left[|x\rangle + i\,|y\rangle\right] \tag{2.26}$$

$$|-\rangle = \frac{1}{\sqrt{2}}\left[|x\rangle - i\,|y\rangle\right]. \tag{2.27}$$

According to Eq. (2.26), a positive circularly polarized state $|+\rangle$ is a superposition of states polarized in the x and y directions. When photons in that state are passed through an x-aligned polarizer, half the photons will pass through, and they will then be polarized along x. Thus we have decomposed the $|+\rangle$ state into $|x\rangle$ and $|y\rangle$ states and projected it onto the $|x\rangle$ state. In this case, the $|x\rangle$ and $|y\rangle$ states serve as basis states. The states $|+\rangle$ and $|-\rangle$ can also serve as basis states, with Eqs. (2.26) and (2.27) representing a transformation between the two sets of basis states. The fact that the physics can be expressed in different sets of basis states and is independent of transformations between the basis states is one of the basic, guiding principles of quantum mechanics.

2.4 Entanglement—"spooky action at a distance"

We now turn to one of the strangest properties of the physical world—the fact that a measurement made at one position in space-time can *appear* have an effect on a measurement at another position that is separated from the first by such a large distance that a signal traveling at the speed of light cannot connect the two measurements. That is, the separation in space Δx between the events satisfies $(\Delta x)^2 > c^2 (\Delta t)^2$, where Δt is the separation in time.

As an example, calcium atoms can be excited in such a way that they emit two photons in succession in opposite directions. These photons are always in the same polarization state, either $|+\rangle$ or $|-\rangle$. Designating the two photons as L and R, (see Fig. 2.13) the state of the two-photon system is written as

$$\frac{1}{\sqrt{2}}\left[\left(|+\rangle_L |+\rangle_R\right) + \left(|-\rangle_L |-\rangle_R\right)\right]. \tag{2.28}$$

This kind of state, which cannot be expressed as a product of the states of the component particles, is called an entangled state. The two photons must be in the same state. What if we send one photon through a linear polarizer? To answer this, we expand the states in Eq. (2.28) in terms of the $|x\rangle$ and $|y\rangle$ basis to obtain

$$\begin{aligned}\frac{1}{\sqrt{2}}\left[(+)_L (+)_R + (-)_L (-)_R\right] &= \frac{1}{\sqrt{2}}\left[|x\rangle + i\,|y\rangle\right]_L \frac{1}{\sqrt{2}}\left[|x\rangle + i\,|y\rangle\right]_R \\ &+ \frac{1}{\sqrt{2}}\left[|x\rangle - i\,|y\rangle\right]_L \frac{1}{\sqrt{2}}\left[|x\rangle - i\,|y\rangle\right]_R \\ &= \frac{1}{\sqrt{2}}\left[|x\rangle_L\,|x\rangle_R - |y\rangle_L\,|y\rangle_R\right].\end{aligned} \tag{2.29}$$

If the left photon L passes through the x polarizer, the right photon, R, for that photon pair must also be polarized along x because by making the first measurement

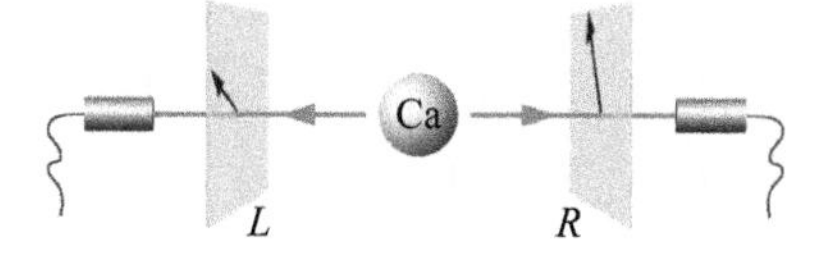

Fig. 2.13 Set up to observe the polarization properties of two photons emitted during the decay of an excited calcium atom. Measurements of the photon count rate as a function of the angle between the L and R polarizer axes show "non-local" quantum behavior.

we projected the state onto the $|x\rangle_L$ state. The R photon is forced into the $|x\rangle_R$ state by the measurement on the L photon. If the R photon is now passed through a linear polarizer with its axis making an angle α with the x-axis, photons will pass through both polarizers with a probability of $(1/2)\cos^2\alpha$. Thus, by choosing to put the L photon through a linear or circular polarizer, we can force the R photon to be either linearly or circularly polarized, even though the two photons are speeding away from each other at the speed of light. Experiments have been done where the choice is made well after the photons have left the atom. In a famous paper, Einstein, Podolsky, and Rosen (EPR) argued that once the L photon has passed through either the linear or circular polarizer, it is certain that the R photon will pass through its similarly set polarizer, so there must be what they called some "element of physical reality" associated with this fact. Since there is no such element (hidden variable) in quantum mechanics, they concluded that the theory must be incomplete. Actually, EPR based their arguments on particle positions. Since the two particles were created in a single reaction, their positions would be correlated, so that measuring the position of one of them would exactly fix the position of the other particle. This is direct evidence of the non-locality of quantum mechanics. Locality means that no action at a given point in space-time can have any influence on points in space-time that are so far away that they cannot be connected by a ray of light (space-like separation). As we will see later, the correlations between the L and R photons predicted by the above formalism have been observed experimentally. John Bell has shown that the photon count rates that have been measured as a function of the angle between the axes of the two polarizers cannot be produced by any local theory (Bell's theorem). In this case, the wave function cannot be interpreted as applying to an ensemble: the correlations arise in individual photon pairs. This is discussed in some detail in Chapter 15.

This non-locality of quantum theory, called "spooky action at a distance" by Einstein, is now realized to be a fact of life having profound philosophical implications. It seems to have been at the heart of Einstein's life-long objections to quantum theory. We now understand these effects as being correlations in the light field that have no classical analogy.

2.5 The Aharonov-Bohm effect and the physical reality of electromagnetic potentials

In classical electromagnetism, the potentials (vector potential $\vec{A}$ and scalar potential ϕ) are introduced as mathematical helpers, while all physical effects are produced by the fields

$$\vec{E} = -\vec{\nabla}\phi - (1/c)\,\partial\vec{A}/\partial t \tag{2.30}$$

$$\vec{B} = \vec{\nabla} \times \vec{A}. \tag{2.31}$$

We know that the total magnetic flux Φ through an area S is given by

$$\Phi = \int \vec{B}\cdot d\vec{S} = \int_S \left(\vec{\nabla}\times\vec{A}\right)\cdot d\vec{S} = \oint \vec{A}\cdot d\vec{l}, \tag{2.32}$$

where the last integral is a line integral taken around the curve that forms the boundary of the area $\vec{S}$. If we have a magnetic field confined to a small ferromagnetic needle, the field outside the needle will be zero but the vector potential $\vec{A}$ cannot be zero according to Eq. (2.32). In fact, the vector potential for this case is, in cylindrical coordinates,

$$A_\phi = \frac{\Phi}{2\pi r} \tag{2.33}$$

$$\left[\vec{\nabla} \times \vec{A}\right]_\phi = \frac{\partial (r A_\phi)}{\partial r} = 0. \tag{2.34}$$

As we saw in Section 2.2.2, de Broglie proposed that massive particles like electrons are associated with a wavelength satisfying

$$k = \frac{2\pi}{\lambda} = \frac{p}{\hbar}, \tag{2.35}$$

where p is the momentum of the particle and $\hbar$ is Planck's constant divided by 2π. We can thus represent a plane wave as $e^{i\frac{\vec{p}}{\hbar}\cdot\vec{r}}$ (or its real part). In classical physics, we can account for the effects of a magnetic field on the motion of a charged particle by replacing the momentum of the particle $\vec{p}$ by (in Gaussian units) $\left(\vec{p} - \frac{e}{c}\vec{A}\right)$, where e is the charge of the particle and c is the velocity of light. In the presence of a magnetic field, our wave will thus be represented by

$$\left(e^{i\frac{\left(\vec{p}-\frac{e}{c}\vec{A}\right)}{\hbar}\cdot\vec{r}}\right) = \left(e^{i\frac{\vec{p}}{\hbar}\cdot\vec{r}}e^{-i\frac{e}{\hbar c}\vec{A}\cdot\vec{r}}\right). \tag{2.36}$$

For a magnetic field varying slowly in space and time, the phase of the second factor accumulates at varying rates as $\vec{r}$ varies, so we can write the second factor as

$$e^{-i\frac{e}{\hbar c}\int \vec{A}\cdot d\vec{r}}. \tag{2.37}$$

In the presence of a vector potential $\vec{A}$, the wave function acquires a phase factor that can result in experimentally observable changes in that particle's behavior, even though $\vec{B} = 0$ throughout the region traversed by the particle. This was first pointed out by Aharonov and Bohm[6] as a case where quantum mechanics predicts a classically anomalous result and requires a new interpretation of the potentials. The effect has since been confirmed experimentally.

We now return to the case of Fig. 2.8, where a ferromagnetic needle was placed between the two slits in an interference experiment. An electron arriving at point P on the screen could have passed through either the top slit ($S1$) or the bottom slit ($S2$). An electron going through the top slit would pick up a phase factor

$$e^{-i\frac{e}{\hbar c}\int_{S1}\vec{A}\cdot d\vec{r}} \tag{2.38}$$

[6]Aharonov, Y., and Bohm, D. *Significance of electromagnetic potentials in the Quantum theory*, Phys. Rev. 113, 485 (1959).

while an electron going through the bottom slit would pick up a phase factor

$$e^{-i\frac{e}{\hbar c}\int_{S2}\vec{A}\cdot d\vec{r}}, \tag{2.39}$$

where the integral is taken along the electron's path. The difference in phase between electrons passing through the two slits will be

$$e^{-i\frac{e}{\hbar c}\left[\int_{S1}\vec{A}\cdot d\vec{r}-\int_{S2}\vec{A}\cdot d\vec{r}\right]} = e^{-i\frac{e}{\hbar c}\oint\vec{A}\cdot d\vec{r}} = e^{-i\frac{e}{\hbar c}\Phi}, \tag{2.40}$$

where Φ is the flux through the area enclosed by the boundary $S1 - S2$. The motion of the electrons has been altered even though they remained in a region where $\vec{B} = 0$ throughout the experiment.

As discussed in Section 2.2.2, we can attempt to use this setup to measure which slit an electron went through. An electron passing the magnetic needle will induce a change of flux in the needle whose direction depends on which side of the needle the electron passes. To determine which slit the electron went through, we must measure the flux in the needle precisely enough to detect the change induced by a single electron. As we will see, a measurement sufficiently precise to determine which slit an electron passed through introduces uncertainties in the flux (and hence the phase difference between electrons passing through the two slits) that are large enough to destroy the interference pattern. The A-B effect ensures that installing the needle to determine which slit a given particle passed through eliminates the interference pattern, confirming the consistency of quantum mechanics.

We now develop a quantitative description of the effect of measuring which slit an electron passed through on the visibility of the interference fringes. Following Furry and Ramsey,[7] we can measure the change in flux within the needle by wrapping an N-turn coil of wire around it. A passing electron will cause a jump in the flux and the changing flux will create a voltage across the coil. If the coil is connected to a capacitor that is part of a voltage measuring device (the gate of a field effect transistor, for example), charge will flow into the device. If we think of the electron moving past the needle as a current flowing in the primary (of 1/2 turn) of a transformer, the maximum input charge impulse will be

$$Q_i = \pm\frac{e}{2N}, \tag{2.41}$$

where the sign depends on which side of the needle the electron passes.

The fluctuation in the magnetic flux, and hence the fluctuation in the charge on the capacitor, can be determined by noting that the coil and capacitor circuit forms a harmonic oscillator with frequency $\omega = 1/\sqrt{LC}$, where L is the inductance of the coil. From the definition of inductance, the flux Φ within the needle is related to the current I in the coil by

$$LI = N\Phi. \tag{2.42}$$

[7]Furry, W.H. and Ramsey, N.F., *Siginificance of potentials in quantum theory*, Phys. Rev. 118, 623 (1960).

The total energy of the system is given by the Hamiltonian, which for a charge Q on the capacitor and current I in the coil is

$$H = \frac{Q^2}{2C} + \frac{LI^2}{2} = \frac{Q^2}{2C} + \frac{(N\Phi)^2}{2L}. \tag{2.43}$$

This is completely analogous to a mechanical harmonic oscillator with position $x \to Q$ and momentum $p \to N\Phi$, when we note that $I = \dot{Q}$. According to the Heisenberg uncertainty principle (Chapter 10), the uncertainties in momentum and position in a mechanical harmonic oscillator must satisfy $\Delta x \Delta p \geq \hbar/2$. Thus we expect fluctuations in the magnetic flux in the needle and the charge in the capacitor to satisfy

$$\Delta Q \Delta(N\Phi) = N \Delta Q \Delta\Phi \geq \frac{\hbar}{2}. \tag{2.44}$$

In cgs units, the magnetic flux requires a factor of $1/c$ to make this relationship dimensionally correct and this factor will be included hereafter.

To detect the sign of the charge induced by the passage of an electron, we need to have $|Q_i| \geq \Delta Q$, implying

$$\frac{e}{2N} > \Delta Q \geq \frac{\hbar c}{2N\Delta\Phi} \tag{2.45}$$

and

$$\Delta\Phi \geq \frac{\hbar c}{e}. \tag{2.46}$$

The phase shift ϕ due to the A-B effect is related to the magnetic flux in the needle by Eq. 2.40:

$$\phi = \frac{e}{\hbar c}\Phi, \tag{2.47}$$

which implies that

$$\Delta\phi = \frac{e}{\hbar c}\Delta\Phi \geq \frac{e}{\hbar c}\frac{\hbar c}{e} = 1. \tag{2.48}$$

This means that adding a coil and a voltage measuring device that would allow us to detect which slit the electron went through results in quantum fluctuations in the magnetic flux that are large enough to wash out the interference pattern. As with light waves, the interference pattern for electron waves can only be observed when waves pass through both slits, so the consistency of the theory requires that any measurement that can determine which slit a particle went through must have some effect on the system that eliminates the interference pattern. The A-B effect provides the means by which such a measurement destroys the interference pattern and is thus necessary to ensure the consistency of quantum mechanics.

2.6 Quantum mechanics and precision measurements

One of the great laments about quantum mechanics is the uncertainty *apparently* built into the theory, due to both the general uncertainty principle, and due to the probabilistic interpretation of the wavefunction. Indeed, popular books on quantum mechanics place so much emphasis on the Heisenberg uncertainty principle that readers

2018 CODATA RECOMMENDED VALUES OF THE FUNDAMENTAL CONSTANTS OF PHYSICS AND CHEMISTRY NIST SP 959 (June 2019)

An extensive constants list is available at physics.nist.gov/constants.

Quantity	Symbol	Numerical value	Unit
*^{133}Cs hyperfine transition frequency	$\Delta\nu_{\text{Cs}}$	9 192 631 770	Hz
*speed of light in vacuum	c	299 792 458	m s^{-1}
*Planck constant	h	$6.626\,070\,15 \times 10^{-34}$	J Hz^{-1}
	$\hbar$	$1.054\,571\,817\ldots \times 10^{-34}$	J s
*elementary charge	e	$1.602\,176\,634 \times 10^{-19}$	C
*Avogadro constant	N_A	$6.022\,140\,76 \times 10^{23}$	mol^{-1}
*Boltzmann constant	k	$1.380\,649 \times 10^{-23}$	J K^{-1}
*luminous efficacy	K_{cd}	683	lm W^{-1}
electron volt (e/C) J	eV	$1.602\,176\,634 \times 10^{-19}$	J
Josephson constant $2e/h$	K_J	$483\,597.848\,4\ldots \times 10^{9}$	Hz V^{-1}
von Klitzing constant $2\pi\hbar/e^2$	R_K	$25\,812.807\,45\ldots$	Ω
molar gas constant $N_A k$	R	$8.314\,462\,618\ldots$	J mol^{-1} K^{-1}
Stefan-Boltzmann const. $\pi^2 k^4/(60\hbar^3 c^2)$	σ	$5.670\,374\,419\ldots \times 10^{-8}$	W m^{-2} K^{-4}

*Defining constants of the International System of Units (SI).

FRONT SIDE

Quantity	Symbol	Numerical value	Unit
(unified) atomic mass unit $\frac{1}{12}m(^{12}\text{C})$	u	$1.660\,539\,066\,60(50) \times 10^{-27}$	kg
Newtonian constant of gravitation	G	$6.674\,30(15) \times 10^{-11}$	m^3 kg^{-1} s^{-2}
fine-structure constant $e^2/(4\pi\epsilon_0\hbar c)$	α	$7.297\,352\,5693(11) \times 10^{-3}$	
inverse fine-structure constant	α^{-1}	$137.035\,999\,084(21)$	
Rydberg frequency $\alpha^2 m_e c^2/(2h)$	cR_∞	$3.289\,841\,960\,2508(64) \times 10^{15}$	Hz
vac. magnetic permeability $4\pi\alpha\hbar/(e^2 c)$	μ_0	$1.256\,637\,062\,12(19) \times 10^{-6}$	N A^{-2}
vac. electric permittivity $1/(\mu_0 c^2)$	ϵ_0	$8.854\,187\,8128(13) \times 10^{-12}$	F m^{-1}
electron mass	m_e	$9.109\,383\,7015(28) \times 10^{-31}$	kg
proton mass	m_p	$1.672\,621\,923\,69(51) \times 10^{-27}$	kg
proton-electron mass ratio	m_p/m_e	$1836.152\,673\,43(11)$	
reduced Compton wavelength $\hbar/(m_e c)$	$\lambda\!\!\!^{-}_C$	$3.861\,592\,6796(12) \times 10^{-13}$	m
Bohr radius $\hbar/(\alpha m_e c)$	a_0	$5.291\,772\,109\,03(80) \times 10^{-11}$	m
Bohr magneton $e\hbar/(2m_e)$	μ_B	$9.274\,010\,0783(28) \times 10^{-24}$	J T^{-1}

The number in parentheses is the one-sigma (1σ) uncertainty in the last two digits of the given value.

CODATA

NIST National Institute of Standards and Technology U.S. Department of Commerce

BACK SIDE

Fig. 2.14 The CODATA fundamental constants Pocket Card; constants on the "Front Side" are now defined (marked with *), or derived from the defined ones (no uncertainty reported, or marked with ...), and thus have no reported uncertainty. Print out your own and become a Card-Carrying Quantum Mechanic. (Missing are those constants mostly associated with high energy and nuclear physics.)(US Gov. Pub., not subject to copyright)

are often left with the impression that quantum phenomena only lead to inaccuracies in physical measurements, and that modern physics is some sort of a sham. However, quantum mechanics brings unprecedented accuracy compared to what is possible with classical systems. Consider the fundamental constants of physics and chemistry, as shown in Fig. 2.14.

The determination of these constants employed quantum mechanical principles, except for one: the value of the Gravitational constant. Its accuracy is thousands of times worse than all other constants; and some constants have no uncertainty at all because they are now defined. These defined and therefored fixed quantities underpin the SI system—as of 2019, this list of defined quantities has expanded and now comprises e, $\hbar$, c, the Cesium ground state hyperfine frequency, Boltzmann's constant k or k_B, etc. The value of G is the only fundamental constant that is not based on some type of quantum system, and is determined by use of a classical measurement apparatus. To date, nobody has figured out how to turn the determination of G into a quantum measurement, which likely would require a quantum theory of gravity, at a minimum. It is interesting to compare the value of G as known in 1929,[8] $G = (6.664 \pm 0.002) \times 10^{-8}$dyne cm^{2}g^{-2}, which has accuracy of 3 parts in 10^4 compared to 2 parts in 10^5 for the currently accepted value, after nearly a century of experimental work. The problems plaguing measurements of G are lack of convergence to a single value, because every measurement is different and requires exact and direct measurements of masses, distances, and forces, in addition to different systematic effects.

The accuracy of measurements of almost all fundamental physical constants that are listed in Fig. 2.14 have improved by factors of about 100,000 during the past nearly 100 years, the exception being G. Although some of the improvements are due to technological and computational advances, most are due to new techniques based on quantum-based measurements. Even though the Heisenberg uncertainty principle often limits measurement accuracy, in most cases the other features of quantum mechanics makes possible vastly improved measurement accuracies. This peculiar situation has gone largely unnoticed and uncommented; Norman Ramsey provides an incisive analysis of the situation, and of the source of precision in quantum mechanics.[9]

Although the principle does imply limitations in a fundamental sense, the uncertainties are often small enough to be unimportant, while other quantum features allow measurements of far greater accuracy than would be otherwise possible. In fact, the momentum recoil of an atom on absorbing a photon , often employed of as a "cause" of momentum-position uncertainty, provides one of the most accurate means of determining h/m. We further note that the time-energy uncertainty relationship is encountered as a matter of course in classical physics as the spread in frequency of the Fourier transform of an oscillatory system based on measurements over a finite time. The degree of accuracy with which a current can be measured is due to "shot noise"generated by the quantization of charge, which is also a fundamental noise source in electronic measurements.

Listed and numbered by Ramsey, op. cit. p. 41, the Seven Features of Quantum Mechanics that have a profound influences on the science of measurement are:

- The existence of discrete quantum states of energy.
- When a system makes a transition between two states, energy is conserved (energy conservation in general).
- Electromagnetic radiation of frequency ν is quantized with energy $h\nu$ per quantum where h is the Planck constant.
- The identity principle.
- The Heisenberg uncertainty principle.
- Addition of probability amplitudes instead of probabilities. The net probability for a superposition of n eigenstates is given by $P = |\sum_{i=1}^{n} a_i\psi_i|^2 \neq \sum_{i=1}^{n} |a_i\psi_i|^2$.
- Wave and coherent phase phenomena.

[8] Raymond T. Birge, *Probable Values of the General Physical Constants*, Rev. Mod. Phys. **1**, 1 (1929). This was the first paper published in this journal, appearing on p. 1.

[9] Norman F. Ramsey, *Quantum Mechanics and Precision Measurements*, Physica Scripta. **T59**, 26–28, 1995.

From the first three of the above features, the emitted photon in a transition from an initial state of energy E_i a final state of energy E_f has the energy hv given by $hv = E_f - E_i$. The existence of discrete frequencies allows the precise measurements required for accurate frequency standards that simply would not be possible if the allowed frequencies had the continuous distribution that characterizes a classical system. Even carefully designed quartz crystal oscillators drift at levels of one part in 10^{10} per day, and can only be fabricated to a part in 10^6 accuracy; of course, such oscillators are useful as a sort of frequency flywheel, but need to be referenced to a fundamentally defined frequency to be generally useful for precision time measurements. The stability of a quartz oscillator is at the limit of the needs for GPS positioning and provides a high-fidelty short time scale time standard, but it must be disciplined against an unchanging fundamental frequency of an atomic transition to be useful in the long term.

The fact that all the atoms of a particular isotope are all exactly identical, and that all elementary particles e, μ, etc. are identical, means that we can define measurements in terms of their properties. The meaning of the identity of two particles in quantum mechanics is far more profound than the normal meaning of the word identical, as we will discuss later. The identity principle has a profound effect on the science of measurement. For example, we know that the frequency of a cesium atom beam standard will be the same for cesium atoms that come from North Korea or from the United States. The identity principle allows for a universality of accurate physical measurements.

Greater precision can be obtained by a better understanding of quantum mechanical limits. As an example, in quantum electrodynamics, the uncertainty principle sets a lower limit to the uncertainty product of the two quadrature phases of an electromagnetic field, $\Delta X \Delta Y$. This does not mean that the uncertainties are necessarily equal. In fact, one of the uncertainties can be greatly reduced if the other is allowed to increase correspondingly, so as to keep the product constant. This technique is used in quantum electronics and optics to form "squeezed photon states," and when applied to mechanical oscillators, serves as the basis of "quantum nondemolition" measurements.

The wave nature of the quantum mechanical state function and the possibilities of coherent effects leads to many different devices for precision measurements, such as interferometers, masers, and lasers. The energy gap that results from electrons forming Cooper pairs in superconductors leads to coherent effects on a macroscopic scale, resulting in zero electrical resistance, and that the magnetic flux enclosed by a superconducting ring must be limited to an integral multiple of the flux quantum $\phi_0 = h/2e = 2.07 \times 10^{-7}$gauss cm^2. Coherence also gives rise to the Josephson effect, due to Cooper pairs tunneling through a thin insulating barrier separating two superconductors. If there is a voltage V across a Josephson junction, the electron pairs will change energy by $2eV$ in crossing the junction and the current through the junction will acquire an oscillatory component with frequency $2eV/h$. Alternatively, if microwaves of frequency ν are coupled to the Josephson junction, abrupt current steps will appear as a function of voltage whenever $n\nu = 2eV/h$ where n is an integer. This relation is apparently exact and permits accurate measurements of voltages in terms of frequencies, with approximately 484 MHz per microvolt, as determined through a combination of fundamental constants.

An outstanding example of a modern precision measurement is the determination of the magnetic moment μ_e of the electron in terms of the Bohr magneton μ_B, by use of a technique originally developed by Dehmelt and his associates.[10] This technique is based on measurements employing a single electron, confined in a Penning trap, and subjected to radiofrequency fields that drive spin flips. Currently, in terms of $g/2 = \mu_e/\mu_b \approx 1$,[11]

$$\frac{g}{2} = \frac{\mu_e}{\mu_B} = 1.00115965218073(028) \tag{2.49}$$

corresponding to an uncertainty of 0.28 parts per trillion, which can be compared to a calculation based on quantum electrodynamics,[12]

$$\frac{\mu_e}{\mu_B} 1.00115965218161(024). \tag{2.50}$$

The agreement is excellent, making QED the most precisely tested theory, and quantum mechanics in general serves as the fundamental basis of the measurement technique and of the calculation.

To quote Ramsey, op. cit., p. 41,

> If we now return to the seven characteristic quantum features we see that only one—the Heisenberg uncertainty principle—provides a limitation to the accuracy of measurement and that this limitation is often so low as not to be serious. The other six characteristics give universality and significance to accurate measurements that are orders of magnitude more accurate than would be possible in a purely classical world. This was vividly brought to my attention by the difficulty I once had in having a slide made which reported some of my measurements with an accuracy of one in 10^{14}. The engineer who was asked to prepare the slide at first refused on the grounds that he had been trained as an engineer and from the point of view of an engineer it was utterly meaningless to express any physical number with such ridiculously high accuracy. As a classical engineer he was quite right. To express the height of the Eiffel Tower to an accuracy of 1 in 10^{14} would be utterly meaningless—its height cannot be defined to that accuracy and it changes more than that from second to second as it blows in the wind and thermally expands. However, in a quantized world such accurate measurements are both meaningful and possible.

2.7 Synopsis

In this chapter we discussed several physical situations where a classical description seems clearly impossible. We showed that particles such as electrons and photons have wave as well as particle properties and must be considered to be in states which can be superposed and decomposed. We saw that certain states of pairs of particles can result in some influence propagating at velocities greater than the speed of light and that the magnetic potential can have observable effects even in regions where the magnetic field is zero. These phenomena represent real aspects of the physical world.

[10] Van Dyck, R. S., Jr., Schwinberg, P. B. and Dehmelt, H.G.., Bull. Am. *Phys. Soc. Electron Magnetic Moment from Geonium Spectra, Atomic Physics 7.* 24, 758 (1979); Atomic Physics 7, 337 and 340 (1981) (Plenum Press).

[11] Hanneke, D.; Fogwell, S.; Gabrielse, G. *New Measurement of the Electron Magnetic Moment and the Fine Structure Constant.* Phys. Rev. Lett. 2008, 100, 120801.

[12] Aoyama, T.; Kinoshita, T.; Nio, M. *Theory of the Anomalous Magnetic Moment of the Electron.* Atoms 2019, 7, 28.

They are not, as some tend to say, the results of quantum mechanics, but features of the world that quantum mechanics must explain. Historically, some of these effects were discovered after the formulation of a complete theory of quantum mechanics, in a series of attempts to challenge the theory by focusing on its most extreme predictions. As we have said, the quantum mechanical predictions have been confirmed in every experiment to date.

In the next chapter we turn to the development of quantum mechanics. This development was a learning process for humanity carried out by a small number of extraordinary individuals based on a massive amount of experimental work. Some knowledge of this development is essential for a proper understanding of the theory.

3 The origin of quantum theory in the crisis of classical physics

As the end of the nineteenth century approached, it was clear that classical mechanics and Maxwell's electromagnetic theory successfully described most observed phenomena, but there were several observations that could not be explained with classical ideas. While some scientists continued to believe that physics was completely understood and these observations would soon be fit into the prevailing classical worldview, this opinion became increasingly less tenable as the non-classical character of these phenomena became clearer. What is particularly striking, from today's perspective, is the vast amount of new phenomena, incompatible with classical ideas and unknown at the time, which have since been discovered as technology advanced and the new quantum theory was developed and elaborated.

From the point of view of the subsequent development of quantum mechanics, the three most important anomalies were:

1. Black body radiation. The intensity of radiation emitted by a body in thermal equilibrium was observed to go to zero at high frequencies while classical physics predicted that it would go to infinity, clearly an impossible result.
2. The photoelectric effect. Exposure of metals to light or radiation resulted in the emission of electrons whose energy was independent of the intensity of the light.
3. Spectral lines. The emission of light from atoms was found to occur at specific discrete wavelengths only.

3.1 Black body radiation

3.1.1 Progress before Planck

During the 1890s, the problem of black body radiation attracted significant attention. Black body radiation is the thermal radiation emitted by a body that is a perfect absorber. Using thermodynamic arguments, Kirchoff had previously shown that electromagnetic radiation at thermal equilibrium at a given temperature must have the same spectrum regardless of the nature of any material it was interacting with. For radiation confined to a closed cavity, the spectrum would be the same regardless of the material properties of the walls or any material in the cavity. Kirchoff showed that this property of equilibrium thermal radiation was a necessary consequence of the second law of thermodynamics, because if the spectrum depended on the nature

The Historical and Physical Foundations of Quantum Mechanics. Robert Golub and Steven K. Lamoreaux, Oxford University Press.
 DOI: 10.1093/oso/9780198822189.003.0003

of the material in contact with the radiation, thermodynamic equilibrium could not exist within a cavity containing different substances. Due to its universal nature, the problem of black body radiation became the focus of much theoretical and experimental research. Stefan proved, again using thermodynamic arguments, that the total energy $E_{\rm tot}$ contained in the radiation varied with the temperature T of the radiation as T^4. Wien then showed that the energy spectrum had to have the form

$$\rho(\omega, T) = \alpha\omega^3 F\left(\frac{\omega}{T}\right), \tag{3.1}$$

where α is a constant, F is an unknown function, and $\rho(\omega, T)\, d\omega$ is the energy density of the radiation in the angular frequency interval between ω and $\omega + d\omega$. The proof of this relation was based on treating the radiation as a gas in a cylinder that is compressed by a moving mirror acting as a piston.[1] Eq. (3.1) leads directly to Stefan's law since

$$E_{tot} = \alpha \int_0^\infty \omega^3 F\left(\frac{\omega}{T}\right) d\omega = \alpha T^4 \int_0^\infty x^3 F(x)\, dx, \tag{3.2}$$

where we have substituted $x = \omega/T$ and the definite integral is a constant whose value depends on the form of $F(x)$. Wien further proposed that $F\left(\frac{\omega}{T}\right)$ should be given by $e^{-\xi\omega/T}$ so that

$$\rho(\omega, T) = \alpha\omega^3 e^{-\frac{\xi\omega}{T}}, \tag{3.3}$$

where α and ξ are experimentally determined proportionality constants. His argument, which assumed that the radiation was in equilibrium with a gas and the interaction depended on the velocity of the gas, was rejected by other physicists, but Eq. (3.3) agreed very well with the existing measurements (which were all in the high frequency region, $\frac{\xi\omega}{T} \gg 1$) and so this formula, called Wien's law, was accepted as empirically correct.

3.1.2 Planck and Wien's law

Max Planck, who was interested in the origin of irreversibility and wanted to show that entropy increase was inevitable according to classical physics, was attracted by the universal nature of the black body spectrum. Since Kirchoff had shown that the nature of the material system in equilibrium with the radiation had no effect on the radiation spectrum, Planck decided to study a set of linear harmonic oscillators in equilibrium with the radiation. The harmonic oscillator is a very simple and completely solvable system. We consider an oscillator with frequency ω that consists of a single particle of mass m and charge e bound by a spring to a fixed center. According to classical electromagnetic theory, an oscillator with a time averaged energy $\bar{\varepsilon}$ would radiate energy at the rate

$$\frac{d\varepsilon}{dt} = -\frac{2e^2}{3mc^3}\omega^2\bar{\varepsilon}. \tag{3.4}$$

[1] Born, M., *Atomic Physics*, 2nd edition, Blackie and Son, London & Glasgow (1937).

On the other hand, an oscillator exposed to a radiation field of energy density ρ_ω will absorb energy at the rate

$$\frac{d\varepsilon}{dt} = \frac{\pi e^2}{3m} 2\pi \rho_\omega. \tag{3.5}$$

In equilibrium the amount of energy absorbed must be equal to the amount of energy radiated, which gives

$$\rho_\omega = \frac{\omega^2 \bar{\varepsilon}}{\pi^2 c^3}. \tag{3.6}$$

The problem of calculating the black body spectrum is then reduced to finding the average energy of an oscillator in thermal equilibrium. Although the work is always presented in a logical forward progression in his published papers, Planck explained in several later writings that he started his treatment of the problem by working backward from Wien's law. What he probably did was to combine Eqs. (3.3) and (3.6) to yield

$$\bar{\varepsilon} = \frac{\pi}{2} c^3 \alpha \omega e^{-\frac{\xi \omega}{T}}, \tag{3.7}$$

rewrite this as

$$\frac{1}{\xi \omega} \ln\left(\frac{2\bar{\varepsilon}}{\pi c^3 \alpha \omega}\right) = -\left(\frac{1}{T}\right), \tag{3.8}$$

and then apply the thermodynamic relation (dropping the bar on $\bar{\varepsilon}$)

$$\begin{aligned} \left(\frac{\partial S}{\partial \varepsilon}\right)_V &= \frac{1}{T} \\ &= -\frac{1}{\xi \omega} \ln\left(\frac{2\varepsilon}{\pi c^3 \alpha \omega}\right). \end{aligned} \tag{3.9}$$

Planck later stated that his key insight was to consider the entropy as a function of ω and T rather than as a function of the energy as other physicists had done. Integrating Eq. (3.9), we find

$$\begin{aligned} S &= -\frac{1}{\xi \omega} \int \ln\left(\frac{2\varepsilon}{\pi c^3 \alpha \omega}\right) d\varepsilon \\ &= -\frac{\varepsilon}{\xi \omega}\left[\ln\left(\frac{2\varepsilon}{\pi c^3 \alpha \omega}\right) - 1\right] = -\frac{\varepsilon}{\xi \omega}\left[\ln\left(\frac{\varepsilon}{b\omega}\right) - 1\right], \end{aligned} \tag{3.10}$$

where $b = \pi^2 c^3 \alpha / 2$. In his published paper[2], Planck defined this as the entropy of an oscillator. Running the argument backward, he then obtained Wien's law for the spectrum of the radiation.

Planck's attention was then drawn to the expression

$$\left(\frac{\partial^2 S}{\partial \varepsilon^2}\right)_V = -\frac{1}{\xi \omega \bar{\varepsilon}}, \tag{3.11}$$

which follows directly from Eq. (3.9). He argued (falsely, as would soon become evident) that this was the only function consistent with the second law of thermodynamics and classical electromagnetic theory so that a violation of Wien's law would amount to a violation of the second law.

[2]Planck, M., *On irreversible radiation processes*, Annalesn der Physik, 1(306), 69, (1900).

3.1.3 The failure of Wien's law and Planck's expression for the black body spectrum

Shortly after Planck published his result, new experimental measurements showed that Wien's law did not hold at low frequencies, for which the intensity was found to be proportional to the temperature T. Planck was informed of these results in advance and thus had time to re-examine the situation so that he was able to give a response at the same meeting where Kurlbaum presented his and Rubens' experimental data[3] for low frequencies.

Planck quickly realized that the connection between Eq. (3.11) and the second law was incorrect. Any negative function of ε would be in agreement with the second law.[4] Then, taking $\varepsilon = CT$ as suggested by the experimental results, he put

$$\left(\frac{\partial S}{\partial \varepsilon}\right)_V = \frac{1}{T} = \frac{C}{\varepsilon}, \tag{3.12}$$

which implies

$$\left(\frac{\partial^2 S}{\partial \varepsilon^2}\right)_V = -\frac{C}{\varepsilon^2}. \tag{3.13}$$

He then saw that Eqs. (3.11) and (3.13) could be considered limiting cases. Eq. (3.11) was valid at high frequencies and small ε, and Eq. (3.13) applied to the case of low frequencies and large ε. If we consider the reciprocals of the rhs of Eqs. (3.11) and (3.13), the obvious way to interpolate between these two limits is to set the reciprocal equal to $\varepsilon\,(\xi\omega + \varepsilon)$, which has the desired limiting values. Then

$$\left(\frac{\partial^2 S}{\partial \varepsilon^2}\right)_V = -\frac{\alpha}{\varepsilon\,(\beta + \varepsilon)}, \tag{3.14}$$

where we have defined $\beta = \alpha\xi\omega$ in order to reproduce Eq. (3.11) for small ε. Integrating (3.14)

$$\left(\frac{\partial S}{\partial \varepsilon}\right)_V = \frac{1}{T} = \frac{\alpha}{\beta}\ln\left(\frac{\beta + \varepsilon}{\varepsilon}\right), \tag{3.15}$$

and solving for ε, we find

$$\bar{\varepsilon} = \frac{\beta}{e^{\beta/\alpha T} - 1}. \tag{3.16}$$

This was Planck's result for the average energy of an oscillator in thermal equilibrium. Using the relation in Eq. (3.6), we find

$$\rho_\omega = \frac{\omega^2}{\pi^2 c^3}\frac{\beta}{e^{\beta/\alpha T} - 1}, \tag{3.17}$$

[3]Rubens, H. and Kurlbaum, F., *Über die Emission langwelliger Wämstrahlen durch den schwarzen Körper bei vershiedenen Temperaturen*, Sitzungberichte der Könglich Preuss. Akad. Wiss. zu Berlin (1900), p. 929.

[4]The easiest way to see that $\left(\frac{\partial^2 S}{\partial \varepsilon^2}\right)_V < 0$ is to recall that S is maximum in equilibrium, so that if ε_0 is the equilibrium energy, for any energy change $\Delta\varepsilon$, $S(\varepsilon_0 + \Delta\varepsilon) < S(\varepsilon_0)$ and $S(\varepsilon_0 - \Delta\varepsilon) < S(\varepsilon_0)$.

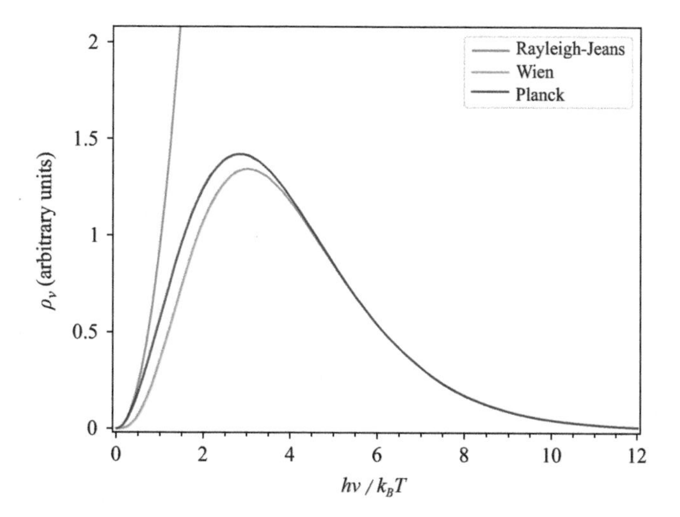

Fig. 3.1 Black body frequency spectrum as a function of the dimensionless variable $h\nu/kT$.

which reduces to Wien's law at high frequencies ($\frac{\xi\omega}{T} \gg 1$). The proportionality constants α and ξ can be subsumed into one constant that we will, for reasons that will become clear later (3.21), call $\hbar$ so that $\beta = \hbar\omega$. This gives the Planck distribution

$$\rho_\omega = \frac{\hbar}{\pi^2 c^3} \frac{\omega^3}{e^{\frac{\hbar\omega}{\alpha T}} - 1}, \tag{3.18}$$

which contains two constants, $\hbar$ and α, to be determined from experiment. Fig. 3.1 shows how the Planck distribution interpolates between the Wien law, which agrees with experimental measurements at high frequencies, and the classically derived Rayleigh-Jeans law, which fits the data at low frequencies, in a way that agrees with the measurements at all frequencies.

3.1.4 An "act of desperation"—the introduction of the quantum

Having arrived at the form of the spectrum given by Eq. (3.18), Planck felt great pressure to find some physical explanation of this result. He was prepared to give up any aspect of classical physics, if necessary, except the first and second laws of thermodynamics. Actually, he used the expression"act of desperation—" to describe the introduction of Eq. (3.14) and did not seem to think the introduction of the energy quantum to be such a radical step.

Integrating Eq. (3.15), we find the entropy of an oscillator to be

$$\begin{aligned} S &= -\frac{\alpha}{\beta}\left[\beta \ln \beta - \varepsilon \ln\left(\beta + \varepsilon\right) + \varepsilon \ln \varepsilon - \beta \ln\left(\beta + \varepsilon\right)\right] \\ &= \alpha\left[\left(1 + \frac{\varepsilon}{\beta}\right) \ln\left(1 + \frac{\varepsilon}{\beta}\right) - \frac{\varepsilon}{\beta} \ln \frac{\varepsilon}{\beta}\right]. \end{aligned} \tag{3.19}$$

This result played an important role in Planck's attempt to find a physical basis for his interpolation.

Eq. (3.19) represents the entropy of a single oscillator. The total entropy for N oscillators is

$$S_N = NS = \alpha\left[\left(N + \frac{U_N}{\beta}\right)\ln\left(N + \frac{U_N}{\beta}\right) - N\ln N - \frac{U_N}{\beta}\ln\frac{U_N}{\beta}\right], \tag{3.20}$$

where we have written the total energy of all N oscillators as $U_N = N\varepsilon$. If U_N/β were an integer, this expression would look like Stirling's formula for large n: $\ln(n!) \sim n\ln n$. Since

$$U_N/\beta = N\varepsilon/\hbar\omega, \tag{3.21}$$

this quantity will be an integer only if the energy ε is always an integer multiple of the energy quantum $\hbar\omega$. If the oscillator energy is quantized in units of $\hbar\omega$, we can write

$$S = \alpha\ln\left[\frac{\left(N + \frac{U_N}{\beta}\right)!}{N!\left(\frac{U_N}{\beta}\right)!}\right] = \alpha\ln\left[\frac{(N+P)!}{N!P!}\right], \tag{3.22}$$

where $P = U_N/\beta$ is the total number of what Planck called energy elements, which were later called quanta or photons, present in the system.

Planck knew that Boltzmann had calculated the entropy of a gas from the expression

$$S = k\ln W, \tag{3.23}$$

where W represents the probability of a given macrostate of the gas, that is, a state with given values of macroscopic variables such as pressure and temperature. W was taken to be proportional to the number of microstates (i.e., states where the coordinates and velocities of the individual molecules are specified) that are compatible with a given macrostate. Actually, Boltzmann never used Eq. (3.23); he always expressed the relation as a proportionality. Planck was the first to introduce and determine the value of the universal constant k that we now call Boltzmann's constant.

In order to calculate W Planck needed to find how many ways we can distribute P packets of energy among N oscillators. As we can see from Fig. 3.2, where the $\times$'s represent energy packets and the spaces between the vertical lines represent the individual oscillators, the problem is reduced to finding how many ways we can arrange the $N-1$ walls and P $+$'s, taking into account that all the elements and walls are identical. Double vertical lines indicate that the oscillator between the lines has $P = 0$ energy quanta. If we have R objects, they can be arranged in $R!$ ways. If L of these objects are identical, they can be arranged in $L!$ ways and the number of different arrangements is $R!/L!$ If in addition another M objects are identical, the number of different arrangements is given by $R!/(L!M!)$. We have a total of $N+P-1$ objects of which $N-1$ are identical to each other and P other objects are also identical. The number of ways these can be arranged is

$$W = \frac{(N+P-1)!}{P!\,(N-1)!}. \tag{3.24}$$

× | × × × || × × × | × × | × × × × × | × | × × × × || ×

Fig. 3.2 Distribution of energy quanta among a group of oscillators. The ×'s represent energy quanta and the vertical lines separate individual oscillators. Double vertical lines indicate that the oscillator between the lines has zero energy quanta.

Using Boltzmann's relation with Eq. (3.24), we find for the entropy

$$S_N = k \ln \left[\frac{(N+P-1)!}{P!\,(N-1)!} \right], \tag{3.25}$$

which in the case of $N, P \gg 1$ reduces to Eq. (3.22) with $\alpha = k$. We can now run the argument backward, differentiating Eq. (3.19) with respect to ε to obtain Eq. (3.15), which leads to the form of ρ_ω given in Eq. (3.18) with $\alpha = k$. Expressed in terms of the frequency $\nu = \omega/2\pi$, the energy density of radiation per unit frequency is

$$\rho_\nu = \frac{8\pi h}{c^3} \frac{\nu^3}{e^{\frac{h\nu}{kT}} - 1}, \tag{3.26}$$

where $h = 2\pi\hbar$ is called Planck's constant.

By comparing this expression with the experimental data, Planck could determine values for h and k. From Boltzmann's calculations of the entropy of an ideal gas, it was known that $k = R/N_A$, where R is the gas constant from the ideal gas equation of state written in terms of moles and N_A is Avogadro's number. Since R was well known by that time, Planck could determine an accurate value of N_A and from this the charge of an electron. In fact, the value of the electron charge calculated in this way turned out to be more accurate than any measured values available at the time.

It is interesting to note that in 1877 Boltzmann also assumed that the energy of a given molecule could only take on discrete values in order to be able to count microstates. If the energy is continuous, the number of different microstates is infinite. Boltzmann assumed that the possible values of a molecule's energy were integer multiples of a fixed amount of energy ζ and later took the limit $\zeta \to 0$ to find that the probability of a state with (continuous) energy E is proportional to $e^{-E/kT}$. He introduced this as a mathematical trick similar to the way an integral is defined as a sum over strips in the limit that the width of the strips goes to zero. Eq. (3.24) appears in Boltzmann's paper.[5] Planck, of course, knew of this work but always felt that it was not quite legitimate. He embraced it, however, when it offered a solution to his problem.

Planck's approach has been criticized many times. His application of the combinatorial arguments was quite different from Boltzmann's, and seemed strange to many of his colleagues. He did not take the limit ($\hbar\omega \to 0$) at the end of the calculation. In deriving Eq. (3.22) he stumbled on the statistics of identical particles, which would only be elucidated much later by Bose and Einstein.

[5] Boltzmann, L., *On the relation between the second law of the mechanical theory of heat and probability calculations with respect to the laws of thermal equilibrium*, Sitz. Ber.Akad.Wiss. (Wien), 76, 373, (1877), reprinted in: Hasenoehrl, F. ed., Boltzmann, L., Wissenschaftliche Abhandlungen, Vol. II paper 42, 164, New York: Chelsea, (1969).

3.1.5 Lord Rayleigh derives the Rayleigh-Jeans law

A few months before Planck presented his law for the radiation spectrum, Lord Rayleigh (John W. Strutt) published a discussion of the black body problem.[6] Instead of considering matter interacting with the radiation field, he concentrated on the radiation field alone. For the radiation field existing in a cubic box of side L with perfectly reflecting walls, the electric field must be zero at the boundary. The solution of the wave equation satisfying the boundary condition is

$$\cos\omega t \sin\left(\frac{n_x\pi}{L}x\right)\sin\left(\frac{n_y\pi}{L}y\right)\sin\left(\frac{n_z\pi}{L}z\right), \tag{3.27}$$

where $n_{x,y,z}$ are integers. For this to be a solution of the wave equation, we must have

$$\omega = c\frac{\pi}{L}\sqrt{n_x^2+n_y^2+n_z^2}. \tag{3.28}$$

We can think of plotting the integers $n_{x,y,z}$ on a three-dimensional lattice. Since there will be one point per unit volume in that space, the number of modes in the frequency interval between ω and $\omega+\,d\omega$ is related to the number of points in the spherical shell between a radius of n_o and n_o+dn, where

$$n_o = \frac{L}{c\pi}\omega = \sqrt{n_{o,x}^2+n_{o,y}^2+n_{o,z}^2} \tag{3.29}$$

$$n_o+dn = \frac{L}{c\pi}\left(\omega+d\omega\right). \tag{3.30}$$

The number of modes is then given by

$$\begin{aligned} 2\frac{4\pi}{8}n_o^2dn &= 2\frac{4\pi}{8}\left(\frac{L}{c\pi}\omega\right)^2\left(\frac{L}{c\pi}d\omega\right) \\ &= L^3\frac{1}{\pi^2c^3}\omega^2d\omega. \end{aligned} \tag{3.31}$$

To obtain the actual number of modes we had to divide the volume of the spherical shell ($4\pi n_o^2dn$) by eight to account for the fact that the integers $n_{x,y,z}$ in Eq. (3.28) take on only positive values (Lord Rayleigh had missed this factor and was corrected by James Jeans) and multiply by two because there are two independent polarizations for every spatial state.

If the average energy of a mode at frequency ω is $\bar{\varepsilon}_\omega$, the energy per unit volume between ω and $\omega+d\omega$ is

$$\rho_\omega d\omega = \frac{\omega^2\bar{\varepsilon}_\omega}{\pi^2c^3}d\omega. \tag{3.32}$$

This is identical to Planck's result in Eq. (3.6) if the average energy of a mode replaces the average energy of Planck's assumed material oscillators. This method of deriving Planck's result was proposed by Debye[7] in 1910.

[6]Lord Rayleigh, *Remarks upon the Law of Complete Radiation*, Phil. Mag. XLIX, 539, (1900).
[7]Debye, P., *Probability in the theory of radiation*, Ann. der Physik 33, 1427, (1910)

According to classical statistical mechanics (equipartition principle), each degree of freedom has an average energy $\bar{\varepsilon}_\omega = kT/2$ and an oscillator has two degrees of freedom (kinetic and potential energy). Inserting this into the expression for ρ_ω, we find the classical result for the black body spectrum:

$$\rho_\omega d\omega = \frac{kT}{\pi^2 c^3}\omega^2 d\omega. \tag{3.33}$$

This result, now known as the Rayleigh-Jeans law, is obviously unsatisfactory. When integrated over all frequencies, it implies that the radiation field contains an infinite energy. Rayleigh took this as an indication that the equipartition principle was not universal and for some unknown reason only applied to the lower frequency modes.

Note what Rayleigh has done. He has calculated the black body spectrum by considering the radiation field on its own, without the need for introducing oscillators interacting with the radiation. In this case, the important quantity is the average energy per mode of the radiation field as opposed to the average energy of the interacting oscillators, which suggests that quantization should be applied not only to the oscillator energies but also to the modes of the radiation field itself.

It is unclear when Planck learned about Rayleigh's work.[8] It is hard to believe that he did not know about what Rayleigh had done when he presented his first, heuristic derivation of his radiation law in October 1900, since the experimenters Kurlbaum and Rubens referred to Rayleigh's result in the work which led Planck to abandon the Wien law. One consequence of the Rayleigh-Jeans law is that the radiation field can absorb an infinite amount of energy, so that in equilibrium between matter and radiation, all the energy would be drained off into the radiation. In a letter to R.W. Wood in 1931, Planck wrote that he had known that classical physics led to the complete loss of energy to the radiation field and hence was incapable of describing the situation. He stated that applying Boltzmann's ideas showed that the transfer of energy to the radiation field could be hindered by the assumption, which he called "purely formal", that the energy is constrained to remain in individual quanta.

After publication, Planck's results were more or less ignored for the following few years. In 1905 Rayleigh wrote that while Planck's expression for ρ_ν "seems to agree very well with the experimental facts," he had not "succeeded in following Planck's reasoning." He could not "understand how another process also based on Boltzmann's ideas can lead to a different result" than the Rayleigh-Jeans law. Einstein stated that Planck's derivation was inconsistent in that it treated the oscillators as classical in establishing the relation between the oscillator energy and the energy density of the radiation field and that the expression that Planck used for W was not really a probability as required by Boltzman's formulation (3.23).

Nevertheless, the introduction of energy quantization by Planck was the breakthrough that eventually led to our present form of quantum mechanics, but the path was not easy and the process took two decades to complete. The year 1905 was also when Einstein, while criticizing Planck's derivation, realized the significance of Planck's idea and showed that several phenomena could be explained by assuming

[8]See the detailed discussion given by Klein, M.J., *Max Planck and the Beginnings of Quantum Theory. Archives for the History of the Exact Sciences*, 1, **5**, 459, (1961).

that the energy of the radiation field itself was quantized. Shortly after this, he gave an alternate derivation of Planck's radiation law.

In a pattern that was often repeated in the development of quantum mechanics, Planck had great difficulty in accepting Einstein's broadening of the field of application of the quantum idea. In 1912 he wrote, critically, that others had made "the assumption of a still more radical nature ... that any radiant energy whatever, even though it travel freely in a vacuum, consists of individual quanta ... " and that this "overstepped" the boundaries of the original proposal. By the time of his Nobel Prize in 1918, however, he was expressing his pride at the many successful applications of the quantum hypothesis. We now turn to this "assumption of a still more radical nature."

3.2 Einstein further develops the quantum idea

3.2.1 Quantization of the radiation field

In his "miracle year" of 1905, Einstein published three papers that changed the course of physics. In fact, in that year he published four papers and 23 reviews of scientific works and submitted his PhD thesis. We turn our attention to his paper on the nature of electromagnetic radiation. He starts with the assumption that, "when a light ray is spreading from a point, the energy ... consists of a finite number of energy quanta that are localized in space, move without dividing and can be absorbed or generated only as a whole."

He began by applying Planck's method to Wien's law in Eq. (3.3) rather than to the oscillator energy extracted from Wien's law. We rewrite Eq. (3.3) in the form

$$\frac{1}{\xi\omega}\ln\frac{\alpha\omega^3}{\rho(\omega,T)} = \frac{1}{T} \tag{3.34}$$

and, defining $E_\omega = \rho(\omega,T)\,V d\omega$ as the energy of radiation between the frequencies ω and $\omega + d\omega$ contained in volume V, we find

$$\frac{1}{\xi\omega}\ln\left(\frac{\alpha\omega^3}{E_\omega}V d\omega\right) = \frac{1}{T} = \frac{dS}{dE_\omega}, \tag{3.35}$$

where S is the entropy of the monochromatic radiation between ω and $\omega + d\omega$. Integrating yields

$$S = \frac{E_\omega}{\xi\omega}\left(\ln\alpha\frac{\omega^3}{E_\omega}V d\omega + 1\right). \tag{3.36}$$

If we consider two systems with volumes V and V_o we find that

$$S - S_o = \frac{E_\omega}{\xi\omega}\ln\frac{V}{V_o} = k\ln\left(\frac{V}{V_o}\right)^{\frac{E_\omega}{k\xi\omega}}. \tag{3.37}$$

If a gas of N point atoms occupies a volume V_o, the probability of finding an atom in the volume V is given by (V/V_o) and the probability of finding all N atoms in the volume V is $(V/V_o)^N$. Boltzmann's formula, Eq. (3.23), then tells us that the entropy of the two states of the gas is related by

$$S - S_o = k \ln (V/V_o)^N . \tag{3.38}$$

This remarkable result indicates that the thermodynamic properties of the radiation are the same as those of a gas of

$$\frac{E_\omega}{k\xi\omega} = \frac{E_\omega}{\hbar\omega} = n_\omega \tag{3.39}$$

number of independent particles (energy quanta) if n_ω is an integer. Here we have used $k\xi = \hbar$. Einstein stated that this argument only applies to cases where $\frac{\hbar\omega}{kT} \gg 1$, where the Wien law is valid, and left open the possibility that the radiation might behave differently in the region where the Rayleigh-Jeans law applies.

3.2.2 The photoelectric effect

The photoelectric effect was discovered by Hertz in 1887.[9] He observed that a spark would jump a gap between two electrodes more easily if the electrodes were exposed to the light of another spark, and showed that the effect was due to ultraviolet light. Subsequent experimenters showed that

1. The effect was due to the emission of electrons.
2. The electrons were emitted with significant kinetic energy.
3. The kinetic energy of the emitted electrons was independent of the intensity of the light and the maximum kinetic energy increased as the frequency of the light increased.
4. The number of electrons emitted per second was proportional to the intensity of the light.

The third point is very hard to understand on the basis of the wave theory of light. As an analogy, think of the waves on a beach moving some pebbles around: clearly, the more intense the waves, the more energy will be transferred to the pebbles.

In the same paper discussed in the previous section, Einstein pointed out that these results could be explained by expanding Planck's hypothesis and assuming that the energy in the light ray was contained in packets (called "quanta"), where each quantum of light of frequency ν contained an energy equal to $h\nu$. If a light quantum can only transfer its entire energy to the electron and the electrons lose some energy when they leave the surface of the metal (this "binding energy" is called the work function of the metal Φ), the kinetic energy (KE) of the electrons leaving the metal would be given by

$$KE = h\nu - \Phi. \tag{3.40}$$

This was Einstein's equation for the photoelectric effect. Millikan spent several years trying to disprove this equation as he felt (rightly) that it was an attack on the classical

[9] Hertz, H. *On the influence of ultra-violet light on electrical disharges*, Annalen der Physik. 267 (8): 983 (1887).

(wave) theory of light. His results showed the opposite of what he had hoped: his data were all in agreement with Eq. (3.40) and constituted an unambiguous experimental verification of this equation.

It seemed that the photon picture was necessary to explain certain phenomena associated with the emission and absorption of light while the well-established results of the wave theory of light—all of classical optics—showed that light could also behave as a wave. We have given a preliminary discussion of the relation between the wave and photon conceptions in the previous chapter and will return to it many times in the sequel. Wave particle duality remains a topic of intense discussion to the present day.

3.2.3 A new derivation of the Planck spectrum

According to statistical mechanics, the probability that a system in thermal equilibrium would be found in a state of energy E is proportional to the Boltzmann factor $e^{-E/kT}$. According to classical theory, E can take on a continuum of values but, in a 1907 paper,[10] Einstein applied the quantization hypothesis to energy, assuming that the energy of an oscillator with resonant frequency ω can only take on values that are integer multiples of a certain energy quantum, that is

$$E_n = n\hbar\omega,$$

where n is an integer. Then the probability of an oscillator having the energy $n\hbar\omega$ is proportional to the Boltzmann factor

$$P(n\hbar\omega) = \frac{e^{-\frac{n\hbar\omega}{kT}}}{\sum_{n=0}^{\infty} e^{-\frac{n\hbar\omega}{kT}}}. \tag{3.41}$$

The denominator ensures that the probability of having some energy is normalized to one. Now

$$\sum_{n=0}^{\infty} e^{-\frac{n\hbar\omega}{kT}} = 1 + e^{-\frac{\hbar\omega}{kT}} + e^{-\frac{2\hbar\omega}{kT}} + \cdots = \frac{1}{1 - e^{-\frac{\hbar\omega}{kT}}} \tag{3.42}$$

because the sum is a geometric series. Differentiating both sides of Eq. (3.42) with respect to $\alpha = \frac{\hbar\omega}{kT}$, we find that the average energy $\bar{E}$ is given by

$$\bar{E} = \frac{\sum_n n\hbar\omega e^{-\frac{n\hbar\omega}{kT}}}{\sum_n e^{-\frac{n\hbar\omega}{kT}}} = \frac{\hbar\omega}{e^{\frac{\hbar\omega}{kT}} - 1}, \tag{3.43}$$

which has the limiting values

$$\bar{E} \approx \hbar\omega e^{-\frac{\hbar\omega}{kT}} \qquad \hbar\omega \gg kT \tag{3.44}$$

$$\bar{E} \approx \frac{\hbar\omega}{\left(1 + \frac{\hbar\omega}{kT}\right)} = kT \qquad \hbar\omega \ll kT. \tag{3.45}$$

Using Rayleigh's expression, Eq. (3.32), for the relation between the spectrum and the average energy per mode, Einstein applied Eq. (3.43) to the modes of the radiation field and rederived Planck's law for the spectrum.

[10]Einstein, A. *Justification of my work: the Planck theory of radiation, etc.*, Annalken der Physik, 22, 800, (1907), received March 8, (1907).

He then considered the atoms bound in a solid, which he represented as a set of oscillators all having the same frequency, with each oscillator having an average energy given by Eq. (3.43). In this way, he obtained an expression for the specific heat of a solid that gave much better agreement with experiment than the classical prediction. In the low temperature limit of Eq. (3.44), the specific heat is strongly suppressed by the exponential, while classical theory predicted that it should remain constant.

3.2.4 A derivation of Planck's law based on interactions between atoms and radiation

In 1913, Bohr introduced his model of the atom (Sec. 3.3), which vastly expanded the field of application of the quantum idea. Afterward Einstein developed another derivation of Planck's law, this time emphasizing the processes of emission and absorption of light, which were not clearly understood in the quantum theory of that time.

We now turn to this later (1916) development by Einstein.[11] He addressed the problem of an atom with two energy levels $E_1 < E_2$ interacting with a radiation field. Both the atom and the field are assumed to be in thermal equilibrium. The question is, what conditions must the radiation field and the probabilities of transition between the two atomic states satisfy for both systems to remain in equilibrium. Taking ρ_ν to be the energy density of the radiation field between the frequencies ν and $\nu + d\nu$, Einstein postulated that the probability for the atom to absorb a photon and make a transition from energy E_1 to energy E_2, per second, is given by

$$P_{1\to 2} = B_{12}\rho_\nu \tag{3.46}$$

while the probability to make a transition from E_2 to E_1 is

$$P_{2\to 1} = A + B_{21}\rho_\nu, \tag{3.47}$$

where the term A represents spontaneous emission by the atom in its excited state and the term $B_{21}\rho_\nu$ represents a new effect called induced or stimulated emission. This effect, intuited by Einstein, would later be verified experimentally and forms the basis for the operation of lasers. It follows that the number of atoms making the transition $1 \to 2$ per second is

$$n_{1\to 2} = B_{12}\rho_\nu n_1, \tag{3.48}$$

while the number of atoms per second making the opposite transition is

$$n_{2\to 1} = (A + B_{21}\rho_\nu)\, n_2, \tag{3.49}$$

where $n_{1,2}$ is the number of atoms in states 1 and 2, respectively. In thermal equilibrium these rates must be equal. Since the population of each state in equilibrium is proportional to the Boltzmann factor, Eq. (3.41),

$$\frac{n_2}{n_1} = e^{-(E_2-E_1)/kT} = e^{-h\nu/kT}. \tag{3.50}$$

The relation $h\nu = E_2 - E_1$ follows from Bohr's hypothesis concerning the relation between the energy change of a radiating system and the frequency of the emitted

[11]Einstein, A., *Quantum theory of radiation*, Mitt. der Phys. Gesell. Zurich, 16, 47, (1916).

radiation field. We will discuss the Bohr model of the atom in Section 3.3. Thus, in equilibrium,

$$B_{12}\rho_\nu = (A + B_{21}\rho_\nu)\, n_2/n_1 = (A + B_{21}\rho_\nu)\, e^{-\frac{h\nu}{kT}}. \tag{3.51}$$

Solving for ρ_ν, we find

$$\rho_\nu = \frac{A/B_{21}}{\frac{B_{12}}{B_{21}} e^{\frac{h\nu}{kT}} - 1}. \tag{3.52}$$

In the limit of high temperatures, $T \to \infty$, we expect $\rho_\nu \to \infty$, which tells us that $B_{12} = B_{21} = B$. For low frequencies, this expression should reduce to the Rayleigh-Jeans law, so we must have $A/B = \frac{8\pi h\nu^3}{c^3}$, which is an important relation between spontaneous and induced emission. With these substitutions, Eq. (3.52) becomes identical to the Planck distribution. This derivation demonstrates the intimate relation between the Planck spectrum and the interaction of radiation with matter.

3.2.5 Fluctuations and the quantization of the energy of the radiation field

In addition to his paper on the quantization of electromagnetic radiation, in 1905 Einstein also published a paper on the theory of Brownian motion. Brownian motion is a random motion of a particle (e.g., a pollen grain) suspended in a fluid due to many collisions with the atoms of the fluid. Einstein's theory of Brownian motion and its subsequent experimental verification provided proof of the existence of atoms, which remained controversial at the time. Based on his understanding of Brownian motion, Einstein reasoned that quantization of the radiation field would affect the fluctuations of the radiation field itself and of the motion of a particle suspended in the field.

In his first approach to the problem of fluctuations (1909), Einstein rewrote Boltzmann's law, Eq. (3.23), as

$$W = e^{\frac{S}{k}}. \tag{3.53}$$

He then assumed that the total volume V of the cavity containing the radiation was divided into a large number of cells of volume V_i, where $V = \sum V_i$, so that the total energy of the field between the angular frequencies ω and $\omega + d\omega$ would be $E_{tot} = V\rho_\omega d\omega$. The entropy and energy of each cell would fluctuate in a random manner so, if ΔE_i is the fluctuation of energy in the ith cell, the entropy in that cell would change to:

$$S_i = S_i\left(E_{io}\right) + \left.\frac{\partial S_i}{\partial E}\right|_o \Delta E_i + \frac{1}{2}\left.\frac{\partial^2 S_i}{\partial E^2}\right|_o \left(\Delta E_i\right)^2 + \ldots, \tag{3.54}$$

where E_{io} is the energy content of cell i at equilibrium and $\left.\frac{\partial S_i}{\partial E}\right|_o = 0$ since the entropy is a maximum at equilibrium. Summing over all the cells and putting the result into Eq. (3.53), we have

$$W \propto \exp\left(\frac{1}{2}\sum_i \left.\frac{\partial^2 S_i}{\partial E_i^2}\right|_o \frac{\left(\Delta E_i\right)^2}{k}\right), \tag{3.55}$$

which is the product of terms

$$W = \prod_i W_i \propto \prod_i \exp\left(\frac{1}{2}\frac{\partial^2 S_i}{\partial E_i^2}\bigg|_o \frac{(\Delta E_i)^2}{k}\right), \tag{3.56}$$

each of which represents the probability of a fluctuation ΔE_i in the cell i. Each W_i is seen to be a Gaussian distribution with a mean square value of ΔE_i given by

$$\overline{(\Delta E_i)^2} = -\frac{k}{\left(\frac{\partial^2 S_i}{\partial E_i^2}\right)}. \tag{3.57}$$

This result relates the energy fluctuations to the inverse of $\frac{\partial^2 S_i}{\partial E_i^2}$, which was Planck's starting point in Eq. (3.14). From Eq. (3.15), we find

$$\frac{1}{T} = \frac{\partial S_i}{\partial E_i} = \frac{k}{\hbar\omega}\ln\left(1 + \frac{\hbar\omega^3}{\pi^2 c^3 \rho_\omega}\right) \tag{3.58}$$

$$= \frac{k}{\hbar\omega}\ln\left(1 + \frac{\hbar\omega^3 V_i d\omega}{\pi^2 c^3 E_i}\right), \tag{3.59}$$

where we used $E_i = V_i \rho_\omega d\omega$. Then

$$\frac{\partial^2 S_i}{\partial E_i^2} = -\frac{k}{\hbar\omega}\frac{1}{\left(1 + \frac{\hbar\omega^3 V_i d\omega}{\pi^2 c^3 E_i}\right)}\frac{\hbar\omega^3 V_i d\omega}{\pi^2 c^3 E_i^2}, \tag{3.60}$$

and (3.57) yields

$$\overline{(\Delta E_i)^2} = \left(\frac{\pi^2 c^3 E_i^2}{\omega^2 V_i d\omega} + \hbar\omega E_i\right). \tag{3.61}$$

The fractional mean square energy fluctuation is then given by

$$\frac{\overline{(\Delta E_i)^2}}{E_i^2} = \left(\frac{\pi^2 c^3}{\omega^2 V_i d\omega} + \frac{\hbar\omega}{E_i}\right). \tag{3.62}$$

Eq. (3.62) consists of two terms, reflecting the wave-particle duality of the radiation. The second (particle-like) term is $1/N_p$ where $N_p = E_i/\hbar\omega$ is the number of photons of $\hbar\omega$ required to form an energy E_i. This term in the fluctuation is seen to be $\overline{(\Delta N)^2}/N^2 = 1/N$ as follows from the Poisson distribution. The first (wave-like) term, which comes from the low frequency part of the spectrum, is expected if the radiation is made up of a sum of waves with random phases.

3.2.6 Photons carry momentum as well as energy

In the year (1917) following the publication of the paper introducing the A and B coefficients, Einstein approached the problem of fluctuations in the radiation field in a new way, closely related to his understanding of Brownian motion.

Einstein considered a gas of atoms (or molecules) in a cavity containing radiation at thermal equilibrium and asked what conditions on the momentum transfer between the atom and the radiation are necessary for the atom to remain in equilibrium with the radiation. If the atom is at rest in the radiation field, the average force exerted on it by the radiation will be zero because of the isotropy of the radiation. The atom would be subject to a series of random kicks coming from all directions. If the atom is moving with velocity v with respect to the radiation, there will be a tendency for more kicks to come from a direction opposed to the motion and it will experience a viscous-like force of Rv, where R is a constant to be determined, due to its interaction with the radiation. This force would eventually bring the atom to rest on average, i.e., the collisions would not transfer any net momentum. This behavior is just what happens in Brownian motion.

During a time interval τ, the momentum of an atom of mass M will change from Mv to $Mv - Rv\tau + \Delta$, where Δ is the random amount of momentum transferred to the atom in time τ by collisions with photons. Obviously the average of Δ is zero. Since the velocity distribution of the atoms must remain constant if equilibrium is to be preserved, we must have

$$\overline{(Mv)^2} = \overline{(Mv - Rv\tau + \Delta)^2} \tag{3.63}$$

$$\overline{(M^2v^2)} = \overline{(M^2v^2)} - 2M\overline{v^2}R\tau + \overline{\Delta^2} \tag{3.64}$$

$$\overline{\Delta^2} = 2M\overline{v^2}R\tau. \tag{3.65}$$

In Eq. (3.64), we have neglected the term $\overline{R^2v^2\tau^2}$ which is negligible for small τ.

Since the atoms are in equilibrium, we must have $M\overline{v^2} = kT$ so that

$$\frac{\overline{\Delta^2}}{\tau} = 2kTR. \tag{3.66}$$

This is an early example of the fluctuation-dissipation theorem; thermally generated fluctuations and dissipation always occur together and there is a quantitative relationship between them. Both effects arise from the same mechanism—the coupling of the system under study to a heat reservoir. Dissipation acts to bring the system to equilibrium; in our example, the dissipative force brings the atom's velocity to the equilibrium value $\bar{v} = 0$ by transferring energy from the atom's motion to the radiation, which we assume can absorb an arbitrary amount of energy. An atom in equilibrium cannot remain completely at rest since each degree of freedom must have an average energy kT; the atom's coupling to the radiation leads to a constant exchange of energy that maintains the equilibrium value $M\overline{v^2} = kT$. Exactly the same phenomenon occurs in Brownian motion, where the Brownian particle's collisions with the molecules of the fluid give rise to a dissipative force that eventually brings its average velocity to zero and also provide constant momentum kicks in random directions that result in the observed erratic motion of the particle.

Einstein went on to study Eq. (3.66) by calculating $\overline{\Delta^2}$ and R separately based on the Planck spectrum. For simplicity he assumed that the atoms could only move in one dimension. To calculate $\overline{\Delta^2}$, he assumed that the atom has two states, k and l, and that the total number of photon absorptions and emissions in the time τ is n, with each absorption or emission event transferring momentum δ_i. Since the net momentum transfer is zero on average, we have $\overline{\sum_i \delta_i} = 0$, and we expect the momentum transferred in each collision to be independent of the other collisions. Then

$$\overline{\Delta^2} = \overline{\sum_{i=1}^{n} \delta_i^2} = n\overline{\delta_i^2}. \tag{3.67}$$

Einstein assumed that each photon absorption or emission event transfers a momentum $(\hbar\omega/c)\cos\phi$, where ϕ is the angle between the directions of motion of the atom and the random direction of the emitted or absorbed photon. Averaging over angles, he found

$$\overline{\delta_i^2} = \frac{1}{3}\left(\frac{\hbar\omega}{c}\right)^2. \tag{3.68}$$

Now n is the total number of absorption and emission events that takes place in time τ. This is twice the number of absorptions given in Eq. (3.51) when the atom makes a transition from state k to state l. We then find

$$n = 2P_k B_{k\to l}\rho_\omega \tau \tag{3.69}$$

$$= 2\left(\frac{e^{-E_k/kT}}{e^{-E_k/kT} + e^{-E_l/kT}}\right) B_{k\to l}\rho_\omega \tau, \tag{3.70}$$

where P_k is the probability of an atom being in the state k. We then obtain $\overline{\Delta^2}$ from Eq. (3.67)

$$\overline{\Delta^2} = \frac{2}{3}\left(\frac{e^{-E_k/kT}}{e^{-E_k/kT} + e^{-E_l/kT}}\right) B_{k\to l}\rho_\omega \tau \left(\frac{\hbar\omega}{c}\right)^2. \tag{3.71}$$

The calculation of R proceeds by transforming the uniformly distributed radiation field in the lab system to the rest system of the moving atom where it will no longer be isotropic. As a result the frequency of the radiation will be changed due to the Doppler shift and the direction of propagation of the radiation will also be altered. Again each absorpion or emission event is associated with a momentum transfer $\frac{\hbar\omega}{c}$. Putting it all together Einstein finds that equation (3.66) is satisfied using the result for R and (3.71) and the Planck formula for ρ_ω. He considered that any viable theory would have to satisfy this criterion of preserving thermal equilibrium.

We see that in order to maintain thermal equilibrium between a gas of atoms and a radiation field, each emission/absorption of a photon has to transfer a momentum $(\hbar\omega/c)$ in a random direction. By considering Eq. (3.51), we see that this applies to absorption and to both spontaneous and stimulated emission. When an atom spontaneously emits a photon of energy $\hbar\omega$, that photon is emitted in a particular direction and the atom suffers a recoil of magnitude $\hbar\omega/c$ in the opposite direction. Einstein considered it a weakness of the theory that it leaves the direction of each individual process to chance. Each emitted photon transfers the momentum $\hbar\omega/c$ directed opposite to the photon's direction, while the classical theory predicts a spherical wave. Einstein's observation that momentum transfers in definite directions are necessary to obtain thermal equilibrium shows that, as he stated, "Outgoing spherical waves do not exist," each photon is emitted in a particular direction. According to the modern theory the classical spherical wave intensity gives the probability that a photon will be emitted in any given direction (which, for spontaneous emission, is equal for all directions). A spherical wave of emitted radiation would result in a spherically symmetric distribution of the recoil velocities of the atoms, which would result in zero average momentum transfer. The atomic wave function must "collapse" into a definite direction in order for the fluctuations to be consistent with thermal equilibrium. Einstein would later use the apparent necessity of wave function collapse as a criticism of quantum mechanics (see Chapter 13).

3.2.7 Summary of Einstein's work on photons, 1905–1917

As we have seen, Einstein returned to the physics of photons at several different times. For clarity we summarize these contributions in chronological order.

1905: Showed that the entropy of radiation obeying Wien's law is the same as that of a gas of independent energy quanta (photons) of energy $\hbar\omega$ and that quantization of the electromagnetic field enabled an understanding of several phenomena such as the photoelectric effect that could not be understood on the basis of the wave theory of light.[12]

1906: Examined Planck's derivation of the black body spectrum and pointed out the contradiction in using classical physics to describe the oscillators and their interaction with the radiation field.[13]

Theory of Brownian motion.[14]

1907: Derived Planck's result for the black body spectrum using the Boltzmann factor for the probability of a given energy state and Planck's assumption $E = \hbar\omega$. He then applied the same ideas to the vibrations of atoms in a solid and solved the long-standing problem that the classical theory did not agree with the observed decrease of the specific heat at low temperatures. Einstein had assumed that all the atoms vibrate at a single frequency, but his idea was further developed by Debye and others, who allowed for a range of oscillation frequencies as in the modern theory of solids.[15]

1909: Calculated the fluctuations in energy present in a small volume in a radiation field at thermal equilibrium and showed that they consist of two types of fluctuations, one characteristic of a gas of particles and the other of a set of waves with random phases. By considering a reflecting surface suspended in a cavity containing black body radiation, he showed that the momentum fluctuations of the mirror that are induced by interaction with the radiation again consist of two terms, one characteristic of interactions with particles and the other of interactions with waves.[16]

[12]Einstein, A., *Generation and conversion of light with regard to a heuristic point of view,* Annalen der Physik, 17 (6) , pp.132-148 (1905).

[13]Einstein, A., *Theory of light production and light absorption,* Annalen der Physik, 20 (6), pp.199-206 (1905).

[14]Einstein, A., *The theory of the Brownian Motion,* Annalen der Physik 19 (2), pp.371-381 (1906).

[15]Einstein, A., *On the validity limit of the thermodynamic equilibrium theorem and the possibility of a new determination of elementary quanta,* Annalen der Physik 22 (3) , pp.569-572, (1907).

[16]Einstein, A., *On the current state of radiation problems.* Phyikalische Zeitschrift, 10, 185-193, (1909) and *On the evolution of our vision on the nature and constitution of radiation,* Phyikalische Zeitschrift, 10, 817-826, (1909).

1916–1917: Derived the Planck distribution by considering spontaneous and induced emission by stipulating that the atom should remain in thermal equilibrium with the radiation. He showed that each spontaneously emitted photon transfers a momentum in a definite direction so that radiation in spherical waves, as expected by classical theory, does not exist. This is indicative of the "wave function collapse" discussed in the previous chapter.[17]

These various interventions had two motivations. The first was to promote the idea that the radiation field was quantized, which had initially encountered considerable resistance, and the second was to investigate the nature of the field quantization with the goal of understanding its relation to the classical wave theory. The latter is a problem that remains controversial even today.

3.3 The Bohr atom

Based on the scattering of alpha particles by heavy atoms, in 1911 Rutherford proposed that atoms consisted of a relatively heavy and small charged nucleus surrounded by a cloud of much lighter electrons with opposite charge to the nucleus. The problem immediately arose as to how such a system could maintain its stability. According to classical theory, an electron orbiting a central charged nucleus would continuously radiate energy because an accelerated charge always radiates, so that classically we would expect that the electron would lose all its energy to radiation and spiral into the nucleus in a matter of nanoseconds.

It had long been observed that atoms emitted radiation in the form of narrow spectral lines at discrete frequencies, whereas classical theory predicted that the electrons would emit radiation at the frequency of their motion, which would vary continuously as they radiated energy, hence the existence of atoms themselves was a puzzle for the classical theory.

Bohr's model of the hydrogen atom provided a solution to the puzzle of spectral lines. Bohr approached the problem with the following assumptions, which represent a profound departure from the prevailing ideas of the time and introduced a new "quantum" method of thinking:

1. Stable orbits exist despite classical electromagnetic theory.
2. Stable orbits are defined by the condition that their angular momentum is *quantized*, i.e., $p_\theta = n\hbar$, where $\hbar = h/2\pi$ and n is an integer.
3. A transition from a higher energy (E_2) to a lower energy (E_1) orbit is accompanied by the emission of light with a frequency

$$h\nu = E_2 - E_1. \tag{3.72}$$

These transitions are not susceptible to analysis by the theory and became known as "quantum jumps." [18]

[17]Einstein, *A Quantum theory of radiation.* Phyikalische Zeitschrift, 18, 121-128, (1917), received March 3, 1917.

[18]Although many theorists, including Heisenberg, took these quantum jumps to be an essential part of the theory, Schrödinger had a deep antipathy to this idea, as we will see.

We begin the analysis with the equations of motion for a particle moving around an attractive center of force exerting a radially directed inverse square force such as the Coulomb force between oppositely charged particles. This gives the equations of motion

$$\frac{d}{dt}\left(mr^2\dot{\theta}\right) = 0$$

$$\frac{d}{dt}(m\dot{r}) - mr\dot{\theta}^2 + \frac{\kappa}{r^2} = 0,$$

where the first equation expresses the conservation of angular momentum, $p_\theta = mr^2\dot{\theta}$. Bohr's quantum condition is then

$$p_\theta = mr^2\dot{\theta} = n\hbar.$$

For a circular orbit, r =const and $\dot{r} = 0$, so the second equation yields

$$\frac{\kappa}{r} = mr^2\dot{\theta}^2 = \frac{p_\theta^2}{mr^2}$$

$$r = \frac{p_\theta^2}{m\kappa} = \frac{(n\hbar)^2}{m\kappa}. \tag{3.73}$$

Using Eq. (3.73), the Hamiltonian is given by

$$H = \frac{p_\theta^2}{2mr^2} - \frac{\kappa}{r} = E_n = -\frac{1}{2}\frac{\kappa}{r}$$

$$= -\frac{1}{2}\left(\frac{m\kappa^2}{p_\theta^2}\right) = -\frac{1}{2}\frac{m\kappa^2}{(n\hbar)^2},$$

so that the energies of the allowed orbits are given in terms of known constants and are $\propto 1/n^2$. From Eq. (3.73), the angular velocity is

$$\dot{\theta} = \frac{m\kappa^2}{(n\hbar)^3}. \tag{3.74}$$

This would be the angular frequency of the emitted radiation according to classical electrodynamics. The radiation frequency would not remain constant as the electron radiated away its energy and the orbit spiraled into the center.

Eq. (3.72) predicts that the frequency emitted when the atom transitions from a state n_1 to a state n_2 is

$$\omega = \frac{1}{2}\frac{m\kappa^2}{\hbar^3}\left(\frac{1}{n_2^2} - \frac{1}{n_1^2}\right). \tag{3.75}$$

The quantitative agreement of this expression with the main features of the observed hydrogen spectrum represented a great success for Bohr's ideas.

One of the most important features of this analysis is that we have been forced to assume that an electron emits radiation at a frequency different from that calculated according to classical electromagnetism in Eq. (3.74). This difference between the frequencies of the orbital motion and the emitted radiation continued to be a puzzle for many years. Bohr pointed out that if we consider emission from a state characterized by a large "quantum number" N making a transition to a state $N-n$, where $n \ll N$, Eq. (3.75) yields

$$\omega = \frac{m\kappa^2}{(\hbar N)^3} n = n\dot{\theta}, \tag{3.76}$$

where we have used Eq. (3.74) for the angular velocity. This agrees with the classical expectation that the emitted radiation frequency would be an integer multiple of the orbital frequency. This result forms the basis of the correspondence principle articulated by Bohr, which says that in the limit of large quantum numbers the results of the quantum theory should agree with the classical predictions. We will see that this played an important role in the further development of the theory.

In his first paper,[19] Bohr gave several other "quantum conditions" that yielded the same results as assuming that stable orbits were defined by quantization of the angular momentum. However, it was the quantization of angular momentum that proved to be the most physically significant.

3.4 Conclusion

In this chapter we have seen how the picture of the world as consisting of continuously distributed matter and energy became untenable. Matter was found to consist of individual atoms and molecules, which were in turn made from smaller particles. Understanding the stability of atoms and the observed properties of electromagnetic radiation required abandoning the classical notion of energy as continuous and assuming that the energies of oscillators and modes of the electromagnetic field could only take on certain discrete values.

However great problems remained. The relation between the highly successful wave theory of light and photons was obscure, to say the least, and discrepancies between the Boht theory and its successors and experimental observations soon became evident.

Nonetheless the early successes of the emerging quantum theory motivated the application of quantum ideas to more complex systems. Bohr's atomic model inspired a wealth of investigations into applying quantum models to various aspects of atomic structure, as we will see in the next chapter.

[19] Bohr, N., *On the constitution of atoms and molecules, Part 1 and Part II, Sustems containing only a single nucleus*, Phil. Mag. (6) 26, pp.1 and 476 (1913).

4

Further steps to quantum mechanics: the old quantum mechanics of Bohr and Sommerfeld

As we have seen, in 1913 Bohr developed a model of the hydrogen atom that agreed very well with the observed spectral lines of hydrogen. He assumed that only a discrete set of the possible electron orbits around the nucleus would be stable against continuously emitting electromagnetic radiation and these orbits, which he took to be circular, were chosen by the condition that their angular momentum p_θ satisfied

$$p_\theta = \frac{nh}{2\pi} = n\hbar \tag{4.1}$$

for some integer n. Radiation of frequency ν was emitted or absorbed when the electron made a transition from one orbit to the other, where ν was determined by the condition

$$h\nu = E_1 - E_2, \tag{4.2}$$

where $E_{1,2}$ are the energies of the orbits involved in the transition and positive (negative) ν means emission (absorption). While the success of this model was impressive, the question arose of how these principles could be applied to more complex systems, for instance to elliptic orbits or systems with more than one electron.

4.1 Quantization conditions

In 1906, well before Bohr's work, Planck was the first to suggest that the proper quantum conditions should be based on the motion of the system in phase space.[1] Phase space is a coordinate system where there is one axis for each spatial coordinate q_k and one for each of the corresponding momenta p_k. For a one-dimensional example, consider the harmonic oscillator (Section A.4.3 in Appendix A). Since the energy is constant, the point (p, x) will lie on an ellipse defined by

$$\begin{aligned} \frac{p^2}{2m} + \frac{kx^2}{2} &= E \\ \frac{p^2}{2mE} + \frac{kx^2}{2E} &= 1 \end{aligned} \tag{4.3}$$

[1]Planck, M., *The dynamics of moving systems*, Sitzungsberichte der Königlich Preussischen Akademie der Wissenschaften, pp. 542-570, (1907).

The Historical and Physical Foundations of Quantum Mechanics. Robert Golub and Steven K. Lamoreaux, Oxford University Press.
 DOI: 10.1093/oso/9780198822189.003.0004

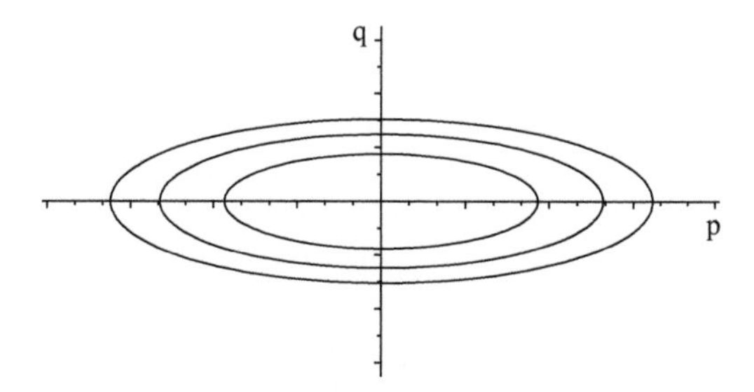

Fig. 4.1 Phase space (p, q) for the linear harmonic oscillator. The ellipses represent the stable quantum states with area $\oint pdq = h$, $2h$ and $3h$, respectively.

with semi-major and semi-minor axes

$$\begin{aligned} a &= \sqrt{2mE} \\ b &= \sqrt{\frac{2E}{k}} \end{aligned} \tag{4.4}$$

and area

$$A = \pi ab = \frac{2\pi E}{\sqrt{k/m}} = \frac{E}{\nu}. \tag{4.5}$$

Planck's original quantization condition on the harmonic oscillator could then be seen to be the requirement that the allowable motions were those which enclose an area of nh in the phase space, as shown in Fig. 4.1. This condition is equivalent to

$$\oint p_i dq_i = n_i h \tag{4.6}$$

with n_i integers, which should be applied to each coordinate-momentum pair in the problem.

4.2 "Old" quantum theory

After Bohr's stunning result, Arnold Sommerfeld and others attempted to apply Eq. (4.6) to more complicated problems. But it was soon recognized that the results of this condition depended on the choice of coordinates in which it was applied. In March 1916, Karl Schwarzschild wrote to Sommerfeld from Brussels, where he was on military service, suggesting that the action variables (Eq. (A.26) in Appendix A) from the Hamilton-Jacobi method were the quantities to be quantized: $J_k = n_k h$. This singled out the coordinate system in which the Hamilton-Jacobi equation could be separated as the system in which the quantum conditions should be applied. The Hamilton-Jacobi theory was known to physicists like Schwarzschild, who had been studying astronomical orbits, but was not known to those like Sommerfeld, who were studying atomic physics. The communication of Hamilton-Jacobi theory to atomic physicists led to a large body of work on applying the quantum conditions to more complex problems in atomic physics.

Ehrenfest pointed out that if a system in a state selected according to a quantum condition was subjected to a weak, slowly varying external influence, it would be impossible for the quantity that was quantized to change by enough so that the quantum number changed by a whole number. The quantized quantity would be frozen at its initial value. The only way to avoid this contradiction was to apply the quantum condition solely to what were called adiabatic invariants, that is, quantities that did not change under adiabatic variations of the system parameters. For a harmonic oscillator, it can be shown that E/ω is an adiabatic invariant, and this is just what is quantized in that case ($E = n\hbar\omega$). In the general case, the action variables $\left(\oint p_i dq_i\right)$ are adiabatic invariants so it is consistent to apply the quantum conditions to them.

When technological advances enabled spectroscopic measurements with higher resolution, each line of the hydrogen spectrum was seen to consist of several closely spaced lines. This effect—called fine structure—was investigated by Sommerfeld, who was able to get good agreement with experiment by calculating the orbits taking account of special relativity. Other phenomena discovered and theoretically studied at this time include the Zeeman effect (the splitting of atomic spectral lines in the presence of a large magnetic field) and the Stark effect (the splitting of the spectral lines in the presence of a strong electric field; experiments were performed with electric fields as high as 10^6 Volts/cm). Much effort was also devoted to understanding the spectra of many-electron atoms. The theoretical treatment of several phenomena, for example, the Stark effect, Zeeman effect, and the fine structure of hydrogen, produced results that agreed with experimental measurements, but there were many cases where the theoretical results disagreed with experiment. We now know that many of these disagreements were due to the existence of electron spin, which was unknown at the time.

4.2.1 Quantization of elliptic orbits in the hydrogen atom

Bohr had confined his analysis to circular orbits. Sommerfeld applied Bohr's ideas to elliptic orbits as well. In Section A.5.2, we have given the action variables for the Kepler problem. To apply this to the hydrogen atom, we put $\kappa = Ze^2$. In Eq. (A.12) in Section A.4.4 of Appendix A, we give the equation of the orbit. It is an ellipse with eccentricity

$$\varepsilon = \sqrt{1 + \frac{2El^2}{m\kappa^2}}. \tag{4.7}$$

We apply the quantum conditions by setting $J_i = n_i h \ (i = r, \theta, \phi)$. From Eq. (A.30) we have $l = \left(J_\theta + J_\phi\right)/2\pi$ and, substituting the energy from Eq. (A.31), we find

$$\varepsilon = \sqrt{1 - \frac{\left(J_\theta + J_\phi\right)^2}{\left(J_r + J_\theta + J_\phi\right)^2}}, \tag{4.8}$$

so that only specific values of the eccentricity are allowed according to the quantum condition. The energy and the principal axis of the ellipse are both functions only of the sum $\left(J_r + J_\theta + J_\phi\right) = \left(n_r + n_\theta + n_\phi\right) h$, which is often written as $n_1 h$, where n_1 is called the principal quantum number. The stable orbits selected by the Bohr-Sommerfeld quantum condition are shown in Fig. 4.2. The orbits corresponding to each

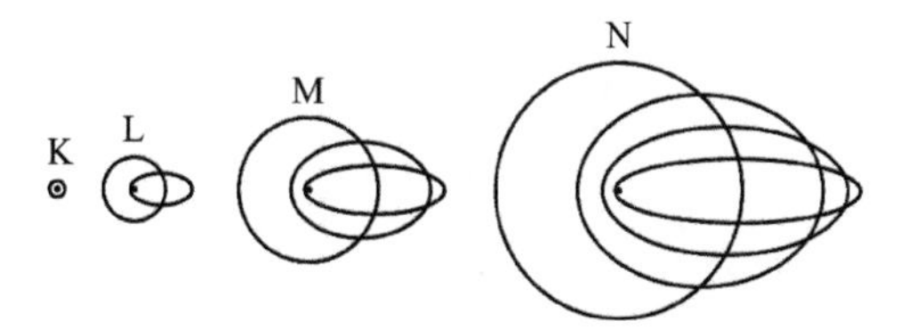

Fig. 4.2 The lowest energy orbits of the hydrogen atom, drawn to scale. The labels correspond to the value of the principal quantum number: K to $n_1 = 1$, L to $n_1 = 2$, and so on.

principal quantum number are to scale but they are all normalized to the principal axis $\left(\sim n_1^2\right)$.

It was the rejection of pictures like these, which represent unobservable objects, that motivated Heisenberg's later development of quantum mechanics.

4.2.2 Spatial quantization

By applying the method discussed above to the Kepler orbit in three dimensions, Sommerfeld discovered another unexpected effect specific to quantum mechanics.[2] We start with Eq. (A.29) in the appendix, which we reproduce below. (Remember l, m are the separation constants of the Hamilton-Jacobi equation (Eq. (A.21) in the appendix). Then

$$J_r = \oint dr \frac{\partial W_r}{\partial r} = \oint dr \sqrt{2m\left(E + \frac{\kappa}{r}\right) - \frac{l^2}{r^2}} = n_r h \tag{4.9}$$

$$\begin{aligned} J_\theta &= \oint d\theta \frac{\partial W_\theta}{\partial \theta} = \oint d\theta \sqrt{l^2 - \frac{m^2}{\sin^2\theta}} = n_\theta h \\ &= 2\pi\left(l - m\right) \quad \text{from Eq. (A.30) in the appendix} \end{aligned} \tag{4.10}$$

$$J_\psi = \oint d\psi \frac{\partial W_\psi}{\partial \psi} = \oint d\psi m = 2\pi m = n_\psi h, \tag{4.11}$$

where we have added the quantization conditions. We have chosen to label the coordinates r, θ, ψ, reserving ϕ for the angle in the plane of the orbit. Combining $2\pi m = n_\psi h$ and $2\pi\left(l - m\right) = n_\theta h$, we find $2\pi l = \left(n_\theta + n_\psi\right) h$. As the electron goes around the orbit, θ will vary between its limits given by the zeroes of the integrand in Eq. (4.10). As seen in Fig. (4.3), the minimum value of θ represents the tilt of the orbital plane with respect to the coordinate axes, the point A in the figure. This minimum is given by

$$\sin\theta = \frac{m}{l} = \frac{n_\psi}{n_\theta + n_\psi}, \tag{4.12}$$

which leads to the quite astonishing result, called spatial quantization, that only a set of discrete orientations of the orbital plane is allowed.

[2]Sommerfeld, A., *Atombau und Spektrallinien (Atomic structure and spectral lines)*, Fred. Vieweg und Sohn. 7th edition, (1950). (English translation of the 1923 edition: Brose, H.L., translator)

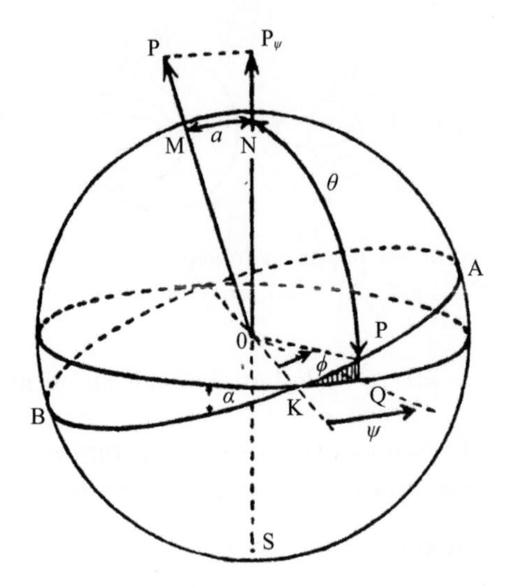

Fig. 4.3 P represents the instantaneous position of the electron on its orbit, which is the ellipse containing P, A, and B. Q is the projection of P in the equatorial plane of the coordinate system with azimuthal angle ψ. p represents the total angular momentum, which is perpendicular to the orbit and makes an angle α with the z-axis. α and the value of θ at the point A are complementary angles.(A. Sommerfeld, op. cit., p. 69, Fig. 28, p.120.)(Public Domain.)

These considerations only apply when a particular orientation in space is specified, for example, due to the presence of an external field. In the general case where no direction is singled out, the z-axis can be chosen to be in any direction, so the above considerations are not relevant. Sommerfeld considered the case where a direction is specified by an applied magnetic field. The above discussion would then apply in the limit of the field going to zero. Sommerfeld had found this spatial quantization effect in a paper studying the Zeeman effect published in 1916.[3]

In 1922 the definitive experiment observing "space quantization" was performed by Stern and Gerlach.[4] They passed a beam of silver atoms through a highly inhomogeneous magnetic field. The results, reproduced in Fig. 4.4, showed that the beam split into two discrete components, indicating that the magnetic moments of the atoms could be oriented in either of only two distinct directions. The two components are seen to come together at the top and bottom of the figure because the magnetic field decreases away from the center of the beam. This result provided dramatic confirmation of the spatial quantization predicted by the Bohr-Sommerfeld model.

[3]Sommerfeld, A., *The Zeeman effect theory of hydrogen lines with a supplement on the power effect*, Physikalische Zeitschrift, 17, 491-507, (1916).

[4]Gerlach, W. and Stern, O., *The experimental evidence of direction quantization in the magnetic field*, Zeitschrift fur Physik 9, 349, (1922).

Abs. Prof. W. Gerlach
Frankfurt a. M.
Physikal. Institut
Universität.

Professor
Niels Bohr
Kopenhagen
Stockholmsgade 37

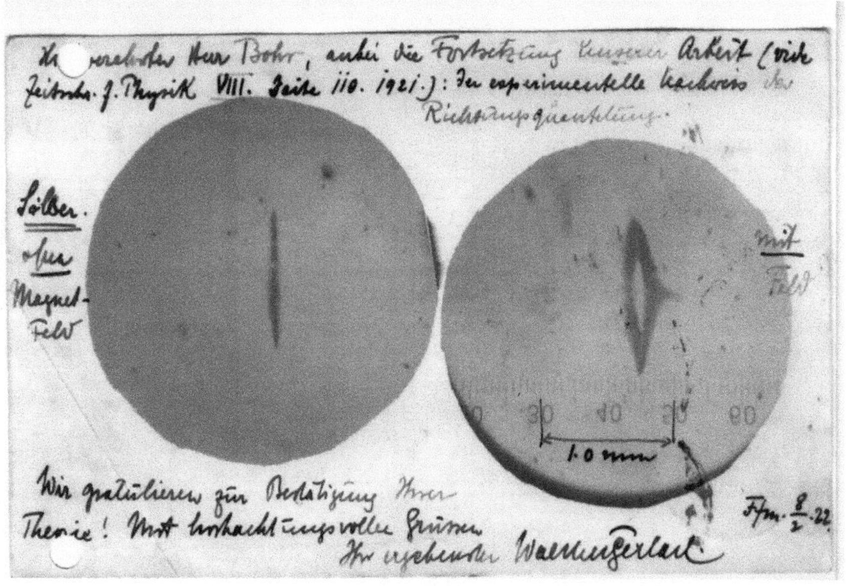

Hochverehrter Herr Bohr, anbei die Fortsetzung unserer Arbeit (vide Zeitschr. f. Physik VIII. Seite 110. 1921.): Den experimentellen Nachweis der Richtungsquantelung.

Silber. ohne Magnet-Feld

mit Feld

1.0 mm

Wir gratulieren zur Bestätigung Ihrer Theorie! Mit hochachtungsvollen Grüßen
Ihr ergebener Walther Gerlach

Ffm. 8/2. 22

Fig. 4.4 The results of the Stern-Gerlach experiment, sent by postcard to Niels Bohr in early 1922. The darkened regions on the photographs of the collection plates show, on the left, the undeflected beam of silver atoms, and on the right, show silver atoms which have passed through a strongly inhomogeneous magnetic field. The fact that the atoms are not distributed uniformly means that their magnetic moments only point in two discrete directions with respect to the field. According to the theory at the time, for an $L = 1$ state, there should be three bands for $m_l = \pm 1\ 0$; the lack of a third (undeflected) stripe for $m_l = 0$ was an unwitting and unnoticed discovery of intrinsic electron spin $s = 1/2$. (Image courtesy of the Niels Bohr Archive, and reproduced here with their permission.)

4.2.3 Fine structure of the hydrogen lines

One of the more difficult problems that Sommerfeld was able to solve using the quantum conditions in Eq. (4.6) with separable variables was the hydrogen atom using special relativistic mechanics (A. Sommerfeld, op. cit., p. 69). We use this as an example of the advanced problems that were solved with this technique. Others were the Stark and Zeeman effects (the splitting of the spectral lines due to external electric or magnetic fields). We start with the relativistic relation between energy (E) and momentum (p) with the potential $V = -\kappa/r$ $\left(\kappa = e^2\right)$ and work in two dimensions because the orbits all remain in a plane. We have

$$(E - V)^2 = (pc)^2 + \left(mc^2\right)^2 \tag{4.13}$$

$$p^2 = \frac{1}{c^2}\left[(E - V)^2 - \left(mc^2\right)^2\right] \tag{4.14}$$

$$\left(\frac{\partial S}{\partial r}\right)^2 + \frac{1}{r^2}\left(\frac{\partial S}{\partial \phi}\right)^2 = \frac{1}{c^2}\left[(E - V)^2 - \left(mc^2\right)^2\right], \tag{4.15}$$

where m is the rest mass of the electron. The separation constant is α_ϕ so that

$$\frac{\partial S}{\partial \phi} = \alpha_\phi = n_\phi \hbar \tag{4.16}$$

by the quantum condition. Then

$$\left(\frac{\partial S}{\partial r}\right) = \sqrt{\frac{E^2}{c^2} - (mc)^2 + \frac{2E}{c^2}\frac{\kappa}{r} + \left[\left(\frac{\kappa}{c}\right)^2 - (n_\phi \hbar)^2\right]\frac{1}{r^2}} \tag{4.17}$$

$$\oint \left(\frac{\partial S}{\partial r}\right) dr = \oint \frac{dr}{r}\sqrt{\left(\frac{E^2}{c^2} - (mc)^2\right) r^2 + \frac{2E}{c^2}\kappa r + \left[\left(\frac{\kappa}{c}\right)^2 - (n_\phi \hbar)^2\right]} \tag{4.18}$$

$$= \oint \frac{dr}{r}\sqrt{Ar^2 + 2Br - C} = n_r h. \tag{4.19}$$

As in the appendix, Section A.5.2 (note that now $A > 0$), we note that the integral

$$\int_{r_1}^{r_2} \frac{dr}{r}\sqrt{(r - r_1)(r - r_2)} = \pi\left(\frac{1}{2}(r_1 + r_2) - \sqrt{r_1 r_2}\right) \tag{4.20}$$

and evaluating the sum and product of the roots of $\left(r^2 + 2Br/A - C/A\right)$ we evaluate the integral as

$$-2\pi\left(\frac{B}{\sqrt{A}} + \sqrt{C}\right), \tag{4.21}$$

so that

$$\frac{-E}{c^2}\kappa\frac{1}{\sqrt{\left(\frac{E^2}{c^2}-(mc)^2\right)}}-\sqrt{(n_\phi\hbar)^2-\left(\frac{\kappa}{c}\right)^2}=n_r\hbar$$

$$-\frac{\alpha}{\sqrt{1-\left(\frac{mc^2}{E}\right)^2}}=n_r+\sqrt{n_\phi^2-\alpha^2}$$

$$E=\frac{mc^2}{\sqrt{1+\frac{\alpha^2}{\left(n_r+\sqrt{n_\phi^2-\alpha^2}\right)^2}}},\tag{4.22}$$

where we have written $\alpha=\frac{\kappa}{\hbar c}=\frac{e^2}{\hbar c}$. This is called the (Sommerfeld) fine structure constant and was introduced by Sommerfeld in the context of this problem. The fact that it is very close to 1/137 (present value 1/137.035999139(31)) led to much speculation at one time. It is a dimensionless number—*the same in all systems of units*—and has a deep significance. It is the strength of the electromagnetic interaction and serves as an expansion parameter for all electromagnetic effects at low momentum transfer.

If we take the lowest order term in α^2, we find

$$E=mc^2\left(1-\frac{1}{2}\frac{\alpha^2}{(n_r+n_\phi)^2}\right),\tag{4.23}$$

which is the rest mass reduced by the classical energy of the bound level specified by (n_r+n_ϕ). At this level of accuracy, the levels are degenerate. The exact relativistic result shows that the levels split: for the same value of n_r+n_ϕ, the energy depends on the values of both n_r and n_ϕ. The complicated line structure is revealed when we consider the frequency condition

$$\hbar\omega=E_{\{n_r,n_\phi\}2}-E_{\{n_r,n_\phi\}1}.\tag{4.24}$$

When fine structure is taken into account, every state of hydrogen with $n_1>1$ splits into multiple states so that each observed spectral line consists of a series of several closely spaced lines. The Lyman-α line ($n_1=2\rightarrow n_1=1$) (see Fig. 9.4), for instance, consists of two lines. The scale of the fine structure splitting compared to the energy differences between states with different n_1 values is $\sim\alpha^2$.

The solution for the electron orbit is a precessing ellipse as shown in Fig. 4.5.[5] It turned out that the fine structure splitting described above agreed with experimental measurements, but that was an amazing stroke of luck. In several other cases (for example, the Zeeman effect in certain energy levels, the second order Stark effect), the results could not be brought into agreement with experiment. The fine structure discussed here can only be completely understood when the electron spin is taken into account.

[5]When applied to the orbit of the planet Mercury, this theory produces a result that is 1/6 of the observed result predicted by general relativity.

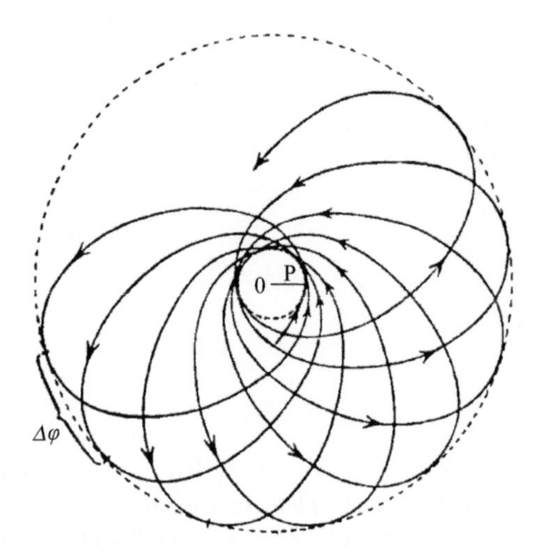

Fig. 4.5 Relativistic Kepler orbits. The center of force (nucleus) is located at 0. The perihelion and the aphelion precess inside the dotted circles. (A. Sommerfeld, op. cit., p. 69, Fig. 67, p. 259.) (Public Domain.)

4.2.4 The Bohr correspondence principle

The old quantum theory as described so far had some modest successes but seemingly had nothing to say about which transitions between different energy levels actually took place (selection rules) and about the intensity and polarization of the emitted radiation. An attempt to deal with these issues was the Bohr correspondence principle, which said that the results of the quantum theory should agree with the classical results in the limit of large quantum numbers. Then the properties of the classical radiation could be applied to cases where the quantum numbers were not too small.

For the harmonic oscillator, the energies are $E_n = n\hbar\omega$. Since classically the only frequency ever emitted is ω, the correspondence principle would indicate that the proper selection rule for allowed transitions is $\Delta n = \pm 1$. To see how the correspondence principle applies in more general cases, we note that the classical frequencies are given by the rates of change of the angle variables and higher harmonics so that the nth harmonic will be

$$n\frac{dw_i}{dt} = \frac{\partial H}{\partial J_i/n} = \nu_{class}. \tag{4.25}$$

In the quantum case J_i would change by $\Delta J_i = \pm nh$ and the frequency will be $\nu_{qu} = \Delta H/h = \Delta H/\left(\Delta J_i/n\right)$ so, in the limit $h \to 0$, the quantum frequency will approach the classical frequency. For the hydrogen atom, we found that the energy can be written as in Eq. (4.23). Neglecting the rest energy, we have

$$H = -\frac{1}{2}\frac{\alpha^2 mc^2}{\left(n_r + n_\phi\right)^2} = -\frac{1}{2}\frac{\alpha^2 mc^2 h^2}{\left(J_r + J_\phi\right)^2}, \tag{4.26}$$

and the fundamental classical frequency will be

$$\nu_{class} = \frac{dw_i}{dt} = \frac{\partial H}{\partial J_i} = \frac{\alpha^2 mc^2 h^2}{(J_r + J_\phi)^3} = \frac{\alpha^2 mc^2}{h\,(n_r + n_\phi)^3}. \tag{4.27}$$

The frequency according to quantum theory would be ($n_1 = n_r + n_\phi$, $\Delta n \ll n_1$)

$$\nu_{qu} = -\frac{1}{2}\alpha^2 \frac{mc^2}{h}\left(\frac{1}{(n_1)^2} - \frac{1}{(n_1 - \Delta n)^2}\right) \tag{4.28}$$

$$\approx \alpha^2 \frac{mc^2}{h}\left(\frac{\Delta n}{n_1^3}\right), \tag{4.29}$$

and we see that for $\Delta n = 1$ the quantum frequency agrees with the fundamental classical frequency and for Δn a larger integer we get the harmonics of the classical frequency.

In recent years these large n_1 levels of hydrogen, called Rydberg states, have been subjected to intense experimental study, but we will need modern quantum mechanics to understand them.

In his Nobel Prize address in 1954,[6] Max Born characterized this method of working by stating "The art of guessing correct formulas, which depart from the classical formulas but pass over into them in the sense of the correspondence principle, was brought to considerable perfection."

4.3 Toward quantum mechanics: classical mechanics as the limit of a wave motion

The function $S(q, t) = \int L dt$ obtained in Appendix A has some very important properties. From Eq. (A.19), valid for time-independent Hamiltonians, we see that $S = W(q) - Et$. At time $t = 0$ we can draw a surface of constant $S = W_1$. At time $t = dt$ later, the surface of constant S will have moved to $S = W_2 - Edt$ with $W_2 = W_1 + Edt$ (see Fig. 4.6).

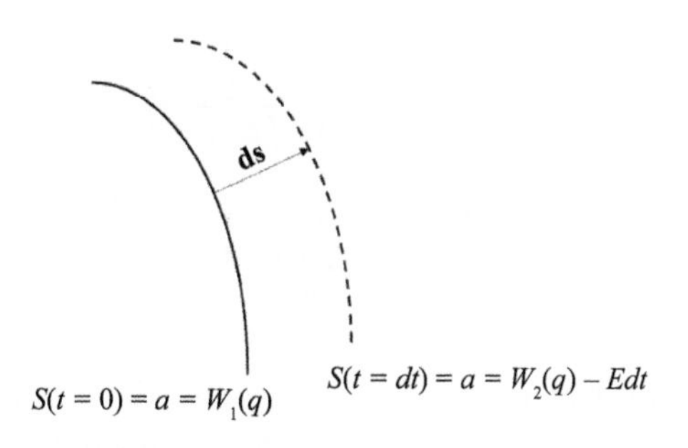

Fig. 4.6 Motion of a surface of constant $S = W(q) - Et$.

[6]Max Born, Nobel Lecture. NobelPrize.org. Nobel Prize Outreach AB 2021. Fri. 22 Oct. 2021, https://www.nobelprize.org/prizes/physics/1954/born/lecture/

We therefore see that the motion of a surface of constant S is given by $dW = Edt$. The distance between the two surfaces, ds, is given by the change in W,

$$dW = W_2 - W_1 = Edt \tag{4.30}$$

divided by the rate of change of W with s, $\partial W/\partial s$:

$$ds = \frac{dW}{\partial W/\partial s} = \frac{Edt}{\partial W/\partial s},$$

so that the surface of constant S moves with a velocity u given by

$$u = \frac{ds}{dt} = \frac{E}{\partial W/\partial s} = \frac{E}{p}.$$

Thus the motion of the surfaces of constant S is analogous to the motion of the surfaces of constant phase of a wave moving with the phase velocity E/p at each point. If there were such a wave, the phase would be proportional to $S = \int Ldt$ and the particles would be moving perpendicular to the surfaces of constant phase along the rays of the wave motion since the momentum $p_i = \partial W/\partial q_i$.

This analogy was already known to Hamilton in the 1830s, who applied these ideas to optics in order to show how geometrical optics followed from a wave equation. However, nobody saw fit to explore this further until de Broglie almost 100 years later, who perhaps was inspired by the successes of the Bohr-Sommerfeld model, as we will see in the next chapter.

For a free particle, $E = mv^2/2 = p^2/2m$ so $u = E/p = p/2m = v/2$. For the more general case of motion in a potential V, we find

$$E = T + V = \frac{p^2}{2m} + V \tag{4.31}$$

$$p = \sqrt{2m\,(E - V)} \tag{4.32}$$

$$u = \frac{E}{p} = \frac{E}{\sqrt{2m\,(E - V)}} \tag{4.33}$$

For a wave, the phase velocity is given by $u = \lambda\nu$, where λ is the wavelength and ν is the frequency of the wave. We can also write this in terms of the wave vector k as $u = \omega/k$, where $\omega = 2\pi\nu$ and $k = 2\pi/\lambda$. For the analogy between classical mechanics and a wave motion to make sense, the energy of motion E would have to be proportional to the frequency of the wave and the wavelength proportional to the inverse of the momentum $(1/p)$ with the same constant of proportionality, i.e., $E = h\nu$ and $p = h/\lambda$ with h an unknown constant. Carrying the analogy further, we usually write the wave amplitude for a wave traveling in the x direction in the form $e^{i\omega(t-x/u)}$ so that the mechanical wave would be written $e^{i(Et-px)/\hbar}$, where we have written $\hbar = h/2\pi$.

However, we shall have to wait a little (not the 100 years that science had to wait) before we see how the development of physics led to speculations of this type.

4.4 Conclusion

In this chapter we have shown how quantum mechanics developed based on the ideas in Bohr's atomic model.

The fundamental ideas of the "old" quantum theory can be stated as follows:

1. Motions are calculated according to classical mechanics, excluding radiation damping but including other electromagnetic interactions. These calculations thus give information on the particle motions as a function of time (electron orbits, for example) which were, and remain, unobservable.
2. A discrete sample of motions is selected from the continuum of classically allowed motions by the quantum conditions.
3. The energies of the selected motions are observable by electron collisions or by absorption and emission of electromagnetic radiation according to the Bohr frequency condition.
4. Concerning the intensity, polarization, and phase of the light, only approximate predictions could be made based on the correspondence principle.

There was an enormous body of work based on this method, dealing with many-electron atoms, the periodic table, chemical properties, and the Zeeman and Stark effects. Many of these calculations ran into the problem that the spin of the electron, which was unknown at the time, significantly influences the effect being studied. In fact, this work eventually led to the discovery of the electron spin (value $1/2$).

The Hamilton-Jacobi theory pointed the way to an unambiguous application of the quantum conditions and showed that classical mechanics can be considered as the limit of a wave motion, analogous to geometric optics. We are getting a little ahead of ourselves here; while the analogy was noticed by Hamilton, it was only after the development of de Broglie's hypothesis that its relevance became clear. We discuss de Broglie's hypothesis in the next chapter.

5

Further steps to quantum mechanics: Louis de Broglie and the world's most important PhD thesis

5.1 Introduction

Louis de Broglie came from a very wealthy family. His 17-year-older brother Maurice had set up his own private physics laboratory and hired several competent experimental physicists. His work was very well known and he was the secretary of the first Solvay Conference on quantum physics held in Brussels in 1911. Louis had access to the texts of the talks and other materials and became interested in the unsolved questions of physics through discussions with his brother. During the First World War, he was in the wireless communication service of the French army and worked with radio communications and vacuum tubes. In 1919 he began his theoretical studies while remaining in close communication with the physicists in his brother's laboratory.

At the time of his initial studies, several difficult points remained in the emerging quantum model. There was the problem of what to quantize: the energy, as was done by Planck for his oscillators interacting with the black body radiation and by Einstein for light itself, or the angular momentum or action, as was done by Bohr, Sommerfeld, and Planck in his later work. For a linear oscillator, the two are equivalent. To de Broglie, the action seemed a very abstract quantity, and the photoelectric effect seemed to imply that it is energy which should be quantized. This was given even more force by Compton's experiments on X-ray scattering in 1923. Compton found that X-rays scattered by electrons had their wavelength changed by an amount that depended on the angle of scattering, which he successfully explained by assuming that X-rays consist of photons with energy hc/λ and momentum h/λ, while von Laue's demonstration that X-rays were diffracted by crystals shows that X-rays also behaved as waves. Compton's equation for the change in wavelength of a photon scattering from an electron is given in the Appendix to this chapter.

Another problem was that the meaning of the quantum conditions remained obscure—they appeared as arbitrary rules. Then there was the question, which had been largely ignored, of the relation between the quanta of light—the photons—and the wave-optical phenomena which had been studied for years. Einstein had addressed this question in 1909. He envisaged the total energy of an electromagnetic wave as being localized in singular points, each of which was surrounded by a force field with the character of a plane wave with its amplitude decreasing with distance from its source.

The Historical and Physical Foundations of Quantum Mechanics. Robert Golub and Steven K. Lamoreaux, Oxford University Press.
 DOI: 10.1093/oso/9780198822189.003.0005

When many points (photons) were close together, the fields would superpose to give something "like" the classical electromagnetic field. This idea was not developed any further, and most of the writings at the time talked about waiting for the "real" theory of quanta to explain the connection between photons and waves.

5.2 De Broglie's contribution

The main results of de Broglie's[1] thinking can be summarized very succinctly:

1. The relation
$$\begin{aligned} \vec{p} &= \hbar \vec{k} \\ k &= 2\pi/\lambda \end{aligned} \tag{5.1}$$
shown by Einstein to apply to photons should apply equally well to massive particles like electrons, implying that there is a wave-like phenomenon associated with massive particles. In this context, the wavelength is called the de Broglie wavelength.
2. The same result can be obtained by noting that the energy and momentum form a four-vector $(E, c\vec{p})$ as do the frequency and wave vector of a light wave $(\omega, c\vec{k})$. Since, for photons, we have a proportionality between the time components, $E = \hbar\omega$, it seems but a small step to ask if we might apply the same to the space components, $\vec{p} = \hbar\vec{k}$. The great leap was, however, to apply this to massive particles.
3. An electron in orbit should be accompanied by a wave whose wavelength is such that the length of the orbit is an integer number of wavelengths; otherwise, the wave would not be single-valued as it moved around the orbit. Thus
$$\oint \frac{ds}{\lambda} = n \tag{5.2}$$
or
$$\oint pds = nh, \tag{5.3}$$
which is the Bohr-Sommerfeld quantization condition.
4. Fermat's principle of least time in optics is equivalent to the principle of least action in classical mechanics so that classical mechanics is the analog of geometrical optics, and there exists a description of mechanics which is the analog of wave optics.

While his work can be summarized so simply, his manner of thinking was much more detailed and has profound echoes in our current understanding. De Broglie was the first to attempt to think through in detail the possible relation between the quanta of light (photons) and the well-known wave-optical phenomena. He began by assigning an extremely small mass to the photon, so that while this implied that the velocity of the light varied with its frequency, the mass was so small that the deviation from c would remain undetectable. This assumption is not really necessary, but we will follow de Broglie in order to see how his thinking developed. In a way that was perhaps

[1] de Broglie, L., *Recherche sur le theorie des quanta*, Annales de Physique, 3, 22 (1925).

fortuitous, this assumption meant that it was just a small step to apply these ideas to massive particles.

5.2.1 Particles accompanied by oscillatory phenomena

A photon of mass m would have a rest energy of mc^2. This should be associated with some kind of periodic vibration of frequency $\nu_o = mc^2/h$. In a frame in which the photon was moving with velocity $v = \beta c$ (lab frame), the energy and frequency are given by the Lorentz transformation:

$$E = \frac{mc^2}{\sqrt{1-\beta^2}} \tag{5.4}$$

$$\nu = \frac{\nu_o}{\sqrt{1-\beta^2}}. \tag{5.5}$$

On the other hand, due to time dilation we would expect something vibrating with frequency ν_o in the rest frame to have a frequency

$$\nu_1 = \nu_o \sqrt{1-\beta^2} \tag{5.6}$$

as seen by an observer in the lab frame. What is going on here? Most people would have concluded at this point that there was something wrong with the whole idea and started working on something else. But not de Broglie. In addition to the internal vibration, he assumed that the particle was also associated with a wave of frequency ν and phase velocity v_p. Thus the wave would be described by a function

$$\sin 2\pi\nu\,(t - x/v_p)\,. \tag{5.7}$$

If the particle (remember we are dealing with photons here) was at $x = 0$ at $t = 0$ and the wave and internal vibration were in phase at that time, at a time t later (in the lab frame), the particle would be at $x = vt$, the wave would be given by

$$\sin 2\pi\nu t\,(1 - v/v_p)\,, \tag{5.8}$$

and the internal vibration would be at a phase $(2\pi\nu_1 t)$. The wave and the internal vibration would always be in phase if

$$\begin{aligned} \nu\,(1 - v/v_p) &= \nu_1 \\ \frac{\nu_o}{\sqrt{1-\beta^2}}\,(1 - v/v_p) &= \nu_o\sqrt{1-\beta^2}, \end{aligned} \tag{5.9}$$

which means that $\beta^2 = v/v_p$ and so $v_p = c/\beta > c$. Thus he concluded that the wave could not transport energy. He initially considered it as a fictional wave associated with the motion and called it a "phase wave."

This was followed by an alternate argument. Imagine a vibration in the rest frame of $\sin 2\pi\nu_o t_o$, where t_o is the proper time. The proper time transforms to the lab frame as

$$t_o = \frac{1}{\sqrt{1-\beta^2}}\left(t - \frac{\beta x}{c}\right), \tag{5.10}$$

so that the vibration as seen from the lab frame will be

$$\sin 2\pi\nu_o\left(\frac{1}{\sqrt{1-\beta^2}}\left(t-\frac{\beta x}{c}\right)\right), \tag{5.11}$$

that is again a wave with frequency ν and phase velocity c/β propagating in the direction of motion. If several waves with nearby frequencies are propagating in the same direction, they will add to form a beat whose peak amplitude will travel with the group velocity. We consider a superposition of plane waves $e^{i\left(\vec{k}\cdot\vec{r}-\omega_k t\right)}$, each determined by its wave vector $\vec{k}$. $\omega_{\vec{k}}$ is the angular dependence which is a function of $\vec{k}$. A typical wave packet can be represented as

$$\psi\left(\vec{r},t\right)=\int A\left(\vec{k}-\vec{k}_o\right)e^{i\left(\vec{k}\cdot\vec{r}-\omega_k t\right)}d^3k \tag{5.12}$$

where $A\left(\vec{k}-\vec{k}_o\right)$ is sharply peaked around $\vec{k}=\vec{k}_o$. Writing $\vec{k}-\vec{k}_o=\delta\vec{k}$ and $\omega_k=\omega_o+\vec{\nabla}_k\omega_k\cdot\delta\vec{k}$ where $\omega_o=\omega\left(k_o\right)$

$$\psi\left(\vec{r},t\right)=e^{i\left(\vec{k_o}\cdot\vec{r}-\omega_o t\right)}\int A\left(\delta\vec{k}\right)e^{i\delta\vec{k}\cdot\left(\vec{r}-\vec{\nabla}_k\omega_k t\right)}d^3\delta k \tag{5.13}$$

The first term is a plane wave $\left(\omega_o,\vec{k_o}\right)$ which is modulated by the second term which is a function moving with the group velocity

$$\vec{v}_g=\vec{\nabla}_k\omega_k. \tag{5.14}$$

In one dimension:

$$v_g=\frac{d\omega}{dk}=\frac{d\nu}{d\left(1/\lambda\right)}=\frac{d\nu}{d\left(\nu/v_p\right)}=\frac{d\nu/d\beta}{d\left(\nu/v_p\right)/d\beta}. \tag{5.15}$$

With ν given by Eq. (5.5) and $v_p=c/\beta$, we have

$$\frac{\nu}{v_p}=\frac{m\beta c}{h\sqrt{1-\beta^2}} \tag{5.16}$$

and we find $v_g=\beta c$, which is the particle velocity. Another way of seeing this is to note that if $H\left(p,q\right)=h\nu=\hbar\omega$ is the total energy, the particle velocity is given by Hamilton's equation:

$$\dot{q}=\frac{\partial H}{\partial p}=\frac{\partial\left(\hbar\omega\right)}{\partial\left(\hbar k\right)}=\frac{\partial\omega}{\partial k}=v_g, \tag{5.17}$$

where we used the de Broglie relation $p=\hbar k$ (5.1).

5.2.2 Relation between the "phase wave" and particle motion

Fermat's principle is based on the following idea, which also plays a big role in modern quantum mechanics. If we assume that light can travel by all possible paths between two points A and B, the contributions of almost all paths will cancel because the light coming from different paths will be out of phase. Since the phase varies with position s as

$$\delta\phi = kds = 2\pi \frac{ds}{\lambda} \tag{5.18}$$

the phase accumulated along a path P_{AB} connecting points A and B will be:

$$\delta\phi_{AB} = 2\pi \int_{P_{AB}} \frac{ds}{\lambda}. \tag{5.19}$$

Here ds is an element of the path length and λ is the wavelength of the light, which may be a function of position if the light is moving through an inhomogeneous medium. This accumulated phase is called the optical length.

If there is a route of minimum optical length, the neighboring routes will differ in optical length by a very small amount of second order in the deviation of the paths. Thus the paths clustered around the path of minimum optical length will interfere constructively, and this will be the route followed by the light. This path will be characterized by the fact that $\int \frac{ds}{\lambda}$ is a minimum or

$$\delta \int \frac{ds}{\lambda} = 0 \tag{5.20}$$

$$\delta \int \frac{\nu}{v_p} ds = 0. \tag{5.21}$$

The second form follows from $\lambda\nu = v_p$, where ν is the frequency and v_p is the phase velocity. For light, the frequency remains constant as λ and v_p vary as the index of refraction varies, $v_p(\overrightarrow{r}) = \lambda(\overrightarrow{r})\,\nu = c/n(\overrightarrow{r})$. In that case, we can take the constant ν out of the integral sign and we have the result that the time of travel of the light $ds/v_p(\overrightarrow{r})$ is a minimum. As we have seen, de Broglie's phase wave has a frequency ν given by Eq. (5.5) and a phase velocity v_p which both vary as the particle velocity varies. Then Eq. (5.21) becomes

$$\delta \int \frac{\nu}{v_p} ds = \delta \int \frac{mc^2}{h\sqrt{1-\beta^2}} \frac{\beta}{c} ds = \frac{1}{h} \delta \int pds = 0. \tag{5.22}$$

The last expression is the principle of least action in classical mechanics. Note that here we have switched from talking about photons with an imaginary, negligible rest mass to particles with finite mass. We thus see that $1/\lambda = p/h$ and that Fermat's principle applied to the phase wave is analogous to the least action principle applied to the particle's motion.

5.2.3 The Bohr-Sommerfeld quantum conditions

From the above, we see that if we consider an electron on a closed orbit in an atom and require that the phase wave be a single-valued function of position, the distance around the orbit would have to be an integer number of wavelengths:

$$\oint \frac{ds}{\lambda} = n. \tag{5.23}$$

The above argument shows that this is equivalent to the quantum condition

$$\oint pds = nh. \tag{5.24}$$

However, de Broglie thought a little more deeply about the problem and argued as follows: Let the electron moving with a constant velocity around the orbit be at a point O at time $t = 0$. The fictitious wave (de Broglie's words) would leave the electron and travel at a velocity c/β, catching up with the particle (after making one or more complete orbits) at a later time τ and at a point O'. In this time, the particle will have traveled a distance $OO' = \beta c\tau$ and the wave would have traveled a distance

$$\frac{c}{\beta}\tau = \beta c\tau + S, \tag{5.25}$$

where $S = \beta c T_{rot}$ is the distance around the orbit and T_{rot} is the rotation period of the electron. Solving this equation, we find

$$\tau = \frac{\beta^2}{1-\beta^2} T_{rot}. \tag{5.26}$$

During the time τ, the phase of the internal vibration will have increased by

$$2\pi\nu_1\tau = 2\pi\nu_o\sqrt{1-\beta^2}\left(\frac{\beta^2}{1-\beta^2}T_{rot}\right) \tag{5.27}$$

$$\nu_1\tau = \frac{mc^2}{h}\frac{\beta^2}{\sqrt{1-\beta^2}}T_{rot}. \tag{5.28}$$

It seems that it is reasonable to expect the orbit to be stable only if the wave is in phase with the internal vibration when it catches up with the particle or that $2\pi\nu_1\tau = 2\pi n$. This gives the condition

$$\frac{mc^2\beta^2}{\sqrt{1-\beta^2}}T_{rot} = nh. \tag{5.29}$$

De Broglie then writes the quantum condition in a form given by Einstein:

$$\oint \vec{p}\cdot\vec{dq} = nh = \oint \frac{m\vec{v}}{\sqrt{1-\beta^2}}\cdot\vec{v}dt = \frac{mc^2\beta^2}{\sqrt{1-\beta^2}}T_{rot}, \tag{5.30}$$

which is the same as the condition that the phase wave and the internal vibration remain in phase.

5.2.4 Quantization of phase space

In Chapter 4, Fig. 4.1 we have seen how Planck had shown that the quantum condition for a harmonic oscillator was equivalent to dividing the phase space into cells of size h, one cell for each quantum state, for a system of one degree of freedom. In Chapter 9 we will discuss how Planck extended this to cells of volume h^N for a system with N degrees of freedom, for calculating the statistical properties of the black body radiation. De Broglie showed that this result followed naturally from the existence of the phase wave.

He started by considering a gas moving in one dimension in a space with length L. It is reasonable to expect that any motions connected to non-stationary states of the phase wave would be automatically eliminated, so that the wavelength should satisfy

$$\begin{aligned} L &= n\lambda = n\left(\frac{h}{mv}\right) \\ n &= \frac{mvL}{h}. \end{aligned} \tag{5.31}$$

Assuming the coordinate space is uniform, we have

$$dn = \frac{m \cdot dx \cdot dv}{h} = \frac{dp \cdot dx}{h}. \tag{5.32}$$

The extension to many dimensions is obvious. This is essentially the modern treatment of this issue.

Plank had originally thought that the kinetic energy of a free particle moving in a straight line cannot be considered as a discrete variable, but de Broglie's expression (5.31) shows that if the particle is freely moving between two walls the momentum (and hence the kinetic energy) is indeed quantized.

5.2.5 De Broglie's ideas on the relation between the phase wave and the particle motion

We have seen that de Broglie thought of particles, both massive particles and photons (which he also considered to have an extremely small mass), as being accompanied by a "phase wave" which could not carry any energy and which he often referred to as "fictitious." This wave had a wavelength and phase velocity which varied from point to point and depended on the velocity the particle would have if it were located at the point in question.

The relation between Fermat's principle for the phase wave and the principle of least action shows that the possible particle trajectories are the rays of the phase wave, that is, the lines perpendicular to surfaces of constant phase at each point in space. This idea was strongly criticized by Pauli at the 1927 Solvay Conference and was then more or less abandoned until, twenty-five years later, David Bohm showed that Pauli had been mistaken and revived the idea. What came to be called the de Broglie-Bohm theory is today considered by many physicists to be a viable interpretation of quantum mechanics. We will discuss this in more detail in Chapter 13.

If waves and particles always moved together, how did de Broglie explain the interference phenomena which had been observed with light for many years? If a free particle following a trajectory fixed by Fermat's principle applied to the phase wave (or the principle of least action) passes through an aperture smaller than the wavelength λ of the wave, the trajectory will bend like the ray of a diffracted wave. This applies to any particles, for example, electrons. The phase wave guides the motion of the energy and the new mechanics is to the old like wave optics is to geometric optics. We note that it was Schrödinger, not de Broglie, who developed this new "wave mechanics" (Chapter 7).

De Broglie then proposed that since light is only observed when it interacts with matter as in the photoelectric effect, the probability of such an interaction is related to the intensity of the phase wave. This is very interesting as de Broglie remained extremely dissatisfied with what came to be the dominant probabilistic interpretation of quantum mechanics for the rest of his life. In some sense, the probabilistic interpretation appears natural but de Broglie began to search for a non-probabilistic interpretation shortly after this view gained widespread acceptance.

For interference to take place, the phases of the waves associated with different particles must have a definite relationship. This could be assured if, in the source, the phase wave of a photon, in passing over neighboring atoms, induced them to emit additional photons with a phase wave identical to the first. This means that a given phase wave can hold many photons. When the number of photons in a single wave was very large, the wave would closely resemble a classical wave. The resemblance to Einstein's prediction of stimulated emission (and modern ideas) is striking, although de Broglie did not mention it.

De Broglie then proposed another model, in which moving particles would be associated with a group of waves with frequencies and wavelengths in a narrow range, and the particle would be a singularity of the wave group which had, at all times, a definite position and velocity. The wave rays would again represent possible trajectories. De Broglie devoted many years to developing these ideas, but they do not seem to have led to anything and we will not discuss them further.

De Broglie pointed out that the best way to test his ideas would be to look for diffraction effects when electrons were scattered by crystals. He failed to persuade the physicists in his brother's laboratory to work on this and it remained to Davisson and Germer, working at Bell Labs in the United States, to provide definitive proof that de Broglie's relation $p = h/\lambda$ applied to electrons. This was not published until 1927, some time after the introduction of modern quantum mechanics, and the authors ascribed the discovery to an accident. A catastrophic failure of a vacuum system had caused their heated polycrystalline nickel sample to become highly oxidized. To remove the oxide, they heated the sample for a long time in hydrogen and vacuum. This had the result of crystallizing the sample so that when the authors continued their electron scattering experiments with the treated sample, it exhibited diffraction peaks.

However, de Broglie's ideas turned out to be very influential even in the absence of experimental verification. Langevin, one of the physicists on de Broglie's PhD committee, sent a copy of the thesis to Einstein, who immediately recognized it as the long-awaited breakthrough. Einstein discussed the ideas with many of his associates. To Born he wrote: "You must read this. It might seem crazy but it is thoroughly sound." Discussions of de Broglie's ideas between Debye and Schrödinger led to the latter's development of a full-fledged wave mechanics based on a wave equation.

It is remarkable how many of the main ideas of the modern theory were already present in de Broglie's initial work.

5.3 Appendix to Chapter 5—Compton scattering

Compton's experiment can be analyzed as follows. An X-ray of wavelength λ is scattered by an electron and deflected through an angle θ. After the scattering, it has a wavelength λ'. As Einstein had shown, a photon carries a momentum $p = h\nu/c = h/\lambda = \hbar k$.

Before the collision, the X-ray had momentum $\vec{p}$ and energy $pc = h\nu$ while the electron (at rest) had momentum $\vec{p}_e = 0$ and energy m_oc^2. After the collision, the photon has momentum $\vec{p}\,'$ and energy $cp' = h\nu'$ and the electron has momentum $\vec{p}\,'_e$ and energy $\sqrt{(m_oc^2)^2 + (p'_ec)^2}$. Then by conservation of momentum

$$\vec{p}\,'_e = \vec{p} - \vec{p}\,', \tag{5.33}$$

and we have

$$\begin{aligned} p_e'^2 &= p^2 + p'^2 - 2pp'\cos\theta \\ (p_e'^2c)^2 &= \left[(h\nu)^2 + (h\nu')^2 - 2h\nu h\nu'\cos\theta\right], \end{aligned} \tag{5.34}$$

where θ is the angle between $\vec{p}$ and $\vec{p}\,'$, i.e., the scattering angle of the photon. Conservation of energy means that

$$\begin{aligned} h\nu + m_oc^2 &= h\nu' + \sqrt{(m_oc^2)^2 + (p'_ec)^2} \\ (p'_ec)^2 &= \left(h\nu + m_oc^2 - h\nu'\right)^2 - \left(m_oc^2\right)^2. \end{aligned} \tag{5.35}$$

Equating the two expressions for $(p'_ec)^2$, we find

$$\lambda' - \lambda = \frac{h}{m_oc}(1 - \cos\theta), \tag{5.36}$$

where (h/m_oc) is called the Compton wavelength of the electron. The agreement of this result with experiment showed in a dramatic fashion that electromagnetic radiation consists of photons. However, some time later Schrödinger gave a purely wave analysis of the Compton effect (see Section 10.6.1).[2]

[2]Schrödinger, E., *Uber den Comptoneffekt*, Ann. d. Phys, 82, 257 (1927) (received November 30, 1926).

6 The invention of quantum mechanics—matrix mechanics

6.1 Introduction

In the years after its introduction, the Bohr-Sommerfeld model was applied to a wide range of problems in atomic physics. The successful calculation of the fine structure using special relativity, outlined in Section 4.2.3, and the calculation of the Stark effect were encouraging. A great deal of effort focused on trying to understand the periodic table of elements, but it was impossible to apply the Bohr-Sommerfeld method to many-electron atoms. In addition, the method gave incorrect results for an atom in crossed electric and magnetic fields, only some instances of the Zeeman effect could be explained, and the case of an atom in periodically varying fields certainly could not be described through the quantum rules.

It was clear that a new idea was needed. Remarkably, two such new ideas appeared in the relatively short time period encompassing the last half of 1925 and the first half of 1926. Both ideas supplied a self-consistent and correct method for dealing with atomic problems but they could not have been more different. One, due to Heisenberg with strong assistance from Born, Jordan, and others, dealt with matrices, and the other, due to Schrödinger building on de Broglie's ideas, dealt with waves (see following chapter). There were some strong criticisms of each method by supporters of the other one, but Schrödinger quickly proved that they were, in fact, identical. Today we know that they are both special cases of a more general theory due to Dirac and von Neumann which we will discuss in chapter 15.

According to classical electromagnetic theory, the frequency of any light emitted by an atomic electron should be equal to a multiple of the frequency of the electron's orbital motion. The actual frequencies radiated by atoms can be described by a difference of terms:

$$\omega\left(n,m\right)=\Omega\left(n\right)-\Omega\left(m\right). \tag{6.1}$$

This relationship, known as the Ritz combination rule, had been deduced empirically from the observation that atomic spectra include lines at frequencies that are either the sum or the difference between the frequencies of two other lines. According to the Bohr theory, the terms in Eq. (6.1) are the energies of the stationary states divided by $\hbar$. The radiation is seen as being emitted during a "quantum jump" from one stationary state to the other.

The Historical and Physical Foundations of Quantum Mechanics. Robert Golub and Steven K. Lamoreaux, Oxford University Press.
 DOI: 10.1093/oso/9780198822189.003.0006

The classical emission frequencies can best be calculated using the action angle-variables (J_i, Θ_i) introduced in Section (A.5) for cases where the Hamiltonian is a function of the J_i alone. The fundamental frequencies of the motion are then given by

$$\dot{\Theta}_i = \frac{\partial H(J)}{\partial J_i} = \nu_i. \tag{6.2}$$

Harmonics $\alpha_i \nu_i$ of these frequencies would also be expected to occur, i.e., frequencies $\nu = \sum_i \alpha_i \nu_i$ where the α_i are integers. The quantum condition is that $J = nh$. For large J, (n) and $\alpha \ll n$, we can write the derivative as a difference

$$\nu_i = \frac{\partial H(J)}{\partial J_i} = \frac{H(nh + \alpha h) - H(nh)}{\alpha h} = \frac{\nu_{\mathrm{Bohr}}}{\alpha}, \tag{6.3}$$

so that the emitted frequencies $\alpha_i \nu_i$ are those given by Bohr's frequency rule. This is an example of Bohr's correspondence principle which says that the result of the quantum theory should agree with classical physics in the limit of large quantum numbers.

6.2 Heisenberg rediscovers matrices

The clash between the frequencies of orbital motion and the actual emitted frequencies given by the Ritz combination rule meant that one had to give up either the relation between the frequencies of the emitted light and the orbital frequencies (as Bohr did) or the classical description of the orbital motion in terms of Fourier series. Heisenberg began by trying to calculate the Fourier series representation of the orbits but found the problem to be intractable. This led him to the conclusion that the pretty ellipses drawn by Sommerfeld had no physical reality and that the theory should be built only on quantities that were observable. The frequencies and intensities of the emitted light were certainly such quantities while the orbital frequencies of the electrons were unobservable. He was convinced that something in the atom had to be vibrating with the frequencies of the emitted light.[1]

He then asked what quantum theoretical object could replace the classical coordinate $x(t)$. He began by comparing the classical and quantum frequencies:

$$\nu(n, \alpha) = \alpha \nu(n) = \alpha \frac{\partial H}{\partial J_n} \qquad \text{classical} \tag{6.4}$$

$$\nu(n, n - \alpha) = \frac{1}{h}\left[H(n) - H(n - \alpha)\right] \qquad \text{quantum.} \tag{6.5}$$

In each case, the frequencies combine as

$$\nu(n, \alpha) + \nu(n, \beta) = \nu(n, \alpha + \beta) \qquad \text{classical} \tag{6.6}$$

$$\nu(n, n - \alpha) + \nu(n - \alpha, n - \alpha - \beta) = \nu(n, n - \alpha - \beta) \qquad \text{quantum.} \tag{6.7}$$

Heisenberg then turned to the amplitudes, which should determine the intensity of the emitted light when substituted into the classical expressions for the power emitted by an accelerating charge. The classical Fourier series expansion for a state n is

$$x(t) = \sum_\alpha A_\alpha(n)\, e^{i\omega(n)\alpha t} \qquad \text{classical.} \tag{6.8}$$

[1] We have seen that several other authors had introduced "virtual oscillators" to play this role.

For non-periodic orbits, the sum could be replaced by a Fourier integral. By an inspired piece of guesswork, Heisenberg proposed a quantum mechanical reinterpretation of this. Noting that the two states n and $n - \alpha$ are equally important in the emission process, he wrote

$$x(t) = \sum_{\alpha} A(n, n-\alpha)\, e^{i\omega(n,n-\alpha)t} \qquad \text{quantum.} \tag{6.9}$$

The question then arose of how to represent $x^2(t)$ or $x(t)\,y(t)$. Classically, we can write

$$x^2(t) = \sum_{\beta} B_{\beta}(n)\, e^{i\omega(n)\beta t} \tag{6.10}$$

with

$$B_{\beta}(n) = \sum_{\alpha} A_{\alpha} A_{\beta-\alpha} \tag{6.11}$$

which leads "almost inevitably," as Heisenberg put it, to

$$B(n, n-\beta) = \sum_{\alpha} A(n, n-\alpha)\, A(n-\alpha, n-\beta)\,. \tag{6.12}$$

According to the combination principle,

$$\omega(n, n-\beta) = \omega(n, n-\alpha) + \omega(n-\alpha, n-\beta)\,, \tag{6.13}$$

so that the multiplication law (6.12) follows from the combination principle. The same principle is then applied to higher powers and to the multiplication of two quantities, $x(t)\,y(t)$, so that

$$y(t) = \sum_{\alpha} B(n, n-\alpha)\, e^{i\omega(n,n-\alpha)t} \tag{6.14}$$

$$x(t)\,y(t) = \sum_{\beta} C(n, n-\beta)\, e^{i\omega(n,n-\beta)t} \tag{6.15}$$

$$C(n, n-\beta) = \sum_{\alpha} A(n, n-\alpha)\, B(n-\alpha, n-\beta)\,. \tag{6.16}$$

Born would later recognize Eqs. (6.12) and (6.16) as the rules of matrix multiplication. Heisenberg realized that this form of multiplication was not commutative, so that there would be situations where $xy \neq yx$. He then introduced the quantum condition as

$$J = nh = \oint p dx = \oint m\dot{x} dx = \oint m\dot{x}^2 dt. \tag{6.17}$$

Using $\dot{x} = \sum_{\alpha} A_{\alpha}(n)\, i\alpha\omega_n e^{i\omega_n \alpha t}$ (note that x Eq. (6.8) is real so that $A_{\alpha}(n) = A^*_{-\alpha}(n)$), we have

$$\oint m\dot{x}^2 dt = m \sum_{\alpha} |A_{\alpha}(n)|^2 \alpha^2 \omega_n^2 \int_0^T dt \tag{6.18}$$

$$= 2\pi m \sum_{\alpha} |A_{\alpha}(n)|^2 \alpha^2 \omega_n = nh \tag{6.19}$$

where we used $T = 2\pi/\omega_n$. By differentiating w.r.t. n, in the spirit of the correspondence principle, he obtained

$$h = 2\pi m \sum_{\alpha} \alpha^2 \frac{d}{dn} \left(|A_\alpha (n)|^2 \omega_n \right). \tag{6.20}$$

After quantum mechanical reformulation, this expression yields

$$h = 2\pi m \sum_{\alpha} \alpha \left[|A(n+\alpha, n)|^2 \omega(n+\alpha, n) - |A(n, n-\alpha)|^2 \omega(n, n-\alpha) \right]. \tag{6.21}$$

Heisenberg's idea was to substitute the quantum mechanical expressions for x and $\dot{x}$ into the classical equations of motion and solve for the amplitudes $A(n, n-\alpha)$ and frequencies $\omega(n, n-\alpha)$. He gave solutions for the nonlinear harmonic oscillator and the rigid rotor. We will return to the methods of solving physical problems by this approach using the notation developed shortly afterward by Born and Jordan.

This work was carried out while Heisenberg was on an island, Helgoland, where he had gone because of a severe hay fever attack. On returning to Göttingen he gave the manuscript to Born and asked his opinion on its suitability for publication. Born was extremely supportive and sent it in for publication[2] (received 7/29/1925). He wrote to Einstein, "Heisenberg's new work looks very mystical but is surely correct and deep."

6.3 The founding of matrix mechanics by Born, Jordan, and Heisenberg

Heisenberg told Born that he felt unable to carry the work forward and asked Born for help. Born, in turn, asked Pauli for help but the latter refused, accusing Born of wanting to spoil Heisenberg's physical ideas with "Göttingen mathematics." However, the final version of the theory will be seen to have a strange kind of beauty. Born then approached Jordan, who agreed to collaborate, and Born and Jordan worked on the problem and submitted a paper[3] (received 9/27/1925). The theory was put in its final form with extended applications in a paper by Born, Jordan, and Heisenberg[4] (received 11/16/1925). In a letter to Pauli written on that day,[5] Heisenberg expressed his reservations about the mathematical content of the theory: "Here I am in an environment which thinks and feels exactly opposed to me and I don't know if I am too dumb to understand mathematics. Göttingen has split into two camps, one like Hilbert ... who speak of the great success due to the introduction of matrices into physics, while the other, like Franck, say that we will never be able to understand matrices." Although Born had learned about matrices as a student, and recognized that Heisenberg's quantities were matrices, it seems that this was pretty rare at that

[2]Heisenberg, W., *On the quantum theoretical reinterpretation of kinematic and mechanical relations*, Z. Phys. 33, 879, (1925).

[3]Born, M., and Jordan, P., *On quantum mechanics*, Zeit. Phys. 34, 858, (1925).

[4]Born, M., Heisenberg, W. and Jordan, P. *On quantum mechanics II*, Zeit. Physik, 36, 557, (1926).

[5] Hermann, A., Meyenn, K. von, and Weisskopf, V. F, eds. Wolfgang Pauli, *Scientific Correspondence With Bohr, Einstein, Heisenberg, A. O. Volume I: 1919-1929* (Springer-Verlag New York. 1979). [105], p. 255.

time—matrices seem to have been considered as exotic mathematics. The contrast with today, when we all learn about matrices in our freshman mathematics courses, could not be starker.

Dirac obtained a copy of Heisenberg's first paper in September. After some effort trying to understand it, he worked out the theory independently, duplicating the results of Born and Jordan. Dirac's paper[6] was received on 11/07/1925.

Born, Jordan, and Heisenberg began by defining the momentum and coordinate of a particle as matrices whose elements give the amplitudes for the corresponding transitions between two different states of the particle and introduced the usual matrix notation. We will use bold letters to designate matrices.

$$\mathbf{q} = q\,(nm)\, e^{i\omega(nm)t} \tag{6.22}$$

$$\mathbf{p} = p\,(nm)\, e^{i\omega(nm)t} \tag{6.23}$$

The matrices are taken to be Hermitian so $q\,(nm) = q\,(mn)^*$ and $\omega\,(nm) = -\omega\,(mn)$. The frequencies are assumed to obey the combination rule so that we can write $\omega\,(nm) = [\Omega\,(n) - \Omega\,(m)]$. We do not assume that $\Omega\,(n)$ is related to the energy. We introduce a diagonal matrix $\mathbf{W} = \hbar\Omega\,(n)\,\delta_{nm}$. Functions $\mathbf{g}\,(\mathbf{p},\mathbf{q})$ of $(\mathbf{p},\mathbf{q})$ can be formed as power series using multiplication and addition. Then

$$\mathbf{g} = g\,(nm)\, e^{i\omega(nm)t} \tag{6.24}$$

$$\dot{\mathbf{g}} = i\omega\,(nm)\, g\,(nm)\, e^{i\omega(nm)t} = i\,[\Omega\,(n) - \Omega\,(m)]\, g\,(nm)\, e^{i\omega(nm)t} \tag{6.25}$$

so that $\dot{\mathbf{g}} = 0$ implies that $\mathbf{g}$ is diagonal $(g\,(nm) \sim \delta_{nm})$. Then

$$\mathbf{Wg} = \sum_k W\,(nk)\, g\,(km) = \sum_k \hbar\Omega\,(n)\,\delta_{nk} g\,(km) = \hbar\Omega\,(n)\, g\,(nm)\,. \tag{6.26}$$

$$\mathbf{gW} = g\,(nm)\,\hbar\Omega\,(m)\,. \tag{6.27}$$

For simplicity, we have left out the factor $e^{i\omega(nm)t}$ in the above equations and we will continue do so in the following discussion. Then

$$\dot{\mathbf{g}} = \frac{i}{\hbar}\,(\mathbf{Wg} - \mathbf{gW}) = \frac{i}{\hbar}\,[\mathbf{W},\mathbf{g}] \tag{6.28}$$

where the square brackets denote the commutator of the two matrices. The quantum condition is now formulated in terms of p and q (using the classical Fourier series representation) as

$$\begin{aligned} J = \oint p dq = \int_0^T p\dot{q} dt = \int_0^T dt \sum_\alpha p_\alpha e^{i\omega\alpha t} \sum_{\alpha'} q_{\alpha'} e^{i\omega\alpha' t} i\alpha'\omega \\ = -2\pi i \sum_\alpha p_\alpha q_{-\alpha}\alpha, \end{aligned} \tag{6.29}$$

[6]Dirac, P.A.M., *The fundamental equations of quantum mechanics*, Proc. Roy. Soc. A109, 642, (1925).

where the period is $T = 2\pi/\omega$ and we have used the fact that the integral is only nonzero for $\alpha + \alpha' = 0$. Differentiating with respect to J, we find

$$1 = -2\pi i \sum_{\alpha} \alpha \frac{\partial}{\partial J} \left(q_{-\alpha} p_{\alpha} \right). \tag{6.30}$$

We substitute $p_{\alpha} \to p\,(n, n-\alpha)$, $q_{-\alpha} \to q(n-\alpha, n)$ so that we can approximate the derivative by a difference ($\Delta J = \alpha h$)

$$1 = -\frac{2\pi i}{h} \sum_{\alpha} \left[p\,(n+\alpha, n)\, q(n, n+\alpha) - p\,(n, n-\alpha)\, q(n-\alpha, n) \right]. \tag{6.31}$$

Since we are summing over all values of α, we can change the sign of α in the first term and obtain

$$1 = \frac{2\pi i}{h} \sum_{\alpha} \left[p\,(n, n-\alpha)\, q(n-\alpha, n) - q(n, n-\alpha) p\,(n-\alpha, n) \right]. \tag{6.32}$$

We can recognize the right-hand side of this equation as the diagonal elements of the commutator $[\mathbf{p}, \mathbf{q}]$. We have shown that the quantum condition implies that the commutation relation

$$\frac{\hbar}{i} = \mathbf{pq} - \mathbf{qp} = [\mathbf{p}, \mathbf{q}] \tag{6.33}$$

holds for the diagonal elements of $[\mathbf{p}, \mathbf{q}]$. Born had difficulty finding the relation for the off-diagonal matrix elements and asked Pauli to help with this problem. At first Born assumed the off-diagonal elements were zero, which will prove to be the case. The solution for the off-diagonal elements was found later by Jordan. This is the only place in the theory where Planck's constant appears. In the limit $\hbar \to 0$, the quantities would commute and we would return to the classical theory.

For a system with many degrees of freedom, the commutation relations are

$$[\mathbf{p}_i, \mathbf{p}_j] = [\mathbf{q}_i, \mathbf{q}_j] = 0 \tag{6.34}$$

$$[\mathbf{p}_i, \mathbf{q}_j] = \frac{\hbar}{i} \mathbf{1} \delta_{ij} \tag{6.35}$$

where $\mathbf{1}$ represents the unit matrix, a diagonal matrix with all the diagonal elements equal to unity, and $\delta_{ij} = 1$ if $i = j$ and 0 otherwise. These commutation relations are the key to quantum mechanics in any representation. They hold for any pair of canonical coordinates. The method for quantizing a dynamical system is to write the classical Lagrangian or Hamiltonian and replace the canonical coordinates by some object satisfying the commutation relations.

An important property of the commutation relations can be seen as follows:

$$\begin{aligned} [\mathbf{p}^2, \mathbf{q}] &= \mathbf{p}^2\mathbf{q} - \mathbf{q}\mathbf{p}^2 = \mathbf{p}^2\mathbf{q} - \mathbf{pqp} + \mathbf{pqp} - \mathbf{q}\mathbf{p}^2 \\ &= \mathbf{p}\,[\mathbf{p}, \mathbf{q}] + [\mathbf{p}, \mathbf{q}]\,\mathbf{p} = 2\frac{\hbar}{i}\mathbf{p}. \end{aligned} \tag{6.36}$$

Using this result, we can calculate the commutator for arbitrary powers:

$$\mathbf{p}^n\mathbf{q} = \mathbf{p}^{n-1}\left(\mathbf{qp}+\frac{\hbar}{i}\right) = \mathbf{p}^{n-2}\left(\left(\mathbf{qp}+\frac{\hbar}{i}\right)\mathbf{p}+\frac{\hbar}{i}\mathbf{p}\right) = \mathbf{p}^{n-2}\left(\mathbf{qp}^2+2\frac{\hbar}{i}\mathbf{p}\right)$$
$$= \mathbf{p}^{n-3}\left(\mathbf{qp}^3+3\frac{\hbar}{i}\mathbf{p}^2\right)\dots = \mathbf{qp}^n + \frac{\hbar}{i}n\mathbf{p}^{n-1}$$
$$[\mathbf{p}^n, \mathbf{q}] = \frac{\hbar}{i}n\mathbf{p}^{n-1} \tag{6.37}$$

and the analogous expression for $\mathbf{q}$ is

$$[\mathbf{q}^n, \mathbf{p}] = -\frac{\hbar}{i}n\mathbf{q}^{n-1}. \tag{6.38}$$

Then for any matrix function $\mathbf{f}(\mathbf{p}, \mathbf{q})$ expressible as a power series, we have

$$[\mathbf{f}, \mathbf{q}] = \frac{\hbar}{i}\frac{\partial \mathbf{f}}{\partial \mathbf{p}} \tag{6.39}$$
$$[\mathbf{f}, \mathbf{p}] = -\frac{\hbar}{i}\frac{\partial \mathbf{f}}{\partial \mathbf{q}}. \tag{6.40}$$

Dirac gave an alternative derivation of Eqs. (6.39) and (6.40). If $\mathbf{f}$ and $\mathbf{g}$ are two functions satisfying these equations, the equations must also hold for $\mathbf{f}+\mathbf{g}$ and for $\mathbf{fg}$. The first is trivial, while the second can be seen as follows:

$$\frac{\partial(\mathbf{fg})}{\partial \mathbf{p}} = \mathbf{f}\frac{\partial \mathbf{g}}{\partial \mathbf{p}} + \frac{\partial \mathbf{f}}{\partial \mathbf{p}}\mathbf{g} = \{\mathbf{f}[\mathbf{g}, \mathbf{q}] + [\mathbf{f}, \mathbf{q}]\mathbf{g}\}\frac{i}{\hbar}$$
$$= \{\mathbf{fgq} - \mathbf{fqg} + \mathbf{fqg} - \mathbf{qfg}\}\frac{i}{\hbar} = [\mathbf{fg}, \mathbf{q}]\frac{i}{\hbar}. \tag{6.41}$$

It therefore follows that Eqs. (6.39) and (6.40) apply to any function composed of sums and products.

Using Eq. (6.28) to express the time derivative as a commutator and applying Eqs. (6.39) and (6.40), we can write the equations of motion as

$$\dot{\mathbf{q}} = \frac{i}{\hbar}[\mathbf{W}, \mathbf{q}] = \frac{\partial \mathbf{H}}{\partial \mathbf{p}} = \frac{i}{\hbar}[\mathbf{H}, \mathbf{q}] \tag{6.42}$$
$$\dot{\mathbf{p}} == \frac{i}{\hbar}[\mathbf{W}, \mathbf{p}] = -\frac{\partial \mathbf{H}}{\partial \mathbf{q}} = \frac{i}{\hbar}[\mathbf{H}, \mathbf{p}]. \tag{6.43}$$

This means that

$$[(\mathbf{W}-\mathbf{H}), \mathbf{q}] = 0$$
$$[(\mathbf{W}-\mathbf{H}), \mathbf{p}] = 0.$$

We see that $(\mathbf{W}-\mathbf{H})$ commutes with $\mathbf{p}$ and $\mathbf{q}$ and thus with every function of $(\mathbf{p}, \mathbf{q})$, including $\mathbf{H}$, which means that

$$[(\mathbf{W}-\mathbf{H}), \mathbf{H}] = 0 \tag{6.44}$$
$$[\mathbf{W}, \mathbf{H}] = 0 \tag{6.45}$$
$$\dot{\mathbf{H}} = 0. \tag{6.46}$$

This means that $\mathbf{H}$ is a diagonal matrix. From Eq. (6.42), $[\mathbf{W},\mathbf{q}] = [\mathbf{H},\mathbf{q}]$, so that

$$q\,(nm)\,(\hbar\Omega\,(n) - \hbar\Omega\,(m)) = q\,(nm)\,(H\,(n) - H\,(m)) \tag{6.47}$$

which means that the frequencies are given by

$$\omega\,(nm) = \frac{(H\,(n) - H\,(m))}{\hbar}. \tag{6.48}$$

In this way, the Bohr frequency condition is seen to follow from the commutation rules. Heisenberg considered this as proof that these rules were correct. Starting with the commutation rules and the Hamiltonian equations of motion, we have found that the Hamiltonian is diagonal. We can also reverse the proof so that the equations of motion follow from a diagonal Hamiltonian and the commutation relations. This means that for a given physical problem, putting the Hamiltonian in a diagonal form is equivalent to solving the equations of motion.

From Eqs. (6.28), (6.42) and (6.43), we also have

$$\dot{\mathbf{f}} = \frac{i}{\hbar}\,[\mathbf{H},\mathbf{f}]\,. \tag{6.49}$$

Classically we have

$$\dot{f}\,(p,q) = \frac{\partial f}{\partial p}\dot{p} + \frac{\partial f}{\partial q}\dot{q} = \frac{\partial f}{\partial q}\frac{\partial H}{\partial p} - \frac{\partial f}{\partial p}\frac{\partial H}{\partial q} = \{f,H\}\,, \tag{6.50}$$

where $\{f,H\}$ is called the Poisson bracket. Thus, as Dirac pointed out, the classical analog of the commutator is $i\hbar$ times the Poisson bracket. Returning to the commutator, we can calculate its time derivative as

$$\frac{d}{dt}\,[\mathbf{p},\mathbf{q}] = \frac{d}{dt}\,[\mathbf{pq} - \mathbf{qp}] = \mathbf{p}\dot{\mathbf{q}} - \dot{\mathbf{q}}\mathbf{p} + \dot{\mathbf{p}}\mathbf{q} - \mathbf{q}\dot{\mathbf{p}} = \left[\mathbf{p},\frac{\partial \mathbf{H}}{\partial \mathbf{p}}\right] - \left[\frac{\partial \mathbf{H}}{\partial \mathbf{q}},\mathbf{q}\right]. \tag{6.51}$$

It is easy to see that this is zero for any Hamiltonian of the form $H = f\,(p) + g\,(q)$, so that $[\mathbf{p},\mathbf{q}]$ is a diagonal matrix as Born had assumed previously.

It is remarkable how the complete theory has emerged almost as if by magic from the starting points: the combination principle for spectral lines, which led to the matrix multiplication rule; the commutation relation, which represents the Bohr-Sommerfeld quantum condition; and the classical Hamiltonian equations of motion.

6.3.1 The simple harmonic oscillator

Before proceeding with further developments of the theory, we show how this theory can be used to solve a real physical problem—the harmonic oscillator. Consider the Hamiltonian

$$\mathbf{H} = \frac{\mathbf{p}^2}{2m} + \frac{m\omega_o^2}{2}\mathbf{q}^2 \tag{6.52}$$

which leads to the equation of motion

$$\ddot{q} + \omega_o^2 q = 0. \tag{6.53}$$

We introduce $\mathbf{q} = q(nm)\, e^{i\omega(nm)t}$ so that

$$\left(-\omega(nm)^2 + \omega_o^2\right) q(nm) = 0. \tag{6.54}$$

The commutation relation is

$$\sum_k \left(p(nk)\, q(km) - q(nk)\, p(km)\right) = \frac{\hbar}{i}\delta_{nm}. \tag{6.55}$$

In order for the left-hand side to be nonzero, there must be at least one $q(nm)$ (and one $p(nm)$) $\neq 0$ for $m \neq n$. According to Eq. (6.54), for those values of (nm), we must have

$$\omega_o^2 - \omega(nm)^2 = 0 = \omega_o^2 - \left(\frac{W_n - W_m}{\hbar}\right)^2. \tag{6.56}$$

(Recall that $\omega(nm) = (W_n - W_m)/\hbar$.) Since this means that $W_n - W_m = \pm\hbar\omega_o$, there are only two possible values of m for a given n and $W_m = W_n \pm \hbar\omega_o$. From the commutation relation, we have

$$\frac{\hbar}{i} = \sum_k \left(p(nk)\, q(kn) - q(nk)\, p(kn)\right) = im \sum_k \begin{pmatrix} \omega(nk)\, q(nk)\, q(kn) - \\ q(nk)\, q(kn)\, \omega(kn) \end{pmatrix} \tag{6.57}$$

$$= 2im \sum_k \omega(nk) \left|q(nk)\right|^2 = 2im \left(\omega(nn') \left|q(nn')\right|^2 + \omega(nn'') \left|q(nn'')\right|^2\right) \tag{6.58}$$

$$= 2im\omega_o \left(\left|q(nn')\right|^2 - \left|q(n''n)\right|^2\right) \tag{6.59}$$

where we have used the fact that the matrices are Hermitian so $q(nn') = q^*(n'n)$. Each state n is connected to states with energy $W_n \pm \hbar\omega_o$ by nonzero matrix elements of $\mathbf{q}$. We label these states by $n' = n \pm 1$ and find

$$\frac{\hbar}{2m\omega_o} = \left|q(n+1, n)\right|^2 - \left|q(n, n-1)\right|^2. \tag{6.60}$$

The numbering and arrangement of the states in the matrices are arbitrary, so we choose to number and arrange them in order of increasing energy. For the harmonic oscillator the energy is always positive, so there must be a state of lowest energy. Calling that state $n = 0$, we require that $q(0, -1) = 0$. It then follows from Eq. (6.60) that

$$\left|q(1,0)\right|^2 = \left(\frac{\hbar}{2m\omega_o}\right); \quad \left|q(2,1)\right|^2 = 2\left(\frac{\hbar}{2m\omega_o}\right); \quad \left|q(3,2)\right|^2 = 3\left(\frac{\hbar}{2m\omega_o}\right) \tag{6.61}$$

$$\left|q(n, n-1)\right|^2 = n\left(\frac{\hbar}{2m\omega_o}\right) \tag{6.62}$$

$$\left|q(n+1, n)\right|^2 = (n+1)\left(\frac{\hbar}{2m\omega_o}\right) = \left|q(n, n+1)\right|^2. \tag{6.63}$$

Thus the matrix **q** is given by (rows and columns are labeled by $0, 1, 2, 3, 4 \ldots$)

$$\frac{\mathbf{q}}{\left(\frac{\hbar}{2m\omega_o}\right)} = \begin{bmatrix} 0 & \sqrt{1} & 0 & 0 & 0 & \rightarrow \\ \sqrt{1} & 0 & \sqrt{2} & 0 & 0 & \rightarrow \\ 0 & \sqrt{2} & 0 & \sqrt{3} & 0 & \rightarrow \\ 0 & 0 & \sqrt{3} & 0 & \sqrt{4} & \rightarrow \\ 0 & 0 & 0 & \sqrt{4} & 0 & \rightarrow \\ \downarrow & \downarrow & \downarrow & \downarrow & \downarrow & \searrow \end{bmatrix} \tag{6.64}$$

The matrix extends to an infinite number of rows and columns. Now that we know the matrix **q**, we can substitute it into the Hamiltonian matrix which should be diagonal. The values of the elements of **H** will then be the energies of the states according to matrix mechanics.

$$\mathbf{H} = \frac{m}{2}\left(\dot{\mathbf{q}}^2 + \omega_o^2 \mathbf{q}^2\right) \tag{6.65}$$

$$H(n,m) = \frac{m}{2}\sum_k \left[\omega_o^2 - \omega(n,k)\,\omega(k,m)\right] q(n,k)\,q(k,m) \tag{6.66}$$

$$H(n,n) = m\omega_o^2 \sum_k q(n,k)\,q(k,n) = m\omega_o^2\left(|q(n,n+1)|^2 + |q(n,n-1)|^2\right)$$

$$= (\hbar\omega_o)\left(n + \frac{1}{2}\right) \tag{6.67}$$

From Eq. (6.66) we see that the only possible values for m are $n, n \pm 2$. In the latter case (off-diagonal terms), the frequency factor in square brackets is seen to be zero, so that the Hamiltonian is indeed diagonal.

The energy levels are equally spaced by $\hbar\omega_o$ and the lowest energy (ground state energy) is not zero. This ground state energy is referred to as the zero-point energy, and it is a purely quantum effect. It does not occur in the classical or Bohr-Sommerfeld theories (see Section **A.5.1**). We will see that the Schrödinger wave mechanics give a physical picture of the origin of this energy.

6.3.2 Canonical transformations and perturbation theory

We have seen in Appendix A how important canonical transformations are in classical mechanics. In matrix mechanics a canonical transformation is one that leaves the commutation relations unchanged. This means that the transformed coordinates $P(p,q)$ and $Q(p,q)$ must satisfy

$$[\mathbf{P},\mathbf{Q}] = \frac{\hbar}{i}; \qquad [\mathbf{Q},\mathbf{Q}] = [\mathbf{P},\mathbf{P}] = 0. \tag{6.68}$$

Born suggested an elegant method of writing such a transformation, namely

$$\mathbf{P} = \mathbf{S}\mathbf{p}\mathbf{S}^{-1} \tag{6.69}$$

$$\mathbf{Q} = \mathbf{S}\mathbf{q}\mathbf{S}^{-1}. \tag{6.70}$$

For example $\mathbf{PQ} = \mathbf{SpS}^{-1}\mathbf{SqS}^{-1} = \mathbf{SpqS}^{-1}$. Then for any function made up of sums of powers of $(\mathbf{p}, \mathbf{q})$, we have

$$\mathbf{f}(\mathbf{P}, \mathbf{Q}) = \mathbf{Sf}(\mathbf{p}, \mathbf{q})\,\mathbf{S}^{-1} \tag{6.71}$$

As we have shown above, given any set of matrices $(\mathbf{p}, \mathbf{q})$ that satisfy the commutation relations, the solution of the equations of motion is equivalent to finding a new set of canonical variables that diagonalize the Hamiltonian. That is, if $(\mathbf{P}, \mathbf{Q})$ are given by Eqs. (6.69) and (6.70), then we require

$$\mathbf{H}(\mathbf{P}, \mathbf{Q}) = \mathbf{SH}(\mathbf{p}, \mathbf{q})\,\mathbf{S}^{-1} = \mathbf{W}, \tag{6.72}$$

where $\mathbf{W}$ is a diagonal matrix. This is the analog of the Hamilton-Jacobi equation with $\mathbf{S}$ playing the role of the action function. We now use this idea to derive a solution to the perturbation problem. Assume we are given a Hamiltonian

$$\mathbf{H}(\mathbf{p}, \mathbf{q}) = \mathbf{H}_0(\mathbf{p}, \mathbf{q}) + \lambda\mathbf{H}_1(\mathbf{p}, \mathbf{q}) \tag{6.73}$$

and that we know the solution to $\mathbf{H}_0$, i.e., we know $(\mathbf{p}_0, \mathbf{q}_0)$ so that $\mathbf{H}_0(\mathbf{p}_0, \mathbf{q}_0) = \mathbf{W}_0$ is diagonal. We want to find

$$\mathbf{H}(\mathbf{P}, \mathbf{Q}) = \mathbf{SH}(\mathbf{p}, \mathbf{q})\,\mathbf{S}^{-1} = \mathbf{W}. \tag{6.74}$$

To do this, we write $\mathbf{W}$ and $\mathbf{S}$ as power series in the small parameter λ:

$$\mathbf{W} = \mathbf{W}_0 + \lambda\mathbf{W}_1 + \lambda^2\mathbf{W}_2 + \ldots \tag{6.75}$$

$$\mathbf{S} = \mathbf{1} + \lambda\mathbf{S}_1 + \lambda^2\mathbf{S}_2 + \ldots \tag{6.76}$$

Then

$$\mathbf{S}^{-1} = \mathbf{1} - \lambda\mathbf{S}_1 + \lambda^2\left(\mathbf{S}_1^2 - \mathbf{S}_2\right) + \ldots \tag{6.77}$$

and

$$\left(\mathbf{1} + \lambda\mathbf{S}_1 + \lambda^2\mathbf{S}_2\right)\left(\mathbf{H}_0 + \lambda\mathbf{H}_1\right)\left(\mathbf{1} - \lambda\mathbf{S}_1 + \lambda^2\left(\mathbf{S}_1^2 - \mathbf{S}_2\right)\right) = \mathbf{W} = \mathbf{W}_0 + \lambda\mathbf{W}_1 + \lambda^2\mathbf{W}_2,$$

where the $\mathbf{W}$ matrices are all diagonal. Collecting terms for each power of λ, we obtain an approximate solution:

$$\mathbf{H}_0 = \mathbf{W}_0 \tag{6.78}$$

$$\mathbf{S}_1\mathbf{H}_0 - \mathbf{H}_0\mathbf{S}_1 + \mathbf{H}_1 = \mathbf{W}_1 \tag{6.79}$$

$$\mathbf{S}_2\mathbf{H}_0 - \mathbf{H}_0\mathbf{S}_2 + \mathbf{H}_0\mathbf{S}_1^2 - \mathbf{S}_1\mathbf{H}_0\mathbf{S}_1 + \mathbf{S}_1\mathbf{H}_1 - \mathbf{H}_1\mathbf{S}_1 = \mathbf{W}_2. \tag{6.80}$$

The first equation holds by assumption. We can write the second equation as ($\sum_k$ understood)

$$S_1(mk)\,H_0(kn) - H_0(mk)\,S_1(kn) + H_1(mn) = W_{1,m}\delta_{mn} \tag{6.81}$$

$$S_1(mn)\left(W_{0,n} - W_{0,m}\right) + H_1(mn) = W_{1,m}\delta_{mn}. \tag{6.82}$$

The diagonal terms give

$$H_1\,(nn) = W_{1,n} \tag{6.83}$$

and the off-diagonal terms give

$$S_1\,(mn) = \frac{H_1\,(mn)\,(\delta_{mn} - 1)}{(W_{0,m} - W_{0,n})} \tag{6.84}$$

in the case where every state has a different energy (no degeneracy). Cases with degeneracy (more than one state with the same energy, i.e., $W_{0,m} - W_{0,n} = 0$ for some $m \neq n$) require special treatment. The third equation yields

$$\mathbf{W}_2 = \mathbf{S}_2\mathbf{H}_0 - \mathbf{H}_0\mathbf{S}_2 + (\mathbf{H}_0\mathbf{S}_1 - \mathbf{S}_1\mathbf{H}_0)\,\mathbf{S}_1 + \mathbf{S}_1\mathbf{H}_1 - \mathbf{H}_1\mathbf{S}_1 \tag{6.85}$$

$$\mathbf{W}_2 = \mathbf{S}_2\mathbf{H}_0 - \mathbf{H}_0\mathbf{S}_2 + (\mathbf{H}_1 - \mathbf{W}_1)\,\mathbf{S}_1 + \mathbf{S}_1\mathbf{H}_1 - \mathbf{H}_1\mathbf{S}_1 \tag{6.86}$$

$$W_{2,m}\delta_{mn} = S_2\,(mn)\,(W_{0,n} - W_{0,m}) + (H_1 - W_1)\,(mk)\,S_1\,(kn) + S_1\,(mk)\,H_1\,(kn) - H_1\,(mk)\,S_1\,(kn) \tag{6.87}$$

$$W_{2,m}\delta_{mn} = S_2\,(mn)\,(W_{0,n} - W_{0,m}) + (H_1 - H_1\delta_{mk})\,(mk)\,S_1\,(kn) + S_1\,(mk)\,H_1\,(kn) - H_1\,(mk)\,S_1\,(kn) \tag{6.88}$$

where we used Eq. (6.83) in the last step. Taking diagonal elements ($n = m$), we find

$$W_{2,m} = (H_1 - H_1\delta_{mk})\,(mk)\,S_1\,(km) + S_1\,(mk)\,H_1\,(km) - H_1\,(mk)\,S_1\,(km)$$

$$W_{2,m} = S_1\,(mk)\,H_1\,(km) = \sum_{k \neq m} \frac{H_1\,(mk)\,H_1\,(km)}{(W_{0,m} - W_{0,k})}. \tag{6.89}$$

This is the extremely useful second order perturbation theory result according to which levels are seen to repel. The terms corresponding to states k with $W_{0,k} > W_{0,m}$ lower the energy of the state m and vice versa. As states which differ in energy by greater amounts have a smaller influence, it is often practical to reduce the infinite matrices to a finite size before diagonalizing the Hamiltonian.

Similarly, we can calculate the first order perturbation to the $\mathbf{q}$ matrix:

$$(\mathbf{1} + \lambda\mathbf{S}_1)\,\mathbf{q}_0\,(\mathbf{1} - \lambda\mathbf{S}_1) = \mathbf{q}_0 + \lambda\mathbf{q}_1 = \mathbf{q}_0 + \lambda\,(\mathbf{S}_1\mathbf{q}_0 - \mathbf{q}_0\mathbf{S}_1) \tag{6.90}$$

$$q_1\,(mn) = \sum_{k \neq m,n} \left[\frac{H_1(mk)}{(W_{0,m} - W_{0,k})} q_0\,(kn) - q_0\,(mk)\,\frac{H_1(kn)}{(W_{0,k} - W_{0,n})} \right] \tag{6.91}$$

6.3.2.1 Case of degeneracy

Equation (6.82) is incompatible with the case $(W_{0,n} - W_{0,m}) = 0$ for $n \neq m$ and $H_1\,(mn) \neq 0$. For all N states belonging to the same energy value, we can write the perturbation as an $N \times N$ matrix.

$$\mathbf{H}_1 = \begin{bmatrix} .. & .. & .. & .. & .. \\ .. & H_1\,(n-1, n-1) & H_1\,(n-1, n) & H_1\,(n-1, n+1) & .. \\ .. & H_1\,(n, n-1) & H_1\,(n, n) & H_1\,(n, n+1) & .. \\ .. & H_1\,(n+1, n-1) & H_1\,(n+1, n) & H_1\,(n+1, n+1) & .. \\ .. & .. & .. & .. & .. \end{bmatrix} \tag{6.92}$$

The N eigenvalues obtained by diagonalizing this matrix are then the first order energy changes of the N degenerate states and the eigenvectors form the transformation matrix $\mathbf{S}_1$ for those states.

As an example of diagonalizing a Hamiltonian, consider a two-level system with unperturbed energy levels E_1 and E_2, so that the unperturbed Hamiltonian is given by

$$\mathbf{H}_0 = \begin{bmatrix} E_2 & 0 \\ 0 & E_1 \end{bmatrix}. \tag{6.93}$$

We assume the perturbation has off-diagonal elements only, so that

$$\mathbf{H}_1 = \begin{bmatrix} 0 & b \\ b^* & 0 \end{bmatrix}. \tag{6.94}$$

Then

$$\mathbf{H} = \mathbf{H}_0 + \mathbf{H}_1 = \begin{bmatrix} E_2 & b \\ b^* & E_1 \end{bmatrix} \tag{6.95}$$

and we have taken account of the fact that $\mathbf{H}_1$ must be Hermitian, so that $\mathbf{H}_{1(mn)} = \mathbf{H}^*_{1(nm)}$. To transform this into a diagonal form, we need to solve Eq. (6.72):

$$\mathbf{H}(\mathbf{p}, \mathbf{q})\,\mathbf{S} = \mathbf{S}\mathbf{W} = \mathbf{S}\begin{bmatrix} \lambda_1 & 0 \\ 0 & \lambda_2 \end{bmatrix}. \tag{6.96}$$

The 2×2 matrix $\mathbf{S}$ can be written as $\begin{bmatrix} \boldsymbol{\Psi}_1 & \boldsymbol{\Psi}_2 \end{bmatrix}$, where $\boldsymbol{\Psi}_{1,2}$ are two-component column vectors, so that Eq. (6.96) can be written as

$$\mathbf{H}\boldsymbol{\Psi}_{1,2} = \lambda_{1,2}\boldsymbol{\Psi}_{1,2} \tag{6.97}$$

$$(\mathbf{H} - \lambda_{1,2}\mathbf{1})\,\boldsymbol{\Psi}_{1,2} = 0. \tag{6.98}$$

These equations have a nonzero solution only if

$$\det(\mathbf{H} - \lambda\mathbf{1}) = 0, \tag{6.99}$$

where $\mathbf{1}$ is the unit matrix. Thus we have to solve

$$(E_2 - \lambda)(E_1 - \lambda) - |b|^2 = 0, \tag{6.100}$$

which yields two solutions for λ :

$$\lambda_{1,2} = \frac{E_1 + E_2}{2} \pm \frac{1}{2}\sqrt{(E_1 - E_2)^2 + 4|b|^2}. \tag{6.101}$$

For simplicity we set the zero of energy halfway between the energies of the two levels, so that $E_1 + E_2 = 0$ and $E_2 - E_1 = 2E_2 = 2E$. Then

$$\lambda_{+,-} = \pm\sqrt{E^2 + |b|^2}. \tag{6.102}$$

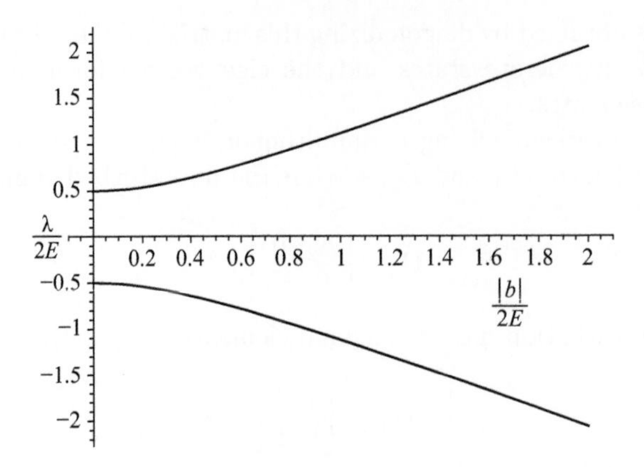

Fig. 6.1 Plot of the eigenvalues $\lambda_{1,2}$ versus |b| in units of $(E_1 - E_2)$.

Substituting these eigenvalues into Eq. (6.97), we find the eigenvectors:

$$\Psi_+ = \begin{bmatrix} \cos \Theta/2 \\ \sin \Theta/2 \end{bmatrix} \tag{6.103}$$

$$\Psi_- = \begin{bmatrix} \sin \Theta/2 \\ -\cos \Theta/2 \end{bmatrix} \tag{6.104}$$

where

$$b/\lambda = \sin \Theta \tag{6.105}$$

$$E/b = \cot \Theta. \tag{6.106}$$

This corresponds to λ being the hypotenuse of a right-angled triangle whose sides are E and b.

As the value of the perturbing matrix element b increases from zero, the energy separation between the states increases quadratically for $|b| \ll E$ and approaches a linear increase for $|b| \gg E$, as shown in Fig. 6.1. Many systems have two energy levels that are much closer to each other than to any other levels. In this case, the matrix elements of the perturbation between these levels and other states will have a negligible effect because the denominator in Eq. (6.89) will be large. We can thus treat the problem by just considering the two neighboring levels and diagonalizing the full Hamiltonian $\mathbf{H}$ within the subspace of those two states, as we have done here, to find the level energies in the presence of the perturbation. This situation occurs for an atom or molecule in an applied electric field E when the electric dipole moment has a nonzero matrix element between the two closely spaced levels. In this case $b = \overrightarrow{p_{12}} \cdot \overrightarrow{E}$, where $\overrightarrow{p_{12}}$ is the matrix element of the electric dipole moment between the states. The energy difference between the states increases as E^2 (quadratic Stark effect) for weak fields and increases linearly for larger fields.

According to Eqs. (6.103) and (6.104), in the limit $|b| \to 0$, Ψ_+ and Ψ_- reduce to the eigenstates of $\mathbf{H}_0$ with eigenvalues E_2 and E_1, respectively. As the perturbation matrix element $|b|$ increases, the eigenstates of $\mathbf{H}$ becomes linear combinations of the

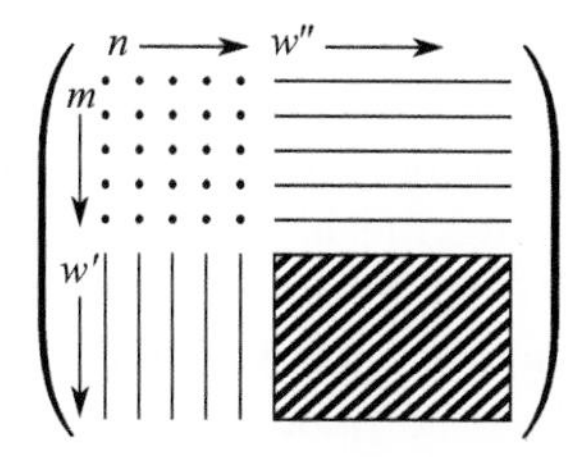

Fig. 6.2 Schematic representation of a matrix including states labeled by both discrete and continuous parameter values. The discrete states are labeled m, n while the continuous states are labeled W', W''.

eigenstates of $\mathbf{H}_0$. When $|b| \gg E$, the eigenstates of $\mathbf{H}$ become equal mixtures of the two unperturbed states. In this limit, the energy shift becomes linear in $|b|$.

6.4 Further developments

One of the features of matrix mechanics was that it allowed the treatment of spectra with both discrete and continuous energy values. There were some mathematical difficulties associated with this, which were finally sorted out in a masterwork by von Neumann (see Chapter 14), but nevertheless the method could be applied to this physically relevant case. To describe a system with states labeled by both discrete and continuous values, such as e.g., the Hydrogen atom with bound (discrete) and unbound (continuous) states, the $\mathbf{p}, \mathbf{q}$ matrices would be represented in the form shown in Fig. 6.2.

As an example, Born, Heisenberg, and Jordan applied the perturbation theory to the anharmonic oscillator. As that played no role in the further development of the theory, we will not discuss it here. They then went on to give a complete description of angular momentum based on the commutation relations. We will reproduce this discussion using more modern notation in a later chapter. In the year following the publication of Born, Heisenberg, and Jordan's paper, their angular momentum results found application to the Zeeman effect and the fine structure of atomic spectra. With the introduction of the electron spin into the theory, these cases could be calculated to great accuracy.

6.5 Conclusion

While matrix mechanics as developed here is still considered to be a correct theory, it is now seen (along with Schrödinger's wave mechanics) as a special case of a more general approach. We will discuss this in the next few chapters. While matrix mechanics and wave mechanics could not appear more different, Schrödinger proved the identity of the two theories within two months of the publication of Born, Heisenberg, and Jordan's paper (op. cit. p. 90). Besides its application to angular momentum, there were not many cases where the program of matrix mechanics, as outlined above, was actually put into practice. The matrix mechanics solution of the hydrogen atom, obtained independently by Dirac and Pauli, was based on special properties of the Kepler problem and used a general vector algebra approach that found no further application. There was a somewhat heated exchange between Heisenberg and Schrödinger over whether the matrix or wave approach was more physically intuitive, which we will return to when we discuss wave mechanics. In a letter to Pauli in 1926,[7] which Heisenberg still quoted as late as 1960, he wrote: "The more I think about the physical part of Schrödinger's theory, the more repulsive I find it. Schrödinger throws overboard everything quantum mechanical as the photoelectric effect, the Franck scattering experiments, the Stern-Gerlach effect, etc. Then it is not hard to make a theory but it disagrees with experience. The great contribution of Schrödinger's theory is the calculation of the matrix elements."

This seems to be the point. Matrix mechanics per se does not offer clear methods for calculating the matrix elements. Many people, including Heisenberg himself initially, had difficulty seeing the physical significance of the matrices, while Heisenberg distrusted the picture offered by the behavior of the wave function in space and time offered by wave mechanics. He felt that the essential physical feature of the theory was the existence of quantum jumps, exemplified in the phenomena mentioned in his letter to Pauli, and which is the basis of his matrix mechanics while being apparently absent from Schrödinger's wave mechanics.

Although matrix mechanics is rarely taught as an independent construct in modern courses, the matrix methods introduced in its development remain a key component of modern quantum theory. The matrix mechanics treatment of angular momentum, for instance, is essentially the same as that used to this day. The angular momentum results were applied to many problems in atomic physics, including the derivation of selection rules that govern which transitions between different atomic states actually take place, which could not be understood previously.

We have seen how starting with the idea of associating a frequency and an amplitude to every possible transition between states in an atom (quantum jumps), rather than with the individual states themselves, Heisenberg was led from the combination principle to the introduction of the matrix multiplication rule, which implied that multiplication was noncommutative. The commutator, the measure of this noncommutative behavior, was then seen to be the key to quantum behavior and replaced the earlier Bohr-Sommerfeld quantization condition. The theory then followed naturally from the properties of the commutator, whose classical analog was shown to be the Poisson bracket.

[7] Wolfgang Pauli, *Scientific correspondence*, op. cit. p. 129, Heisenberg to Pauli,[136], p.328. June 8, (1926).

7 Schrödinger and the development of wave mechanics

7.1 Ideas leading to wave mechanics

7.1.1 Introduction

We have seen how matrix mechanics was developed to give a consistent theory capable of calculating the hydrogen atom spectrum and treating problems like the harmonic oscillator and the rigid rotor. The theory gave a procedure for calculating the energy levels by diagonalization of the Hamiltonian matrix and emphasized the transitions between different levels (quantum jumps). The intensity of a transition was proportional to the square of the matrix element q_{nm}, and the time dependence of the matrix element corresponded to the frequency ω_{nm} of the light emitted (or absorbed) as a result of the transition. However, methods for calculating the matrix elements were not always evident, and the method of calculation (discovered independently by Pauli and Dirac) of the hydrogen atom spectrum was not obviously applicable to other problems. The completion of the theory of matrix mechanics was marked by the publication of Born, Heisenberg, and Jordan's paper[1] (received in November 1925) and Pauli's work on the hydrogen spectrum (received in January 1926[2]). Many problems, such as the fine structure and the Zeeman effect, could not be understood until the concept of electron spin was incorporated into the theory. In March 1926, Heisenberg and Jordan published a matrix analysis of the Zeeman effect incorporating spin.[3]

During this period (November 1925 to June 1926) another form of quantum mechanics based on de Broglie's ideas of particles being accompanied by "phase waves" was being developed by Erwin Schrödinger, an Austrian physicist who had been appointed to a chair in Zurich.

In his famous physics lectures, Feynman said the following concerning Schrödinger's equation: "Where did we get that from? Nowhere. It's not possible to derive it from anything you know. It came out of the mind of Schrödinger, invented in his struggle to find an understanding of the experimental observations of the real world."[4]

[1]Op cit, Chapter 6.

[2]Pauli, W., *On the hydrogen spectrum from the viewpoint of the new quantim mechanics*, Zeit. Phys. 36, 336 (1926).

[3]Heisenberg, W. and Jordan, P., *Application of quantum mechanics to the problem of the anomalous Zeeman effect*, Zeit. Physik 37,263, (1926).

[4]Feynman, R.P., Leighton, R.B., and Sands, M., *The Feynman Lectures on Physics*, (Basic Books, NY, 2011) Vol.3 p.16-12.

The Historical and Physical Foundations of Quantum Mechanics. Robert Golub and Steven K. Lamoreaux, Oxford University Press.
 DOI: 10.1093/oso/9780198822189.003.0007

It seems likely that Schrödinger found the wave equation after about two months of work.

Several different strands of thought converged in Schrödinger's work leading to his development of wave mechanics, and the study of them can help us to a deeper understanding of quantum mechanics. There is some controversy among historians of science as to the relative importance of the various ideas that influenced Schrödinger's thinking. The development of wave mechanics by Schrödinger has been and continues to be one of the most intensely studied episodes in the history of science. We will attempt to discuss the various ideas in what seems to be chronological order but this too is controversial.

7.1.2 First glimmers of a relationship between phase and the quantum condition

While the Bohr-Sommerfeld quantization procedure had some successes (and some failures), the question arose as to whether the quantization rules could be understood on a more fundamental basis. It was hoped that a deeper understanding would lead to an improved theory.

In 1922, a year or two before de Broglie started developing his wave-particle ideas, Schrödinger was working on a theory that had been proposed by Hermann Weyl in an attempt to extend Einstein's theory of general relativity. Weyl's modification of general relativity was based on the idea that there was a scale length (called "gauge" by Weyl) associated with each particle and that this gauge altered by a factor

$$e^{-\frac{e}{\gamma}\int \phi\, dt} \tag{7.1}$$

for a particle of charge e moving in an electric potential ϕ. The factor γ was a constant with dimensions of action that was to be determined. Schrödinger considered applying this to an electron in a Bohr orbit with orbital period τ. He noticed that the exponent could be rewritten

$$-\frac{e}{\gamma}\int_0^{\tau} \phi\, dt = -\frac{\overline{V}\tau}{\gamma} = \frac{2\tau\overline{T}}{\gamma}, \tag{7.2}$$

where $\overline{V}$ is the average potential energy and $\overline{T} = -\frac{1}{2}\overline{V}$ is the average kinetic energy. Now

$$2\tau\overline{T} = \int_0^{\tau} \frac{p^2}{m} dt = \int_0^{\tau} p\dot{q} dt = \oint pdq, \tag{7.3}$$

so the Weyl gauge factor can be written

$$e^{\frac{1}{\gamma}\oint pdq}. \tag{7.4}$$

For a Bohr-Sommerfeld stable state $\oint pdq = nh$ (n integer), so if $\gamma = \hbar/i$, the Weyl factor would become $e^{2\pi i n} = 1$ for a periodic motion allowed by the quantum conditions, and the gauge length would be the same after every period. Schrödinger called this a "noteworthy property of the quantum trajectory of a single electron."[5] He stated that it was "hard to believe that this was a random mathematical result without deeper meaning," but was unable to carry the idea any further. However, when Schrödinger read de Broglie's thesis three years later (both he and Einstein were impressed by de

[5]Schrödinger, E., *On a noteworthy property of the quantum trajectory of a single electron*, Zeit. Physik, **22**, 13, (1922), received Oct. 5 1922.

Broglie's explanation of the quantum condition), he recognized an affinity and noted in a letter to Einstein that[6]

"de Broglie's interpretation of the quantum condition seems to be related in some way to my note concerning the Weyl gauge factor. The mathematical situation is, as far as I can see, the same. Naturally de Broglie's consideration in the framework of his broad theory is of far greater value than my single statement, which I did not know what to make of, at first."

Thus his early work on the Weyl theory would have predisposed Schrödinger to pay more attention to de Broglie's ideas. On the other hand there is a story that Schrödinger's first reaction to de Broglie's thesis was that it was "rubbish" and Ehrenfest persuaded him to give it a careful reading.

7.1.3 The relationship between particles and waves in the quantum theory of the monatomic ideal gas

In 1924 Satyendra Nath Bose, a physicist from India, sent Einstein a paper with a new derivation of Planck's radiation law which considered the radiation as consisting of a gas of identical photons, and which Einstein then forwarded to be published.[7] Between the end of 1924 and the beginning of 1925, Einstein presented a two-part paper[8] to the Prussian Academy of Science in which he applied Bose's ideas to an ideal monatomic gas and demonstrated the existence of "Bose-Einstein" condensation. At thermal equilibrium at low enough temperatures, many gas atoms will "condense" into the ground state of the system. This effect is at the root of superfluidity and superconductivity and is actively studied in gases of ultracold atoms.

Just as in the case of the Planck distribution, Einstein found that the fluctuations in this gas consist of two terms, one associated with particle fluctuations and the other with waves. Einstein wrote: "We have to associate with the gas a radiation process and calculate its fluctuations due to interference. De Broglie showed how a scalar wave field can be associated with a system of particles."

Impressed with this work, Schrödinger obtained a copy of de Broglie's thesis and approached the gas problem from a different direction, taking "seriously the de Broglie-Einstein wave theory of moving particles according to which the particles are nothing more than a type of white foam on the wave crests of an underlying wave motion." We see here the beginning of his thinking about the wave-particle relation. Schrödinger noted that the Planck law can be derived (as we have shown in Section 3.2.3, Eqs. (3.42) and (3.43)) by applying what he called "natural statistics" (i.e., the Boltzmann factor), to the modes of the radiation field. The photons are regarded as being the excitations of the modes of vibration of the radiation field.[9] Vibrations in a solid were treated in the same way in calculations of the specific heat by Einstein[10] and later by Debye.

[6]Schröedinger to Einstein, Nov. 3, 1925, cited in Hanle, P.A., *The Schröedinger-Einstein correspondence and the sources of wave mechanics*, Am. J. Phys. 47, 644, (1979)

[7]Bose, S.N., *Planck's law and the light quantum hypothesis*, Zeit. Physik **26**, 176 (1924) received July 2, (1924).

[8]Einstein, A., *Quantum theory of the monoatomic gas 1 & 2*, Sitz. Ber. Preuss. Akad. Wiss. p. 262, (July 10, 1924) and p.3 (Jan. 8, 1925)

[9]Schroedinger, E., *On Einstien's theory of gases*, Phyik. Zeit. 27, 95 (1926) received Dec. 15, (1925).

[10]Einstein, A., *The Planck theory of radiation and the specific heat*, Ann. d. Phy., 22, 180 (1906).

Schrödinger proposed that the atoms in a gas have a series of available energy levels, $\varepsilon_1, \varepsilon_2, \ldots \varepsilon_n$. Arbitrary numbers of atoms could be in each level, which would then have the energy $n_i\varepsilon_i$, where n_i is the number of atoms in the level with energy ε_i. Each level (labeled by i) then behaves like a one-dimensional oscillator with energy levels $0, \varepsilon_i, 2\varepsilon_i, \ldots n_i\varepsilon_i$, and the probability of a state specified by the set of occupation numbers n_i is proportional to the Boltzmann factor

$$e^{-\frac{1}{k_BT}[n_1\varepsilon_1+n_2\varepsilon_2+\ldots]} \tag{7.5}$$

where k_B is Boltzmann's constant. The idea of considering atoms as excitations of a normal mode is the basis of second quantization, which played a great role in the further development of quantum theory (Chapter 12). The fact that both de Broglie and Schrödinger proposed this model prior to the development of wave mechanics is an indication of their deep physical insight.

Schrödinger went on to calculate the energy levels using an argument that had been applied previously by Jeans in connection with radiation and which would have wide-ranging applications. Consider a wave confined in a cube with sides of length L. The boundary condition means that the wave must vanish at the boundaries of the volume $(x, y, z = 0, L)$. Using $k = 2\pi/\lambda$, and taking the wave to be $\sim \sin kx$ means that we must put $kL = \pi n$ (n an integer) or $n = kL/\pi$. Consider a three-dimensional grid of points labeled by (n_x, n_y, n_z). The number of points dN in a shell between $n = \sqrt{n_x^2 + n_y^2 + n_z^2}$ and $n + dn$ is given by the volume of the grid, divided by eight to account for the fact that n can take only positive values, so that

$$dN = 4\pi n^2 dn = \frac{\pi}{2}\frac{L^3k^2}{\pi^3}dk, \tag{7.6}$$

and the number of states N having wave vector less than k is

$$N = \int_0^k dN = \frac{1}{2\pi^2}\frac{Vk^3}{3} \tag{7.7}$$

where $V = L^3$ is the volume of the cube. This means that k can be expressed as

$$k = \left(6\pi^2\frac{N}{V}\right)^{1/3}. \tag{7.8}$$

The energy of the Nth level is thus

$$E = \frac{\hbar^2k^2}{2m} = \frac{\hbar^2}{2m}\left(6\pi^2\frac{N}{V}\right)^{2/3} = \frac{h^2}{2m}\left(\frac{3N}{4\pi V}\right)^{2/3} \tag{7.9}$$

which agrees with Einstein's result. Using some formidable mathematics, Schrödinger, using an explicit wave picture, then showed that the distribution and fluctuations of the atoms in the gas were the same as had been found by Einstein (Section 9.1.5).

Schrödinger concluded the paper with a remarkable discussion of "the possibility to represent molecules or light quanta through the interference of plane waves." If we superpose a great number of plane waves with wave vectors which vary over a small range of magnitudes and directions clustered around some central wave vector, the interference of the different waves will produce a function that is only nonzero in a small region of space (which is still large compared to the central wavelength). The peak of this function will move with the group velocity which, as de Broglie had shown, is the classical particle velocity. The amplitude of this wave can be written as

$$\psi\left(\vec{x}\right) = \int A\left(\vec{k}\right) e^{i\left(\vec{k}\cdot\vec{r}-\omega(k)t\right)} d^3k, \tag{7.10}$$

where $A\left(\vec{k}\right)$ is nonzero only in a small region around the central wavevector, $\vec{k} = \vec{k}_o$. Writing $\vec{k} = \vec{k}_o + \overrightarrow{\Delta k}$, we have

$$\begin{aligned}\psi\left(\vec{x}\right) &= \int A\left(\overrightarrow{\Delta k}\right) e^{i\left(\vec{k_o}\cdot\vec{r}-\omega(k_o)t\right)} e^{i\left(\overrightarrow{\Delta k}\cdot\vec{r}-\vec{\nabla}_k\omega(k)\cdot\overrightarrow{\Delta k}t\right)} d^3\left(\Delta k\right) \\ &= e^{i\left(\vec{k_o}\cdot\vec{r}-\omega(k_o)t\right)} \int A\left(\overrightarrow{\Delta k}\right) e^{i\overrightarrow{\Delta k}\cdot\left(\vec{r}-\vec{\nabla}_k\omega(k)t\right)} d^3\left(\Delta k\right),\end{aligned} \tag{7.11}$$

where $\vec{\nabla}_k\omega\left(k\right) = \hat{\imath}\frac{\partial\omega_k}{\partial k_x} + \hat{\jmath}\frac{\partial\omega_k}{\partial k_y} + \hat{k}\frac{\partial\omega_k}{\partial k_z}$ ($\hat{\imath}$, $\hat{\jmath}$, and $\hat{k}$ are unit vectors in the x, y, and z directions) and we take $A\left(\overrightarrow{\Delta k}\right)$ to have a maximum at $\overrightarrow{\Delta k} = 0$. This represents a plane wave $\left(\vec{k_o}, \omega\left(k_o\right)\right)$ modulated by a function moving with the group velocity, $\vec{v} = \vec{\nabla}_k\omega\left(k\right)$. Schrödinger noted that such a wave packet will tend to spread and expressed the hope that future quantum theory would modify this behavior. This was not to be, and Schrödinger's future attempts to support this model eventually failed because of this spreading, which we will discuss in Chapter 10.

In studying the behavior of a quantum gas, Schrödinger wholeheartedly embraced de Broglie's idea that the motion of particles (photons and molecules) was accompanied by a wave motion. In fact, Schrödinger took the wave concept much further in that he regarded particles as a product of the wave motion. For de Broglie, the wave accompanied the particle. They were separate entities. For Schrödinger, at this stage, the particle was a wave packet.

7.1.4 First appearance of a wave equation

In Schrödinger's notebooks,[11] there are a few pages headed "H-Atom Eigenvibrations," which seem to contain the first appearance of a wave equation for the de Broglie waves (see Fig. 7.1). In these notes, Schrödinger writes a relativistic wave equation for a wave with the phase velocity given by de Broglie. A standard wave equation is given by

$$\nabla^2\psi - \frac{1}{u^2}\frac{\partial^2\psi}{\partial t^2} = 0, \tag{7.12}$$

where u is the phase velocity. (This first use of ψ in this context persists to this day.) Taking out the time dependence in the usual way by writing

$$\psi = \psi' e^{i\omega t}, \tag{7.13}$$

[11]Archive for the History of Quantum Physics, 1898-1950, American Philosophical Society, Philadelphia, PA, http://www.amphilsoc.org/library.

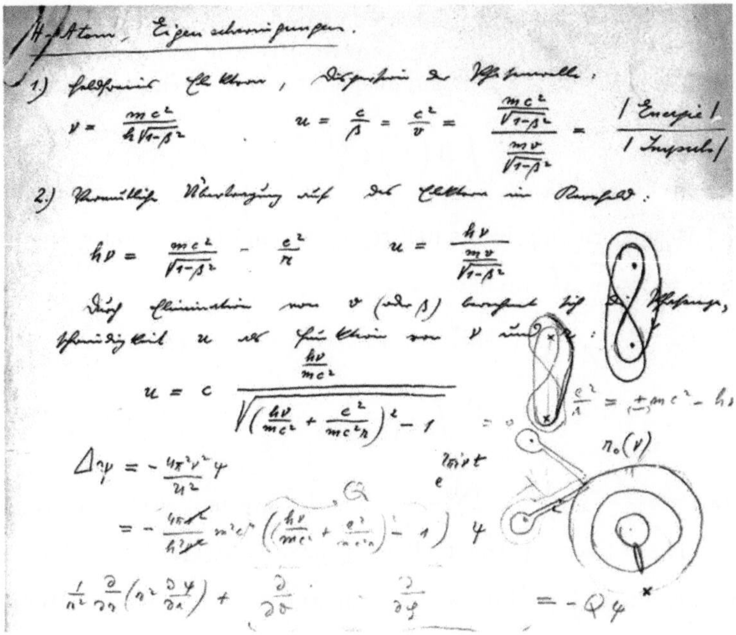

Fig. 7.1 The page from Schrödinger's notebook containing the first appearance of the wave equation for the hydrogen atom. The wave equation is just visible under the expression for u. The loop diagrams with intricate loops around two points refer to the path for a contour integration used to solve second-order differential equations by a method not in general use today. The lower doodle seems to be concerned with jumps between Bohr orbits. (Used by permission: Earle E. Spamer Reference Archivist, Coordinator for Reference and Library Programs, The American Philosophical Society Library, Archive for the History of Quantum Physics, "Rough notes, H-Atom, Eigenschwingungen," Film 40, Reel-Frame 5.)

we get

$$\nabla^2\psi + \frac{\omega^2}{u^2}\psi = 0, \tag{7.14}$$

where we have dropped the prime in ψ'. This equation is often used in studying optics problems. In the presence of a potential V, the relativistic energy of a massive particle is given by (with $\beta = v/c$)

$$E - V = \frac{mc^2}{\sqrt{1-\beta^2}} = \hbar\omega - V, \tag{7.15}$$

where we used $E = \hbar\omega$. Defining γ by

$$\gamma = \frac{\hbar\omega - V}{mc^2} = \frac{1}{\sqrt{1-\beta^2}} \tag{7.16}$$

$$\beta = \sqrt{1 - \frac{1}{\gamma^2}} \tag{7.17}$$

we find for the de Broglie phase velocity

$$u = \frac{E}{p} = \frac{\hbar\omega}{m\beta c/\sqrt{1-\beta^2}} \tag{7.18}$$

$$\frac{\omega^2}{u^2} = \left(\frac{mc}{\hbar}\right)^2 \left(\gamma^2 - 1\right). \tag{7.19}$$

Then for an electron in a hydrogen atom ($V = -e^2/r$), the relativistic wave equation becomes

$$\nabla^2\psi + \left(\frac{mc}{\hbar}\right)^2 \left(\left(\frac{\hbar\omega + e^2/r}{mc^2}\right)^2 - 1\right)\psi = 0. \tag{7.20}$$

Using the non-relativistic relation for the de Broglie phase velocity,

$$\frac{\omega^2}{u^2} = \frac{E^2}{\hbar^2}\frac{p^2}{E^2} = \frac{p^2}{\hbar^2} = k^2 = \frac{2m}{\hbar^2}\left(E - V\right), \tag{7.21}$$

we obtain

$$\nabla^2\psi + \frac{2m}{\hbar^2}\left(E - V\right)\psi = 0$$

$$-\frac{\hbar^2}{2m}\nabla^2\psi + V\psi = E\psi. \tag{7.22}$$

This is the most straightforward way of arriving at the time-independent non-relativistic Schrödinger equation, but we should keep in mind Feynman's remark and avoid the tendency to trivialize the enormous breakthrough that this represented. The relativistic equation, Eq. (7.20), is now called the Klein-Gordon equation and is known to be the correct relativistic wave equation for spin-zero particles. It turns out that relativistic single-particle wave equations have only a limited usefulness because of the possibility of pair production. Schrödinger worked on the solution of this relativistic equation for the hydrogen atom and was terribly disappointed because it gave half-integer quantum numbers and the results for the energy levels did not agree with experiment. This is now seen to be due to the absence of electron spin. However, Schrödinger soon found that the non-relativistic equation gave reasonable results.

Both the relativistic and non-relativistic equations for the hydrogen problem can be solved by the same method, the only difference between the equations being the value of the constants involved. We will give the solution to the non-relativistic wave equation for several physical problems in Section 7.3.

7.1.5 "Quantization as an eigenvalue problem"

This was the title of Schrödinger's series of four papers[12] presenting the non-relativistic equation and its many ramifications. Differential equations like Eq. (7.22) are known to have solutions that satisfy specific boundary conditions such as

[12]Schrödinger's papers on wave mecahnics: *Quantization as an eigen-value problem*, Parts 1-4, Ann, der Physik, 79, 361 (1926), 79, 489, (1926), 80, 437 (1926), 81, 109 (1926) as well as: *On the relation between the Heisenberg -Born-Jordan quantum mechanics and mine*, Ann, der Physik, 79, 734 (1926) are reprinted in English in Schrödinger, E., *Collected Papers On Wave Mechanics*, Minkowski Institute Press, 2020, and in Ludwig, G., *Wave Mechanics*, Pergamon Press, 1968, among others.

$$\lim_{r \to \infty} \psi(\vec{r}) = 0 \tag{7.23}$$

only for certain values of a parameter in the equation such as E. These special values are called eigenvalues after the German word "eigen" which means "own." The corresponding solutions are called eigenfunctions. In Chapter 12 we will show how this works in one dimension for the case of a vibrating string, such as a violin chord that is fixed at both ends. We see that the boundary conditions force the eigenvalue k to take on only discrete values as given by Eq. (12.74). A vibrating membrane would provide a two-dimensional eigenvalue problem where the solution would depend on the shape and the size of the membrane. Schrödinger expressed the idea as follows: "I wish to show that the customary quantum conditions can be replaced by another postulate, in which the notion of integer numbers, as such, is not introduced. Rather when integers do appear, they arise in the same natural way as the node numbers on a vibrating string. The new conception is capable of generalization, and strikes, I believe, very deeply at the true nature of the quantum rules." This is yet another train of thought leading to the wave equation.

7.1.6 Peter Debye

Schrödinger was Debye's successor at the University of Zurich after Debye had moved on to ETH, (Communal Technical University) Zurich, and they organized a colloquium together. Since Schrödinger's work on the ideal gas problem was based on de Broglie's thesis, Debye asked Schrödinger to give a talk on de Broglie's ideas. After the talk Debye remarked that he found de Broglie's treatment to be childish. He had learned from Sommerfeld that the proper way to discuss waves was with a wave equation. A few weeks later Schrödinger gave another talk that he began by saying that Debye had asked for a wave equation and he had now found such an equation. After this colloquium, Debye remarked that all the most important theories in physics had been expressed in the form of a variational principle that encompassed the content of the theory. This would mean that Schrödinger's first "derivation" of the wave equation did not involve a variational principle.

7.1.7 Summary of Schrödinger's work leading to the wave equation

As we have seen, several of the topics that Schrödinger worked on are closely related to his discovery of the wave equation:

1. Early work on the possible relation of Weyl's gauge factor to the Bohr-Sommerfeld quantization condition.
2. Application of the wave picture to the statistics of a gas (Bose-Einstein).
3. The eigenvalue problem as a natural way of obtaining discrete values.
4. Direct substitution of the de Broglie phase velocity into a wave equation.

In his published work, Schrödinger introduced two additional methods for obtaining the equation: via a variational principle and by using the analogy between geometric and wave optics to derive a wave equation from the H-J equation. In the second publication he denounced the argument of the first publication, which is based on a variational principle, as "incomprehensible." From this it might seem that this first method was constructed after the fact in order to justify the result he had obtained using the fourth method listed above. However, there is some evidence that he indeed considered the variational method as the foundation of the theory.

When presenting a new result, a researcher often does not attempt to describe their own original path to the result, but selects an argument they feel will help the reader best understand the concept. There has been some speculation that Schrödinger was reluctant to present the approach based directly on de Broglie's ideas because these ideas were either not known or were regarded as unrealistic by most physicists working in the field. In addition, Debye's remarks might have brought the idea of a variational principle to the front of his mind.

7.2 The development of wave mechanics as presented in Schrödinger's publications

7.2.1 Derivation of the wave equation from a variational principle

In Schrödinger's first paper on wave mechanics, "Quantization as an eigenvalue problem, First Communication," he presented an argument justifying the form he chose for the wave equation and then showed how solving the equation for the hydrogen atom led to the correct Balmer spectrum. We will discuss that solution in Section 7.3.4.

Schrödinger introduced the argument by saying that in his work the usual quantization rules are replaced by another requirement. The argument starts with the Hamilton-Jacobi (H-J) equation:

$$H\left(q, p = \frac{\partial S}{\partial q}\right) = E. \tag{7.24}$$

As we have seen, this equation can be separated into independent equations by writing, e.g., (as we have seen the system of coordinates in which the equation can be separated depends on the potential)

$$S(x, y, z) = X(x) + Y(y) + Z(z). \tag{7.25}$$

For systems of more than one electron, the process is repeated for the coordinates of every electron. Schrödinger then introduced a new function Ψ such that

$$S = K \ln \Psi, \tag{7.26}$$

and the H-J equation is written

$$H\left(q, p = \frac{K}{\Psi}\frac{\partial \Psi}{\partial q}\right) - E = F(q, \Psi) = 0. \tag{7.27}$$

Instead of looking for a solution and then applying the quantum conditions, he replaced the quantum condition by a variational problem. He required that Ψ be single-valued, finite, and twice differentiable over the entire configuration space, and that the integral of $F(q, \Psi)$ over the entire space be an extremum (maximum or minimum). Substituting $H = p^2/2m + V(q)$,[13] the variational principle (in rectangular coordinates) is

$$\delta \int \left[\frac{1}{2m} \left[\left(\frac{K}{\Psi} \frac{\partial \Psi}{\partial x} \right)^2 + \left(\frac{K}{\Psi} \frac{\partial \Psi}{\partial y} \right)^2 + \ldots \right] + V - E \right] d\tau = 0, \tag{7.28}$$

where $d\tau$ represents the volume of the whole configuration space $d^3r_1 d^3r_2 \ldots . d^3r_n$. We can rewrite this as

$$\delta \int \left[\left[\left(\frac{\partial \Psi}{\partial x} \right)^2 + \left(\frac{\partial \Psi}{\partial y} \right)^2 + \ldots \right] + \frac{2m}{K^2} (V - E) \Psi^2 \right] d\tau = 0. \tag{7.29}$$

The procedure for calculating the variation of the integral is to replace Ψ by $\Psi + \delta\Psi$ and take the terms that are linear in $\delta\Psi(q)$ in the difference $[F(\Psi + \delta\Psi) - F(\Psi)]$, where $\delta\Psi$ is considered to be an infinitesimal function of the coordinates. The variation is thus given by

$$\int \left[2 \left\{ \left(\frac{\partial \Psi}{\partial x} \right) \frac{\partial \delta\Psi}{\partial x} + \left(\frac{\partial \Psi}{\partial y} \right) \frac{\partial \delta\Psi}{\partial y} + \ldots \right\} + \frac{4m}{K^2} (V - E) \Psi \delta\Psi \right] d\tau. \tag{7.30}$$

We then integrate the terms containing the partial derivatives by parts to obtain

$$\int \left[- \left[2 \left\{ \left(\frac{\partial^2 \Psi}{\partial x^2} \right) + \left(\frac{\partial^2 \Psi}{\partial y^2} \right) + \ldots \right\} \right] + \frac{4m}{K^2} (V - E) \Psi \right] \delta\Psi d\tau + \int \frac{\partial \Psi}{\partial n} \delta\Psi d^2S. \tag{7.31}$$

The last term is the boundary term from the integration by parts: d^2S is an element of the surface surrounding the volume of integration and $\frac{\partial \Psi}{\partial n}$ is the derivative of Ψ with respect to the normal to this surface. If the surface S is taken to be infinitely far away, the second term will vanish since $\delta\Psi$ must satisfy the boundary condition that it vanishes at infinity. We can write Eq. (7.30) in a more modern notation as

$$\int \left[2\vec{\nabla}\Psi \cdot \vec{\nabla}\delta\Psi + \frac{4m}{K^2} (V - E) \Psi \delta\Psi \right] d\tau, \tag{7.32}$$

and the integration by parts yields

$$\int \frac{\partial \Psi}{\partial n} \delta\Psi d^2S + \int \left[-2 \left(\nabla^2 \Psi \right) + \frac{4m}{K^2} (V - E) \Psi \right] \delta\Psi d\tau. \tag{7.33}$$

As argued above, the first term vanishes. In order for the second term to be zero for all $\delta\Psi$, the expression in square brackets must be zero, so that

$$- \left(\nabla^2 \Psi \right) + 2 \frac{m}{K^2} (V - E) \Psi = 0. \tag{7.34}$$

Using modern notation, this expression becomes

$$- \frac{\hbar^2}{2m} \left(\nabla^2 \Psi \right) + V\Psi = E\Psi. \tag{7.35}$$

This is known as the time-independent Schrödinger equation. Note that going from the variational principle of Eq. (7.29) to the transformed expression Eq. (7.31) has the

[13] Here q stands for all the coordinates of all the electrons: $x_1, y_1, z_1, x_2, y_2, z_2 \ldots x_n, y_n, z_n$ for an n-particle system.

effect of changing the square of the partial derivative $\left(\frac{\partial\Psi}{\partial x_i}\right)^2$ to the second derivative $\left(\frac{\partial^2\Psi}{\partial x_i^2}\right)$, which is what is needed for a wave equation.

In his second communication Schrödinger gave another justification for the wave equation based on the relation between geometrical optics and classical point mechanics, which we will explain in Section 7.2.3. The second communication began with the statement that he had previously (in the first communication described above) used what he called the "incomprehensible" transformation of Eq. (7.26) and the "equally incomprehensible" transfer from setting an expression equal to zero as in the H-J equation to the requirement that the volume integral of the same expression should be stationary. He explained that "this was only for a temporary fast orientation concerning the relation between the wave equation and the H-J equation." This would seem to indicate that the variational principle argument was only introduced as a sort of "marketing" measure, perhaps in response to Debye's remark concerning the importance of variational principles in physics, or due to a reluctance to base the theory on de Broglie's ideas, which did not have general recognition at the time.

However, in a letter to Planck sent a few days after the second communication containing the above remarks was received by the journal, Schrödinger framed the entire problem in terms of the variational principle argument. He wrote: "The stationary values of the Hamiltonian integral

$$\int \left[\frac{K}{2m}\left[\left(\frac{\partial\Psi}{\partial x}\right)^2 + \left(\frac{\partial\Psi}{\partial y}\right)^2 + \ldots\right] + V\Psi^2\right] d\tau \tag{7.36}$$

imposing the condition $\int \Psi^2 d\tau = 1$, are the quantum values of the energy for the mechanical system under consideration, if we set the constant $K = \hbar^2$. Isn't that very remarkable?"[14] So even though he devoted all his attention to solving the wave equation, Schrödinger continued to take the more general point of view that he was, in fact, solving the variational problem.

7.2.2 Applications of the variational principle

Integrating the partial derivatives in (7.36) by parts gives

$$\int \left(\frac{\partial\Psi}{\partial x}\right)^2 d\tau = -\int \Psi\frac{\partial^2\Psi}{\partial x^2} d\tau \tag{7.37}$$

plus a surface term which vanishes because of the boundary conditions. Thus the variational principle can be written

$$\delta\int \Psi\left[-\frac{\hbar^2}{2m}\left(\nabla^2\Psi\right) + (V)\,\Psi\right] d\tau = 0. \tag{7.38}$$

In this form it is used as an approximation method to find the ground state of systems that are not solvable by other methods. A form of the wave function is guessed that

[14]Schrdinger to Planck Feb. 26, 1926, quoted in Mehra, J. and Rechenberg, H., *The Historical Development of Quantum Mechanics, Vol.5, Part2*, p. 623.

depends on some parameters to be determined and then the integral is minimized as a function of these parameters. The resulting parameter values are then used to calculate an estimate of the energy. The true energy is always less than the calculated value since it represents the true minimum of the integral. As an example, we can apply this method to the linear harmonic oscillator, for which the variational principle is written as

$$\delta \int \Psi \left[-\frac{\hbar^2}{2m} \left(\frac{\partial^2}{\partial x^2} \Psi \right) + \frac{1}{2} m\omega^2 x^2 \Psi \right] dx = 0, \tag{7.39}$$

and we will take $\Psi = \frac{\sqrt{\alpha}}{\sqrt[4]{\pi}} e^{-\alpha^2 x^2/2}$ for a normalized trial wave function (with some hindsight, of course). The Hamilton integral is

$$\frac{\alpha}{\sqrt{\pi}} \int_{-\infty}^{\infty} e^{-\alpha^2 x^2} \left[\frac{\hbar^2}{2m} \alpha^2 + \left(\frac{1}{2} m\omega^2 - \frac{\hbar^2}{2m} \alpha^4 \right) x^2 \right] dx = \frac{1}{4} \left(\frac{\hbar^2}{m} \alpha^2 + m\omega^2 \frac{1}{\alpha^2} \right) = E. \tag{7.40}$$

Minimizing with respect to the parameter α, we find

$$\begin{aligned} \frac{d}{d\alpha} \left(\frac{\hbar^2}{m} \alpha^2 + m\omega^2 \frac{1}{\alpha^2} \right) &= 0 \\ \alpha^2 &= \frac{m\omega}{\hbar} \\ E &= \frac{1}{4} \left(\frac{\hbar^2}{m} \alpha^2 + m\omega^2 \frac{1}{\alpha^2} \right) = \frac{1}{2} \omega\hbar, \end{aligned} \tag{7.41}$$

which as we know is the ground state energy. By choosing the correct form of the wave function, we obtained the exact ground state energy as the minimum value of the integral.

In a footnote, Schrödinger pointed out that the variational principle of Eq. (7.38) could be used to obtain the wave equation in any coordinate system, thus avoiding the sometimes-cumbersome calculation of ∇^2. We will demonstrate this in spherical coordinates defined by the element of distance given by

$$ds^2 = dx^2 + dy^2 + dz^2 = dr^2 + r^2 d\theta^2 + r^2 \sin^2\theta d\phi^2, \tag{7.42}$$

so that the kinetic energy is

$$T(q, \dot{q}) = \frac{m}{2} \left(\left(\frac{dr}{dt} \right)^2 + r^2 \left(\frac{d\theta}{dt} \right)^2 + r^2 \sin^2\theta \left(\frac{d\phi}{dt} \right)^2 \right), \tag{7.43}$$

and the momenta $\left(p_i = \frac{\partial T}{\partial \dot{q}_i} \right)$ are

$$p_r = m\frac{dr}{dt}, \quad p_\theta = mr^2 \frac{d\theta}{dt}, \quad p_\phi = mr^2 \sin^2\theta \frac{d\phi}{dt}. \tag{7.44}$$

Then

$$T(q,p) = \frac{p_r^2}{2m} + \frac{p_\theta^2}{2mr^2} + \frac{p_\phi^2}{2mr^2\sin^2\theta} \tag{7.45}$$

$$T\left(q, p_i = \hbar\frac{\partial\psi}{\partial q_i}\right) = \frac{\hbar^2}{2m}\left[\left(\frac{\partial\psi}{\partial r}\right)^2 + \frac{1}{r^2}\left(\frac{\partial\psi}{\partial\theta}\right)^2 + \frac{1}{r^2\sin^2\theta}\left(\frac{\partial\psi}{\partial\phi}\right)^2\right]. \tag{7.46}$$

The variational principle is then

$$\delta\int\left[T\left(q, p_i = \hbar\frac{\partial\psi}{\partial q_i}\right) + V(q)\,\psi^2\right]dq = 0 \tag{7.47}$$

$$\delta\int\left[\frac{\hbar^2}{2m}\left[\left(\frac{\partial\psi}{\partial r}\right)^2 + \frac{1}{r^2}\left(\frac{\partial\psi}{\partial\theta}\right)^2 + \frac{1}{r^2\sin^2\theta}\left(\frac{\partial\psi}{\partial\phi}\right)^2\right] + V(q)\,\psi^2\right] r^2\sin\theta d\theta d\phi dr = 0$$

$$\int\left[\frac{\hbar^2}{2m}\left[\frac{\partial\psi}{\partial r}\frac{\partial\delta\psi}{\partial r} + \frac{1}{r^2}\frac{\partial\psi}{\partial\theta}\frac{\partial\delta\psi}{\partial\theta} + \frac{1}{r^2\sin^2\theta}\frac{\partial\psi}{\partial\phi}\frac{\partial\delta\psi}{\partial\phi}\right] + V(q)\,\psi\delta\psi\right] r^2\sin\theta d\theta d\phi dr = 0$$

where $\delta\psi(r,\theta,\phi)$ is an arbitrary small function of position. In integrating by parts, we have to take the factor from the volume element into account, so that we find

$$\int\left[\left\{-\frac{\hbar^2}{2m}\left[\sin\theta\frac{\partial}{\partial r}r^2\frac{\partial\psi}{\partial r} + \frac{\partial}{\partial\theta}\sin\theta\frac{\partial\psi}{\partial\theta} + \frac{1}{\sin\theta}\frac{\partial^2\psi}{\partial^2\phi}\right] + V(q)\,r^2\sin\theta\psi\right\}\delta\psi\right] d\theta d\phi dr = 0.$$

As this holds for arbitrary $\delta\psi$, we have

$$-\frac{\hbar^2}{2m}\left[\sin\theta\frac{\partial}{\partial r}r^2\frac{\partial\psi}{\partial r} + \frac{\partial}{\partial\theta}\sin\theta\frac{\partial\psi}{\partial\theta} + \frac{1}{\sin\theta}\frac{\partial^2\psi}{\partial^2\phi}\right] + V(q)\,r^2\sin\theta\psi = 0$$

$$-\frac{\hbar^2}{2m}\left[\frac{1}{r^2}\frac{\partial}{\partial r}r^2\frac{\partial\psi}{\partial r} + \frac{1}{r^2\sin\theta}\frac{\partial}{\partial\theta}\sin\theta\frac{\partial\psi}{\partial\theta} + \frac{1}{r^2\sin^2\theta}\frac{\partial^2\psi}{\partial^2\phi}\right] + V(q)\,\psi = 0, \tag{7.48}$$

and the expression in square brackets is $\nabla^2\psi$ in spherical coordinates.

7.2.3 Derivation of the wave equation using Hamilton's analogy between point mechanics and geometric optics

In his second communication, Schrödinger gave a more intuitive derivation of the wave equation. He again started with the Hamilton-Jacobi equation, this time in its time-dependent form:

$$\frac{\partial S}{\partial t} + H\left(q, p = \frac{\partial S}{\partial q}\right) = \frac{\partial S}{\partial t} + T\left(q, p = \frac{\partial S}{\partial q}\right) + V(q) = 0, \tag{7.49}$$

where (p,q) represent all the coordinates and momenta in the problem, $H(q,p)$ is the Hamiltonian, $T(q,p)$ is the kinetic energy and $V(q)$ is the potential energy. Writing

$$S = W(q) - Et, \tag{7.50}$$

we obtain

$$T\left(q, p = \frac{\partial W}{\partial q}\right) = E - V, \tag{7.51}$$

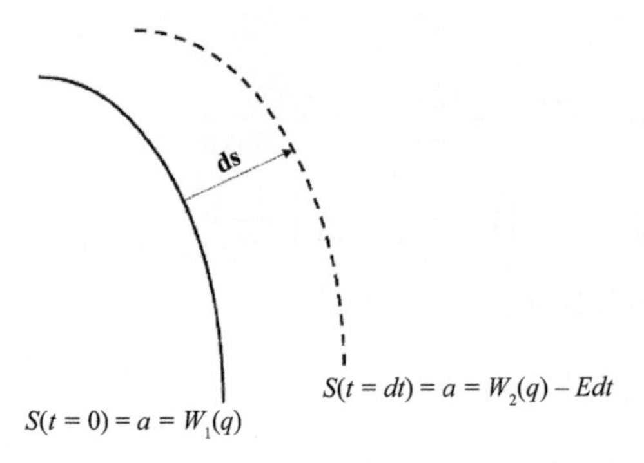

Fig. 7.2 Motion of a surface of constant S.

or for $T = p^2/2m$

$$\left|\vec{\nabla} W\right| = \left|\vec{\nabla} S\right| = \sqrt{2m\,(E - V)}. \tag{7.52}$$

From Eq. (7.50), valid for time-independent Hamiltonians, we see that at time $t = 0$ we can draw a surface of constant $S_1 = W_1$. At time $t = dt$ later, the surface of constant S will have moved to $S_2 = W_2 - Edt$ with $W_2 = W_1 + Edt$ (see Fig. 7.2). We see that the surface of constant S moves from the curve $W_1\,(q,t)$ in Fig. 7.2 to the curve $W_2\,(q,t)$, so that $W_2 - W_1 = dW = Edt$. The perpendicular distance between the two surfaces dq is then given by dW divided by the rate of change of W with q, $\left|\vec{\nabla} W\right|$:

$$dq = \frac{dW}{\left|\vec{\nabla} W\right|} = \frac{Edt}{\left|\vec{\nabla} W\right|} \tag{7.53}$$

so that the surface of constant S moves with a velocity u given by

$$u = \frac{dq}{dt} = \frac{E}{\left|\vec{\nabla} W\right|} = \frac{E}{p} = \frac{E}{\sqrt{2m\,(E - V)}}. \tag{7.54}$$

Thus the motion of the surfaces of constant S is analogous to the motion of the surfaces of constant phase of a wave moving with the phase velocity E/p (normal to the surface) at each point. If there were such a wave, the phase would be proportional to $S = \int Ldt$ and the particle trajectories would be moving perpendicular to the surfaces of constant phase along the rays of the wave motion, since the momentum $p_i = \left|\vec{\nabla}_i W\right|$. The propagation from one surface to the next can also be described by Huygens' principle.

Using Eq. (7.54), Fermat's principle of least time becomes

$$\delta \int dt = \delta \int \frac{ds}{u} = \delta \int \frac{ds}{E/p} = \frac{1}{E}\,\delta \int pds \propto \delta \int p\dot{q}dt \propto \delta \int 2Tdt = 0 \tag{7.55}$$

where the last integral represents the action. Fermat's principle is thus equivalent to the principle of least action, an analogy that had been noticed earlier by de Broglie.

This analogy was also known to Hamilton in the 1830s. However, there was absolutely no experimental reason to explore this further until almost 100 years later.

For a wave, the phase velocity is given by $u = \lambda\nu$ where λ is the wavelength and ν is the frequency of the wave. We can also write $u = \omega/k$ where $\omega = 2\pi\nu$ and $k = 2\pi/\lambda$. So for the analogy between classical mechanics and a wave motion to make sense, we would need the energy E of motion proportional to the frequency of the wave and the wavelength λ to be proportional to the inverse of the momentum $(1/p)$ with the same constant of proportionality (i.e., $E = h\nu$ and $p = h/\lambda$ with h an unknown constant). Schrödinger acknowledged that these relations had already been found by de Broglie for the "phase wave" under special relativity and he thanked de Broglie for the stimulation of his own work. Carrying the analogy further, we usually write the amplitude for a wave traveling in the x direction in the form $e^{i\omega(t-x/u)}$ so that the mechanical wave would be written $e^{i(Et-px)/\hbar}$ where we have written $\hbar$ for the unknown constant h divided by 2π.

Schrödinger pointed out that the particle velocity is $v = \sqrt{\frac{2}{m}(E-V)}$, so the point particle will not remain on a surface of constant S (constant phase) but the particle trajectories will lie on the rays of the wave. The particle will have to be represented by a wave packet, a superposition of waves of different (ω, k), whose maximum, as we have shown, (Eq. 7.11) moves with the group velocity

$$v = \frac{d\omega}{dk} = \frac{dE}{dp} = \frac{p}{m}. \tag{7.56}$$

The packet can represent a particle as long as the size of the packet is small enough. When the trajectory curvature is on the order of a wavelength, classical mechanics is just as useless for describing the particle motion as geometric optics is for explaining diffraction. In that case we must give up the idea of a trajectory. The true laws of quantum mechanics are not applied to single trajectories but to a whole set of trajectories. The motion of a point particle in mechanics is thus equivalent to geometric optics in a region with a spatially varying index of refraction, a description that is valid when the wavelength is small compared to the size of obstacles and the curvature of the rays. Schrödinger suggested that the failure of classical mechanics at very small dimensions is fully analogous to the failure of geometric optics and that we therefore need to search for a "wave mechanics."

For the relation between geometric and wave optics, Schrödinger referred to a work by Sommerfeld and Runge from 1911[15] in which they, following a suggestion by Debye, showed how geometric optics arises in the short-wavelength limit of wave optics. Sommerfeld and Runge started with the time-independent wave equation obtained by substituting $A = ue^{i\omega t}$ with $\omega = ck$ into the usual wave equation:

$$\nabla^2 u + k^2 u = 0. \tag{7.57}$$

[15]Sommerfeld, A. and Runge, J., *The application of vector calculations on the basis of geometric optics*, Ann, der Physik, 35, 277, June (1911), received April 17, 1911.

Writing $u = u_o e^{i\frac{k}{n}W}$, we have

$$\vec{\nabla} u = e^{i\frac{k}{n}W}\left(i\frac{k}{n}u_o\vec{\nabla} W + \vec{\nabla} u_o\right) \tag{7.58}$$

$$\nabla^2 u = e^{i\frac{k}{n}W}\left(-\frac{k^2}{n^2}u_o\left|\vec{\nabla} W\right|^2 + i\frac{k}{n}u_o\nabla^2 W + 2i\frac{k}{n}\vec{\nabla} W\cdot\vec{\nabla} u_o + \nabla^2 u_o\right), \tag{7.59}$$

where we have assumed that $\left(\vec{\nabla} n/n \ll \vec{\nabla} W/W\right)$. Then Eq. (7.57) becomes

$$\frac{k^2 u_o}{n^2}\left(-\left|\vec{\nabla} W\right|^2 + i\frac{n}{k}\nabla^2 W + 2i\frac{n}{k u_o}\vec{\nabla} W\cdot\vec{\nabla} u_o + \frac{n^2}{k^2 u_o}\nabla^2 u_o + n^2\right) = 0. \tag{7.60}$$

In the limit $k \to \infty$ ($\lambda \to 0$), we find

$$\left|\vec{\nabla} W\right| = n, \tag{7.61}$$

which is the same form as Eq. (7.52) with $n \sim \sqrt{2m\,(E-V)}$.

Schrödinger's problem was to go in the opposite direction, from Eq. (7.52) to a true wave equation. The phase velocity is not enough to determine the form of the wave equation, but for simplicity he assumed that it was a second-order equation. Starting with

$$\nabla^2\Psi - \frac{1}{u^2}\frac{\partial^2\Psi}{\partial t^2} = 0,$$

substituting $\Psi = \psi e^{\frac{i}{\hbar}Et}$ and using Eq. (7.54) for the phase velocity, we obtain

$$\nabla^2\psi + \frac{2m}{\hbar^2}\left(E - V\right)\psi = 0. \tag{7.62}$$

Schrödinger then noted that the function ψ can be visualized in three-dimensional space only for the one-body problem. In many-body problems, ψ is a function of all the coordinates ($3N$ in the case of N particles) and so cannot be easily visualized. At this point he acknowledged the work of Heisenberg, Born, and Jordan as containing at least a part of the truth. Their advantage was that they could calculate line intensities, something that Schrödinger did not see how to do using wave mechanics at the time.

7.3 First applications of the wave equation

In his first paper, in which he derived the wave equation from a variational principle, Schrödinger solved the equation for the hydrogen atom, as this was essential to establish the credibility of the theory. In the second paper, he gave the solutions for the harmonic oscillator, the rigid rotor with a free axis, with a fixed axis and a diatomic molecule with rotation and vibration.

7.3.1 The harmonic oscillator

We will reverse the order of presentation chosen by Schrödinger and will first present the solution to the harmonic oscillator for which the potential energy can be written

$V(x) = \frac{1}{2}m\omega_o^2 x^2$, where $\omega_o^2 = k/m$ and k is the force constant, $F = kx$. In this case the wave equation is

$$\frac{\partial^2 \psi}{\partial x^2} + \frac{2m}{\hbar^2}\left(E - \frac{1}{2}m\omega_o^2 x^2\right)\psi = 0. \tag{7.63}$$

In the limit of very large x, the solution will be proportional to a Gaussian, as we can check by differentiating:

$$\frac{\partial^2 e^{-\frac{1}{2}\frac{m}{\hbar}\omega_o x^2}}{\partial x^2} = e^{-\frac{1}{2}\frac{m}{\hbar}\omega_o x^2}\left(\frac{m^2\omega_o^2 x^2}{\hbar^2} - \frac{m\omega_o}{\hbar}\right). \tag{7.64}$$

The Gaussian function thus satisfies the equation in leading order, and we are led to try a solution of the form

$$\psi = f(x)\, e^{-\frac{1}{2}\frac{m}{\hbar}\omega_o x^2}. \tag{7.65}$$

Substituting this into Eq. (7.63), we obtain

$$\frac{d}{dx} f e^{-\frac{1}{2}\frac{m}{\hbar}\omega_o x^2} = -\frac{m}{\hbar}\omega_o x e^{-\frac{1}{2}\frac{m}{\hbar}\omega_o x^2} f + f' e^{-\frac{1}{2}\frac{m}{\hbar}\omega_o x^2} \tag{7.66}$$

$$\frac{d^2 f}{dx^2} - 2\frac{m\omega_o}{\hbar}x\frac{df}{dx} + \frac{m}{\hbar^2}(2E - \hbar\omega_o) f = 0. \tag{7.67}$$

We look for a solution in terms of a power series

$$f = a_o + a_1 x + \ldots a_n x^n. \tag{7.68}$$

Substituting the nth term $a_n x^n$ into this equation, we find

$$n(n-1) a_n x^{n-2} - 2\frac{m\omega_o}{\hbar} a_n n x^n + \frac{m}{\hbar^2}(2E - \hbar\omega_o) a_n x^n + \ldots = 0. \tag{7.69}$$

Since this is to hold for all values of x, the coefficient of each power of x must separately be equal to zero. Taking the coefficient of x^n we find

$$(n+2)(n+1) a_{n+2} + \frac{m}{\hbar^2}\left[2E - \hbar\omega_o(2n+1)\right] a_n = 0. \tag{7.70}$$

For large n we see that $a_{n+2}/a_n \sim \frac{2m\omega_o}{\hbar n}$, which is the same as for the series expansion of $e^{\frac{2m\omega_o}{\hbar}x^2}$. This means that in general the solution will diverge as $x \to \infty$ unless the series cuts off at some n. This will occur if

$$2E - \hbar\omega_o(2n+1) = 0 \tag{7.71}$$

$$E_n = \hbar\omega_o\left(n + \frac{1}{2}\right). \tag{7.72}$$

The normalized polynomials determined according to Eq. (7.70) are called Hermite polynomials. They are orthogonal to each other and there is a rich body of mathematical techniques for using them. The first few (unnormalized) eigenfunctions are

n	$\psi(x)$	$\frac{E}{\hbar\omega_o}$
0	$e^{-\frac{1}{2}\frac{m}{\hbar}\omega_o x^2}$	$\frac{1}{2}$
1	$xe^{-\frac{1}{2}\frac{m}{\hbar}\omega_o x^2}$	$\frac{3}{2}$
2	$\left(\frac{2m\omega_0}{\hbar}x^2 - 1\right)e^{-\frac{1}{2}\frac{m}{\hbar}\omega_o x^2}$	$\frac{5}{2}$
3	$\left(\frac{2m\omega_0}{3\hbar}x^3 - x\right)e^{-\frac{1}{2}\frac{m}{\hbar}\omega_o x^2}$	$\frac{7}{2}$

Figure 7.3 shows the first four solutions (eigenfunctions) of the harmonic oscillator. Besides determining the energy eigenvalue, the quantum number n also determines the number of zero crossings.

After completing the calculation, Schrödinger asked himself if there were any special properties associated with the classical turning points x_n given by

$$E_n = \frac{1}{2}m\omega_o^2 x_n^2 = \left(n + \frac{1}{2}\right)\hbar\omega_o \tag{7.73}$$

$$\frac{x_n}{\sqrt{\frac{\hbar}{m\omega_o}}} = \sqrt{2n+1}. \tag{7.74}$$

As seen in Figure 7.3, nothing dramatic happens to the wave function at these classical turning points. According to Eq. (7.63), these are inflection points where the first derivative is at an extremum. Schrödinger then noted that the phase velocity in Eq. (7.54) becomes infinite at these points and imaginary at larger values of x. One

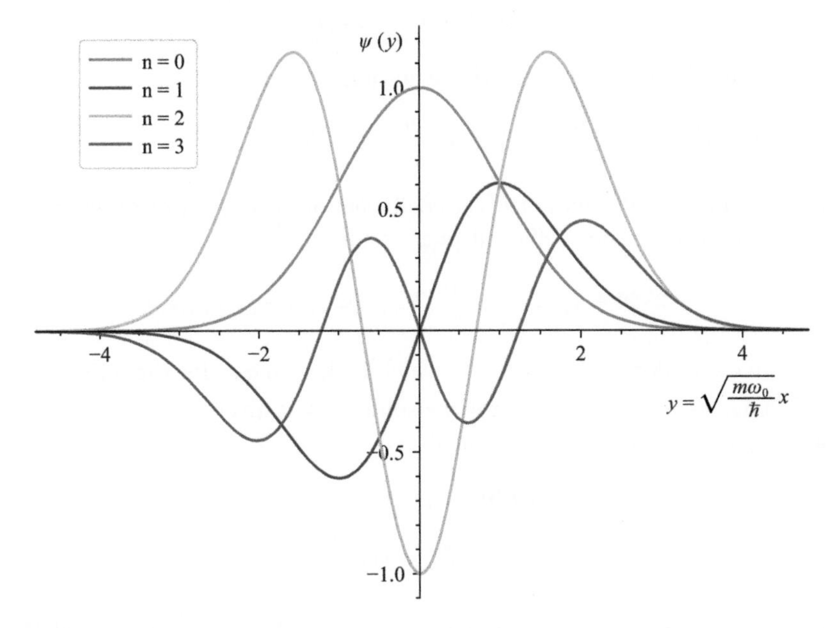

Fig. 7.3 The lowest-energy eigenfunctions of the harmonic oscillator plotted as a function of $y = \sqrt{m\omega_o/\hbar}x$.

remarkable point is that the wave function is nonzero in this region (where $E < V$), which is forbidden to the classical particle. Schrödinger then remarked that this is the reason that the simple requirement that $\lim_{x\to\infty} \psi(x) = 0$ leads to discrete eigenvalues. This applies in general, not only to the harmonic oscillator. Since a wave given by $\sim e^{i\omega(t-x/u)}$ becomes $e^{i\omega t}e^{\pm\omega x/u}$ when the phase velocity is imaginary, a very skillful choice of the energy value is required to avoid the positive exponential growth. This is more clearly seen in the next example, which was not published by Schrödinger.

7.3.2 Square well potential box

We consider a one-dimensional potential given by

$$V(x) = \begin{matrix} V_o & |x| > L \\ 0 & |x| < L \end{matrix} \tag{7.75}$$

which is illustrated in Fig. 7.4. The wave equation is then

$$\begin{aligned} \frac{\partial^2\psi}{\partial x^2} + \frac{2m}{\hbar^2}E\psi = 0 \quad |x| < L \\ \frac{\partial^2\psi}{\partial x^2} + \frac{2m}{\hbar^2}(E - V_o)\psi = 0 \quad |x| > L. \end{aligned} \tag{7.76}$$

We seek a solution for bound states, for which $E < V_o$, using trial solutions of the form

$$\begin{aligned} \psi = Ae^{ikx} + Be^{-ikx} \quad |x| < L \\ \psi = Ce^{\kappa x} + De^{-\kappa x} \quad |x| > L \end{aligned} \tag{7.77}$$

with $k = \sqrt{\frac{2m}{\hbar^2}E}$ and $\kappa = \sqrt{\frac{2m}{\hbar^2}(V_o - E)}$, where κ corresponds to an imaginary phase velocity or imaginary k. This demonstrates Schrödinger's point that it is the appearance of an imaginary phase velocity that results in the requirement for discrete eigenvalues.

The function ψ and its derivative $d\psi/dx$ must be continuous everywhere in order for $d^2\psi/dx^2$ to exist at every point. Thus we require that

$$\left[Ae^{ikx} + Be^{-ikx} = Ce^{\kappa x} + De^{-\kappa x}\right]_{x=L} \tag{7.78}$$

$$\left[ik\left(Ae^{ikx} - Be^{-ikx}\right) = \kappa\left(Ce^{\kappa x} - De^{-\kappa x}\right)\right]_{x=L} \tag{7.79}$$

and the same condition will apply at $x = -L$.

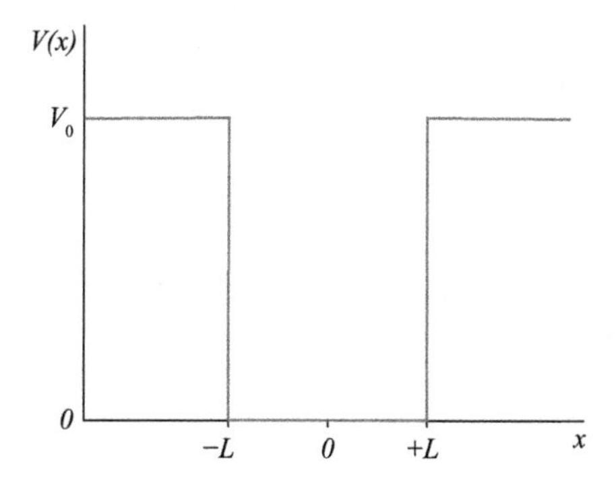

Fig. 7.4 Potential energy as a function of x for the square well potential.

Eq. (7.75) is an even function of x, which means that for every solution $u(x)$ of Eq. (7.76) that corresponds to an energy E, $u(-x)$ is also a solution with the same energy. If the system is non-degenerate, there is only one solution for each value of E, which means that the two solutions must differ only by a constant:

$$u(x) = \lambda u(-x). \tag{7.80}$$

If we substitute $x \to -x$, this equation becomes

$$u(-x) = \lambda u(x) \tag{7.81}$$

so that

$$u(x) = \lambda^2 u(x) \tag{7.82}$$

$$\lambda = \pm 1. \tag{7.83}$$

For a potential symmetric in x, the solutions are thus either even or odd functions of x and we need only search for a solution for $x > 0$. We start with an even solution. Then $A = B$ and we have

$$2A \cos kL = Ce^{\kappa L} + De^{-\kappa L} \tag{7.84}$$

$$-2Ak \sin kL = \kappa \left(Ce^{\kappa L} - De^{-\kappa L}\right). \tag{7.85}$$

Now we must have $C = 0$ for ψ to remain finite as $x \to \infty$. Eliminating D from the above equations, we find

$$A\left(\cos kL - \frac{k}{\kappa} \sin kL\right) = Ce^{\kappa L} \tag{7.86}$$

and the only way to have $C = 0$ is to put

$$k \tan kL = \kappa \tag{7.87}$$

$$x \tan x = y \tag{7.88}$$

with $x = kL,\ y = \kappa L$. Further we have $k^2 + \kappa^2 = \frac{2m}{\hbar^2} V_o$ or

$$x^2 + y^2 = \frac{2m}{\hbar^2} V_o L^2. \tag{7.89}$$

Eqs. (7.88) and (7.89) can be solved to yield the energy eigenvalues of the bound states. For the odd solutions we obtain

$$k \cot kL = -\kappa. \tag{7.90}$$

Eq. (7.86) demonstrates the crucial point that only certain specific values of k and κ, and therefore of the energy E, allow the wave function to remain finite at infinity.

7.3.3 Rigid rotor with a free axis

Among other problems which Schrödinger solved in his second paper is the rigid rotor with a free axis. This problem serves as an introduction to angular momentum in quantum mechanics, so we will present it now.

The rotor is taken to be a point mass m rigidly fixed at a distance R from a fixed point. The direction of the axis, which is free to move, is specified by the polar coordinates (θ, ϕ). The angular dependent parts of ∇^2 in spherical coordinates are (7.48)

$$\nabla^2 \psi = \frac{1}{\sin\theta}\frac{\partial}{\partial\theta}\left(\sin\theta\frac{\partial\psi}{\partial\theta}\right) + \frac{1}{\sin^2\theta}\frac{\partial^2\psi}{\partial\phi^2} \tag{7.91}$$

and the Schrödinger equation becomes

$$\frac{1}{\sin\theta}\frac{\partial}{\partial\theta}\left(\sin\theta\frac{\partial\psi}{\partial\theta}\right) + \frac{1}{\sin^2\theta}\frac{\partial^2\psi}{\partial\phi^2} + \frac{2mR^2}{\hbar^2}E\psi = 0. \tag{7.92}$$

We use the technique of separation of variables by writing

$$\psi = \Theta(\theta)\,\Phi(\phi). \tag{7.93}$$

Substituting and dividing by ψ, we have

$$\frac{1}{\Theta(\theta)\sin\theta}\frac{\partial}{\partial\theta}\left(\sin\theta\frac{\partial\Theta(\theta)}{\partial\theta}\right) + \frac{1}{\sin^2\theta\Phi(\phi)}\frac{\partial^2\Phi(\phi)}{\partial\phi^2} + \frac{2mR^2}{\hbar^2}E = 0 \tag{7.94}$$

$$\frac{\sin\theta}{\Theta(\theta)}\frac{\partial}{\partial\theta}\left(\sin\theta\frac{\partial\Theta(\theta)}{\partial\theta}\right) + \frac{2mR^2}{\hbar^2}E\sin^2\theta = -\frac{1}{\Phi(\phi)}\frac{\partial^2\Phi(\phi)}{\partial\phi^2}. \tag{7.95}$$

Since the left-hand side is a function of θ only and the right-hand side is a function of ϕ only, they must both be equal to a constant which we will call m^2. Then

$$\frac{\partial^2\Phi(\phi)}{\partial\phi^2} = -m^2\Phi(\phi) \tag{7.96}$$

$$\Phi(\phi) = \frac{1}{\sqrt{2\pi}}e^{\pm im\phi} \tag{7.97}$$

where we have chosen to normalize $\Phi(\phi)$. Then

$$\frac{1}{\sin\theta}\frac{\partial}{\partial\theta}\left(\sin\theta\frac{\partial\Theta(\theta)}{\partial\theta}\right) + \left(\Gamma - \frac{m^2}{\sin^2\theta}\right)\Theta(\theta) = 0, \tag{7.98}$$

where $\Gamma = \frac{2mR^2}{\hbar^2}E$. We begin by substituting $z = \cos\theta$. Then $dz = -\sin\theta d\theta$, $d\theta = -dz/\sqrt{1-z^2}$, $\partial/\partial\theta = -\sqrt{1-z^2}\partial/\partial z$, $\frac{1}{\sin\theta}\frac{\partial}{\partial\theta} = -\partial/\partial z$, so that Eq. (7.98) becomes

$$\frac{\partial}{\partial z}\left(\left(1-z^2\right)\frac{\partial\Theta(z)}{\partial z}\right) + \left(\Gamma - \frac{m^2}{1-z^2}\right)\Theta(z) = 0. \tag{7.99}$$

Now the trick is to substitute

$$\Theta(z) = \left(1 - z^2\right)^{m/2} P(z). \tag{7.100}$$

In general, m can take on positive or negative values, and $|m|$ should appear in the formula. We consider this as understood to simplify the equations. Then

$$\frac{\partial \Theta(z)}{\partial z} = -m\left(1 - z^2\right)^{m/2-1} zP(z) + \left(1 - z^2\right)^{m/2} \frac{dP(z)}{dz}$$

$$\frac{\partial}{\partial z}\left(\left(1 - z^2\right)\frac{\partial \Theta(z)}{\partial z}\right) = \frac{\partial}{\partial z}\left[\left(-m\left(1 - z^2\right)^{m/2} zP(z)\right) + \left(1 - z^2\right)^{m/2+1}\frac{dP(z)}{dz}\right] =$$

$$= \left(1 - z^2\right)^{m/2-1}\left(-m + \left(m^2 + m\right)z^2\right)P(z) - m\left(1 - z^2\right)^{m/2} z\frac{dP(z)}{dz} +$$
$$\left(1 - z^2\right)^{m/2+1}\frac{d^2P(z)}{dz^2} - 2(m/2 + 1)z\left(1 - z^2\right)^{m/2}\frac{dP(z)}{dz}$$

$$\frac{\partial}{\partial z}\left(\left(1 - z^2\right)\frac{\partial \Theta(z)}{\partial z}\right) + \left(\Gamma - \frac{m^2}{1 - z^2}\right)\Theta(z)$$
$$= \left(1 - z^2\right)^{m/2}\left[\begin{array}{c}\left(1 - z^2\right)\frac{d^2P(z)}{dz^2} - \frac{dP(z)}{dz}2z(m + 1) + \\ P(z)(\Gamma - m(m + 1))\end{array}\right] = 0$$

$$\left(1 - z^2\right)\frac{d^2P(z)}{dz^2} - \frac{dP(z)}{dz}2z(m + 1) + P(z)(\Gamma - m(m + 1)) = 0. \tag{7.101}$$

We again look for a solution in terms of a power series:

$$P(z) = \sum_{n=0} a_n z^n. \tag{7.102}$$

Substituting into Eq. (7.101), we obtain

$$\sum_n \left(1 - z^2\right)a_n n(n - 1)z^{n-2} - 2(m + 1)na_n z^n + a_n z^n(\Gamma - m(m + 1)) = 0$$
$$\sum_n a_n n(n - 1)z^{n-2} + a_n z^n\left[\Gamma - m(m + 1) - 2(m + 1)n - n(n - 1)\right] = 0.$$

Since the equation should hold for all z, the coefficient of each power of z should separately be equal to zero, so that

$$a_{n+2}(n + 2)(n + 1) + a_n\left[(\Gamma - m(m + 1)) - 2(m + 1)n - n(n - 1)\right] = 0$$
$$a_{n+2}(n + 2)(n + 1) + a_n\left[\Gamma - (n + |m|)(n + |m| + 1)\right] = 0$$

$$a_{n+2} = \frac{a_n\left[(n + |m|)(n + |m| + 1) - \Gamma\right]}{(n + 2)(n + 1)}. \tag{7.103}$$

The boundary condition is that the function should remain finite for all $|z| \leq 1$. We see that for large n $\lim_{n\to\infty} a_{n+2}/a_n = 1$ so that for $|z| = 1$ the series in Eq. (7.102) will diverge unless it is cut off for some n. That means that

$$\Gamma = [(n + |m|)(n + |m| + 1)] \tag{7.104}$$

$$= l(l+1), \tag{7.105}$$

where we have redefined $(n + |m|) = l$ so that l can take on the values $|m|, |m| + 1, |m| + 2, \ldots$ For a fixed l, the maximum value of $|m| = l$. Since m can be positive or negative, the number of m values for each l is $(2l + 1)$. Returning to Eqs. (7.98) and (7.94), we see that

$$\frac{1}{\sin\theta}\frac{\partial}{\partial\theta}\left(\sin\theta\frac{\partial\Theta(\theta)}{\partial\theta}\right) + l(l+1)\Theta(\theta) - \frac{m^2}{\sin^2\theta}\Theta(\theta) = 0 \tag{7.106}$$

$$\frac{1}{\sin\theta}\frac{\partial}{\partial\theta}\left(\sin\theta\frac{\partial\Theta(\theta)}{\partial\theta}\right)\Phi(\phi) + \frac{1}{\sin^2\theta}\frac{\partial^2\Phi(\phi)}{\partial\phi^2}\Theta(\theta) = -l(l+1)\Theta(\theta)\Phi(\phi). \tag{7.107}$$

From the definition of $\Gamma = \frac{2mR^2}{\hbar^2}E$, we find

$$E = \frac{\hbar^2}{2mR^2}l(l+1), \tag{7.108}$$

and each value of l corresponds to $(2l + 1)$ degenerate states. The eigenfunctions of Eq. (7.92) are then

$$\psi = \Theta(\theta)\Phi(\phi) = e^{\pm im\phi}\left[\left(1 - z^2\right)^{|m|/2}\sum_n a_n z^n\right]. \tag{7.109}$$

The functions defined by the finite series in the square brackets are called the associated Legendre polynomials $P_l^{|m|}(\cos\theta)$. When normalized, these eigenfunctions are referred to as spherical harmonics $Y_l^{\pm m}(\theta,\phi) = N_{l|m|}P_l^{|m|}(\cos\theta)\,e^{\pm im\phi}$ where $N_{l|m|}$ is a normalizing factor. These functions are of great importance in the theory of angular momentum, which we will discuss in Chapter 18.

7.3.4 The hydrogen atom

In his first paper, Schrödinger applied the wave equation to the case of the hydrogen atom. He considered this the essential demonstration that his method had at least some connection with reality. Taking $V = -e^2/r$, the Coulomb potential of the nucleus, he wrote the wave equation as

$$\nabla^2\psi + \frac{2m}{\hbar^2}\left(E + e^2/r\right)\psi = 0. \tag{7.110}$$

This is actually valid only in the center of mass system of the electron and proton with m equal to the reduced mass, $m = m_e m_p/(m_e + m_p)$. Substituting for ∇^2 in spherical coordinates (7.48), this equation takes the form

$$\frac{1}{r^2}\frac{\partial}{\partial r}\left(r^2\frac{\partial \psi}{\partial r}\right)+\frac{1}{r^2}\left(\frac{1}{\sin\theta}\frac{\partial}{\partial\theta}\sin\theta\frac{\partial\psi}{\partial\theta}+\frac{1}{\sin^2\theta}\frac{\partial^2\psi}{\partial\phi^2}\right)+\frac{2m}{\hbar^2}\left(E+e^2/r\right)\psi=0.$$

We now separate the variables with $\psi = R(r)\,\Theta(\theta)\,\Phi(\phi)$:

$$\frac{1}{R}\frac{\partial}{\partial r}\left(r^2\frac{\partial R}{\partial r}\right)+\frac{1}{\Theta(\theta)\,\Phi(\phi)}\left(\frac{1}{\sin\theta}\frac{\partial}{\partial\theta}\sin\theta\frac{\partial\Theta(\theta)\,\Phi(\phi)}{\partial\theta}+\frac{1}{\sin^2\theta}\frac{\partial^2\Theta(\theta)\,\Phi(\phi)}{\partial\phi^2}\right)$$
$$+\frac{2m}{\hbar^2}\left(E+e^2/r\right)r^2=0$$

$$\frac{1}{R}\frac{\partial}{\partial r}\left(r^2\frac{\partial R}{\partial r}\right)+\frac{2m}{\hbar^2}\left(E+e^2/r\right)r^2=-\frac{1}{\Theta(\theta)\,\Phi(\phi)}\left\{\frac{1}{\sin\theta}\frac{\partial}{\partial\theta}\sin\theta\frac{\partial\Theta(\theta)\,\Phi(\phi)}{\partial\theta}\right.$$
$$\left.+\frac{1}{\sin^2\theta}\frac{\partial^2\Theta(\theta)\,\Phi(\phi)}{\partial\phi^2}\right\}.$$

From Eq. (7.106), we know that the only solutions for $\Theta(\theta)\,\Phi(\phi)$ that remain finite for all values of θ are $Y_l^m(\theta,\phi)$, which give $-l(l+1)\,Y_l^m(\theta,\phi)$ for the expression in the curly brackets. For bound states ($E<0$), the radial equation is thus

$$\frac{1}{r^2}\frac{\partial}{\partial r}\left(r^2\frac{\partial R}{\partial r}\right)+\left(\frac{2m}{\hbar^2}\left(-|E|+e^2/r\right)-\frac{l(l+1)}{r^2}\right)R=0. \tag{7.111}$$

Schrödinger solved this equation using a method that is no longer taught. It seems that at this time he did not know about the book *Methods of Mathematical Physics* by Courant and Hilbert (published in 1924), which he referred to in later work.

To solve this equation using the modern technique, we note that for very large r we have

$$\frac{\partial^2 R}{\partial r^2}=\frac{2m}{\hbar^2}|E|\,R \tag{7.112}$$

which has solutions $R=\exp\left(\pm\sqrt{\frac{2m}{\hbar^2}|E|}r\right)$. The positive exponential is unacceptable, so we look for solutions of the form

$$R=e^{-\sqrt{\frac{2m|E|}{\hbar^2}}r}F(r)=e^{-\alpha r}F(r). \tag{7.113}$$

This defines $\alpha=\sqrt{2m|E|/\hbar^2}$, which has the dimensions of an inverse length.

$$\frac{1}{r^2}\frac{\partial}{\partial r}\left(r^2\frac{\partial R}{\partial r}\right)=\left(\frac{2}{r}\frac{\partial R}{\partial r}+\frac{\partial^2 R}{\partial r^2}\right)$$
$$=\begin{bmatrix}\frac{2}{r}\left(-\alpha F(r)+F'(r)\right)+\\ \alpha^2F(r)-\alpha F'(r)-\alpha F'(r)+F''(r)\end{bmatrix}e^{-\alpha r}$$
$$=\left[F''(r)+F'(r)\left(\frac{2}{r}-2\alpha\right)+F(r)\left(\alpha^2-\frac{2}{r}\alpha\right)\right]e^{-\alpha r}$$

The radial wave equation is then

$$F''(r)+F'(r)\left(\frac{2}{r}-2\alpha\right)+F(r)\left(\left(\frac{2}{a_0}-2\alpha\right)\frac{1}{r}-\frac{l(l+1)}{r^2}\right)=0,$$

where $a_0=\frac{\hbar^2}{me^2}$ is the Bohr radius.

As in previous cases, we look for a solution in the form $F = \sum_{n=0} a_n r^n$. Then substituting the nth term gives

$$\sum_n \left[a_n n (n-1) r^{n-2} + a_n n r^{n-1} \left(\frac{2}{r} - 2\alpha \right) + a_n r^n \left(\left(\frac{2}{a_0} - 2\alpha \right) \frac{1}{r} - \frac{l(l+1)}{r^2} \right) \right] = 0. \tag{7.114}$$

The coefficient of each power of r must vanish individually. Taking the coefficient of r^{n-1}, we have

$$a_{n+1} \left[(n+2)(n+1) - l(l+1) \right] + a_n 2 \left[\frac{1}{a_0} - (n+1)\alpha \right] = 0 \tag{7.115}$$

$$a_{n+1} = \frac{2\left[(n+1)\alpha - \frac{1}{a_0} \right]}{(n+2)(n+1) - l(l+1)} a_n. \tag{7.116}$$

Since n and l are both integers we must be careful to avoid the case where $(n+2)(n+1) - l(l+1) = 0$ or $n = l-1$. For a given l, if any values of $n < l-1$ had a_n that were nonzero, then in general a_{l-1} would be nonzero and a_l would diverge. Therefore the series must start at $n = l$. For large n, we see that $a_{n+1}/a_n \sim 1/n$ so that the series will diverge unless it is cut off at some finite n value. So for some $n' \geq l$ we must have

$$(n'+1) = \frac{1}{a_0 \alpha} = \frac{me^2}{\hbar^2} \sqrt{\frac{\hbar^2}{2m|E|}} \tag{7.117}$$

$$|E| = \frac{e^4 m}{2\hbar^2 n^2} \tag{7.118}$$

where $n = n' + 1 \geq l + 1$. Thus the condition that the solutions to the wave equation be finite leads to the fact that the energy must be equal to one of the terms in the Balmer spectrum in agreement with the Bohr-Sommerfeld model and with Heisenberg, Born, and Jordan's matrix mechanics calculation. Thus the wave equation passed its first, most crucial test.

It is conventional to write the function $F(r)$ in terms of the dimensionless variable $\rho = \alpha r$:

$$F_{n,l}(\rho) = \rho^l \sum_{i=0}^{n-l-1} a_i \rho^i. \tag{7.119}$$

When normalized, the functions defined by this series are called the associated Laguerre polynomials $L_{n,l}(\rho)$. They are discussed in great detail in the literature. The complete solutions for the hydrogen atom problem are thus written as

$$\psi_{nlm}(r,\theta,\phi) = \rho^l e^{-\rho} L_{n-l-1}^{2l+1}(2\rho)\, Y_l^m(\theta,\phi) = R_{nl}(\rho)\, Y_l^m(\theta,\phi), \tag{7.120}$$

where $\rho = \frac{r}{a_0 n}$. Figure 7.5 shows the probability distribution $r^2 R_{nl}(r)$ as a function of r for the lowest-energy eigenfunctions.

7.4 The relation between matrix and wave mechanics

The differences between matrix and wave mechanics could not be more pronounced. In the one, we have these somewhat mysterious matrices consisting of lists of numbers, and in the other, we have continuously varying solutions of a partial differential equation. Schrödinger remarked that the two theories departed from the classical mechanics of point particles in opposite directions. Matrix mechanics was more discontinuous while wave mechanics dealt with a continuum. In matrix mechanics, continuous classical variables are replaced by discrete objects—matrices—satisfying algebraic equations. Its authors called it a "true discontinuous theory." Wave mechanics does the exact opposite. It replaces classical mechanics, which has many dependent variables satisfying a system of ordinary differential equations, with a continuous field-like process in configuration space (the $3N$-dimensional space of the coordinates of N particles), determined by a single partial differential equation derivable from a variational principle. This variational principle or partial differential equation replaces both the equations of motion and the quantum conditions of the old quantum theory.

In fact, the particles seemed to have almost disappeared from wave mechanics. Matrix mechanics provided a method for calculating intensities and selection rules for the observed atomic spectra, which wave mechanics at that time could not do. However, in his work proving the identity of the two theories, Schrödinger found that wave mechanics could also calculate these quantities. On the other hand, wave mechanics seemed to be applicable to a wider range of problems. Pauli claimed that he noticed the

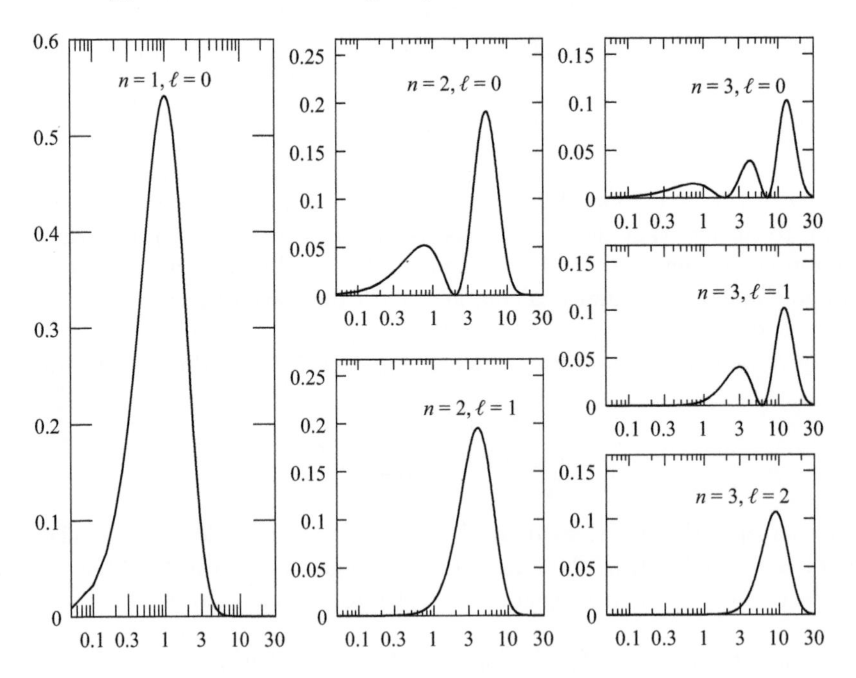

Fig. 7.5 Probability density $4\pi r^2 R_{nl}^2(r)$ for the lowest-energy radial eigenfunctions of the hydrogen atom. The x-axis is in units of r/a, where $a = \frac{\hbar^2}{me^2}$ is the Bohr radius.

identity of the two theories as soon as he had seen Schrödinger's first paper. In a letter to Jordan written between the date when Schrödinger had submitted his publication on the relation between the two theories and the date it was published,[16] he essentially outlined Schrödinger's proof, which we will now discuss.

Schrödinger wrote that, naturally, he had known about Heisenberg's theory when he was formulating wave mechanics, "but I was scared away, if not repulsed by the transcendental algebra and lack of visualizability." He began his demonstration of the identity of the two approaches by remarking that matrices can be constructed from a complete, arbitrary, orthonormal set of functions of the ($3N$-dimensional) configuration space q, not the ($6N$-dimensional) phase space (p, q). When the orthonormal system of functions are the eigenfunctions of the Schrödinger equation, the matrices satisfy the Heisenberg equations of motion. Actually it was Lanczos, in a paper submitted before the first publication on wave mechanics, who had noticed that the matrix elements could be calculated on the basis of a set of eigenfunctions, but he chose the eigenfunctions of an integral equation. Schrödinger pointed out that Lanczos' model was not related to his model because the kernel of Lanczos' integral equation could not be the Green's function of his wave equation. Modern quantum mechanics makes use of both the differential and integral equation formulations, as we will discuss in detail in Section 7.4.2.

An orthonormal set of eigenfunctions $\psi_n(q)$, where q can represent one or several variables, satisfies

$$\int \psi_m^*(q)\,\psi_n(q)\,dq = \delta_{mn} = \begin{matrix} 1 & m = n \\ 0 & m \neq n. \end{matrix} \tag{7.121}$$

The eigenfunctions of the Schrödinger wave equation share this property, which can be seen as follows. We start with

$$\vec{\nabla}\cdot\left(\psi_n^*\vec{\nabla}\psi_m - \psi_m\vec{\nabla}\psi_n^*\right) = \left(\psi_n^*\nabla^2\psi_m - \psi_m\nabla^2\psi_n^*\right). \tag{7.122}$$

The divergence theorem relates the volume integral to a surface integral:

$$\int \left(\psi_n^*\nabla^2\psi_m - \psi_m\nabla^2\psi_n^*\right) d\tau = \int d\vec{S}\cdot\left(\psi_n^*\vec{\nabla}\psi_m - \psi_m\vec{\nabla}\psi_n^*\right) = 0. \tag{7.123}$$

Here $d\tau$ is an element of volume and $d\vec{S}$ is an element of the surface bounding that volume. The surface integral is zero because the surface S can be considered to be infinitely far away and the eigenfunctions must go to zero sufficiently rapidly at infinity. Substituting from the wave equation and its complex conjugate, we have

$$\int \left(\psi_n^*(V - E_m)\,\psi_m - \psi_m(V - E_n)\,\psi_n^*\right) d\tau = (E_n - E_m)\int \psi_n^*\psi_m d\tau = 0, \tag{7.124}$$

which proves Eq. (7.121) for $E_n \neq E_m$ (with normalization of the eigenfunctions).

[16] Pauli to Jordan April 12, 1926, in: Hermann, A., Meyenn, K. von and Weisskopf, V.W., eds., *Wolfgang Pauli: Scientific Correspondence with Bohr, Einstein, Heisenberg A.O. Vol. 1 1919-1929*, (Springer Verlag 1979)

This means that a function $f(q)$ can be expanded as follows:

$$f(q) = \sum_k f_k \psi_k(q), \tag{7.125}$$

where the f_k are numbers. Using Eq. (7.121), we find

$$\int \psi_m^*(q) f(q)\, dq = \sum_k f_k \int \psi_m^*(q) \psi_k(q)\, dq = f_m. \tag{7.126}$$

Substituting this result into Eq. (7.125) we have

$$f(q) = \sum_k \int \psi_k^*(q') f(q')\, dq' \psi_k(q) = \int f(q') \sum_k \psi_k^*(q') \psi_k(q)\, dq', \tag{7.127}$$

which can hold only if

$$\sum_k \psi_k^*(q') \psi_k(q) = \delta(q - q'), \tag{7.128}$$

where $\delta(x)$ is the Dirac delta function defined by

$$\delta(x) = 0 \quad \text{for } x \neq 0 \tag{7.129}$$

$$\int \delta(x) f(x)\, dx = f(0). \tag{7.130}$$

Eq. (7.128) is called the completeness relation and both it and Eq. (7.121) can be shown to hold for the eigenfunctions of the wave equation. This is discussed in more detail in Chapter 11.

Schrödinger started his demonstration by noting that if the momentum component p_i is replaced by the operator $K\frac{\partial}{\partial q_i}$, where K is a universal constant which we will see is equal to $\hbar/i$, the commutation relation $[p_i, q_j] = (\hbar/i)\,\delta_{ij}$ will hold. Namely

$$\left(\frac{\partial}{\partial q_i} q_i - q_i \frac{\partial}{\partial q_i}\right)\psi = \psi. \tag{7.131}$$

Then an arbitrary function can be written

$$F(p, q) = F\left(\frac{\hbar}{i}\frac{\partial}{\partial q_i}, q_i\right), \tag{7.132}$$

where the function is interpreted as a power series. The objects p_i and q_i are operators which can be applied to arbitrary functions of (p, q). The operator q represents the operation of multiplying by the numerical value of the coordinate q. Using our orthonormal set of functions, we define a matrix element of the operator F as

$$F_{lm} = \int \psi_l^*(q) F \psi_m(q)\, dq. \tag{7.133}$$

Then the product of F with a similarly defined matrix G is

$$\sum_l F_{kl}G_{lm} = \sum_l \int \psi_k^*(q)\,F\psi_l(q)\,dq \int \psi_l^*(q')\,G\psi_m(q')\,dq' \tag{7.134}$$

$$= \int\int dqdq'\psi_k^*(q)\,F\sum_l [\psi_l(q)\,\psi_l^*(q')]\,G\psi_m(q') \tag{7.135}$$

$$= \int dq\psi_k^*(q)\,FG\psi_m(q) = (FG)_{km}, \tag{7.136}$$

where we used the completeness relation of Eq. (7.128). Thus the matrix elements defined by Eq. (7.133) satisfy the rule of matrix multiplication. The matrix elements of the commutation relation in Eq. (7.131) are then

$$[p_iq_j - q_jp_i]_{mn} = \int u_m^*(q)\,[p_iq_j - q_jp_i]\,u_n(q)\,dq$$
$$= \frac{\hbar}{i}\delta_{ij}\int u_m^*(q)\,u_n(q)\,dq = \frac{\hbar}{i}\delta_{ij}\delta_{mn}. \tag{7.137}$$

This shows that $[p, q]$ is a diagonal matrix.

As we showed in Chapter 6, the relations

$$\left(\frac{\partial F}{\partial q_i}\right) = \frac{i}{\hbar}[p_iF - Fp_i] \tag{7.138}$$

$$\left(\frac{\partial F}{\partial p_i}\right) = \frac{i}{\hbar}[Fq_i - q_iF] \tag{7.139}$$

follow from the commutation relations. Schrödinger explained the first relation as follows: when we apply the operator $\left(\frac{\partial F}{\partial q_i}\right)$ to a function $\psi(q)$, the differentiation would apply not only to F but also to $\psi(q)$ so that $\frac{\partial F}{\partial q_i}\psi = \psi\frac{\partial F}{\partial q_i} + F\frac{\partial \psi}{\partial q_i}$. The second unwanted term is eliminated by the second term in Eq. (7.138). Now if we let $H\left(p = \frac{\hbar}{i}\frac{\partial}{\partial q_i}, q\right)$ represent the Hamiltonian as an operator, the equations of motion for the matrix elements are

$$\left(\frac{\partial q_j}{\partial t}\right)_{mn} = \left(\frac{\partial H}{\partial p_j}\right)_{mn} = \frac{i}{\hbar}[Hq_j - q_jH]_{mn} \tag{7.140}$$

$$\left(\frac{\partial p_j}{\partial t}\right)_{mn} = -\left(\frac{\partial H}{\partial q_j}\right)_{mn} = \frac{i}{\hbar}[Hp_j - p_jH]_{mn}. \tag{7.141}$$

If the basis states are now chosen to be eigenfunctions of the Hamiltonian, so that

$$H\psi_n = E_n\psi_n \tag{7.142}$$

$$H_{mn} = \int \psi_m^* H\psi_n dq = E_n\delta_{mn}, \tag{7.143}$$

then

$$[Hq_j - q_jH]_{mn} = \sum_l \left[H_{ml}(q_j)_{\text{ln}} - (q_j)_{ml}H_{\text{ln}}\right] \tag{7.144}$$

$$= (E_m - E_n)(q_j)_{mn}, \tag{7.145}$$

so that Eq. (7.140) is

$$\left(\frac{\partial q_j}{\partial t}\right)_{mn} = \frac{i}{\hbar}\left(E_m - E_n\right)\left(q_j\right)_{mn}, \tag{7.146}$$

which is just the equation satisfied by the Heisenberg matrices. Thus Schrödinger had shown that it is possible to produce the Heisenberg matrices from a set of eigenfunctions of the Hamiltonian for the problem at hand. To prove the equivalence from the reverse direction, we start with the matrix elements q_{ik}

$$q_{ik} = \int \psi_i^* q \psi_k dq. \tag{7.147}$$

Given this we can calculate, by matrix multiplication:

$$\left(q^n\right)_{ik} = \int \psi_i^* q^n \psi_k dq \tag{7.148}$$

so that we can calculate all the moments of the function $\psi_i^*(q)\,\psi_k(q)$, which is enough to determine the function for a wide class of functions. For $i = k$, we can calculate $\left|\psi_k(q)\right|^2$. This is not sufficient to uniquely determine the functions $\psi_k(q)$ since $\psi_k(q)$ can be negative or even complex. The introduction of a complex wave function in association with the time-dependent wave equation means that knowledge of $\left|\psi_k(q)\right|^2$ is only sufficient to determine the function $\psi_k(q)$ for a limited class of functional forms of $\psi_k(q)$.

These and similar arguments have been advanced against Schrödinger's proof of the equivalence of matrix and wave mechanics but all the conclusions were shown to be valid in a mathematically rigorous way by von Neumann. Remarkably, Eckart,[17] working independently at Caltech, derived the wave equation from matrix mechanics, thus demonstrating the equivalence in the reverse way.

Wave mechanics also differed from matrix mechanics in its applicability to various problems. For example, wave mechanics predicted that the amount of charge in a given volume Δ is $e\int_\Delta |\psi|^2\, dq$, while matrix mechanics made no such statement. Wave mechanics was also capable of describing time-dependent processes such as scattering (whose treatment we will discuss in Section 10.3 and Chapter 20), something matrix mechanics could not do.

7.4.1 First speculations on the emission of radiation according to wave mechanics

In matrix mechanics, the intensity of the light emitted in a transition $n \to m$ was calculated by substituting eq_{nm} into the classical expression for the energy emitted by an oscillating dipole, so that the intensity is proportional to $\left|q_{nm}\right|^2$. Schrödinger began by thinking of $e\psi\frac{\partial\psi^*}{\partial t}$ as a charge density but he soon realized that $e\psi\psi^*$ was a

[17]Eckart, C., *Operator calculus and the solution of the equations of quantum dynamics*, Phys. Rev. 28, 711 (1926), received June 7, 1926.

better choice. We consider a hydrogen atom in a superposition state[18] with the wave function given by

$$\psi = \sum_n c_n u_n(q)\, e^{iE_n t/\hbar} \tag{7.149}$$

$$\psi\psi^* = \sum_{n,m} c_n c_m^* u_m^*(q)\, u_n(q)\, e^{i(E_n - E_m)t/\hbar} \tag{7.150}$$

where $u_n(q)$ are the eigenfunctions of the Hamiltonian. If we multiply this density by a component of the electric dipole moment eq_i and integrate over all space we get a dipole moment given by

$$M_i = e \sum_{n,m} c_n c_m^* \left[q_i\right]_{mn} e^{i(E_n - E_m)t/\hbar}. \tag{7.151}$$

We have here an expression of the dipole moment as a sum of all the frequencies in the emission spectrum. The intensity of each line will then be proportional to the square of the matrix elements q_{nm} of q, in agreement with matrix mechanics.

For a transition between two excited states, both states must be excited in order for emission of radiation to occur. Consider the energy that must be transferred to the atom to excite both states and thus produce the radiation. This energy can be provided, for example, by an electron collision. If the two excited states have energies $E_2 > E_1$ with respect to the ground state, we would expect that the electron would have to give up an amount of energy equal to $E_1 + E_2$ to excite both states simultaneously, but experimentally an energy exchange of E_2 is sufficient so the above picture cannot be complete. In the modern theory, we know that the amplitude for the state E_2 builds up over time due to perturbation of the atom by the radiation field in its ground state, as we shall see.

7.4.2 Relation to integral equations

In a remarkable paper submitted before Schrödinger's first publication on wave mechanics,[19] Lanczos showed that he could base the matrix mechanics of Heisenberg, Born and Jordan on a set of eigenfunctions ϕ_n defined by an integral equation:

$$\phi_n(s) = \lambda_n \int K(s, \sigma)\, \phi_n(\sigma)\, d\sigma. \tag{7.152}$$

Here s and σ represent points in the configuration space and $K(s, \sigma)$ is a function of the position of two points called the kernel. He showed that matrices based on these eigenfunctions had all the properties of the Heisenberg matrices as long as

$$\frac{1}{\lambda_n} = E_n \tag{7.153}$$

with E_n the energy eigenvalues. There is some uncertainty in Lanczos' paper as to the meaning of the kernel. At several points, he states that the kernel is proportional

[18] Note that if ψ_1 and ψ_2 are solutions of the wave equation then so is $\psi_1 + \psi_2$ as the equation is linear.

[19] Lanczos, C., *On a field-like model of the new quantum mechanics*, Zeit. Phys. 35, 812 (1926), received Dec. 22, 1925.

to the Hamiltonian, but in another place he says that his method requires knowledge of the kernel in addition to the Hamiltonian. It appears that this latter remark is incorrect. Lanczos pointed out the possibility that the kernel could be the given by a differential equation. At a symposium honoring Dirac's seventieth birthday in 1972, van der Waerden[20] stated, incorrectly, that Lanczos had actually used the Green's function for the Schrödinger equation as the kernel for his integral equation. We now turn to a discussion of Lanczos' work.

We can see the connection between the wave equation and an integral equation as follows. Write the wave equation using the differential operator H as

$$H\left(\frac{\hbar}{i}\frac{\partial}{\partial q}, q\right)\psi = E\psi. \tag{7.154}$$

Now assume we have a solution to the equation

$$H\left(\frac{\hbar}{i}\frac{\partial}{\partial q}, q\right)G(q,q') = \delta\left(q-q'\right), \tag{7.155}$$

where $\delta(q)$ is the $3N$-dimensional Dirac delta function. $G(q,q')$ is called the Green's function of the differential equation. The method of Green's functions is a very fruitful method of solving differential equations. One way of finding the Green's function is to Fourier transform Eq. (7.155). Then

$$\psi(q) = E\int G(q,q')\psi(q')\,dq' \tag{7.156}$$

since

$$\begin{aligned} H\left(\frac{\hbar}{i}\frac{\partial}{\partial q}, q\right)\psi(q) &= E\int H\left(\frac{\hbar}{i}\frac{\partial}{\partial q}, q\right)G(q,q')\psi(q')\,dq' \\ &= E\int \delta\left(q-q'\right)\psi(q')\,dq' = E\psi(q). \end{aligned} \tag{7.157}$$

Eq. (7.156) is an integral equation for $\psi(q)$, which appears on both sides of the equation. It only has solutions for certain values of $E = E_n$ which are the same as the eigenvalues of the wave equation, Eq. (7.154), and it is equivalent to that equation. The eigenfunctions $\psi(q)$ are seen to be the same as the eigenfunctions of Eq. (7.154). In general, the eigenfunctions of such an integral equation form an orthogonal set and can be normalized to form an orthonormal set. These methods are applied to scattering problems and lead to the path integral formulation of quantum mechanics as described in Chapter 20.

Lanczos chose his eigenfunctions to be determined by the equation

$$\phi_i(q) = \lambda_i\int K(q,q')\,\phi_i(q')\,dq' \tag{7.158}$$

[20]Waerden, B. L. van der, *From matrix mechanics to wave mechanics to unified quantum mechanics, The Physicist's Conception of Nature,* Mehra, J., ed., (D. Reidel Pub. Co., Dordrect, 1973) reprinted in Notices of the AMS **44**, 323 (1997).

and defined the matrix element of a general function of two positions as

$$F(q,q') = \sum_{m,n} F_{mn}\phi_m(q)\,\phi_n^*(q') \tag{7.159}$$

or

$$F_{mn} = \int F(q,q')\,\phi_m^*(q)\,\phi_n(q')\,dq dq', \tag{7.160}$$

as can be seen from the orthonormality condition. The matrix multiplication rule follows from

$$\begin{aligned} A(q,q') &= \int B(q,q'')\,C(q'',q')\,dq'' = \\ &= \int \sum_{m,n} B_{mn}\phi_m(q)\,\phi_n^*(q'') \sum_{i,j} C_{ij}\phi_i(q'')\,\phi_j^*(q')\,dq'' \\ &= \sum_{m,n,j} B_{mn}C_{nj}\phi_m(q)\,\phi_j^*(q') = \sum_{m,j} A_{mj}\phi_m(q)\,\phi_j^*(q') \end{aligned} \tag{7.161}$$

with

$$A_{mj} = \sum_n B_{mn}C_{nj}. \tag{7.162}$$

At this point, we step back from Lanczos' treatment in order to give a broader, modern view of the situation.

We use ψ_i for the eigenfunctions of the Hamiltonian in a differential equation and the corresponding Green's function and ϕ_i for the eigenfunctions of Lanczos' kernel (which is the Hamiltonian) in the integral equation, and treat physical quantities such as the Hamiltonian H as two-point functions so that we can write

$$\int H(q,q')\,\psi_i(q')\,dq' = E_i\psi_i(q). \tag{7.163}$$

We can then find the matrix elements of H as follows:

$$\int \sum_{m,n} H_{mn}\psi_m(q)\,\psi_n^*(q')\,\psi_i(q')\,dq' = \sum_{m,n} H_{mn}\psi_m(q)\,\delta_{ni} \tag{7.164}$$

$$= \sum_m H_{mi}\psi_m(q) = E_i\psi_i(q) \tag{7.165}$$

$$H_{ii} = E_i \tag{7.166}$$

The last step follows by using the orthonormality of the eigenfunctions. Since H is diagonal according to the Heisenberg prescription, these are the only nonzero matrix elements. Then we can reconstruct H:

$$H(q,q') = \sum_{m,n} H_{mn}\psi_m(q)\,\psi_n^*(q') = \sum_i H_{ii}\psi_i(q)\,\psi_i^*(q') = \sum_i E_i\psi_i(q)\,\psi_i^*(q'). \tag{7.167}$$

The Green's function for this Hamiltonian satisfies

$$\int H(q,q')\,G(q',q_0)\,dq' = \delta(q-q_0) = \sum_i \psi_i(q)\,\psi_i^*(q_0), \tag{7.168}$$

where $\delta(q-q_0)$ is the Dirac delta function introduced previously. This will be discussed in more detail in Chapter 11. Then

$$\int \sum_{m,n} H_{mn}\psi_m(q)\,\psi_n^*(q') \sum_{i,j} G_{ij}\psi_i(q')\,\psi_j^*(q_0)\,dq' = \sum_{m,n,j} H_{mn}G_{nj}\psi_m(q)\,\psi_j^*(q_0)$$

$$= \sum_{m,n,j} E_m\delta_{mn}G_{nj}\psi_m(q)\,\psi_j^*(q_0) = \sum_{m,j} E_mG_{mj}\psi_m(q)\,\psi_j^*(q_0) = \sum_i \psi_i(q)\,\psi_i^*(q_0).$$

By multiplying by $\psi_k(q_0)$ and integrating with respect to dq_0 and then repeating this procedure with $\psi_k^*(q)$ and dq, we find

$$E_kG_{kk} = 1 \tag{7.169}$$

$$G_{kk} = \frac{1}{E_k} \tag{7.170}$$

$$G(q,q') = \sum_k \frac{\psi_k(q)\,\psi_k^*(q')}{E_k}, \tag{7.171}$$

which is the well-known expression for the eigenfunction expansion of the Green's function. We can rewrite the integral equation for the Schrödinger eigenfunctions (7.156) as

$$\begin{aligned}\psi_i(q) &= E_i \int G(q,q')\,\psi_i(q')\,dq' \\ &= E_i \int \sum_{k,j} G_{kj}\psi_k(q)\,\psi_j^*(q')\,\psi_i(q')\,dq' \\ &= E_i \sum_{k,j} \frac{\delta_{kj}}{E_k}\psi_k(q)\,\delta_{ij} = \psi_i(q)\,.\end{aligned}$$

Now Lanczos' integral equation is

$$\begin{aligned}\phi_i(q) &= \lambda_i \int H(q,q')\,\phi_i(q')\,dq' = \lambda_i \int \sum_{k,j} H_{kj}\phi_k(q)\,\phi_j^*(q')\,\phi_i(q')\,dq' \\ &= \lambda_i \int \sum_{k,j} E_k\delta_{kj}\phi_k(q)\,\delta_{ji} = \lambda_iE_i\phi_i(q) \end{aligned} \tag{7.172}$$

$$\lambda_i = \frac{1}{E_i}, \tag{7.173}$$

as obtained by Lanczos. Thus we see that, contrary to van der Waerden's assertion, the kernel used by Lanczos was not the Green's function for the wave equation and the eigenvalues and eigenfunctions obtained in this way were not those obtained by Schrödinger. Nevertheless the Lanczos treatment is valid, as far as it went, and his great insight—relating the quantum mechanical matrices to continuous functions—was very prophetic.

Since matrix mechanics could be brought into full agreement with this theory involving continuous fields, Lanczos concluded that the difference between quantum mechanics and classical mechanics is not based on the opposition between continuous and discontinuous functions and that the solution of the "quantum secrets" cannot have anything to do with a "quantum-type" reinterpretation of geometry or calculus.

8

Further developments of wave mechanics by Schrödinger

8.1 Introduction

A little less than two months after the submission of his demonstration of the equivalence of wave and matrix mechanics, Schrödinger submitted another paper[1] in which he presented perturbation theory and applied it to the Stark effect (the shift in the frequency of the light emitted by atoms due to the presence of an external electric field) and gave an exhaustive comparison of his results with experiments. About six weeks after that, he submitted another paper[2] where he introduced the time-dependent wave equation and time-dependent perturbation theory, applied it to the interaction of atoms with radiation, introduced interactions with an external magnetic field, and discussed the physical meaning of the wave function.

8.2 Perturbation theory

As in classical physics, perturbation theory addresses the problem where the Hamiltonian consists of a part H_0 for which an exact solution is already known and one or more additional parts which represent much smaller energies than H_0.

Schrödinger began with a general type of differential equation called the Sturm-Liouville equation. This is written as

$$L\left(y\right)+E\rho\left(x\right)y=0 \tag{8.1}$$

$$L\left(y\right)=p\left(x\right)y''+p'y'-q\left(x\right)y, \tag{8.2}$$

where $p\left(x\right)$, $\rho\left(x\right)$, and $q\left(x\right)$ are all arbitrary functions of the independent variable x. We will discuss Sturm-Liouville theory in Chapter 11. Here we present Schrödinger's perturbation theory, which was based on a theory that Lord Rayleigh had worked out for acoustics adapted to the specialized case of the Schrödinger equation.

[1]Schrödinger, E., *Quantization as an eigenvalue problem (Third communication), Perturbation theory with application to the Stark effect and Balmer lines*, Ann.. der Physik, 80, 437, (1926), received May 10, 1926.

[2]Schrödinger, E., *Quantization as an eigenvalue problem (Fourth communication)*, Ann. der Physik, 81,109 (1926) received June, 21, 1926.

The Historical and Physical Foundations of Quantum Mechanics. Robert Golub and Steven K. Lamoreaux, Oxford University Press.
 DOI: 10.1093/oso/9780198822189.003.0008

We consider the case where the Hamiltonian is of the form $H\left(p=\frac{\partial}{\partial q},q\right) = H_0\left(p=\frac{\partial}{\partial q},q\right) + \lambda H_1\left(p=\frac{\partial}{\partial q},q\right)$, where H_0 is a Hamiltonian for which we already know the eigenfunctions[3] and λ is a small parameter. In general we want to find the eigenfunctions and eigenvalues for H as a power series in λ, but for now we will restrict ourselves to the first order in λ. We look for solutions in the form

$$\phi_k = \phi_k^0 + \lambda v_k \tag{8.3}$$

with eigenvalues

$$E_k^* = E_k + \lambda\varepsilon_k. \tag{8.4}$$

Then

$$H\left(\phi_k^0 + \lambda v_k\right) = \left(E_k + \lambda\varepsilon_k\right)\left(\phi_k^0 + \lambda v_k\right) = H_0\left(\phi_k^0 + \lambda v_k\right) + \lambda H_1\left(\phi_k^0 + \lambda v_k\right), \tag{8.5}$$

which means that

$$H_0 v_k + H_1\phi_k^0 = E_k v_k + \varepsilon_k\phi_k^0 \tag{8.6}$$

$$\left(H_0 - E_k\right) v_k = \left(\varepsilon_k - H_1\right)\phi_k^0. \tag{8.7}$$

Since the ϕ_k^0 are a complete set, we can expand the v_k as

$$v_k = \sum_i \gamma_{ki}\phi_i^0. \tag{8.8}$$

Then

$$\left(H_0 - E_k\right)\sum_i \gamma_{ki}\phi_i^0 = \sum_i \gamma_{ki}\left(E_i - E_k\right)\phi_i^0 = \left(\varepsilon_k - H_1\right)\phi_k^0. \tag{8.9}$$

Multiplying by $\left(\phi_m^0\right)^*$ and integrating over the whole space, we find

$$\varepsilon_k\delta_{km} - \int^* \phi_m^{0*}(q)\, H_1\phi_k^0(q)\, dq = \gamma_{km}\left(E_m - E_k\right). \tag{8.10}$$

Taking $m = k$, we have

$$\varepsilon_k = \int \phi_k^{0*}(q)\, H_1\phi_k^0 k(q)\, dq = \left(H_1\right)_{kk} \tag{8.11}$$

and for $m \neq k$

$$\gamma_{km} = \frac{\int \phi_m^{o*}(q)\, H_i\phi_k^o(q)\, dq}{E_k - E_m} = \frac{\left(H_1\right)_{mk}}{E_k - E_m} \tag{8.12}$$

where we have substituted the matrix element notation. The first order perturbed energies and wave functions are thus

[3] $H_0\phi_k^0 = E_k\phi_k^0$ and the functions ϕ_k^0 form a complete orthonormal set.

$$E_k^* = E_k + \lambda (H_1)_{kk} \tag{8.13}$$

$$\phi_k = \phi_k^o + \lambda \sum_{m \neq k} \frac{(H_1)_{mk}}{E_k - E_m} \phi_m^o. \tag{8.14}$$

The case of degeneracy, when $E_k = E_m$ for some $m \neq k$, will be discussed in a later chapter. In the case of a partially continuous spectrum, the sum over m is replaced by an integral for states in the continuum.

8.3 The time-dependent Schrödinger equation

We recall that what we have called the Schrödinger equation,

$$-\frac{\hbar^2}{2m} \nabla^2 \psi + (V - E) \psi = 0, \tag{8.15}$$

was obtained from the wave equation for a wave with the phase velocity u given by de Broglie's result, $u = E/\sqrt{2m(E - V)}$:

$$\nabla^2 \psi - \frac{2m(E - V)}{E^2} \frac{\partial^2 \psi}{\partial t^2} = 0. \tag{8.16}$$

Assuming a time dependence

$$\psi \sim e^{\pm i \frac{Et}{\hbar}} \tag{8.17}$$

based on the Bohr frequency condition, $E = \hbar\omega$, gives

$$\frac{\partial^2 \psi}{\partial t^2} = -\frac{E^2}{\hbar^2} \psi, \tag{8.18}$$

so that Eq. (8.16) reduces to Eq. (8.15). Equations (8.15) and (8.16) are only valid for a fixed value of the energy (wave frequency); thus, they are not really wave equations but rather equations for the amplitude of the wave. Eq. (8.15) is well known in optics and is called the Helmholtz equation. It is perfectly adequate for situations when $V(q)$ is not a function of time. To treat problems where $V(q, t)$ is a function of both position and time, as in in the case of atomic collisions or an atom interacting with a light wave, we need a wave equation that is not restricted to a fixed value of energy.

Schrödinger noted, after some discussion, that equation (8.15) would also follow from Eq. (8.16) if, instead of considering the second derivative as in Eq. (8.18), he considered the first derivative:

$$\frac{\partial \psi}{\partial t} = \pm i \frac{E}{\hbar} \psi. \tag{8.19}$$

Using this to eliminate E from Eq. (8.15), we find

$$-\frac{\hbar^2}{2m} \nabla^2 \psi + V\psi = \pm i\hbar \frac{\partial \psi}{\partial t}. \tag{8.20}$$

The complex wave function ψ should obey this equation with either the plus or minus sign. Then the complex conjugate ψ^* will obey the equation with the opposite sign. It is conventional to choose the plus sign for the equation for ψ.

We should not overlook the profound implications of the inclusion of the factor i in this equation. Without it we would have the diffusion equation, whose solutions do not show any oscillatory behavior and which does not allow any time-independent stationary state solutions. In addition, we now know that the factor of i is responsible for the interference and diffraction effects that are characteristic of quantum mechanics. The introduction of a complex wave function ψ along with an equation that is first order in the time, means that we only have to specify ψ at a given time, say $t = 0$, in order to specify the complete future development of the wave function. If we want to work with a real wave function, we can define it by the Fourier integral

$$\psi\left(\overrightarrow{q},t\right) = \text{Re}\left[\frac{1}{2\pi}\int A\left(\overrightarrow{k}\right)e^{i\left(\overrightarrow{k}\cdot\overrightarrow{q}-\omega t\right)}d^{3N}k\right]. \tag{8.21}$$

In the above $\overrightarrow{k}$ represents all the the wave vectors along all $3N$ coordinate directions in the problem and $d^{3N}k$ represents the product of their differentials. The frequency ω is a known function of k. Then at $t = 0$, we have

$$\psi\left(\overrightarrow{q},0\right) = \text{Re}\frac{1}{2\pi}\int A\left(\overrightarrow{k}\right)e^{i\left(\overrightarrow{k}\cdot\overrightarrow{q}\right)}d^{3N}k \tag{8.22}$$

$$= \frac{1}{4\pi}\left[\int A\left(\overrightarrow{k}\right)e^{i\left(\overrightarrow{k}\cdot\overrightarrow{q}\right)}d^{3N}k + \int A^*\left(\overrightarrow{k}\right)e^{-i\left(\overrightarrow{k}\cdot\overrightarrow{q}\right)}d^{3N}k\right] \tag{8.23}$$

$$= \frac{1}{4\pi}\left[\int\left(A\left(\overrightarrow{k}\right) + A^*\left(-\overrightarrow{k}\right)\right)e^{i\left(\overrightarrow{k}\cdot\overrightarrow{q}\right)}d^{3N}k\right] \tag{8.24}$$

or

$$\frac{1}{2}\left[A\left(\overrightarrow{k}\right) + A^*\left(-\overrightarrow{k}\right)\right] = \int\psi\left(\overrightarrow{q},0\right)e^{-i\left(\overrightarrow{k}\cdot\overrightarrow{q}\right)}d^{3N}q. \tag{8.25}$$

Given $\psi\left(\overrightarrow{q},0\right)$, we cannot uniquely determine $A\left(\overrightarrow{k}\right)$ since Eq. (8.25) is compatible with

$$A\left(\overrightarrow{k}\right) = \int\psi\left(\overrightarrow{q},0\right)e^{-i\left(\overrightarrow{k}\cdot\overrightarrow{q}\right)}d^{3N}q + C\left(\overrightarrow{k}\right) \tag{8.26}$$

if $C\left(\overrightarrow{k}\right) = -C^*\left(-\overrightarrow{k}\right)$. In order to determine $\psi\left(\overrightarrow{q},t\right)$ for times $t > 0$, we need additional information such as the value of $\left.\frac{\partial\psi\left(\overrightarrow{q},t\right)}{\partial t}\right|_{t=0}$ as would be expected from the form of Eq. (8.16). With complex ψ, we can write

$$\psi\left(\overrightarrow{q},t\right) = \frac{1}{2\pi}\int A\left(\overrightarrow{k}\right)e^{i\left(\overrightarrow{k}\cdot\overrightarrow{q}-\omega t\right)}d^{3N}k \tag{8.27}$$

so that

$$A\left(\overrightarrow{k}\right) = \int\psi\left(\overrightarrow{q},0\right)e^{-i\left(\overrightarrow{k}\cdot\overrightarrow{q}\right)}d^{3N}q \tag{8.28}$$

is completely determined by $\psi\left(\overrightarrow{q},0\right)$. Thus the introduction of a complex wave function in Eq. (8.20) by means of Eq. (8.19) means that we only have to specify the wave function at a given time to determine all future behavior. The state of the system is completely specified by the wave function ψ alone.

8.3.1 Time-dependent perturbation theory: interaction of light with an atom

In the presence of a light wave, the atom will be exposed to oscillating electric and magnetic fields. As a first approach to the problem, we neglect the effect of the magnetic field and consider the interaction between the atom and an electric field $E(q,t) = F_o \cos\omega t$ that is linearly polarized in the z direction, with an interaction energy

$$V(q,t) = -eF_o \sum_i z_i \cos\omega t \equiv A(q)\cos\omega t, \tag{8.29}$$

where the sum is over all the electrons in the atom. As this interaction energy is small compared to the Coulomb interaction $V_c(q)$, we can treat the problem by perturbation theory. The time-dependent Schrödinger equation, Eq. (8.20), is then

$$-\frac{\hbar^2}{2m}\nabla^2\psi + (V_c + A\cos\omega t)\psi - i\hbar\frac{\partial\psi}{\partial t} = 0. \tag{8.30}$$

For $A = 0$, the equation reverts to the time-independent equation, Eq. (8.15), by substituting $\psi = u(q)\, e^{-i\frac{Et}{\hbar}}$. We regard the solutions for the eigenfunctions $u_n(q)$ and eigenvalues E_n of the unperturbed problem as previously known. We assume these to be non-degenerate and discrete (no continuous spectrum) and look for a solution for the perturbed wave function that is close to the unperturbed state $u_n(q)$ in the form

$$\psi(q,t) = u_n(q)\, e^{-i\frac{E_n t}{\hbar}} + \varepsilon(q,t), \tag{8.31}$$

where $\varepsilon(q,t)$ is considered small so that its product with A can be neglected. Substituting in Eq. (8.30), we find

$$-\frac{\hbar^2}{2m}\left(\nabla^2 u_n e^{-i\frac{E_n t}{\hbar}} + \nabla^2\varepsilon\right) + V_c\left(u_n e^{-i\frac{E_n t}{\hbar}} + \varepsilon\right) + A\cos\omega t u_n e^{-i\frac{E_n t}{\hbar}}$$
$$- i\left(-iE_n u_n e^{-i\frac{E_n t}{\hbar}} + \hbar\dot{\varepsilon}\right) = 0. \tag{8.32}$$

Since the u_n are eigenfunctions with eigenvalues E_n,

$$-\frac{\hbar^2}{2m}\nabla^2 u_n + V_c u_n - E_n u_n = 0, \tag{8.33}$$

and

$$\frac{\hbar^2}{m}\nabla^2\varepsilon - 2V_c\varepsilon + 2i\hbar\dot{\varepsilon} = 2A(\cos\omega t)\, u_n e^{-i\frac{E_n t}{\hbar}} \tag{8.34}$$

$$= Au_n\left(e^{i(\omega-\omega_n)t} + e^{-i(\omega+\omega_n)t}\right), \tag{8.35}$$

where $\omega_n = \frac{E_n}{\hbar}$. Introducing

$$\varepsilon = \varepsilon_+ e^{i(\omega-\omega_n)t} + \varepsilon_- e^{-i(\omega+\omega_n)t} \tag{8.36}$$

we find

$$\frac{\hbar^2}{m}\nabla^2\varepsilon_\pm - 2V_c\varepsilon_\pm + 2\hbar\varepsilon_\pm(\omega_n \mp \omega) = A(q)\,u_n. \tag{8.37}$$

Now we expand both $\varepsilon_\pm$ and $A(q)\,u_n$ in terms of the eigenfunctions u_i of the unperturbed problem:

$$\varepsilon_\pm = \sum_i \gamma_i^\pm u_i \tag{8.38}$$

$$A(q)\,u_n = F_o\sum_i V_{ni}u_i. \tag{8.39}$$

Using the orthonormal properties of the u_n, we see that V_{ni} is given by $V_{ni} = \frac{1}{F_o}\int u_i^* A(q)u_n dq = \int u_i^*\left(-e\sum_i z_i\right)u_n dq$ and is the matrix element of the electric dipole moment. Then

$$\sum_i \gamma_i^\pm\left(\frac{\hbar^2}{m}\nabla^2 u_i - 2V_c u_i + 2\hbar(\omega_n \mp \omega)\,u_i\right) = F_o\sum_i V_{ni}u_i \tag{8.40}$$

$$\sum_i 2\gamma_i^\pm\left(-E_i + \hbar(\omega_n \mp \omega)\right)u_i = F_o\sum_i V_{ni}u_i, \tag{8.41}$$

which means that

$$\gamma_i^\pm = \frac{F_o V_{ni}}{2\,(E_n - E_i \mp \hbar\omega)}. \tag{8.42}$$

The correction to the state u_n due to the perturbation is then

$$\varepsilon(q,t) = \frac{F_o}{2}\sum_i V_{ni}u_i(q)\left(\frac{e^{i(\omega-\omega_n)t}}{(E_n - E_i - \hbar\omega)} + \frac{e^{-i(\omega+\omega_n)t}}{(E_n - E_i + \hbar\omega)}\right). \tag{8.43}$$

Under the influence of the perturbation, the wave function for the unperturbed state u_n becomes a superposition of all the other unperturbed states u_i, with the amplitude of each state being proportional to the electric dipole moment matrix element between the states u_n and u_i.

To further explore the physical meaning of this result, we calculate $\psi\psi^*$ using Eq. (8.31) and assuming for simplicity that the u_n are real:

$$\begin{aligned}\psi\psi^* &= u_n(q)^2 + \frac{F_o}{2}\sum_i V_{ni}u_i(q)\,u_n(q)\left\{\begin{array}{l} e^{i\omega_n t}\left(\frac{e^{i(\omega-\omega_n)t}}{(E_n-E_i-\hbar\omega)} + \frac{e^{-i(\omega+\omega_n)t}}{(E_n-E_i+\hbar\omega)}\right) + \\ e^{-i\omega_n t}\left(\frac{e^{-i(\omega-\omega_n)t}}{(E_n-E_i-\hbar\omega)} + \frac{e^{i(\omega+\omega_n)t}}{(E_n-E_i+\hbar\omega)}\right)\end{array}\right\} \\ &= u_n(q)^2 + 2\cos\omega t F_o\sum_i V_{ni}u_i(q)\,u_n(q)\left\{\frac{E_n - E_i}{\left((E_n - E_i)^2 - (\hbar\omega)^2\right)}\right\}. \end{aligned} \tag{8.44}$$

Schrödinger wanted to interpret $-e\psi\psi^*$ as the charge density in the case of a single electron. We see that this quantity oscillates at the frequency of the perturbing electric field and the influence of the other states depends in a resonance-like fashion on the energy difference between the states. Of course the above treatment breaks down in the case of exact resonance when the changes in the wave function can no longer be treated as small perturbations. Also Schrödinger's interpretation of $e\psi\psi^*$ as a charge density turned out to be untenable, due to the spreading of the wave packet, as we shall soon see.

To calculate the scattered radiation—the radiation emitted by the oscillating charge distribution—we need the electric dipole moment of the charge distribution as a function of t. We consider the y component of the electric dipole moment, $M_y = -\sum_i ey_i$. The matrix elements of M_y with respect to the unperturbed states are

$$M_{kn}^{(y)} = \int u_k(q)\, M_y u_n(q)\, dq. \tag{8.45}$$

Recall that the incoming field is given by $F_z = F_o \cos\omega t$, and the interaction with the atom is given by $A(q) = F_o M_z \cos\omega t$ with $M_z = -\sum_i ez_i$. The intensity of the scattered radiation polarized along the y direction will be proportional to the square of

$$\int \psi M_y \psi^* dq = 2\cos\omega t F_o \sum_i V_{ni} M_{in}^{(y)} \left\{ \frac{E_n - E_i}{\left((E_n - E_i)^2 - (\hbar\omega)^2\right)} \right\}. \tag{8.46}$$

This reproduces a by then well-known result previously obtained by Kramers. An important point is the presence of terms with $E_i < E_n$, that is, the possibility of a transition to a lower energy state. This possibility had previously been deduced by Kramers on the basis of the correspondence principle.

Schrödinger then mentioned that the Heisenberg matrix theory does not say anything about the fact that the atom can find itself in different energy states at different times. Matrix mechanics seems to treat the atom as a timeless whole. He then referred to a publication where Heisenberg states that in the matrix theory, the question of the time development of a process has "no direct meaning":[4]

"...in the theory (*matrix mechanics*), up unitl now the question of the time development of events has no direct meaning and the terms 'sooner' or 'later' can not be exactly defined... because of the nature of atomic systems such difficulties were certainly to be expected."

In wave mechanics, the emission of radiation will eventually result in the higher states disappearing as the atom decays to the ground state. This is the equivalent of the classical radiation reaction force. We will see below that the value of $|\psi|^2$ is a constant so that the total wave function remains normalized.

8.3.2 First discussion of the physical meaning of the wave function

For a single-particle system, Schrödinger considered $e\psi\psi^*$ to be the electric charge density and the point particle is replaced by this charge density. For a system containing N particles, ψ is a function of the $3N$ coordinates of all the particles and cannot be visualized in ordinary three-dimensional space. Classically, the several-particle system

[4]Heisenberg, W., *On quantum mechanical kinematics and mechanics*, Mathematische Annalen, 95, 683, (1926) received, Dec. 21, 1925.

is represented by a point in a $3N$-dimensional space (configuration space) whose movement indicates the motion of all the particles in the system. $\psi\psi^*$ is a kind of "weight function" (Schrödinger's term) in the configuration space. The wave-mechanical configuration of the system is a superposition of all possible point configurations, where each point in the configuration space contributes with a certain weight ($\psi\psi^*$) to the overall configuration. One could say that the system finds itself simultaneously in all possible point configurations but each point configuration has a different weight. Of highest interest is the change of the distribution with time and the resulting variations in charge density. The importance of the wave function is that it allows all of these fluctuations to be described by a single partial differential equation.

To go from the configuration space to a charge density in ordinary space, Schrödinger proposed to calculate the total charge at a point $\vec{r}$ as (using a modern notation)

$$\rho(\vec{r}) = -e\sum_{i=1}^{N}\int \delta^{(3)}(\vec{r} - \vec{r}_i)\,\psi\psi^* d^{3N}r, \tag{8.47}$$

where $\vec{r}_i$ are the coordinates of each of the particles in the system and $d^{3N}r$ is the volume element in the configuration space.

For all this to make sense physically, the integral of the weight function over the entire configuration space must remain constant. We investigate this question by calculating

$$\frac{d}{dt}\int \psi\psi^* d^{3N}r = \int\left(\psi\frac{d\psi^*}{dt} + \psi^*\frac{d\psi}{dt}\right)d^{3N}r. \tag{8.48}$$

From the time-dependent wave equation,

$$-\frac{\hbar^2}{2m}\nabla^2\psi + V\psi = i\hbar\frac{\partial\psi}{\partial t}, \tag{8.49}$$

where ∇^2 is understood to act separately on all $3N$ coordinates, $\nabla^2 = \sum_{i=1}^{3N}\frac{\partial^2}{\partial x_i^2}$, we obtain

$$\frac{d}{dt}\int \psi\psi^* d^{3N}r = \frac{1}{i\hbar}\left[\psi\left(\frac{\hbar^2}{2m}\nabla^2\psi^* - V^*\psi^*\right) + \psi^*\left(-\frac{\hbar^2}{2m}\nabla^2\psi + V\psi\right)\right] \tag{8.50}$$

$$= \frac{1}{i\hbar}\int\left[\frac{\hbar^2}{2m}\left(\psi\nabla^2\psi^* - \psi^*\nabla^2\psi\right) + (V - V^*)\,\psi\psi^*\right]d^{3N}r. \tag{8.51}$$

For the usual case of V real, the last term vanishes, and we have

$$\frac{d}{dt}\int \psi\psi^* d^{3N}r = \frac{\hbar}{2mi}\int\left[\left(\psi\nabla^2\psi^* - \psi^*\nabla^2\psi\right)\right]d^{3N}r \tag{8.52}$$

$$= \frac{\hbar}{2mi}\int\vec{\nabla}\cdot\left[\left(\psi\vec{\nabla}\psi^* - \psi^*\vec{\nabla}\psi\right)\right]d^{3N}r \tag{8.53}$$

$$= \frac{\hbar}{2mi}\int\left[\left(\psi\vec{\nabla}\psi^* - \psi^*\vec{\nabla}\psi\right)\right]\cdot d\vec{S}, \tag{8.54}$$

where $d\vec{S}$ is the $3N-1$ dimensional surface surrounding a volume of the configuration space and we have used the $3N$-dimensional divergence theorem. If ψ vanishes fast enough as we go to infinite distances, the right-hand side is zero and the total "weight function" is conserved. For finite distances, we have

$$\frac{\partial}{\partial t}(\psi\psi^*) = \vec{\nabla} \cdot \vec{J}, \tag{8.55}$$

where

$$\vec{J} = \frac{\hbar}{2mi}\left(\psi\vec{\nabla}\psi^* - \psi^*\vec{\nabla}\psi\right) \tag{8.56}$$

is the wave-mechanical current density. This shows that the choice of $\psi\psi^*$ as a density is consistent.

In the case of a complex potential V, we see that

$$\frac{\partial}{\partial t}(\psi\psi^*) = \vec{\nabla} \cdot \vec{J} + \frac{2}{\hbar}\,\mathrm{Im}\,(V)\,\psi\psi^*, \tag{8.57}$$

so an imaginary part of the potential represents a phenomenological loss of weight function (particles) and is often used in nuclear physics (e.g., the optical model) to represent the absorption of a particle (e.g., a neutron) when it is not necessary to follow through the details of the entire reaction.

In the final paragraph of the five-paper series in which he developed wave mechanics,[5] Schrödinger discussed the meaning of the fact that he was led to a complex wave function. If this were fundamentally unavoidable, it would mean that in reality two functions were necessary to determine the state of a system. He preferred the idea that eventually the theory would be developed in such a manner that a single real function and its time derivative would determine the state of a system. He thought that the real wave function would probably satisfy a fourth-order differential equation, but he was unable to find such a formulation for the nonconservative case.

This is in strong opposition to the current idea that the wave function alone should determine the state of the system. This idea led Dirac to the formulation of the relativistic equation for spin-1/2 particles, which is a first order equation. In that case the wave function consists of four real functions, a fact that may be related to Schrödinger's speculations about the possibility of describing quantum systems purely in terms of real functions.

8.3.3 Modern treatment of time-dependent perturbation theory

Another approach to time-dependent perturbation theory is to expand the entire wave function, not just the perturbed part, in eigenfunctions of the unperturbed Hamiltonian. Writing the total Hamiltonian as

$$H = H_0 + V(q,t), \tag{8.58}$$

where $V(q,t)$ is a small, time-dependent perturbation, we write the wave function as

$$\psi = \sum_m c_m(t)\,u_m(q)\,e^{-iE_m t/\hbar} \tag{8.59}$$

[5]Schrödinger, E., *Quantization as an eigenvalue problem*, (Fourth communication) Ann. der Physik, 81, 109, (1926), op. cit. received June 21, 1926.

where $H_0 u_m = E_m u_m$. Substituting this into the time-dependent wave equation gives

$$i\hbar\frac{\partial\psi}{\partial t} = \sum_m \left(i\hbar\dot{c}_m(t) + E_m c_m(t)\right) u_m(q)\, e^{-iE_m t/\hbar} \tag{8.60}$$

$$= \sum_m \left(E_m + V(q,t)\right) c_m(t)\, u_m(q)\, e^{-iE_m t/\hbar} \tag{8.61}$$

$$\sum_m i\hbar\dot{c}_m(t)\, u_m(q)\, e^{-iE_m t/\hbar} = \sum_m c_m(t)\, V(q,t)\, u_m(q)\, e^{-iE_m t/\hbar} \tag{8.62}$$

$$i\hbar\dot{c}_k(t) = \sum_m c_m(t)\, V_{km}(t)\, e^{i(E_k - E_m)t/\hbar}. \tag{8.63}$$

The last step follows by multiplying by $u_k^*(q)$, integrating over the total coordinate volume, and making use of the orthonormality of the eigenfunctions. Here $V_{km}(t) = \int u_k^*(q)\, V(q,t)\, u_m(q)\, dq$ is the matrix element of the potential. Equation (8.63) is exact. To proceed, we assume the system is initially (at $t = 0$) in a given state u_n so that $c_n(0) = 1$, $c_k(0) = 0$ for $k \neq n$, and we assume the c_k remain small throughout and that c_n remains close to 1. Then

$$i\hbar\dot{c}_k(t) = V_{kn}(t)\, e^{i(E_k - E_n)t/\hbar} \tag{8.64}$$

$$c_k(t) = \frac{1}{i\hbar}\int_0^t V_{kn}(t)\, e^{i(E_k - E_n)t/\hbar} dt. \tag{8.65}$$

If the potential is turned on at a given time (taken here as $t = 0$) and switched off at a later time T, the limits of integration can be extended to $t = -\infty$ and $t = +\infty$, and we see that the amplitude of the state k is given for all times later than T by the Fourier transform of the matrix element of the perturbation between the initial state n and the final state k, evaluated at the Bohr frequency between the two states, $\omega_{kn} = (E_k - E_n)/\hbar$. There are many interesting applications of these results which will be addressed in later chapters.

8.4 Conclusion

The development we have outlined in the last three chapters took place during the period between the submission of Heisenberg's first paper outlining the ideas of matrix mechanics (July 29, 1925) and the publication of Schrödinger's last paper in his series (September 5, 1926). The publication of Schrödinger's "first communication" on March 13, 1926, set off an extraordinary activity among the physicists that were capable of keeping up with the rapid developments.

In the calendar year 1926, there were at least 104 publications on quantum mechanics, 48 devoted to wave mechanics (10 by Schrödinger himself) and eight to matrix mechanics. By 1927, the number of papers on quantum mechanics had dropped to 35 (21 on wave mechanics). This shows the extremely rapid acceptance of wave mechanics and was an indication of the extremely broad reach that the theory would develop in future years.

Turning again to the strange fact that matrix and wave mechanics could be so different yet have essentially the same physical content, we note that both theories started from the Bohr-Planck frequency condition. In matrix mechanics the classical Fourier coefficients were modified so that the frequencies obeyed the combination principle and the time dependence of the matrix elements reflected this condition, while wave mechanics resulted from applying de Broglie's relations to a wave equation. Both theories integrated Planck's nonclassical relation with the basic concepts of classical mechanics. Matrix mechanics dealt with discrete quantities satisfying algebraic equations while wave mechanics dealt with a continuous, wave-like process in configuration space that was described by a partial differential equation.

Both Heisenberg and Schrödinger were highly critical of the other's work. Schrödinger wrote[6] that he was "scared away if not repelled" by the "apparently very difficult transcendental algebra and lack of visualizability" of matrix mechanics. At one stage Lorentz asked, "Can you imagine me to be nothing but a matrix? It is hard to believe all this is real." Heisenberg loathed the idea of a visual model of unobservable processes; his starting point was to include only observable elements in the theory. In his view, the matrix elements $|q_{mn}|^2$ give the intensities of the radiation emitted as a result of an atomic transition and any association with the actual position of an electron should be avoided. He expressed his opinion of wave mechanics in a letter to Pauli:

"The more I think about the physical part of Schrödinger's theory, the more horrible I find it...I find it to be 'Mist' (polite translation - 'rubbish'). ...The great accomplishment of Schrödinger's theory is the calculation of matrix elements"[7]

Schrödinger, on the other hand felt that

"Now the damned Göttingen people use my beautiful wave mechanics to calculate their 'scheiss' (polite translation 'trashy') matrix elements."[8]

The key issue was that Heisenberg's theory emphasized the discrete "quantum jumps" between energy levels, whereas in Schrödinger's theory, the jumps were replaced by gradual transfers of "weight" from one state to another. In fact, the electron had essentially disappeared in wave mechanics, being replaced, in Schrödinger's unsuccessful interpretation, by a charge cloud. We will see that the questions raised by this dichotomy were to be the subject of much controversy and, to some extent, remain controversial today. We now know that matrix and wave mechanics, different ways of looking at the same model, are based on a generalized Hilbert space where the coordinate axes represent the complete set of expansion functions. It is as if the two theories had grabbed on to different parts of the elephant.

[6]Schrödinger, E., *On the relation between the Heisenberg-Born-Jordan quantum mechanics and mine*, Ann. der Physik, 79, 734, (1926), received March 18, 1926.

[7]Heisenberg to Pauli, June 8, 1926 in Hermann, A. et al., op. cit. 129.

[8]Quoted in Mehra, J., and Rechenberg, H., *The Historical Development of Quantum Theory, Vol. 5, Erwin Schrödinger and the Rise of Wave Mechanics, Part 2,* footnote 292, p. 822.

9 Quantum statistics and the origin of wave mechanics

9.1 Bose-Einstein statistics

9.1.1 Introduction

The idea that energy might not be a continuous variable in some systems first arose in the consideration of black body radiation (Chapter 3). In order to derive a spectral distribution that agreed with experiment, Planck started by assuming that in order for radiation to reach thermal equilibrium, it had to be in contact with some matter at a given temperature. As it was known on thermodynamic grounds that the spectrum of the radiation must be the same regardless of the nature of the matter, Planck took the simplest case and assumed the matter to be a system of harmonic oscillators. In order to obtain a spectrum that interpolated between the Rayleigh-Jeans result at low frequencies and Wien's law at high frequencies, it was necessary to assume that the energies of the material oscillators were limited to integer multiples of a basic energy unit that was proportional to the frequency of the oscillators. This was followed by Einstein's suggestion that it was the energy of the radiation field that was quantized, an idea that Planck opposed for many years

Given the fundamental importance of Planck's radiation law and the ongoing lack of progress in understanding the relation between the quantized energy of the radiation field and the wide range of optical phenomena that could only be explained by a wave theory of radiation, it is understandable that many physicists revisited the derivation of this law in search of a more fundamental understanding. In Chapter 3, we saw how Einstein derived the Planck law in two entirely different ways. Other physicists who offered derivations were Debye (1910),[1] Planck himself (1911),[2] Nernst (1911),[3] and Natanson (1911).[4] Most of the derivations used the classical electromagnetic (wave) theory to calculate the number of modes of the radiation field in a given frequency

[1]Debye, P., *The role of probability in the theory of radiation*, Ann. der Phys. 33, 1427, (1910), received Oct. 12, 1910.

[2]Planck, M., *The law of black body radiation and the hypothesis of the elementary quantum of radiation* (in French), in Langevin, P. and de Broglie, L., eds La Theorie du Rayonnement et les Quanta, Réunion tenue à Bruxelles, du 30 octobre au 3 novembre 1911, (first Solvay Conference) Gauthier-Villars, 1912.

[3]Nernst, W., *On new problems of the theory of heat*, Sitz. ber. Preuss. Akad. Wiss. (Berlin) (1911), p. 65 Jan. 26, 1911.

[4]Natanson, L., *On the statistical theory of radiation*, Phys. Zeits, 12, 659. (1911), received April 29, 1911.

The Historical and Physical Foundations of Quantum Mechanics. Robert Golub and Steven K. Lamoreaux, Oxford University Press.
 DOI: 10.1093/oso/9780198822189.003.0009

range and/or the relation between the average energy of a Planck oscillator and the energy density of the radiation field. The one exception seems to have been Planck. At the 1911 Solvay Conference (the first of the Solvay Physics Conferences), he proposed dividing the six-dimensional phase space into cells of size h^3 (h being the "elementary quantum of action" that is now called Planck's constant) and considering each such cell a state of the system. He then stated that using this idea and calculating the entropy in the usual way leads to the Planck spectral distribution for the black body radiation.

Bose and Einstein seem not to have known of this work. In 1924 Bose sent a paper to Einstein, who translated it into German and arranged for it to be published. In this paper, Bose claimed that all previous calculations of the black body spectrum (except for Einstein's derivation based on the A and B coefficients, Section 3.2.5) had relied on a classical derivation of the density of states, and he then presented a derivation based on the idea of phase space cells. One new feature of Bose's calculation was that he considered the radiation as being composed of photons ("quanta" or "atoms of light" as they were called at the time), and Einstein realized very quickly that the same argument would apply to a gas of material atoms. A week after Bose's paper[5] was received by the journal *Zeitschrift für Physik* (July 2, 1924), Einstein presented a paper[6] to the Berlin Academy in which he derived the velocity distribution for a gas of material particles (now called the Bose-Einstein distribution) and found that there was a tendency for the atoms to collect in the lowest energy state at low temperatures. This phenomenon is now known as Bose-Einstein condensation. In a second work presented about six months later[7] (January 8, 1925), Einstein calculated the fluctuations in energy for a small volume of the gas. Just as in the case of black body radiation, the fluctuations consisted of two parts: one that would be expected in a gas of massive particles and another that would be produced by the interference of waves. In the case of the Planck distribution, the fluctuations pointed to a particle nature of the radiation, while in the case of a gas of material particles, the fluctuations indicated a wave nature of the massive particles. Einstein referred to de Broglie's thesis, which showed how a wave could be connected with the motion of massive particles. This idea was taken up by Schrödinger, who gave his own derivation of the Bose-Einstein result[8] based on quantizing the amplitude of the de Broglie waves, thus anticipating second quantization.

9.1.2 Planck

At the first Solvay Conference on Physics (1911), Planck presented a survey of derivations of the black body radiation spectrum. He began by considering two systems in thermal equilibrium. $W_{1,2}$ were the probabilities that system $(1, 2)$ had energies $E_{1,2}$.

[5]Bose, S.N., *Planck's law and the light quantum hypothesis*, Zeit. Phys. 26, 178, (1924), received July 2, 1924.

[6]Einstein, A., *Quantum theory of the mono-atomic ideal gas*, Sitz, Ber, Preuss. Akad. Wiss., meeting of July 10, (1924), p. 261.

[7]Einstein, A., *Quantum theory of the mono-atomic ideal gas*, Sitz, Ber, Preuss. Akad. Wiss., meeting of Jan. 8, (1925), p. 3.

[8]Schrödinger, E., *On Einstein's gas theory*, Phys. Zeit., 27, 95 (1926), received Dec. 15 1925.

Then the joint probability of system 1 having energy E_1 and system 2 having energy E_2 would be $W_1 \cdot W_2$. The condition that this be a maximum is

$$d\left(W_1 W_2\right) = 0 \tag{9.1}$$

$$\frac{dW_1}{W_1} + \frac{dW_2}{W_2} = 0\,. \tag{9.2}$$

Since the total energy must be conserved during the variation,

$$dE_1 + dE_2 = 0\,, \tag{9.3}$$

so that

$$\left(\frac{1}{W_1}\frac{dW_1}{dE_1} - \frac{1}{W_2}\frac{dW_2}{dE_2}\right) = 0 \tag{9.4}$$

$$\frac{1}{W_1}\frac{dW_1}{dE_1} = \frac{1}{W_2}\frac{dW_2}{dE_2} = \text{const} = f\left(T\right). \tag{9.5}$$

We can define

$$\frac{d\ln W}{dE} = \frac{1}{k_B T}\,, \tag{9.6}$$

where k_B is Boltzmann's constant.

Planck then defined the probability W as the number of possible realizations (microstates) compatible with a given macrostate of the gas. N oscillators with a total energy E_N would contain a total number $P = E_N/h\nu$ of quanta (photons). Then there would be $(N+P)!$ ways of arranging the P photons among the N oscillators of which $N!P!$ arrangements would be identical. This takes into account only the number of quanta, treating them as indistinguishable particles. Thus

$$W = \frac{(N+P)!}{N!\,P!}. \tag{9.7}$$

Using Eq. (9.6) with Stirling's approximation in the form $\ln(N!) = N \ln N$ gives

$$\frac{1}{k_B T} = \frac{1}{h\nu}\frac{d\ln W}{dP} = \frac{1}{h\nu}\left(\ln\left(N+P\right) + 1 - \ln P - 1\right) \tag{9.8}$$

$$\frac{h\nu}{k_B T} = \ln\left(1 + \frac{N}{P}\right) \tag{9.9}$$

$$P = \frac{N}{e^{\frac{h\nu}{k_B T}} - 1}, \quad E_N = \frac{Nh\nu}{e^{\frac{h\nu}{k_B T}} - 1}. \tag{9.10}$$

Based on a linear oscillator which traces out an ellipse in the two-dimensional phase space (p,q), Planck showed that setting the area of the ellipse equal to a multiple of h,

$$\oint pdq = nh, \tag{9.11}$$

leads to the energy levels being quantized ($E_n = nh\nu$), as shown in Section 4.1. Each additional quantum of energy $h\nu$ increases the area by Planck's constant h for a one-dimensional oscillator or h^3 for a three-dimensional oscillator. So the number of cells in a phase space volume ($d\Omega$) corresponding to an energy interval dE is given by

$$\frac{1}{h^3}\int_{dE} d\Omega = \frac{1}{h^3}\int_{dE} dx\,dy\,dz\,dp_x\,dp_y\,dp_z. \tag{9.12}$$

It seems that Planck used this idea of applying quantization in phase space only to show that the oscillator energy was quantized, since he then went on to use the relation between the energy density of the radiation and the average oscillator energy based on the classical theory of electromagnetism. It was Bose who first (1924) used the idea of quantizing phase space to eliminate classical considerations from the derivation.

Planck presented two other methods of calculating the average energy of an oscillator. In the most interesting derivation, he characterizes the system by the number of oscillators in a given energy state, so that N_j oscillators have the energy $jh\nu$. Then the number of microstates for a system of N oscillators is given by

$$W = \frac{N!}{\prod_{j=0}^{\infty} N_j!}. \tag{9.13}$$

Using the method of Lagrange multipliers to maximize $S = k_B \ln W$ with respect to N_j with the constraints[9]

$$\sum_j jN_j h\nu = E_N \tag{9.14}$$

$$\sum_j N_j = N, \tag{9.15}$$

we find

$$-\ln N_j - \lambda - \beta j h\nu = 0 \tag{9.16}$$

$$e^{-\lambda}e^{-\beta j h\nu} = N_j. \tag{9.17}$$

Then Eq. (9.15) yields

$$e^{-\lambda}\sum_j e^{-\beta j h\nu} = N \tag{9.18}$$

$$e^{-\lambda} = \left(1 - e^{-\beta h\nu}\right) N. \tag{9.19}$$

From Eq. (9.14), the energy is

$$\begin{aligned} E_N &= \sum_j jN_j h\nu = N\left(1 - e^{-\beta h\nu}\right)\sum_j jh\nu e^{-\beta j h\nu} \\ &= N\left(1 - e^{-\beta h\nu}\right)\left(-\frac{\partial}{\partial\beta}\sum_j e^{-\beta j h\nu}\right) = N\frac{h\nu}{e^{\beta h\nu} - 1} \end{aligned} \tag{9.20}$$

where β is identified as $\beta = 1/k_B T$.

[9]Note that here we are dealing with the number of oscillators, which is fixed, unlike the number of photons.

Planck finished his paper by expressing his disquiet with the idea of quantizing the radiation field. His work had been concerned with the material oscillators taken to be interacting with the field. He proposed that a minimum assumption compatible with observations was that only the energy emitted by an oscillator was quantized, thus solving the problem of reconciling the existence of photons with the observed wave properties of light.

After Planck concluded, Einstein[10] found it "a little shocking" that Planck had introduced a probability W without giving a physical definition of it. Even though Planck had defined the probability so that the entropy derived from it agreed with the experiment, Einstein remarked that "we cannot conclude that the theory is correct."

9.1.3 Bose

Satyendra Nath Bose[11] was an Indian physicist who had been educated in Calcutta. At that time the universities were controlled by the colonial government and most of the teaching positions had previously been held by Scotsmen. In 1914 a new University College of Science was founded in Calcutta with financial support from a group of rich Indians who stipulated that all the professors should be Indians. Bose and his friend M. N. Saha were appointed as lecturers in 1916. Together they published two papers on statistical mechanics in the *Philosophical Magazine* of London.

In June 1924, Bose sent a paper (in English) to Einstein asking him for his opinion of the work and, if that was favorable, to arrange for its publication in *Zeitschrift für Physik*. Bose reminded Einstein that he had previously asked Einstein's permission to translate his book on general relativity into English and that the translation had been published. Bose did not feel his German was good enough to do the translation in the other direction. The paper had previously been submitted to *Philosophical Magazine* but had been rejected. Einstein was immediately impressed and sent the German translation to *Zeitschrift für Physik*[12] about a week after receiving the letter. He only made one small change, which we will discuss presently, and added a comment at the end to the effect that the work represented real progress and that the method also yields the quantum theory of the ideal gas of massive particles. About a week after submitting the translated paper, he presented the results for the ideal gas to the Berlin Academy (July 10, 1924).[13] Bose sent a second paper to Einstein which Einstein again translated and submitted for publication. Einstein added an appendix to Bose's paper in which he strongly disagreed with one of Bose's conclusions[14].

[10]Einstein, A. *Discussion of the report of Mr. Planck, in Langevin and de Broglie*, op cit, p. 115.

[11]P. Ghose, "Bose Statistics: A Historical Perspective", in *Satyendra Nath Bose—His Life and Times*, edited by K.C. Wali (Singapore, World Scientific, 2009), pp. 296–331; W. A. Blanpied, *Satyendranath Bose- Co-Founder of Quantum Statistics*, Am. J. Phys., **40** 1212 (1972). This volume contains English translations of Bose's papers.

[12]Bose, S.N., *Planck's law and the light quantum hypothesis*, Zeit. Phys, 26, 178, (1924), received July 2, 1924.

[13]Einstein, A., *Quantum theory of the monoatomic ideal gas*, Sitz.ber. Preuss. Akad. Wiss. p. 261, July 10, (1924).

[14]Bose, S.N., *Thermal Equilibrium in the Radiation Field in the Presence of Matter*, Zeit. Phys. 27, 384, (1924) received July 10, 1924.

Although Bose spent two years in Europe, one year in Paris and one in Berlin (1925–26), he only had one brief meeting with Einstein.

Bose's original manuscript is not extant but an English (re)translation has been published.[15] Bose begins with the statement that every previous derivation of the Planck spectrum relied on the relation

$$\rho_\nu d\nu = 8\pi \frac{\nu^2 d\nu}{c^3} \bar{E}\,, \tag{9.21}$$

where $\rho_\nu d\nu$ is the energy density of the radiation between frequencies ν and $\nu + d\nu$ and $\bar{E}$ is the average energy of an oscillator interacting with the radiation field in this frequency range.

Eq. (9.21), derived from the classical electromagnetic theory, relates the energy spectrum of the radiation and the average energy of the oscillators. The only derivation of the Planck spectrum that did not use (9.21) was Einstein's derivation using the A and B coefficients, which Bose pointed out used the correspondence principle and the Wien law, which was based on classical theory. In a postcard to Bose[16] (July 2, 1924), Einstein replied that he did not use the correspondence principle and the Wien law is not based on the wave theory, but that it was of no importance since Bose was the first to have derived the factor from quantum theory, even if not completely, because of the "polarization factor" of two which Bose had introduced. Bose later claimed that Einstein had altered his paper on this point and that he (Bose) had proposed that the photons existed in two states with an angular momentum parallel or antiparallel to the direction of motion.

Bose considered the radiation to be enclosed in a cavity of volume V and to consist of photons of various frequencies. If N_s is the number of photons having energy $h\nu_s$, the total energy is given by

$$E = \sum_s N_s h\nu_s = V \int \rho_\nu d\nu\,. \tag{9.22}$$

The photons are considered to have a definite position and momentum $h\nu_s/c$ in the direction of propagation. The phase space volume associated with the interval $d\nu$ is given by

$$\Phi = \int d^3x \int d^3p = V 4\pi p^2 dp = V 4\pi \left(\frac{h\nu_s}{c}\right)^2 \frac{h}{c} d\nu_s. \tag{9.23}$$

If the phase space is divided into cells of volume h^3, the number of cells associated with $d\nu_s$ is given by

$$A_s = V 8\pi \frac{\nu_s^2}{c^3} d\nu_s\,, \tag{9.24}$$

where the result has been multiplied by two to take into account the two possible polarizations of the photons. In the published translation, Einstein stated that this factor of two that accounts for the two possible polarizations of the radiation was

[15] Op. cit. p. 153 Wali, pp. 26–29.

[16] Einstein's postcard is reproduced in Wali (op. cit., p. 153, p. xxi).

the only remnant of classical theory in Bose's derivation. There is some evidence that Bose's original manuscript contained the proposal that the photons had a spin with two different possible orientations. This would have been a purely quantum origin of the factor of two.

The argument proceeds by letting p_r^s be the number of cells, associated with frequencies between ν_s and $\nu_s + d\nu_s$, that contain r photons. The total number of photons N_s is then given by

$$N_s = \sum_r r p_r^s \tag{9.25}$$

and the total number of cells is

$$A_s = \sum_r p_r^s \,. \tag{9.26}$$

The thermodynamic probability W (the number of microstates comprising the same macrostate) is given by the number of ways of choosing the A_s cells ($A_s!$) divided by the number of ways of arranging the cells with the same number of photons

$$W = \prod_s \left[\frac{A_s!}{\prod_r p_r^s!} \right] \tag{9.27}$$

and the equilibrium distribution is the set of values p_r^s which maximizes W subject to the constraint (9.26) and

$$E = \sum_s N_s h\nu_s \tag{9.28}$$

with N_s given by Eq. (9.25). Then, using Stirling's approximation, we have

$$\ln W = \sum_s A_s \ln A_s - \sum_{s,r} p_r^s \ln p_r^s \,. \tag{9.29}$$

We find the maximum, with respect to variations in p_r^s , of

$$\ln W - \lambda_s \sum_r p_r^s - \beta \sum_s h\nu_s \sum_r r p_r^s \tag{9.30}$$

using the method of Lagrange multipliers. This gives

$$\ln p_r^s + 1 + \lambda_s + \beta h\nu_s r = 0 \tag{9.31}$$

$$p_r^s = B_s e^{-\beta h\nu_s r} \tag{9.32}$$

where we put $B_s = e^{-(1+\lambda_s)}$. From Eq. (9.26),

$$B_s = A_s \left(1 - e^{-\beta h\nu_s}\right) \tag{9.33}$$

and then from (9.25)

$$N_s = \sum_r r p_r^s = A_s \left(1 - e^{-\beta h\nu_s}\right) \sum_r r e^{-\beta h\nu_s r} \tag{9.34}$$

$$= \frac{A_s}{\left(e^{\beta h\nu_s} - 1\right)}, \tag{9.35}$$

where we used the fact that

$$\sum_r r e^{-zr} = -\frac{\partial}{\partial z} \sum_r e^{-zr} = -\frac{\partial}{\partial z} \frac{1}{1 - e^{-z}} = \frac{e^{-z}}{\left(1 - e^{-z}\right)^2}. \tag{9.36}$$

Finally Eq. (9.22) gives

$$E = \sum_s N_s h\nu_s = \sum_s hV 8\pi \frac{\nu_s^3}{c^3} \frac{d\nu_s}{\left(e^{\beta h\nu_s} - 1\right)}. \tag{9.37}$$

We can find the entropy from Eq. (9.29):

$$\frac{S}{k_B} = \ln W = \sum_s A_s \ln A_s - \sum_{s,r} p_r^s \ln p_r^s$$

$$\sum_{s,r} p_r^s \ln p_r^s = \sum_{s,r} p_r^s \ln \left(A_s \left(1 - e^{-\beta h\nu_s}\right) e^{-\beta h\nu_s r}\right)$$

$$= \sum_s A_s \left(1 - e^{-\beta h\nu_s}\right) \sum_r e^{-\beta h\nu_s r} \left[\ln A_s + \ln\left(1 - e^{-\beta h\nu_s}\right) - \beta h\nu_s r\right]$$

$$= \sum_s A_s \left[\ln A_s + \ln\left(1 - e^{-\beta h\nu_s}\right)\right] - \beta \sum_s h\nu_s A_s \left(1 - e^{-\beta h\nu_s}\right) \sum_r r e^{-\beta h\nu_s r}$$

$$= \sum_s A_s \left[\ln A_s + \ln\left(1 - e^{-\beta h\nu_s}\right)\right] - \beta \sum_s h\nu_s A_s \frac{1}{\left(e^{\beta h\nu_s} - 1\right)}$$

$$\frac{S}{k_B} = \beta E - \sum_s A_s \ln\left(1 - e^{-\beta h\nu_s}\right), \tag{9.38}$$

where we used Eqs. (9.35), (9.36), and (9.37). Then we have

$$\frac{\partial S}{\partial E} = \frac{1}{T} = k_B \beta, \tag{9.39}$$

so that $\beta = 1/k_B T$ and Eq. (9.37) gives the Planck distribution.

In his second paper, Bose turned to Einstein's derivation of the Planck spectrum by considering the radiation field as interacting with a Bohr atom and introducing the A and B coefficients. The calculation yielded the Planck distribution when the atomic energy levels were occupied according to the Boltzmann distribution and the emission and absorption probabilities satisfied certain relations. Bose proposed a more general method, independent of any assumptions as to the elementary processes of emission and absorption. He considered the thermodynamic probability of the whole system of radiation and material particles and maximized it subject to the usual constraints.

For the radiation field characterized by the number of cells A_ν in a frequency interval $d\nu$ (given by Eq. (9.24)) and the number of photons N_ν in that interval, the thermodynamic probability is

$$W_{rad} = \prod_\nu \frac{(A_\nu + N_\nu)!}{N_\nu!\, A_\nu!}. \tag{9.40}$$

For the material particles, we again divide the phase space into cells and consider that every cell i has a probability of occupancy g_i. For free particles these are all equal, but for the Bohr atom they represent the degeneracy of the states. With n_i the number of atoms in the ith cell and a total number of atoms N, the thermodynamic probability is

$$W_a = N! \prod_i \frac{g_i^{n_i}}{n_i!}. \tag{9.41}$$

Thus he treats the atoms classically (i.e., as distinguishable), and this expression for W yields the ordinary Boltzmann distribution. For fixed total energy and number of particles, W_a is a maximum for $n_i = Cg_i e^{-\epsilon_i/k_BT}$ with ϵ_i being the energy of state i. For the whole system of radiation and particles, the probability is

$$W = \prod_\nu \frac{(A_\nu + N_\nu)!}{N_\nu!\, A_\nu!} \prod_i \frac{g_i^{n_i}}{n_i!} N! \tag{9.42}$$

The constraints are

$$\sum_s N_\nu E_\nu + \sum_i n_i E_i = E \tag{9.43}$$

$$\sum_i n_i = N. \tag{9.44}$$

Now consider an elementary process in which an atom goes from the state r to the state s, while a photon of frequency ν is absorbed and one of frequency ν' is emitted. Thus $n_r \to n_r - 1$ and $n_s \to n_s + 1$, while $N_\nu \to n_\nu - 1$ and $N_{\nu'} \to N_{\nu'} + 1$. The probability W_0 of the original state, given by Eq. (9.42), contains factors

$$\frac{g_r^{n_r}}{n_r!} \frac{g_s^{n_s}}{n_s!} \frac{(A_\nu + N_\nu)!\,(A_{\nu'} + N_{\nu'})!}{N_\nu! A_\nu! N_{\nu'}! A_{\nu'}!}, \tag{9.45}$$

while the probability W_1 of the new state contains factors

$$\frac{g_r^{n_r-1}}{(n_r-1)!} \frac{g_s^{n_s+1}}{(n_s+1)!} \frac{(A_\nu + N_\nu - 1)!\,(A_{\nu'} + N_{\nu'} + 1)!}{(N_\nu - 1)! A_\nu!\,(N_{\nu'} + 1)! A_{\nu'}!}, \tag{9.46}$$

so we can write

$$W_1 = W_0 \frac{n_r}{g_r} \frac{g_s}{(n_s+1)} \frac{N_\nu}{(A_\nu + N_\nu)} \frac{(A_{\nu'} + N_{\nu'} + 1)}{(N_{\nu'} + 1)}. \tag{9.47}$$

At equilibrium W must be a maximum, so $W_1 = W_0$, which means that

$$\frac{n_r}{g_r} \frac{N_\nu}{(A_\nu + N_\nu)} = \frac{(n_s + 1)}{g_s} \frac{(N_{\nu'} + 1)}{(A_{\nu'} + N_{\nu'} + 1)} \tag{9.48}$$

$$\cong \frac{n_s}{g_s} \frac{N_{\nu'}}{(A_{\nu'} + N_{\nu'})}. \tag{9.49}$$

Using $N_\nu h\nu = V\rho_\nu d\nu$, $A_\nu = V8\pi\frac{\nu^2}{c^3}d\nu$, $N_\nu = (V\rho_\nu/h\nu)\, d\nu = B_\nu\rho_\nu d\nu$, we can write Eq. (9.49) as

$$\frac{n_r}{g_r} \frac{B_\nu\rho_\nu}{(A_\nu + B_\nu\rho_\nu)} = \frac{n_s}{g_s} \frac{B_{\nu'}\rho_{\nu'}}{(A_{\nu'} + B_{\nu'}\rho_{\nu'})} \tag{9.50}$$

where

$$\frac{A_\nu}{B_\nu} = 8\pi h\frac{\nu^3}{c^3}. \tag{9.51}$$

For the case of a Bohr atom considered by Einstein where a single photon is absorbed so that $N_\nu \to N_\nu - 1$ and the atom goes from a state r to a state s, with no change in $N_{\nu'}$ Eq. (9.49) reduces to

$$\frac{n_r}{g_r} \frac{N_\nu}{(A_\nu + N_\nu)} = \frac{n_r}{g_r} \frac{B_\nu\rho_\nu}{(A_\nu + B_\nu\rho_\nu)} = \frac{n_s}{g_s}, \tag{9.52}$$

which Bose interpreted as saying that the absorption probability (left-hand side $r \to s$) was equal to the emission probability (right-hand side $s \to r$).

Bose compared this to Einstein's result that the probabilities of absorption and emission must be equal in order to attain thermal equilibrium, so that

$$n_r B'_\nu\rho_\nu = n_s\left(A'_\nu + B'_\nu\rho_\nu\right). \tag{9.53}$$

Bose remarked how Einstein had to assume the presence of an induced emission (proportional to the intensity of the radiation) as well as spontaneous emission (independent of the state of the field). Additional assumptions as to the relations between A' and B' were also necessary to obtain the Planck spectrum. Bose felt that these assumptions were unfounded and favored his interpretation, according to which the absorption probability was given by

$$P_{abs} = \frac{N_\nu}{(A_\nu + N_\nu)}, \tag{9.54}$$

which he noted was not simply proportional to the number of photons present. He favored this because it did not require additional assumptions.

In an addendum to Bose's paper, Einstein pointed out that thermal equilibrium only required that the transition probabilities were in the ratio

$$\frac{g_s}{g_r} \frac{N_\nu}{(A_\nu + N_\nu)} : 1 \tag{9.55}$$

and that according to Bose's interpretation the emission would be independent of the incident radiation, which contradicts the correspondence principle that the classical theory must represent a limit of the quantum theory. Bose wrote a third paper, which he sent to Einstein, in which he accepted the existence of stimulated emission while rejecting spontaneous emission. The paper was never published and no copy of it has been found, but the cover letter sent to Einstein is available. Bose seemed to feel that emission was a single process and should not be due to two separate mechanisms. Thus he gave up induced emission in his second paper, and then, in response to Einstein's criticisms, he evidently produced a theory with only induced emission, which he described in his third paper.

An undated two-page manuscript in Bose's handwriting was found in Einstein's papers.[17] It contains a very interesting calculation of the fluctuations in the number of particles in a small volume of a gas by direct calculation of the probability. The gas is characterized by N particles having velocities between v and $v+dv$ and the number of phase space cells in this region is A. The volume is divided into two parts, one with p cells and the other with q cells. Thus $p+q=A$. The number of particles in the p cells is n_p and the number in the q cells is n_q and $n_p+n_q=N$. Then the thermodynamic probability is given by[18]

$$W_0\left(n_p, n_q\right)=\frac{\left(p+n_p-1\right)!}{(p-1)!\,n_p!}\frac{\left(q+n_q-1\right)!}{(q-1)!\,n_q!}. \tag{9.56}$$

This form was introduced by Planck (Eq. (3.24) in Chapter 3). The (-1) plays no role but its origin can be seen as follows. We indicate a given arrangement of the n_p particles among p cells by the following symbol

$$\{\times\times\mid\times\times\times\times\mid\times\mid\mid\times\times\times\mid.....\}, \tag{9.57}$$

where the $\times$'s represent particles and the vertical lines indicate cell walls. This symbol means there are two particles in the first cell, four in the second cell, one in the third, no particles in the fourth cell, etc. Thus we need to calculate the number of arrangements of $(p-1)$ cell walls and n_p particles, which is given by the p-dependent part of Eq. (9.56).

If a particle were to move from the region labeled q to the one labeled p, W would change to

$$W\left(n_p+1, n_q-1\right)=\frac{\left(p+n_p\right)!}{(p-1)!\left(n_p+1\right)!}\frac{\left(q+n_q-2\right)!}{(q-1)!\left(n_q-1\right)!} \tag{9.58}$$

$$=W_0\left(n_p, n_q\right)\left[\frac{\left(p+n_p\right)}{\left(n_p+1\right)}\frac{n_q}{\left(q+n_q-1\right)}\right]. \tag{9.59}$$

Since W is a maximum at equilibrium, we must have $W\left(n_p+1, n_q-1\right)=W\left(n_p, n_q\right)$, which means that

$$\left(p+n_p\right)n_q=\left(n_p+1\right)\left(q+n_q-1\right). \tag{9.60}$$

[17] P. Ghose, "Bose Statistics: A Historical Perspective", in *Satyendra Nath Bose—His Life and Times*, edited by K.C. Wali (Singapore: World Scientific, 2009), p. 314.

[18] Note that here we are considering the number of particles distributed among the cells whereas previously we considered the number of cells with a given number of particles.

For a fluctuation of ε particles, we see that

$$W(n_p+\varepsilon, n_q-\varepsilon) = W_0 \frac{(p+n_p)(p+n_p+1)\dots(p+n_p+\varepsilon-1)}{(n_p+1)\dots(n_p+\varepsilon)} \times \\ \times \frac{n_q(n_q-1)..(n_q-(\varepsilon-1))}{(q+n_q-1)(q+n_q-2)\dots(q+n_q-\varepsilon)}. \tag{9.61}$$

Writing this as

$$\frac{W(n_p+\varepsilon, n_q-\varepsilon)}{W_0} = \frac{A_p}{B_p}\frac{C_q}{D_q}, \tag{9.62}$$

we calculate

$$\begin{aligned} \ln A_p &= \ln(p+n_p) + \ln(p+n_p+1) + \dots \ln(p+n_p+\varepsilon-1) \\ &= \ln(p+n_p) + \ln\left[(p+n_p)\left(1+\frac{1}{p+n_p}\right)\right] + \dots \ln\left[(p+n_p)\left(1+\frac{\varepsilon-1}{p+n_p}\right)\right] \\ &= \varepsilon\ln(p+n_p) + \frac{1}{p+n_p} + \dots\dots\frac{\varepsilon-1}{p+n_p} \\ &= \varepsilon\ln(p+n_p) + \frac{1}{p+n_p}\frac{\varepsilon(\varepsilon-1)}{2}. \end{aligned} \tag{9.63}$$

Similarly, using $\varepsilon \ll n_p, n_q$, we have

$$\ln B_p = \varepsilon\ln(n_p+1) + \frac{\varepsilon(\varepsilon-1)}{2}\frac{1}{(n_p+1)} \tag{9.64}$$

$$\ln\frac{A_p}{B_p} = \varepsilon\ln\frac{(p+n_p)}{(n_p+1)} + \frac{\varepsilon(\varepsilon-1)}{2}\left[\frac{1-p}{(p+n_p)(n_p+1)}\right] \tag{9.65}$$

$$\ln C_q = \varepsilon\ln n_q - \frac{\varepsilon(\varepsilon-1)}{2}\frac{1}{n_q} \tag{9.66}$$

$$\ln D_q = \varepsilon\ln(n_q+q-1) - \frac{\varepsilon(\varepsilon-1)}{2}\frac{1}{(n_q+q-1)} \tag{9.67}$$

$$\ln\frac{C_q}{D_q} = \varepsilon\ln\frac{n_q}{n_q+q-1} + \frac{\varepsilon(\varepsilon-1)}{2}\left[\frac{1-q}{(n_q+q-1)n_q}\right]. \tag{9.68}$$

So finally

$$\ln\frac{W(n_p+\varepsilon, n_q-\varepsilon)}{W_0} = \varepsilon\ln\left[\frac{p+n_p}{n_p+1}\frac{n_q}{n_q+q-1}\right] + \frac{\varepsilon^2}{2}\left[\frac{1-p}{(p+n_p)(n_p+1)} + \frac{1-q}{n_q(n_q+q-1)}\right] \tag{9.69}$$

using $\varepsilon \gg 1$. The first term is zero due to Eq. (9.60). For $n_p, n_q \gg 1$, we can neglect the 1 in the denominator, so that

$$\ln\frac{W(n_p+\varepsilon.n_q-\varepsilon)}{W_0} = \frac{\varepsilon^2}{2}\left[\frac{1-p}{(p+n_p)n_p} + \frac{1-q}{n_q(n_q+q)}\right]. \tag{9.70}$$

We see that $(p+n_p), (n_q+q)$ are proportional to the respective volumes $V_{p,q}$. Taking $V_p \ll V_q$, we can neglect the second term and find

$$\frac{W(n_p+\varepsilon, n_q-\varepsilon)}{W_0} = \exp\left[-\frac{\varepsilon^2}{2}\left[\frac{p-1}{(p+n_p)\,n_p}\right]\right]. \tag{9.71}$$

The fractional (mean square) particle number fluctuations are thus

$$\frac{\overline{\Delta\varepsilon^2}}{n_p^2} = \frac{(p+n_p)\,n_p}{pn_p^2} = \frac{1}{p} + \frac{1}{n_p}. \tag{9.72}$$

In the case of a photon gas where $n_p = E_p/h\nu$, with E_p the energy of the gas in volume V_p, and $p = A = 8\pi V\nu^2 d\nu/c^3$, this is identical to Einstein's result (Section 3.2.4), and shows that the fluctuations in a gas of material particles also consist of a wave-type and a particle-type contribution. This means that the wave-particle duality is already contained in the form of Eq. (9.56).

As this manuscript is undated, it is unknown if this work was done before or after Einstein's calculation of the same quantity using the Bose statistics. We now turn to Einstein's application of Bose's ideas to an ideal gas of material particles.

9.1.4 Einstein

At the end of his translation of Bose's paper, Einstein wrote:

"In my opinion Bose's derivation of the Planck formula constitutes important progress. The method used also yields the quantum theory of the ideal gas, which I will work out in another place."

Einstein presented his results to the Berlin Academy a week after his translation of Bose's paper had been received by the publisher.

Since Bose had treated the radiation field as a gas of particles, Einstein realized that the only change necessary to apply the calculation to an ideal gas of massive point particles was to add a constraint fixing the number of particles. Einstein began by reviewing the elements of Bose's calculation:

1. The phase space of an atom was divided into cells of size h^3.
2. The thermodynamically important feature is the distribution of the atoms over the cells.
3. The probability W of a macrostate is the number of different microstates consistent with the macrostate.
4. The entropy is given by Boltzmann's law

$$S = k_B \ln W. \tag{9.73}$$

For a gas of atoms of mass m confined in a volume V, the phase space volume for a single atom is given by

$$\Phi = \int dx\, dy\, dz\, dp_x\, dp_y\, dp_z \tag{9.74}$$

and the volume occupied by atoms with energy less than $E_0 = \left(p_{0x}^2 + p_{0y}^2 + p_{0z}^2\right)/2m$ is the volume of a sphere in momentum space:

$$\Phi = V\frac{4\pi}{3}p_0^3 = V\frac{4\pi}{3}\left(2mE_0\right)^{3/2}. \tag{9.75}$$

The number of cells Δs corresponding to an energy between E and $E + \Delta E$ is clearly

$$\Delta s = V\frac{4\pi}{h^3}p_o^2 dp = V\frac{2\pi}{h^3}\left(2m\right)^{3/2}\sqrt{E}dE. \tag{9.76}$$

V can always be taken large enough so that $\Delta s \gg 1$. Slightly changing Bose's definition, Einstein took p_r^s to be the fraction of cells in Δs that contain r atoms. Thus $\sum_r p_r^s = 1$ and the number of possible distributions of Δn atoms over Δs cells is

$$W_s = \frac{\Delta s!}{\prod_{r=0}^{r=\infty}\left(p_r^s \Delta s\right)!} \tag{9.77}$$

in agreement with Eq. (9.27). Using Stirling's approximation,

$$\begin{aligned}\ln W_s &= \Delta s \ln \Delta s - \Delta s \sum_r p_r^s \left(\ln p_r^s + \ln \Delta s\right) \\ &= -\Delta s \sum_r p_r^s \ln p_r^s \end{aligned} \tag{9.78}$$

$$W_s = \prod_{r=0}^{r=\infty} \frac{1}{\left[\left(p_r^s\right)^{p_r^s}\right]^{\Delta s}} \tag{9.79}$$

and

$$W = \prod_s W_s = \prod_{s,r} \frac{1}{\left(p_r^s\right)^{p_r^s}}\,. \tag{9.80}$$

since the Δs factors in Eq. (9.79) are included in the product over all s. The result is then

$$S = k_B \ln W = k_B \sum_s \ln W_s = -k_B \sum_{r,s} p_r^s \ln p_r^s\,. \tag{9.81}$$

In thermodynamic equilibrium, this is a maximum subject to the constraints

$$n = \sum_{r,s} r p_r^s \tag{9.82}$$

$$E = \sum_{r,s} E_s r p_r^s \tag{9.83}$$

$$\sum_r p_r^s = 1. \tag{9.84}$$

Dividing (9.75) by h^3 gives us $s = \Phi/h^3$—the number of cells corresponding to energies $\leq E_s$. The energy is thus

$$E_s = s^{2/3}\left(\frac{3h^3}{4\pi V}\right)^{2/3}\frac{1}{2m} = cs^{2/3}. \tag{9.85}$$

S is a maximum subject to the constraints when

$$\ln p_r^s + 1 + \gamma + \lambda r + \beta E_s r = 0 \tag{9.86}$$

$$p_r^s = e^{-(1+\gamma)} e^{-r(\lambda+\beta E_s)} \tag{9.87}$$

which we write as

$$p_r^s = \left(1 - e^{-\alpha_s}\right) e^{-\alpha_s r} \tag{9.88}$$

with

$$\alpha_s = \lambda + \beta E_s = A + B s^{2/3} \tag{9.89}$$

so as to satisfy Eq. (9.84). Then

$$n_s = \sum_r r p_r^s = \left(1 - e^{-\alpha_s}\right) \sum_r r e^{-\alpha_s r} = \frac{1}{e^{\alpha_s} - 1} \tag{9.90}$$

and

$$\bar{E}_s = \frac{c s^{2/3}}{e^{\alpha_s} - 1}. \tag{9.91}$$

The entropy is given by

$$S = -k_B \sum_{r,s} p_r^s \ln p_r^s = -k_B \left[\sum_{r,s} p_r^s \left(\ln\left(1 - e^{-\alpha_s}\right) - \alpha_s r \right) \right] \tag{9.92}$$

$$= -k_B \left[\sum_s \ln\left(1 - e^{-\alpha_s}\right) - An - \frac{B}{c} E \right] \tag{9.93}$$

and then

$$\frac{1}{T} = \frac{\partial S}{\partial E} = \frac{B}{c} k_B \tag{9.94}$$

$$\alpha_s = \lambda + \beta E_s = A + E_s / k_B T, \tag{9.95}$$

so that

$$TS = -k_B T \sum_s \ln\left(1 - e^{-\alpha_s}\right) + A n k_B T + E\,. \tag{9.96}$$

The Helmholtz free energy is

$$F = E - TS = k_B T \left[\sum_s \ln\left(1 - e^{-\alpha_s}\right) - An \right]. \tag{9.97}$$

We can find the equation of state using $p = -\frac{\partial F}{\partial V}$ and Eq. (9.85):

$$p = -\sum_s \frac{k_B T e^{-\alpha_s}}{1 - e^{-\alpha_s}} \frac{\partial \alpha_s}{\partial V} = -k_B T \sum_s n_s \frac{\partial \alpha_s}{\partial V} = -\sum_s n_s \frac{\partial E_s}{\partial V}$$

$$= \sum_s n_s \frac{2}{3} \frac{E_s}{V} = \frac{2}{3} \frac{E}{V} = \frac{2}{3V} \sum_s \frac{E_s}{e^A e^{E_s / k_B T} - 1}. \tag{9.98}$$

To summarize,

$$n = \sum_s \frac{1}{e^A e^{E_s/kT} - 1} \tag{9.99}$$

$$E = \sum_s \frac{E_s}{e^A e^{E_s/kT} - 1}. \tag{9.100}$$

Comparing Eq. (9.86) to Eq. (9.31) for the Bose case we see that the difference—the term λr—comes from the additional constraint fixing the total number of particles, and this term results in the term e^A in Eqs. (9.99) and (9.100). As we will see, it is the presence of this term which results in Bose-Einstein condensation.

About six months later (January 8, 1925) Einstein read a second paper[19] to the Berlin Academy in which he explored the consequences of the e^A term and showed this would lead to what is now called Bose-Einstein condensation. Following publication of this paper, Schrödinger wrote a letter to Einstein (February 28, 1925) in which he criticized the approach taken above and stated that p_r^s should be given by a Poisson distribution. Einstein replied that his calculation was correct, and the difference between the Poisson distribution and the Bose statistics is that in the former the particles are treated as independent (i.e., distinguishable), while in the latter the particles are not independent. He expressed regret that he had not clearly emphasized the fact that he and Bose had applied a special type of statistics which could only be justified by the experimental results. According to this statistics, the atoms had a preference to be in a cell with other atoms.

In his reply to Schrödinger, Einstein considered a system of two particles in two cells as an example. For independent particles labeled I and II, the possible distributions are

Case	Cell 1	Cell 2
1	I II	-
2	I	II
3	II	I
4	-	I II

while in the Bose case of indistinguishable particles, the possible distributions are

Case	Cell 1	Cell 2
1	++	-
2	+	+
3	-	++

so that in this case the molecules are in the same cell more often (two-thirds of the time) than in the distinguishable particle case (half of the time).

[19] EInstein, A., *Quantum theory of the monoatomic gas*, Second communication, Sitz.ber. Preuss. Akad. WIss..p. 3 Jan. 8, (1925)

In his second paper Einstein continued the discussion of degeneracy. Recognizing that $A > 0$ in (9.99) so that the result for n is not negative,[20] he rewrote Eq. (9.99) using Eq. (9.85):

$$n = \sum_s \frac{1}{e^A e^{E_s/kT} - 1} = \sum_s \frac{\lambda e^{-E_s/kT}}{1 - \lambda e^{-E_s/kT}} = \sum_{s,r=1} \left(\lambda e^{-E_s/kT}\right)^r$$
$$= \sum_{s,r=1} \lambda^r e^{-cs^{2/3}r/kT}. \tag{9.101}$$

Replacing the sum over s by an integral, we find

$$n = \sum_{r=1} \lambda^r \int_0^\infty e^{-cs^{2/3}r/kT} ds\,. \tag{9.102}$$

The integral can be evaluated to obtain

$$\int_0^\infty e^{-a_r s^{2/3}} ds = \frac{3}{2a_r^{3/2}} \int_0^\infty \sqrt{y} e^{-y} dy = \frac{3\sqrt{\pi}}{4a_r^{3/2}} \tag{9.103}$$

$$n = \frac{V}{h^3} (2\pi m k_B T)^{3/2} \sum_{r=1} \frac{\lambda^r}{r^{3/2}}\,. \tag{9.104}$$

Thus n will be a maximum when $\lambda = 1$. In this case, the density is

$$\frac{n_{\max}}{V} = \frac{(2\pi m k_B T)^{3/2}}{h^3} \sum_{r=1} \frac{1}{r^{3/2}} = \frac{(2\pi m k_B T)^{3/2}}{h^3} \zeta\left(\frac{3}{2}\right) = 2.612 \frac{(2\pi m k_B T)^{3/2}}{h^3}, \tag{9.105}$$

where $\zeta(x)$ is the Riemann zeta function.

What happens if we start with a gas at this density and temperature and compress it isothermally? Einstein suggested that the excess of atoms over that given by Eq. (9.105) would go into the lowest energy state characterized by zero kinetic energy, and the remainder would be distributed according to the distribution with $\lambda = 1$. This is similar to what happens to a saturated vapor in contact with its liquid. Isothermal compression of the vapor leads to atoms condensing out into the liquid state. As in that case, equilibrium is determined by the requirement that the Gibbs free energy G is the same for both phases.[21] For the degenerate case $(A = 0, \lambda = 1)$, the entropy in Eq. (9.93) is given by

$$S = -k_B \left[\sum_s \ln\left(1 - e^{-\alpha_s}\right) - \frac{E}{k_B T} \right]. \tag{9.106}$$

[20] This implies that the "degeneracy parameter" defined by Einstein as $\lambda = e^{-A}$ must satisfy $0 < \lambda < 1$.

[21] At the time, $-G/T$ was called the "Planck function".

Replacing the sum by an integral and integrating by parts, we have

$$\sum_s \ln\left(1-e^{-\alpha_s}\right) = \int \ln\left(1-e^{-cs^{2/3}/k_BT}\right) ds \tag{9.107}$$

$$= -\int s \frac{e^{-cs^{2/3}/k_BT}}{\left(1-e^{-cs^{2/3}/k_BT}\right)} \frac{2}{3} \frac{cs^{-1/3}}{k_BT} ds \tag{9.108}$$

$$= -\frac{2}{3}\int \frac{E_s}{k_BT} n_s ds = -\frac{2}{3}\frac{E}{k_BT} = -\frac{pV}{k_BT} \tag{9.109}$$

so that

$$S = \frac{pV+E}{T} \tag{9.110}$$

and the Gibbs free energy for the saturated gas is found to be

$$G = E + pV - TS = 0, \tag{9.111}$$

in agreement with that of the "condensed phase" where $S = E = p = 0$.

We can see the process of condensation more clearly by rewriting Eq. (9.99) and separating the ground state ($E_s = 0$) from the sum ($N =$ total number of particles):

$$N = \frac{1}{e^A - 1} + \sum_s \frac{1}{e^A e^{E_s/kT} - 1}. \tag{9.112}$$

We can rewrite the sum over s as an integral using (9.85):

$$E_s = \frac{s^{2/3}}{2m}\left(\frac{3h^3}{4\pi V}\right)^{2/3}, \quad s = (2mE_s)^{3/2}\left(\frac{4\pi V}{3h^3}\right) \tag{9.113}$$

$$ds = (2m)^{3/2} E_s^{1/2}\left(\frac{2\pi V}{h^3}\right) dE_s \tag{9.114}$$

$$N = \frac{1}{e^A - 1} + \left(\frac{2\pi V}{h^3}\right)(k_BT)^{3/2}(2m)^{3/2}\int \frac{x^{1/2}}{e^A e^x - 1} dx, \tag{9.115}$$

where we put $E_s = xk_BT$. As before, we require $A > 0$ so that the integral remains finite. The first term is the number of atoms in the ground state, and this term can accommodate as many particles as necessary as $A \to 0$. For large numbers of atoms in the ground state, we can approximate the integral by its value at $A = 0$: $\int_0^\infty \frac{x^{1/2}}{e^x - 1} dx = \frac{1}{2}\sqrt{\pi}\zeta\left(\frac{3}{2}\right) = 2.3152 = K_0$. We define the critical temperature T_c as that temperature where all the atoms fit into the excited states, i.e., just before atoms start accumulating (condensing) into the ground state while cooling, so that

$$N = \left(\frac{2\pi V}{h^3}\right)(k_BT_c)^{3/2}(2m)^{3/2} K_0\,. \tag{9.116}$$

Then we can rewrite Eq. (9.115) as

$$N = n_0 + N\left(\frac{T}{T_c}\right)^{3/2} \tag{9.117}$$

where $n_0 = \left(e^A - 1\right)^{-1}$ is the number of atoms in the ground state and

$$T_c = \left(\frac{h^3 N}{2\pi V K_0}\right)^{2/3} \frac{1}{2mk_B}. \tag{9.118}$$

We thus find

$$n_0 = N\left[1 - \left(\frac{T}{T_c}\right)^{3/2}\right], \tag{9.119}$$

the characteristic behavior of Bose condensation, which is illustrated in Fig. 9.1.

Bose's method of counting cells rather than particles tended to obscure the fundamental physical property that was at play. After the publication of Einstein's first paper on the Bose gas, Ehrenfest and others, including Schrödinger in his correspondence with Einstein, criticized this "Bose statistics" as being incompatible with the usual Boltzmann statistics as already mentioned above. Einstein apologized to Schrödinger for not making the matter clearer. As indicated in his little two-particle, two-cell model, Einstein was well aware that the key point was that the particles were indistinguishable. This was implicit in Planck's case, as he was considering the energy of excitation of harmonic oscillators. In order to make the matter completely clear, Einstein gave a direct comparison of the two methods.

He took the number of cells in a region ΔE as Eq. (9.76)

$$z_\nu = \frac{2\pi V}{h^3}(2m)^{3/2} E^{1/2}\Delta E \tag{9.120}$$

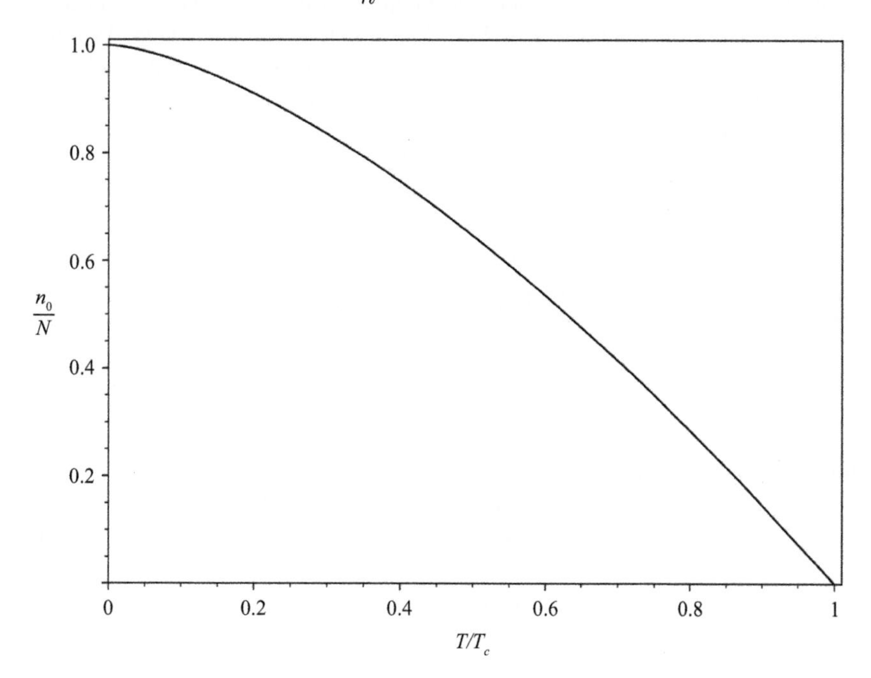

Fig. 9.1 Fraction of atoms in the ground state as a function of temperature.

and the number of microstates per macrostate as given by Bose (Eq. (9.56)):

$$W = \prod_{\nu} \frac{(n_\nu + z_\nu - 1)!}{n_\nu! z_\nu!}, \tag{9.121}$$

where n_ν is the number of particles contained in ΔE, so that the entropy is

$$S_{BE} = k_B \sum_{\nu} \left[(n_\nu + z_\nu) \ln (n_\nu + z_\nu) - n_\nu \ln n_\nu - z_\nu \ln z_\nu \right]. \tag{9.122}$$

As we will see below, this is different from the case in which the particles are distributed independently among the cells. Since it deviates from the independent distribution, the Bose-Einstein distribution indirectly expresses a certain hypothesis concerning a "mutual influence of the particles of a completely puzzling nature."

For the case of completely independent (distinguishable) particles, we can distribute n_ν atoms over z_ν cells in $(z_\nu)^{n_\nu}$ different ways and there can be

$$n! \prod_{\nu} \frac{(z_\nu)^{n_\nu}}{n_\nu!} \tag{9.123}$$

possible assignments of atoms to the regions specified by n_ν. Then

$$S_{\text{Boltz}} = k_B \left[n \ln n + \sum_{\nu} \left[n_\nu \ln z_\nu - n_\nu \ln n_\nu \right] \right]. \tag{9.124}$$

Einstein used these expressions to calculate the mean square fluctuations of the number of atoms in a given volume just as he had done for the radiation field. We discussed this calculation in Section 3.2.4. Taking n_ν as the average number of atoms in a volume V and energy interval ΔE, and $n_\nu + \Delta_\nu$ as the instantaneous number of atoms, we can rewrite Eq. (3.54) as

$$\overline{(\Delta_\nu)^2} = \frac{-k_B}{\left(\frac{\partial^2 S}{\partial \Delta_\nu^2} \right)}. \tag{9.125}$$

For the Bose-Einstein case, we obtain

$$\frac{\partial^2 S_{BE}}{\partial \Delta_\nu^2} = -\frac{k_B z_\nu}{n_\nu (z_\nu + n_\nu)} \tag{9.126}$$

$$\overline{\left(\frac{\Delta_\nu}{n_\nu} \right)^2_{BE}} = \frac{1}{n_\nu} + \frac{1}{z_\nu}, \tag{9.127}$$

while for the Boltzmann case, we have

$$\frac{\partial^2 S_{\text{Boltz}}}{\partial \Delta_\nu^2} = -\frac{k_B}{n_\nu} \tag{9.128}$$

$$\overline{\left(\frac{\Delta_\nu}{n_\nu} \right)^2_{\text{Boltz}}} = \frac{1}{n_\nu}. \tag{9.129}$$

The first term in Eq. (9.127) represents the fluctuations in a gas of independent particles while the second term is independent of the atomic density and corresponds to interference fluctuations. Just as for the black body radiation, the fluctuations of the Bose-Einstein gas consist of a particle and a wave term, which implies wave-like properties for the massive particles just as it had implied particle-like properties for the radiation. Einstein stated:

"I believe this is more than a simple analogy. de Broglie has shown how to associate a (scalar) wave field with material particles in a very noteworthy paper.[22] It seems that every motion is coupled to an oscillating field whose physical nature remains in the dark, but must be observable in principle."

Einstein then considered some possibilities for observing the wave-like behavior. The achievable wavelengths seemed to be too small to observe diffraction with a man-made slit but the diffraction should exert an observable influence on the scattering of atoms by atoms and would be expected to affect the temperature dependence of viscosity. Application to the free electron gas in conductors raised some difficulties, which we now understand in terms of the Fermi-Dirac statistics to be discussed in the next section. Direct observation of Bose-Einstein condensation was precluded by the fact that physical gases were not ideal and helium, for example, reached its critical point at a density about five times lower than that required for the onset of condensation.

Today the superfluidity of ^{4}He is explained as being due to condensation of the Bose-Einstein type modified by the interaction between the atoms. The fraction of atoms in the lowest state has been measured by neutron scattering. In recent years much work has been devoted to creating Bose-Einstein condensation in gases at extremely low temperatures and many interesting experiments have been and are being carried out.

9.1.5 Schrödinger

Following the exchange of letters with Einstein, Schrödinger submitted a paper to *Physikalische Zeitschrift*[23] in which he approached the problem of quantizing the ideal gas in another way. This was right after the publication of the Heisenberg, Born, and Jordan paper and was just before the Christmas holiday during which Schrödinger developed the wave equation for the hydrogen atom. Schrödinger's starting point was the fact that Einstein's work was based on Bose's statistics considered as a primary hypothesis that did not follow from anything else in the theory and was justified only by comparison with experiment. This contradicted Schrödinger's natural feeling and seemed to be hiding a certain mutual dependence or interaction between the gas molecules. He expected to gain a deeper insight from making the necessary changes at a more fundamental level, so that he was not required to give up known and verified methods.

Einstein had applied the Bose statistics, which were known to lead to the Planck law for photons, to gas molecules. But Schrödinger remembered that the Plank law could also be derived by applying "natural" (Boltzmann) statistics, not to the assumed Planck resonators, as we did in Section 3.2.3, but to the degrees of freedom of the

[22] L. de Broglie, *Recherches sur la théorie des quanta* [Research on quantum theory] (PhD thesis, Paris, 1924); *Ann. de Physique* (10) **3**, 22 (1925).

[23] Schrödinger, E., *On the Einstein theory of gases*, Phys. Zeit. 27, 95, (1926), received Dec.15, 1925.

radiation field itself—what he called the "ether resonators." The photons then appear only as the energy levels of these resonators (degrees of freedom). Now he pictured the gas as a "cavity radiation" of de Broglie waves and, using Planck's method of the partition function (Eq. (3.42)), arrived at the same result as the Bose statistics.

The true meaning of Einstein's gas theory is that the gas is a wave system with normal modes that behave as linear oscillators like the radiation in a cavity or vibrating atoms in a crystal. In the case of radiation, we have an infinite number of normal modes without any limits on the quantum numbers; for a solid we have a finite number of modes without a limit on the quantum numbers; while a gas has infinitely many modes but the sum of the quantum numbers is fixed.

As soon as experiment requires that the Bose statistics be applied to a system, we must conclude that this class of systems consists, not of individual particles, but of energetic excited states, so that the Bose statistics can be replaced by Boltzmann statistics applied to another class of objects.

This picture of particles as the energy levels of an underlying wave field is precisely the view of second quantization (see Chapter 12). In this formulation, the problem of a $3N$-dimensional space does not arise because the normal modes (degrees of freedom) of the de Broglie waves in a cavity are always expressed in three dimensions.

We start with n atoms in a volume V. The atoms have a discrete series of energy levels, $\varepsilon_1, ...\varepsilon_n...\varepsilon_s$. At any instant an arbitrary number of atoms can be in the same state. A more realistic description is that every state corresponds to a degree of freedom (normal mode) of a wave field and the sth mode oscillates with the energy $n_s\varepsilon_s$; that is, it behaves as an oscillator with possible energies $0, \varepsilon_s, 2\varepsilon_s...n_s\varepsilon_s$. The partition function is defined as the sum of Boltzmann factors for all states of the system:

$$Z = \sum_{n_1, n_2 ...} e^{-\frac{1}{k_BT}(n_1\varepsilon_1 + n_2\varepsilon_2 + ... n_s\varepsilon_s)} \tag{9.130}$$

which can be rewritten as

$$Z(x_s) = \prod_s \sum_{n_s=0}^{\infty} e^{-\frac{n_s\varepsilon_s}{k_BT}} = \prod_s \frac{1}{1 - e^{-\frac{\varepsilon_s}{k_BT}}} = \prod_s \frac{1}{1 - x_s} \tag{9.131}$$

where $x_s = e^{-\frac{\varepsilon_s}{k_BT}}$. This leads to the Planck spectrum. In the case of the gas, we must have

$$\sum_s n_s = n. \tag{9.132}$$

We can satisfy Eq. (9.132) by selecting the terms in (9.131) which are of order n in all the x_s.[24] This can be accomplished by replacing x_s by zx_s and treating z as a complex variable. This method was originally introduced by Darwin and Fowler (1922).[25] Using the fact that the integral around a closed path enclosing the origin is

[24]This is easiest to see by considering the first form of Eq. (9.131) with the sum over n_s. In the product over s, each term will be of the form $x_1^{k_1}x_2^{k_2}x_3^{k_3}...$ (s runs from 1 to inf), and we must select those terms where $\sum_s k_s = n$.

[25]Darwin, C.G. and Fowler, R.H., *On the partition of energy*, Philos. Mag. **44**:261, 450-479 (1922).

$$\frac{1}{2\pi i}\oint \frac{dz}{z^k} = \delta_{k1}, \tag{9.133}$$

we obtain the terms conforming to the constraint by calculating

$$\frac{1}{2\pi i}\oint_0 \frac{dz}{z^{n+1}} Z(z) = \frac{1}{2\pi i}\oint \frac{dz}{z^{n+1}} \prod_s \frac{1}{1 - zx_s}. \tag{9.134}$$

This can be evaluated by the method of steepest descent. The integrand has a pole at the origin and poles at $z_s = 1/x_s = e^{\varepsilon_s/k_B T}$ and a minimum on the real axis between the origin and the smallest z_s pole. Taking the first derivative of the integrand,

$$\frac{d}{dz}\left(\frac{1}{z^{n+1}}\prod_s \frac{1}{1 - zx_s}\right) = -\frac{(n+1)}{z^{n+2}}\prod_s \frac{1}{1 - zx_s} + \frac{1}{z^{n+1}}\prod_s \frac{1}{1 - zx_s}\sum_s \frac{x_s}{1 - zx_s}, \tag{9.135}$$

we find that the integrand is stationary at $z = r$ with r given by

$$-\frac{n+1}{r} + \sum_s \frac{x_s}{(1 - rx_s)} = 0. \tag{9.136}$$

We will choose the contour as a circle centered at the origin and passing through the point $z = r$. This point $(z = r)$ is a minimum along the x-axis. We will call the integrand $I(z)$. Since this function is analytic, the Cauchy-Riemann equations tell us that $\partial^2 I/\partial x^2 = -\partial^2 I/\partial y^2$, and the point $z = r$ is a maximum along the y direction (saddle point), as shown in Fig. 9.2.

Since the decrease in $I(z)$ as we go away from $z = r$ in the y direction is very rapid, we can expand $I(z)$ in a series in y, keeping only terms of second order (saddle point method). We have

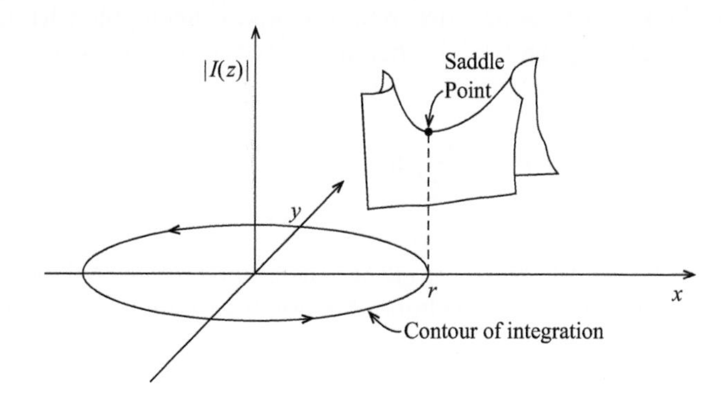

Fig. 9.2 The integrand $I(z)$ of Eq. (9.134) in the x-y plane.

$$\ln(I(z)) = -(n+1)\ln z - \sum_s \ln(1 - zx_s) \tag{9.137}$$

$$\left.\frac{d\ln(I)}{dz}\right|_{z=r} = -\frac{n+1}{r} + \sum_s \frac{x_s}{(1-rx_s)} = 0 \tag{9.138}$$

$$\begin{aligned}\left.\frac{d^2\ln(I)}{dz^2}\right|_{z=r} &= \frac{n+1}{r^2} + \sum_s \frac{x_s^2}{(1-rx_s)^2} \\ &= \frac{1}{r}\sum_s \frac{x_s}{(1-rx_s)} + \sum_s \frac{x_s^2}{(1-rx_s)^2} \\ &= \sum_s \frac{\frac{x_s}{r}(1-rx_s) + x_s^2}{(1-rx_s)^2} = \frac{1}{r}\sum_s \frac{x_s}{(1-rx_s)^2}.\end{aligned} \tag{9.139}$$

We can expand $\ln(I)$ around $z = r$ with $dz = dy = ri\phi$ for ϕ small:

$$\ln(I(z)) = \ln\left(\frac{1}{r^{n+1}}\prod_s \frac{1}{1-rx_s}\right) - \frac{1}{2}\phi^2 r \sum_s \frac{x_s}{(1-rx_s)^2}$$

$$I(z) = \left(\frac{1}{r^{n+1}}\prod_s \frac{1}{1-rx_s}\right) e^{-\phi^2 \frac{r}{2}\sum_s \frac{x_s}{(1-rx_s)^2}}$$

Then we find, from (9.134)

$$Z = \int_0^{2\pi} I(z)\,d\phi \approx \int_{-\infty}^{\infty} I(z)\,d\phi = \left(\frac{1}{r^{n+1}}\prod_s \frac{1}{1-rx_s}\right)\left(\frac{r}{2\pi}\sum_s \frac{x_s}{(1-rx_s)^2}\right)^{-1/2} \tag{9.140}$$

where we extended the ϕ integral from $-\infty$ to ∞ because the function falls off very fast with increasing ϕ.

Then we can calculate the Helmholtz free energy

$$F = k_B T \ln Z = -k_B T\left[\sum_s \ln(1-rx_s) + (n+1)\ln r + \frac{1}{2}\ln\left(\frac{r}{2\pi}\sum_s \frac{x_s}{(1-rx_s)^2}\right)\right]. \tag{9.141}$$

Due to (9.136), the last term is of order $\ln(n+1)$ and is negligible with respect to the first two terms. Neglecting the trivial difference between n and $n+1$, this is identical with Einstein's result in (9.97):

$$F = k_B T\left[\sum_s \ln\left(1 - e^{-\alpha_s}\right) - An\right], \tag{9.142}$$

with $r = e^{-A}$ and $\alpha_s = A + \varepsilon_s / k_B T$.

Schrödinger then used the known result for the number of normal modes per energy (frequency) interval in an enclosed volume (Section 3.1.5) and showed that this led to Einstein's result:

$$\varepsilon_s = cs^{2/3}. \tag{9.143}$$

This method avoids all the questions about the choice of the Boltzmann probability W and led Schrödinger to understand the crucial role played by second quantization (regarding the particles as excited, quantized states of a wave field), where the particle numbers are given by the quantum numbers of the energy states of the de Broglie waves. This convinced him of the reality of the de Broglie phase waves, whose guiding equation he would discover a few weeks later. About a month earlier (November 16, 1925) Jordan had added to the "three man work" (Born, Heisenberg, and Jordan) a section on the quantization of the radiation field, which was certainly known to Schrödinger.

Both Einstein and Schrödinger came to quantum mechanics from statistical considerations, whereas the Copenhagen-Göttingen group was involved mostly with problems of spectroscopy. This may be partly responsible for their later divergent views toward the theory.

9.1.6 Summary

Bose's method of counting cells rather than particles tended to obscure the crucial role played by the indistinguishability of identical particles. Einstein understood this but neglected to mention it in his papers, applying Bose statistics to a gas of material particles. The fact that the calculated fluctuations for the gas of massive particles showed wave-like as well as particle-like behavior, just like the fluctuations of the black body radiation, convinced Einstein that de Broglie's idea that the motion of massive particles was accompanied by a wave motion had to be taken seriously.

Schrödinger, whose first impression of Einstein's work was that there was an error in the counting method, turned to a more detailed consideration of the nature of the waves accompanying the gas particles. Carrying the analogy with the radiation field further, he showed that considering the particles to be the excited states of the normal modes of the de Broglie wave field would lead directly to the Bose-Einstein behavior. By nature these excitations would be indistinguishable. This view is precisely that of second quantization.

Quantization of the radiation field had been proposed by Jordan in the famous Born, Heisenberg, and Jordan paper,[26] which was received one month earlier than Schrödinger's paper but was only published six weeks after Schrödinger's work, and was not mentioned by Schrödinger. Two weeks after the submission of his gas paper, Schrödinger was in a skiing resort developing the non-relativistic wave equation for the hydrogen atom.

9.2 Fermi-Dirac statistics

9.2.1 Introduction

We have seen that the key feature of the Bose-Einstein statistics, the condensation of atoms into the ground state, was unobservable at the time because the interactions between atoms in the most likely candidate gases, hydrogen and helium, were so strong that they turned into liquids before the conditions for the condensation could

[26]M. Born, W. Heisenberg, and P. Jordan, *Zur Quantenmechanik II* [On quantum mechanics II], *Z. Phys.* **36**, 557 (1926).

be reached. Except for their application to the black body spectrum, there was no experimental evidence to support these ideas.

In January 1925 Pauli was studying the relation between atomic spectra and the periodic table of the elements, when he first articulated the idea that two or more electrons could never occur in the same state (all quantum numbers the same), which was in direct contradiction to the predictions of the Bose-Einstein condensation. In February 1926, Fermi realized that this would imply a new type of statistics and calculated the energy distribution of an ideal gas of particles obeying Pauli's exclusion principle. A few months later (August 1926) Dirac related the type of statistics to the symmetry of the multi-particle wave function under exchange of particles. Since the exchange of identical particles could have no physical effects, $|\psi|^2$ must remain unchanged and ψ can either be unchanged (symmetric) or change sign (antisymmetric) under the interchange of identical particles. In the antisymmetric case, no two electrons could occupy the same state, so that particles represented by an antisymmetric wave function (now called fermions) obey the Pauli exclusion principle while particles represented by a symmetric wave function (now called bosons) obey Bose-Einstein statistics. Since electrons were known to obey the Pauli principle, Dirac assumed this would also hold for the massive molecules of a gas and went on to derive the distribution previously derived by Fermi.

9.2.2 The physics of multi-electron atoms and the Pauli exclusion principle

We start with a rough summary of the work on atomic spectroscopy which led to the exclusion principle.[27]

The introduction of the Bohr atomic model in 1913 stimulated a large body of work trying to extend the results for the hydrogen atom to the rest of the elements. According to this model, the electrons moved in orbits around the positively charged nucleus, with the number of electrons equal to the charge of the nucleus for a neutral atom. The orbits were normal, classical orbits selected according to the quantization rules.

Light was emitted during transitions between the allowed orbits and the frequency of the light was determined by the Bohr frequency rule applied to the energy difference of the initial and final states. The quantization rules resulted in three quantum numbers for every electron (one per degree of freedom). For hydrogen, the quantum numbers n_r, n_θ, n_ϕ were used to define the quantum numbers $n = n_r + n_\theta + n_\phi$, (the principal quantum number), $k_1 = n_\theta + n_\phi$ (the azimuthal (inner) quantum number), and $m = n_\phi$ (the magnetic quantum number). In hydrogen the principal quantum number determined the major axis of the ellipse and hence the energy of the orbit, the inner quantum number determined the eccentricity of the orbit, and the magnetic quantum number determined the orientation of the orbit with respect to an axis fixed in space (spatial quantization). Fig. 9.3 shows the set of orbits for $n = 4$.

The magnetic quantum number was important for the Zeeman effect, the change in spectral lines due to the application of an external magnetic field. This was treated

[27] D. ter Haar , *The Old Quantum Theory* (London, Pergamon Press, 1967).

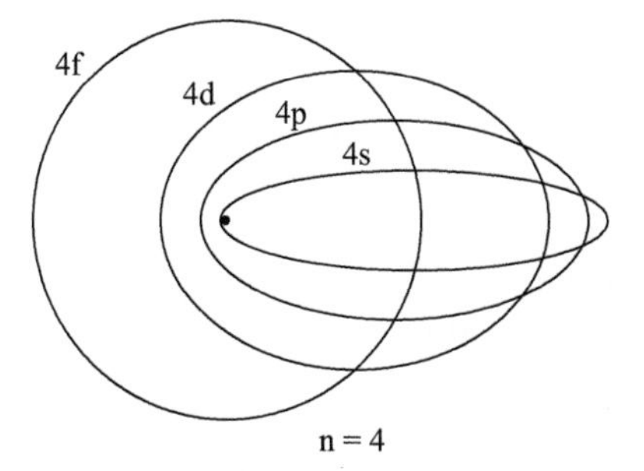

Fig. 9.3 Orbits for $n = 4$ in hydrogen. The lower k_1 orbits are more eccentric and penetrate closer to the nucleus. The labels s, p, d, f denote $k_1 = 1, 2, 3, 4$, respectively. The orbital angular momentum quantum number in the modern notation is given by $l = k_1 - 1$.

in the Bohr-Sommerfeld theory as follows. The Hamiltonian for an electron in the presence of a magnetic field is given by

$$H = \frac{\left(\vec{p} - \frac{e}{c}\vec{A}\right)^2}{2m_e} + V(\vec{r}) \tag{9.144}$$

and the first order contribution of the field is given by

$$H' = -e\frac{\vec{p} \cdot \vec{A}}{m_e c}. \tag{9.145}$$

For a fixed magnetic field $\vec{A} = \frac{1}{2}\left(\vec{B} \times \vec{r}\right)$ so that

$$-H' = \frac{e}{2m_e c}(\vec{r} \times \vec{p}) \cdot \vec{B} = \frac{e}{2m_e c}\left(\vec{L} \cdot \vec{B}\right) = \frac{e}{2m_e c}L_z B = \frac{e}{2m_e c}p_\phi B, \tag{9.146}$$

where $\vec{L} = \vec{r} \times \vec{p}$ is the orbital angular momentum with z component $L_z = p_\phi$, the momentum conjugate to the spherical coordinate ϕ, and B is taken to be along the z direction. The momentum p_ϕ is determined by the quantum condition $p_\phi = m_\phi \hbar$. Treating H' as a perturbation, we find that the energies are shifted by

$$-\frac{e\hbar}{2m_e c}m_\phi B = -\mu_B m_\phi B. \tag{9.147}$$

$\mu_B = \frac{e\hbar}{2m_e c}$ is called the Bohr magneton and is the natural unit for magnetic moments. This is why m_ϕ is called the magnetic quantum number. During transitions it can only change by $\Delta m_\phi = 0, \pm 1$. This is an example of a selection rule, fixing the amount by which the different quantum numbers can change during a transition. These selection rules were derived from the correspondence principle. This orbital magnetic moment is produced by the current consisting of the electron going around its orbit.

The observed frequencies had to be translated into energy levels called "terms" and the selection rules were deduced from the observations but were then shown to follow from the correspondence principle. Thus applying a magnetic field would split a spectral line into three lines, one of unchanged frequency and two lines separated from the original frequency by $\pm\mu_B B/\hbar$. This is the normal Zeeman effect. In many cases the pattern induced by a magnetic field is more complex. These cases are called the anomalous Zeeman effect. The inability of the Bohr-Sommerfeld theory to account for this was one of the great failures of that theory. The concept of electron spin (see Chapter 19) is crucial to understanding this effect.

According to the Bohr rule, the frequency of a spectral line was given by the difference between the terms. By careful study of the spectrum, it was possible to assign a term structure to each atom. Figures 9.4–9.6 (called Grotrian diagrams) show the term structure for the alkali metals hydrogen, lithium, and sodium.

Based on their chemical properties and X-ray spectra, the elements were arranged in a periodic table (Fig. 9.7). The periodicity of the chemical properties led to the concept of closed shells but there was no explanation for the observed periodicities or for the closing of the shells. Columns in the periodic table correspond to 1,2,3,4... electrons outside a closed shell. These numbers determine the chemical properties, which vary with them until the shell is filled, starting another period. In a multi-electron atom, the electrons move in the field due to the atomic nucleus as well as that due to the other electrons.

The similarities in the spectra shown in the figures combined with the position of these elements in the same column of the periodic table (all had valence 1) indicated that these spectra were due to a single electron. Thus the radial quantum condition for a valence electron is given by

$$J_r = n_r h = \oint dr \sqrt{2m\left(E - V\left(r\right)\right) - \frac{l^2}{r^2}} \tag{9.148}$$

where $V(r)$ is the potential seen by the valence electron. At large r the potential is $\left(-e^2/r\right)$, the field of the nucleus shielded by the other electrons, and at small r the potential is $\left(-Ze^2/r\right)$, the field of the bare nucleus. Let the difference between this integral with $V(r) = -Ze^2/r$ and the corresponding one for $V(r) = -e^2/r$ be αh. Then

$$\left(n_r - \alpha\right) h = \oint dr \sqrt{2m\left(E + \frac{e^2}{r}\right) - \frac{l^2}{r^2}}. \tag{9.149}$$

Evaluating the integral as described in Chapter 4 leads to

$$E = -\frac{me^4}{2\hbar^2\left(n - \alpha\right)^2} \tag{9.150}$$

where $n = n_r + n_\theta + n_\phi$. α is called the quantum defect, which depends on both n and l, and is found empirically to be on the order of one for lower l states and close to zero for higher l states.

This theory was remarkably successful in explaining the general features of a large number of spectra, but it had several notable failures. It was unable to calculate the spectrum for helium and unable to explain the anomalous Zeeman effect. Additionally,

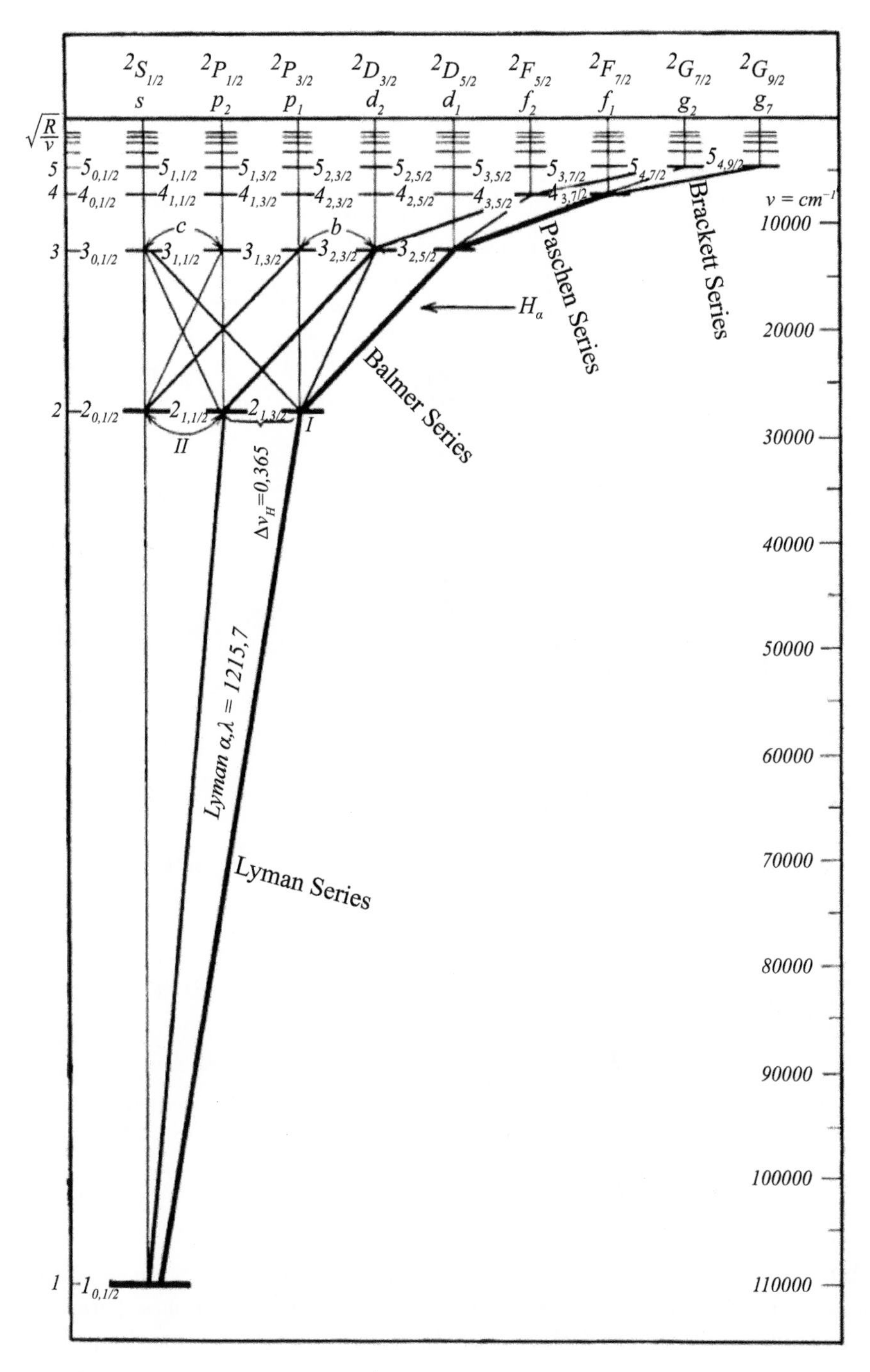

Fig. 9.4 Term diagram for hydrogen. (Reprinted by permission from Springer Nature Customer Service Centre GmbH: Springer Nature, "Einzelne Spektren in ihrer Serienauflösung und Niveauschemata einzelner Spektren" by Dr. W. Grotrian ©1928. Fig. 4, p. 5.)

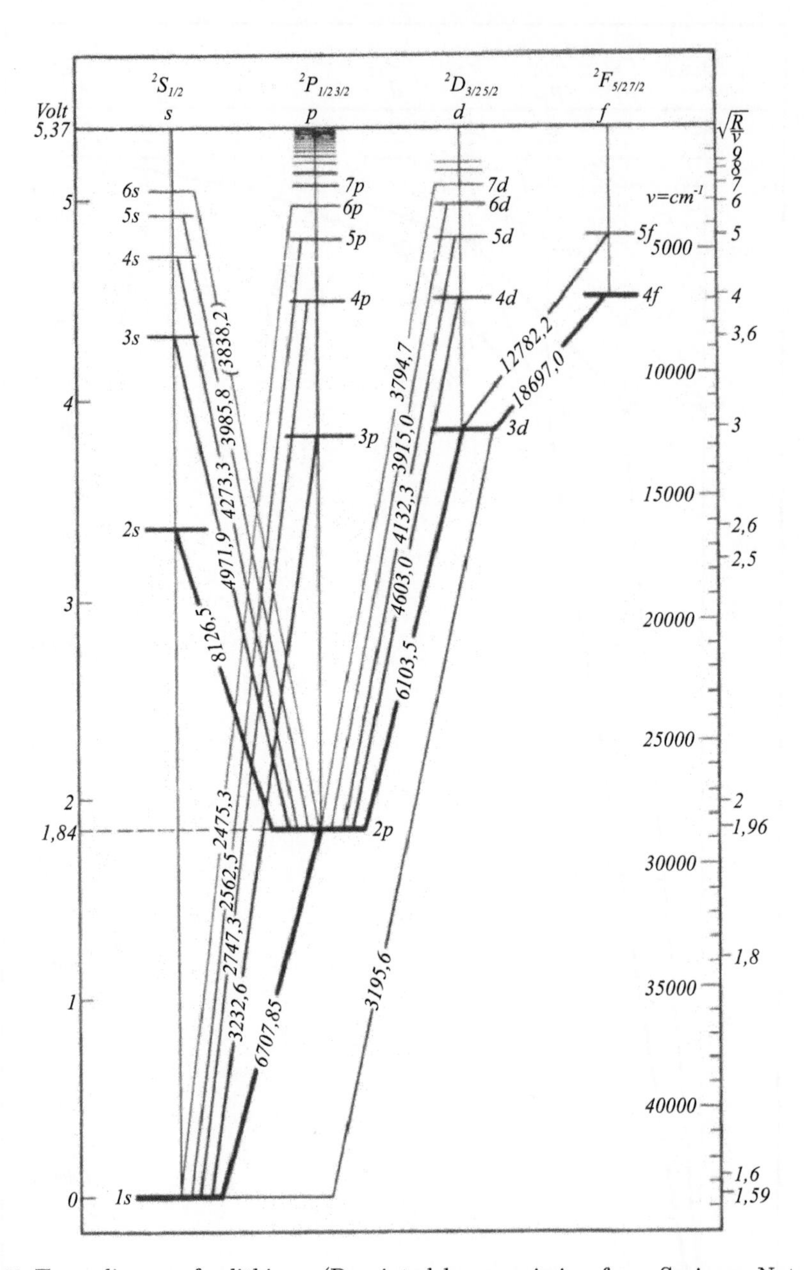

Fig. 9.5 Term diagram for lithium. (Reprinted by permission from Springer Nature Customer Service Centre GmbH: Springer Nature, "Einzelne Spektren in ihrer Serienauflösung und Niveauschemata einzelner Spektren" by Dr. W. Grotrian ©1928. Fig. 12, p. 15.)

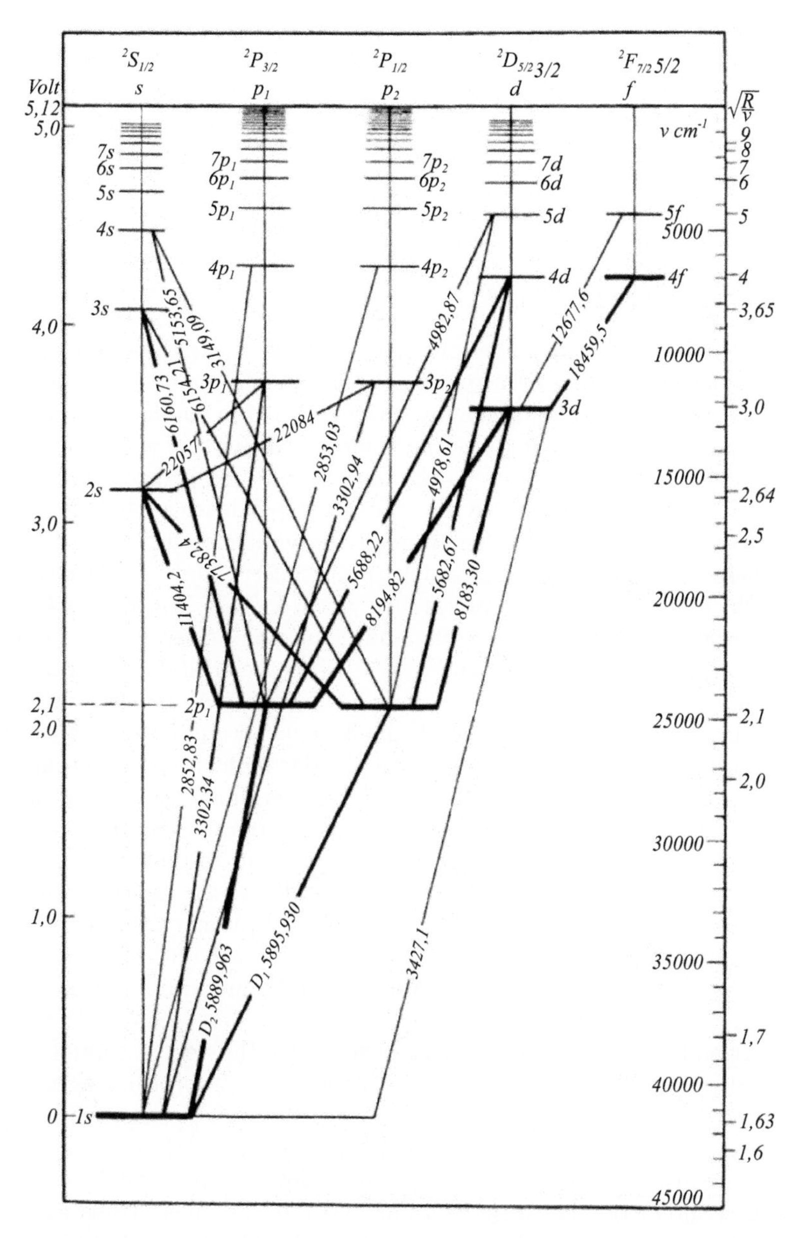

Fig. 9.6 Term diagram for sodium. (Reprinted by permission from Springer Nature Customer Service Centre GmbH: Springer Nature, "Einzelne Spektren in ihrer Serienauflösung und Niveauschemata einzelner Spektren" by Dr. W. Grotrian ©1928. Fig. 18, p. 21.)

when atomic spectra were observed at high resolution, many lines were found to be split into multiplets of two or three closely spaced lines, which were called singlets (no splitting), doublets, and triplets. A significant amount of work was devoted to understanding these effects, and the continued failures led to a sense of despair in 1923 as to the validity of the theory. Bohr and other physicists felt it would be years, or maybe generations, before a solution was found. In November 1924 Born wrote:

"I strongly share with the Copenhagen researchers, the conviction that we are still very far from a 'final' quantum mechanics."

As we saw in Chapter 4, Sommerfeld achieved a great success in calculating the fine structure splitting in hydrogen using a relativistic Hamiltonian. This turned out to be a lucky accident since we now know the spin plays a crucial role in these splittings, which are caused by the interaction of the spin magnetic moment with the orbital magnetic moment. A cancellation of two effects in hydrogen resulted in Sommerfeld's calculation giving the correct answer. The origin of the Sommerfeld splitting was that orbits with higher eccentricity penetrated closer to the nucleus and had higher velocities, which led to a detectable effect of the relativistic velocity dependence of the mass.

It was necessary to introduce a fourth quantum number to describe the multiplet splitting. There were several ways of doing this. Stoner[28] (July 1924) introduced a number k_2 which had two possible values for a given k_1: $k_2 = k_1$ or $k_2 = k_1 - 1$. For each set of values k_1, k_2 there were $2k_2$ electrons. Thus for a given k_1 there were $2k_1 + 2(k_1 - 1) = 2(2k_1 - 1)$ electrons. With $l = k_1 - 1$ (l is the orbital angular momentum quantum number), we have $2(2l + 1)$ electrons, which is the number of states for a given l according to the modern theory of angular momentum. We now know that k_2 corresponds to $(j + 1/2)$, where j is the total angular momentum quantum number. The selection rule for k_2 was $\Delta k_2 = 0, \pm 1$ and k_2 determined the penetration of the orbit into the core and thus the size of the relativistic correction. Since $n_r = n - k_1$, k_1 can run from 1 to n, which means that l runs from 0 to $n - 1$. The state with $k_1 = 0$ does not occur because the corresponding orbit would pass through the origin.

One of the cornerstones of the model for many-electron atoms was Bohr's "build-up" (Aufbau) principle according to which the quantum numbers of the existing electrons remained unchanged when new electrons were added to produce an atom with more electrons.

Heisenberg and Landé devised a model which divided the electrons into two groups, the valence electrons responsible for light emission and absorption, as well as the chemical properties, and a core or rump consisting of the other electrons. The rump was thought to have angular momentum and a magnetic moment, these being attributed to the two innermost K(1s) electrons. The multiplet structure would then be due to the magnetic interaction between the rump and the valence electrons varying with different relative orientations of the magnetic moments. The deviation from the simple Larmor triplet in the anomalous Zeeman effect would come from the fact that the ratio of magnetic moment to angular momentum (the gyromagnetic ratio γ) was twice as

[28]Stoner, E.C., *The distribution of electrons among atomic levels*, Phil. Mag. 48, 719, 1924, (July, 1924).

large for the core as for the valence electrons, a sign that the core was a replacement for what we now know is the electron spin.

Another funny issue with the rump model was that if an electron were removed to produce an ionized atom, the angular momentum of the rump changed by $\hbar/2$. This, as well as the duality of k_2 discussed above, is another sign that the rump was a substitute for the electron spin.

Another problem with the core model was that the angular momentum assigned to the core was different for the alkali atoms (one valence electron) and the alkaline earths (two valence electrons). Bohr attributed this to a "mechanical constraint" (Zwang).

The structure of the optical spectrum of many-electron atoms and the fine structure of single-electron atoms, along with that of the X-ray spectra, were thought to have different causes, the former being attributed to the rump model and the latter to relativistic effects. Landé showed that both models predicted splittings of the same order of magnitude which differed in their dependence on the quantum numbers, with the relativistic splitting being proportional to Z^4 while the magnetic splitting went as Z^3. Millikan and Landé showed that the optical doublets could also be calculated by a relativistic formula. This should have been enough to kill the rump model but people (including Landé) were reluctant to give it up.

In December 1924,[29] Pauli pointed out that the relativistic change of mass with velocity should be significant for the electrons in the $n = 1$ shell (called K electrons) in high-Z atoms, and he calculated that this would result in an observable dependence of the anomalous Zeeman effect on Z, which was contradicted by experiment. This was a significant blow to the rump model and Pauli went on to propose a new model where the rump electrons contributed nothing: zero angular momentum and zero magnetic moment. In the alkali atoms (single valence electron outside a closed shell), the angular momentum and magnetic energy changes were exclusively due to the valence electron, which should also be the location of the gyromagnetic anomaly. This required what Pauli called a classically indescribable duality of the quantum properties of the valence electron (which we now call the spin).

In a second paper (January 1925),[30] Pauli presented an analysis of the problems of atomic spectra, emphasizing his new model but pointing out that it by no means solved all the outstanding problems. He then turned to the problem of the closing of the groups in the periodic table.

He was convinced that this problem was related to the multiplet structure. A remark in Stoner's paper provided the essential clue. Stoner had observed that the number of terms (i.e., the number of energy levels) in the Zeeman effect of the alkali metals in a strong external magnetic field was equal to the number of electrons in a completed subgroup corresponding to the same value of l. This led Pauli to the general principle:

"There can never be, in an atom, two or more equivalent electrons for which, in strong fields, the values of all the quantum numbers (n, k_1, k_2, m) agree. If there is an electron in the atom

[29] Pauli, W., Jr., *On the influence of the velocity dependence of the electron mass on the Zeeman effect*, Zeit. Phys. **31**, 373 (1925). Received Dec. 2, 1924.

[30] Pauli, W., Jr., *On the connection between the completion of electron groups in an atom with the complex structure of its spectrum*, Zeit. Phys. **31**, 765, (1925) received Jan. 16, 1925, English translation in ter Haar, op. cit. p. 174.

for which these quantum numbers, in an external field, have definite values, then this state is 'occupied'. We cannot give a more detailed basis for this rule, however it seems to be very natural."

Pauli then went on to say that the fact that the argument is based on strong fields is not really a restriction since, on thermodynamic grounds, the statistical weight of a state should be an adiabatic invariant and the number of stationary states should be the same in weak fields. Thus we see that for $l = 0, 1, 2, 3$ there are $2, 6, 10, 14$ states, respectively, and for each n there are $2n^2$ states, giving periods of $2, 8, 18, 32$ for $n = 1-4$. In this way the periodic table is built up. When an electron is added to an existing atom in order to produce the next atom in the table, the new electron occupies the next unfilled state (taking account of the exclusion principle) with the lowest energy. In the lighter elements, the lower energy states are those with the lower principal quantum number n, so the groups with increasing n are successively filled in the first rows of the periodic table. However, in the heavier atoms, it often happens that states with a greater n have a lower energy than certain states with smaller n. This leads to irregularities in the table. So it can happen that a shell with a given n begins to be occupied before a shell with a lower n is completely filled. To understand this, we need to calculate the energies of the different states. These energies are determined by the shielding of the nuclear charge by the other electrons. In the periodic table shown in Fig. 9.7, the symbols to the left of each row represent the principal quantum number and the orbital angular momentum quantum number (designated by a letter) of the states which are filled in for the elements to the right. The letters $i =$ s,p,d,f signify $l = 0, 1, 2, 3$, respectively. For the rare earths, the $4f$ shell fills before the $5d$ shell is complete.[31] we see that the $6s$ states fill before the $4f$ states, which are first occupied in the rare earths. The $4f$ wave functions are closer to the nucleus, so that they do not affect the chemical properties, which remain mainly unchanged as the $4f$ shell fills up.

9.2.3 Fermi

Pauli's statement of the exclusion principle was quickly accepted by his colleagues. Enrico Fermi was impressed by the application of the exclusion principle to the explanation of the shell structure and the analysis of the spectra of many-electron atoms.[32] He seems to have been the first to realize that this represented a challenge to the Bose-Einstein treatment of an ideal gas. Since the deviations of the Bose-Einstein gas from the classical behavior were unobservable because they were masked by the deviations from the ideal gas behavior due to interactions between the molecules, and the theory was based on many statistical assumptions, Fermi felt justified in proposing an alternative. As the exclusion principle was so successful when applied to the electrons in atoms, he attempted to apply it to an ideal monatomic gas. In order to quantize the motion of the atoms using the Bohr-Sommerfeld rule, the atoms need to be held in a

[31] Robert B. Leighton, *Principles of Modern Physics* (1959, out of print, https://archive.org/details/principlesofmode00leig, Digitized by the Kahle/Austin Foundation) p. 251.

[32] An English translation of Fermi's original paper, E. Fermi, *Sulla quantizzazione del gas perfetto monoatomico* ('On the quantization of the ideal monoatomic gas'), *Rend. Lincei* **3**, 145 (1926) is available at arXiv:cond-mat/9912229. Also published as Fermi, E., Zeit. Phys. **36**, 902 (1926), received Mar. 24, 1926.

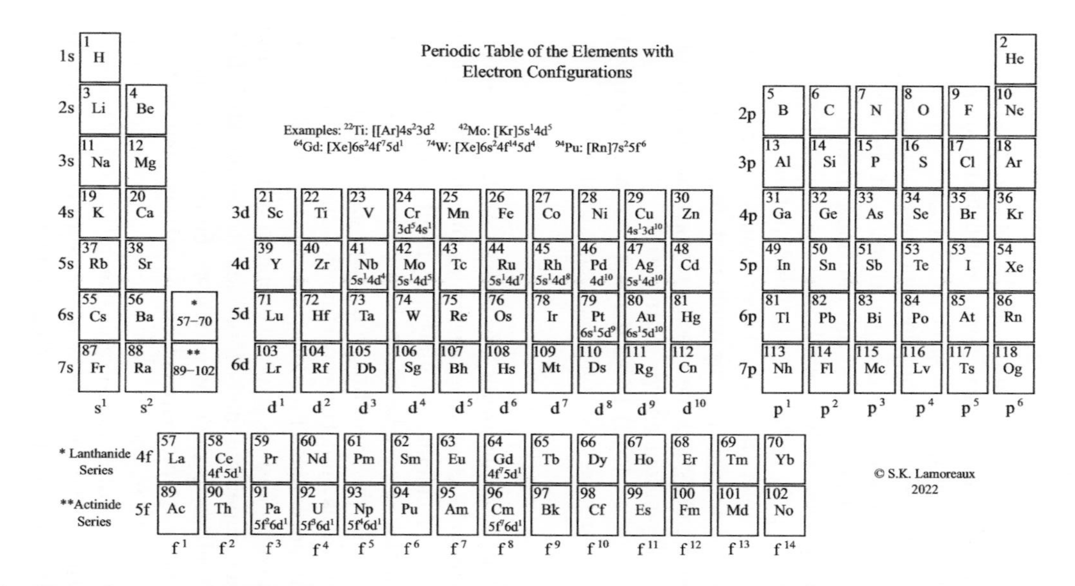

Fig. 9.7 Periodic table of the elements. The symbols to the left of each row represent the principal quantum number and the orbital angular momentum quantum number (designated by a letter) of the states which are filled in the elements to the right. The letters $i = $ s,p,d,f signify $l = 0, 1, 2, 3$, respectively. For the rare earths, the $4f$ shell fills before the $5d$ shell is complete.

container so that their motion is periodic. While the type of container has no effect on the results, and it would be possible to consider the atoms as being contained in a box with rigid walls, Fermi chose to consider the atoms as being harmonically bound to a fixed point in space so that each atom was part of a degenerate, three-dimensional harmonic oscillator. The kinetic energies would then be given by $E = \hbar\omega_o\,(n_x + n_y + n_z)$, where ω_o is the oscillator frequency and $n_{x,y,z}$ are integers. He felt that this would make the calculations easier. Although modern experiments with degenerate Fermi-Dirac gases (as they are now called) do use a harmonic confinement,[33] the result of this choice was that the paper was difficult to understand and so did not have a large resonance among the physics community. Fermi published a more detailed description of the calculation in *Zeitschrift für Physik* (in German) some weeks later (March 1926). As the linear dimensions of the container increase, the energy levels get closer together, so that for macroscopic vessels the quantization of the energy levels can have no real effect. This indicates that we need an additional rule in order to obtain the expected degeneracy of the gas. As we stated, this was the exclusion principle, which Fermi asserted should also apply to the translational motion of gas atoms, so that for each state designated by a triplet (n_x, n_y, n_z) there could be at most one atom. Fermi applied the usual method of maximizing the number of microstates corresponding to a macrostate subject to constraints and then used a classical treatment of the oscillatory motion to find the space-dependent energy distribution.

A few months later (August 1926) Dirac derived the same energy distribution for the case of the more familiar box-type container using Schrödinger's wave mechanics.

9.2.4 Dirac

In this paper, "On the Theory of Quantum Mechanics"[34], Dirac reviewed Schrödinger's work, emphasizing the equivalence of wave mechanics and Heisenberg's matrix mechanics as well as the method of calculating matrix elements from the eigenfunctions of Schrödinger's theory. In this paper, he showed how the statistical behavior of a gas depends on the symmetry of the wave function under exchange of identical particles. He demonstrated that an antisymmetric wave function implies the Pauli principle and presented the statistical mechanics of a monatomic ideal gas subject to the exclusion principle. He did not refer to Fermi's paper, stating in a much later interview (1963) that he had read Fermi's paper but had "forgotten it completely." In October 1926, Fermi wrote to Dirac, calling Dirac's attention to Fermi's previous work: "Since I suppose that you have not seen my paper, I beg to attract your attention on it."

Dirac praised Heisenberg's method of only calculating physically observable quantities while saying nothing about unmeasurable theoretical constructions. He reviewed the method of calculating matrix elements by expanding the result of applying an operator **a** to an eigenfunction as a series of eigenfunctions

$$\mathbf{a}\psi_n = \sum_m \psi_m a_{nm}\,, \tag{9.151}$$

[33] B. DeMarco and D. S. Jin, *Onset of Fermi Degeneracy in a Trapped Atomic Gas*, *Science* 285, 1703, (1999).

[34] *Proc. R. Soc. A* **112**, 661 (1926) received Aug. 26, 1926.

in which case the a_{nm} are the matrix elements, which obey the Heisenberg equations.

He then turned to systems of more than one electron. In a two-electron system where the electrons do not interact, the states can be designated as (mn) which indicates electron (1) is in the state m and electron (2) is in state n. The question then arises as to whether or not (mn) and (nm), which are physically indistinguishable, are to be regarded as the same state. This is equivalent to asking whether the matrix needed to represent this situation contains one or two rows and columns. If (mn) and (nm) are different states, there will be a separate transition amplitude for the transitions $(mn) \to (m'n')$ and $(mn) \to (n'm')$. Since $(m'n')$ and $(n'm')$ are indistinguishable experimentally, only the sum of these two transition amplitudes can be observed. In this situation, the theory would enable the calculation of an experimentally unobservable quantity. If (mn) and (nm) are the same state, we must have

$$x_1 \left(mn, m'n'\right) = x_2 \left(nm, n'm'\right), \tag{9.152}$$

where $x_{1,2}$ are the coordinates of the two electrons. If (mn) and (nm) define the same row and column of the matrices, each element of the matrix x_1 is equal to the corresponding element of x_2, so that we have $x_1 = x_2$, which is impossible. This means that functions of the coordinates (and momenta) that are not symmetric with respect to exchange of the two particles cannot be represented by matrices. For a matrix function $\mathbf{A}$ symmetric with respect to particle exchange, the matrix elements will be given by

$$\mathbf{A} \left|mn\right\rangle = \mathbf{A}\psi_{mn} = \sum_{m'n'} \psi_{m'n'} A_{mn,m'n'}, \tag{9.153}$$

where $\psi_{mn} = (mn) = \psi_m(1)\,\psi_n(2)\,,\ (nm) = \psi_m(2)\,\psi_n(1)$. Since these represent the same state, we must find a set of eigenfunctions of the form

$$\psi_{mn} = a_{mn}\psi_m(1)\,\psi_n(2) + b_{mn}\psi_m(2)\,\psi_n(1) \tag{9.154}$$

and must allow the determination of the matrix for any symmetric function $\mathbf{A}$ of the coordinates. We must thus choose $a_{mn} = \pm b_{mn}$ in order for ψ_{mn} to have a definite symmetry with respect to particle exchange. Depending on the choice of sign, the left-hand side of Eq. (9.153) will be either symmetric or antisymmetric. Only functions of the same symmetry will be required for the expansion on the right-hand side of Eq. (9.153) and either type of function gives a complete solution of the problem.

Another form of the argument is to note that if P_{ij} is an operator that exchanges the electron in state i with that in state j, P_{ij} must commute with the Hamiltonian, which must be symmetric under the exchange of identical particles. This means that eigenstates of the Hamiltonian can always be written as eigenstates of P_{ij},

$$P_{ij}\psi = \pi_{ij}\psi, \tag{9.155}$$

where π_{ij} is a c-number (ordinary classical number according to Dirac). Since applying P_{ij} twice in succession must leave the state unchanged, we have $\pi_{ij}^2 = 1, \pi_{ij} = \pm 1$.

Extending this to a large number of electrons, the antisymmetric wave function can be written as a determinant

$$\psi(1,2....K) = \begin{vmatrix} \psi_{n_1}(1) & \psi_{n_1}(2) & ... & \psi_{n_1}(K) \\ \psi_{n_2}(1) & \psi_{n_2}(2) & ... & \psi_{n_2}(K) \\ ... & ... & ... & ... \\ \psi_{n_K}(1) & \psi_{n_K}(2) & ... & \psi_{n_K}(K) \end{vmatrix} \tag{9.156}$$

Since the determinant vanishes when two of the electrons are in the same state ($n_i = n_j,\ i \neq j$), this wave function only allows one electron per state. The symmetrical solution would allow any number of electrons to be in the same state. Dirac then assumed that all stationary states (represented by one eigenfunction) have equal *a priori* probability. Then the symmetric wave function would yield the Bose-Einstein distribution. This distribution is correct for photons but the antisymmetric case was considered more likely to apply to gas molecules since they would be expected to "resemble electrons more closely than light quanta." Similar ideas regarding the importance of the symmetry of the wave function had been advanced by Heisenberg (June 1926).[35]

Now consider a gas of particles with an antisymmetric wave function, where no more than one particle can be associated with each of the normal modes of the de Broglie waves in a box. As we saw in Equation (9.76), the number of phase space cells A_s in the energy interval between E_s and $E_s + dE_s$ is given by

$$A_s = \frac{V}{(2\pi)^2}\left(\frac{\sqrt{2m}}{\hbar}\right)^3 E_s^{1/2} dE_s\,. \tag{9.157}$$

The calculation proceeds as for the Bose-Einstein case. The number of ways of assigning N_s particles to A_s cells with a maximum one particle per cell is the number of ways of choosing N_s cells out of A_s cells, which is given by

$$\binom{A_s}{N_s} = \frac{A_s!}{N_s!\,(A_s - N_s)!}\,. \tag{9.158}$$

For the whole gas, we have

$$W = \prod_s \frac{A_s!}{N_s!\,(A_s - N_s)!}\,, \tag{9.159}$$

so the entropy is

$$S = k_B \ln W = k_B \sum_s \left[A_s(\ln A_s - 1) - N_s(\ln N_s - 1) - (A_s - N_s)(\ln(A_s - N_s) - 1)\right] \tag{9.160}$$

$$\delta S = k_B \sum_s \left[-\ln N_s + \ln(A_s - N_s)\right]\delta N_s\,, \tag{9.161}$$

[35] Heisenberg, W., *The many body problem and resonance in quantum mechanics*, Zeit. Phys. **38**, 411 (1926). Received June 11, 1926.

which must be a maximum subject to the conditions of fixed particle number and total energy:

$$N = \sum_s N_s, \; E = \sum_s E_s N_s \,. \tag{9.162}$$

Introducing Lagrange multipliers, we have

$$\delta S = k_B \sum_s (\alpha + \beta E_s) \, \delta N_s \tag{9.163}$$

$$\ln \left(\frac{A_s}{N_s} - 1 \right) = \alpha + \beta E_s \tag{9.164}$$

$$N_s = \frac{A_s}{e^{\alpha + \beta E_s} + 1}. \tag{9.165}$$

By letting the energy vary at fixed N ($\sum_s \delta N_s = 0, \; \delta E = \sum_s E_s \delta N_s$), Eq. (9.163) becomes

$$\delta S = k_B \beta \cdot \delta E \tag{9.166}$$

$$\frac{\delta S}{\delta E} = \frac{1}{T} = k_B \beta \tag{9.167}$$

Then, writing $\alpha = -E_F/k_B T$, the number of particles in the energy interval is

$$dN_s = \frac{V}{(2\pi)^2} \left(\frac{\sqrt{2m}}{\hbar} \right)^3 \frac{E_s^{1/2} dE_s}{e^{(E_s - E_F)/k_B T} + 1}. \tag{9.168}$$

The Fermi energy E_F is determined by setting the integral of Eq. (9.168) equal to the total number of particles N. Fig. 9.8 shows the occupation probability

$$f(x, T') = \frac{1}{e^{(x-1)/T'} + 1} \tag{9.169}$$

where $x = E/E_F$ and $T' = k_B T/E_F$.

For $T = 0$, the energy levels are all occupied until the number of levels exhausts the number of particles—this condition determines the Fermi energy. From Eq. (9.168) at $T = 0$ K, we have

$$N = \frac{V}{(2\pi)^2} \left(\frac{\sqrt{2m}}{\hbar} \right)^3 \int_0^{E_F} E_s^{1/2} dE_s = \frac{V}{6\pi^2} \left(\frac{\sqrt{2m}}{\hbar} \right)^3 E_F^{3/2} \tag{9.170}$$

$$E_F = \left(6\pi^2\right)^{2/3} \left(\frac{\hbar^2}{2m} \right) \left(\frac{N}{V} \right)^{2/3}. \tag{9.171}$$

The Fermi energy thus depends on the density as $(N/V)^{2/3}$.

As the temperature is increased, particles are excited from states just below the Fermi level to states just above it. The width of the affected region is on the order of $k_B T$, so that higher temperatures excite particles from deeper inside the distribution. These effects are important in a wide range of physical situations ranging from computer chips to neutron stars.

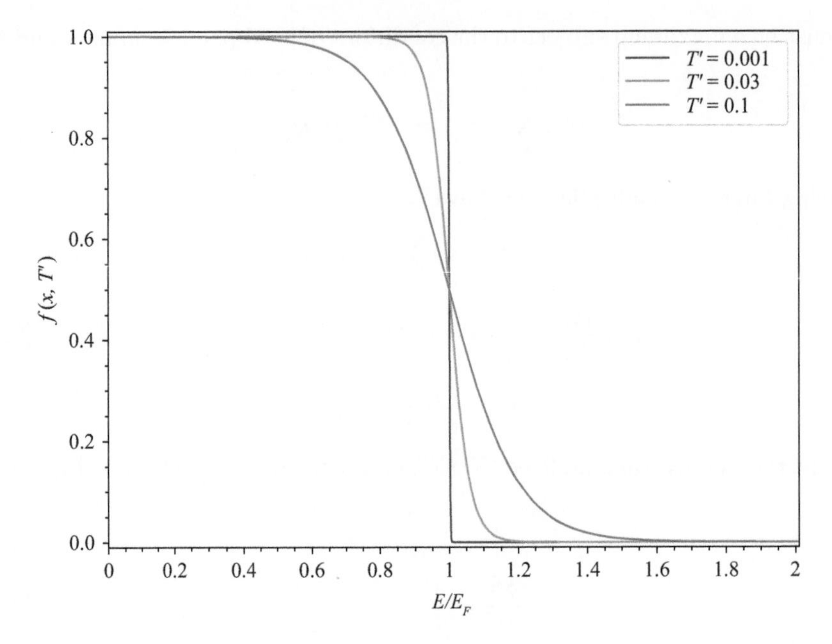

Fig. 9.8 The occupation probability for a Fermi gas as a function of energy $x = E/E_F$ and temperature $T' = k_B T/E_F$: Blue $T' = 0.001$ (essentially $T = 0$); Green $T' = 0.03$; Red $T' = 0.1$.

9.2.5 Early applications of Fermi-Dirac statistics

9.2.5.1 Fowler and dense stars

In one of the first applications of Fermi-Dirac statistics, in December 1926 Fowler pointed out that in extremely dense stars, the kinetic energies would be so great that the atoms would be ionized and the electrons would form a degenerate Fermi gas.[36] Neglecting the Coulomb energy, he then applied Eq. (9.171) to show that, as the star radiated energy, a time would come when the temperature was effectively zero but the energy and hence the pressure of the star would remain significant. This degeneracy pressure of the electrons is what stabilizes white dwarf stars against collapsing due to their own gravity.

9.2.5.2 Pauli paramagnetism

Another early application of the Fermi-Dirac statistics was to the electrons in a metal. This problem had been studied for many years without much success. One of the approaches was to consider that there were two types of electrons in a metal: those that were bound to the nuclei, which in turn were bound to the crystal lattice, and a group of electrons (the valence or conduction electrons) which were free to move throughout the metal. These were modeled as an ideal gas. The densities of the electrons, assuming one electron per atom, were high enough that the Fermi energy was much higher than

[36] Fowler, R.H., *On Dense Matter*, *Mon. Not. R. Astron. Soc.*, **87**, 114 (1926).

$k_B T$, so that the energy distribution of the assumed free electron gas would correspond to the $T = 0$ plot in Fig. 9.8. Thus it is no surprise that efforts to treat the electron gas as a classical gas were unsuccessful. With the advent of Fermi-Dirac statistics, great progress was achieved in understanding the properties of electrons in metals. In December 1926, Pauli was the first to apply these statistics to the free electron gas.[37] Much later (1956), Pauli explained that he had written to Dirac in the fall of 1926, asking if he (Dirac) knew how the electron spin (proposed by Uhlenbeck and Goudsmit in October 1925) would affect the results of his paper on the statistics of an ideal gas. He also mentioned Fermi's paper. Pauli recounts that Dirac replied that he had never considered the question of spins and that "Fermi's paper was entirely new to him."[38] Pauli then started to work on the problem himself. In an attempt to understand the choice between the Bose-Einstein and Fermi-Dirac statistics, Pauli turned to exploring further applications of the Fermi-Dirac statistics.

If the electrons had a spin and a related magnetic moment

$$\mu_e = -g\left(\frac{e\hbar}{2m_e c}\right) m_s = -g\mu_B m_s, \tag{9.172}$$

where $m_s = \pm 1/2$ is the component of spin angular momentum along an axis fixed in space, m_e is the electron mass and $g = 2$ (approximately) for an electron, why then do most metals exhibit such a weak paramagnetism? Comparing Eq. (9.172) to Eq. (9.147), where m_ϕ is an integer, we see that the electron spin is associated with a magnetic moment that is twice as large with respect to its angular momentum as the magnetic moment associated with the orbital angular momentum. This fact was initially found experimentally. Its explanation was one of the great triumphs of Dirac's relativistic wave equation.

Pauli started with a review of Fermi's calculation of the properties of an ideal gas obeying the exclusion principle. One of the new things he found was that the fluctuations of the number of atoms n_s in a fixed volume element in a fixed energy range differed significantly from those predicted for a Bose-Einstein gas. For a gas obeying the exclusion principle, $n_s = 0, 1$, which means that $n_s^2 = 0, 1$, so that $\langle n_s^2 \rangle = \langle n_s \rangle$ and

$$\left\langle (\Delta n_s)^2 \right\rangle = \langle n_s^2 \rangle - \langle n_s \rangle^2 = \langle n_s \rangle - \langle n_s \rangle^2. \tag{9.173}$$

Let Z be the number of phase space cells in a region of energies between E and $E + dE$. Then the average number of electrons in this region is $\langle N_Z \rangle = Z \langle n_s \rangle$ and the mean square of its fluctuations $\Delta_Z = (N_Z - \langle N_Z \rangle)$ is

$$\langle \Delta_Z^2 \rangle = Z \left\langle (\Delta n_s)^2 \right\rangle \tag{9.174}$$

since we assume the fluctuations in each state are independent of each other. Then

$$\langle \Delta_Z^2 \rangle = Z \langle n_s \rangle - Z \langle n_s \rangle^2 = \langle N_Z \rangle - \frac{\langle N_Z \rangle^2}{Z} \tag{9.175}$$

[37] Pauli, W., Jr., *On gas degeneracy and paramagnetism*, Zeit. Phys. **41**, 81, (1927) received Dec. 10, 1926.

[38] J. Mehra and H. Rechenberg *The Historical Development of Quantum Mechanics*, vol. 1, The Completion of Quantum Mechanics: 1926-1941, part 1 (New York: Springer-Verlag, 2000).

and the fractional mean square fluctuation is

$$\frac{\langle\Delta_Z^2\rangle}{\langle N_Z\rangle^2} = \frac{1}{\langle N_Z\rangle} - \frac{1}{Z}. \tag{9.176}$$

This differs from the Bose-Einstein result (9.72) by the difference in sign. Remember that the first term describes the fluctuations expected for a classical ideal gas of statistically independent particles, while the second term describes the fluctuations due to interference of the different de Broglie waves. In the Fermi-Dirac case, the contribution of the wave interference serves to reduce the particle number fluctuations, which are thus less than those of a classical gas. Since Einstein's demonstration that the second term corresponded to wave interference required that the partial waves had random phases, Pauli speculated that the Fermi-Dirac behavior might indicate specific phase relations between the different normal modes of the de Broglie waves and may point the way to a physical explanation of the basis of the Fermi statistics.

For a Fermi gas of particles with spin and magnetic moment in the presence of a magnetic field, the total energy of an electron is given by $E = E_s \pm E_m$ with $E_m = -\mu_e B$ and μ_e given by (9.172). This would replace E_s in Eq. (9.162) since the total energy must be held fixed in the variation. The phase space factor A_s will still be determined by the kinetic energy E_s; however, the equilibrium distribution will be altered compared to Eq. (9.168) to

$$dN_s = \frac{V}{(2\pi)^2}\left(\frac{\sqrt{2m}}{\hbar}\right)^3 \frac{E_s^{1/2} dE_s}{e^{(E_s \pm E_m - E_F)/k_B T} + 1} \tag{9.177}$$

and E_F in Eqs. (9.170) and (9.171) should be replaced by $E_F \mp E_m$. The presence of spin, with its two possible states ($m_s = \pm 1/2$), means that two electrons can inhabit each phase space cell or be represented by the same normal mode of the de Broglie wave field, so that Eq. (9.170) should be multiplied by two. The magnetic moments in filled cells cancel out.

Electrons in the two spin states behave as two separate gases. In the absence of a magnetic field, the two gases contain the same number of electrons. In the presence of a magnetic field, the Fermi energies of the two gases shift in opposite directions. Since the two Fermi levels must be the same in equilibrium, a number dN_s of electrons has to move from one group to the other, resulting in a net magnetization. As the Fermi energy (on the order of 2-10 eV) is much greater than $k_B T$ (25 meV at room temperature), the gas is fully degenerate and we can use the $T = 0$ result. From Eq. (9.170),

$$dN_s = \frac{3}{2}\frac{N'}{E_F} dE_F, \tag{9.178}$$

where $N' = N/2$ is the number of electrons in one spin state. The magnetization of the gas is then given by (note $dE_F = \mp dE_m$)

$$M = \frac{dN_s^{(+)} - dN_s^{(-)}}{V}\mu_e = \frac{3}{2}\mu_e \frac{N}{V}\left(\frac{-E_m}{E_F}\right) = \frac{3}{2}\mu_e \frac{N}{V}\left(\frac{\mu_e B}{E_F}\right), \tag{9.179}$$

so the magnetic moment per electron is reduced by the factor $\left(\frac{\mu_e B}{E_F}\right)$.

9.2.5.3 The Thomas-Fermi statistical model of the electron distribution in a many-electron atom

Fermi's first approach to the theory of a gas of particles obeying the exclusion principle was to consider the gas as confined by a harmonic oscillator potential. When he turned his attention to problems of atomic structure about two years later (February 1928)[39], he had the idea of considering an atom as consisting of an electron gas trapped by the Coulomb potential of the nucleus—a kind of atmosphere surrounding the nucleus. L.H. Thomas had published a similar idea more than a year earlier (November 1926).[40] Thomas was interested in calculating the distribution of electric fields inside an atom for the purpose of calculating certain atomic constants. The idea of this model is to consider the space around the nucleus as being divided up into cells that are so small that the potential energy $V(r) = -\phi(r)$ of a given electron, due to the nucleus and all the other electrons, does not vary significantly over the cell. We further assume that the system is spherically symmetric. Since the maximum energy is zero for a bound electron, the momentum of the electrons at a given distance r from the nucleus ranges between 0 and $p_F = \sqrt{2m\phi(r)}$, and the number of phase space cells at that position is given by

$$d^3r \frac{4\pi}{3h^3} p_F^3 \tag{9.180}$$

with d^3r the volume of the cell. If all cells are filled, which corresponds to a completely degenerate gas, there will be $n(r)$ electrons per unit volume:

$$n(r) = \frac{8\pi}{3h^3} p_F^3 = \frac{8\pi}{3h^3} (2m\phi(r))^{3/2}, \tag{9.181}$$

where we have multiplied by two to account for the two spin states. This charge density will produce an electric field given by a potential energy

$$\nabla^2\phi(r) = \frac{1}{r^2}\frac{d}{dr}\left(r^2\frac{d\phi}{dr}\right) = 4\pi e^2 n(r) = e^2\frac{32\pi^2}{3h^3}(2m\phi(r))^{3/2},$$

where e is the charge of an electron. Defining

$$\phi(r) = \frac{Ze^2}{r}\psi(r) \tag{9.182}$$

$$r = x/x_o, \tag{9.183}$$

with

$$x_o = \left(\frac{32\pi^2}{3}\right)^{2/3} \frac{2me^2Z^{1/3}}{h^2}, \tag{9.184}$$

[39] Fermi, E., *A statistical Method for Determining some Properties of the Atoms and its Application to the Theory of the periodic Table of Elements*, Zeit. Phys. **48**, 73 (1928) Received Feb. 23, 1928.

[40] Thomas, L.H., *The calculation of atomic fields*, Math. Proc Cambridge Phil. Soc. **23**, 542 (1927) Received Nov. 6, 1926.

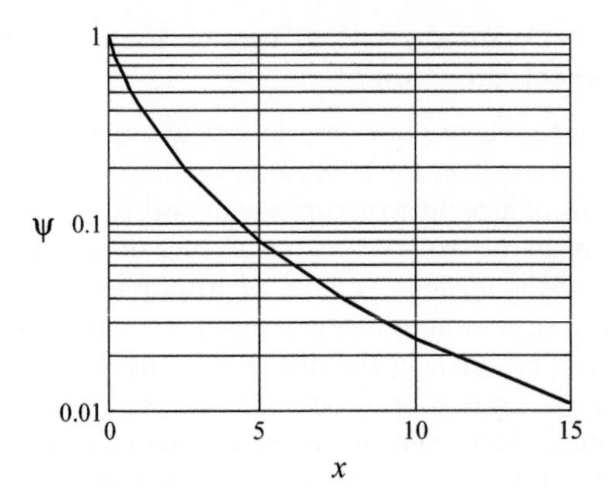

Fig. 9.9 The Thomas-Fermi function $\Psi(x)$.

we obtain the dimensionless form of the equation:

$$\frac{d^2\psi}{dx^2} = \frac{\psi^{3/2}}{x^{1/2}}. \tag{9.185}$$

Since at small r the potential is that of the unshielded nucleus, ψ must satisfy the boundary condition $\psi(0) = 1$. The other boundary condition follows from the requirement

$$\int n(r)\, d^3r = Z. \tag{9.186}$$

Using Eq. (9.181) and integration by parts, this condition reduces to $\psi(r \to \infty) = 0$. The solution can be obtained numerically and is shown in Fig. 9.9.

It is interesting to apply this technique to the original situation addressed by Fermi: a gas confined by a harmonic oscillator potential $V(r) = Kr^2$. We will only consider the degenerate case with $T = 0$. The energy states will fill up to an energy E_F determined by the number of particles present. The number of states per unit volume will again be given by

$$n(r) = \frac{4}{3h^3}\pi p_F^3(r) = \frac{4}{3h^3}\pi\left(2m\left(E_F - Kr^2\right)\right)^{3/2} \tag{9.187}$$

and the total number of particles is

$$\begin{aligned} N = 4\pi \int_0^{r_F} n(r)\, r^2 dr &= (4\pi)^2 \frac{(2m)^{3/2}}{3h^3} \int_0^{r_F} \left(E_F - Kr^2\right)^{3/2} r^2 dr \\ &= (4\pi)^2 \frac{(2m)^{3/2}}{3h^3} \frac{E_F^3}{K^{3/2}} \int_0^1 \left(1 - x^2\right)^{3/2} x^2 dx \end{aligned} \tag{9.188}$$

where $r_F = (E_F/K)^{1/2}$ and $x = r/r_F$. The dimensionless integral is given by $\pi/32$ so

$$N = \left(\frac{\pi}{h}\right)^3 \frac{(2m)^{3/2}}{6} \frac{E_F^3}{K^{3/2}} \tag{9.189}$$

$$E_F = N^{1/3} \frac{h}{\pi} \frac{6^{1/3} K^{1/2}}{(2m)^{1/2}}. \tag{9.190}$$

Then

$$n(r) = \frac{4\pi}{3h^3}(2mE_F)^{3/2}\left(1 - \frac{r^2}{r_F^2}\right)^{3/2} \tag{9.191}$$

which is quite different from the result obtained by Fermi, who treated the motion in the harmonic potential using the Bohr-Sommerfeld theory, as classical motion restricted by the quantization rule.

9.3 Conclusion

As the first phenomenon to demonstrate the wave-particle duality that is now known to be central to quantum mechanics, the Planck spectrum for black body radiation was crucial to the development of quantum theory. Due to its importance, a great deal of work was devoted to developing different methods of deriving the Planck distribution in an attempt to gain insight into the underlying physics.

S. Bose found a new way of deriving the Planck distribution by dividing the phase space up into cells of volume h^3. He arrived at the Planck spectrum by treating the radiation field as a collection of photons and considering the number of cells that held a given number of photons. He sent a manuscript describing his work to Einstein, who immediately realized that the same treatment could be applied to atoms in a gas and derived the distribution for a gas using Bose's method. Based on these results Einstein demonstrated that, at a low enough temperature and high enough density, the gas would "condense," with an increasing fraction of atoms in the ground state. This phenomenon is connected with superfluidity and superconductivity, although the interaction between the particles significantly complicates the physics involved in these effects. Einstein derived an expression for the particle number fluctuations in a small volume of the gas. These fluctuations consisted of two contributions: one that would occur in a classical gas of independent particles and another that arises from the interference of waves. Thus the material gas showed wave-like properties just as the radiation field described by the Planck distribution had shown particle-like properties. This led Einstein to conclude that de Broglie's proposal of some kind of wave associated with the material particles had to be taken seriously.

Schrödinger, who at first doubted the Bose-Einstein approach, showed that the Bose-Einstein result could be obtained by considering the particles as excitations of the normal modes (standing waves) of the de Broglie wave field in a cavity and applying the standard Boltzmann statistics. This idea would later be formulated as "second quantization" (Chapter 12), which has the advantage of being able to treat many-body systems in three dimensions.

After Pauli deduced, based on the periodic table and atomic spectroscopy, that no two electrons could occupy the same quantum state,[41] Fermi realized that this was in direct contradiction to the Bose-Einstein statistics and proceeded to calculate the distribution function for a gas of massive particles satisfying the exclusion principle. He considered the molecules as being bound to a fixed point in space by a harmonic force but treated the motion classically. Some months later Dirac showed that matrix

[41] This was originally called the "Pauli prohibition" (Verbot) and is now called the Pauli exclusion principle.

elements could only be assigned in a meaningful way to functions with a well-defined symmetry under the exchange of identical particles. This implied that the Schrödinger wave function had to be either symmetric or antisymmetric under the exchange and that an antisymmetric wave function would automatically satisfy the Pauli principle. He then, like Schrödinger, regarded the gas as consisting of a set of waves where either zero or one atoms were assigned to each wave. The results of Fermi's and Dirac's approaches were identical and showed the phenomenon of degeneracy at low temperatures.

Thomas and Fermi independently modeled many-electron atoms as a nucleus surrounded by an "atmosphere" consisting of an electron gas that obeyed the Pauli principle with two electrons per cell (volume h^3) of phase space. This found application as an approximate way to treat the spectra of complex atoms, but this method was quickly replaced by more detailed methods using the many-electron ($3N$-dimensional) wave function developed by Hartree and Fock, where N is the number of electrons, so that $N = Z$ for a neutral atom.

To conclude, we list the different forms for the thermodynamic probability used in this chapter:

- **Maxwell-Boltzmann** (distinguishable particles):
  N = total number of particles, n_i = number of particles in cell i, g_i = the *a priori* probability that a cell is occupied:

$$W = N! \prod_i \frac{g_i^{n_i}}{n_i!} \tag{9.192}$$

- **Planck** (photons):
  P = number of photons, N = number of oscillators (a similar form was used by Debye and Natanson):

$$W = \frac{(N+P)!}{P!\,(N-1)!} \tag{9.193}$$

  * N oscillators with N_j oscillators having an energy $jh\nu$:

$$W = \frac{N!}{\prod_j N_j!} \tag{9.194}$$

- **Bose** (photons):
  N_s = number of cells (phase space volume h^3) in shell s, defined by $E_s = h\nu_s$, p_i^s = number of cells in the shell s that contain i photons:

$$W = \prod_s \frac{N_s!}{\prod_i p_i^s} \tag{9.195}$$

  * N_s = number of cells in shell s, P_s = the number of photons in that shell:

$$W = \prod_s \frac{(N_s+P_s)!}{N_s!P_s!} \tag{9.196}$$

- **Bose-Einstein** (indistinguishable massive particles):
  N_s = number of cells in shell s, n_s = number of particles in shell s:

$$W = \prod_s \frac{(N_s + n_s - 1)!}{(N_s - 1)!n_s!} \tag{9.197}$$

- **Fermi-Dirac** (indistinguishable massive particles):
  N_s = number of cells in shell s, n_s = number of particles in shell s:

$$W = \prod_s \frac{N_s!}{n_s!\,(N_s - n_s)!} \tag{9.198}$$

10 Early attempts at interpretation of the theory

10.1 Introduction

As we have seen, the first successes of the new quantum theory left open the question of the physical meaning of the basic quantities: the wave function $\psi(q,t)$ and the matrices A_{mn}.

Schrödinger's first idea was that the squared magnitude of the wave function $|\psi(q,t)|^2$ multiplied by the charge e gave the charge density of the particle, so that classical point particles would be replaced by wave packets—charge clouds smeared out in space. However, this idea was found to be insupportable when it became clear that in most cases the wave packets would spread with time. The harmonic oscillator, the example picked by Schrödinger, turned out to be a special case where the wave packets do not spread due to the fact that the energy levels are all equally separated.

Unlike Schödinger, Heisenberg was strongly opposed to any intuitive picture, insisting that the matrix elements provided all the information that was experimentally accessible—namely the frequency, amplitude, and polarization of the emitted radiation—and that the quantum jumps were intrinsically unknowable.

A major contribution to the physical understanding of the theory came from Born, who studied scattering phenomena using Schrödinger's wave mechanics. These studies led him to the probabilistic interpretation of the wave function that remains the standard interpretation today.

10.2 Schrödinger and the spreading of wave packets

Schrödinger's first idea concerning the meaning of the wave function was that point particles did not exist: instead, particles were instances of spread out mass and charge density described by wave packets—superpositions of Schrödinger eigenstates. As an example, he considered the one-dimensional harmonic oscillator. A general wave function, a solution of the time-dependent Schrödinger equation, given by a superposition of the eigenstates is:

$$\psi(x,t) = \sum_n a_n u_n(x) e^{-\frac{iE_n t}{\hbar}}, \tag{10.1}$$

The Historical and Physical Foundations of Quantum Mechanics. Robert Golub and Steven K. Lamoreaux, Oxford University Press.
 DOI: 10.1093/oso/9780198822189.003.0010

where $u_n(x), E_n$ are the eigenfunctions and eigenvalues of the harmonic oscillator Hamiltonian H_o ($H_o u_n = E_n u_n$), a_n are amplitudes, and the energy eigenvalues are

$$E_n = \left(n + \frac{1}{2}\right)\hbar\omega_o \tag{10.2}$$

with ω_o the frequency of the oscillator. At $t = 0$, the wave function is $\psi(x, 0) = \sum_n a_n u_n(x)$, so that if we choose

$$a_n = \int u_n^*(x) f(x)\, dx, \tag{10.3}$$

we can fix $\psi(x, 0)$ to be any arbitrary function $f(x)$. Schrödinger chose $f(x)$ to be a narrow Gaussian and found that this packet oscillated back and forth like a classical particle without any spread of the wave function. He thought the wave function did not spread because the energy levels were discrete and thus expected that the same should hold for wave packets made of discrete eigenfunctions for any system. But it is easy to see that the wave packets do not spread in this case because the energy levels are equally spaced, so that Eq. (10.1) is a Fourier series and $\psi(x, t)$ is periodic.

To show explicitly that in other discrete systems the wave packets do indeed spread, we consider a particle moving in one dimension between impenetrable walls located at $x = 0, L$ so that $\psi(x = 0) = \psi(x = L) = 0$. This example was used by Einstein[1] in 1953 in an attempt to show that quantum mechanics was incompatible with the classical behavior of a single particle and could only be applied to an ensemble (a set of many identically prepared systems), with the wave function only giving information concerning quantities averaged over the whole ensemble. According to this view, quantum mechanics was incomplete because it provided no information on individual particles.

Born pointed out that Einstein had based his argument on considering a single eigenstate. He then proceeded to show that, if the initial state is taken to be a wave packet (a suitably chosen superposition of eigenstates), the quantum result approaches the classical one in a smooth fashion. The observed spreading of the wave packet would also occur classically if we allow for the fact that there is always some uncertainty in the value of the initial velocity. Born worked out this case in great detail.[2] His treatment was later embellished by Pauli[3] in an exercise in his lectures on quantum mechanics.

10.2.1 Wave packets for a particle in a box

The eigenstates for the particle in a box are given by

$$u_n(x) = \sqrt{\frac{2}{L}} \sin(k_n x), \tag{10.4}$$

[1]Einstein, A., "Elementary Considerations on the Foundations of Quantum Mechanics," in *Scientific Papers Presented to Max Born on his Retirement from the Tate Chair of Natural Philosophy in the University of Edinburgh* (Oliver and Boyd, 1953).

[2]Born, M., *Continuity, Determinism and Reality*, Kgl. Danske Vidensk. Selskab, Math-Fys. Medd. 30, No. 2 (1955) received April 15, 1955.

[3]Pauli, in Enz, C.P., ed. *Pauli Lectures on Physics, vol. 5 Wave Mechanics*, Dover, (2000).

and the general solution is given by

$$\psi(x,t) = \sqrt{\frac{2}{L}} \sum_n a_n \sin(k_n x)\, e^{-i\frac{E_n t}{\hbar}} \tag{10.5}$$

with $k_n = n\pi/L$. This satisfies the wave equation with $E_n = \frac{1}{2m}\left(\frac{\hbar\pi n}{L}\right)^2 = \hbar\alpha n^2$, where m is the mass of the particle and $\alpha = \frac{\hbar}{2m}\left(\frac{\pi}{L}\right)^2$. Following Born and Pauli, We will discuss this problem by means of the Green's function. This is the solution to the wave equation with the initial condition

$$\psi(x, t=0) = \delta(x - x_o), \tag{10.6}$$

where $\delta(x)$ is the Dirac delta function. Thus the particle is certain to be at x_o at $t = 0$. It is easy to see that this function is given by

$$G(x - x_o, t) = \sum_n u_n(x)\, u_n^*(x_o)\, e^{-i\frac{E_n t}{\hbar}}. \tag{10.7}$$

This is clearly a solution of the wave equation in (x,t), and at $t = 0$ we find

$$G(x - x_o, t = 0) = \sum_n u_n(x)\, u_n^*(x_o) = \delta(x - x_o) \tag{10.8}$$

using the completeness relation for the eigenfunctions (Eq. (7.128) in Chapter 7). Any solution of the problem can be built up as a sum of such solutions:

$$\psi(x,t) = \int G(x - x', t)\, \psi_o(x')\, dx', \tag{10.9}$$

where $\psi_o(x')$ is the value of the wave function at $t = 0$, as can be seen by substituting Eq. (10.3) into (10.1) and using (10.8). Thus if $G(x,t)$ spreads with time, any wave packet will also spread.

In our case,

$$G(x - x_o, t) = \frac{2}{L}\sum_n \sin\left(\frac{n\pi x}{L}\right) \sin\left(\frac{n\pi x_o}{L}\right) e^{-i\frac{E_n t}{\hbar}} \tag{10.10}$$

$$= \frac{1}{L}\sum_{n=0}\left[\cos\frac{n\pi}{L}(x - x_o) - \cos\frac{n\pi}{L}(x + x_o)\right] e^{-i\alpha n^2 t} \tag{10.11}$$

$$= \frac{1}{L}\left\{\begin{matrix} 1 + \sum_{n=1}\cos\frac{n\pi}{L}(x - x_o) - \\ (1 + \sum_{n=1}\cos\frac{n\pi}{L}(x + x_o)) \end{matrix}\right\} e^{-i\alpha n^2 t} \tag{10.12}$$

$$= \frac{1}{2L}\left\{\begin{matrix} (1 + 2\sum_{n=1}\cos\frac{n\pi}{L}(x - x_o)) - \\ (1 + 2\sum_{n=1}\cos\frac{n\pi}{L}(x + x_o)) \end{matrix}\right\} e^{-i\alpha n^2 t}. \tag{10.13}$$

We can write this in terms of the Jacobi theta function $\vartheta_3(z|\tau)$, which is defined as

$$\vartheta_3(z|\tau) = 1 + 2\sum_{n=1}\cos(2nz)\, e^{i\pi n^2 \tau}, \tag{10.14}$$

so that $G(x - x_o, t)$ can be written as

$$G(x - x_o, t) = \frac{1}{2L}\left\{\vartheta_3\left(\frac{\pi}{2L}(x - x_o)\middle| -\frac{\alpha t}{\pi}\right) - \vartheta_3\left(\frac{\pi}{2L}(x + x_o)\middle| -\frac{\alpha t}{\pi}\right)\right\}. \quad (10.15)$$

Now $\vartheta_3(z|\tau)$ satisfies a relation known as Jacobi's imaginary transformation,[4]

$$\vartheta_3(z|\tau) = \frac{1}{(-i\tau)^{1/2}} e^{-i\frac{z^2}{\pi\tau}} \vartheta_3\left(-\frac{z}{\tau}\middle| -\frac{1}{\tau}\right), \quad (10.16)$$

so that

$$G(x - x_o, t) = \frac{1}{2L}\frac{1}{(i\alpha t/\pi)^{1/2}}\left\{\begin{array}{c} e^{i\frac{1}{\alpha t}\left[\frac{\pi}{2L}(x-x_o)\right]^2}\vartheta_3\left(\frac{\pi^2}{2L}\frac{(x-x_o)}{\alpha t}\middle|\frac{\pi}{\alpha t}\right) \\ -e^{i\frac{1}{\alpha t}\left[\frac{\pi}{2L}(x+x_o)\right]^2}\vartheta_3\left(\frac{\pi^2}{2L}\frac{(x+x_o)}{\alpha t}\middle|\frac{\pi}{\alpha t}\right)\end{array}\right\}. \quad (10.17)$$

Reapplying Eq. (10.14), we find

$$\begin{aligned}
G(x - x_o, t) &= \frac{1}{2L}\frac{1}{(i\alpha t/\pi)^{1/2}}\left\{\begin{array}{c} e^{i\frac{1}{\alpha t}\left[\frac{\pi}{2L}(x-x_o)\right]^2}\left(1 + 2\sum_{n=1}\cos\left(n\frac{\pi^2}{L}\frac{(x-x_o)}{\alpha t}\right)e^{i\frac{\pi^2 n^2}{\alpha t}}\right) \\ -e^{i\frac{1}{\alpha t}\left[\frac{\pi}{2L}(x+x_o)\right]^2}\left(1 + 2\sum_{n=1}\cos\left(n\frac{\pi^2}{L}\frac{(x+x_o)}{\alpha t}\right)e^{i\frac{\pi^2 n^2}{\alpha t}}\right)\end{array}\right\} \\
&= \frac{1}{2L}\frac{1}{(i\alpha t/\pi)^{1/2}}\left\{\begin{array}{c} \left(\sum_{n=-\infty}^{\infty}\exp i\frac{\pi^2}{\alpha t}\left(\left[\frac{1}{2L}(x - x_o)\right]^2 + n\frac{(x-x_o)}{L} + n^2\right)\right) \\ -\left(\sum_{n=-\infty}^{\infty}\exp i\frac{\pi^2}{\alpha t}\left(\left[\frac{1}{2L}(x + x_o)\right]^2 + n\frac{(x+x_o)}{L} + n^2\right)\right)\end{array}\right\} \\
&= \frac{1}{2L}\frac{1}{(i\alpha t/\pi)^{1/2}}\sum_{n=-\infty}^{\infty}\left[\begin{array}{c}\exp i\left\{\frac{\pi^2}{\alpha t 4L^2}\left((x - x_o) + 2Ln\right)^2\right\} - \\ \exp i\left\{\frac{\pi^2}{\alpha t 4L^2}\left((x + x_o) + 2Ln\right)^2\right\}\end{array}\right] \\
&= \left(\frac{m}{it\pi 2\hbar}\right)^{1/2}\sum_{n=-\infty}^{\infty}\left[\exp i\left\{\frac{m}{2\hbar t}\left((x - x_o) + 2Ln\right)^2\right\}\right. \\
&\quad \left. - \exp i\left\{\frac{m}{2\hbar t}\left((x + x_o) + 2Ln\right)^2\right\}\right]. \quad (10.18)
\end{aligned}$$

This function spreads with time as $\sim (1/t)^{1/2} f(x/t)$, so that any wave packet made up of a sum of such functions will always expand with time despite the fact that $G(x - x_o, t)$ is a sum of eigenfunctions with discrete energies. Note that the two terms are functions of $(x \pm x_o + 2Ln)$ corresponding to sources located at $(x = \mp x_o + 2Ln)$, i.e., the $n = 0$ term in Eq. (10.18) is $G_o(x - x_o, t)$, the Green's function for a free particle with the physical source at x_o, while the other terms represent free particle Green's functions emanating from all the mirror images of the physical source. Including the mirror images as sources is necessary to satisfy the boundary conditions at the walls $(x = 0, L)$. To further illustrate the point, Fig. 10.1 shows the first 100 terms in Eq. (10.7) in units of x/L and t/t_o, where $t_o = \frac{2m}{\hbar}\left(\frac{L}{\pi}\right)^2$ and $x_o = L/2$.

The spreading with time is obvious, thus explicitly showing that wave packets consisting of superpositions of eigenfunctions with discrete energies do indeed spread.

[4] J. Sondow and E. W. Weisstein, '*Jacobi's Imaginary Transformation.*' From MathWorld—A Wolfram Web Resource. http://mathworld.wolfram.com/JacobisImaginaryTransformation.html

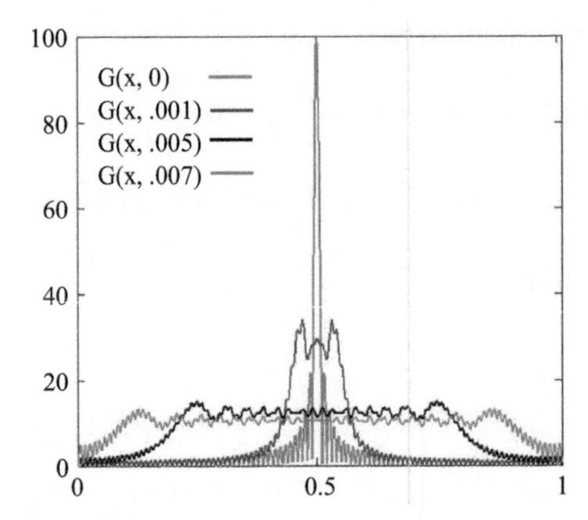

Fig. 10.1 $G(x - x_o, t)$ for a particle in a one-dimensional box, approximated by the first 100 terms in the eigenfunction expansion (10.7). The particle is initially at $x_o = L/2$. Red: $t/t_o = 0$; Blue: $t/t_o = 0.001$; Black: $t/t_o = 0.005$; Magenta: $t/t_o = 0.007$.

10.3 Born's insight and the loss of determinacy in physics

Max Born, who we have seen played a central role in the development of matrix mechanics, became interested in the question of how to describe non-periodic processes in quantum mechanics. Matrix mechanics started from reinterpreting the coefficients of Fourier series for the positions and momenta of the particles involved and so was intrinsically applicable to periodic motions only. Born realized that Schrödinger's wave mechanics could be applied to nonperiodic processes such as scattering and could give a description of the transition process, which was absent in the matrix theory. As scattering experiments (e.g., The Frank-Hertz experiment, 1914) had furnished definitive, direct proof for the existence of stationary states in atoms, Born considered wave mechanics "the deepest version of the theory."

Born[5] considered scattering from an atom which is taken, for simplicity, to be fixed. At large distances from the atom, the interaction potential is negligible and the solution of the wave equation describing the scattered particle must go over to the free particle solution. The free particle solutions of the time-independent wave equation can be written as

$$\psi_k = Ce^{i\vec{k}\cdot\vec{r}}, \tag{10.19}$$

which represents a momentum eigenstate with momentum $\vec{p}\psi_k = \frac{\hbar}{i}\vec{\nabla}\psi_k = \hbar\vec{k}\psi_k$. The direction of propagation is given by $\hat{k}$, the unit vector in the $\vec{k}$ direction. A particle moving in the z direction would be represented by the plane wave $\psi_k(z) \sim e^{ikz}$. In order to normalize these states, we consider the one-dimensional case:

$$\int dk\, \psi_k^*(z')\, \psi_k(z) = |C|^2 \int dk\, e^{ik(z-z')} = |C|^2 \lim_{\Gamma\to\infty} \int_{-\Gamma}^{\Gamma} dk\, e^{ik(z-z')} \tag{10.20}$$

$$= |C|^2 \lim_{\Gamma\to\infty} 2\frac{\sin[\Gamma(z-z')]}{(z-z')} = |C|^2\, 2\pi\delta(z-z'). \tag{10.21}$$

[5]Born, M., *Quantum mechanics of collision processes*, Zeit. der Phys. **37**, 863 (1926) received June 25, 1926 and **38**, 803 (1926) received July 21, 1926.

The last step follows from the fact that the limit vanishes for $(z - z') \neq 0$ and

$$\int_{-\infty}^{\infty} \frac{\sin \Gamma z}{z} dz = \pi, \tag{10.22}$$

so that $\lim_{\Gamma \to \infty} \int_{-\Gamma}^{\Gamma} \frac{\sin \Gamma z}{\pi z} dz$ has all the properties of the Dirac delta function, namely $\delta(x) = 0$ for $x \neq 0$ and $\int_{-\varepsilon}^{\varepsilon} \delta(x) \, dx = 1$ for $\varepsilon > 0$, . Thus we must set $C = \frac{1}{\sqrt{2\pi}}$.

10.3.1 Elastic scattering of a particle by an atom

For scattering of a particle by an atom when the atom is fixed at the origin of coordinates and does not change state (elastic scattering), the scattered particle, say an electron, sees the atom as a fixed potential $V(\vec{r})$. The wave equation for the electron can then be written in the form

$$\nabla^2 \psi + k^2 \psi = U(\vec{r}) \, \psi(\vec{r}) \tag{10.23}$$

with $U(\vec{r}) = \frac{2m}{\hbar^2} V(\vec{r})$. We solve this by the method of Green's functions. The Green's function satisfies

$$\nabla^2 G(\vec{r}) + k^2 G(\vec{r}) = \delta^{(3)}(\vec{r}), \tag{10.24}$$

where $\delta^{(3)}(\vec{r}) = \delta(x) \, \delta(y) \, \delta(z)$ is the three-dimensional delta function. Given $G(\vec{r})$, the solution to Eq. (10.23) is

$$\psi(\vec{r}) = \int d^3r' G(\vec{r} - \vec{r'}) U(\vec{r'}) \psi(\vec{r'}). \tag{10.25}$$

While this equation is exact, it is not really a solution since the unknown $\psi(\vec{r})$ appears on both sides of the equation. As discussed in Section 7.4.2, Eq. (10.25) represents the integral equation equivalent of Eq. (10.23). We proceed by iteration, writing

$$\psi = \psi_o + \psi_1 + \psi_2 + ... \tag{10.26}$$

where $\psi_o = \frac{1}{(2\pi)^{3/2}} e^{i\vec{k_o} \cdot \vec{r}}$ is the solution of Eq. (10.23) with the right-hand side set equal to zero (free particle, plane wave). The initial state is specified by the value of $\vec{k}$. Then

$$\psi_n = \int d^3r' G(\vec{r} - \vec{r'}) U(\vec{r'}) \psi_{n-1}(\vec{r'}). \tag{10.27}$$

Terminating the sum at ψ_1 is called the first Born approximation. To calculate this, we need to solve Eq. (10.24) for the Green's function. We do this by expanding $G(\vec{r})$ and $\delta^{(3)}(\vec{r})$ in the free particle eigenfunctions (Fourier transform):

$$G(\vec{r}) = \frac{1}{(2\pi)^{3/2}} \int d^3q \, g(\vec{q}) \, e^{i\vec{q} \cdot \vec{r}} \tag{10.28}$$

$$\delta^{(3)}(\vec{r}) = \frac{1}{(2\pi)^3} \int d^3q \, e^{i\vec{q} \cdot \vec{r}}. \tag{10.29}$$

Substituting into Eq. (10.24) we find

$$\frac{1}{(2\pi)^{3/2}}\int d^3q\, e^{i\vec{q}\cdot\vec{r}}\left[\left(-q^2+k^2\right)g\left(\vec{q}\right)-\frac{1}{(2\pi)^{3/2}}\right]=0. \tag{10.30}$$

Since this must hold for all values of $\vec{r}$, we must have

$$g\left(\vec{q}\right)=\frac{1}{(2\pi)^{3/2}\left(-q^2+k^2\right)}. \tag{10.31}$$

Then

$$G\left(\vec{r}\right)=\frac{1}{(2\pi)^3}\int d^3q\frac{e^{i\vec{q}\cdot\vec{r}}}{\left(-q^2+k^2\right)} \tag{10.32}$$

$$=\frac{1}{(2\pi)^3}\int\frac{e^{iqr\cos\theta}}{\left(-q^2+k^2\right)}\sin\theta\, d\theta\, d\phi\, q^2dq\,, \tag{10.33}$$

where we introduced the spherical coordinates for $\vec{q}$ and θ is the angle between $\vec{q}$ and $\vec{r}$. Then, defining $z=\cos\theta$, we have

$$G\left(\vec{r}\right)=\frac{1}{(2\pi)^2}\int_0^\infty\left[\int_{-1}^{1}e^{iqrz}dz\right]\frac{q^2dq}{\left(-q^2+k^2\right)} \tag{10.34}$$

$$=\frac{1}{(2\pi)^2\, ir}\int_0^\infty\left[\frac{e^{iqr}-e^{-iqr}}{\left(-q^2+k^2\right)}\right]q\,dq \tag{10.35}$$

$$=\frac{1}{2\,(2\pi)^2\, ir}\int_{-\infty}^\infty\left[\frac{e^{iqr}-e^{-iqr}}{\left(-q^2+k^2\right)}\right]q\,dq. \tag{10.36}$$

We can evaluate this by contour integration. The integrand has poles at $q=\pm k$, and we must choose the contours differently for the two terms. In order for the integrals to converge, we must choose the contour for the first term in the upper half q plane and the contour for the second term in the lower half q plane. We choose the contours such that the result contains only outgoing spherical waves e^{ikr} and not incoming spherical waves e^{-ikr}, which are not applicable for scattering problems. Thus the contour for the first term encloses only the pole at $q=k$, while the contour for the second term encloses only the pole at $q=-k$. Then we find

$$G\left(\vec{r}\right)=\frac{-1}{4\pi}\frac{e^{ikr}}{r}, \tag{10.37}$$

and the first-order solution to our problem is

$$\psi_1=\int d^3r'\frac{1}{4\pi}\frac{e^{ik|\vec{r}-\vec{r'}|}}{|\vec{r}-\vec{r'}|}U(\vec{r'})\frac{e^{i\vec{k_o}\cdot\vec{r'}}}{(2\pi)^{3/2}}, \tag{10.38}$$

where the incident plane wave has the wave vector $\vec{k_o}$. Since the distance r to the point where the scattered particle is observed is much larger than the values of r' over which $U(\vec{r'})$ is nonnegligible, we can write $|\vec{r}-\vec{r'}|\approx r-(\hat{r}\cdot\vec{r'})$ and

$$\psi_1 = \frac{1}{2(2\pi)^{5/2}} \frac{e^{ikr}}{r} \int d^3r' e^{-ik\left(\hat{r}\cdot\vec{r'}\right)} U\left(\vec{r'}\right) e^{i\vec{k_o}\cdot\vec{r'}} \tag{10.39}$$

$$= \frac{1}{2(2\pi)^{5/2}} \frac{e^{ikr}}{r} \int d^3r' U\left(\vec{r'}\right) e^{i\vec{Q}\cdot\vec{r'}} \tag{10.40}$$

$$= \frac{1}{(2\pi)^{3/2}} \frac{e^{ikr}}{r} \Phi\left(\vec{Q}\right) \tag{10.41}$$

$$\Phi\left(\vec{Q}\right) = \frac{1}{4\pi} \int d^3r' U\left(\vec{r'}\right) e^{i\vec{Q}\cdot\vec{r'}} \tag{10.42}$$

where $\vec{Q} = \vec{k_o} - \vec{k}$ and $\vec{k} = k\hat{r}$ points in the direction of observation $\hat{r}$. Since energy conservation requires that $|\vec{k_o}| = |\vec{k}|$, the momentum transfer is $\hbar|\vec{Q}| = 2\hbar k \sin(\theta/2)$, where θ is the scattering angle.

The wave function for the system, correct to first order, is

$$\psi = \psi_o + \psi_1 = \frac{1}{(2\pi)^{3/2}} \left[e^{i\vec{k_o}\cdot\vec{r}} + \frac{e^{ikr}}{r} \Phi\left(\vec{Q}\right) \right]. \tag{10.43}$$

The first-order change in the wave function is a spherical wave $\left(\frac{e^{ikr}}{r}\right)$ with an amplitude given by the Fourier transform of the scattering potential evaluated at the momentum transfer divided by $\hbar$. The radial current density at large r is given by

$$\vec{j}\left(\vec{r}\right) = \frac{\hbar}{2mi} \left(\psi^* \vec{\nabla}_r \psi - \psi \vec{\nabla}_r \psi^*\right) \tag{10.44}$$

$$= \frac{1}{(2\pi)^3} \frac{\hbar}{m} \frac{\vec{k}}{r^2} \left|\Phi\left(\vec{Q}\right)\right|^2 \tag{10.45}$$

and the flux through an area $r^2 d\Omega$ corresponding to a solid angle $d\Omega$ is $\frac{1}{(2\pi)^3} \frac{\hbar}{m} k \left|\Phi\left(\vec{Q}\right)\right|^2 d\Omega$. Dividing by the incident flux, $\frac{1}{(2\pi)^3} \frac{\hbar}{m} k$, we find the cross section for scattering into $d\Omega$:

$$d\sigma(\theta) = \left|\Phi\left(\vec{Q}\right)\right|^2 d\Omega. \tag{10.46}$$

Note that in applying Eq. (10.44) we have only considered the second term in (10.43). The first term would yield the incident current and the cross terms would imply an interference between the scattered and incident waves, which cannot occur in practice because the incident beam is always collimated to a size small compared to the distance to the detector.

The plane wave is an idealization. To confine the beam to a small region, we construct a wave packet consisting of small values of $\vec{k}$ transverse to the beam. For a macroscopic beam width, this range of $\vec{k}$ values is very small, much smaller than the resolution of the apparatus.

10.3.2 Inelastic scattering of a particle by a fixed atom

Born now turned to the problem of a scattering event where the energy state of the atom is changed as a result of the collision with the incident particle. To describe this situation, we need a wave function that describes the state of the atom. We take $\psi_n(q)$ to be the nth eigenfunction of the Hamiltonian for the atom $H_a(q)$, where q represents all the coordinates of the electrons in the atom, and we have

$$H_a(q)\,\psi_n(q) = E_n\psi_n(q)\,. \tag{10.47}$$

When it is located at large distances from the atom, the particle is described by the free particle eigenstates $\psi_k(\vec{r}) = e^{i\vec{k}\cdot\vec{r}}/(2\pi)^{3/2}$. In the absence of an interaction, the eigenstates of the whole system are given by $\psi_{n\vec{k}} = \psi_n(q)\,\psi_{\vec{k}}(\vec{r})$. The time-independent Schrödinger equation for the problem is

$$\left[H_a(q) + H_e(\vec{r}) + \lambda V(q,\vec{r})\right]\psi(q,\vec{r}) = W\psi(q,\vec{r}) \tag{10.48}$$

where $V(q,\vec{r})$ is the interaction potential between the atomic nucleus and electrons and the scattering particle and $H_e(\vec{r}) = -\frac{\hbar^2}{2m}\nabla_r^2$ is the Hamiltonian for the free particle. We introduced the parameter λ to keep track of the order of the calculation. Working to first order in λ, we write $\psi(q,\vec{r}) = \psi_o(q,\vec{r}) + \lambda\psi_1(q,\vec{r})$ where $\psi_o(q,\vec{r}) = \psi_n(q)\,\psi_{k_o}(\vec{r})$ is the initial state of the system (i.e., a particle incident from the $-z$ direction and the atom in the eigenstate ψ_n). The zero order energy is $W_o = E_n + \frac{\hbar^2 k_o^2}{2m}$. Then the zero order equation is satisfied identically, and the first order terms are

$$\left[H_a(q) - \frac{\hbar^2}{2m}\nabla_r^2 - \left(E_n + \frac{\hbar^2 k_o^2}{2m}\right)\right]\psi_1 = -V\psi_o\,. \tag{10.49}$$

Now we expand $\psi_{1(nk)}(q,\vec{r})$ and $V\psi_o$ in the unperturbed atomic eigenstates:

$$\psi_{1(nk)}(q,\vec{r}) = \sum_m c_{nm}(\vec{r})\,\psi_m(q) \tag{10.50}$$

$$V\psi_o = V(q,\vec{r})\,\psi_n(q)\,\psi_{k_o}(\vec{r}) = \sum_m V_{mn}(\vec{r})\,\psi_m(q)\,\psi_{k_o}(\vec{r})\,, \tag{10.51}$$

where $V_{mn}(\vec{r}) = \int \psi_m^*(q)\,V(q,\vec{r})\,\psi_n(q)\,dq$. We find

$$\sum_m \left\{c_{nm}(\vec{r})\left((E_n - E_m)\frac{2m}{\hbar^2} + k_o^2\right) + \nabla_r^2 c_{nm}(\vec{r}) - \frac{2m}{\hbar^2}V_{mn}(\vec{r})\,\psi_{k_o}(\vec{r})\right\} = 0. \tag{10.52}$$

Since this holds for all values of $\vec{r}$, we must have

$$\nabla_r^2 c_{nm}(\vec{r}) + k_{nm}^2 c_{nm}(\vec{r}) = \frac{2m}{\hbar^2}V_{mn}(\vec{r})\,\psi_{k_o}(\vec{r}) \tag{10.53}$$

with

$$k_{nm}^2 = (E_n - E_m)\frac{2m}{\hbar^2} + k_o^2\,. \tag{10.54}$$

We know the Green's function for this equation is given by (10.37), so we can write down the solution:

$$c_{nm}\left(\vec{r}\right) = \int d^3r' G\left(\vec{r} - \vec{r'}\right)\left(\frac{2m}{\hbar^2}\right) V_{mn}\left(\vec{r'}\right) \psi_{k_o}\left(\vec{r'}\right) \tag{10.55}$$

$$= \int d^3r' \frac{1}{4\pi} \frac{e^{ik_{nm}\left|\vec{r} - \vec{r'}\right|}}{\left|\vec{r} - \vec{r'}\right|} \left[\frac{2m}{\hbar^2} V_{mn}\left(\vec{r'}\right)\right] e^{i\vec{k_o}\cdot\vec{r'}} \tag{10.56}$$

$$= \frac{e^{ik_{nm}r}}{4\pi r} \int d^3r' \left[\frac{2m}{\hbar^2} V_{mn}\left(\vec{r'}\right)\right] e^{i\vec{Q}\cdot\vec{r'}} \tag{10.57}$$

where $\vec{Q} = \vec{k_o} - \vec{k}_{nm}$ and the direction of $\vec{k}_{nm}$ is given by the direction of observation $\hat{r}$. As in the previous section, we have made use of the limit $r \gg r'$, and we have also dropped the factor $1/(2\pi)^{3/2}$ since this has no effect on the cross section. The solution for the first-order change in the wave function is then

$$\psi_{1(nk)}\left(q, \vec{r}\right) = \sum_m c_{nm}\left(\vec{r}\right) \psi_m(q) = \sum_m \psi_m(q) \frac{e^{ik_{nm}r}}{r} \Phi_{nm}\left(\vec{Q}\right) \tag{10.58}$$

$$\Phi_{nm}\left(\vec{Q}\right) = \frac{1}{4\pi} \int d^3r' \left[\frac{2m}{\hbar^2} V_{mn}\left(\vec{r'}\right)\right] e^{i\vec{Q}\cdot\vec{r'}}. \tag{10.59}$$

This is a sum over the eigenfunctions of the atom, each multiplied by an outgoing spherical wave with a wave vector corresponding to the conservation of energy, as expressed in Eq. (10.54). As in the previous section, the cross section for each transition $(\psi_n \to \psi_m)$ will be given by $\frac{d\sigma}{d\Omega} = \left|\Phi_{nm}\left(\vec{Q}\right)\right|^2$.

In this way Born laid the basis for scattering theory in quantum mechanics. These results are very general and applicable to the scattering of all types of waves. With a proper choice of momentum variables, the results are also applicable to relativistic particle collisions. The relation between the scattering amplitude and the Fourier transform of the interaction potential is universal for the scattering of any wave.

10.3.3 Born's interpretation of the wave function

Born pointed out that, if we want to understand the results of the last section in a particle picture, there is only one interpretation possible: $\Phi_{nm}\left(\vec{Q}\right)$ (in a footnote he amended this, due to "more exact considerations," to $\left|\Phi_{nm}\left(\vec{Q}\right)\right|^2$) "determines the probability that a particle coming from the $-z$ direction will be scattered into the direction given by $\vec{k}_{nm}$ with an energy change of the atom given by $(E_n - E_m)$." So we have an exact answer but no causal relation. This raises the problem of determinism. In quantum mechanics there is no quantity that determines the effect of a collision in single cases. Experimentally we also have no indication that there is an inner property of the atom that determines a definite outcome. "Should we hope to discover such properties in the future and determine their values in single events, or should we believe that the agreement between theory and experiment in their inability to give conditions for a causal process represents a harmony which rests on the nonexistence of such quantities?" Born himself "leaned to giving up determinism in the atomic world." We see that the question of the possible existence of what we now call "hidden variables" was raised at the very introduction of indeterminism into physics.

In the second paper cited above, Born characterized Heisenberg's matrix mechanics as starting from the idea that an exact description of processes in space and time is impossible. The theory is satisfied with relations between observable quantities. Schrödinger, on the other hand, saw the de Broglie waves as being as real as light waves and tried to build up wave packets which should directly represent the moving particle. Born went on to reject both viewpoints and offer a third viewpoint, the probability interpretation that has proved its usefulness in the collision problems discussed above.

He mentioned (without citation) a remark by Einstein to the effect that the waves are there only to show the way for the light quanta. He called them "ghost waves" that determine the probability that the light quanta, the carriers of energy and momentum, follow a given path. The field itself carries no energy or momentum. This is a remarkable statement given Einstein's life-long objections to quantum theory, which we will discuss later.

Born applied this idea to the electron, in which case the de Broglie-Schrödinger waves are seen as the "ghost wavess" or, better, "guiding waves." The basic idea is that $\psi(q,t)$ spreads according to the Schrödinger equation and momentum and energy are transferred as if particles are really flying around. The trajectories are only determined insofar as required by energy-momentum conservation. The probability of a given, definite trajectory is determined by $|\psi|^2$. The particle motion proceeds according to probabilistic laws while the probability spreads according to causal laws, which means that knowledge of the conditions at all points at a given time fixes the distribution of the state at all later times.

This wave-particle duality, along with the dichotomy between causal and probabilistic laws, is one of the features (not the only one) of quantum mechanics that is at the root of the difficulties of interpretation that persist, to some extent, to the present time.

10.4 Heisenberg's uncertainty principle

Possibly as a reaction to Schrödinger's comment that matrix mechanics was repulsively nonintuitive, Heisenberg published a paper with the title "On the intuitive content of the quantum theoretical kinematics and mechanics."[6] In that paper Heisenberg gave a detailed description of his intuitive view of quantum mechanics as it had recently been generalized by Dirac's transformation theory, which we discuss in Chapter 11. Knowledge of this is not necessary for Heisenberg's arguments. He quoted Schrödinger's remark, and stated his opinion that "the popular intuitiveness of wave mechanics has led away from the straight path that had been laid out through the work of Einstein and de Broglie on the one hand and the work of Bohr and the quantum mechanics on the other hand."

[6]W. Heisenberg, *Über den anschaulichen Inhalt der quantentheoretischen Kinematik und Mechanik*, *Z. Phys.* **43**, 172-98 (1927).

The quantum condition $[p, q] = \hbar/i$ means that we have to be very suspicious of applying the concepts of location and velocity. For example, if we wanted to measure the position of an electron at a given time, we might try to shine some light on it and look for the scattered photons. The accuracy of the measurement will be limited by the wavelength of the light used. During the interaction of the electron with a photon, we cannot avoid the transfer of a certain momentum to the electron, as in the Compton effect. If we try to increase the accuracy of the position measurement by decreasing the wavelength of the light, we cannot avoid increasing the momentum of the photons $(p = h/\lambda)$.

We could measure the electron's position by looking at it under a microscope using light with wavenumber k. In a microscope subtending an angle θ at the position of the electron, the diffraction-limited resolution results in a minimum uncertainty Δx in the electron's position, which is given by

$$k\Delta x \sin\theta \sim \pi \tag{10.60}$$

$$\Delta x \sim \frac{\pi}{k\sin\theta} = \frac{\lambda}{2\sin\theta}. \tag{10.61}$$

A photon scattered into the microscope will be distributed somewhere within the cone fixed by the angle θ, so that the uncertainty in the momentum of the photon, and hence of the electron, will be

$$\Delta p \sim p_\gamma \sin\theta \tag{10.62}$$

where $p_\gamma = h/\lambda$ is the momentum of the photon. Thus

$$\Delta p \Delta x \sim \frac{1}{2}h. \tag{10.63}$$

The same argument can be presented in many other ways. Consider a parallel beam of electrons, with wave number k going through a slit of width Δx. After passing through the slit, the beam will be spread by diffraction over an angular range given by

$$k\Delta x \sin\theta = \pi. \tag{10.64}$$

After the slit, the transverse momentum of the beam will be between $\pm p\sin\theta$ so that $\Delta p \sim 2p\sin\theta$ with $p = \hbar k$ the momentum of the particles in the beam. Thus

$$\Delta p \Delta x \sim 2\hbar k \frac{\pi}{k} \sim h. \tag{10.65}$$

We see that the wave-particle duality is always involved. In fact, this relation—called the Heisenberg uncertainty principle—follows directly from the properties of the Fourier transform and the de Broglie relation between momentum and wavelength. If we want to build up a wave packet consisting of the sum of a number of plane waves (e^{ikx}) that is only nonzero for a restricted region of space, we can write (we work in one dimension for simplicity)

$$\psi(x) = \int A(k)\, e^{ikx} dk. \tag{10.66}$$

Thus $\psi(x)$ is the Fourier transform of $A(k)$ and we can write the inverse transform as

$$A(k) = \frac{1}{2\pi}\int \psi(x)\, e^{-ikx} dx\,. \tag{10.67}$$

Taking $\psi(x)$ to be constant in the region $-\Delta x/2 < x < \Delta x/2$ and zero elsewhere, we have

$$A(k) = \frac{1}{2\pi}\int_{-\Delta x/2}^{\Delta x/2} e^{-ikx} dx = \frac{\Delta x}{2\pi}\frac{\sin\frac{k\Delta x}{2}}{k\Delta x/2}\,, \tag{10.68}$$

which has its first zero at $k\Delta x = 2\pi$ so that $\Delta k \sim 4\pi/\Delta x$. This last relation was well known from the theory of optical instruments, especially Rayleigh's discussion of the resolution of spectrometers. We write $\Delta k \sim 4\pi/\Delta x = \Delta p/\hbar$. Thus

$$\Delta p \Delta x \sim 2h. \tag{10.69}$$

A more rigorous argument starts by defining $(\Delta x)^2 = \left\langle (x - \langle x\rangle)^2 \right\rangle$ and $(\Delta p)^2 = \left\langle (p - \langle p\rangle)^2 \right\rangle$, where $\langle x\rangle$ denotes the average value of x. Since, according to Born, the probability of being in a region between x and $x + dx$ is given by $|\psi(x)|^2\, dx$, we have

$$\langle \Delta x^2 \rangle = \int \psi^*(x)\,(\Delta x)^2\,\psi(x)\,dx = \int |\Delta x\,\psi(x)|^2\,dx. \tag{10.70}$$

Similarly,

$$\langle \Delta p^2 \rangle = \int |\Delta p\,\psi(x)|^2\,dx, \tag{10.71}$$

so that

$$\langle \Delta p^2 \rangle \langle \Delta x^2 \rangle = \int |\Delta x\,\psi(x)|^2\,dx \int |\Delta p\,\psi(x)|^2\,dx. \tag{10.72}$$

According to the Schwarz inequality,

$$\int |f|^2\,dx \int |g|^2\,dx \geq \left| \int f^* g\,dx \right|^2, \tag{10.73}$$

where the equality holds only if $f = \gamma g$ for some constant γ. Then we have

$$\langle \Delta p^2 \rangle \langle \Delta x^2 \rangle \geq \left| \int (\Delta x\,\psi)^* (\Delta p\,\psi)\,dx \right|^2 = \left| \int \psi^*\,(\Delta x)\,(\Delta p)\,\psi dx \right|^2 \tag{10.74}$$

$$= |\langle \Delta x \Delta p\rangle|^2. \tag{10.75}$$

Now

$$\Delta x \Delta p = \frac{1}{2}[\Delta x, \Delta p] + \frac{1}{2}(\Delta x \Delta p + \Delta p \Delta x) \tag{10.76}$$

$$= \frac{i\hbar}{2} + \frac{1}{2}(\Delta x \Delta p + \Delta p \Delta x)\,. \tag{10.77}$$

Since $(AB)^\dagger = BA$ for two Hermitian operators A and B, the second term on the right is Hermitian, and its average is real. Since the first term is imaginary, we can write

$$\left\langle \Delta p^2 \right\rangle \left\langle \Delta x^2 \right\rangle \geq \left| \left\langle \Delta x \Delta p \right\rangle \right|^2 = \frac{\hbar^2}{4} + \left| \frac{\left\langle \Delta x \Delta p + \Delta p \Delta x \right\rangle}{2} \right|^2 \tag{10.78}$$

$$\left\langle \Delta p^2 \right\rangle \left\langle \Delta x^2 \right\rangle \geq \frac{\hbar^2}{4}. \tag{10.79}$$

This is a rigorous statement of the uncertainty principle.

10.4.1 The minimum uncertainty wave packet

It is interesting to see under what conditions the equality in (10.79) holds. For the Schwarz inequality to be an equality, we require

$$\Delta x \, \psi = \gamma \Delta p \, \psi. \tag{10.80}$$

To minimize $\left\langle \Delta p^2 \right\rangle \left\langle \Delta x^2 \right\rangle$, we also need

$$0 = \int \psi^* \left(\Delta x \Delta p + \Delta p \Delta x \right) \psi \, dx \tag{10.81}$$

$$= \int \left[\left(\Delta x \psi \right)^* \Delta p \, \psi + \psi^* \Delta p \Delta x \psi \right] dx \tag{10.82}$$

$$= \int \left[\left(\gamma \Delta p \, \psi \right)^* \Delta p \, \psi + \psi^* \left(\Delta p \right)^2 \gamma \psi \right] dx \tag{10.83}$$

$$= \left(\gamma^* + \gamma \right) \int \psi^* \left(\Delta p \right)^2 \psi dx, \tag{10.84}$$

so that we must have $\gamma = \pm i\beta$. Then we can write Eq. (10.80) as

$$\left(x - \langle x \rangle \right) \psi = -i\beta \left(p - \langle p \rangle \right) \psi \tag{10.85}$$

$$= -i\beta \left(\frac{\hbar}{i} \frac{\partial \psi}{\partial x} - \langle p \rangle \psi \right), \tag{10.86}$$

which yields

$$\frac{1}{\psi} \frac{\partial \psi}{\partial x} = -\frac{\left(x - \langle x \rangle - i\beta \langle p \rangle \right)}{\beta \hbar} = \frac{d \ln \psi}{dx} \tag{10.87}$$

$$\ln \psi = -\frac{\left(x - \langle x \rangle - i\beta \langle p \rangle \right)^2}{2\beta\hbar} \tag{10.88}$$

$$\psi = \exp \left[-\frac{\left(x - \langle x \rangle - i\beta \langle p \rangle \right)^2}{2\beta\hbar} \right] \tag{10.89}$$

$$= \exp \left[-\frac{\left(x - \langle x \rangle \right)^2}{2\beta\hbar} + \frac{\beta \langle p \rangle^2}{2\hbar} + i \frac{\langle p \rangle}{\hbar} \left(x - \langle x \rangle \right) \right]. \tag{10.90}$$

The constant terms will be removed by normalizing this function, so the minimum uncertainty wave packet is given by

$$\psi(x) = \left(\frac{1}{2\pi\sigma^2}\right)^{1/4} \exp\left[-\frac{(x - \langle x\rangle)^2}{4\sigma^2} + i\frac{\langle p\rangle}{\hbar}x\right]. \tag{10.91}$$

That is a plane wave with wave vector corresponding to the average momentum modulated by a Gaussian function. Note this has a minimum uncertainty product regardless of the value of σ. If σ increases, the momentum distribution, which is given by the Fourier transform of this wave packet and is itself a Gaussian, will narrow.

10.4.2 Spreading of the minimum uncertainty wave packet

Taking this minimum uncertainty packet as the wave function at a given time, say $t = 0$, we will find that the wave function spreads with time. We begin the calculation by writing the general solution for the free particle Schrödinger equation:

$$\psi(x,t) = \int A(k)\, e^{i(kx - E_k t/\hbar)} dk \tag{10.92}$$

with $E_k = (\hbar k)^2/2m$ and

$$A(k) = \frac{1}{2\pi}\int \psi(x,0)\, e^{-ikx} dx, \tag{10.93}$$

so that

$$\psi(x,t) = \frac{1}{2\pi}\int dx'\psi(x',0)\int e^{ik(x-x') - i\alpha k^2 t} dk \tag{10.94}$$

with $\alpha = \hbar/2m$. The integral over k is the Fourier transform of the free particle Green's function $G(x - x', t)$. Working out the integral by contour integration, we find

$$G(x - x', t) = \left(\frac{-im}{2\pi\hbar t}\right)^{1/2} \exp\left[i(x - x')^2 \frac{m}{2\hbar t}\right], \tag{10.95}$$

which is a spreading function of time. This indicates that any free particle wave packet will spread.

It is instructive to carry out the calculation for the minimum uncertainty packet in detail. Taking $\langle x\rangle = \langle p\rangle = 0$ for simplicity and substituting Eq. (10.91) into Eq. (10.94), we have

$$\psi(x,t) = \left(\frac{1}{2\pi\sigma^2}\right)^{1/4} \frac{1}{2\pi}\int e^{ikx - i\alpha k^2 t} dk \int dx' \exp\left[-\frac{x'^2}{4\sigma^2} - ikx'\right] \tag{10.96}$$

$$= \left(\frac{\sigma^2}{2\pi^3}\right)^{1/4} \int e^{ikx - i\alpha k^2 t} e^{-k^2\sigma^2} dk \tag{10.97}$$

$$= \left(\frac{1}{2\pi}\right)^{1/4} \left(\sigma + i\frac{\alpha t}{\sigma}\right)^{-1/2} \exp\left[-\frac{x^2}{4\sigma^2 + 4i\alpha t}\right], \tag{10.98}$$

which yields

$$|\psi(x,t)|^2 = \left[2\pi\left(\sigma^2 + \frac{\hbar^2 t^2}{4m^2\sigma^2}\right)\right]^{-1/2} \exp\left[-\frac{x^2}{2\left(\sigma^2 + \frac{\hbar^2 t^2}{4m^2\sigma^2}\right)}\right] \tag{10.99}$$

for the probability density. The wave packet maintains its initial Gaussian form, but the width increases with time. If we note that $\hbar/2\sigma = \Delta p$ we can write the width as $\sigma_{eff}^2(t) = \sigma^2 + (\Delta p)^2 t^2/m^2$. For large times, the influence of the initial width σ becomes negligible and the excess width is $\sigma = \Delta vt$, exactly what we would have for a classical particle starting with an uncertain velocity. Thus classical mechanics with the initial conditions given by a probability distribution would behave probabilistically in a way similar to quantum mechanics. At any time, there would be a probability distribution of position and velocity, and making a measurement would collapse the probability. However, in the classical system, measurements could be made without disturbing the system and increasing the uncertainty of the variable canonical to the measured one.

10.4.3 Heisenberg's interpretation of quantum mechanics as presented in his 1927 paper "On the intuitive content of the quantum theoretical kinematics and mechanics"

Heisenberg's paper,[7] which we discussed above, begins by pointing out that quantum mechanics began with the attempt to break with the usual kinematical variables and find relations between concrete, experimentally given numbers. In the case of the electron in a hydrogen atom, the concept of trajectory has no meaning. To measure the trajectory, we would have to use light with a wavelength less than 1 Å (10^{-8} cm) in order to localize the electron to sufficient precision. This corresponds to a photon energy greater than 10 keV, so that a single photon would be able to knock an electron completely out of its orbit. Thus we can only hope to measure a single location in the trajectory, so that the term "trajectory" has no meaning. Repeated measurements of position would yield the probability $|\psi_n(\vec{r})|^2$ of a given location when the atom is in the state n. This corresponds to an average of the classical trajectories over all phases of the electron in the orbit.

Because of the commutation relations, canonically conjugate variables satisfy an uncertainty principle and can only be simultaneously determined to a certain accuracy. This inexactness is, according to Heisenberg, the real reason for the appearance of statistical relations in quantum mechanics. However, Dirac felt that the statistics was introduced through the act of experimental measurement, a view shared by Bohr. In classical mechanics, there is also a probability connected with the unknown phases. Although classically we can think of determining the phases, in practice this is impossible without destroying the atom.

One method of determining the energy state of an atom is by means of a Stern-Gerlach experiment. In these experiments, beams of neutral atoms are sent

[7] Heisenberg, W., *On the intuitive content of the quantum theoretical kinematics and mechanics*, Zeit. Phys. 43, 172, 1927, received March 23, 1927, English translation: in Wheeler, J.A. and Zurek, W.H., *Quantum Theory and Measurement*, Princeton, 1983, p.62.

through inhomogeneous (electric or magnetic) fields. Atoms passing through an inhomogeneous magnetic field will be subject to a force $\vec{\nabla}\left(\vec{\mu}\cdot\vec{B}\right)$ transverse to the beam direction and will thus be deflected by an amount proportional to $\vec{\mu}\cdot\vec{B}$ ($\vec{\mu}$=magnetic moment of the particle). In cases where $\vec{\mu}\cdot\vec{B}$ differs with energy state, the different energy states will be separated into different beams. For a beam of width d, the force on the particle is given by the gradient of the potential energy $\Delta V/d$, where ΔV is the change in potential energy across the beam width d. The deflection will then be

$$\Delta\theta = \frac{\Delta V}{d}\frac{T}{p}, \tag{10.100}$$

where T is the time that the force acts on the beam and p is the momentum of the beam. Since a beam going through a slit of width d will be diffracted through an angle $\Delta\theta \geq \lambda/d$ (the de Broglie wavelength is $\lambda = h/p$), we find

$$\frac{\Delta V}{d}\frac{T}{p} \geq \Delta\theta \geq \frac{h}{pd} \tag{10.101}$$

$$(\Delta V)\,T \geq h. \tag{10.102}$$

In order to measure the energy of the state in question, we must have $\Delta V \ll \Delta E$, the desired accuracy of the measurement of the energy of the state. Thus $(\Delta E)\,T \geq h$, and we see how a given accuracy of an energy measurement can only be achieved when a corresponding time interval is available for the measurement. If we were to consider the action of the field as occurring suddenly (quantum jump) then T would be the uncertainty in the time of that jump. This equation of T with an uncertainty only applies when we are analyzing processes that can be considered as quantum jumps. A more rigorous proof proceeds by use of the Fourier transform and the relation $E = \hbar\omega$, just as was done for the momentum–position uncertainty relation.

Another experiment starts by sending a beam of atoms in a known energy state through a field with a very strong gradient in the direction of the beam so that the atoms see a very quickly varying external field which contains a wide range of frequencies capable of inducing transitions between the initial state and a large number of other energy states. The initial state can be written as $\psi(1) = \psi_n e^{-iE_n(t+\beta_n)/\hbar}$ with β_n a random phase. After passing through the strong gradient (F_1), the state will be given by

$$\psi(2) = \sum_m c_{nm}\psi_m e^{-iE_m(t+\beta_m)/\hbar} \tag{10.103}$$

If now we performed a transverse Stern-Gerlach experiment to determine the state of the system, we would find that the probability of finding the state m is given by $|c_{nm}|^2$. For further investigation, we take $\psi_m e^{-iE_m(t+\beta_m)/\hbar}$ as the initial state. Having determined the state to be m, if we then pass the beam through another very high gradient field (F_2), separated by the distance L from the first, the state at the output would be

$$\psi(3) = \sum_l d_{ml}\psi_l e^{-iE_l(t+\beta_l)/\hbar}. \tag{10.104}$$

If we had not made any measurement on the state $\psi(2)$, we would have

$$\psi(1) = \psi_n e^{-iE_n(t+\beta_n)/\hbar} \underset{F_1}{\longrightarrow} \psi(2) \underset{F_2}{\longrightarrow} \sum_{m,l} c_{nm} d_{ml} \psi_l e^{-iE_l(t+\beta_l)/\hbar}$$
$$= \sum_l e_{nl} \psi_l e^{-iE_l(t+\beta_l)/\hbar} \tag{10.105}$$

with

$$e_{nl} = \sum_m c_{nm} d_{ml} \, . \tag{10.106}$$

Measurement of the state behind the field F_2 will give the state l with the probability

$$|e_{nl}|^2 = \sum_{m,k} (c_{nm} d_{ml}) (c^*_{nk} d^*_{kl}). \tag{10.107}$$

If we repeat the experiment with the Stern-Gerlach measurement in the region L many times, we will find the state l after the field F_2 with the probability

$$Z_{nl} = \sum_m |c_{nm}|^2 |d_{ml}|^2 \neq |e_{nl}|^2. \tag{10.108}$$

The intermediate measurement alters the state by changing the phase of the atoms in a random way so that only the terms with $m = k$ contribute to the sum in Eq. (10.107), and we obtain Z_{nl} in this case.

All this is only possible because of the linearity of the Schrödinger equation, which means that the sum of solutions is also a solution. It is because of this property that the Schrödinger equation can be understood as an equation for waves in phase space. Thus, Heisenberg considered every attempt to replace this equation with a non-linear equation (e.g., in the relativistic many-electron problem) to be hopeless.

A hydrogen atom in a highly excited state, say $n = 10^3$, is large enough for the position of its electron to be measured with sufficient accuracy using long wavelength light. After the measurement, a range of states, say between $n = 950$ and $n = 1050$ will be excited. These will form a wave packet with dimensions comparable to the wavelength of light used, which as we have seen will spread with time. A further measurement of the same type will again yield a wave packet of the same size but with its position randomly chosen from the spreading wave packet produced by the previous measurement. The process can be repeated. Every position measurement reduces the wave packet to its original size given by the wavelength of the light used in the measurement.

Heisenberg concluded his paper with the remark that causal laws hold in quantum mechanics as long as all quantities can be exactly determined. In the statement, "When we know the present state exactly we can calculate the future," the assumption is false. We cannot measure exactly all the variables determining the present state, so any observation entails a choice, fixing the values of some variables while increasing the uncertainty in others, which sets a limit on our ability to predict the future evolution of the system.

Since the statistical character of quantum mechanics is so closely related to the uncertainty principle, could we think that behind the statistically experienced world there is a hidden, real world in which causal laws are valid? This seemed to Heisenberg to be unfruitful and meaningless. Physics should only formally describe relations between observable quantities. The true situation can be better characterized as follows: because all experiments obey the laws of quantum mechanics and therefore the uncertainty principle, the invalidity of the causal laws is definitively established by quantum mechanics.

In a note added in proof, Heisenberg explained Bohr's view that the origin of the uncertainties is in the need to simultaneously account for the somewhat contradictory effects associated with the wave and particle theories, in other words from the wave-particle duality, as we will see in the next section.

10.5 Niels Bohr and complementarity: the Copenhagen interpretation of quantum mechanics

During 1926 Heisenberg was a lecturer at Bohr's institute in Copenhagen. Schrödinger visited the institute in October to lecture on his wave mechanics. Schrödinger had previously given a lecture at Munich which Heisenberg had attended. At that lecture Heisenberg tried to raise some objections to Schrödinger's ideas, but he found little support in the audience. Heisenberg and Bohr had been having intense discussions for some months about the physical meaning of the new theory, and the discussions continued after Schrödinger left. Heisenberg worked out the results on the uncertainty principle that were discussed in the last section while Bohr was away on a skiing vacation. When Bohr returned, he and Heisenberg disagreed on how best to present these results. Heisenberg wanted to base the discussion on the transformation theory of Jordan and Dirac, while Bohr wanted to give more emphasis to the wave mechanical picture. In fact, the transformation function used by Heisenberg, which specified the transformation between matrix elements based on energy eigenstates and those based on position eigenstates, was the Schrödinger wave function ψ, and so the differences between Heisenberg's and Bohr's positions were mostly superficial. We discuss the transformation theory in Chapter 11. On his return from vacation, Bohr started to put his thoughts on the relationship between the competing interpretations of quantum mechanics into writing, and he presented the results at a conference held in honor of Volta at Lake Como, Italy, in September 1927.

Bohr[8] expressed the hope that his work would bring about an agreement between the "seemingly contradictory views of different physicists." In this we see that, even at the beginning, quantum mechanics raised difficulties of interpretation. We will see that there are physicists who continue to hold contradictory views even at the present time. Taking Heisenberg's uncertainty relations as a basis, Bohr developed a physical understanding of their meaning. According to classical mechanics, measurements can be carried out without disturbing the system being measured, but in reality, due to the quantization of action, the unavoidable interaction between the system and the

[8]Bohr, N., *The quantum postulate and the recent development of atomic theory, Atti del Congresso Internazionale del Fisici 11-20 Septembere 1927, Como-Pavia-Roma*, Nicola Zanichelli, Vol.2, 565, also Nature 121, 580, 1928

measuring instrument introduces a random (Bohr called it "irrational") disturbance of the system. Bohr framed this in terms of the "quantum postulate," "which attributes to any atomic process an essential discontinuity or rather individuality, completely foreign to the classical theories and symbolized by Planck's quantum of action." Otherwise said—transitions occur instantly at random times. In classical physics, a measurement does not influence the system being measured, but in atomic physics, the unavoidable interaction between the instrument and the system means that neither the phenomenon nor the instrument can be ascribed an independent reality. Bohr was aware that the cut between what we consider the system and what we consider the measuring instrument was arbitrary, and this feature continues to play a major role in contemporary discussions of interpretations of quantum mechanics as we shall see.

Bohr pointed out that there are two mutually exclusive ways of describing observations. For example, processes involving light can be described in terms of propagation in space-time (i.e., the wave description) or in terms of what he calls causality, or the conservation of energy-momentum (i.e., the photon picture). These concepts cannot both be applied independently to the same problem. The two viewpoints on the nature of light are two different attempts to reconcile the experimental results to our usual (classical) ways of thinking. The limits on classical concepts are expressed in complementary ways. The two views are not contradictory but complementary ways of viewing events, and only together can they offer a realistic generalization of the classical methods of description.

These concepts apply not only to light but also to material particles. In light of the de Broglie-Schrödinger theory and the recent experiments on electron diffraction from crystal planes (Davisson-Germer), it had become clear that the behavior of electrons is characterized by the same type of wave-particle duality.

The concepts of radiation in empty space or isolated particles are abstractions because their properties can only be determined by interactions with other systems obeying the quantum postulate. Nevertheless, the abstractions are a necessary means of understanding the results of experiments using our usual ideas and concepts. In order to apply the new methods to physical problems without generating contradictions, Heisenberg's reciprocal uncertainties must be applied to every atomic measurement.

The wave picture with a definite frequency and wavelength implies unbounded sinusoidal waves. Only by superposition (interference in a group of elementary sinusoidal waves) can we achieve a limit on the space-time extension. This results in the Fourier transform derivation of the uncertainty relations given above. Instead of a well-defined event at a point in space-time, we can only speak of the collision of inexactly defined individuals (spread in energy and momentum) inside a finite region of space-time.

Bohr then pointed out that any attempt to violate the uncertainty principle by measuring the change in momentum of the microscope in the Heisenberg thought experiment is impossible, as the uncertainty principle has to be applied to the microscope as well, so that if we determined its momentum accurately enough, its position, and hence the position of the observed particle, would increase in uncertainty. This and similar arguments formed the centerpiece of the debate between Bohr and Einstein that took place at the Solvay Conference one month after the Como conference at which Bohr presented this paper. We discuss these arguments in Chapter 13.

To summarize, according to complementarity, both the wave and particle pictures are valid, and whether the wave or particle aspect is observed in any given experiment depends on the nature of the measuring apparatus. Measurements are accompanied by an uncontrollable disturbance of the measured system due to the fact that action is quantized (i.e., there is a minimum amount that can be transferred). In Bohr's view, transitions (energy transfers) occur instantaneously at random times.

10.6 Conflicting views on quantum jumps

One of the fundamental tenets of the original Bohr theory of the atom was that the electrons could find themselves in stationary states, which were visualized as classical orbits. For those states satisfying the Bohr-Sommerfeld quantum conditions, the laws of classical physics were suspended, so that even though the electrons were being accelerated as they circulated around their orbits, they did not radiate.

Radiation occurred only when the electron made a transition to another energy level. These transitions (quantum jumps) were intrinsically unpredictable and instantaneous. Einstein had treated the occurrence of these jumps as random events in his derivation of Planck's law.

Heisenberg and Schrödinger disagreed on several points, including on the nature of these types of transitions. Schrödinger was put off by the nonintuitive aspect of Heisenberg's theory, and Heisenberg countered with an intuitive discussion of the meaning of his theory, which we presented above. In addition, Heisenberg claimed that Schrödinger's theory disagreed with experiment because it could not account for observations like the Compton effect, the photoelectric effect and Planck's radiation law, where quantum jumps seemed to be emphasized. Each of these phenomena was eventually described in wave mechanics.

10.6.1 The Compton effect as a wave phenomenon

As an example, Schrödinger[9] discussed the Compton effect by pointing out the analogy with the known classical effect of the diffraction of light from a sound wave when the density variation caused by the sound wave changes the index of refraction. Following de Broglie, the energy and momentum of the accompanying particles are given by $E = \hbar\omega$ and $\overrightarrow{p} = \hbar\overrightarrow{k}$. In the relativistic case, we have $c\left|\overrightarrow{k}\right| = \sqrt{\omega^2 - \omega_o^2}$ for a plane wave with $\hbar\omega_o = mc^2$, and $\left(\omega, c\overrightarrow{k}\right) = k_\mu$ form a four-vector. For radiation we put $\omega_o = 0$. If ψ is a plane wave, $|\psi|^2$ is constant in space and time. If we take a superposition of two plane waves with four-vectors Q_μ and Q'_μ, $|\psi|^2$ (which Schrödinger interprets as a charge density) will vary as $\exp(iD_\mu x^\mu)$ with the four-vector $D_\mu = Q_\mu - Q'_\mu$. This charge density wave will diffract a colliding light wave just as the sound wave would in the classical case. We can use Lorentz invariance to avoid calculating the diffraction on the moving lattice by going into the frame where the density wave is at rest, $D_\mu = \left(0, \overrightarrow{D}\right)$. In this case the frequency of the light (the time component of Q_μ) does not change and the Bragg condition is given by

[9] Schrödinger, E., *On the Compton effect*, Ann. de Phys. **82**, 257 (1927) received Nov. 30, 1926.

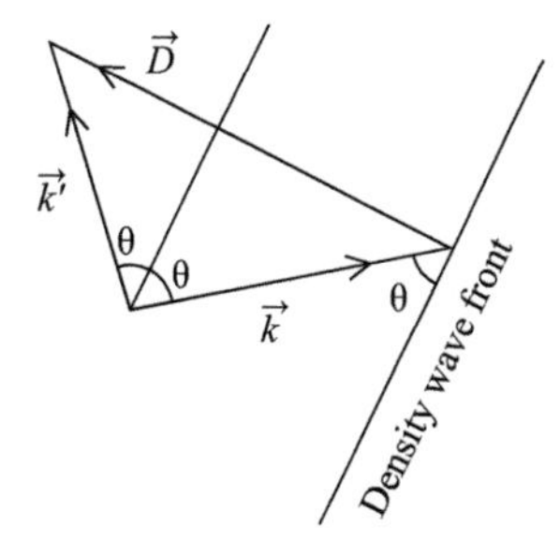

Fig. 10.2 Relations between the $\vec{k}$ vectors of the photon in the Compton effect in the rest frame of the density wave.

$$k_\mu + D_\mu = k'_\mu \tag{10.109}$$

$$k_\mu + Q_\mu = k'_\mu + Q'_\mu, \tag{10.110}$$

where k_μ, k'_μ are the four-vectors of the incident and reflected light, respectively. Fig. 10.2 shows the relationship between the space components of the four-vectors involved. As these are four-vectors, the same relation holds in the lab frame. Multiplying this last expression by $\hbar$ gives the conservation of energy and momentum of the Compton effect, here derived from purely wave considerations. Schrödinger noted that he considered the problem as a steady state situation, as opposed to starting with an initial state containing only two incident waves.

10.6.2 Transitions without quantum jumps

Schrödinger[10] started by reviewing a more concise form of the time-dependent perturbation theory given by Heisenberg and Jordan. Given a potential $V(q)$ with known eigenfunctions and eigenvalues (ψ_k, E_k),

$$-\frac{\hbar^2}{2m}\nabla^2\psi_k + V(q)\,\psi_k = E_k\psi_k\,, \tag{10.111}$$

we seek a solution of the time-dependent problem with an additional time-independent potential $\varepsilon(q)$. This could be treated in time-independent perturbation theory which would yield the new eigenvalues and eigenstates of the potential $(V+\varepsilon)$. But we want to consider a problem where the system starts in eigenstates of the unperturbed system and study how these states, which are not stationary in the presence of the perturbation, evolve with time. This is a problem with many applications. Thus we seek a solution of

$$-\frac{\hbar^2}{2m}\nabla^2\psi + [V(q)+\varepsilon(q)]\,\psi = i\hbar\frac{\partial\psi}{\partial t}\,. \tag{10.112}$$

We start by expanding the wave function in eigenfunctions of the unperturbed problem,

$$\psi = \sum_k c_k\psi_k(q)\,e^{-iE_kt/\hbar}, \tag{10.113}$$

[10]Schrödinger, E., *Energy exchange according to wave mechanics*, Ann. der Physik **83**, 956 (1927) received June 10, 1927.

and find

$$\sum_k c_k \varepsilon(q)\,\psi_k(q)\,e^{-iE_k t/\hbar} = i\hbar \sum_k \dot{c}_k \psi_k(q)\,e^{-iE_k t/\hbar}. \tag{10.114}$$

Multiplying by ψ_m^* and integrating over all q using the fact that the ψ_k are orthonormal, we find

$$i\hbar \dot{c}_m = \sum_k c_k \varepsilon_{mk} e^{i(E_m - E_k)t/\hbar} \tag{10.115}$$

with

$$\varepsilon_{mk} = \int \psi_m^* \varepsilon(q)\,\psi_k dq \tag{10.116}$$

the matrix elements of the perturbation in the basis of the unperturbed states.

This is exact up to this point. Now we consider the case where the frequencies $\omega_{mk} = (E_m - E_k)/\hbar$ are large compared to $\varepsilon_{mk}/\hbar$, so the terms in the sum with $k \neq m$ are oscillating rapidly and will not contribute significantly to the result.

Then the leading term is $\dot{c}_m = \varepsilon_{mm} c_m / i\hbar$, which leads to $c_m = c_m^{(0)} e^{-i\varepsilon_{mm} t/\hbar}$. This means that only the phase of c_m varies and $|c_m|$ is constant in time. In the case of degeneracy, all the terms in Eq. (10.115) that have $E_m = E_k$ will be non-oscillating, and we have to keep all of them. Thus, in the case of α degenerate states,

$$i\hbar \dot{c}_m = \sum_{k=1}^{\alpha} c_k \varepsilon_{mk}\,. \tag{10.117}$$

Multiplying by c_m^* and summing over m yields

$$\sum_{m=1}^{\alpha} c_m^* \dot{c}_m = \frac{1}{i\hbar} \sum_{k=1}^{\alpha} \sum_{m=1}^{\alpha} c_m^* c_k \varepsilon_{mk}. \tag{10.118}$$

Since ε is Hermitian, we have $\varepsilon_{mk}^* = \varepsilon_{km}$. We can now find the time derivative of $\sum_{m=1}^{\alpha} c_m^* c_m$:

$$\frac{d}{dt} \sum_{m=1}^{\alpha} c_m^* c_m = \frac{1}{i\hbar} \sum_{k=1}^{\alpha} \sum_{m=1}^{\alpha} \left(c_m^* c_k \varepsilon_{mk} - c_m c_k^* \varepsilon_{km}\right) = 0 \tag{10.119}$$

$$\sum_{m=1}^{\alpha} c_m^* c_m = \text{const} \tag{10.120}$$

This means that the probability of being in some combination of the degenerate states remains constant, but the occupation probability can be redistributed among these states.

We now consider two different systems described by different coordinates (q_1, q_2), the first with eigenfunctions and eigenvalues $\psi_k(q_1), E_k$ and the second with $\Phi_m(q_2), F_m,$

each satisfying their own time-dependent Schrödinger equation. Imagine the two systems joined into a single system with no coupling between them. The wave equation for the combined system is then

$$-\frac{\hbar^2}{2m}\left(\nabla_1^2+\nabla_2^2\right)\Psi+\left(V\left(q_1\right)+V\left(q_2\right)\right)\Psi=i\hbar\frac{\partial\Psi}{\partial t} \tag{10.121}$$

with eigenfunctions and eigenvalues $\Psi=\psi_k\left(q_1\right)\Phi_m\left(q_2\right),\left(E_k+F_m\right)$. We will now add a small coupling term, $\varepsilon\left(q_1,q_2\right)$, which depends on both sets of coordinates. If there is no degeneracy in the combined system, all the terms in (10.115) will be oscillating and there will be no significant mutual influence between the systems. Now assume, for simplicity, that there is a two-fold degeneracy between the states $\Psi_1=\psi_k\Phi_{l'}$ (red) and $\Psi_2=\psi_{k'}\Phi_l$ (blue) with energies $E_k+F_{l'}=E_{k'}+F_l$ or $E_k-E_{k'}=F_l-F_{l'}$, as shown in Fig. 10.3. Note that if one of the systems has continuous eigenvalues, say a free electron or a photon, there will always be such a degeneracy. Under these conditions the equations (10.117) become

$$i\hbar\dot{c}_1=\varepsilon_{11}c_1+\varepsilon_{12}c_2 \tag{10.122}$$

$$i\hbar\dot{c}_2=\varepsilon_{21}c_1+\varepsilon_{22}c_2 \tag{10.123}$$

If at time $t=0$ we start in the state Ψ_1 so that $c_1\left(0\right)=1,c_2\left(0\right)=0$, the amplitude $c_2\left(t\right)$ will increase with time while $c_1\left(t\right)$ will decrease, as required by Eq. (10.120).

This has wide physical applications. The system ψ_k could be a free electron with Φ_l an atom. Then the transition from Ψ_1 to Ψ_2 would be the excitation of the atom accompanied by the loss of electron energy as in the Franck-Hertz experiment. Or ψ_k could be the radiation field and Φ_l an electron making a transition from a bound state in a metal to a free state as in the photoelectric effect. In every case we have a smooth transition in time between one state and another. There is no need to introduce energy at all. The eigenvalues are frequencies and we have a simple resonance phenomenon analogous to two coupled pendulums, where the excitation energy oscillates back and forth between the two pendulums.

Without the quantum postulate of transitions between levels with Bohr's frequency condition, we find that the wave system behaves completely *as if* the quantum postulate was correct. Schrödinger suggested that this is ground to mistrust the quantum postulate, to which he attributed an "axiomatic incomprehensibility." Once the terms in atomic spectra are viewed as discrete energy levels, every newly discovered exchange process (e.g., the Franck-Hertz experiment) will appear to confirm this idea, even if, in reality, we have nothing more than a resonance phenomenon between different wave fields. Schrödinger had a seemingly instinctual abhorrence of the idea of quantum jumps. As Heisenberg recalled the discussions between Bohr and Schrödinger

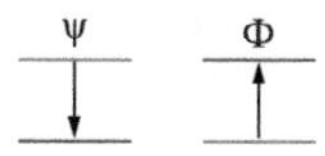

Fig. 10.3 Energy levels for two systems ψ,Φ showing a two-fold degeneracy. The initial state Ψ_1 is in red while the final state Ψ_2 is in blue.

in Copenhagen,[11] Schrödinger had argued "...the electron is said to jump from one orbit to the next and to emit radiation. Is this jump supposed to be gradual or sudden? If it is gradual, the orbital frequency and energy of the electron must change gradually as well. But in that case, how do you explain the persistence of narrow spectral lines? On the other hand, if the jump is sudden ... we must ask ourselves how precisely the electron behaves during the jump. Why does it not emit a continuous spectrum as electromagnetic theory demands? ... the whole idea of quantum jumps is sheer fantasy."

Bohr replied that the contradictions were not removed by Schrödinger's picture but simply swept aside. The closed system, consisting of the two subsystems isolated from the rest of the world, is itself an abstraction. For us to learn anything about it, the system must interact with an external object (e.g., a detector), and this introduces discontinuity and randomness in agreement with the quantum postulate. He insisted that Planck's radiation law could only be derived with the photon picture as per Einstein's derivation.

As the resonance picture and the quantum postulate constitute two different explanations for the same thing, one or, as Schrödinger politely suggested, both must be false. In fact, we have here a nice example of Bohr's complementarity between what he called the space-time (wave description) and causality (conservation of energy-momentum or particle description) as discussed in his paper, which only appeared some months after Schrödinger's work on transitions in the wave picture.

According to Bohr, an instantaneous transfer of energy-momentum between particles occurs at a random time and place, while Schrödinger sees the same phenomenon as a continuous transfer of energy from one system to another. Schrödinger's description is an example of the shifting of the cut between the system and the measuring instrument, and actual measurements show discrete particles at randomly distributed positions and times, in accordance with Bohr's description. To describe the experimental results, we must turn to Born's probabilistic interpretation of the waves, and thus randomness (Bohr's "irrationality") returns.

10.7 Chronology of Bohr–Heisenberg–Schrödinger discussions

July 1926: Schrödinger lectures in Munich, attended by Heisenberg.

October 1926: Schrödinger in Copenhagen.

November 1926: Schrödinger's paper on the Compton effect.

February 1927: Bohr goes on vacation; Heisenberg writes paper explaining the uncertainty principle.

March 1927: Submission of Heisenberg's paper.

June 1927: Schrödinger's paper on the resonance picture as an alternative to quantum jumps.

September 1927: Bohr's paper on complementarity presented to Como Conference.

October 1927: Solvay Conference, which we discuss in Chapter 13.

[11]Heisenberg, W., *Physics and Beyond, Encounters and Conversations* (Harper and Row, 1971).

11

The final synthesis of quantum mechanics: the "transformation theory" and Dirac notation

11.1 Introduction

The early development of quantum mechanics progressed at an alarmingly rapid rate. The mathematics was apparently left as assumed to be understood by the readers of the brisk flow of publications. Slightly before the accelerated development of quantum mechanics, mathematics had its own epiphany. This chapter is an interlude where we will describe the mathematics behind the theory in more detail than we have done until now, and give a brief overview of the development of mathematics in the years before, paralleling the period of rapid progress in quantum theory.

The entire development of quantum mechanics followed largely from abductive reasoning, which is regarded as possibly the only method of generating new fundamental knowledge. As described by Eco, quoting Pierce,[1] "Abduction is a case of synthetic inference 'where we find some very curious circumstances, which would be explained by the supposition that it was a case of a certain general rule, and thereupon adopt that supposition.'"

By late 1926, quantum mechanics had been developed in two different forms: matrix mechanics/operator calculus/q-number theory and wave mechanics.[2] At this time, the theory bore little resemblance to its modern form, and the relationships between the various forms were murky. However, after the realization that matrix and wave mechanics were equivalent, it became evident that all forms might be different aspects of a more general structure. The first step in the development of this general structure was provided by F. London in a paper received on September 19, 1926, where he introduced the concept of a quantum state vector in Hilbert space, with quantum dynamics described by the motion of this vector.[3] A principal goal of this chapter is to explain exactly what this means. The formulation of the transformation theory, which refers to the fact that the theory can be transformed to any basis, was the last significant abductive step in the development of quantum mechanics.

[1] U. Eco, *A Theory of Semiotics* (Bloomington, IN: Indiana University Press, 1979), pp. 131-132.

[2] A. Duncan and M. Janssen, *From canonical transformations to transformation theory, 1926-1927: The road to Jordan's Neue Begründung*, *Stud. Hist. Philos. Mod. Phys.* **40**, 352 (2009).

[3] F. London, *Winkelvariable und kanonische Transformationen in der Undulationsmechanik* [Angle variables and canonical transformations in wave mechanics], *Z. Phys.* **40**, 193 (1926).

The Historical and Physical Foundations of Quantum Mechanics. Robert Golub and Steven K. Lamoreaux, Oxford University Press.
 DOI: 10.1093/oso/9780198822189.003.0011

After two subsequent papers by Dirac[4] and by Jordan,[5] which were received on December 2 and December 18 1926, respectively, quantum mechanics became fully understood as a linear transformation theory, for which a notational system was developed by Dirac that is still in use today. Although Dirac's notation went through an evolutionary process, as can be seen in the three subsequent editions of his book (1930, 1935, and 1947),[6] the conceptual foundations presented in that context remain unchanged. According to von Neumann, Dirac's representation of quantum mechanics is "scarcely to be surpassed in its brevity and elegance."[7] Through a compactification of superfluous information, this notation allows the deeper questions and their solutions to become self-evident: There is an old adage, "A good notation proves its own theorems."

In the years following 1926, frenzied developments led to the essentially complete modern formulation of quantum theory, including the relativistic theory of the electron formulated by Dirac in 1928. In 1935 Condon and Shortley wrote *The Theory of Atomic Spectra*,[8] in which essentially every atomic phenomenon observed to that date was explained through application of quantum mechanics. When first published, a reviewer for *Nature* stated, "Its power and thoroughness leave the general impression of a work of the first rank, which successfully unifies the existing state of our knowledge, and will prove for many years a starting point for further researches and an inspiration to those who may undertake them." The nearly complete understanding of atomic phenomena at this time is demonstrated by the fact that, according to the authors' preface to the 1963 edition, the reissue of the book included only a few corrections, with the original notation being left intact. This book remains a useful atomic physics reference, even though some of the notation is slightly dated.

Dirac's notation was not without its detractors. In particular, the need for a continuum of momentum eigenstates $\vec{p}$ and position (coordinate) eigenstates $\vec{x}$ led to the introduction of the so-called "delta function," which had no apparent fundamental mathematical basis.[9] In the margins of the 1935 edition of Dirac's book retrieved from the Yale Library (reproduced in Fig. 11.1), there are handwritten comments, in the ink and penmanship style of the era, regarding his introduction and use of the delta function: "Extreme foolhardiness," "Baloney," "Mathematical suicide," and "Shoddy opportunism." It was von Neumann, in his 1932 book, who showed that Dirac's "clear and unified" representation can be cast in a mathematically rigorous way by use of the Hilbert theory of operators. In this case, the delta function is viewed as a functional

[4]Dirac, P. A. M. *On the theory of quantum mechanics*, *Proc. R. Soc. A* **113**, 621 (1927).

[5]Jordan, P. *Über eine neue Begründung der Quantenmechanik* [On a new foundation of quantum mechanics], *Z. Phys.* **40**, 809 (1927).

[6]Dirac, P.A.M., *The Principles of Quantum Mechanics*, Oxford, 1930, second edition, 1935, third edition, 1947

[7]Neumann, J. von , *Mathematical Foundations of Quantum Mechanics*, trans. R. T. Beyer (Princeton: Princeton University Press, 1955). Originally published as *Mathematische Grundlagen der Quantenmechanik* (Berlin: Springer, 1932).

[8]Condon, E.U. and Shortley, G.H., *The Theory of Atomic Spectra* (Cambridge, 1935) 1st Edition, still in print.

[9]The delta function of the Dirac form has been used in different contexts since the 1820s.

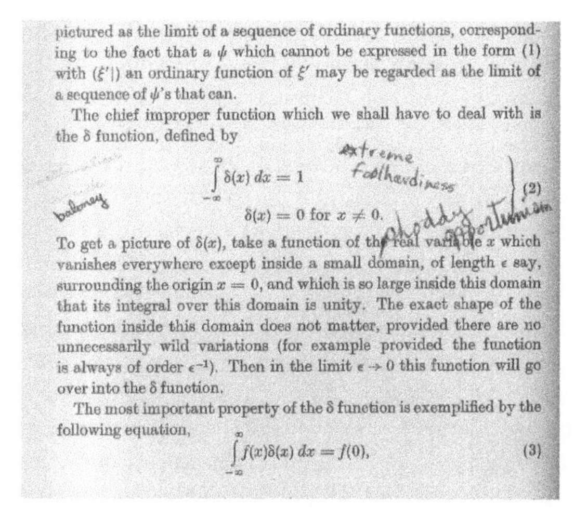

pictured as the limit of a sequence of ordinary functions, corresponding to the fact that a ψ which cannot be expressed in the form (1) with $(\xi'|)$ an ordinary function of ξ' may be regarded as the limit of a sequence of ψ's that can.

The chief improper function which we shall have to deal with is the δ function, defined by

$$\int_{-\infty}^{\infty} \delta(x)\,dx = 1$$
$$\delta(x) = 0 \text{ for } x \neq 0. \quad (2)$$

To get a picture of $\delta(x)$, take a function of the real variable x which vanishes everywhere except inside a small domain, of length ϵ say, surrounding the origin $x = 0$, and which is so large inside this domain that its integral over this domain is unity. The exact shape of the function inside this domain does not matter, provided there are no unnecessarily wild variations (for example provided the function is always of order ϵ^{-1}). Then in the limit $\epsilon \to 0$ this function will go over into the δ function.

The most important property of the δ function is exemplified by the following equation,

$$\int_{-\infty}^{\infty} f(x)\delta(x)\,dx = f(0), \quad (3)$$

Fig. 11.1 Handwritten annotation (graffiti), by a person or persons unknown, in the second edition (1935) of Dirac's *The Principles of Quantum Mechanics*, p. 72, as obtained from the Yale Library. The handwriting style and the ink used in writing the notations suggests that they were written before the δ-function had been established on a firm mathematical foundation; before, the use of such was considered, at best, sloppy, in a mathematical sense. (Fair use—annotations of historical interest.)

to be interpreted in the context of the integration of a function. Von Neumann was well prepared for this task, as he had worked closely with Hilbert for several years.

(It was not until the work of Laurent Schwartz on the theory of distributions that full mathematical rigor was brought to the subject, for which he received the Fields Medal in 1950.)[10]

On the other hand, quantum mechanics is a physical theory, not a mathematical theory. The architects of modern physics in the early twentieth century cannot be faulted for not knowing the necessary mathematical techniques, which were largely developed in the early-to-mid-nineteenth century. The development of physical theories during this time was in general hindered because the required mathematics was not universally at hand. For example, it is well known that Hilbert nearly "scooped" Einstein on general relativity, after Einstein gave a seminar at Göttingen describing the physical theory and his difficulties with its full mathematical description, which Hilbert and his students accomplished on their own in a few days of work. As Hilbert commented, "Every boy in the streets of Göttingen understands more about four-dimensional geometry than Einstein. Yet, in spite of that, Einstein did the work and not the mathematicians."[11]

Similarly, there was surprise when it was realized that the fundamental mathematics of quantum mechanics followed directly from the theory of vibrating drumheads and musical instruments, along with Fourier's analysis of heat flow. The initial principal

[10] J. Lützen, *The Prehistory of the Distribution Function* (New York: Springer-Verlag, 1982).

[11] Reid, C., *Hilbert* (New York: Springer-Verlag, 1996), p. 142.

success of the Schrödinger equation was in its application to fundamental atomic processes that exhibit sharp resonances. However, new problems and ideas were introduced with this equation: the world was left to figure out what different aspects of the theory, for example the wave function, really represent. Schrödinger's development of the wave equation is another example of abductive analysis. This *modus operandi* in the development of a physical theory was largely unprecedented in the history of science.[12]

In contrast to the abductive method used by the founders of quantum mechanics, Einstein formulated the mathematical descriptions that transformed his mental pictures into a physical theory using the method of *reductio ad absurdum* via *gedankenexperiments.* This mathematical description, whether arrived at by abductive reasoning or by other methods, could then be used to explain other phenomena or predict new ones yet to be observed in experiments. Looking for these predicted effects is referred to as "testing the theory." Indeed, those original questions brought forward in 1926 in many ways persist to this day; in particular, the question of the "completeness" of quantum mechanics (in the sense that all phenomena are adequately described by the theory) is still discussed.

With that said, students are often left with the impression that quantum mechanics is a mere exercise in differential equations that have characteristics we like to think are required in a "good quantum theory," with that impression fueled by the treatments presented in many elementary textbooks on the subject. Our goal is to bring the focus back to the underlying physical principles, and to show that an accurate description of the physical world flows directly from those principles. This chapter will repeat some aspects of the earlier chapters; however, in our opinion this recapitulation is important when we now recast the theory into its transformational form.

11.2 Sturm-Liouville theory, Hilbert space, and linear operators

The Schrödinger equation turned out to be a special case of the more general Sturm-Liouville theory[13] that deals with a specific form of second-order, real, linear differential equations. These equations follow the form

$$-\frac{\mathrm{d}}{\mathrm{d}x}\left[p(x)\frac{\mathrm{d}y}{\mathrm{d}x}\right] + q(x)y = \lambda w(x)y \tag{11.1}$$

for an unknown (possibly complex) eigenfunction y and its eigenvalue λ on an interval $[a, b]$, where $p(x)$ (not to be confused with the momentum), $q(x)$, and the weighting function $w(x)$ are given real functions of x. The boundary values for $y \equiv y(x)$ or $y' \equiv y'(x)$, or a linear combination of them, must be specified at a and b:

$$\begin{aligned} \alpha y(a) + \alpha' y'(a) &= 0 \\ \beta y(b) + \beta' y'(b) &= 0 \end{aligned} \tag{11.2}$$

[12]This method is much more accepted and used today: String theory is a prominent example of theories developed in this manner.

[13]The theory was developed by Jacques Charles François Sturm (1803–1855) and Joseph Liouville (1809–1882).

where $|\alpha| + |\alpha'| > 0$ and $|\beta| + |\beta'| > 0$. For $\alpha' = \beta' = 0$, these are called Dirichlet boundary conditions, while $\alpha = \beta = 0$ corresponds to Neumann boundary conditions. The general case is referred to as mixed boundary conditions.

As we will discuss, the eigenvalues λ_n are real, are countably infinite in number, and can be ordered as $\lambda_0 < \lambda_1 < \lambda_2 ... < \lambda_n < \lambda_{n+1} ...$. The eigenfunction $y_n(x)$ is square integrable on the interval $[a, b]$ and has $n-1$ zeros on (a, b). (The number of zeros on the closed interval $[a, b]$ of course depends on the boundary conditions.) The eigenfunctions y_n are unique up to a multiplicative constant. We will show that the eigenfunctions are orthogonal, which means that the inner product $(y_n, y_{n'})$ satisfies

$$(y_n, y_{n'}) \equiv \int_a^b dx\, w(x)\, y_n^*(x)\, y_{n'}(x) = \delta_{nn'}, \tag{11.3}$$

assuming that the y_n are normalized to unity. The Kronecker delta function is defined as

$$\delta_{nn'} = \begin{cases} 0 & \text{if } n \neq n' \\ 1 & \text{if } n = n' \end{cases} \tag{11.4}$$

We will choose the multiplicative constant for each y_n so that the eigenfunctions are normalized to unity. These normalized eigenfunctions form the orthonormal basis for a closed subspace of a general Hilbert space $L^2([a, b], w(x), dx)$, where the square indicates that the norms are defined as the sum of squared magnitudes, $[a, b]$ is the range over which the functions are defined, and $w(x)$ specifies the weighting function. In general, a, b, and x can be multidimensional.

11.2.1 The Sturm-Liouville operator is self-adjoint

The differential equation (11.1) can be rewritten using the operator

$$L = \frac{1}{w(x)} \left(-\frac{\mathrm{d}}{\mathrm{d}x} \left[p(x) \frac{\mathrm{d}}{\mathrm{d}x} \right] + q(x) \right), \tag{11.5}$$

so that, for an eigenfunction y_n with eigenvalue λ_n,

$$L y_n = \lambda_n y_n. \tag{11.6}$$

An operator L is defined as self-adjoint (or Hermitian) if the inner product, as defined in Eq. (11.3), satisfies $(Lf, g) = (f, Lg)$ for two arbitrary, possibly complex, functions $f^* \equiv f^*(x)$ and $g \equiv g(x)$ defined on the interval $[a, b]$ and which satisfy the boundary conditions of Eq. (11.2). This can be shown to be true for L as defined by Eq. (11.5) as follows.

First, because all the terms in Eq. (11.5) are real, $L^* = L$. We can then write $[Lf]^* = Lf^*$ and similarly for g. Then

$$(Lf, g) - (f, Lg) = \int_a^b dx w(x)[Lf^*]g - \int_a^b dx w(x) f^*[Lg] \tag{11.7}$$

$$= \int_a^b dx\, g\left(-\frac{\mathrm{d}}{\mathrm{d}x}\left[p(x)\frac{\mathrm{d}}{\mathrm{d}x}\right] + q(x)\right) f^* - ... \tag{11.8}$$

$$- \int_a^b dx f^*\left(-\frac{\mathrm{d}}{\mathrm{d}x}\left[p(x)\frac{\mathrm{d}}{\mathrm{d}x}\right] + q(x)\right) g \tag{11.9}$$

$$= \int_a^b dx g\left(-\frac{\mathrm{d}}{\mathrm{d}x}\left[p(x)\frac{\mathrm{d}}{\mathrm{d}x}\right]\right) f^* - \int_a^b dx f^*\left(-\frac{\mathrm{d}}{\mathrm{d}x}\left[p(x)\frac{\mathrm{d}}{\mathrm{d}x}\right]\right) g, \tag{11.10}$$

where the terms with $q(x)$ cancel each other. This equation can be integrated by parts to give

$$\begin{aligned}(Lf, g) - (f, Lg) &= p(a)g(a)f^{*\prime}(a) - p(b)g(b)f^{*\prime}(b) + p(b)f^*(b)g'(b) - p(a)f^*(a)g'(a) \\ &= p(a)[g(a)f^{*\prime}(a) - f^*(a)g'(a)] + p(b)[f^*(b)g'(b) - f^{*\prime}(b)g(b)],\end{aligned}$$

where we use the prime notation for the derivatives. The boundary conditions given by the set of linear equations (11.2) apply to both f^* and g, and we can write them as a matrix equation:

$$\begin{pmatrix} f^*(a) & f^{*\prime}(a) \\ g(a) & g'(a) \end{pmatrix}\begin{pmatrix} \alpha \\ \alpha' \end{pmatrix} = \begin{pmatrix} 0 \\ 0 \end{pmatrix} \tag{11.11}$$

and similarly for the boundary conditions at b. Since at least some of the coefficients α, α' and β, β' are nonzero, the determinant of the matrix must vanish, which means that

$$\det\begin{pmatrix} f^*(a) & f^{*\prime}(a) \\ g(a) & g'(a) \end{pmatrix} = f^*(a)g'(a) - f^{*\prime}(a)g(a) = 0 \tag{11.12}$$

$$\det\begin{pmatrix} f^*(b) & f^{*\prime}(b) \\ g(b) & g'(b) \end{pmatrix} = f^*(b)g'(b) - f^{*\prime}(b)g(b) = 0. \tag{11.13}$$

This immediately shows that the boundary terms from the integration by parts are zero. Therefore

$$(Lf, g) - (f, Lg) = 0, \tag{11.14}$$

which proves that the Sturm-Liouville operator is self-adjoint.

11.2.2 The eigenvalues are real

Consider an eigenfunction y_n so that $Ly_n = \lambda_n y_n$. Because L is self-adjoint,

$$(Ly_n, y_n) = (y_n, Ly_n) = (y_n, \lambda_n y_n) = \lambda_n(y_n, y_n). \tag{11.15}$$

Next we take the complex conjugate of Ly_n,

$$[Ly_n]^* = [\lambda_n y_n]^* = \lambda_n^* y_n^*, \tag{11.16}$$

and use Eq. (11.3) to calculate (Ly_n, y_n):

$$(Ly_n, y_n) = \int_a^b dx\, w(x)\, [Ly_n(x)]^*\, y_n(x) = \int_a^b dx\, w(x)\, \lambda_n^* y_n(x)^* y_n(x) = \lambda_n^*(y_n, y_n). \tag{11.17}$$

Since L is self-adjoint, this must equal $\lambda_n(y_n, y_n)$, which means that

$$\lambda_n = \lambda_n^*, \tag{11.18}$$

so that the eigenvalues of a self-adjoint operator are real.

11.2.3 The eigenfunctions are orthogonal

Consider two distinct eigenfunctions y_m and y_n of L with eigenvalues λ_m and λ_n. Because L is self-adjoint,

$$(Ly_m, y_n) = \lambda_m(y_m, y_n) = (y_m, Ly_n) = \lambda_n(y_m, y_n), \tag{11.19}$$

which implies that either

$$\lambda_m = \lambda_n \quad \text{or} \quad (y_m, y_n) = 0 \tag{11.20}$$

must be true for all m, n with $m \neq n$. If the eigenfunctions y_m and y_n are not degenerate (which means they have the same eigenvalue, $\lambda_m = \lambda_n$), we conclude that $(y_m, y_n) = 0$ for $m \neq n$, which means that y_m and y_n are orthogonal functions.

If two distinct eigenfunctions have the same eigenvalue, we can always form two mutually orthogonal functions from them using the Gram-Schmidt orthogonalization procedure. Given two degenerate eigenfunctions v_1 and v_2 that are normalized so that $(v_1, v_1) = (v_2, v_2) = 1$, let

$$u_1 = v_1; \quad u_2 = v_2 - (u_1, v_2)u_1. \tag{11.21}$$

Then

$$(u_1, u_2) = (v_1, v_2 - (v_1, v_2)v_1) = (v_1, v_2) - (v_1, v_2)(v_1, v_1) = 0, \tag{11.22}$$

and similarly, noting that $(v_1, v_2) = (v_2, v_1)^*$ is a (possibly complex) number, we find

$$(u_2, u_1) = (v_2 - (v_1, v_2)v_1, v1) = (v_2, v_1) - (v_1, v_2)^*(v_1, v_1) = 0. \tag{11.23}$$

This process can be extended to an arbitrary number of functions by defining a projection operator,

$$\text{proj}_u(v) = \frac{(u, v)}{(u, u)} u. \tag{11.24}$$

Then we can form a series:

$$\begin{aligned}
u_1 &= v_1, \\
u_2 &= v_2 - \mathrm{proj}_{u_1}(v_2), \\
u_3 &= v_3 - \mathrm{proj}_{u_1}(v_3) - \mathrm{proj}_{u_2}(v_3), \\
&\vdots \\
u_n &= v_n - \sum_{k=1}^{n-1} \mathrm{proj}_{u_k}(v_k).
\end{aligned}$$

It should be noted that there is an infinite number of ways to construct an orthogonal set of functions. It is easily seen that one could shuffle the v_k in the above series and form a different set of u_k. The functions u_k produced using this procedure are not normalized, but that is easily accomplished. And of course, orthogonality implies linear independence among a set of eigenfuctions.

11.2.4 The eigenvalues form an ascending series

The eigenvalues can be ordered as $\lambda_0 < \lambda_1 < \lambda_2 ... < \lambda_n < \lambda_{n+1} ...$ with a minimum eigenvalue λ_0 and $\lambda_n \to \infty$ for $n \to \infty$. To prove the existence of a minimum eigenvalue, we note that the Sturm-Liouville equation arises from a variational equation. Consider the functional integral[14]

$$\Omega[f] = \int_a^b F(x, f, f') = \int_a^b \left[p(x)(f')^2 + q(x)f^2\right] dx \tag{11.25}$$

subject to the constraint (which we take as unit normalization to simplify the calculation)

$$\int_a^b G(x, f, f')dx = \int_a^b fw(x)dx = 1, \tag{11.26}$$

where $f = f(x)$ is some arbitrary real function on $[a, b]$ and the real functions $p(x)$, $q(x)$ and $w(x)$ are as defined for the Sturm-Liouville equation. (Note that the eigenfunctions can always be cast into a form that they are real; this is not always the most convenient form but that simplifies the following.)

Defining $\mathcal{L} = F - \lambda G$, the functional $\Omega[f]$ is minimized when $\mathcal{L}$ satisfies the Euler-Lagrange equation:

$$\frac{\partial \mathcal{L}}{\partial f} - \frac{\mathrm{d}}{\mathrm{d}x}\frac{\partial \mathcal{L}}{\partial f'} = 0. \tag{11.27}$$

Here we accounted for the constraint (11.26) using Lagrange's method of undetermined multipliers by including the $-\lambda G$ term. The Euler-Lagrange equation yields

$$-\frac{\mathrm{d}}{\mathrm{d}x}\left[p(x)\frac{\mathrm{d}f}{\mathrm{d}x}\right] + q(x)f - \lambda w(x)f = 0, \tag{11.28}$$

which is equivalent to Eq. (11.1) for the function f.

[14] P. M. Morse and H. Feshbach, *Methods of Theoretical Physics* (New York: McGraw-Hill, 1957), pp. 736-737.

Let us now consider the functional integral

$$\Omega[f] = \int_a^b \left[p(f')^2 + qf^2\right] dx \tag{11.29}$$

and vary f, which can be thought of as a trial function that is normalized and satisfies the boundary conditions, until the lowest value of $\Omega[f]$ is obtained. Let us call the function for which this minimization occurs f_0. As we saw above, f_0 must be a solution of the Sturm-Liouville equation in order for $\Omega[f]$ to be stationary. Integrating the functional integral by parts ($u = pf'$ and $dv = f'$)

$$\min \Omega[f] = \Omega[f_0] = \int_a^b \left[p(f_0')^2 + qf_0^2\right] dx \tag{11.30}$$

$$= \left[pf_0f_0'\right]_a^b + \int_a^b f_0 \left[-\frac{\mathrm{d}}{\mathrm{d}x}\left[p(x)\frac{\mathrm{d}f_0'}{\mathrm{d}x}\right] + q(x)f_0\right] dx. \tag{11.31}$$

The boundary term is zero because of the Sturm-Liouville boundary conditions Eq. (11.2). The second term can be determined using Eq. (11.28). Substituting this equation into Eq. (11.31) yields

$$\Omega[f_0] = \int_a^b \lambda w(x) f_0^2 dx = \lambda \int_a^b w(x) f_0^2 dx = \lambda \tag{11.32}$$

where we have assumed that f_0 is normalized. Because this is the global minimum, we can immediately identify $\lambda = \lambda_0$ and $f_0 = y_0$, the lowest eigenvalue and its corresponding eigenfunction, respectively.

From the foregoing, we immediately see the motivation for the minimization procedure used in Chapter 7 to find the ground state energy. We want to minimize the Rayleigh quotient,

$$R(f) = \frac{(f, Lf)}{(f, f)}, \tag{11.33}$$

where we have relaxed the condition that f is normalized. Note that f is any real function on the interval that has a finite number of discontinuities, that is square integrable, and that satisfies the boundary conditions. We are not assuming *a priori* that f contains only a linear combination of the eigenfunctions of the Sturm-Liouville operator L; f can be any imaginable function subject to the aforementioned criteria.

However, using the results of Eqs. (11.30) and (11.32) and assuming uniqueness of the eigensolutions of L, the lowest eigenvalue always satisfies

$$\lambda_0 \leq R(f). \tag{11.34}$$

If the function f is varied until a global minimum is found, we can conclude that

$$\lambda_0 = \min R(f), \tag{11.35}$$

in which case $f \propto y_0$. This is in essence the variational method that is often used to estimate the ground state energy of a system.

Technically, the variational method described above can find either a global minimum or a global maximum; this amounts to a sign change on L. For the Hamiltonian formulation of quantum mechanics, it is a global minimum. It is always true that

$$|\lambda_0| = |\min R(f)| \tag{11.36}$$

so we can have a descending (toward more negative) or an ascending (toward more positive) spectrum of eignenvalues, depending on the sign of L. We will choose the case for the spectrum to be ascending, which applies to quantum mechanics.

It is possible to generate the non-descending series (allowing for degeneracies) of eigenvalues by forcing f to be orthogonal to a subset of all lower eigenfunctions $y_0, y_1, ...y_N$. This can be accomplished by defining

$$f_N = \sum_{n=0}^{N} c_n y_n \quad \text{with} \quad c_n = (y_n, f) \tag{11.37}$$

and

$$h_N = f - f_N. \tag{11.38}$$

The function h_N is orthogonal to all y_n for $n \leq N$, which can be easily seen by direct calculation:

$$(y_n, h_N) = (y_n, f) - (y_n, \sum_{n=0}^{N} c_n y_n) = c_n - c_n \quad \text{for } n \leq N. \tag{11.39}$$

Because we have constructed h_N to be orthogonal to all eigenfunctions with $n \leq N$, none of those eigenvalues (except in the case of degeneracies) can be obtained from the Rayleigh quotient of h_N. Therefore

$$\lambda_{N+1} = \min R(h_N), \tag{11.40}$$

for which $f = y_{N+1}$. The overall conclusion is that, given a minimum eigenvalue and corresponding eigenfunction, an ascending series of eigenfunctions can be constructed, with corresponding eigenvalues that have no upper limit.

11.2.5 The eigenfunctions form a complete set

Let us now consider h_N for an arbitrary function f. We can think of h_N as being the error between the eigenfunction expansion of f and f itself, and can see how its squared magnitude $|h_N|^2 = (h_N, h_N)$ varies in the limit of $N \to \infty$. If $|h_N|^2$ goes to zero in the limit of $N \to \infty$, we can expand any function in terms of the eigenfunctions, and therefore they form a complete set.

Using the result (11.40) for the Rayleigh quotient and the definition of the inner product, we find

$$|h_N|^2 \leq \frac{1}{\lambda_{N+1}}(h_N, Lh_N). \tag{11.41}$$

The right-hand side can be expanded as

$$
\begin{aligned}
(h_N, Lh_N) &= (f, Lf) - (f, L\sum_{n=0}^{N} c_n y_n) - (\sum_{n=0}^{N} c_n y_n, Lf) + (\sum_{m=0}^{N} c_m y_m, L\sum_{n=0}^{N} c_n y_n) \\
&= (f, Lf) - \sum_{n=0}^{N} c_n \lambda_n (f, y_n) - \sum_{n=0}^{N} c_n^* (y_n, Lf) + \sum_{n=0}^{N} |c_n|^2 \lambda_n \\
&= (f, Lf) - \sum_{n=0}^{N} c_n \lambda_n (f, y_n) - \sum_{n=0}^{N} c_n^* (Ly_n, f) + \sum_{n=0}^{N} |c_n|^2 \lambda_n \\
&= (f, Lf) - \sum_{n=0}^{N} c_n \lambda_n (f, y_n) - \sum_{n=0}^{N} c_n^* \lambda_n (y_n, f) + \sum_{n=0}^{N} |c_n|^2 \lambda_n \\
&= (f, Lf) - \sum_{n=0}^{N} c_n c_n^* \lambda_n - \sum_{n=0}^{N} c_n^* c_n \lambda_n + \sum_{n=0}^{N} |c_n|^2 \lambda_n \\
&= (f, Lf) - \sum_{n=0}^{N} |c_n|^2 \lambda_n,
\end{aligned}
\tag{11.42}
$$

where we have used the definition of c_n, the orthogonality of the eigenfunctions, and the self-adjoint property of L to reduce the sum.

Let us divide the sum into negative and nonnegative eigenvalues by assuming the largest negative eigenvalue λ_n has $n = N_m < N$,

$$
\begin{aligned}
(h_N, Lh_N) = (f, Lf) - \sum_{n=0}^{N} |c_n|^2 \lambda_n &= (f, Lf) + \sum_{n=0}^{N_m} |c_n|^2 |\lambda_n| - \sum_{n=N_m+1}^{N} |c_n|^2 \lambda_n \\
&\leq (f, Lf) + \sum_{n=0}^{N_m} |c_n|^2 |\lambda_n|
\end{aligned}
\tag{11.43}
$$

because the weighted sum over positive eigenvalues is greater than zero, and not subtracting that term makes the right-hand side larger.[15] Therefore,

$$
|h_N|^2 \leq \frac{1}{\lambda_{N+1}} (h_N, Lh_N) \leq \frac{1}{\lambda_{N+1}} \left[(f, Lf) + \sum_{n=0}^{N_m} |c_n|^2 |\lambda_n| \right]. \tag{11.45}
$$

The term in the brackets on the right-hand side is constant. Since $\lim_{N\to\infty} \lambda_{N+1} = \infty$, we conclude that $|h_{N\to\infty}|^2 = 0$ and therefore the set of eigenfunctions y_n is complete, so that any function f defined on $[a, b]$ can be expanded in terms of that set.

[15]Note that if there is an infinite number of negative eigenvalues, as in the case of the hydrogen atom radial wave function, a positive constant K can be added to the eigenvalues in the sum to form a maximum negative $\lambda_n + K$ for some $n = N_m < N$. The proof then proceeds in a similar manner except that there will be an additional term on the right-hand side of Eq. (11.43) that is also constant as $N \to \infty$:

$$
(h_N, Lh_N) \leq (f, Lf) + K(f, f) + \sum_{n=0}^{N_m} |c_n|^2 |K + \lambda_n|. \tag{11.44}
$$

The set of functions $\{y_n(x)\}$ form a complete set in that any square integrable, piecewise continuous, and piecewise differentiable function f on the domain $[a, b]$ can be written as a linear combination of the y_n. In this sense, the set $\{y_n(x)\}$ are a generalization beyond the well-known functions $\sin nx$ and $\cos nx$ used in the Fourier series. That is, we can write some arbitrary function $f(x)$ defined on $[a, b]$ as

$$f(x) = c_1 y_1(x) + c_2 y_2(x) + c_3 y_3(x) + \ldots = \sum_n c_n y_n(x) \tag{11.46}$$

where the $\{c_n\}$ are determined using Eq. (11.3):

$$c_n = \int_a^b dx\ w(x)\ y_n^*(x) f(x). \tag{11.47}$$

The set of eigenfunctions y_n serve as a basis of a $L^2([a, b], w(x), dx)$ closed subspace of a general Hilbert space. The function f can be thought of as a vector in this infinite dimensional space. This vector is referred to as the quantum state vector, which we will describe in detail later in this chapter.

11.2.6 Delta function and completeness

Let us consider the eigenfunction expansion of an arbitrary function f that satisfies the boundary conditions, is square integrable, and is piecewise continuous and piecewise differentiable. With the coefficients c_n given by Eq. (11.47), we expand in the complete set, $y_n(x)$:

$$f(x) = \sum_n c_n y_n(x) = \sum_n y_n(x) \int_a^b y_n^*(x') f(x') dx' = \sum_n \int_a^b y_n(x) y_n^*(x') f(x') dx' \tag{11.48}$$

$$\equiv \int_a^b \delta(x - x') f(x') dx' = f(x) \tag{11.49}$$

with

$$\delta(x - x') = \sum_n y_n(x) y_n^*(x'). \tag{11.50}$$

Thus the δ function is given by (11.50) as long as the $y_n(x)$ are a complete set.

This defines the δ-function and provides an auxiliary definition of completeness. This form of the completeness relation was proved in Eq. (7.128).

Any function that is peaked and symmetric around $x = 0$ can be used to approximate the δ-function. A simple example is a rectangular window function $W(x - x_0)$ of height $1/2\epsilon$ with width 2ϵ centered at x_0 and zero elsewhere. When used in an integral, this function yields

$$\int_{-\infty}^{\infty} W(x - x_0) f(x) dx = \int_{x_0-\epsilon}^{x_0+\epsilon} \frac{1}{2\epsilon} f(x) dx \approx f(x_0), \tag{11.51}$$

and the approximation becomes exact when $\epsilon \to 0$.

Another form of the δ-function that goes back to Cauchy and Poisson as their "selector" function is

$$\delta(x) = \lim_{\epsilon \to 0} \frac{\epsilon}{\pi(x^2 + \epsilon^2)}. \tag{11.52}$$

A very useful form employs the sampling function $\operatorname{sinc}(x) = \sin(x)/x$, for which $\int_{-\infty}^{\infty} \operatorname{sinc}(x)dx = \pi$, so that

$$\delta(x) = \lim_{\epsilon \to 0} \frac{\operatorname{sinc}(x/\epsilon)}{\pi\epsilon} = \lim_{\epsilon \to 0} \frac{\sin(x/\epsilon)}{\pi x}. \tag{11.53}$$

This can be rewritten as

$$\delta(x) = \lim_{L \to \infty} \frac{\sin(Lx)}{\pi x} = \lim_{L \to \infty} \frac{1}{2\pi} \int_{-L}^{L} dk\ e^{ikx} \equiv \frac{1}{2\pi} \int_{-\infty}^{\infty} dk\ e^{ikx}. \tag{11.54}$$

This integral is not absolutely convergent; if it is taken as a definition of the δ-function, it should be rather considered as a functional, as suggested by von Neumann, for it alone is not well-defined but only makes sense when included in an integral. From the above considerations, the δ-function can be informally thought of as the Fourier transform of the constant function $f(x) = 1$.

The δ-function was introduced to quantum mechanics by Dirac as an extension of the Kronecker delta to a continuum of states (coordinate position x and momentum p). However, the function itself has a long history; for example, Heaviside (circa 1895) introduced it as the derivative of his step function, and it is used as the impulse function in electrical and mechanical engineering. Von Neumann put the δ-function on a firmer mathematical foundation but showed that quantum mechanics could be developed without it. However, it was not until the work of Laurent Schwartz on the theory of distributions that full mathematical rigor was brought to the subject.

11.2.7 Applications to quantum mechanics via the Schrödinger equation

To put the one-dimensional Schödinger equation with a potential $V(x)$ in Sturm-Liouville form, we set $p(x) = \hbar^2/2m$, $q(x) = V(x)$, and $w(x) = 1$.[16] Since the Hamiltonian is a Sturm-Liouville operator, the results of the previous subsections apply to its eigenvalues and eigenfunctions; in particular, this means that an arbitrary state can be expanded as a linear combination of the eigenstates of the Hamiltonian.

In von Neumann's mathematically rigorous formulation of quantum mechanics, quantum states are represented by vectors in a Hilbert space, which is referred to as a state space. The set of coefficients in the eigenfunction expansion of a general wave function define the state vector, and the dynamics of a quantum system are described by the motion of this vector in Hilbert space. In fact, in almost all situations, the wave function itself is only needed to calculate matrix elements, with the system dynamics being fully parameterized by the expansion coefficients $\{c_n(t)\}$, where a possible time dependence is now indicated.

[16] Other forms are used when we use separation of variables to disentangle the angular and radial dependence of the three-dimensional Schrödinger equation.

The eigenfunctions $y_n(x)$ can be left in abstract form, without reference to the coordinate x, leaving the possibility of using different representations (e.g., momentum or position). The (infinite) set of all x can itself be thought of as a Hilbert space basis onto which the state vector is projected, with the eingenfunction $y_n(x)$ being the projection of the state $|n\rangle$ onto this basis. This will become clearer in the next section where Dirac's notation is introduced. A most important point is that quantum mechanics is a linear theory, with linear operators that reorient the state vector or select components.

The interpretation of the physical state as a sum of eigenfunctions requires certain postulates or axioms of quantum mechanics. These postulates are actually hypotheses, albeit now well-tested ones, so perhaps the term "principles of quantum mechanics" is more apt. After introducing Dirac notation, we will reiterate these principles in terms of that notation.

11.3 Dirac's bra-ket notation

The quantum state of a physical system can be expressed as a sum of eigenstates of a Hamiltonian of the form (11.1). In this context, the eigenstates are usually denoted by Greek symbols such as $\psi_n(x)$. For example, two different physical states can be written in terms of the same set of eigenfunctions as

$$\eta(x) = \sum_n b_n \psi_n(x), \qquad \phi(x) = \sum_m c_m \psi_m(x), \tag{11.55}$$

where b_n and c_m are complex numbers that are called the probability amplitudes, so that, for example, the probability for state ϕ to be in the eigenstate ψ_m is given by $|c_m|^2 \equiv c_m^* c_m$. We require that

$$\sum_m |c_m|^2 = 1 \Longrightarrow \int \phi^* \phi dx = 1. \tag{11.56}$$

We will develop this notation for a one-dimensional coordinate system; the generalization to multiple dimensions is straightforward. Furthermore, we will first consider the case where the eigenvalues form a discrete spectrum, but we will later generalize to the case of a continuous spectrum. The inner product of these two states is

$$(\phi(x), \psi(x)) = \int dx\ \phi^*(x)\psi(x) \equiv \langle\phi|\psi\rangle, \tag{11.57}$$

where the integral is over the domain for which the eigenfunctions are defined. The symbol $\langle ...|...\rangle$ is called a Dirac bracket; the left side $\langle\phi|$ is referred to as the "bra" while the right side $|\psi\rangle$ is referred to as the "ket" (what happened to the c remains a mystery), with

$$\langle\phi| \cdot |\psi\rangle \equiv \langle\phi|\psi\rangle. \tag{11.58}$$

Let us investigate the features of this notation. First, we have not specified the meaning of the variable x—at this point it could represent position or momentum—we will return to this point in the next section. Second, calculation of the inner product using Eq. (11.55), the expansion in terms of eigenstates, yields

$$\langle\phi|\psi\rangle = \int dx\, \phi^*(x)\psi(x) = \int dx \sum_{m,n} b_m^* c_n \psi_m^*(x)\psi_n(x) = \sum_{m,n} b_m^* c_n \delta_{mn}$$

$$= \sum_n b_n^* c_n = b_1^* c_1 + b_2^* c_2 + b_3^* c_3 + ... \tag{11.59}$$

where we have used the orthogonality condition on the eigenstates ψ_m, ψ_n. This immediately suggests that the bra can be represented as

$$\langle\phi| = \begin{pmatrix} b_1^* & b_2^* & b_3^* \ldots \end{pmatrix} \tag{11.60}$$

and the ket can be represented as

$$|\psi\rangle = \begin{pmatrix} c_1 \\ c_2 \\ c_3 \\ \cdot \\ \cdot \\ \cdot \end{pmatrix} \tag{11.61}$$

so that taking the matrix product yields a sum equivalent to the inner product between states. This is the basis of matrix mechanics, and this illustrates our previous assertion that the eigenfunctions need not be directly employed to define a physical system. This might seem surprising, but it is the quintessential feature of the transformation theory.

It is useful to introduce a shorthand notation for the eigenstates so that $|\phi_n\rangle \equiv |n\rangle$. At this stage, the eigenfunction can be completely abstract. The expansion of a general state is

$$|\psi\rangle = \sum_n c_n |n\rangle. \tag{11.62}$$

Such a state is referred to as a superposition state. Completeness is expressed through the identity operator I, defined as

$$I = \sum_k |k\rangle\langle k|, \tag{11.63}$$

where k extends over all eigenvalues. Operating on a general state yields

$$I|\psi\rangle = \sum_k |k\rangle\langle k| \sum_n c_n |n\rangle = \sum_{k,n} c_n |k\rangle\langle k|n\rangle = \sum_{k,n} \delta_{kn} c_n |k\rangle = \sum_n c_n |n\rangle = |\psi\rangle. \tag{11.64}$$

The operator that projects a general state onto a specific eigenfunction is

$$P_k = |k\rangle\langle k|. \tag{11.65}$$

If P_k operates on a general state, it extracts the k component of the wave function:

$$P_k \sum_n c_n|n\rangle = \sum_n c_n|k\rangle\langle k|n\rangle = |k\rangle \sum_n c_n \delta_{kn} = c_k|k\rangle. \tag{11.66}$$

We can immediately see that

$$\langle\psi|P_k|\psi\rangle = \left[\sum_m c_m^* \langle m|\right] c_k|k\rangle = c_k^* c_k = |c_k|^2, \tag{11.67}$$

which is the probability for the general state to be in the eigenstate k.

11.3.1 Operators

Life would be boring if we could only define quantum states; we can also do something to these states by acting on them with specific types of operators. Let us define an operator $\mathbf{A}$ that acts on a ket and transforms it into a new ket:

$$|\psi'\rangle = \mathbf{A}|\psi\rangle. \tag{11.68}$$

The expectation value of such an operator for a given arbitrary state $|\psi\rangle$ is $\langle\psi|\mathbf{A}|\psi\rangle = \langle\psi|\psi'\rangle$. The expectation value can thus be thought of as the probability that the system remains in its original state after the action of $\mathbf{A}$. We define the matrix elements as

$$A_{mn} = \langle m|\mathbf{A}|n\rangle = \int dx\, \psi_m^*(x) A(x) \psi_n(x). \tag{11.69}$$

We can therefore write an operator in matrix form as

$$\mathbf{A} = \sum_{mn} A_{mn}|\psi_m\rangle\langle\psi_n|. \tag{11.70}$$

The new state $\mathbf{A}|\psi\rangle = |\psi'\rangle$ generated by $\mathbf{A}$ acting on an arbitrary state $|\psi\rangle$ can be expanded in terms of the eigenfunctions denoted by $|m\rangle$ with coefficients c'_m given by

$$c'_m = \langle m|\psi'\rangle = \langle m| \left[\sum_{m'n'} A_{m'n'}|m'\rangle\langle n'|\right] \sum_n c_n|n\rangle = \sum A_{mn} c_n, \tag{11.71}$$

where c_n are the expansion coefficients of the original state $|\psi\rangle$ in terms of the same eigenfunctions. The orthogonality relations for the eigenfunctions were used to collapse the sums in Eq. (11.71). This result implies that $\mathbf{A}$ can be interpreted as a matrix that converts a state vector $|\psi\rangle$ to $|\psi'\rangle$:

$$\begin{pmatrix} c'_1 \\ c'_2 \\ c'_3 \\ \cdot \\ \cdot \\ \cdot \end{pmatrix} = \begin{pmatrix} A_{11} & A_{12} & A_{13} & \cdots \\ A_{21} & A_{22} & A_{23} & \cdots \\ A_{31} & A_{32} & A_{33} & \cdots \\ \cdot & \cdot & \cdot & \cdots \\ \cdot & \cdot & \cdot & \cdots \\ \cdot & \cdot & \cdot & \cdots \end{pmatrix} \begin{pmatrix} c_1 \\ c_2 \\ c_3 \\ \cdot \\ \cdot \\ \cdot \end{pmatrix} \tag{11.72}$$

The bra $\langle\psi'|$ can be determined by taking the complex conjugate and transpose (which converts a ket into a bra) of Eq. (11.72):

$$|\psi'\rangle = \mathbf{A}|\psi\rangle \rightarrow \langle\psi'| = \langle\psi|\mathbf{A}^\dagger \tag{11.73}$$

where we have used the rule $(\mathbf{ab})^T = \mathbf{b}^T\mathbf{a}^T$ for any two matrices $\mathbf{a}$ and $\mathbf{b}$. This rule can be readily proved by considering the i, j elements of the matrix product of $\mathbf{A}$, an $m \times n$ dimension matrix, and $\mathbf{B}$, an $n \times m'$ dimension matrix:

$$(\mathbf{AB})_{ij} = \sum_{k=1}^{n} a_{ik}b_{kj}. \tag{11.74}$$

Therefore

$$(\mathbf{AB})^T_{ij} = (\mathbf{AB})_{ji} = \sum_{k=1}^{n} a_{jk}b_{ki}. \tag{11.75}$$

However

$$(\mathbf{B}^T\mathbf{A}^T)_{ij} = \sum_{k=1}^{n} b^T_{ik}a^T_{kj} = \sum_{k=1}^{n} b_{ki}a_{jk} = \sum_{k=1}^{n} a_{jk}b_{ki} \equiv (\mathbf{AB})^T_{ij}. \tag{11.76}$$

The adjoint of a matrix is therefore defined by

$$\mathbf{A}^\dagger = (\mathbf{A}^T)^* \rightarrow A^*_{nm} = (A_{mn})^\dagger. \tag{11.77}$$

An operator is self-adjoint (or Hermitian) if $\mathbf{A}^\dagger = \mathbf{A}$. Hermitian operators are of particular interest here since, as we showed in Section 11.2, their eigenvalues are real and their eigenvectors form an orthonormal basis for the state space. Observables in quantum mechanics are represented by Hermitian projection operators that project superposition states into eigenstates of that operator, with the corresponding eigenvalues representing the possible measurement outcomes. For example, if the observable corresponding to operator $\mathbf{A}$ is measured to be a', this means that the state of the system after measurement is $|a'\rangle$, the eigenstate of $\mathbf{A}$ corresponding to eigenvalue a'. Since the eigenvectors of $\mathbf{A}$ form a basis for the state space, we can expand the initial state $|\psi\rangle$ as

$$|\psi\rangle = \sum_a c_a|a\rangle. \tag{11.78}$$

The probability of measuring this observable to be a' is then given by $|c_{a'}|^2$.

A special subgroup of Hermitian operators is operators that preserve the inner product. If $\mathbf{A}$ is such an operator and $\mathbf{A}|\psi\rangle = |\psi'\rangle$ and $\mathbf{A}|\phi\rangle = |\phi'\rangle$, the inner product satisfies

$$\langle\psi'|\phi'\rangle = \langle\psi|\mathbf{A}^\dagger\mathbf{A}|\phi\rangle = \langle\psi|\phi\rangle \tag{11.79}$$

for any states $|\psi\rangle$ and $|\phi\rangle$. In particular, this means that these operators preserve the normalization of the state. As seen above, for an operator to preserve the inner product, it must satisfy $A^\dagger A = I$. Operators satisfying this condition are called unitary.[17]

[17]Sometimes the notation $\langle\psi|\mathbf{A}^\dagger = \langle\mathbf{A}\psi|$ is employed as a shorthand; however, caution is required in its use. For example, for an eigenfunction of position multiplied by a number a, $a|x\rangle \neq |ax\rangle$ since $|ax\rangle$ represents a translation. This will be further explained later in this chapter.

Unitary operators include rotations, spatial translations, and translations in time. Among the most important unitary operators is the time translation operator $U(t, t_0)$, which describes how the state evolves with time. From the time-dependent Schrödinger equation, we know that

$$i\hbar \frac{\partial \psi(x,t)}{\partial t} = H\psi(x,t). \tag{11.80}$$

The evolution of a state $\psi(x, t_0)$ for an infinitesimal time step dt is

$$\psi(x, t_0{+}dt) = U(t_0{+}dt, t_0)\psi(x, t_0) = \psi(x, t_0){+}dt\, \frac{\partial}{\partial t}\psi(x,t)\bigg|_{t_0} = \left[1 - \frac{i\,dt}{\hbar}H\right]\psi(x, t_0). \tag{11.81}$$

If we divide a finite time translation $\Delta t = t - t_0$ into N steps with $N \to \infty$, then

$$\psi(x, t + t_0) = \left[1 - \frac{\Delta t}{N}\frac{i}{\hbar}H\right]^N \psi(x, t_0) \to e^{-iH\Delta t/\hbar}\psi(x, t_0). \tag{11.82}$$

This suggests writing the time translation operator as

$$U(t, t_0) = e^{-iH(t-t_0)/\hbar} \tag{11.83}$$

so that

$$\psi(x, t + t_0) = U(t, t_0)\psi(x, t_0). \tag{11.84}$$

This operator is unitary because $H^\dagger = H$, so that $U(t, t_0)^\dagger U(t, t_0)$ $= e^{iH(t-t_0)/\hbar}e^{-iH(t-t_0)/\hbar} = 1$. Therefore, in the limit $N \to \infty$,

$$\langle\psi(t_0)|U^\dagger(t, t_0)U(t, t_0)|\psi(t_0)\rangle = \langle\psi(t_0)|\left[1 + \frac{\Delta t}{N}\frac{i}{\hbar}H\right]^N \left[1 - \frac{\Delta t}{N}\frac{i}{\hbar}H\right]^N |\psi(t_0)\rangle =$$

$$= \langle\psi(t_0)|\left[1 + \frac{\Delta t^2}{N^2}\frac{1}{\hbar}H\right]^N |\psi(t_0)\rangle \to \langle\psi(t_0)|\psi(t_0)\rangle. \tag{11.85}$$

To see the importance of this requirement, consider expanding $|\psi(t_0)\rangle$ and $|\psi(t)\rangle$ in terms of the eigenstates of some observable as

$$|\psi(t_0)\rangle = \sum_k c_k(t_0)|k\rangle \tag{11.86}$$

$$|\psi(t)\rangle = \sum_k c_k(t)|k\rangle. \tag{11.87}$$

Since the time evolution operator preserves normalization, we have $\langle\psi(t_0)|\psi(t_0)\rangle =$ $\langle\psi(t)|\psi(t)\rangle$, which means that

$$\sum_k |c_k(t_0)|^2 = \sum_k |c_k(t)|^2. \tag{11.88}$$

This expresses the conservation of probability.

Other unitary operators include spatial rotations, which will be developed in a subsequent chapter, and position translation, which will be briefly described in the next section.

11.3.2 Continuous spectra

So far we have only considered eigenfunctions of operators with discrete spectra of eigenvalues λ_n. We know that position x and momentum p are continuous variables, so we expect there to be eigenfunctions of the x (p) operator for each value of position (momentum). Let us first consider eigenstates of position $\mathbf{x}$. We can think of the $x-$axis as an infinite-dimensional vector of the form $...x - 2\Delta, x - \Delta, x, x + \Delta, x + 2\Delta$ in the limit of $\Delta \to 0$. Each entry represents a dimension in an infinite Hilbert space. We therefore need to determine eigenfunctions of the position operator $\mathbf{x}$ as

$$\mathbf{x}|x\rangle = x|x\rangle, \tag{11.89}$$

where x is a real number representing coordinate position. Consider the general matrix element,

$$\langle x'|\mathbf{x}|x\rangle = x\langle x'|x\rangle = 0 \ \text{ unless } x' = x. \tag{11.90}$$

We also require that a general state can be converted from a coordinate basis to its abstract form by superposition of the $|x\rangle$ eigen- (or basis) states, which in the continuum limit becomes

$$|\psi\rangle = \sum \psi(x_k)|x_k\rangle \to \int dx\, \psi(x)|x\rangle. \tag{11.91}$$

This implies that

$$\langle x'|\psi\rangle = \int dx\, \psi(x)\langle x'|x\rangle = \psi(x') \tag{11.92}$$

and therefore

$$\langle x'|x\rangle = \delta(x' - x), \tag{11.93}$$

which is the analog of $\langle n'|n\rangle = \delta_{n'n}$ in the discrete case, where the δ-function is of the Kronecker type. In addition, this relationship proves immediately that the $|x\rangle$ states form a complete set. Therefore, the identity operator is

$$I = \int dx|x\rangle\langle x| \tag{11.94}$$

which can be verified as

$$I|x'\rangle = \int dx|x\rangle\langle x|x'\rangle = \int dx|x\rangle\delta(x - x') = |x'\rangle.$$

We further note that writing $a|x\rangle = |ax\rangle$, where a is a real constant, results in

$$a\langle x|x'\rangle = a\delta(x - x') \neq \langle ax|x'\rangle = \delta(ax - x'), \tag{11.95}$$

which illustrates the danger of moving operators into a bra or ket.

We can be tempted to determine the coordinate representation of the $|x\rangle$ state. Let

$$\psi_y(x) = \delta(x-y). \tag{11.96}$$

We should then have

$$\int \psi_y(x)\psi_{y'}(x)dx = \int \delta(x-y)\delta(x-y')dx \stackrel{?}{=} \delta(y-y'). \tag{11.97}$$

This query can be answered by use of a test function:

$$\int f(y)\left[\int \delta(x-y)\delta(x-y')dx\right]dy = \int \delta(x-y')\left[\int f(y)\delta(x-y)dy\right]dx =$$

$$= \int f(x)\delta(x-y')dx = f(y'), \tag{11.98}$$

which shows that the final equality in Eq. (11.97) holds.

Next, consider the eigenfunctions of the momentum operator

$$\mathbf{p}_x = \frac{\hbar}{i}\frac{\partial}{\partial x}. \tag{11.99}$$

This operator is Hermitian because

$$(\phi(x), \mathbf{p}_x\psi(x)) = \int_{-\infty}^{\infty} \phi^*(x)\frac{\hbar}{i}\frac{\partial}{\partial x}\psi(x)dx =$$

$$= \phi^*(x)\psi(x)|_{-\infty}^{\infty} + \int_{-\infty}^{\infty}\left[\frac{\hbar}{-i}\frac{\partial}{\partial x}\phi^*(x)\right]\psi(x)dx = (\mathbf{p}_x\phi(x), \psi(x)), \tag{11.100}$$

where we have taken the functions to satisfy $\phi(x), \psi(x) \to 0$ for $x \to \infty$. We therefore know that the eigenvalues of $\mathbf{p}$ are real, and we can determine the eigenfunctions from the requirement that[18]

$$\mathbf{p}\,\psi_p(x) = \frac{\hbar}{i}\frac{\partial}{\partial x}\psi_p(x) = p\,\psi_p(x), \tag{11.101}$$

which has general solution

$$\psi_p(x) = C_p e^{ipx/\hbar}, \tag{11.102}$$

where C_p is a possibly complex constant and p is a real number, with $x \in (-\infty, \infty)$.

Defining $k = p/\hbar$, let us take the orthogonality relation, Eq. (11.20), to the continuum case,

$$\int \psi_{p'}^*(x)\psi_p(x)\,dx = C_{k'}^* C_k \int e^{i(k-k')x}dx = 2\pi|C_k|^2\delta(k-k'), \tag{11.103}$$

[18]We are considering one dimension only so that $\mathbf{p} = \mathbf{p}_x$, momentum in the x direction, however, this can be readily generalized to all three dimensions.

where we have used Eq. (11.54). This result implies that if we choose $C_k = 1/\sqrt{2\pi}$ then

$$\langle k'|k\rangle = \delta(k - k') \tag{11.104}$$

with

$$\psi_k(x) = \langle x|k\rangle = \frac{1}{\sqrt{2\pi}}e^{ikx}. \tag{11.105}$$

We can define an identity operator I_k similarly to I_x, as

$$I_k = \int dk\ |k\rangle\langle k|. \tag{11.106}$$

11.3.3 Momentum space wave functions

Because the $|k\rangle$ states form a complete set, we can expand any abstract state vector (function) terms of them in the same way that we can expand in the $|x\rangle$ states:

$$|\psi\rangle = \int dk\, \psi(k)|k\rangle.$$

We point out that any state $|\psi\rangle$ expanded in terms of a general basis $|q\rangle$ can be expanded in terms of any other basis $|\ell\rangle$ with expansion coefficients $\langle \ell|q\rangle\langle q|\psi\rangle$, and the Fourier transform is an example of this. Completeness allows us to project the abstract function $|\psi\rangle$ onto a coordinate representation as

$$\psi(x) = \langle x|\psi\rangle = \langle x| \int \psi(k)|k\rangle dk = \int dk\, \psi(k)\langle x|k\rangle = \int\ dk\, \psi(k)\frac{1}{\sqrt{2\pi}}e^{ikx}. \tag{11.107}$$

Then $\psi(k)$ can be determined from

$$\psi(k') \equiv \langle k'|\psi\rangle = \int dx\, \langle k'|x\rangle\langle x|\psi\rangle = \int dx\, \frac{1}{\sqrt{2\pi}}e^{-ik'x}\psi(x), \tag{11.108}$$

which is the Fourier transform, or spectral amplitude, of $\psi(x)$. We can further check that this is consistent by writing $\psi(x)$ in the form of the integral over k:

$$\int dx\, \frac{1}{\sqrt{2\pi}}e^{-ik'x}\psi(x) = \int dx\, \frac{1}{\sqrt{2\pi}}e^{-ik'x} \int dk\, \psi(k)\frac{1}{\sqrt{2\pi}}e^{ikx} =$$

$$= \frac{1}{2\pi}\int dk \int dx\, \psi(k)e^{-ik'x}e^{ikx} = \frac{1}{2\pi}\int dk\, 2\pi\psi(k)\delta(k - k') = \psi(k'). \tag{11.109}$$

The function $\psi(k')$ is the wave function $|\psi\rangle$ in the k basis, and is the Fourier transform of the original function $\psi(x)$.

As an aside, we note that the momentum operator is proportional to the generator of displacements. For an infinitesimal displacement dx, we have

$$\psi(x + dx) = \psi(x) + dx\frac{\partial\psi(x)}{\partial x} = \psi(x) + dx\frac{i}{\hbar}\mathbf{p}_x\psi(x). \tag{11.110}$$

The position translation operator for finite displacements can be developed by considering a finite translation as a sequence of infinitesimal steps, in the same way we did for the time translation operator. This yields

$$\hat{T}_x(x, x_0) = e^{\frac{i}{\hbar}\mathbf{p}_x(x-x_0)}. \tag{11.111}$$

11.3.3.1 Not every bra has a ket

It might be tempting to take $|x\rangle$ or $|k\rangle$ as possible quantum state kets; however, both $\langle x|x\rangle$ and $\langle k|k\rangle$ diverge and therefore cannot be normalized. It is best to consider the bra $\langle x|$ as a projection operator. Above we considered $\langle \psi_{k'}|x\rangle$, which we can define as $[\langle x|\psi_{k'}\rangle]^{\dagger}$.

In the case of the $|k\rangle$ kets, we saw above that if we have a continuous superposition of those states, we can define a perfectly reasonable, normalizable state function. Another method of handling the divergent normalization is to make the $|k\rangle$ state discrete by enclosing the system in a large box of dimension L along each axis, with the boundary condition that the wave function is zero on the boundaries. Then each state is quantized as $k_n = n\pi/L$ and can be normalized. Calculations can be performed using these states, and at the end the limit $L \to \infty$ can be taken. This is called normalization in a box.

In the spirit of the Dirac notation, it is usual to ignore these issues and simply use the delta function normalization as defined in Eq. (11.104).

11.4 General features of the theory and Dirac notation

Let us recapitulate the useful features of the Dirac notation in quantum mechanics.

- Not all physical states arise from solutions of the Schrödinger equation and have a coordinate or momentum representation. An example is the intrinsic spin of a particle. Dirac notation allows us to represent these states as kets.
- The abstract states do not rely on a specific coordinate system or representation (for example, position or momentum) but can be expanded in terms of the eigenfunctions of any Hermitian operator (which, as we have seen, form a basis for the space). Transformations of this basis are coordinate transformations in the Hilbert space. Both the state vectors and the operators are changed by these transformations.
- The notation allows a simple reformulation of quantum mechanics that only involves operators.
- Use of the abstract states and their operators allows many problems to be solved by algebraic methods without resort to the spatial coordinate or momentum basis wave functions. This will be explored in a later chapter.
- It is simple to change between Schrödinger style wave mechanics notation and Dirac notation. In fact, the Schrödinger wave function $\psi(x)$ is given by $\psi(x) = \langle x|\psi\rangle$.

11.4.1 The rules of quantum mechanics

Over the years, the rules or axioms of quantum mechanics have been expounded many times, perhaps first by Jordan in 1927.[19] The first attempts at forming an axiomatic theory were cumbersome. Today the rules are most easily cast in the context of the Hilbert space formulation of quantum mechanics.

There is no official set of rules for quantum mechanics as there is for chess or for poker; however, the following guidelines, which represent the nearly universally accepted interpretation of the Hilbert space formulation, allow any problem in the non-relativistic theory to be uniquely solved, assuming a Hamiltonian H with suitable boundary conditions, as well as operators that act on the eigenstates of H, can be defined. The postulates, axioms, or rules outlined here are for one-particle systems but these results can be generalized to multi-particle systems and carry over, largely unmodified, to the relativistic theory, where the conservation of probability is modified because of the possibility of pair creation.

1. The state of a microscopic system is given by a linear (coherent) superposition of a complete set $|n\rangle$ of eigenstates of any Hermitian operator as $|\psi\rangle = \sum_n c_n|\psi_n\rangle$. These eigenstates can be internal quantized states (intrinsic spin) that do not explicitly depend on coordinates or momentum, or they can be space- and time-dependent eigenstates that are solutions to the Schrödinger equation with a potential defined over some region of space, or they can be products of the two (for example, spin and spatial wave functions for atomic electrons).
2. The probability of being in a given eigenstate is determined by $|c_n|^2$ (which can be a continuous function of n) and $\sum_{\text{all } n} |c_n|^2 = 1$ (normalization).
3. The time evolution of a state is given by the time displacement operator $e^{-iHt/\hbar}$ for a time-independent Hamiltonian. For an eigenstate for which $H|\psi_n\rangle = E_n|\psi_n\rangle$, time evolution is simply given by the phase factor $e^{-iE_nt/\hbar}$. The case of a time-varying Hamiltonian can be treated by straightforward generalizations.
4. States are evolved by rotations, displacements in time or space, or other actions that operate on general states $|\psi\rangle$. These operators are unitary, so that they preserve normalization.
5. A measurement is described by a Hermitian operator $\mathbf{A}$. The results of a measurement can only be eigenvalues of $\mathbf{A}$. Expanding the state in the eigenfunctions of $\mathbf{A}$ gives the relative probability of obtaining each of the different eigenvalues when a measurement is made. At some point in the measurement process, the initial superposition state collapses to a single eigenstate, with the probability of ending up in $|n\rangle$ given by $|c_n|^2$. If the state collapses to $|n\rangle$, measurement of $\mathbf{A}$ will return the corresponding eigenvalue a_n. This is sometimes referred to as the "collapse of the wave function". The collapse is irreversible and occurs at the *Heisenberg Cut*, which lies perhaps within the measurement apparatus; until that collapse, the state remains in a superposition.

[19] Jordan, P. *Über eine neue Begründung der Quantenmechanik II* [On a new foundation of quantum mechanics II], Z. Phys. **44**, 1 (1927). See also Lacki, J., *The early axiomatizations of quantum mechanics: Jordan, von Neumann and the continuation of Hilbert's program*, Arch. Hist. Exact Sci. **54**, 279 (2000).

6. As discussed in Chapter 9, a system that contains identical particles has explicit symmetry requirements.
7. Planck's constant h is the same for all fields and particles and does not depend on energy.

In essence, these rules explain how the Hilbert space functions associated with some Hamiltonian are to be physically interpreted. It must also be emphasized that these interpretations of the Hilbert space functions are based on experimental evidence, and hence really cannot be thought of as axioms in the classical mathematical sense. To date, no violations of these rules have been discovered. As is usual in any logical system, we cannot use that system itself to decide on its completeness and ultimate validity in describing the physical world. We rely on empirical evidence, and we will return to this question in Chapter 14.

One complication with the Born interpretation, that the probability to be in a particular quantum eigenstate is given by the squared magnitude of that state's amplitude, is that it applies to a so-called *pure state*. Although this is the basis of the statistitical nature of quantum mechanics, treating an ensemble of particles with slightly different quantum states, or calculating the effects of random perturbations, are fraught with difficulties. Such effects are expected in experimental situations, and not accounting for the fact that a system can be in a *mixed state* (a blend of different single particle pure states) leads to paradoxes such as Schrödinger's cat. The density matrix was introduced to address this next level of statistitical variation within a system.

In later chapters, we will demonstrate how these rules are applied to a variety of problems and will touch again on their foundational understanding. In particular, we will review von Neumann's treatment of the subject and will discuss the density matrix non-locality effects and hidden variables in some detail.

12

Dirac and Jordan commit "sin squared": second quantization and the beginning of quantum field theory

12.1 Introduction

Heisenberg and some other physicists were initially annoyed when Schrödinger's work started to appear. They felt that they had a complete theory and were annoyed at the intrusion and hence the necessity to think differently about the problem and learn new techniques. Schrödinger then added insult to injury by criticizing matrix mechanics as being horribly nonintuitive. His goal was to produce an intuitive physical model. However, when Schrödinger and then the Dirac-Jordan transformation theory showed that both models were instances of a more general theory, this was accepted by all sides.

One somewhat unattractive feature of Schrödinger's wave mechanics was the fact that the wave equation for an N-particle system was written in a $3N$ dimensional space. This detracted severely from the intuitiveness of Schrödinger's method and had played a role in the argument between Schrödinger and Heisenberg over whose method was more intuitive. As both models were seen to make exactly the same predictions for experimental results, there was no objective criterion to allow selection between them, and so intuitiveness appeared as such a criterion. Heisenberg countered Schrödinger's charge that matrix mechanics was horribly abstract with the countercharge that $3N$ dimensional space was even more abstract.

The German word used, "anschaulich," means something like "intuitive"—something that can be visualized.

Heisenberg asserted that nothing could be less "anschaulich" than a wave in $3N$ dimensional space and published his paper[1] introducing the uncertainty principle. He started by defining an "intuitive" theory as a theory where "in all simple cases the results of the theory can be understood qualitatively, and we can be certain that the application of the theory never leads to any inner contradictions." He went on to say that the (then) intuitive meaning of quantum mechanics was still full of inner contradictions: the relationship between the continuous evolution of the wave function and discontinuous quantum jumps, as well as the connection between the wave and particle

[1] Heisenberg, W., (1927), op. cit.

The Historical and Physical Foundations of Quantum Mechanics. Robert Golub and Steven K. Lamoreaux, Oxford University Press.
 DOI: 10.1093/oso/9780198822189.003.0012

descriptions, had yet to be clarified. He concluded that it is not possible to understand quantum mechanics (by which he meant matrix mechanics) with the usual kinematical and mechanical quantities. He then, by means of numerous examples, outlined the results of the uncertainty principle, stating that the statistical nature of the theory was the result of the necessary uncertainties involved in measuring canonically conjugate quantities. In his conclusion he quoted Schrödinger's characterization of quantum mechanics as "horribly, even repulsively non-intuitive and abstract" and replied that in his opinion "regarding the main physical questions, the popular intuitive nature of wave mechanics leads away from the straight path which was laid out by the work of Einstein and de Broglie, on the one hand, and Bohr and the quantum (matrix) mechanics on the other."

This question of intuitiveness became politically important in the 1930s, when a group of physicists led by Stark and Lenard tried to promote what they called "Deutsche" (German) physics with the argument that relativity and quantum mechanics were too abstract and nonintuitive. This was strongly opposed by Heisenberg and Jordan: the latter even published a book called *Anschaulich Quantum Mechanics* (1936). Jordan's position was rather complicated. He defended relativity and quantum mechanics against the proponents of "Deutsche" physics, but as a member and avid supporter of the Nazi Party, concerned to preserve (or improve) his position, he used the practical argument that the physics of Einstein and Copenhagen worked and their application would contribute to the power and prestige of the Third Reich.

Jordan, who was a big contributor to the development of matrix mechanics, had the flexibility to take up the newer forms of the theory and make advances in other directions. As we will now see, further development of the quantum theory, starting with Dirac's and Jordan's extraction of the essence of the theory by considering operators and vectors in Hilbert space and the introduction of what was called second quantization, provided a way for the N-particle systems to be described in ordinary three-dimensional space.

On hearing of the latter development Einstein, expressing his long-held unhappiness with quantum mechanics as it had developed, characterized second quantization as "sin squared," although in fact the method was based on his earlier method of treating the ideal gas as a system of oscillating modes.

12.2 Dirac's q-numbers, operators, and the quantum mechanics of Dirac, Jordan, and von Neumann

In this section we will use Dirac's[2] bra-ket notation (developed in 1939 and first published in the third edition of his book, 1947) which we have discussed in Chapter 11. While introducing nothing new physically, this notation makes the discussion much easier to understand.

Some time after receiving a copy of Heisenberg's original paper, Dirac realized, as had Born and Jordan, that the key feature was the non-commutativity of the canonical variables (p, q) and that any quantities satisfying the canonical commutation relations

[2]Dirac, P.A.M., *On the theory of quantum mechanics*, Proc, Roy, Soc, (London), A112, 661, (1926), received Aug. 26, 1926 and Jordan, P., *On a new foundation of quantum mechanics*, Zeit. Phys. 40, 809, (1927) received Dec. 18, 1926.

could be used to build a theory. He called such noncommuting quantities, q (quantum) numbers as opposed to ordinary numbers, which he called c−numbers. q−numbers obeyed all the rules of ordinary arithmetic except that pairs of canonical variables satisfied the commutation relation

$$[p, x] = px - xp = \frac{\hbar}{i}. \tag{12.1}$$

Thus in quantum mechanics the classical canonical variables become quantities that do not satisfy the commutation rules of ordinary arithmetic.

One realization of q-numbers was the representation

$$p = \frac{\hbar}{i}\frac{\partial}{\partial x}, \tag{12.2}$$

which satisfies (12.1) because

$$\frac{\hbar}{i}\frac{\partial}{\partial x}(x\psi(x)) - x\frac{\hbar}{i}\frac{\partial}{\partial x}\psi(x) = \frac{\hbar}{i}\psi(x). \tag{12.3}$$

Thus q-numbers morphed into operators and the term "q-number" does not appear in modern usage. As we have emphasized earlier the essence of the physics is in the commutation relations, that is the physics is in the operators. Using Schrödinger's theory these are differential operators acting on the wave function. In the Heisenberg, Born, and Jordan matrix mechanics, the operators are matrices calculated in the basis of the eigenstates of the Hamiltonian. In general, the operators can act on "states" of the system turning, say, a given energy eigenstate into a sum of several such states. Several years after the completion of the theory Dirac invented a notation which is so powerful it makes the mathematical relations obvious. In this notation states are represented by $|n\rangle$, which means that the state is an eigenfunction of some operator with eigenvalue designated by the quantum number n. The states can be considered as vectors in a (Hilbert) space whose axes are taken as representing an orthonormal set of functions, and the components of the vector along a given axis represent the amplitude, a_n, of that basis function in the superposition representing the state, $\psi(x)$:

$$\psi(x) = \sum_n a_n\psi_n(x), \tag{12.4}$$

where the $\psi_n(x)$ are the orthonormal basis functions. This is called a Hilbert space. Different sets of functions can serve as the coordinates in this space. One set of coordinates could be the energy eigenfunctions, another could be the position eigenfunctions as shown in the previous chapter.

Like ordinary vectors the state vectors can be expanded in components relative to another coordinate system. So the energy eigenstates could be represented by components along the direction of the position eigenstates. Using the expression $\langle a|\, b\rangle$ to express the dot product in Hilbert space, the expansion of an energy eigenstate in position eigenstates would be given by

$$|n\rangle = \sum_x |x\rangle\,\langle x|\, n\rangle, \tag{12.5}$$

which is equivalent to the usual vector relation:

$$\vec{A} = \sum_{i=1,2,3} \vec{x}_i \left(\vec{x}_i \cdot \vec{A}\right), \tag{12.6}$$

where the $\vec{x}_i$ are a set of normalized, orthogonal vectors spanning the ordinary three dimensional space.

The component along a given x eigenstate direction $\langle x|\, n\rangle$ is just the Schrödinger eigenfunction $\psi_n(x) = \langle x|\, n\rangle$ and its conjugate is $\psi_n^*(x) = \langle n|\, x\rangle$.

Thus the orthonormality of the eigenstates is expressed as

$$\delta_{mn} = \int dx \psi_m^*(x)\, \psi_n(x) = \int dx\, \langle m|\, x\rangle\, \langle x|\, n\rangle = \langle m|\, n\rangle. \tag{12.7}$$

The last step is based on the completeness relation $\sum_j |j\rangle\, \langle j| = 1$ which also follows from (12.5). In this expression the summation sign represents a sum over discrete values and an integral over continuous values of the index j. We can see the correctness of this relation as follows. The usual form of the completeness relation is written

$$\sum_n \psi_n^*(x')\, \psi_n(x) = \delta(x - x'), \tag{12.8}$$

which in Dirac notation is

$$\sum_n \langle x|\, n\rangle\, \langle n|\, x'\rangle = \langle x|\, x'\rangle = \delta(x - x'). \tag{12.9}$$

This is an example of the notation proving the theorems. We gave a more rigorous discussion of these questions in Chapter 11.

The average value of the position of a particle in the state $\psi_n(x)$ is

$$\langle x\rangle = \langle n|\, \mathbf{x}\, |n\rangle = \sum_{x'',x'} \langle n|\, x''\rangle\, \langle x''|\, \mathbf{x}\, |x'\rangle\, \langle x'\, |n\rangle = \int dx' \psi_n^*(x') x' \psi_n(x') = \int dx\, |\psi(x)|^2\, x$$

so that the probability of finding the value of x to be between x and $x+dx$ is $|\psi(x)|^2\, dx$. Here we have used $\langle x''|\, \mathbf{x}\, |x'\rangle = x'\delta(x' - x'')$ as well as completeness, $\sum_{x'} |x'\rangle\, \langle x'| = 1$.

The basis could be chosen to be the eigenstates of the Hamiltonian, the angular momentum or any other physical quantity such as the position or momentum of a particle. An operator $\mathbf{A}$ operating on the state $|n\rangle$ gives a new state $\mathbf{A}\, |n\rangle$ which, as any state, can be expanded in the original eigenstates:

$$\mathbf{A}\, |n\rangle = \sum_m a_{mn}\, |m\rangle = \sum_m |m\rangle\, \langle m|\, A\, |n\rangle \tag{12.10}$$

$$\langle k|\, A\, |n\rangle = \sum_m \langle k|\, m\rangle\, \langle m|\, A\, |n\rangle = \sum_m \langle k|\, m\rangle\, a_{mn} = a_{kn}. \tag{12.11}$$

The usual method of expressing this with Schrödinger eigenfunctions is

$$\langle k|\, A\, |n\rangle = \sum_{x,x'} \langle k|\, x'\rangle \langle x'|\, A\, |x\rangle \langle x|\, n\rangle = \int\int dx dx' \psi_k^* (x') A(x) \delta(x - x') \psi_n(x) \tag{12.12}$$

$$= \int dx \psi_k^* (x) A(x) \psi_n(x), \tag{12.13}$$

where $\langle x'|\, A\, |x\rangle = A(x)\, \delta(x - x')$ are the matrix elements of $\mathbf{A}$ in the position basis. This demonstrates the way the matrix elements in one basis, $|x\rangle$, are transformed into those in another basis $|m\rangle$. Dirac considered this "transformation theory" as the characteristic feature of quantum mechanics. In Hilbert space, the dot product is Hermitian. This means that $\langle k|\, m\rangle = \langle m|\, k\rangle^* = \langle k|\, m\rangle^\dagger$. (The symbol † stands for Hermitian conjugate.) Then the matrix element $\langle k|\, A\, |n\rangle = \langle k|\, (A\, |n\rangle) = ((\langle n|\, A)\, |k\rangle^* = \langle n|\, A\, |k\rangle^*$ or $A_{kn} = A_{nk}^* = A_{kn}^\dagger$. Operators are thus represented by Hermitian matrices.

Using this notation, we can describe physical systems with operators operating on states or vectors in Hilbert space. The operators transform vectors (states) into other states, we can express the eigenfunctions of a given Hermitian operator in terms of the eigenfunctions of another such operator, and the time dependence of the system will be described by the time dependence of the state vector. This is called the Schrödinger picture. An equivalent picture (the Heisenberg picture) puts the time dependence in the operators (Chapter 18).

Observables are represented by Hermitian operators, an observable "O" corresponding to an operator, $\mathbf{O}$. The eigenfunctions of $\mathbf{O}$ are given by

$$\mathbf{O}\, |o_i\rangle = o_i\, |o_i\rangle \tag{12.14}$$

One of the main features of quantum mechanics is that any state of the system in question, $|n\rangle$, can be expanded as a series in the eigenfunctions of any Hermitian operator.

$$|n\rangle = \sum_i c_i\, |o_i\rangle \tag{12.15}$$

Then a measurement of "O" on the system when it is in state $|n\rangle$ will yield one of the eigenvalues of $\mathbf{O}$ with probability $|c_i|^2$ and following the measurement the system will be in the state $|o_i\rangle$. This is sometimes called the "collapse of the wave function." If many measurements are made the average value (expectation value) of these measurements will be given by

$$\langle n|\, \mathbf{O}\, |n\rangle = \sum_{i,j} c_i c_j^* \langle o_j|\, \mathbf{O}\, |o_i\rangle = \sum_i |c_i|^2\, o_i. \tag{12.16}$$

The fact that there is no way to predict the result of a single measurement is one of the characteristics of quantum mechanics that was the most disturbing to the classical physicists.

As an example consider the eigenstates of momentum $|p_o\rangle$, so that $\mathbf{p}\, |p_o\rangle = p_o\, |p_o\rangle$. We write this in the x representation, where $\langle x\, |n\rangle = \psi_n(x)$ as

$$\sum_{x'} \langle x| \mathbf{p} |x'\rangle \langle x' |p_o\rangle = p_o \langle x |p_o\rangle \tag{12.17}$$

$$\int dx' \delta(x - x') \frac{\hbar}{i} \frac{\partial}{\partial x'} \langle x' |p_o\rangle = \frac{\hbar}{i} \frac{\partial}{\partial x} \psi_{p_o}(x) = p_o \psi_{p_o}(x) \tag{12.18}$$

$$\psi_{p_o}(x) = \frac{1}{\sqrt{2\pi}} e^{\frac{i}{\hbar} p_o x} = \frac{1}{\sqrt{2\pi}} e^{ik_o x} = \langle x |p_o\rangle \tag{12.19}$$

where $k_o = p_o/\hbar$ and we have chosen the normalization so that

$$\sum_{x} \langle p' |x\rangle \langle x |p\rangle = \frac{1}{2\pi} \int dx e^{\frac{-i}{\hbar}(p'-p)x} = \delta(p' - p) = \langle p' |p\rangle. \tag{12.20}$$

Then for say eigenfunctions of the Hamiltonian $\psi_n(x) = \langle x |n\rangle$ we have

$$\sum_{x} \langle p |x\rangle \langle x |n\rangle = \langle p |n\rangle = \psi_n(p) \tag{12.21}$$

$$= \int dx \frac{1}{\sqrt{2\pi}} e^{\frac{-i}{\hbar} px} \psi_n(x) \tag{12.22}$$

so eigenfunctions of the momentum operator are plane waves and the wave functions in p representation and in x representation are related by the Fourier transform.

The Schrödinger equation in the $|x\rangle$ representation is

$$i\hbar \frac{\partial \psi_n(x)}{\partial t} = H \psi_n(x), \tag{12.23}$$

or

$$i\hbar \frac{\partial \langle x |n\rangle}{\partial t} = \sum_{x'} \langle x| \mathbf{H} |x'\rangle \langle x' |n\rangle, \tag{12.24}$$

with $\langle x| \mathbf{H} |x'\rangle = H(x) \delta(x - x')$. The generalized form of this is (remember $\sum_{x'} |x'\rangle \langle x'| = 1$)

$$i\hbar \frac{\partial |n\rangle}{\partial t} = \mathbf{H} |n\rangle, \tag{12.25}$$

as can be seen by multiplying from the left with $\langle x|$ which yields:

$$i\hbar \frac{\partial \langle x |n\rangle}{\partial t} = \langle x| \mathbf{H} |n\rangle \tag{12.26}$$

$$i\hbar \frac{\partial \psi_n(x)}{\partial t} = \sum_{x'} \langle x| \mathbf{H} |x'\rangle \langle x'| n\rangle = \int dx' H(x) \delta(x - x') \psi_n(x') \tag{12.27}$$

$$= H(x) \psi_n(x). \tag{12.28}$$

12.2.1 Some additional properties of noncommuting operators

In Section 6.3 we saw that the commutation relation

$$[\mathbf{p}, \mathbf{q}] = \frac{\hbar}{i} \tag{12.29}$$

implies

$$[f(\mathbf{p},\mathbf{q}),\mathbf{q}] = \frac{\hbar}{i}\frac{\partial f(\mathbf{p},\mathbf{q})}{\partial \mathbf{p}} \tag{12.30}$$

$$[f(\mathbf{p},\mathbf{q}),\mathbf{p}] = -\frac{\hbar}{i}\frac{\partial f(\mathbf{p},\mathbf{q})}{\partial \mathbf{q}}. \tag{12.31}$$

These will apply to any pair of canonical variables satisfying (12.29). Applying these relations, we find

$$\left[e^{i\alpha\mathbf{q}},\mathbf{p}\right] = -\hbar\alpha e^{i\alpha\mathbf{q}} \tag{12.32}$$

$$e^{i\alpha\mathbf{q}}\mathbf{p} = (\mathbf{p}-\hbar\alpha)\,e^{i\alpha\mathbf{q}} \tag{12.33}$$

$$\left[e^{i\alpha\mathbf{p}},\mathbf{q}\right] = \hbar\alpha e^{i\alpha\mathbf{p}} \tag{12.34}$$

$$e^{i\alpha\mathbf{p}}\mathbf{q} = (\mathbf{q}+\hbar\alpha)\,e^{i\alpha\mathbf{p}} \tag{12.35}$$

For a general function $f(\mathbf{p},\mathbf{q})$ we write, tentatively,

$$e^{i\alpha\mathbf{q}}f(\mathbf{p},\mathbf{q}) = f(\mathbf{p}-\hbar\alpha,\mathbf{q})\,e^{i\alpha\mathbf{q}} \tag{12.36}$$

$$e^{i\alpha\mathbf{p}}f(\mathbf{p},\mathbf{q}) = f(\mathbf{p},\mathbf{q}+\hbar\alpha)\,e^{i\alpha\mathbf{q}}. \tag{12.37}$$

If this holds for two functions f_1, f_2, it clearly holds for (f_1+f_2). It also holds for the product $f_1 f_2$ as can be seen as follows

$$e^{i\alpha\mathbf{q}}f_1(\mathbf{p},\mathbf{q}) = f_1(\mathbf{p}-\hbar\alpha,\mathbf{q})\,e^{i\alpha\mathbf{q}} \tag{12.38}$$

$$e^{i\alpha\mathbf{q}}f_2(\mathbf{p},\mathbf{q}) = f_2(\mathbf{p}-\hbar\alpha,\mathbf{q})\,e^{i\alpha\mathbf{q}} \tag{12.39}$$

$$f_2(\mathbf{p},\mathbf{q}) = e^{-i\alpha\mathbf{q}}f_2(\mathbf{p}-\hbar\alpha,\mathbf{q})\,e^{i\alpha\mathbf{q}} \tag{12.40}$$

$$e^{i\alpha\mathbf{q}}f_1(\mathbf{p},\mathbf{q})\,f_2(\mathbf{p},\mathbf{q}) = f_1(\mathbf{p}-\hbar\alpha,\mathbf{q})\,e^{i\alpha\mathbf{q}}e^{-i\alpha\mathbf{q}}f_2(\mathbf{p}-\hbar\alpha,\mathbf{q})\,e^{i\alpha\mathbf{q}} \tag{12.41}$$

$$= f_1(\mathbf{p}-\hbar\alpha,\mathbf{q})\,f_2(\mathbf{p}-\hbar\alpha,\mathbf{q})\,e^{i\alpha\mathbf{q}}. \tag{12.42}$$

Since this relation holds for $\mathbf{p}$ and any power of $\mathbf{p}$, it will hold for any function of $\mathbf{p}$ which is expandable in a power series, so (12.36) is seen to be valid for the usual functions we encounter in physics. Then $e^{i\alpha\mathbf{q}}$ is a displacement operator that shifts the momentum, $\mathbf{p}\to\mathbf{p}-\hbar\alpha$. The same relations hold for any pair of canonical variables satisfying the commutation relations (12.29).

12.2.2 Solution of the one dimensional harmonic oscillator by the operator method

The Hamiltonian is given by

$$H = \frac{p^2}{2\mu} + \frac{1}{2}\mu\omega_o^2 x^2 \tag{12.43}$$

where $\omega_o^2 = k/\mu$ with k the force constant and μ the mass. The energy eigenstates $|m\rangle$ satisfy $H|m\rangle = E|m\rangle$. The Schrödinger equation can be formulated as

$$H|m\rangle = E|m\rangle \tag{12.44}$$

$$\langle x|H|m\rangle = E\langle x|m\rangle \tag{12.45}$$

$$\sum_{x'} \langle x|H|x'\rangle\langle x'|m\rangle = E\langle x|m\rangle. \tag{12.46}$$

In the coordinate representation, this is

$$-\frac{\hbar^2}{2\mu}\frac{\partial^2}{\partial x^2}\psi + \frac{1}{2}\mu\omega_o^2 x^2\psi = E\psi \tag{12.47}$$

where $\langle x|m\rangle = \psi_m(x)$, $\langle x|H|x'\rangle = \delta(x-x')H\left(p = \frac{\hbar}{i}\frac{\partial}{\partial x}, x\right)$. We proceed by introducing the operators

$$a = \sqrt{\frac{\mu\omega_o}{2\hbar}}x + i\sqrt{\frac{1}{2\mu\hbar\omega_o}}p \tag{12.48}$$

$$a^\dagger = \sqrt{\frac{\mu\omega_o}{2\hbar}}x - i\sqrt{\frac{1}{2\mu\hbar\omega_o}}p. \tag{12.49}$$

Then

$$\left[a, a^\dagger\right] = 2\sqrt{\frac{\mu\omega_o}{2\hbar}}\sqrt{\frac{1}{2\mu\hbar\omega_o}}i\,[p,x] = 1 \tag{12.50}$$

since $[p,x] = \hbar/i$. Then

$$x = \left(a + a^\dagger\right)\frac{1}{2}\sqrt{\frac{2\hbar}{\mu\omega_o}} \tag{12.51}$$

$$p = \left(a - a^\dagger\right)\frac{1}{2i}\sqrt{2\mu\hbar\omega_o} \tag{12.52}$$

and the Hamiltonian (12.43) becomes

$$H = \left[-\left(a - a^\dagger\right)^2 + \left(a + a^\dagger\right)^2\right]\frac{1}{4}\hbar\omega_o \tag{12.53}$$

$$= \left[\left(aa^\dagger + a^\dagger a\right)\right]\frac{1}{2}\hbar\omega_o = \left(a^\dagger a + \frac{1}{2}\right)\hbar\omega_o, \tag{12.54}$$

so that the eigenstates of H are identical with the eigenstates of $a^\dagger a$. This operator is called the number operator $N = a^\dagger a$, for reasons given below. Assume we have such an eigenstate $a^\dagger a|j\rangle = j|j\rangle$ where j is some number. Now operate on that state with the operator a to generate the state $a|j\rangle$. Is this state also an eigenstate of $a^\dagger a$? We calculate

$$a^\dagger a\left(a|j\rangle\right) = \left(aa^\dagger - 1\right)a|j\rangle = a\left(a^\dagger a - 1\right)|j\rangle = (j-1)\,a|j\rangle. \tag{12.55}$$

Thus if $|j\rangle$ is an eigenstate of $N = a^\dagger a$ with eigenvalue j, then the state $a|j\rangle$ is an eigenstate of the same operator with eigenvalue $(j-1)$. According to (12.54), the state $|j-1\rangle$ has an energy eigenvalue $(j-1/2)\,\hbar\omega_o$. We can continue applying the operator a, thus obtaining eigenstates with lower and lower energies. In order to avoid this leading to states with an infinite negative energy, the process must be cut off, which means there must be a state of lowest energy, $|j=0\rangle$ (called the ground state) for which $a\,|j=0\rangle = 0$. This means that the j's must be integers so that successive applications of a will eventually lead to $j=1$ and then to $j=0$, which will designate the ground state. This state will still have a nonzero energy $\frac{1}{2}\hbar\omega_o$. We have seen (in the application of the Schrödinger equation to this problem) that this zero-point energy is an indication of the fluctuations in x and p in the ground state.

Given an eigenstate $|j\rangle$, we can form the state $a^\dagger\,|j\rangle$. Applying the operator N to this, we have

$$\begin{aligned}\left(a^{+}a\right)a^{\dagger}\,|j\rangle &= a^{\dagger}\left(1+a^{\dagger}a\right)|j\rangle \\ &= (1+j)\,a^{\dagger}\,|j\rangle \end{aligned} \tag{12.56}$$

so that $a^\dagger\,|j\rangle$ is an eigenstate of H with eigenvalue $(E_{j+1}+\hbar\omega_o/2)$. Now if we apply $a^\dagger$ to the ground state ψ_o we obtain a state with energy $\hbar\omega_o\left(1+\frac{1}{2}\right)$. Repeating this procedure n times we obtain a state with energy $E_n = \hbar\omega\left(n+\frac{1}{2}\right)$. Thus the only allowable energies of the harmonic oscillator differ from each other by an integer multiple of $\hbar\omega_o$, and we have deduced the quantization of the energy as well as the existence of the zero-point energy from the commutation relation.

We can easily find the ground state in the $\mathbf{x}$ representation as follows.

The ground state satisfies (using Eq. 12.48)

$$\mathbf{a}\,|n_o\rangle = 0 = \sum_{x'}\langle x|\,\mathbf{a}\,|x'\rangle\,\langle x'\,|n_o\rangle \tag{12.57}$$

$$= \int dx'\delta\left(x-x'\right)\left(\sqrt{\frac{\mu\omega_o}{2\hbar}}x' + \sqrt{\frac{\hbar}{2\mu\omega_o}}\frac{\partial}{\partial x'}\right)\psi_o\left(x'\right) \tag{12.58}$$

$$= \left(\sqrt{\frac{\mu\omega_o}{2\hbar}}x + \sqrt{\frac{\hbar}{2\mu\omega_o}}\frac{\partial}{\partial x}\right)\psi_o\left(x\right). \tag{12.59}$$

Thus

$$\frac{\partial}{\partial x}\psi_o\left(x\right) = -\frac{\mu\omega_o}{\hbar}x\psi_o\left(x\right) \tag{12.60}$$

$$\psi_o\left(x\right) = Ne^{-\frac{\mu\omega_o}{2\hbar}x^2}. \tag{12.61}$$

In this case we see that while the average value of x is zero, $\langle n_o|\,\mathbf{x}\,|n_o\rangle = 0$, the average value of $\mathbf{x}^2$ is nonzero so there is a fluctuation in the position of the oscillator. Since the representation of the ground state in the momentum basis is given by the Fourier transform of $\psi_o\left(x\right)$, there will be a nonzero average of $\mathbf{p}^2 = \langle n_o|\,\mathbf{p}^2\,|n_o\rangle$ so there will be a fluctuation in the momentum as well. This effect is called zero-point fluctuations and results in the nonzero ground state energy $E_o = \hbar\omega_o/2$.

We can obtain the matrix elements of $a, a^\dagger$ as follows:

$$\langle k| a |j\rangle \neq 0 \quad \text{only for } k = j-1 \tag{12.62}$$

$$\langle k| a^\dagger |j\rangle \neq 0 \quad \text{only for } k = j+1 \tag{12.63}$$

$$\langle j| a^\dagger a |j\rangle = j = \sum_k \langle j| a^\dagger |k\rangle \langle k| a |j\rangle = \sum_k |\langle k| a |j\rangle|^2 = \sum_k \left|\langle j| a^\dagger |k\rangle\right|^2 \tag{12.64}$$

where only the value $k = j-1$ occurs in the implied sum over k. Thus

$$\langle j-1| a |j\rangle = \sqrt{j}. \tag{12.65}$$

Also $j = \left|\langle j| a^\dagger |j-1\rangle\right|^2$ so $\langle j| a^\dagger |j-1\rangle = \sqrt{j}$ or

$$\langle j+1| a^\dagger |j\rangle = \sqrt{j+1}. \tag{12.66}$$

Substituting these values into Eq. (12.51), we find the matrix elements for position x, in agreement with those found using matrix mechanics in Section 6.3.1.

12.3 The beginning of quantum field theory

In the second paper ever written on matrix mechanics,[3] where Born and Jordan explained and expanded on Heisenberg's original idea (July 1925, op. cit.), Jordan added a section at the end where he suggested that the electromagnetic field should be represented by matrices, i.e., noncommuting objects which, according to the later Dirac-Jordan generalization of the theory, would be considered operators. Born and Jordan pointed out that the radiation field in a cavity, which had an infinite number of degrees of freedom, could be considered as a system of an infinite number of uncoupled independent oscillators, but they did not say what the commutation relations might be.

In their next paper, the famous "three-man paper" of Born, Heisenberg, and Jordan,[4] Jordan again contributed a last section dealing with quantizing the electromagnetic field. He again pointed out that the radiation in a cavity could be considered as a system of harmonic oscillators, and this time, taking a vibrating string as a simplified model of the radiation field, he applied the quantum conditions to the oscillators. Using this method, he obtained the correct expression for the energy fluctuations containing both wave and particle contributions that had been originally obtained by Einstein. In the next section we show how a continuous system such as a field is treated according to quantum mechanics, taking the vibrating string as the simplest example.

12.3.1 The vibrating string as an example of a continuous field with an infinite number of degrees of freedom

We begin with a review of the application of the Lagrangian and Hamiltonian formalism to continuous fields.[5] The key point is that in systems of many particles the Lagrangian is given as a function of the positions and velocities of all the particles as

[3]M. Born and P. Jordan, *Zur Quantenmechanik* [On quantum mechanics], *Zeit. Phys.* **34**, 858 (1925) received September 27, 1925.

[4]Born, M, Heisenberg, W. and Jordan, P., *Zur Quantenmechanik II* [On quantum mechanics II], *Z. Phys.* **36**, 557 (1926) received November 16, 1925.

[5]See Appendix E for more details.

$$L = L(\dot{q}_i, q_i) = \sum_i \frac{m\dot{q}_i^2}{2} - V\left(q_1, q_2......\right), \tag{12.67}$$

where i runs over all the coordinates of all the particles, and we have assumed Cartesian coordinates. For the case of the vibrating string or a field, the dynamical variables depend on position as well as time $y\left(x,t\right)$, so that the continuous variable x serves as the discrete index i and sums over i are replaced by integrals over x.

Thus, we have shown how the equation of motion for the vibrating string,

$$\rho \frac{\partial^2 y\left(x,t\right)}{\partial t^2} - \tau \frac{\partial^2 y\left(x,t\right)}{\partial x^2} = 0, \tag{12.68}$$

can be derived from the Lagrangian

$$L = \frac{1}{2}\int dx \left[\rho\left(\dot{y}\left(x,t\right)\right)^2 - \tau\left(\frac{\partial y\left(x,t\right)}{\partial x}\right)^2\right] = \int \mathcal{L}\left(\dot{y}, \frac{\partial y}{\partial x}\right) dx, \tag{12.69}$$

where $\mathcal{L}$ is the Lagrangian density. Here ρ is the density of the string per unit length and τ is the tension in the string.

With the field momentum defined by $\Pi = \partial\mathcal{L}/\partial\dot{y} = \rho\,\dot{y}$, the Hamiltonian is given by $\left(y' = \partial y\left(x,t\right)/\partial x\right)$

$$H = \int dx \left(\frac{\partial\mathcal{L}}{\partial\dot{y}}\dot{y} - \mathcal{L}\right) \equiv \int dx\,\mathcal{H} \tag{12.70}$$

$$\mathcal{H} = \frac{\partial\mathcal{L}}{\partial\dot{y}}\dot{y} - \mathcal{L}\left(\dot{y}, y'\right). \tag{12.71}$$

For the vibrating string, we find

$$\mathcal{H} = \rho\dot{y}^2 - \frac{1}{2}\left(\rho\dot{y}^2 - \tau\left(\frac{\partial y}{\partial x}\right)^2\right) \tag{12.72}$$

$$= \frac{1}{2}\left(\frac{\Pi^2}{\rho} + \tau\left(\frac{\partial y}{\partial x}\right)^2\right). \tag{12.73}$$

For a string fixed at $x = 0, L$, the boundary conditions are $y\left(x=0\right) = y\left(x=L\right) = 0$ and the solution to (12.68) can be written in the form

$$y = \sum_k q_k\left(t\right)\sin\left(\frac{k\pi}{L}x\right) \tag{12.74}$$

with k an integer. Substituting into (12.68), we find

$$\sum_k \left(\rho\ddot{q}_k + \tau\left(\frac{k\pi}{L}\right)^2 q_k\right)\sin\left(\frac{k\pi}{L}x\right) = 0. \tag{12.75}$$

This can only hold if for each individual k we have

$$\ddot{q}_k + \frac{\tau}{\rho}\left(\frac{k\pi}{L}\right)^2 q_k = \ddot{q}_k + \left(\frac{k\pi}{L}v\right)^2 q_k \tag{12.76}$$

$$= \ddot{q}_k + \omega_k^2 q_k = 0\,, \tag{12.77}$$

where we have put $\tau/\rho = v^2$, with v the phase velocity of the wave motion on the string and $\omega_k = k\,(\pi v/L)$. Substituting Eq. (12.74) into the Hamiltonian (12.73) and using $(\Pi_k = \rho\,\dot{q}_k)$, we find

$$H = \int dx\,\mathcal{H} = \frac{1}{2}\int dx\left(\frac{\Pi^2}{\rho} + \tau\left(\frac{\partial y}{\partial x}\right)^2\right) \tag{12.78}$$

$$= \frac{1}{2}\int dx\left[\sum_k\left(\rho\,\dot{q}_k^2\sin^2\left(\frac{k\pi}{L}x\right) + \tau q_k^2\left(\frac{k\pi}{L}\right)^2\cos^2\left(\frac{k\pi}{L}x\right)\right)\right] \tag{12.79}$$

$$= \frac{L}{2}\left[\sum_k\left(\frac{\Pi_k^2}{2\rho} + \frac{\rho}{2}\omega_k^2 q_k^2\right)\right], \tag{12.80}$$

where in the last step we used the fact that the cross terms $\sin\left(\frac{k\pi}{L}x\right)\sin\left(\frac{m\pi}{L}x\right)|_{k\neq m}$ vanish on integration. This is just a sum of harmonic oscillator Hamiltonians which we treat, as in the previous section, by defining

$$P_k = \frac{p_k}{\sqrt{2\rho\hbar\omega_k}}, \quad Q_k = q_k\sqrt{\frac{\rho\omega_k}{2\hbar}}\,. \tag{12.81}$$

Then

$$H = \frac{L}{2}\sum_k (P_k^2 + Q_k^2)\,\hbar\omega_k\,, \tag{12.82}$$

where L is the length of the string. The commutation relations are

$$[P_k, Q_k] = \frac{1}{2\hbar}\,[p_k, q_k] = \frac{1}{2i}\,. \tag{12.83}$$

Defining

$$a_k = Q_k + iP_k \tag{12.84}$$

$$a_k^\dagger = Q_k - iP_k, \tag{12.85}$$

We find that

$$\left[a_k, a_k^\dagger\right] = 1, \tag{12.86}$$

so that

$$(P_k^2 + Q_k^2) = \left(a_k^\dagger a_k + \frac{1}{2}\right) \tag{12.87}$$

and

$$H = \frac{L}{2}\sum_k \hbar\omega_k\left(a_k^\dagger a_k + \frac{1}{2}\right). \tag{12.88}$$

The eigenfunctions of the Hamiltonian are the eigenfunctions of the number operator $\mathbf{N}_k = a_k^\dagger a_k$ which must be integers $n_k \geq 0$.

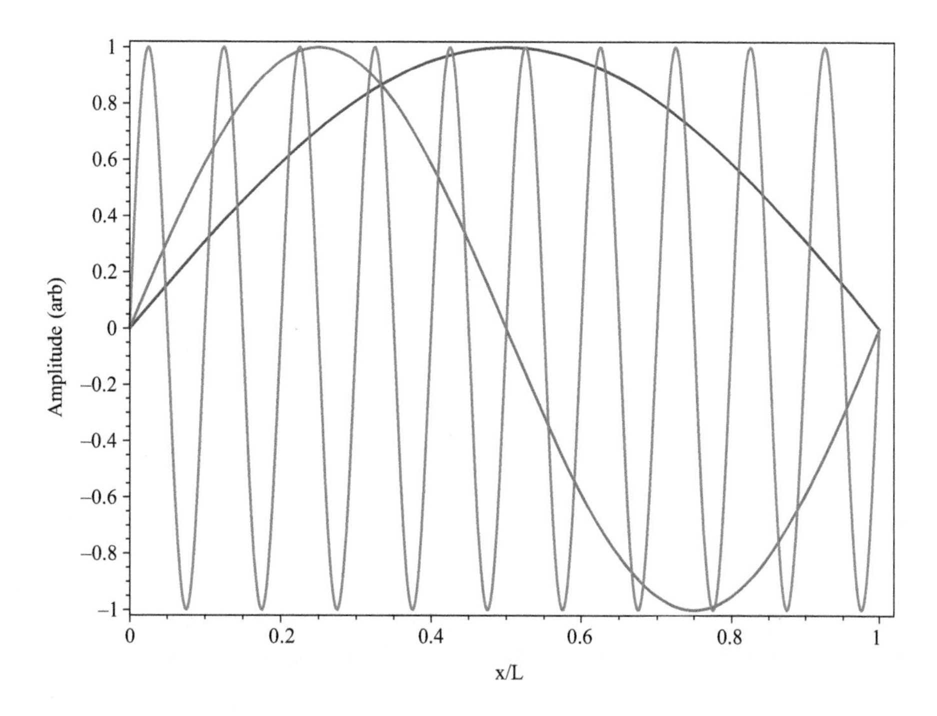

Fig. 12.1 Modes of the vibrating string for $k = 1$ (blue), $k = 2$ (green), and $k = 20$ (red) plotted as a function of x/L.

Figure 12.1 shows some of the modes ($k = 1, 2, 20$) of the vibrating string. As explained in Appendix E, these are normal modes: once such a mode is set in motion, its amplitude remains constant, and there is no transfer of energy between the modes. Classically these modes can be excited with an amplitude which can be varied continuously. The result of the "quantization" of the string (i.e., the assignment of noncommuting operators to the canonical variables describing the motion) is that each mode can only have a discrete set of amplitudes corresponding to the energies $E_k = (n_k + 1/2)\,\hbar\omega_k$ where $\omega_k = k\,(\pi v/L)$.

The state of the string is specified by the set of integers n_k. The energy of each mode can only change in units of $\hbar\omega_k$, so that it appears that discrete particles are being absorbed or emitted by the vibrating string. The operators $a_k, a_k^\dagger$, or p_k, q_k change the number n_k by ± 1 so that an interaction depending on these quantities will result in the increase ($n_k \to n_k + 1$) or decrease ($n_k \to n_k - 1$) of the energy of the mode by $\hbar\omega_k$. These considerations apply to any continuous vibrating system. We speak of the state of the mode specified by n_k as containing n_k "xxx-ons" where xxx designates the type of field we are studying. For vibrations of a solid where the modes are acoustic waves, the excitations of the wave are called "phonons"; in the case of the electromagnetic field, they are called "photons"; in the present case of a vibrating string, we might call them "vibrons". So when an electron, say, loses (gains) energy while interacting with the vibrations in a crystal, we say a phonon was created (annihilated) by the interaction and this is described by the operator $a_k^\dagger$ (a_k).

12.3.2 Dirac shows how to quantize the electromagnetic field

12.3.2.1 The electromagnetic field in free space

The suggestion of Born, Heisenberg, and Jordan that the electromagnetic field should be treated as a mechanical system to be "quantized" was taken up by Dirac a little over a year later (1927).[6] He approached the problem in two ways: first by considering the interaction of an atom with a system of massless particles (photons) and second by quantizing the field directly, similarly to the vibrating string example. We will describe the second method using a modern notation. This was based on Einstein's idea, which had also occurred to de Broglie, that an ideal gas of material particles could be considered as quantized waves in ordinary three-dimensional space.

Dirac began by treating the radiation field as a dynamical system whose interaction with atoms is described by a Hamiltonian. We consider the radiation field to be confined in a finite-sized cavity with boundary conditions such that the field goes to zero on the boundary, so that the modes of the cavity have discrete k values like those of the vibrating string with fixed ends. We could equally well use what are called periodic boundary conditions, namely that the field has a period equal to the size of the cavity. In either case, the Fourier transform variable $\vec{k}$ is restricted to discrete values $k_n^{(i)} = n\pi/L_i$ where $i = x, y, z$ and L_i is the length of the cavity in the i direction.[7]

We start with the classical Maxwell equations for free space with no sources:

$$\vec{\nabla} \times \vec{B} - \frac{1}{c}\frac{\partial \vec{E}}{\partial t} = \frac{4\pi}{c}\vec{j} = 0 \text{ (free field)} \tag{12.89}$$

$$\vec{\nabla} \times \vec{E} + \frac{1}{c}\frac{\partial \vec{B}}{\partial t} = 0. \tag{12.90}$$

The fields can be expressed in terms of a vector potential $\vec{A}$ and scalar potential ϕ as

$$\vec{E} = -\vec{\nabla}\phi - \frac{1}{c}\frac{\partial \vec{A}}{\partial t}. \tag{12.91}$$

$$\vec{B} = \vec{\nabla} \times \vec{A} \tag{12.92}$$

We will work in the Coulomb gauge, defined by $\vec{\nabla} \cdot \vec{A} = 0$. In this gauge, $\phi = 0$ when there are no charges present. Then the first Maxwell equation takes the form

$$\vec{\nabla} \times \vec{\nabla} \times \vec{A} + \frac{1}{c^2}\frac{\partial^2 \vec{A}}{\partial t^2} = 0 \tag{12.93}$$

$$\vec{\nabla}\left(\vec{\nabla} \cdot \vec{A}\right) - \nabla^2 \vec{A} + \frac{1}{c^2}\frac{\partial^2 \vec{A}}{\partial t^2} = -\nabla^2 \vec{A} + \frac{1}{c^2}\frac{\partial^2 \vec{A}}{\partial t^2} = 0. \tag{12.94}$$

Thus, in the Coulomb gauge, $\vec{A}$ satisfies a wave equation. Equation (12.93) can be derived from the Lagrangian density

[6] Dirac, P.A.M., *The quantum theory of the absorption and emission of radiation* Proc. Roy. Soc. **A114**, 243 (1927) received April 4, 1927.

[7] Similar ideas had been used by Einstein in his discussion of Planck's radiation law.

$$\mathcal{L} = \frac{1}{8\pi}\left[\left(\frac{1}{c}\frac{\partial \vec{A}}{\partial t}\right)^2 - \left(\vec{\nabla}\times\vec{A}\right)^2\right]. \tag{12.95}$$

Variation of $\vec{A}(\vec{r},t) \to \vec{A} + \delta\vec{A}(\vec{r},t)$ yields, using integration by parts,

$$\delta\int L\,dt = \frac{1}{4\pi}\int dt\int d^3r\left[\frac{1}{c^2}\frac{\partial\vec{A}}{\partial t}\cdot\frac{\partial\delta\vec{A}}{\partial t} - \left(\vec{\nabla}\times\vec{A}\right)\cdot\left(\vec{\nabla}\times\delta\vec{A}\right)\right] \tag{12.96}$$

$$= -\frac{1}{4\pi}\int dt\int d^3r\left[\frac{1}{c^2}\frac{\partial^2\vec{A}}{\partial t^2} + \vec{\nabla}\times\left(\vec{\nabla}\times\vec{A}\right)\right]\delta\vec{A}(\vec{r},t) = 0\,, \tag{12.97}$$

which yields (12.93). Since $\vec{A}$ is the canonical coordinate here, the conjugate momentum is given by

$$\vec{P} = \frac{\partial\mathcal{L}}{\partial\left(\frac{\partial\vec{A}}{\partial t}\right)} = \frac{1}{4\pi c^2}\frac{\partial\vec{A}}{\partial t} = -\frac{1}{4\pi c}\vec{E}. \tag{12.98}$$

Then the Hamiltonian density is found to be

$$\mathcal{H} = \vec{P}\cdot\left(\frac{\partial\vec{A}}{\partial t}\right) - \mathcal{L} = \frac{1}{8\pi}\left((4\pi)^2c^2\left|\vec{P}\right|^2 + \left(\vec{\nabla}\times\vec{A}\right)^2\right) \tag{12.99}$$

$$= \frac{1}{8\pi}\left(\left|\vec{E}\right|^2 + \left|\vec{B}\right|^2\right). \tag{12.100}$$

We then expand $\vec{A}$ in a Fourier series

$$\vec{A}(\vec{x},t) = \frac{1}{L^{3/2}}\sum_{k,\lambda}\widehat{\varepsilon}_{k\lambda}q_k(t)\,e^{i\vec{k}\cdot\vec{x}} \tag{12.101}$$

$$\vec{P} = \frac{1}{L^{3/2}}\sum_{k,\lambda}\widehat{\varepsilon}_{k\lambda}p_{k\lambda}(t)\,e^{i\vec{k}\cdot\vec{x}} \tag{12.102}$$

where $\lambda = 1,2$ indicates two different polarizations in the directions given by the unit vectors $\widehat{\varepsilon}_{k,\lambda}$. In order for $\vec{A}(\vec{x},t)$ to be a real quantity, we require $q_{-k,\lambda} = q^*_{k,\lambda}$. $\vec{\nabla}\cdot\vec{A} = 0$ means that $\widehat{\varepsilon}_{k,\lambda}\perp\vec{k}$. Then the Hamiltonian density becomes

$$\mathcal{H} = \left(2\pi c^2\left|\vec{P}\right|^2 + \frac{1}{8\pi}\left(\vec{\nabla}\times\vec{A}\right)^2\right) \tag{12.103}$$

$$= \frac{2\pi c^2}{L^3}\left|\sum_{k,\lambda}\widehat{\varepsilon}_k p_k(t)\,e^{i\vec{k}\cdot\vec{x}}\right|^2 + \frac{1}{8\pi L^3}\left|\sum_{k,\lambda}\left(\vec{k}\times\widehat{\varepsilon}_k\right)q_k(t)\,e^{i\vec{k}\cdot\vec{x}}\right|^2 \tag{12.104}$$

and the Hamiltonian $H = \int\mathcal{H}\,d^3x$ is

$$H = 2\pi c^2\sum_k p_k p_k^* + \frac{1}{8\pi}\sum_k k^2 q_k q_k^* \tag{12.105}$$

where we used $\int e^{i\left(\vec{k}-\vec{k}'\right)\cdot\vec{x}}d^3x = L^3\delta^{(3)}\left(\vec{k}-\vec{k'}\right)$. Equation (12.105) is a sum of harmonic oscillator (HO) Hamiltonians. As is usual for the HO we now introduce noncommuting operators $a_k, a_k^\dagger$:

$$q_k = \sqrt{\frac{2\pi\hbar c}{k}}\left(a_k + a_k^\dagger\right) \tag{12.106}$$

$$p_k = \sqrt{\frac{\hbar k}{8\pi c}}\left(a_k^\dagger - a_k\right)i, \tag{12.107}$$

so that p_k, q_k are Hermitian operators and we find (putting $\omega_k = ck$)

$$H = 2\pi c^2 \sum_k p_k p_k^\dagger + \frac{1}{8\pi}\sum_k k^2 q_k q_k^\dagger \tag{12.108}$$

$$= \sum_k \frac{\hbar\omega_k}{2}\left(a_k^\dagger a_k + a_k a_k^\dagger\right). \tag{12.109}$$

Applying the commutation rules to the canonical variables p_k, q_k, we find

$$[p_k, q_k] = \frac{\hbar}{i} = \frac{\hbar i}{2}\left\{\left[a_k^\dagger, a_k\right] - \left[a_k, a_k^\dagger\right]\right\} \tag{12.110}$$

$$= \frac{\hbar}{i}\left[a_k, a_k^\dagger\right]. \tag{12.111}$$

Thus we require

$$\left[a_k, a_k^\dagger\right] = 1 \tag{12.112}$$

and all other operator pairs commute. Then

$$H = \sum_k \frac{\hbar\omega_k}{2}\left(a_k^\dagger a_k + a_k a_k^\dagger\right) = \sum_k \hbar\omega_k 2\left(a_k^\dagger a_k + \frac{1}{2}\right). \tag{12.113}$$

As with the harmonic oscillator, the number operator is $n_k = a_k^\dagger a_k$, $a_k^\dagger, a_k$ are raising and lowering operators and there is a lowest energy state $|n_o\rangle$ so that $a\,|n_o\rangle = 0$. Then

$$E_{tot} = \sum_k \hbar\omega_k\left(n_k + 1/2\right) \tag{12.114}$$

$$a\,|n\rangle = \sqrt{n}\,|n-1\rangle, \qquad a_k^\dagger\,|n\rangle = \sqrt{n+1}\,|n+1\rangle. \tag{12.115}$$

In the ground (vacuum) state, the square of the electric field amplitude $\langle n_o|\,\vec{E}^2\,|n_o\rangle \neq 0$ since

$$E_{tot} = \frac{1}{8\pi}\int\left(E^2 + B^2\right)d^3r = \frac{1}{4\pi}\int\left(E^2\right)d^3r \tag{12.116}$$

$$= \langle n_o|\sum_k \frac{\hbar\omega_k}{2}\,|n_o\rangle. \tag{12.117}$$

This is called the "zero-point energy" and the mean square fluctuations of the field, $\langle n_o | \left(\vec{E} - \left\langle \vec{E} \right\rangle \right)^2 | n_o \rangle = \langle n_o | \vec{E}^2 | n_o \rangle \neq 0$ are called "zero-point fluctuations." Substituting (12.101) into the wave equation (12.94) we have

$$\sum_{k,\lambda} \widehat{\varepsilon}_{k,\lambda} \left(\frac{1}{c^2} \ddot{q}_{k,\lambda} + k^2 q_{k,\lambda} \right) e^{i \vec{k} \cdot \vec{x}} = 0, \tag{12.118}$$

which is a more direct demonstration that the classical quantities $q_{k,\lambda}, q^*_{k,\lambda}$ satisfy the equation of motion of a simple harmonic oscillator:

$$\ddot{q}_{k,\lambda} + \omega_k^2 q_{k,\lambda} = 0. \tag{12.119}$$

Dirac actually worked with a different pair of noncommuting operators. He introduced the number operator and its canonical conjugate, which is the phase θ of the wave, and put

$$a_k = (N_k + 1)^{1/2} e^{i\theta_k/\hbar} = e^{i\theta_k/\hbar} N_k^{1/2} \tag{12.120}$$

$$a_k^\dagger = N_k^{1/2} e^{-i\theta_k/\hbar} = e^{-i\theta_k/\hbar} (N_k + 1)^{1/2}, \tag{12.121}$$

where the equality of the two expressions for $a_k, a_k^\dagger$ follows from Eq. (12.37). Substituting (12.120) and (12.121) into (12.112), we have

$$\left[a_k, a_k^\dagger \right] = 1 = N_k + 1 - N_k \tag{12.122}$$

$$= N_k + 1 - e^{-i\theta_k/\hbar} (N_k + 1) e^{i\theta_k/\hbar} \tag{12.123}$$

$$= N_k - e^{-i\theta_k/\hbar} N_k e^{i\theta_k/\hbar} \tag{12.124}$$

$$\left[e^{i\theta_k/\hbar}, N_k \right] = e^{i\theta_k/\hbar}. \tag{12.125}$$

The Hermitian adjoint of Eq. (12.125) yields

$$N_k e^{-i\theta_k/\hbar} - e^{-i\theta_k/\hbar} N_k = e^{-i\theta_k/\hbar}. \tag{12.126}$$

Applying (12.125) to an eigenstate of N_k ($N_k |n_k\rangle = n_k |n_k\rangle$), we find

$$N_k e^{i\theta_k/\hbar} |n_k\rangle - e^{i\theta_k/\hbar} N_k |n_k\rangle = -e^{i\theta_k/\hbar} |n_k\rangle \tag{12.127}$$

$$N_k e^{i\theta_k/\hbar} |n_k\rangle = (n_k - 1) e^{i\theta_k/\hbar} |n_k\rangle. \tag{12.128}$$

Similarly, using (12.126) gives

$$N_k e^{-i\theta_k/\hbar} |n_k\rangle - e^{-i\theta_k/\hbar} N_k |n_k\rangle = e^{-i\theta_k/\hbar} |n_k\rangle \tag{12.129}$$

$$N_k e^{-i\theta_k/\hbar} |n_k\rangle = (n_k + 1) e^{-i\theta_k/\hbar} |n_k\rangle. \tag{12.130}$$

So $e^{\pm i\theta_k/\hbar} |n_k\rangle$ is an eigenstate of N_k with eigenvalue $(n_k \mp 1)$. Thus we confirm that the different forms in (12.120, 12.121) are equivalent. For example the first form for

a_k, applied to a state $|n_k\rangle$ will produce the state $|n_k - 1\rangle$ multiplied by $\sqrt{n_k}$ while the second form produces exactly the same state.

Eq. (12.125) is equivalent to

$$[\theta_k, N_k] = \hbar/i \tag{12.131}$$

since this latter implies

$$[f(\theta_k), N_k] = \frac{\hbar}{i}\frac{\partial f}{\partial \theta_k} \tag{12.132}$$

which yields (12.125) when applied to $f(\theta_k) = e^{i\theta_k/\hbar}$. Thus the photon number and the phase satisfy an uncertainty principle. States of the radiation field with a definite number of photons correspond to a completely undetermined phase or a chaotic wave motion. On the other hand, states with a large number of photons can have a well-determined phase; that is, the electromagnetic field can be in a "coherent" state such as that produced by a laser. This illustrates the wave-particle duality or complementarity: the particle nature (N_k) and the wave nature (θ_k) are seen as mutually incompatible in the same way as the position and momentum of a point particle. Depending on the state of the system, either the wave or particle aspect can dominate, or there can be a mixed state corresponding to different degrees of incoherence of the wave.

However, it turns out that the operator θ_k introduced by Dirac is not Hermitian and the definition of a proper phase operator is more complex. This has been pointed out by several modern authors.[8] Dirac abandoned this approach in the third edition (1947) of his book on quantum mechanics, stating in the preface that the approach using the $a, a^\dagger$ operators, originally introduced by Fock,[9] was "simpler and more powerful."

12.3.2.2 Interaction of a charged particle with the radiation field

The interaction between a charged particle and a general electromagnetic field is given in classical mechanics by replacing $\vec{p}$ by $\left(\vec{p} - \frac{e}{c}\vec{A}\right)$ so that the Hamiltonian becomes[10]

$$\begin{aligned} H &= \frac{\left(\vec{p} - \frac{e}{c}\vec{A}\right)^2}{2m} + V(\vec{r}) \\ &= \frac{p^2}{2m} - \frac{e}{mc}\vec{p}\cdot\vec{A} + \frac{e^2}{2mc^2}A^2 + V(\vec{r}). \end{aligned} \tag{12.133}$$

Taking, for simplicity, $V(\vec{r}) = 0$, Hamilton's equations become [note that the total time derivative is $\frac{d}{dt} = \left(\frac{\partial}{\partial t} + \vec{v}\cdot\vec{\nabla}\right)$]

[8] P. Carruthers and M. M. Nieto, *Phase and Angle Variables in Quantum Mechanics*, Rev. Mod. Phys. **40**, 411 (1968).

[9] Fock,V., *Configuration space and second quantization*, Zeit. Phys. **75**, 622 (1932) received March 10, 1932.

[10] Note that $\vec{p}\cdot\vec{A} = \vec{A}\cdot\vec{p}$ in the gauge $\vec{\nabla}\cdot\vec{A} = 0$.

$$\dot{\vec{r}} = \frac{\partial H}{\partial \vec{p}} = \frac{\left(\vec{p} - \frac{e}{c}\vec{A}\right)}{m} = \vec{v} \tag{12.134}$$

$$\vec{p} = m\dot{\vec{r}} + \frac{e}{c}\vec{A} \tag{12.135}$$

$$\dot{\vec{p}} = -\frac{\partial H}{\partial \vec{r}} = m\ddot{\vec{r}} + \frac{e}{c}\frac{d}{dt}\vec{A} = m\ddot{\vec{r}} + \frac{e}{c}\left(\frac{\partial \vec{A}}{\partial t} + \left(\vec{v}\cdot\vec{\nabla}\right)\vec{A}\right) \tag{12.136}$$

$$= \frac{\left(\vec{p} - \frac{e}{c}\vec{A}\right)}{m} \cdot \left[\left(\hat{\imath}\frac{\partial}{\partial x} + \hat{\jmath}\frac{\partial}{\partial y} + \hat{k}\frac{\partial}{\partial z}\right)\frac{e}{c}\vec{A}\right] = \frac{e}{c}\vec{\nabla}\left(\vec{v}\cdot\vec{A}\right) \tag{12.137}$$

$$m\ddot{\vec{r}} = \frac{e}{c}\left(-\frac{\partial \vec{A}}{\partial t} - \left(\vec{v}\cdot\vec{\nabla}\right)\vec{A} + \vec{\nabla}\left(\vec{v}\cdot\vec{A}\right)\right) = \tag{12.138}$$

$$= e\vec{E} + \frac{e}{c}\vec{v}\times\left(\vec{\nabla}\times\vec{A}\right) = e\left(\vec{E} + \frac{\vec{v}}{c}\times\vec{B}\right) \tag{12.139}$$

where we used the identity

$$\vec{v}\times\left(\vec{\nabla}\times\vec{A}\right) = \vec{\nabla}\left(\vec{v}\cdot\vec{A}\right) - \left(\vec{v}\cdot\vec{\nabla}\right)\vec{A}. \tag{12.140}$$

An alternative method is to apply the relation $\vec{\nabla}\left(\vec{u}\cdot\vec{u}\right) = 2\left(\vec{u}\cdot\vec{\nabla}\right)\vec{u} + 2\vec{u}\times\left(\vec{\nabla}\times\vec{u}\right)$ to $\vec{u} = \left(\vec{p} - \frac{e}{c}\vec{A}\right)$. This shows that the Hamiltonian (12.133) is correct for a charged particle moving in an electromagnetic field.

The interaction between the particle and the field is then given by

$$V_i = -\frac{e}{mc}\vec{p}\cdot\vec{A} + \frac{e^2}{2mc^2}A^2. \tag{12.141}$$

The second term is of higher order, so it can be neglected for the problems we are going to discuss in this section. Quantum mechanically, $\vec{p}$ is an operator on the states of the electron and $\vec{A}$ operates on the field states. Using (12.101) and (12.106), we can write the first term as

$$V_i = -\frac{e}{m}\frac{1}{L^{3/2}}\sum_{k,\lambda}\vec{p}\cdot\hat{\varepsilon}_{k,\lambda}\sqrt{\frac{2\pi\hbar}{ck}}\left(a_{k,\lambda} + a^{\dagger}_{k,\lambda}\right)e^{i\vec{k}\cdot\vec{x}}. \tag{12.142}$$

We will apply this to a physical problem below.

12.3.2.3 Time-dependent perturbation theory

We consider a system consisting of an electron in an atom with energy levels given by $|N\rangle_A$ and an electromagnetic field with states $|n_k\rangle$. The states of the field are specified by giving an integer n_k for each of the modes of the field, $|n_1, n_2, n_3, ...n_k, n_{k+1}....\rangle$. We will only display the nonzero photon numbers. A state of the complete system can be written as $|\psi_{Nn}\rangle = |N\rangle_A |n_{k_1}, n_{k_2}\rangle$ depending on how many modes of the field are occupied. The energy will be $E_N = E_{NA} + \sum_k n_k \hbar\omega_k$, suppressing the zero-point energy of each mode. A general state can be represented as

$$|\psi\rangle = \sum_{nN} b_{Nn}(t) |\psi_{Nn}\rangle. \tag{12.143}$$

The Hamiltonian is $H = H_o + V_i$, where $H_o = H_{\text{atom}} + H_{\text{field}}$ and V_i is given by (12.142). Then the generalized Schrödinger equation is

$$i\hbar\frac{\partial}{\partial t}|\psi\rangle = H|\psi\rangle \tag{12.144}$$

$$i\hbar\sum_{Nn}\dot{b}_{Nn}(t)|\psi_{Nn}\rangle = \sum_{nN} b_{Nn}(t)[E_N + V_i]|\psi_{Nn}\rangle. \tag{12.145}$$

Multiplying on the left with $\langle\psi_{Mm}|$ and using orthonormality, we obtain

$$i\hbar\dot{b}_{Mm}(t) = b_{Mm}(t)E_M + \sum_{nN} b_{Nn}(t)\langle\psi_{Mm}|V_i|\psi_{Nn}\rangle \tag{12.146}$$

$$i\hbar\dot{c}_{Mm}(t) = \sum_{nN} c_{Nn}(t)\langle\psi_{Mm}|V_i|\psi_{Nn}\rangle e^{i(E_M - E_N t/\hbar)} \tag{12.147}$$

where we substituted $b_{Mm}(t) = c_{Mm}(t)e^{-iE_M t/\hbar}$. So far this equation is exact. If we assume all of the c_{Mm} only change from their initial values by a small amount, we can substitute $c_{Nn}(0)$ on the right-hand side and obtain a solution accurate to first order in V_i.

$$i\hbar\dot{c}_{Mm}(t) = \sum_{nN} c_{Nn}(0)\langle\psi_{Mm}|V_i|\psi_{Nn}\rangle e^{i(E_M - E_N)t/\hbar} \tag{12.148}$$

$$c_{Mm}(t) = c_{Mm}(0) + \sum_{nN} c_{Nn}(0)\frac{\langle\psi_{Mm}|V_i|\psi_{Nn}\rangle}{(E_M - E_N)}\left(1 - e^{i(E_M - E_N)t/\hbar}\right) \tag{12.149}$$

$$c_{Mm}(t) = \frac{\langle\psi_{Mm}|V_i|\psi_{Nn}\rangle}{E_M - E_N}\left(1 - e^{i(E_M - E_N)t/\hbar}\right). \tag{12.150}$$

The last result applies when the system is initially in a single state $|\Psi_{Nn}\rangle$, i.e., $c_{Mm}(0) = \delta_{Mm,Nn}$. The amplitude for transitions to another state is periodic with an amplitude proportional to V_i, unless the final state has an energy close to the initial state, $E_M \approx E_N$. For states with $E_M = E_N$ the amplitude increases linearly with time (probability $\sim t^2$). If we consider the modes of the electromagnetic field to be discrete due to the system being confined in a cavity, the photon energies will be very close together, so that the number of states satisfying $(E_M = E_N)$ exactly is negligible compared to the number of states in the vicinity, and we need to calculate the integral of the probability over a small range of energy values about the initial energy E_N.

The probability of finding the system in the state $|\psi_{Mm}\rangle$ if it was initially in the state $|\psi_{Nn}\rangle$ is then given by

$$P_{Nn\to Mm} = |c_{Mm}(t)|^2 = 4|\langle\psi_{Mm}|V_i|\psi_{Nn}\rangle|^2\frac{\sin^2[(E_M - E_N)t/2\hbar]}{(E_M - E_N)^2} \tag{12.151}$$

Considering this as a function of final energy E_M, we see that the peak of the function (at $E_M - E_N$) grows as t^2 and its width decreases as $1/t$. Thus the area of the peak, proportional to the total transition probability, grows as t (time-proportional transition probability). Summing over all states M, we can replace the sum by an integral with a weighting function $\rho(E)$, where $\rho(E)\,dE$ is the number of states with energies between E and $E+dE$, and find for the total transition probability W_{NM}:

$$W_{NM} = 4\left|\langle\psi_{Mm}|\,V_i\,|\psi_{Nn}\rangle\right|^2 \int \frac{\sin^2\left[(E_M - E_N)\,t/2\hbar\right]}{(E_M - E_N)^2}\rho(E_M)\,dE_M. \qquad (12.152)$$

If $\rho(E_M)$ and the matrix element $\langle\psi_{Mm}|\,V_i\,|\psi_{Nn}\rangle$ are slowly varying functions of E_M, we can remove them from under the integral sign and find

$$W_{NM} = 4\left|\langle\psi_{Mm}|\,V_i\,|\psi_{Nn}\rangle\right|^2 \rho(E_N) \int \frac{\sin^2\left[(E_M - E_N)\,t/2\hbar\right]}{(E_M - E_N)^2}dE_M \qquad (12.153)$$

$$= \frac{2\pi}{\hbar}\left|\langle\psi_{Mm}|\,V_i\,|\psi_{Nn}\rangle\right|^2 \rho(E_N)\,t. \qquad (12.154)$$

Here we made use of the result $\int_{-\infty}^{\infty}\frac{\sin^2(x/2)}{x^2}dx = \frac{1}{2}\pi$. Conservation of energy $(E_M = E_N)$ follows automatically from the procedure. Thus, under these conditions the transition probability will grow linearly with time. This result is called "Fermi's golden rule," although it was derived by Dirac using a somewhat more obscure notation. The Dirac notation used here was first introduced in the third edition of his book, *Principles of Quantum Mechanics* (1947). For any system, the number of states in a given phase space volume is equal to the (phase space volume)$/h^3$ with the phase space volume given by L^3d^3p:

$$dN = L^3\frac{d^3p}{h^3} = L^3\frac{d^3k}{(2\pi)^3} = \rho(E_M)\,dE_M. \qquad (12.155)$$

For photons $E_k = \hbar ck = \hbar\omega_k$, and multiplying by two for the two states of polarization, the number of states within energy interval dE_k and solid angle $d\Omega_k$ is given by

$$dN = \frac{L^3}{(2\pi)^3}\frac{\omega_k^2}{\hbar c^3}d\Omega_k\,dE_k. \qquad (12.156)$$

12.3.2.4 Emission and absorption of radiation by an atom

Applying this to a system consisting of an atom interacting with the radiation field, Dirac considered two states of the entire system with the same energy $|\psi_1\rangle = |1\rangle_A\,|n_k+1\rangle$ and $|\psi_2\rangle = |2\rangle_A\,|n_k\rangle$ with energies

$$E_1 = E_{1A} + \hbar\omega_k\,(n_k+1) = E_2 = E_{2A} + \hbar\omega_k n_k \qquad (12.157)$$

where the atomic states $|1\rangle_A$ and $|2\rangle_A$ have energies E_{1A}, E_{2A} with $E_{2A} = E_{1A} + \hbar\omega_k$. Then a transition from state $|1\rangle_A$ to $|2\rangle_A$ corresponds to absorption of a photon and the transition in the opposite direction to emission of a photon. Using the perturbation (12.142), the matrix element is

$$\langle\psi_{Mm}|\,V_i\,|\psi_{Nn}\rangle = -\frac{e}{m}\frac{1}{L^{3/2}}\sum_k\sqrt{\frac{2\pi\hbar}{ck}}\,\langle\psi_{\text{final}}|\,e^{i\vec{k}\cdot\vec{x}}\vec{p}\cdot\hat{\varepsilon}_k\left(a_k + a_k^\dagger\right)|\psi_{\text{initial}}\rangle \tag{12.158}$$

$$= -\frac{e}{m}\frac{1}{L^{3/2}}\sum_k\sqrt{\frac{2\pi\hbar}{ck}}\,\langle\psi_{f,A}|\,e^{i\vec{k}\cdot\vec{x}}\vec{p}\cdot\hat{\varepsilon}_k\,|\psi_{in,A}\rangle\,\langle n_{k,\text{final}}|\left(a_k + a_k^\dagger\right)|n_{k,\text{initial}}\rangle \tag{12.159}$$

$$= -\frac{e}{m}\frac{1}{L^{3/2}}\sum_k\sqrt{\frac{2\pi\hbar}{ck}}\,\langle\psi_{f,A}|\,e^{i\vec{k}\cdot\vec{x}}\vec{p}\cdot\hat{\varepsilon}_k\,|\psi_{in,A}\rangle\begin{Bmatrix}\sqrt{n_k} & \text{absorption}\\ \sqrt{n_k+1} & \text{emission}\end{Bmatrix}. \tag{12.160}$$

The photon operators a_k and $a_k^\dagger$ only have nonzero matrix elements between states $|n_k\rangle$ and $|n_k \pm 1\rangle$. The matrix element involving the atomic operator between the atomic states can be calculated in the x representation by the usual volume integration.

In the usual optical transitions, we have $\vec{k}\cdot\vec{x} \ll 1$ so that in first approximation we can set $e^{i\vec{k}\cdot\vec{x}} = 1$. This is called the electric dipole approximation.[11] The next terms in the expansion of $e^{i\vec{k}\cdot\vec{x}}$ lead to the magnetic dipole and electric quadrupole approximations.

The transition probability per unit time is then given by Eq. (12.151):

$$P_{Nn\to Mm} = 4\left|\langle\psi_{Mm}|\,V_i\,|\psi_{Nn}\rangle\right|^2\frac{\sin^2\left[(E_M - E_N)\,t/2\hbar\right]}{(E_M - E_N)^2} \tag{12.161}$$

$$P_{Nn\to Mm} = 4\left(\frac{e}{m}\right)^2\frac{1}{L^3}\sum_k\frac{2\pi\hbar}{ck}\left|(\vec{p}\cdot\hat{\varepsilon}_k)_{fi}\right|^2\begin{Bmatrix}n_k & \text{absorption}\\ n_k+1 & \text{emission}\end{Bmatrix}$$
$$\frac{\sin^2\left[(E_M - E_N)\,t/2\hbar\right]}{(E_M - E_N)^2}. \tag{12.162}$$

From Hamilton's equations in matrix form, we have

$$\vec{p}_{fi} = m\dot{\vec{r}}_{fA,iA} = \frac{m}{i\hbar}[\vec{r}, H]_{fA,iA} = m\vec{r}_{fA,iA}\left(E_{fA} - E_{iA}\right)\frac{i}{\hbar} = i\omega_{fA,iA}m\vec{r}_{fA,iA}. \tag{12.163}$$

We have seen that $\omega_{fA,iA} = \pm\omega_k$, since $E_1 = E_2$ and then

$$P_{Nn\to Mm} = 4e^2\left|(\vec{r}\cdot\hat{\varepsilon}_k)_{fi}\right|^2\frac{1}{L^3}\sum_k\frac{2\pi\hbar}{ck}\omega_{fA,iA}^2\begin{Bmatrix}n_k & \text{absorption}\\ n_k+1 & \text{emission}\end{Bmatrix}$$
$$\frac{\sin^2\left[(E_M - E_N)\,t/2\hbar\right]}{(E_M - E_N)^2} \tag{12.164}$$

where n_k is the number of photons present in the initial state.

[11] For optical transitions, $\vec{k}\cdot\vec{x} \ll 1$ because $k = 2\pi/\lambda$ and $x \sim R_{\text{atom}}$, so that $\vec{k}\cdot\vec{x} \sim R_{\text{atom}}/\lambda$. A typical atomic radius is about 1 Å$(10^{-10}$ m), so for a typical wavelength for visible light, say 500 nm, $\vec{k}\cdot\vec{x} \sim 10^{-3}$.

We see that, in agreement with Einstein's original assumptions, the absorption is proportional to the number of photons n_k present in the initial state of the field (Einstein's $B\rho_\nu$ term) while there are two types of emission: induced emission, proportional to $B\rho_\nu \sim n_k$, and spontaneous emission, which exists even when the field is in its ground state, $(n_k = 0)$. Spontaneous emission is considered to be emission induced by the zero-point fluctuations associated with the zero-point energy. Both the spontaneous emission and zero-point fluctuations have their origin in the fact that the operators $a_k, a_k^\dagger$ do not commute.

We work in the limit where L is large, so the discrete $\overrightarrow{k}$ values are very close together, and we can replace the sum $\sum_k$ by an integral. As each mode contributes an amount $n_k\hbar\omega_k/L^3$ to the energy density, we see that it contributes an amount $I(\omega_k) = n_k\hbar c\omega_k/L^3$ to the radiation intensity (power/area). Then we can replace $\sum_k n_k$ by $L^3 \int \frac{I(\omega_k)}{\hbar^2 c\omega_k} dE_k$. For the spontaneous emission term we replace (see Eq. (12.156))

$$\sum_k \to \int dN = \frac{L^3}{(2\pi)^3} \frac{\omega_k^2}{\hbar c^3} d\Omega_k dE_k. \tag{12.165}$$

Then

$$\begin{aligned} P_{Nn\to Mm} &= 8\pi e^2 \left|\langle\psi_{f,A}|\overrightarrow{r}|\psi_{in,A}\rangle \cdot \widehat{\varepsilon}_k\right|^2 \left\{ \begin{matrix} \int \frac{I(\omega_k)}{\hbar^2 c\omega_k} dE_k & \text{absorption} \\ \int \frac{I(\omega_k)}{\hbar^2 c\omega_k} dE_k + \int\int \frac{1}{(2\pi)^3} \frac{\omega_k^2}{\hbar c^3} d\Omega_k dE_k & \text{emission} \end{matrix} \right\} \\ &\times \frac{\hbar}{ck} \omega_{fA,iA}^2 \left[\frac{\sin^2\left[(E_M - E_N)t/2\hbar\right]}{(E_M - E_N)^2} \right]. \end{aligned} \tag{12.166}$$

As the term in the square brackets gets very narrow in energy as time increases, all the other energy-dependent terms will be slowly varying in comparison and can be taken out of the integral, as in the previous section, so that

$$\begin{aligned} P_{Nn\to Mm} &= 8\pi \frac{e^2}{\hbar c} \left|\langle\psi_{f,A}|\overrightarrow{r}|\psi_{in,A}\rangle \cdot \widehat{\varepsilon}_k\right|^2 \left\{ \begin{matrix} \frac{I(\omega_k)}{\hbar\omega_k^2} & \text{absorption} \\ \frac{I(\omega_k)}{\hbar\omega_k^2} + \frac{1}{(2\pi)^3} \int \frac{\omega_k^2}{c^2} d\Omega_k & \text{emission} \end{matrix} \right\} \\ &\times \frac{\hbar}{ck} \omega_{fA,iA}^2 \int dE_k \left[\frac{\sin^2\left[(E_M - E_N)t/2\hbar\right]}{(E_M - E_N)^2} \right] \end{aligned} \tag{12.167}$$

$$= 4\pi^2 \frac{e^2}{\hbar c} \left|\langle\psi_{f,A}|\overrightarrow{\mathbf{r}}|\psi_{in,A}\rangle \cdot \widehat{\varepsilon}_k\right|^2 \left\{ \begin{matrix} \frac{I(\omega_k)}{\hbar} & \text{absorption} \\ \frac{I(\omega_k)}{\hbar} + \frac{1}{(2\pi)^3} \int \frac{\omega_k^3}{c^2} d\Omega_k & \text{emission} \end{matrix} \right\} t \tag{12.168}$$

and the transition probability is seen to grow proportionally to time. There are two independent directions for $\widehat{\varepsilon}_k$, both perpendicular to the direction of propagation given by the $\overrightarrow{k}$ vector. We can choose one of these in the plane determined by $\overrightarrow{k}$ and the matrix element of $\overrightarrow{\mathbf{r}}$, with the second polarization vector being perpendicular to that plane, so that it does not contribute to the transition probability. Then

$|\langle\psi_{f,A}|\vec{r}|\psi_{in,A}\rangle\cdot\widehat{\epsilon}_k|^2 = |\langle\psi_{f,A}|\vec{r}|\psi_{in,A}\rangle|^2\sin^2\theta$ where θ is the angle between $\vec{k}$ and the vector matrix element. The spontaneous emission term is evaluated as:

$$\frac{e^2}{\hbar c}|\langle\psi_{f,A}|\vec{r}|\psi_{in,A}\rangle\cdot\widehat{\varepsilon}_k|^2\frac{1}{(2\pi)}\int\frac{\omega_k^3}{c^2}d\Omega_k = \frac{e^2}{\hbar c}|\vec{r}_{f,in}|^2\frac{\omega_k^3}{c^2}\int_0^\pi\sin^3\theta d\theta \tag{12.169}$$

$$= \frac{4}{3}\frac{e^2}{\hbar c}|\vec{r}_{f,in}|^2\frac{\omega_k^3}{c^2}. \tag{12.170}$$

This is Einstein's A coefficient giving the rate of spontaneous transitions. His B coefficient is given by the absorption/induced emission term divided by $\rho(\omega) = I(\omega)/c$.

The theory of the quantized radiation field yields both the spontaneous and induced emission processes, shown by Einstein to be necessary to maintain thermal equilibrium between matter and black body radiation, without any auxiliary assumptions.

12.3.3 The width of spectral lines: the Weisskopf–Wigner theory[12]

12.3.3.1 Width of the final state distribution

In deriving Fermi's golden rule above, we considered a system described by a Hamiltonian consisting of a large part H_o whose solutions are known and a smaller time-independent part V that is considered as a perturbation. If we expand the wave function for such a system in terms of the eigenfunctions ψ_n of the unperturbed Hamiltonian, with eigenvalue E_n, we have

$$\psi(t) = \sum_n c_n(t)\,e^{-iE_nt/\hbar}\psi_n. \tag{12.171}$$

The $c_n(t)$ satisfy the following equations

$$i\hbar\frac{\partial c_n}{\partial t} = \sum_m c_m(t)\,e^{i(E_m-E_n)t/\hbar}\langle n|V|m\rangle. \tag{12.172}$$

The golden rule follows from assuming that the perturbation is sufficiently weak that the values of the c_n only vary from their initial values by a small amount and we can replace the $c_m(t)$ on the right-hand side of (12.172) by their initial values $c_m(0)$. Clearly the amplitude of the initial state will eventually go to zero, so that its decrease must be included in the model if we want to know something about the distribution of the final states.

To that end, we again assume that initially the system is in a single state $|i\rangle$, so that

$$c_i(0) = 1,\ \text{all other } c_n(0) = 0. \tag{12.173}$$

This is obviously only valid for relatively short times in which the system does not change its state by a significant amount. In reality the probability of being in the initial state will decrease with time. To proceed beyond this approximation, Weisskopf and

[12] V. Weisskopf and E. Wigner, *Berechnung der natürlichen Linienbreite auf Grund der Diracschen Lichttheorie* (Calculation of the natural linewidth on the basis of Dirac's theory of light), Z. Phys. **63**, 54 (1930).

Wigner started with the system in a single state $|i\rangle$ as in Eq. (12.173). Then (12.172) becomes

$$i\hbar\frac{\partial c_i}{\partial t} = \sum_f c_f(t)\, e^{i(E_f - E_i)t/\hbar}\langle i|\, V\,|f\rangle \tag{12.174}$$

$$i\hbar\frac{\partial c_f}{\partial t} = \sum_{f'} c_{f'}(t)\, e^{i(E_{f'} - E_f)t/\hbar}\langle f|\, V\,|f'\rangle. \tag{12.175}$$

Since $c_f(0) = 0$ for all $f \neq i$ and we assume that these amplitudes remain small (of first order in V), we can write (12.175) as

$$i\hbar\frac{\partial c_f}{\partial t} = c_i(t)\, e^{i(E_i - E_f)t/\hbar}\langle f|\, V\,|i\rangle. \tag{12.176}$$

To proceed further, we assume $c_i(t) = e^{-\Gamma t}$. Then a simple integration yields ($\hbar\Omega_{i,f} = E_{i,f}$)

$$c_f(t) = -\frac{1}{\hbar}\langle f|\, V\,|i\rangle\,\frac{e^{i(\Omega_i - \Omega_f + i\Gamma)t} - 1}{(\Omega_i - \Omega_f + i\Gamma)}, \tag{12.177}$$

which satisfies the initial condition. Substituting this solution into (12.174), we find

$$i\hbar\frac{\partial c_i}{\partial t} = -\frac{1}{\hbar}\sum_f |\langle i|\, V\,|f\rangle|^2\,\frac{e^{-\Gamma t} - e^{i(\Omega_f - \Omega_i)t}}{(\Omega_i - \Omega_f + i\Gamma)}. \tag{12.178}$$

We are going to apply this result to the emission of radiation by an atom. If the atom is located in a very large cavity, the allowed photon energies will be so close together that we can consider them as forming a continuum and the sum can be written as an integral. Then

$$i\hbar\frac{\partial c_i}{\partial t} = -\int |\langle i|\, V\,|f\rangle|^2\,\frac{e^{-\Gamma t} - e^{i(\Omega_f - \Omega_i)t}}{(\Omega_i - \Omega_f + i\Gamma)\,\hbar}\rho(E_f)\, dE_f \tag{12.179}$$

where $\rho(E_f)\, dE_f$ is the number of states having energies between E_f and $E_f + dE_f$.

The magnitude of the denominator has a minimum when $\Omega_i = \Omega_f$ and doubles when $\Omega_i - \Omega_f = \pm\Gamma$. In atomic spectroscopy, it is observed that the line widths are much smaller than the frequencies themselves so it is reasonable to assume here that $\Gamma \ll \Omega_{i,f}$ and only a narrow band of final state energies will contribute to the integral. We expect that $|\langle i|\, V\,|f\rangle|^2\rho(E_f)$ will vary only slightly over this narrow band, so that they can be taken out of the integral sign and the limits of the integral can be extended to $\pm\infty$. Thus

$$i\hbar\frac{\partial c_i}{\partial t} = \left[|\langle i|\, V\,|f\rangle|^2\rho(E_f)\right]_{E_f = E_i}\int_{-\infty}^{\infty}\frac{e^{-\Gamma t} - e^{i(\Omega_f - \Omega_i)t}}{(\Omega_f - \Omega_i - i\Gamma)}d\Omega_f. \tag{12.180}$$

The first integral

$$\begin{aligned} e^{-\Gamma t}\lim_{M\to\infty}\int_{-M}^{M}\frac{1}{(\Omega_f - \Omega_i - i\Gamma)}d\Omega_f &= e^{-\Gamma t}\lim_{M\to\infty}\ln\left[\frac{M - \Omega_i - i\Gamma}{-M - \Omega_i - i\Gamma}\right] \\ &= e^{-\Gamma t} i\pi. \end{aligned} \tag{12.181}$$

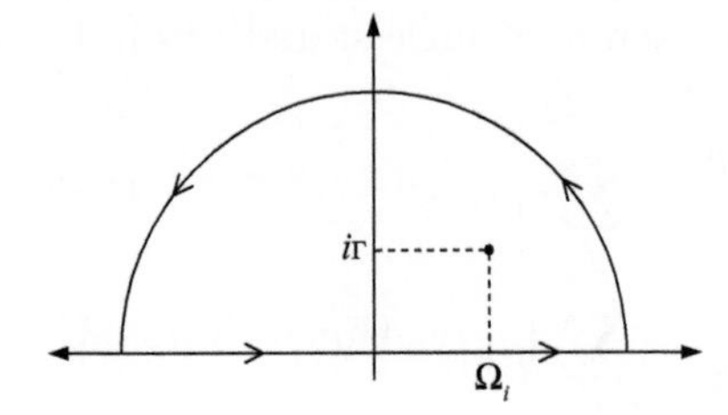

Fig. 12.2 Contour of integration for the second term in Eq. (12.180). The arrows show the direction of integration.

The second integral can be evaluated by contour integration around a semicircle in the upper half plane (for $t > 0$), as shown in Fig. 12.2. Since there is a simple pole at $\Omega_f = \Omega_i + i\Gamma$,

$$-\int_{-\infty}^{\infty} \frac{e^{i(\Omega_f - \Omega_i)t}}{(\Omega_f - \Omega_i - i\Gamma)} d\Omega_f = -2\pi i e^{-\Gamma t}. \tag{12.182}$$

Then (12.180) becomes

$$\frac{\partial c_i}{\partial t} = -\frac{\pi}{\hbar}\left[\left|\langle i| V |f\rangle\right|^2 \rho(E_f)\right]_{E_f = E_i} e^{-\Gamma t}, \tag{12.183}$$

so that

$$\Gamma = \frac{\pi}{\hbar}\left[\left|\langle i| V |f\rangle\right|^2 \rho(E_f)\right]_{E_f = E_i}. \tag{12.184}$$

Using (12.177),

$$\left|c_f(t)\right|^2 = \left|\frac{1}{\hbar}\langle f| V |i\rangle \frac{e^{i(\Omega_i - \Omega_f)t} e^{-\Gamma t} - 1}{(\Omega_i - \Omega_f + i\Gamma)}\right|^2 \tag{12.185}$$

$$= \left|\frac{1}{\hbar}\langle f| V |i\rangle\right|^2 \left(\frac{e^{-2\Gamma t} + 1 - 2e^{-\Gamma t}\cos\left[(\Omega_i - \Omega_f)t\right]}{(\Omega_i - \Omega_f)^2 + \Gamma^2}\right) \tag{12.186}$$

$$= \left|\frac{1}{\hbar}\langle f| V |i\rangle\right|^2 \frac{1}{(\Omega_i - \Omega_f)^2 + \Gamma^2} \quad \text{for } t \to \infty \tag{12.187}$$

$$\sum_f \left|c_f(t \to \infty)\right|^2 = \left|\frac{1}{\hbar}\langle f| V |i\rangle\right|^2 \rho(E_f) \int \frac{1}{(\Omega_i - \Omega_f)^2 + \Gamma^2} \hbar\, d\Omega_f$$

$$= \frac{\pi}{\hbar}\left|\langle f| V |i\rangle\right|^2 \rho(E_f) \frac{1}{\Gamma} = 1, \tag{12.188}$$

where we used

$$\int_{-\infty}^{\infty} \frac{dx}{x^2 + a^2} = \frac{\pi}{a}. \tag{12.189}$$

Equation (12.187) gives the distribution of final state energies, which is seen to be a Lorentzian with a full width at half maximum (FWHM) of 2Γ. We now sum equation (12.186) over frequencies at a finite time:

$$\sum_f |c_f(t)|^2 = \frac{1}{\hbar} |\langle f| V |i\rangle|^2 \rho(E_f) \int \left(\frac{e^{-2\Gamma t} + 1 - 2e^{-\Gamma t} \cos[(\Omega_i - \Omega_f) t]}{(\Omega_i - \Omega_f)^2 + \Gamma^2} \right) d\Omega_f. \tag{12.190}$$

Using (12.189) and

$$\int_{-\infty}^{\infty} \frac{\cos xt}{x^2 + \Gamma^2} dx = \frac{\pi}{\Gamma} \left(e^{-\Gamma t}\right) \tag{12.191}$$

we find

$$\sum_f |c_f(t)|^2 = \frac{1}{\hbar} |\langle f| V |i\rangle|^2 \rho(E_f) \left(1 - e^{-2\Gamma t}\right) \frac{\pi}{\Gamma} \tag{12.192}$$

and

$$\frac{d}{dt} \sum_f |c_f(t)|^2 = \frac{2\pi}{\hbar} |\langle f| V |i\rangle|^2 \rho(E_f) e^{-2\Gamma t} = 2\Gamma e^{-2\Gamma t}. \tag{12.193}$$

Recalling that $c_i(t) = e^{-\Gamma t}$, we see that probability is conserved:

$$\frac{d}{dt} |c_i(t)|^2 + \frac{d}{dt} \sum_f |c_f(t)|^2 = 0. \tag{12.194}$$

If the system consists of just two discrete, degenerate states, (12.172) becomes

$$i\hbar \frac{\partial c_1}{\partial t} = c_2(t) \langle 1| V |2\rangle \tag{12.195}$$

$$i\hbar \frac{\partial c_2}{\partial t} = c_1(t) \langle 2| V |1\rangle \tag{12.196}$$

which has the solution

$$c_1 = \cos \omega_o t \tag{12.197}$$

$$c_2 = \sin \omega_o t \tag{12.198}$$

$$\omega_o = \frac{|\langle 1| V |2\rangle|}{\hbar}. \tag{12.199}$$

This shows that in this case the transition is completely reversible. After going over completely into state $|2\rangle$ at $\omega_o t = \pi/2$, the system returns to state $|1\rangle$ at $\omega_o t = \pi$ and so on. This is in agreement with the fact that the Schrödinger equation is symmetric under time reversal ($t \to -t$, $\psi \to \psi^*$). However, Eq. (12.180) shows an irreversible decay of the probability of the system being in the initial state. The reason for this is that the final state consists of a superposition of states with a continuous energy distribution, so that the reverse transitions from each of these states cancel out. This idea is somewhat controversial: the method of taking the limit to the continuum distribution has been questioned by some mathematicians, who feel that an equation symmetric under time reversal should always have time-reversal symmetric solutions. This might be considered a weakness of the theory, but the results presented here are in agreement with experiment. The problem of the existence of effects violating time reversal in a time reversal invariant theory also appears in classical mechanics, as illustrated by the energetic debates over Boltzmann's H theorem.

12.3.3.2 Frequency shift of the final state distribution

According to the above approximation, the final state energy distribution will have a finite width. A slightly better approximation will show that the center of the energy distribution will also be shifted slightly. To see this, we start by writing the solution to (12.176) as

$$c_f = \frac{1}{i\hbar} \langle f|\, V\, |i\rangle \int_0^t c_i\,(t')\, e^{i(E_i - E_f)t'/\hbar} dt' \tag{12.200}$$

and substitute this into (12.174):

$$i\hbar \frac{\partial c_i}{\partial t} = \sum_f c_f\,(t)\, e^{i(E_f - E_i)t/\hbar} \langle i|\, V\, |f\rangle \tag{12.201}$$

$$\frac{\partial c_i}{\partial t} = -\frac{1}{\hbar^2} \sum_f |\langle i|\, V\, |f\rangle|^2\, e^{i(E_f - E_i)t/\hbar} \int_0^t c_i\,(t')\, e^{i(E_i - E_f)t'/\hbar} dt'. \tag{12.202}$$

This equation is exact. We will solve it using the method of Laplace transforms,[13] which are an extension of the Fourier transform to complex frequencies. The Laplace transform of a function $f\,(t)$ is given by (note it only applies to $t > 0$)

$$F\,(s) = \mathfrak{L}\,[f\,(t)] = \int_0^\infty e^{-st} f\,(t)\, dt\,. \tag{12.203}$$

The inverse transform is

$$f\,(t) = \frac{1}{2\pi i} \int_{\varepsilon - i\infty}^{\varepsilon + i\infty} F\,(s)\, e^{st} ds \tag{12.204}$$

where the contour is closed in the left half plane $(x < 0)$. We see that if $F\,(s)$ has a pole at $s = x + iy$, $f\,(t)$ will have a term $e^{(x+iy)t}$, i.e., a damped imaginary exponential $(x < 0)$. If $F\,(s)$ is the Laplace transform of $f\,(t)$, the transform of df/dt is $(sF\,(s) - f\,(0))$. Applying the transform to a differential equation transforms it into an algebraic equation for $F\,(s)$. We will apply the transform to Eq. (12.202). The right-hand side will be proportional to

$$\begin{aligned} \int_0^\infty e^{-st} dt \int_0^t c_i\,(t')\, e^{i(\Omega_{if})(t'-t)} dt' &= \int_0^\infty c_i\,(t')\, e^{i(\Omega_{if})t'} dt' \int_{t'}^\infty e^{-st} e^{-i\Omega_{if}t} dt \\ &= \int_0^\infty \frac{c_i\,(t')\, e^{-st'}}{s + i\Omega_{if}} dt' = \frac{C_i\,(s)}{s + i\Omega_{if}} \end{aligned} \tag{12.205}$$

with $\Omega_{if} = (E_i - E_f)\,/\hbar$. On changing the order of integration we have to be careful to cover the same region of the $t - t'$ plane, namely the region where $t > t'$. Obviously $C_i\,(s)$ is the Laplace transform of $c_i\,(t)$. In this way Eq. (12.202) is transformed into

[13] W. H. Louisell, *Quantum Statistical Properties of Radiation* (New York: Wiley, 1973).

$$(sC_i(s) - 1) = -\frac{1}{\hbar^2}\sum_f |\langle i| V |f\rangle|^2 \frac{C_i(s)}{s + i\Omega_{if}} \tag{12.206}$$

$$C_i(s) = \frac{1}{s + \frac{1}{\hbar^2}\sum_f |\langle i| V |f\rangle|^2 \frac{1}{s+i\Omega_{if}}}. \tag{12.207}$$

This is again an exact relation. It often happens that solving differential equations by Laplace transform yields a result which has no inverse transform; that is, the differential equation has no analytical solution. In the present case, we look for an approximate solution by taking the limit of the sum as $s \to 0^+$. As we have seen, the solution $c_i(t)$ will be determined by the pole s_o of $C_i(s)$:

$$s_o = -\frac{i}{\hbar^2}\sum_f |\langle i| V |f\rangle|^2 \lim_{s\to 0^+} \frac{1}{is - \Omega_{if}}. \tag{12.208}$$

Now

$$\lim_{s\to 0^+} \frac{1}{is - \Omega_{if}} = -\lim_{s\to 0^+} \frac{(\Omega_{if} + is)}{s^2 + \Omega_{if}^2} = -\left[\frac{1}{\Omega_{if}} + i\pi\delta(\Omega_{if})\right] \tag{12.209}$$

since

$$\lim_{s\to 0^+} \frac{s}{s^2 + \Omega_{if}^2} = \begin{matrix} 0 \text{ for } \Omega_{if} \neq 0 \\ \infty \text{ for } \Omega_{if} = 0 \end{matrix} \tag{12.210}$$

and

$$s\int_{-\infty}^{\infty} \frac{dx}{s^2 + x^2} = \pi. \tag{12.211}$$

Then

$$s_o = \frac{1}{\hbar^2}\sum_f |\langle i| V |f\rangle|^2 \left[\frac{i}{\Omega_{if}} - \pi\delta(\Omega_{if})\right]. \tag{12.212}$$

According to (12.204), $c_i(t)$ is given by

$$c_i(t) = e^{s_o t} = \exp[i\Delta - \Gamma]\,t \tag{12.213}$$

where the decay constant

$$\Gamma = \frac{\pi}{\hbar^2}\sum_f |\langle i| V |f\rangle|^2 \delta(\Omega_{if}) \tag{12.214}$$

is the total transition rate out of the initial state $|i\rangle$. Substituting this solution into (12.200)

$$c_f(t) = \frac{1}{i\hbar}\langle f| V |i\rangle \int_0^t e^{[i(\Delta+\Omega_{if})-\Gamma]t'}dt' \tag{12.215}$$

$$= \frac{1}{i\hbar}\langle f| V |i\rangle \frac{e^{[i(\Delta+\Omega_{if})-\Gamma]t} - 1}{i(\Delta + \Omega_{if}) - \Gamma} \tag{12.216}$$

$$|c_f(t\to\infty)|^2 = \left|\frac{1}{\hbar}\langle f| V |i\rangle\right|^2 \frac{1}{(\Delta + \Omega_{if})^2 + \Gamma^2} \tag{12.217}$$

so the distribution of final states is again a Lorentzian with a FWHM of 2Γ, but its peak is shifted from $\Omega_{if} = 0$ to $\Omega_{if} = -\Delta$ with

$$\Delta = \frac{1}{\hbar^2} \sum_f |\langle i| V |f\rangle|^2 \frac{1}{\Omega_{if}}. \tag{12.218}$$

Note that, unlike the expression for Γ, the expression for Δ does not require conservation of energy between the initial state and the final states included in the sum. So in the case where the final state includes a photon, the sum extends over all photon energies. Because of the increase in the density of states with photon energy, the sum does not converge. This is one of many divergences that appear in calculations in quantum electrodynamics (QED). Physicists fought with this problem for about twenty years. The solution consists of recognizing that similar divergences occur for free particles, so that, according to the theory, electrons would have an infinite mass and charge. Since we observe finite values for these quantities, we consider that the mass, say, is composed of a "bare" mass in addition to the infinite mass calculated by the theory, the sum of these two infinite quantities being the finite observed mass. Strange as it seems, this procedure—called "renormalization"—is the basis for all modern particle theories.

12.3.3.3 Applications to the quantized electromagnetic field

While the above considerations apply to any system with a continuum of energy levels, one of the main applications is to the problem of an atom interacting with the electromagnetic field. In that case, the state $|i\rangle$ is taken to consist of the atom in an excited state $|e\rangle_A$ and the electromagnetic field in its ground vacuum state $|0\rangle_k$ (no photons present). In the final state, the atom is in a lower energy state $|f\rangle_A$ with one photon present $|1\rangle_k$. The matrix element of the interaction is then given by Eq. (12.160)

$$\langle f| V |i\rangle = -\frac{e}{m} \frac{1}{L^{3/2}} \sqrt{\frac{2\pi\hbar}{\omega_k}}\ {}_A\langle f| \vec{p} \cdot \vec{\varepsilon}_k |i\rangle_A \tag{12.219}$$

$$= -\frac{e}{m} \frac{1}{L^{3/2}} \sqrt{\frac{2\pi\hbar}{\omega_k}} i\omega_{fA,eA}\, m_A \langle f| \vec{r} \cdot \vec{\varepsilon}_k |i\rangle_A, \tag{12.220}$$

where we have used the electric dipole approximation in the second step. Replacing the sum over final states by $\int \rho(E_k)\, d\Omega_k dE_k$ and using Eqs. (12.214) and (12.165), we find

$$\Gamma = \frac{1}{4\pi} \left(\frac{e^2}{\hbar c}\right) \int |{}_A\langle f| \vec{r} \cdot \vec{\varepsilon}_k |e\rangle_A|^2 \frac{\omega_k^3}{c^2} d\Omega_k \tag{12.221}$$

$$2\Gamma = \left(\frac{e^2}{\hbar c}\right) |{}_A\langle f| \vec{r} |e\rangle_A|^2 \frac{\omega_k^3}{c^2} \int_0^\pi \sin^3\theta d\theta \tag{12.222}$$

$$= \frac{4}{3} \left(\frac{e^2}{\hbar c}\right) |{}_A\langle f| \vec{r} |e\rangle_A|^2 \frac{\omega_k^3}{c^2} \tag{12.223}$$

in agreement with our previous result for the rate of decay of an atomic energy level (Einstein's A coefficient, Eq. (12.170)).

12.3.4 Discussion: wave-particle duality

This treatment of the electromagnetic field throws a new light on the nature of photons and allows some insight into the nature of Bohr's idea of complementarity between the wave and particle pictures. If the system is described as containing $n_{\vec{k},\hat{\varepsilon}}$ photons with wave vector $\vec{k}$ and polarization $\hat{\varepsilon}$ then the $(\vec{k},\hat{\varepsilon})$ oscillator (representing the amplitude of the plane wave denoted by $(\vec{k},\hat{\varepsilon})$) is in the $(n_{\vec{k},\hat{\varepsilon}})^{\text{th}}$ excited state. Creation (destruction) of a photon means that the corresponding oscillator makes a transition to a lower (higher) state. This shows the intimate relation between the wave and particle pictures.

States corresponding to a definite value of the field (coherent states), consist of superpositions of states with different values of $n_{\vec{k},\hat{\varepsilon}}$, and states with different spatial dependence can be formed by superpositions of states with different $\vec{k}$ values.

We can see how classical optics is contained in the above field quantization model by considering the example of a generic optical system consisting of a single atom in an excited state with energy E_1 which serves as a source of light when it emits a photon while making a transition to the ground state with energy E_o (Fig. 12.3).[14] The detector consists of another atom with an ionization energy less than E_1, so that it can be ionized by absorption of the photon emitted by the first atom. In traveling from the source atom to the detector atom, the photon travels through a generic optical system which may consist of a diaphragm containing two slits or reflection from a diffraction grating or any system exhibiting wave optical phenomena.

We will calculate the amplitude for the system to go from the state $|E_1\rangle_S\,|E_o\rangle_D\,|0\rangle_k$ to the state given by $|E_o\rangle_S\,|E_i\rangle_D\,|0\rangle_k$, where S, D, k refer to the source atom, detector atom, and electromagnetic field, respectively. Since the interaction between the atom and the field is $\sim \vec{p}\cdot\vec{A}$, it can only change the photon number by one (to lowest order), which means that in order to describe emission of a photon (from atom S) and absorption of a photon (by atom D), we need to consider the perturbation to second order.

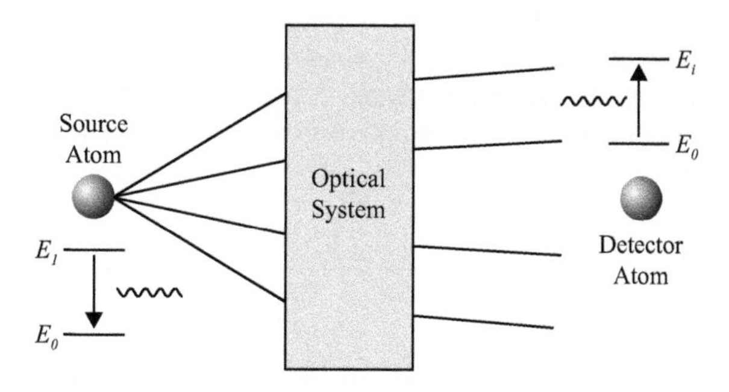

Fig. 12.3 A source atom excites the electromagnetic field by making a transition to its ground state and a detector atom, separated from the source by an optical system, is ionized.

[14]This argument was first presented by Schiff, L. I., *Quantum Mechanics*, second edition, section 50, McGraw Hill, (1955).

12.3.4.1 Second-order perturbation theory

We obtained the first-order solution by substituting $c_{Nn}(t) = c_{Nn}(0)$ in the right side of Eq. (12.147). To obtain the second-order solution, we substitute instead the first-order solution (12.147)

$$i\hbar\dot{c}^{(2)}_{Mm}(t) = \sum_{nN} c^{(1)}_{Nn}(t)\langle\psi_{Mm}|V_i|\psi_{Nn}\rangle e^{i(E_M-E_N)t/\hbar} \tag{12.224}$$

$$c^{(1)}_{Nn}(t) = c_{Jj}(0)\frac{\langle\psi_{Nn}|V_i|\psi_{Jj}\rangle}{E_N-E_J}\left(1-e^{i(E_N-E_J)t/\hbar}\right) \tag{12.225}$$

(12.226)

$$i\hbar\dot{c}^{(2)}_{Mm}(t) = \sum_{nN} c_{Jj}(0)\frac{\langle\psi_{Mm}|V_i|\psi_{Nn}\rangle\langle\psi_{Nn}|V_i|\psi_{Jj}\rangle}{E_N-E_J}\left(e^{i(E_M-E_N)t/\hbar}-e^{i(E_M-E_J)t/\hbar}\right)$$

(12.227)

where we assumed the system was in the state Jj at $t = 0$. Performing the time integration as before, we find

$$c^{(2)}_{Mm}(t) = \sum_{nN} c_{Jj}(0)\frac{\langle\psi_{Mm}|V_i|\psi_{Nn}\rangle\langle\psi_{Nn}|V_i|\psi_{Jj}\rangle}{E_N-E_J}\times\ldots \tag{12.228}$$

$$\ldots\left[\frac{e^{i(E_M-E_J)t/\hbar}-1}{E_M-E_j}-\frac{e^{i(E_M-E_N)t/\hbar}-1}{E_M-E_N}\right]. \tag{12.229}$$

The second term in the brackets is a result of the 1 in the expression for $c^{(1)}_{Nn}(t)$ and this in turn is the result of the initial condition, which in effect corresponds to a sudden turning on of the perturbation at $t = 0$. This term can thus be considered an artifact and neglected. The remaining term is treated just as we did in the first order case and forces energy conservation between the initial (Jj) and final (Mm) states. If there is an intermediate state (Nn) with the same energy as the initial state and there is a nonzero matrix element between these two states, we have to interpret any integrals over energy as the principal value of a contour integral. Thus we can use this expression as the matrix element in our previous result for the time-proportional transition probability:

$$\sum_{nN}\frac{\langle\psi_{Mm}|V_i|\psi_{Nn}\rangle\langle\psi_{Nn}|V_i|\psi_{Jj}\rangle}{E_N-E_J}\delta(E_M-E_J). \tag{12.230}$$

12.3.4.2 Amplitude for a photon to be detected after traversing an optical system

Previously we expanded the vector potential in terms of plane waves. The amplitudes of the individual plane waves were then taken to be noncommuting operators operating on occupation number eigenstates. The occupation numbers referred to the various plane wave states. In the present problem, we will assume that the entire system—

source, detector, and optical system—is surrounded by a reflecting boundary. We will expand the vector potential in eigenfunctions $\vec{\alpha}_k(\vec{r})$ of

$$\nabla^2 \vec{\alpha}_k(\vec{r}) + k^2 \vec{\alpha}_k(\vec{r}) = 0 \tag{12.231}$$

that satisfy the boundary conditions imposed by the optical system (say two slits in a screen) and the distant reflecting boundary which serves to ensure that the eigenvalues k will be discrete. The expansion amplitudes will again become noncommuting operators but now the occupation numbers will refer to these eigenstates rather than to plane wave states.

There are two types of intermediate states by which the system can go from the initial state $|E_1\rangle_S|E_0\rangle_D|0\rangle_k$ to the final state $|E_0\rangle_S|E_i\rangle_D|0\rangle_k$. The first is the state where the source atom emits a photon and jumps to its ground state: $|E_0\rangle_S\,|E_0\rangle_D\,|1\rangle_k$. The second-order matrix element for this case is then

$$M_{fi}^{(1)} = \sum_k \frac{\langle E_{0S}, E_{iD}, 0_k|\, V_D\, |E_{0S}, E_{0D}, 1_k\rangle\, \langle E_{0S}, E_{0D}, 1_k|\, V_S\, |E_{1S}, E_{0D}, 0_k\rangle}{\hbar\omega_k - (E_{1S} - E_{0S})} \tag{12.232}$$

The second type of intermediate state occurs when the detector atom makes a transition to its ionized state while *emitting* a photon which is absorbed by the source atom when it jumps to its ground state. While this state violates conservation of energy (it is called a virtual state), it must be included since we need to include all possible states in the summation over intermediate states. The matrix element for this process is

$$M_{fi}^{(2)} = \sum_k \frac{\langle E_{0S}, E_{iD}, 0_k|\, V_S\, |E_{1S}, E_{iD}, 1_k\rangle\, \langle E_{1S}, E_{iD}, 1_k|\, V_D\, |E_{1S}, E_{0D}, 0_k\rangle}{\hbar\omega_k + (E_{1S} - E_{0S})} \tag{12.233}$$

where we made use of the overall conservation of energy

$$E_{1S} + E_{0D} = E_{0S} + E_{iD}. \tag{12.234}$$

To evaluate the matrix elements, we modify Eq. (12.142) to reflect the expansion of $\vec{A}$ in the eigenfunctions $\vec{\alpha}_k(\vec{r})$ (which can be taken as real in the case of non-absorbing boundaries) rather than plane waves

$$V_i = -\frac{e}{m}\frac{1}{L^{3/2}}\vec{p}\cdot\sum_k \sqrt{\frac{2\pi\hbar}{\omega_k}}\,\vec{\alpha}_k(\vec{r})\left(a_k + a_k^\dagger\right). \tag{12.235}$$

Then the matrix element

$$\langle E_{0S}, E_{iD}, 1_k | V_D | E_{0S}, E_{0D}, 0_k \rangle = \langle E_{iD}, 1_k | V_D | E_{0D}, 0_k \rangle \tag{12.236}$$

$$= -\frac{e}{m} \frac{1}{L^{3/2}} \sum_k \sqrt{\frac{2\pi\hbar}{\omega_k}} \langle E_{iD} | \overrightarrow{p} \cdot \overrightarrow{\alpha}_k (\overrightarrow{r}_D) | E_{0D} \rangle \tag{12.237}$$

$$= -\frac{e}{m} \frac{1}{L^{3/2}} \sum_k \sqrt{\frac{2\pi\hbar}{\omega_k}} \times .. \tag{12.238}$$

$$\int \psi_i^* (\overrightarrow{r}_D) [\overrightarrow{p} \cdot \overrightarrow{\alpha}_k (\overrightarrow{r}_D)] \psi_0 (\overrightarrow{r}_D) d^3 r_D. \tag{12.239}$$

We see that the matrix elements for $M_{fi}^{(1)}$ and $M_{fi}^{(2)}$ are the same, so we can add the two terms together by combining the denominators and obtain the amplitude for our process of a photon being emitted from the source and absorbed by the detector:

$$M_{fi} = \left(\frac{e}{m}\right)^2 \frac{4\pi}{L^3} \sum_k \sum_{i.j=x,y,z} \int \int \psi_i^* (\overrightarrow{r}_D) [p_i] \psi_0 (\overrightarrow{r}_D) \psi_0^* (\overrightarrow{r}_S) [p_j] \psi_1 (\overrightarrow{r}_S) \times \left[\frac{\alpha_{k,i} (\overrightarrow{r}_D) \alpha_{k.j} (\overrightarrow{r}_S)}{(\omega_k)^2 - (E_{1S} - E_{0S})^2 / \hbar^2}\right] d^3 r_S d^3 r_D. \tag{12.240}$$

The atomic wave functions are nonzero only for positions very close to the atomic nuclei, which are separated by a macroscopic distance so that we can take the expression in square brackets out of the integrations

$$M_{fi} = \left(\frac{e}{m}\right)^2 \frac{4\pi}{L^3} \sum_k \sum_{i.j=x,y,z} \left[\frac{\alpha_{k,i} (\overrightarrow{r}_D) \alpha_{k.j} (\overrightarrow{r}_S)}{(\omega_k)^2 - (E_{1S} - E_{0S})^2 / \hbar^2}\right] \times .. \langle E_{iD} | p_i | E_{oD} \rangle \langle E_{oS} | p_j | E_{1S} \rangle \tag{12.241}$$

and the position dependence of the amplitude is contained in the sums of the terms in square brackets. The probability of the process is found by substituting this expression for the matrix element in Eq. (12.154).

12.3.4.3 Classical optics solution of the problem

To address the problem, we need to calculate the field amplitude at the position of the detector atom when there is an oscillating current at the position of the source atom. The vector potential satisfies the wave equation:

$$\nabla^2 \overrightarrow{A} (\overrightarrow{r}, t) - \frac{1}{c^2} \frac{\partial^2}{\partial t^2} \overrightarrow{A} (\overrightarrow{r}, t) = -\frac{4\pi}{c} e^{i\omega t} \overrightarrow{J} (\overrightarrow{r}) \tag{12.242}$$

Letting $\overrightarrow{A} (\overrightarrow{r}, t) = \overrightarrow{A} (\overrightarrow{r}) e^{i\omega t}$ we find

$$\nabla^2 \overrightarrow{A} (\overrightarrow{r}) + \frac{\omega^2}{c^2} \overrightarrow{A} (\overrightarrow{r}) = -\frac{4\pi}{c} \overrightarrow{J} (\overrightarrow{r}). \tag{12.243}$$

Again we will expand $\vec{A}(\vec{r})$ in a complete orthonormal set of eigenfunctions satisfying the boundary conditions imposed by the optical system and the distant boundaries. As these boundary conditions are the same as in the quantum case discussed above, the functions $\vec{\alpha}_k(\vec{r})$ will be the same as in that problem. We write

$$\vec{A}(\vec{r}) = \sum_k g_k \vec{\alpha}_k(\vec{r}) \tag{12.244}$$

and find

$$\sum_k g_k \left[-k^2 \vec{\alpha}_k(\vec{r}) + \frac{\omega^2}{c^2} \vec{\alpha}_k(\vec{r}) \right] = -\frac{4\pi}{c} \vec{J}(\vec{r}). \tag{12.245}$$

Taking the dot product with $\vec{\alpha}_n(\vec{r})$ and integrating over $\mathrm{d}^3 r$ we obtain

$$g_n \left[-n^2 + \frac{\omega^2}{c^2} \right] = -\frac{4\pi}{c} \int \vec{\alpha}_n(\vec{r}) \cdot \vec{J}(\vec{r})\, d^3 r \tag{12.246}$$

$$g_k = -\frac{4\pi}{c} \frac{\int \vec{\alpha}_k(\vec{r}) \cdot \vec{J}(\vec{r})\, d^3 r}{\frac{\omega^2}{c^2} - k^2} \tag{12.247}$$

$$A_j(\vec{r}_D) = -\frac{4\pi}{c} \int \sum_{i=x,y,z} J_i(\vec{r}) \sum_k \left[\frac{\alpha_{k,i}(\vec{r})\, \alpha_{k,j}(\vec{r}_D)}{\frac{\omega^2}{c^2} - k^2} \right] d^3 r. \tag{12.248}$$

The term in square brackets agrees with the term in square brackets in equation (12.241). The sum of the terms in the square brackets is the eigenfunction expansion of the Green's function as can be seen by writing

$$A_j(\vec{r}_D) = \int \sum_{i=x,y,z} J_i(\vec{r})\, G_{ij}(\vec{r}_D, \vec{r})\, d^3 r \tag{12.249}$$

with

$$G_{ij}(\vec{r}_D, \vec{r}) = -\frac{4\pi}{c} \sum_k \left[\frac{\alpha_{k,i}(\vec{r})\, \alpha_{k,j}(\vec{r}_D)}{\frac{\omega^2}{c^2} - k^2} \right]. \tag{12.250}$$

The intensity of radiation at the detector will be proportional to $\left| \vec{A}(\vec{r}_D) \right|^2$, that is, proportional to the quantum mechanical probability calculated above. The usual optical effects—interference, diffraction, etc.—are contained in the dependence of the result on $\vec{r}_D$.

This result shows how the wave behavior is contained in the quantized field picture. In making a transition, the source atom excites a set of modes of the radiation field and these modes, extended in space, excite the detector atom. The idea of trajectories, in fact the particle concept itself, seems to play no role here. By inverse Fourier transforming the eigenfunction expansion of the Green's function, we obtain a time-dependent picture showing how the wave spreads out from the source as a function of time. The discrete, statistical behavior is associated with the absorption and emission of energy by the atoms.

The mathematics seems to force us to the viewpoint that what we would tend to think of as the emission of a photon and its transmission through a system until it induces a transition in a detector is really the excitation of some modes of the field, which in turn excite the atom. The discreteness comes in when we measure the state of the detector atom and get an either/or (excited/unexcited) result. This is what Bohr meant in his discussions with Schrödinger when he said that Schrödinger's attempt to give a continuous description of physical processes did not go far enough: the discreteness was just pushed down the line.

All of this gives us a deeper insight into the wave-particle duality and Bohr's complementarity principle.

12.3.5 Following Dirac, Jordan commits "sin squared" on his own

The Schrödinger form of quantum mechanics was constructed by considering the mechanics of particles as analogous to optics. Classical mechanics was assumed to correspond to geometrical optics and Schrödinger's wave mechanics was constructed to correspond to wave optics. However, in view of the above, the analogy seems incomplete. To account for the existence of photons, the electromagnetic field must be quantized; that is, the field amplitudes must be replaced by noncommuting operators. Thus it is tempting, in order to account for the existence of electrons, to try to similarly replace the Schrödinger wave amplitudes by noncommuting operators. Jordan, who saw the need to quantize the electromagnetic field as early as 1925, later claimed that he understood the need to quantize the Schrödinger field as soon as he read Schrödinger's first paper. Dirac felt that to obtain a proper quantum theory you had to start with a known, classical system, while Jordan believed any physical system could be considered as a starting point, in particular, the Schrödinger equation considered as a classical wave equation. Dirac argued against this idea, saying that while photons can be created and destroyed as described by the quantization formalism discussed above, this did not apply to electrons, and thus the quantization procedure was inappropriate in that case.

Jordan responded that this meant that, indeed, electrons could be created and destroyed and went ahead with his quantization idea, providing the foundation for modern particle theory. As we will see below, the non-relativistic theory as applied to electrons conserves the total particle number. The interactions merely move electrons from one state to another.

In relativistic field theories, particle creation and destruction are commonplace. In 1928 Dirac discovered the relativistic wave equation for electrons, which allows for the existence of the positron, and thus led the way to the relativistic theory that allows for electron-positron pair creation and destruction. After positron creation and annihilation were observed, Dirac became a supporter of the second quantization idea.

Another advantage of second quantization is that the states are defined by basis functions in three-dimensional space. This is in contrast to Schrödinger's wave mechanics, where the wave functions for an N-particle system is a function in $3N$-dimensional space.

As shown above, the state of the electromagnetic field is specified by the set of occupation numbers (suppressing $\vec{\varepsilon}$ for simplicity) $n_{\vec{k}_i}$ for every possible mode $\vec{k}_i$, and there is no restriction on the allowed values of $n_{\vec{k}_i}$. The exchange of two excitations has no effect since packets of excitation energy are indistinguishable. Einstein had shown that the Bose-Einstein distribution of atoms in a gas could be obtained by representing the atoms by quantized waves in ordinary three-dimensional space. The introduction of noncommuting operators, which imposed no restrictions on $n_{\vec{k}_i}$ and resulted in states symmetric under particle exchange, was thus not compatible with the behavior of fermions which obey the Pauli exclusion principle, and thus can only be described by a set of occupation numbers limited to the values $n = 0, 1$. This was a major stumbling block to the idea of second quantization, which Jordan solved in his first paper devoted to this topic.[15] These results were further elaborated by Jordan and Wigner.[16]

We start with a system in a state described by a Hamiltonian H_0 and a wave function $\psi(\vec{r})$ and expand this wave function in a complete set of (orthonormal) eigenstates $u_k(\vec{r})$ of H_0, where $H_0 u_k(\vec{r}) = E_k u_k(\vec{r})$

$$\psi(\vec{r}) = \sum_k a_k u_k(\vec{r}). \tag{12.251}$$

When normalized to one, $\rho(\vec{r}) = \psi^*(\vec{r})\psi(\vec{r})$ represents the probability of finding a particle at the position $\vec{r}$ (between $\vec{r}$ and $\vec{r} + \vec{dr}$). For an ensemble of N identically prepared systems, there would be $N\rho(\vec{r})$ systems where the particle was found at $\vec{r}$ and we can normalize $\psi(\vec{r})$ so that

$$\int \psi^*(\vec{r})\psi(\vec{r})\,d^3r = N = \sum_{k,j} a_k^* a_j \int u_k^*(\vec{r})\,u_j(\vec{r})\,d^3r = \sum_k a_k^* a_k\,. \tag{12.252}$$

Then the energy of the state $\psi(\vec{r})$ will be given by

$$E = \int \psi^*(\vec{r})\,H_0\psi(\vec{r})\,d^3r = \sum_{k,j} a_k^* a_j \int u_k^*(\vec{r})\,H_0 u_j(\vec{r})\,d^3r \tag{12.253}$$

$$= \sum_j a_j^* a_j E_j = \sum_j N_j E_j. \tag{12.254}$$

Considering a_j and $p_j = i\hbar a_j^*$ as canonical variables, we see that they satisfy the equations of motion with Hamiltonian $H_j = a_j p_j E_j / i\hbar$:

$$\dot{a}_j = \frac{\partial H_j}{\partial p_j} = a_j \frac{E_j}{i\hbar} \qquad a_j = b_j e^{E_j t/i\hbar} = b_j e^{-i\omega_j t} \tag{12.255}$$

$$\dot{p}_j = -\frac{\partial H_j}{\partial a_j} = i\hbar\dot{a}_j^* = -a_j^* E_j \qquad a_j^* = b_j^* e^{-E_j t/i\hbar} = b_j^* e^{i\omega_j t}. \tag{12.256}$$

[15]Jordan, P., *Quantum Mechanics of Gas Degeneracy*, Zeit. Phys., 44, 473, (1927), received July 7, 1927.

[16]Jordan, P., and Wigner, E.P., *On the Pauli Exclusion Principle*, Zeit. Phys., 47, 631, (1928), received Jan. 26, 1928.

This shows that a_j and a_j^* behave like the canonical coordinates of a harmonic oscillator. Considering this as a classical system, we would quantize it by replacing the coordinates (a, a^*) by noncommuting operators $(a, a^\dagger)$ satisfying

$$\left[a_j, a_k^\dagger\right] = \delta_{jk} \tag{12.257}$$

$$[a_j, a_k] = \left[a_j^\dagger, a_k^\dagger\right] = 0. \tag{12.258}$$

We can see the connection with a harmonic oscillator more explicitly by writing

$$a_k = \frac{1}{\sqrt{2}}\left(q_k + ip_k\right) \tag{12.259}$$

$$a_k^\dagger = \frac{1}{\sqrt{2}}\left(q_k - ip_k\right), \tag{12.260}$$

where (q_k, p_k) are Hermitian. Substituting into (12.257), we see that the (q_k, p_k) satisfy

$$[p_k, q_j] = -i\delta_{kj}. \tag{12.261}$$

We also find

$$H_j = a_j^\dagger a_j E_j = \frac{\hbar\omega_j}{2}\left(p_j^2 + q_j^2 - i\left[p_j, q_j\right]\right) = \frac{\hbar\omega_j}{2}\left(p_j^2 + q_j^2 - 1\right). \tag{12.262}$$

Thus the $\left(a_j^\dagger, a_j\right)$ are the raising and lowering operators acting on eigenstates of N_j. As we have seen in our discussion of the simple harmonic oscillator, this follows directly from the commutation relations.

Because the a_k are now noncommuting operators, the $\psi\left(\vec{r}\right), \psi^\dagger\left(\vec{r}\right)$ must also be operators:

$$\left[\psi\left(\vec{r}\right), \psi^\dagger\left(\vec{r'}\right)\right] = \left[\sum_j a_j u_j\left(\vec{r}\right), \sum_k a_k^\dagger u_k^*\left(\vec{r'}\right)\right] = \sum_{j,k} u_j\left(\vec{r}\right) u_k^*\left(\vec{r'}\right)\left[a_j, a_k^\dagger\right] \tag{12.263}$$

$$= \sum_k u_k\left(\vec{r}\right) u_k^*\left(\vec{r'}\right) = \delta^{(3)}\left(\vec{r} - \vec{r'}\right), \tag{12.264}$$

where the last step makes use of the completeness of the basis functions.

We can also choose another complete set of functions as a basis for the expansion (12.251), say the momentum eigenstates of a free particle. In that case the Hamiltonian (12.253) would be written

$$E = \int \psi^*\left(\vec{r}\right) H_0 \psi\left(\vec{r}\right) d^3r = \sum_{k,j} a_k^\dagger a_j \int u_k^*\left(\vec{r}\right) H_0 u_j\left(\vec{r}\right) d^3r \tag{12.265}$$

$$= \sum_{k,j} a_k^\dagger a_j H_{kj} \tag{12.266}$$

where we introduced the matrix element for the Hamiltonian operator. We see that this is an operator which moves a particle from the state described by the jth basis function to that described by the kth function. If the Hamiltonian contains a two-particle interaction

$$H = H_0\left(\vec{r}\right) + V\left(\vec{r} - \vec{r'}\right), \tag{12.267}$$

then we have

$$E = \sum_{k,j} a_k^\dagger a_j H_{kj} + \int\int d^3r d^3r' \psi^*\left(\vec{r'}\right)\psi\left(\vec{r'}\right) V\left(\vec{r} - \vec{r'}\right)\psi^*\left(\vec{r}\right)\psi\left(\vec{r}\right) \tag{12.268}$$

$$= \sum_{k,j} a_k^\dagger a_j H_{kj} + \int\int d^3r d^3r' \sum_k a_k^\dagger u_k^*\left(\vec{r'}\right)\sum_j a_j u_j\left(\vec{r'}\right) V\left(\vec{r} - \vec{r'}\right) \times ..$$

$$.. \sum_l a_l^\dagger u_l^*\left(\vec{r}\right)\sum_m a_m u_m\left(\vec{r}\right) \tag{12.269}$$

$$= \sum_{k,j} a_k^\dagger a_j H_{kj} + \sum_{k,j,l,m} a_k^\dagger a_j a_l^\dagger a_m \int\int d^3r d^3r' u_k^*\left(\vec{r'}\right) u_l^*\left(\vec{r}\right) V\left(\vec{r} - \vec{r'}\right)$$

$$u_j\left(\vec{r'}\right) u_m\left(\vec{r}\right)$$

$$= \sum_{k,j} a_k^\dagger a_j H_{kj} + \sum_{k,j,l,m} a_k^\dagger a_j a_l^\dagger a_m V_{klmj} \tag{12.270}$$

where the second term is an operator which moves two particles from the states u_j, u_m to the states u_k, u_l, that is, a two-particle scattering event.

The second term in (12.268),

$$\int\int d^3r d^3r' \psi^*\left(\vec{r'}\right)\left[\psi\left(\vec{r}'\right)\psi^*\left(\vec{r}\right)\right]\psi\left(\vec{r}\right) V\left(\vec{r} - \vec{r}'\right), \tag{12.271}$$

can be rewritten using the commutation relation for the second-quantized wave function as

$$= \int\int d^3r d^3r' \psi^*\left(\vec{r}'\right)\left[\psi^*\left(\vec{r}\right)\psi\left(\vec{r}'\right) + \delta^{(3)}\left(\vec{r} - \vec{r}'\right)\right]\psi\left(\vec{r}\right) V\left(\vec{r} - \vec{r}'\right) \tag{12.272}$$

$$= \int\int d^3r d^3r' \psi^*\left(\vec{r}'\right)\psi^*\left(\vec{r}\right)\psi\left(\vec{r}'\right)\psi\left(\vec{r}\right) V\left(\vec{r} - \vec{r}'\right)$$

$$+ \int d^3r \psi^*\left(\vec{r}\right)\psi\left(\vec{r}\right) V\left(\vec{r} - \vec{r}'\right), \tag{12.273}$$

where the last term will cancel the interaction of the electrons with themselves, which has not been included in (12.267). Then we can rewrite (12.270) as

$$E = \sum_{k,j} a_k^\dagger a_j H_{kj} + \sum_{k,j,l,m} a_k^\dagger a_l^\dagger a_j a_m V_{klmj} \tag{12.274}$$

where the *normal ordering*, all creation operators on the left, removes the singular self-energy term.

12.3.5.1 Equivalence to the Schrödinger equation in $3N$-dimensional space[17]

We consider a system of N particles and choose, as above, an orthonormal complete set of single-particle functions $\phi_n(x)$ satisfying the appropriate boundary conditions as a basis. We will expand the wave function of the complete N-particle system in products of the single-particle functions

$$\phi_{n_1}(x_1)..\phi_{n_k}(x_k)..\phi_{n_N}(x_N) \tag{12.275}$$

where the kth particle is in the state labeled by n_k, etc.

As the particles are identical, the wave function must be symmetric (for bosons) or antisymmetric (for fermions) under exchange of any two particles and can only depend on the number of particles in each state (i.e., the occupation numbers N_n, where N_n equals the number of atoms in the state $\phi_n(x)$). For fermions, the only possible values are $N_n = 0$ or $N_n = 1$. In this section, we will consider a system of bosons, for which the N_n can range from zero to the total number of particles.

We will use the notation

$$(\phi_n)^{N_n} = \phi_n(x_1)\,\phi_n(x_2)....\phi_n(x_{N_n}), \tag{12.276}$$

which represents N_n atoms in the state labeled by the quantum numbers n. Noting that the number of ways that N_1 particles can be placed in the first state, N_2 in the second, etc. is given by (with $N = \sum_k N_k$ the total number of particles)

$$Z^2 = \frac{N!}{N_1!...N_m!}, \tag{12.277}$$

we write the wave function for the system as

$$\Psi_{n_1,...n_N}(x_1,....x_N) = \frac{1}{\sqrt{N!}} \sum_{\substack{\text{all perm}\\ 1,2,...N}} \frac{(\phi_{n_1})^{N_{n_1}}}{\sqrt{N_{n_1}!}} \frac{(\phi_{n_2})^{N_{n_2}}}{\sqrt{N_{n_2}!}} \cdots \frac{(\phi_{n_k})^{N_{n_k}}}{\sqrt{N_{n_k}!}} \tag{12.278}$$

where n_i represents the set of quantum numbers designating the ith state and N_{n_i} is the number of particles in that state. The sum is over the $N!$ permutations of the particle indices $1, 2, ...N$. Summing over the permutations produces a state symmetric under the exchange of any two particles. We designate the state as $\Psi_{[N]}$ where $[N]$ is the set of occupation numbers N_k. These states are orthogonal and normalized.

Now we operate on $\Psi_{[N]}$ with the one-particle Hamiltonian $\sum_i H_o(x_i)$ using

$$H_o\,|n_i\rangle = \sum_m |n_m\rangle \langle n_m|\, H_o\,|n_i\rangle \tag{12.279}$$

where $|n_i\rangle$ is the state occupied by the ith particle. Then (note that $\phi_{n_i}(x) = \langle x|\, n_i\rangle$)

$$H_o(x_i)\,(\phi_{n_i})^{N_{n_i}} = N_{n_i} \sum_m \langle n_m|\, H_o\,|n_i\rangle\, \phi_m(x_i)\,(\phi_{n_i})^{N_{n_i}-1}, \tag{12.280}$$

[17] See also Heisenberg, W., *The Physical Principles of the Quantum Theory,* (Dover, 1949).

remembering that $H_o(x_i)$ is a sum of individual particle operators, so that there will be N_{n_i} identical terms. Then

$$\sum_i H_o(x_i)\,\Psi_{[N_1,\dots N_i\dots N_m\dots.N_k]} = \frac{1}{\sqrt{N!}}\sum_{\substack{\text{all perm}\\1,2,\dots N}}\sum_{m,i}\langle n_m|\,H_o\,|n_i\rangle\,\frac{(\phi_{n_1})^{N_{n_1}}}{\sqrt{N_{n_1}!}}\times .. \tag{12.281}$$

$$..N_{n_i}\frac{(\phi_{n_i})^{N_{n_i}-1}}{\sqrt{N_{n_i}!}}\dots\frac{(\phi_{n_m})^{N_{nm}+1}}{\sqrt{N_{n_m}!}}\dots\frac{(\phi_{n_k})^{N_{n_k}}}{\sqrt{N_{n_k}!}} \tag{12.282}$$

$$= \frac{1}{\sqrt{N!}}\sum_{\substack{\text{all perm}\\1,2,\dots N}}\sum_{m,i}\langle n_m|\,H_o\,|n_i\rangle\,\frac{(\phi_{n_1})^{N_{n_1}}}{\sqrt{N_{n_1}!}}\times .. \tag{12.283}$$

$$..\sqrt{N_{n_i}}\frac{(\phi_{n_i})^{N_{n_i}-1}}{\sqrt{(N_{n_i}-1)!}}..\sqrt{N_{n_m}+1}\frac{(\phi_{n_m})^{N_{nm}+1}}{\sqrt{(N_{n_m}+1)!}}\dots\frac{(\phi_{n_k})^{N_{n_k}}}{\sqrt{N_{n_k}!}} \tag{12.284}$$

$$= \sum_{n_i} N_{n_i}\langle n_i|\,H_o\,|n_i\rangle\,\Psi_{[N_1,\dots N_i\dots N_m\dots.N_k]} + .. \tag{12.285}$$

$$\sum_{n_m\neq n_i}\langle n_m|\,H_o\,|n_i\rangle\,\sqrt{N_{n_i}}\sqrt{N_{n_m}+1}\Psi_{\left[N_1\dots\left(N_{n_i}-1\right)\!,..(N_{n_m}+1)..N_k\right]} \tag{12.286}$$

where we made use of the identities

$$\frac{N}{\sqrt{N!}} = \frac{\sqrt{N}}{\sqrt{(N-1)!}} \tag{12.287}$$

$$\frac{1}{\sqrt{N!}} = \frac{\sqrt{N+1}}{\sqrt{(N+1)!}} \tag{12.288}$$

We know that the operators $\left(a, a^\dagger\right)$ satisfying the commutation relations (12.112) have the following effect when applied to an n-particle state $|N\rangle$

$$a\,|N\rangle = \sqrt{N}\,|N-1\rangle \tag{12.289}$$

$$a^\dagger\,|N\rangle = \sqrt{N+1}\,|N+1\rangle \tag{12.290}$$

so that we can write

$$\sum_i H_o(x_i)\,\Psi_{[N_1,\dots N_i\dots N_m\dots.N_k]} = \sum_{n_m,n_i}\langle n_m|\,H_o\,|n_i\rangle\,a^\dagger_{n_m}a_{n_i}\Psi_{\left[N_1\dots\left(N_{n_i}\right)\!,..\left(N_{nm}\right)\!..N_k\right]} \tag{12.291}$$

in agreement with equation (12.270). For the two-particle operator in (12.267), we find by a similar argument the result in (12.270).

Thus we see that by using the occupation number eigenstates, the many-body Schrödinger equation can be transformed to basis functions in ordinary three-dimensional space. This is an argument for the second quantized theory and is an answer to those who felt uncomfortable with the wave equation in $3N$-dimensional space. In addition, requiring the wave amplitudes to satisfy the canonical commutation relations is seen to lead to the existence of particles, or particle-like excitations of the wave field, and sheds a new light on the question of wave-particle duality. This formalism is widely used in the study of condensed matter.[18]

12.3.5.2 Application to fermions satisfying Pauli's exclusion principle with states antisymmetric under exchange of identical particles

Using Jordan's prescribed quantization method of replacing the Schrödinger wave function with quantities (operators, matrices) which satisfy the commutation relations (12.112) leads to wave functions symmetric under exchange of any two particles and to the fact that the individual particle states can contain any number of particles ranging from $N_k = 0$ to $N_k \to \infty$. This led to some confusion at first, as it was known that only one electron is allowed to occupy a given state, as formulated in the Pauli exclusion principle which is necessary to understand the shell structure in atoms.

At first it was proposed that second quantization based on operators satisfying the commutation relations might only apply to a photon gas (black body radiation) and that all matter would behave like electrons. In that case the second quantization discussed above would not apply to matter. If it applied to electrons, all the electrons in an atom would be expected to eventually fall into the ground state, clearly an unphysical result.

We now know that second quantization using commutators applies only to particles with integer spin (bosons), while for particles with half-integer spins (fermions), anticommutators must be used instead.

In a paper submitted in July 1927,[19] Jordan realized that it was not really necessary to impose the commutation relations, and he proposed using operators with the property that only $N_k = 0, 1$ were allowed. The condition that these operators had to satisfy was remarkably simple but with very deep consequences. Instead of (12.112), Jordan and Wigner[20] showed that all that is needed is that the operators be made to satisfy so-called anticommutation rules:

$$b_k b_{k'}^\dagger + b_{k'}^\dagger b_k = \left\{ b_k, b_{k'}^\dagger \right\} = \delta_{kk'} \tag{12.292}$$

$$\{b_k, b_{k'}\} = 0 \tag{12.293}$$

$$\left\{ b_{k'}^\dagger, b_k^\dagger \right\} = 0. \tag{12.294}$$

[18] Mattuck, R.D., *A Guide to Feynman Diagrams in the Many-Body Problem*, second edition, (Dover, 1992), see also Abrikosov, A.A., Gor'kov, L.P. and Dzyaloshinskii, I.Ye., *Quantum Field Theoretical Methods in Statistical Physics*, second edition (Pergamon Press, 1965).

[19] Jordan, P., *Zur Quantenmechanik der Gasentartung* [On the quantum mechanics of gas degeneracy], *Z. Phys.* **44** 473 (1927) received July 7, 1927.

[20] Jordan, P. and WIgner, E.P., *Über das Paulische Äquivalenzverbot* [On the Pauli exclusion principle], *Z. Phys.* **47**, 631 (1928) received January 26, 1928.

Given this, the operator $N_k = b_k^\dagger b_k$ is seen to satisfy

$$N^2 = b_k^\dagger b_k b_k^\dagger b_k = b_k^\dagger \left(1 - b_k^\dagger b_k\right) b_k = b_k^\dagger b_k = N, \tag{12.295}$$

so the only possible eigenvalues are $N = 0, 1$. Here we used the fact that $b_k b_k = 0$ according to the anticommutation relations. We can also use the anticommutation relations to show that the commutator

$$[N_k, N_m] = b_k^\dagger b_k b_m^\dagger b_m - b_m^\dagger b_m b_k^\dagger b_k = 0, \tag{12.296}$$

so that N_k, N_m can be simultaneously diagonalized.

Defining the state with $N_k = 0$ as $|0\rangle_k$ where $b_k^\dagger b_k |0\rangle_k = 0$ we find that the state $b_k^\dagger |0\rangle_k$ is characterized by $N_k = 1$ since

$$N_k b_k^\dagger |0\rangle_k = b_k^\dagger b_k b_k^\dagger |0\rangle_k = b_k^\dagger \left(1 - b_k^\dagger b_k\right) |0\rangle_k = 1 b_k^\dagger |0\rangle_k \tag{12.297}$$

$$b_k^\dagger |0\rangle_k = |1\rangle_k . \tag{12.298}$$

The anticommutation relations also yield $b_k^\dagger b_k^\dagger = 0$, which means that $b_k^\dagger |1\rangle_k = b_k^\dagger \left(b_k^\dagger |0\rangle_k\right) = 0$, confirming that only one particle can occupy a given state. Further,

$$b_k |1\rangle_k = b_k b_k^\dagger |0\rangle_k = \left(1 - b_k^\dagger b_k\right) |0\rangle_k = |0\rangle_k . \tag{12.299}$$

A state given by the set of occupation numbers $\{N\} = \{n_1, n_2, ...n_N\}$ is then written as

$$|\{N\}\rangle = |n_1, n_2, ...n_N\rangle = \left(b_1^\dagger\right)^{n_1} \left(b_2^\dagger\right)^{n_2} ... \left(b_N^\dagger\right)^{n_N} |0, 0...0\rangle. \tag{12.300}$$

Operating on this state with $b_k^\dagger$ yields

$$b_k^\dagger |n_1, n_2, ...n_k...n_N\rangle = (-1)^{\Sigma_k} |n_1, n_2, ...n_k + 1, ...n_N\rangle \text{ when } n_k = 0 \tag{12.301}$$

$$= 0 \text{ when } n_k = 1 \tag{12.302}$$

and similarly

$$b_k |n_1, n_2, .n_k...n_N\rangle = (-1)^{\Sigma_k} |n_1, n_2, ...n_k - 1, ...n_N\rangle \text{ when } n_k = 1 \tag{12.303}$$

$$= 0 \text{ when } n_k = 0 \tag{12.304}$$

$$b_k^\dagger b_k |n_1, n_2, ...n_k...n_N\rangle = n_k |n_1, n_2, ...n_k, ...n_N\rangle \tag{12.305}$$

with $\Sigma_k = \Sigma_{i=1}^{i=k-1} n_i$. That is, due to the anticommutation relations, the sign of the state changes every time $b_k^\dagger, b_k$ passes an occupied state as it moves from left to right. This results in the sign of the state changing when two particles are exchanged, but the number operator does not introduce any phase change.

In summarizing the method, Jordan explained:

With second quantization light and matter can be simultaneously considered as waves interacting in three-dimensional space. The major feature of electron theory, the existence of discrete electrically charged particles turns out to be a characteristic quantum phenomenon, namely the fact that matter waves appear only in discrete quantum states. Thus the non-commuting of the wave amplitudes in three-dimensional space is responsible for the existence of particles and the validity of the exclusion principle.

12.3.5.3 Wave mechanical behavior of material particles

As in our discussion of photon optics, we again consider a generic optical system consisting of various optical elements (e.g., slits, diffraction gratings, etc.) and bounded by a reflecting boundary a large distance away. We will expand our wave function in energy eigenfunctions as in Eq. (12.251), with the eigenfunctions chosen to satisfy the boundary conditions on all the boundaries. We ask for the amplitude for a particle, which is located at $\vec{r}_S$ at a time t_S, to be found at a position $\vec{r}_D$ at time t_D. Let $|0\rangle$ represent the vacuum state where all the eigenstates are unoccupied. Then our initial state is given by

$$\Psi^\dagger\left(\vec{r}_S, t_S\right)|0\rangle \tag{12.306}$$

where the operator $\Psi^\dagger\left(\vec{r}_S, t_S\right)$ creates a particle at $\left(\vec{r}_S, t_S\right)$ and the final state is given by

$$\Psi^\dagger\left(\vec{r}_D, t_D\right)|0\rangle \tag{12.307}$$

so the amplitude we are seeking is then

$$M_{fi} = \langle 0|\,\Psi\left(\vec{r}_D, t_D\right)\Psi^\dagger\left(\vec{r}_S, t_S\right)|0\rangle. \tag{12.308}$$

Introducing the eigenfunction expansions

$$\begin{aligned} M_{fi} &= \langle 0|\sum_j a_j u_j\left(\vec{r}_D\right) e^{-\frac{i}{\hbar}\varepsilon_j t_D} \sum_k u_k^*\left(\vec{r}_S\right) e^{\frac{i}{\hbar}\varepsilon_k t_S} a_k^\dagger\,|0\rangle \\ &= \sum_k\sum_j u_j\left(\vec{r}_D\right) u_k^*\left(\vec{r}_S\right) e^{-\frac{i}{\hbar}\varepsilon_j t_D} e^{\frac{i}{\hbar}\varepsilon_k t_S}\,\langle 0|\,a_j a_k^\dagger\,|0\rangle \\ &= \sum_k u_k\left(\vec{r}_D\right) u_k^*\left(\vec{r}_S\right) e^{-\frac{i}{\hbar}\varepsilon_k\left(t_D - t_S\right)}\Theta\left(t - t_S\right) \end{aligned} \tag{12.309}$$

where the function $\Theta(t)$ ($\Theta(t) = 1$ for $t \geq 0$ and $\Theta(t) = 0$ otherwise) guarantees causality (i.e., that a particle cannot be detected before it is emitted).

The "classical" Green's function for a free particle satisfies the Schrödinger equation:

$$i\hbar\frac{\partial G\left(\vec{r}, t\right)}{\partial t} + \frac{\hbar^2}{2m}\nabla^2 G\left(\vec{r}, t\right) = i\hbar\,\delta^{(3)}\left(\vec{r} - \vec{r}_S\right)\delta\left(t - t_S\right). \tag{12.310}$$

We expand $G\left(\vec{r}, t\right)$ as

$$G\left(\vec{r}, t\right) = \sum_k g_k(t)\, u_k\left(\vec{r}\right), \tag{12.311}$$

where the eigenfunctions satisfy $\nabla^2 u_k + k^2 u_k = 0$ as well as the boundary conditions. We find

$$\sum_n \left(i\hbar \frac{\partial g_n(t)}{\partial t} - \frac{\hbar^2 k^2}{2m} g_n(t) \right) u_n(\vec{r}) = i\hbar\, \delta^{(3)}(\vec{r} - \vec{r}_S)\, \delta(t - t_S) \tag{12.312}$$

$$i\hbar \frac{\partial g_k(t)}{\partial t} - \frac{\hbar^2 k^2}{2m} g_k(t) = i\hbar\, u_k^*(\vec{r}_S)\, \delta(t - t_S) \tag{12.313}$$

$$g_k(t) = u_k^*(\vec{r}_S)\, e^{-\frac{i}{\hbar}\varepsilon_k(t-t_S)} \Theta(t - t_S) \tag{12.314}$$

$$G(\vec{r}, t) = \sum_k u_k(\vec{r})\, u_k^*(\vec{r}_S)\, e^{-\frac{i}{\hbar}\varepsilon_k(t-t_S)} \Theta(t - t_S) \tag{12.315}$$

where we have multiplied the first equation by $u_k^*(\vec{r})$ and integrated over all values of $\vec{r}$ and $\varepsilon_k = \hbar^2 k^2/2m$.

Since (12.309) is identical to the Green's function of the Schrödinger equation (12.315), we see that the second quantized description reproduces the "classical" Schrödinger wave optics result. Just as in the case of the electromagnetic field, the amplitude for a particle created at $(\vec{r}_S, t_S)$ to be detected at $(\vec{r}_D, t_D)$ after traversing an optical system is the same as that predicted by the Schrödinger equation, which the second quantization procedure treats as a classical field.

12.3.5.4 Superposition of states in second quantization

Until now we have spoken of a number of particles in different states. The state given by

$$|.\rangle = \left(a_i^\dagger\right)^{N_i} \left(a_j^\dagger\right)^{N_j} |0\rangle, \tag{12.316}$$

where $|0\rangle$ is the vacuum, represents N_i particles in the state $|i\rangle$ and N_j particles in the state $|j\rangle$. It is an eigenstate of the operators $\hat{N}_i, \hat{N}_j$ and $\hat{N}_i + \hat{N}_j$.

On the other hand, the state

$$|.\rangle = \left(\alpha a_i^\dagger + \beta a_j^\dagger\right) |0\rangle \tag{12.317}$$

represents a superposition. It is not an eigenstate of $\hat{N}_i$ or $\hat{N}_j$ but is an eigenstate of $\hat{N}_i + \hat{N}_j$ with eigenvalue 1. The expectation value of $\hat{N}_i \left(\hat{N}_j\right)$ is $|\alpha|^2 \left(|\beta|^2\right)$ so α, β are

the probability amplitudes of the states $|i\rangle, |j\rangle$ being occupied.

12.3.5.5 Summary

We have seen that the commutation and anticommutation rules guarantee all the required symmetry properties of the multi-particle states of bosons and fermions, respectively, without any additional measures being necessary. Thus we have found a method of representing an N-body system in three-dimensional space. In addition, second quantization (applying the (anti) commutation rules) can be considered as being responsible for the existence of particles and, as we have seen, gives new insight into the wave-particle duality.

The Schrödinger equation is continuous just like Maxwell's equations, and it does not, by itself, contain any particle-like discrete element. In the words of Jordan and Wigner, the second quantization process

provides a formulation of the quantum mechanics of material particles which avoids the representation of waves in the abstract $3N$-dimensional configuration space in favor of a representation by means of quantum mechanical waves in the usual three-dimensional space, one which attempts to explain the existence of material particles in a similar fashion to the explanation of the existence of light quanta by the quantization of the electromagnetic waves.

It was Jordan who had the insight to see the importance of this feature of second quantization. Although he worked with several collaborators—Klein, Pauli, and Wigner were coauthors on papers developing these concepts—many other physicists seemed to be rather underwhelmed by the new theory. Dirac did not like the feature that particles could be created and destroyed in the theory, but he became a strong supporter after the discovery of positron creation and annihilation. Note that the theory does conserve particle number; the physically meaningful operators simply remove particles from one state and add them to another. The method of second quantization is the basis of modern many-particle theory and the quantum field theory of elementary particles.

12.4 Ehrenfest's theorem and the classical limit of quantum mechanics

In September 1927, shortly before the Solvay Conference (see following chapter), Ehrenfest submitted a paper[21] where he discussed the question of how we can "interpret Newton's fundamental equations of classical mechanics from the standpoint of quantum mechanics" and called attention to an "elementary relation, which follows exactly, without any approximation from the Schrödinger equation and shows the relation between wave mechanics and classical mechanics in an obvious way." Ehrenfest used the Schrödinger equation in its original form but we will give the argument based on the generalized theory. A state $|\psi\rangle$ satisfies the generalized Schrödinger equation:

$$i\hbar\frac{\partial}{\partial t}|\psi\rangle = \mathbf{H}|\psi\rangle = \left(\frac{\mathbf{p}^2}{2m} + V\right)|\psi\rangle. \tag{12.318}$$

The rate of change of the average momentum is then given by

$$\frac{d}{dt}\langle\psi|\,\mathbf{p}\,|\psi\rangle = \frac{1}{i\hbar}\langle\psi|\,\mathbf{pH} - \mathbf{Hp}\,|\psi\rangle = -\langle\psi|\,\frac{\partial}{\partial x}V\,|\psi\rangle, \tag{12.319}$$

so that the time rate of change of the expectation value of momentum is equal to the expectation value of the Newtonian force. This follows directly from the fundamental equations of matrix mechanics, where the matrix operators were postulated to satisfy Hamilton's equations, and taking the expectation values leads to Ehrenfest's results. This means that a narrow wave packet would move according to Newton's laws just

[21] Ehrenfest, P., *Notes on the approximate validity of classical mechanics in quantum mechanics*, Zeit. Phys. 45, 455, (1927), received Sept. 5, 1927.

like a classical particle. The spreading of the wave packet, which Heisenberg cited in rejecting Schrödinger's wave packet model, also occurs in classical mechanics given an initial uncertainty in the particle's momentum. As Heisenberg said, the "system trajectories of the classical theory will spread just as the wave packet." The spreading of a wave packet of initial width Δq is given by $(\Delta p/m)\,t$ with $\Delta p = \hbar/\Delta q$.

One place where Ehrenfest's theorem does not apply is in a Stern-Gerlach type experiment where a superposition of energy states is split into a series of macroscopically separated beams. In this case the expectation value of position has no physical meaning. However, it can still be applied to each state separately in cases where the atoms in the beam are statistically divided between the two states (mixed state).

12.5 Stability of matter—second quantization

In concluding this chapter, we point out that perhaps the most amazing aspect of quantum mechanics, with absolutely no known classical analog, is second quantization, which requires that the wavefunction of a system of identical particles be symmetric with particle exchange (bosons) or antisymmetric (fermions—the Pauli exclusion principle or PEP). This is sometimes referred to as the Symmetrization Postulate (SP). The SP is one of the most important but least understood aspects of quantum mechanics. Specifically, without the PEP, according to Dyson, "Not only individual atoms, but matter in bulk, would collapse into a condensed high density phase. The assembly of any two macroscopic objects would release energy comparable to that of an atomic bomb."[22] The observed stability of matter is a balance between the PEP and the Coulomb interaction, and has been discussed at length.[23]

Perhaps surprisingly, the need for second quantization and the SP apparently cannot be derived from fundamental concepts without invoking unproven assumptions. Pauli's paper on the subject is assumed to be a proof, but if fact it only shows that the symmetrization postulate produces a consistent framework for quantum mechanics in the relativistic limit.[24] Feynman commented that he was unable to find a simple explanation of the SP and therefore we really do not understand it.[25] A derivation of the PEP in the framework of quantum field theory relies on the individual fermion states having a well-defined parity and on the 4π symmetry of spinor rotation (see Section 19.3.2).[26] However, Feyman does point out that if we *mechanically* exchange two electrons (fermions), they undergo a *relative* 2π rotation, hence the sign of the wavefunction of one of the electrons changes relative to the other, and, consequently, so does the sign of the product state. Of course we have neglected the overall phase

[22] Dyson, F.J., *Ground-State Energy of a Finite System of Charged Particles*, J. Math. Phys. **8**, 1538 (1967), received Dec. 13, 1966.

[23] Lieb, E. H., *The stability of matter*, Rev. Mod. Phys. 48, 553 (1976).

[24] Pauli, W. *The connection between spin and statistics*, Phys. Rev. 58, 716, (1940), received Aug. 19, 1940.

[25] Feynman, R.P., Leighton, R. B., and Sands, M.L. *The Feynman Lectures on Physics:Commemorative Issue* (Addison-Wesley, Redwood City, California, 1989), Vol. 1, p. 12.

[26] Berestetskii, V. B., Lifshitz, E.M., and . Pitaevskii, L.P., *Quantum Electrodynamics* (Pergamon, Oxford, 1982).

change such an operation would impose, however, this picture at least suggests the importance of the spinor rotation symmetry.[27] It is difficult to imagine symmetry having so much power as to stabilize all the matter in the universe, so the exchange operation might be a shadow of a much deeper process that we cannot directly observe, but can be implemented with infinite aplomb.

The fundamental aspect of the SP is that all particles of a given type are identical, in the mathematical sense of perfectly equal in all ways—even more identical than ants in a colony. What this means is that a lump of ^{12}C, for example, taken from as far away in the universe as can be imagined, comprises identical atoms to a similar lump collected up on Earth. There has been speculation that the fundamental constants of physics might have varied with time, meaning the masses and other properties of particles might have been different in the distant past. Such issues are generally dealt with by the assumption that variations would be due to a scalar field. This field has the property that otherwise identical particles with a different time history that are brought together at a point in space-time have properties not determined by their paths, but determined only by the end point where they meet.

[27]Feynman, R.P., in *Elementary Particles and the Laws of Physics. The 1986 Dirac Memorial Lectures* (Cambridge University Press, Cambridge, 1987) pp. 29-30.

13

The "completion of quantum mechanics"—the fifth Solvay Conference on Physics, October 1927

13.1 Introduction

The two years following Heisenberg's first paper on matrix mechanics[1] saw an amazing output of creative activity on the part of the small group of physicists who were actively engaged in developing the new ideas. In this period, we see the development of matrix mechanics, wave mechanics, the proof of the equivalence of the two, perturbation theory, the probabilistic interpretation of the wave function, the generalized form of the theory involving operators and transformations of the basis (Hilbert space), quantization of the electromagnetic field, and second quantization of matter waves.

Ernst Solvay had invented a process for manufacturing soda (sodium carbonate, Na_2CO_3) in the 1860s and in the following fifty years had built an industrial empire. He had a keen interest in science and established a foundation that had been supporting a series of separate physics and chemistry conferences—the Solvay Conferences—since 1911.[2] The fifth conference, held in October 1927 under the scientific leadership of Lorentz, was titled "Electrons and photons" and was devoted to a discussion of the present state of quantum theory.

Attendance was by invitation only. All the main contributors to the rapid development of the field were there, with the notable exception of Pascual Jordan. Both his main collaborators, Born and Heisenberg, were there. The participants can be seen in this widely circulated conference photo (Fig. 13.1).

The conference consisted of a series of reports, and speakers were asked to circulate written versions of their talks before the conference began. Each report was followed by discussion, with a general discussion at the end. The proceedings were published and are available online in French. Many books and articles have been written discussing the conference and analyzing the arguments. The proceedings have also been published

[1]Heisenberg, W., *Uber die quantentheoretische Umdeutung kinematischer und mechanischer Beziehungen* [On the quantum theoretical reinterpretation of kinematic and mechanical relations], *Z. Phys.* **33**, 879 (1925) received July 29, 1925.

[2]Mehra, J., *The Solvay Conferences on Physics, Aspects of the Development of Physics since 1911*, (D. Reidel, 1975).

The Historical and Physical Foundations of Quantum Mechanics. Robert Golub and Steven K. Lamoreaux, Oxford University Press.
 DOI: 10.1093/oso/9780198822189.003.0013

Fig. 13.1 Solvay Conference 1927. (Public domain.)

in English, along with a summary of the historical background and an analysis of the arguments.[3]

The oral reports were

1. W. L. Bragg: The intensity of X-ray reflection.
2. A. H. Compton: Disagreements between experiment and electromagnetic theory.
3. L. de Broglie: The new dynamics of quanta.
4. M. Born and W. Heisenberg: Quantum mechanics.
5. E. Schrödinger: Wave mechanics.

The Born-Heisenberg talk included a discussion of the generalized transformation theory, but only a brief mention of Dirac's quantization of the electromagnetic field and Jordan's idea of second quantization of the Schrödinger wave function.

After the conference, participants were asked to submit written versions of their contributions to the discussion, so the published discussion represents positions that

[3] G. Bacciagaluppi and A. Valentini, *Quantum Theory at the Crossroads: Reconsidering the 1927 Solvay Conference* (Cambridge: Cambridge University Press, 2009).

were carefully thought through in the light of what other speakers had said at the conference.

This is generally considered to be one of the most important physics conferences ever held and has been presented as the place where the ideas of quantum mechanics were crystallized and the objections of some physicists were decisively answered. Notable among these were arguments produced by Einstein, who remained unhappy with the theory, as it developed, for the rest of his life. There was an evidently exciting debate between Einstein and Bohr during the meeting, but very little of this appears in the conference proceedings. While the grounds of the disagreement are clearly seen in the proceedings, details of the "off-stage" discussions were only published by Bohr more than 20 years later in a joint work in which Einstein published a rather restrained reply.[4] As we will see, recent work has questioned Bohr's presentation of the discussion and illuminated Einstein's arguments.

Writing in 1929, Heisenberg described the conference as follows:

> This congress has, due to the possibility of discussion between representatives of the different research directions, contributed extraordinarily to the clarification of the physical foundations of the quantum theory. It forms, so to say, officially, the completion of the quantum theory, which can now be applied, without hesitation, as a complete, consistent theory to all problems of atomic physics and which forms a basis for further investigations.[5]

Another opinion was voiced by Langevin: the conference was "where the confusion of ideas reached its peak."[6]

There were four main lines of research in quantum mechanics, all of which were represented in the formal reports:

1. De Broglie's pilot wave theory.
2. Heisenberg, Born, and Jordan's matrix mechanics.
3. Schrödinger's wave mechanics.
4. Dirac and Jordan's demonstration that 2. and 3. were special cases of a more general transformation theory.

The main issues discussed at the conference were:

1. The collapse of the wave function and its meaning—the measurement problem.
2. Wave-particle duality.
3. Bohr's complementarity interpretation.
4. de Broglie's proposal of a deterministic "pilot wave" model.
5. Einstein's objections to quantum mechanics, centered on the non-locality of the theory.
6. Need for the use of "configuration" space: for an N-particle system, this is the $3N$-dimensional space of the coordinates (x_i, y_i, z_i) of all the particles, where i runs from 1 to N.

[4] P. A. Schilpp, ed., *Albert Einstein, Philosopher-Scientist*, vol.1 (New York: Harper, 1949).

[5] Heisenberg, W., *Die Entwicklung der Quantentheorie*, 1918-1928, (The Development of Quantum Theory, 1918-1928), Die Naturwissenshaftenn 17, 490, (1929)

[6] Cited in Bacciagaluppi and Valentini, op. cit., p. 294.

Many of these questions, which were recognized so soon after the introduction of the theory, are still being discussed today, 90 years later.

13.2 The collapse of the wave function and its meaning—the measurement problem

13.2.1 Born and Heisenberg's discussion of superposition

According to Bohr's original ideas, a quantum system could only be in one state (energy eigenstate) at a time.

However, the quantum theory in all its forms is linear, so a sum of solutions is also a solution. We have seen that, in the presence of a time-dependent perturbation (e.g., an applied external field or interaction with the quantized electromagnetic field), the wave function evolves into a superposition

$$\psi(t) = \sum_n c_n(t)\,\psi_n^o(q) \tag{13.1}$$

where the ψ_n^o are the eigenstates of the unperturbed Hamiltonian ($H_o\psi_n^o(q) = E_n^o\psi_n^o(q)$). In the case where the system is initially in the state ψ_k^o ($c_k(0) = 1,\ c_n(0) = 0$ for $n \neq k$), we can write $c_n(t) = S_{nk}(t)\,c_k(0)$ where $S_{nk}(t)$ are determined by the perturbation $V(t)$. Thus if we wish to maintain the idea that a system can only be in a single state at any one time, we are forced to interpret (13.1) in a statistical fashion. The average energy in a state $\psi = \sum_n a_n\psi_n^o(q)$ is given by

$$\begin{aligned}\langle E\rangle = \langle\psi|\,H_o\,|\psi\rangle &= \int dq\,\psi^* H_o\psi = \int dq\left[\left(\sum_m a_m^*\psi_m^{o*}(q)\right)H_o\left(\sum_n a_n\psi_n^o(q)\right)\right]\\ &= \int dq\left[\left(\sum_m a_m^*\psi_m^{o*}(q)\right)\left(\sum_n E_n^o a_n\psi_n^o(q)\right)\right] = \sum_n |a_n|^2 E_n^o,\end{aligned} \tag{13.2}$$

where we made use of the orthogonality of the eigenfunctions. This leads us to interpret $|c_n(t)|^2$ as being the probability of finding the system in the state ψ_n^o at time t and $|S_{nk}(t)|^2$ as the probability of having made a transition from the original state k to the final state n between $t = 0$ and t. If we had started with a system in the superposition (13.1) at $t = 0$, we would have a new state designated by the coefficients $d_n(t)$:

$$d_n(t) = \sum_m S_{nm}(t)\,c_m(0), \tag{13.3}$$

so that

$$|d_n(t)|^2 = \left|\sum_m S_{nm}(t)\,c_m(0)\right|^2, \tag{13.4}$$

which is not the same as

$$|d_n(t)|^2 = \sum_m |S_{nm}(t)|^2\,|c_m(0)|^2 \tag{13.5}$$

because of the existence of cross terms between different m terms in (13.4).

Knowledge of the $c_m(0)$ requires knowledge of a phase in addition to the occupation probability of the state m, $c_m(0) = |c_m(0)| e^{i\gamma_m}$. Born and Heisenberg point out that a measurement of $c_m(0)$ would result in a randomization of the phases, so that (13.4) would go over into (13.5) if the distribution $|c_m(0)|^2$ was measured. The fact that an energy measurement randomizes the phases of the quantum amplitudes was shown by Bohr in the special case of a Stern-Gerlach measurement of energy[7] (Section 13.3.1). In his 1951 textbook,[8] Bohm showed that it holds in general.

According to the modern interpretation, a measurement of energy E_n^o would result in the system being left in the state $\psi_n^o(q)$, but Born and Heisenberg imply the system is left in the state (13.1) with randomized phases for the $c_n(0)$. This makes sense if we consider an ensemble of N identical systems. Then $N|c_m(0)|^2$ systems will be found in the state m, and averaging over all the possible outcomes, m, in the complete ensemble will yield (13.5). In the case where no measurement is made, (13.4) would remain valid and there would be interference.

13.2.2 Wave function collapse as seen by Dirac and Heisenberg

In the general discussion at the end of the conference, Dirac again brought up the case of a wave function of the form

$$\psi(t) = \sum_n c_n \psi_n. \tag{13.6}$$

Dirac did not mention, in this context, that the eigenfunctions of any operator could be used for this expansion; the selection of the appropriate set is determined by the measuring apparatus used. This point was emphasized by Bohr with his idea of complementarity (see Section 13.3.1).

The wave function normally varies according to a causal law (e.g., the Schrödinger equation) so that its initial value determines its value at all later times. Now according to Dirac, if at time t_o the functions are of such a nature that they cannot interfere, then at later times the system will be described by only one of the ψ_n. It is as if nature chooses the particular ψ_n since the only information that the theory provides is that the probability of ψ_n is given by $|c_n|^2$. The choice, once it is made, is irrevocable and will influence the future state of the world. The results of the choice made by nature can be determined by experiment and the results of any experiment are numbers describing such choices of nature. Taking as an example a scattering experiment where the incident wave packet is scattered in all directions (Fig. 13.2 (a)), we should take for the wave function after the scattering a wave packet traveling in a single direction, since from the results of an experiment we can work backward through a chain of causally connected events and determine in which direction the electron was scattered and conclude that nature has chosen that direction. On the other hand, if we introduce a mirror (Fig. 13.2 (b)) so that electrons scattered in direction d_1 are reflected and can interfere with electrons scattered in direction d_2, we would not be able to distinguish

[7] Bohr, N., *The Quantum Postulate and the Recent Development of Atomic Theory*, Nature, 121, 580, April 14, 1928,. (Bohr's lecture at the Como conference, Sept. 17, 1927).

[8] Bohm, D., *Quantum Theory*, (Prentice-Hall, 1951).

between the two scattering directions and would not be able to follow back the causal chain as far as in the previous case, so that we cannot say that nature has chosen a direction immediately after the scattering event. It is only later that nature will choose the region where the electron will appear. The interference forces nature to delay its choice. Thus Dirac raised for the first time the problem of wave function collapse. This would be addressed in detail by von Neumann, (see Chapter 14) whose then unpublished work was referred to by Born immediately after Dirac's contribution.

Heisenberg responded that he disagreed completely with Dirac's formulation. No matter how far away from the scatterer you choose to place two mirrors (Fig. 13.2 (c)), you can always obtain an interference. Evidently the choice of nature could never be known before the decisive experiment has been done. Then the expression "nature makes a choice" has no physical consequences. It is better to say that the observer himself makes the choice, because it is only at the moment of observation that the "choice" becomes a physical reality and the phase relation between the waves—the ability to produce an interference pattern—is destroyed. Again he is equating a measurement with the randomization of the phases (see below). Surely Heisenberg must have meant that the observer brings about the choice.

The statement that at the moment of observation the "choice" becomes a physical reality implies that some irreversibility is necessary, as indicated by Dirac with his emphasis on the irrevocability of the choice. What determines the "moment of observation" remained a matter of debate. Does it need a human consciousness to observe the result, or is it enough if there is a mark on a recording tape or a number written

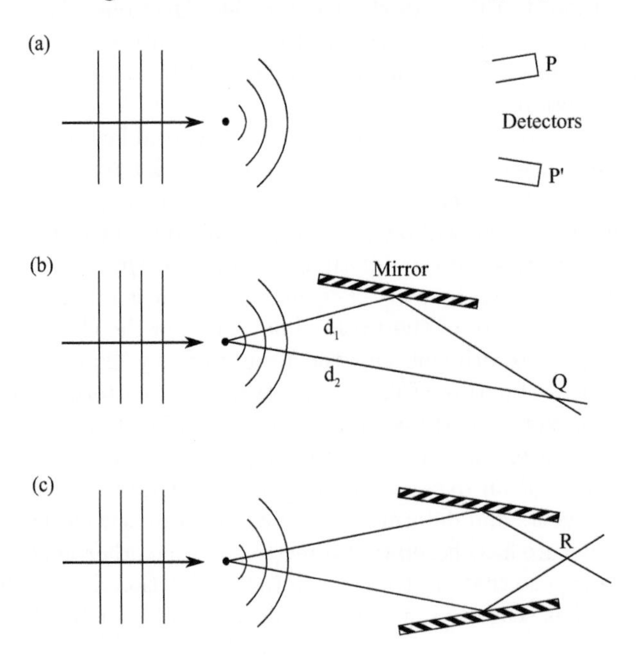

Fig. 13.2 Electron interference and complementarity as discussed by Dirac and Heisenberg (op. cit., p. 294).

into a computer memory in order for the possibility of interference to be destroyed? The idea that the "collapse" occurs when a macroscopic object is involved is fraught with difficulties. Because of conservation of energy and momentum, the uncertainties will be transferred when a microscopic system interacts with a macroscopic one, which of course must happen in every measurement. We will see that Bohr, in his discussions with Einstein, found it necessary to apply the uncertainty principle (i.e., quantum mechanics) to macroscopic instruments in order to preserve the consistency of quantum mechanics.

Equally problematic is the idea that a conscious observer is needed to observe the result.

"Was the world wave function waiting to jump for thousands of millions of years until a single-celled living creature appeared? Or did it have to wait a little longer for some more qualified measurer with a Ph.D.[9]"

The exact place in the causal chain involved in a microscopic measurement where the collapse is considered to occur (i.e., the phases are randomized), called the von Neumann cut, does not have any physical consequences. As Bell wrote,

...you are not told where the division comes—on which side of my spectacles for example—or at which end of my optic nerve.

13.3 Wave-particle duality

As summarized by Compton in his report, the difficult challenges faced by classical electromagnetic theory were:

1. The need for an ether to sustain the oscillations of the electromagnetic field. Its existence, however, contradicts experiment and the special theory of relativity.
2. Classically, electromagnetic waves are produced by sources oscillating at the same frequency as the radiated waves. Atoms, however, radiate frequencies quite different from the possible orbital frequencies of the electrons.
3. The photoelectric effect: the necessity for energy concentration in the case of very weak illumination and experiments with pulsed radiation.
4. Scattering of radiation by single electrons: the Compton effect.
5. The fact that energy and momentum are conserved in individual scattering processes.

Points 2–5 indicate the existence of photons of energy $E = \hbar\omega$ as the carriers of the energy of the electromagnetic field. Nonetheless, the existence of interference and diffraction effects means that the wave theory cannot be discarded. With the confirmation of de Broglie's proposal of a wave being associated with massive particles by the electron diffraction experiments of Davisson and Germer (1925), it became clear

[9]Bell, J.S., '*Quantum Mechanics for Cosmologists*', in Isham, C., Penrose, R. and Sciama, D., eds., *Quantum Gravity 2*, (Oxford, 1981), reprinted in Bell, J. S., *Speakable and Unspeakable in Quantum Mechanics*, (Cambridge, 1987).

that the phenomenon of wave-particle duality applies not only to photons but also to electrons and other massive particles.

The problem of how to reconcile these facts was (and is) very difficult, and many of the founders of quantum mechanics had grappled with it for years. Einstein had speculated at one time that each photon was surrounded by a wave field falling off with distance from the photon; in the case of multiple photons, these waves would combine in some unspecified way.

Bohr had attempted to eliminate photons entirely. In January 1924 he, Kramers, and Slater proposed what is known as the BKS theory, where each atom is accompanied by a set of virtual oscillators corresponding to the possible transition frequencies connecting to the current state of the atom. These oscillators were considered to emit virtual radiation, which would determine the transition probability of the emitting atom as well as distant atoms that could absorb radiation. As both emission and absorption were probabilistic, it was possible for a distant atom to make a transition to a higher state (absorption of radiation) before the initiating atom had made a transition to a lower state (emission), so that energy and momentum would only be conserved statistically but not in individual events. This theory had a wide influence until it was disproved in April 1925 by an experiment by Bothe and Geiger which showed, by coincidence measurements on the Compton effect, that energy and momentum were indeed conserved in each individual event, confirming the reality of photons. Slater later said that he had originally wanted to include photons in the theory but Bohr persuaded him otherwise.

In a comment on Compton's report at the conference, Bohr stated that the frequency change observed in the Compton effect is measured by instruments described by wave theory and the waves must be extended over finite regions in space and time in order to have any meaning. On the other hand, the change in energy of the electron is considered as occurring at a sharply defined point in space-time. To resolve the conflict between these two incompatible representations of the same event, BKS theory had rejected the idea of photons, which would mean that the conservation of energy and momentum was only valid in a statistical sense. The same considerations applied in this theory to electrons and their accompanying de Broglie waves. After experiment forced the abandonment of the BKS theory, Bohr developed his ideas of complementarity as a resolution of the wave-particle duality puzzle. We will return to this in the next section.

Schrödinger also felt that particles were not a necessary part of the theory. His initial idea was that particles could be represented by wave packets, but this had the problem that in most circumstances the wave packets would spread with time. Born pointed out that the spreading also occurs in classical mechanics if the initial momentum is not precisely known, so the rejection of the wave packet model for this reason is unjustified. In the case of an electron in an atom, the wave would be spread throughout the atom and the particle concept would have no meaning.

Born's proposal for the reconciliation of the wave and particle properties was to interpret the square of the magnitude of the wave function at any given point in space and time, $|\psi(\vec{x},t)|^2\, d^3x$, as the probability of finding the particle at time t in a region d^3x centered on $\vec{x}$.

13.3.1 Bohr and complementarity

Bohr did not turn in a written version of his contributions to the discussions at the conference. Instead he asked that a paper he had published after the conference be included in the proceedings. This paper was based on a lecture he had given at another physics conference held at Lake Como in Italy a month before the Solvay Conference. He had worked on this paper for months, and it presented a detailed statement of his views at the time. Evidently Bohr had some difficulties formulating his ideas for publication. As late as March 1928, Pauli, who had been critically proofreading the various drafts of the paper, refused—perhaps tongue in cheek—to visit Bohr in Copenhagen until the latter had certified that the final version of the paper had been sent to the publisher, so that they could discuss other topics during his visit. The paper published in the Solvay proceedings was the French translation of the Como paper originally published in German. In addition, an English translation was published in *Nature*.[10]

By the quantum postulate Bohr meant that every atomic process had a discontinuity or individuality completely foreign to the classical theory and characterized by Planck's quantum of action. This means that every observation involves a non-negligible reciprocal action between the observed object and the measuring instrument, so that we cannot attribute an independent physical reality to either the object or the instrument. The choice of which parts of the complete system we consider as the object and which as the measuring instrument is somewhat arbitrary. Every measurement is eventually carried out by our sense organs but, since we always need a theoretical representation to interpret the observations, it is a matter of convenience at what point we introduce the notion of observation and its irrational (random) action specified by the quantum postulate. This is the choice of where the wave function collapses (cut).

The usual definition of a state of a system assumes the system is completely isolated from all external influences, but in that case there is no possibility of making observations since the quantum postulate says that any observation involves a non-negligible effect on the observed system. A space-time (wave-like) description of a definite state is then impossible. If, in order to make an observation, we introduce an interaction with an external instrument, we can no longer define the state of the system and there can no longer be causality due to the unpredictable element introduced by the interaction (collapse of the wavefunction). Representation in space-time (wave view) and causality (particle view, which requires energy-momentum conservation in individual events) which are considered as simultaneously valid in classical theory are, according to the quantum theory, complementary but mutually exclusive descriptions of experience.

Consider the relations

$$\frac{E}{\omega} = \frac{p}{k} = \hbar \,. \tag{13.7}$$

In this expression, energy E and momentum p refer to a particle with implied definite space-time coordinates, and frequency ω and wave number k refer to plane waves with

[10] Bohr, N., *The quantum postulate and the recent development of atomic theory,* Nature (supplement) **121**, 580 (1928).

an infinite extent in space and time. To obtain a wave train of a finite extent in space and time, we need to superpose waves of different wave number and frequency. The uncertainty relations result from the requirement that we need a certain minimum spread in (ω, k) to achieve a given limitation in space-time, as discussed in the proof of the uncertainty principle in Chapter 10.

In tracing observation back to our sensations, we have to take into account the fact that the quantum postulate refers to the instrument as well as to the observed system and that the cut between them is arbitrary. We might hope to determine the momentum transferred to a particle during a position measurement by measuring the recoil momentum of the microscope. However, that would result in the loss of knowledge of the position of the microscope to sufficient accuracy. Note that here Bohr is applying quantum mechanics to macroscopic objects, and this argument reflects the discussions with Einstein during the conference as we shall see.

Bohr discusses critically Schrödinger's suggestion that the quantum jumps—the "irrational element" required by the quantum postulate—could be eliminated by considering the time development of the wave function. The case considered by Schödinger is a closed system, not accessible to observation. In order to detect a transition, we have to detect a photon, which is a random discontinuous event. This argument was already advanced during Schrödinger's visit to Copenhagen in 1926 (Chapter 10).

After any observation allowing us to determine the stationary state occupied by a system, we can disregard the previous history because the phase of the state is randomized. Bohr then refers to an argument by Heisenberg concerning the Stern-Gerlach experiment. In order to observe the splitting between the two (in the case of spin-1/2) energy states, the angular deflection $\Delta\theta_B$ due to the inhomogeneous field must be greater than the spread $\Delta\theta_d$ of the beam due to diffraction caused by passing through the slit of width a: $\Delta\theta_d = 2\pi/ka$. Assuming the force $F = \mu G$, where μ is the magnetic moment of the particle and G is the field gradient, acts for a time t, we find

$$\Delta\theta_B = \frac{\Delta p}{p} = \frac{Ft}{\hbar k} = \frac{\mu G t}{\hbar k} \gg \frac{2\pi}{ka} \tag{13.8}$$

$$\mu G t a \gg h. \tag{13.9}$$

Force is the gradient of the potential, so $F = \Delta E/a$ with ΔE the difference in potential energy across the beam and a the beam width. Then

$$\Delta E \cdot t \gg h \tag{13.10}$$

$$\Delta\nu \cdot t \gg 1 \tag{13.11}$$

This means that the phase will be washed out if the two energy states are separated. In summary, the concept of complementarity means that because of the unavoidable, uncontrollable action of a measuring instrument on the system being measured, the results depend on the properties of the measuring instrument as much as on those of the measured system. An apparatus to measure the position of a particle is very different from an apparatus to measure its momentum. Use of the one disturbs the system so much that the other quantity is no longer measurable.

13.3.2 De Broglie's proposal of a pilot wave

Contrary to these approaches to the problem of wave-particle duality, de Broglie sought for models in which both the wave and the particle had a simultaneous existence. His first idea, which he called the theory of the "double solution," was that the wave equation had two solutions: the ordinary solution found by Schrödinger and a singular solution representing the particle. Finding this singular solution proved to be mathematically difficult, so at the conference de Broglie presented a simpler model, in which the particle and wave had a separate existence but the wave guided the motion of the particle according to given mathematical rules. De Broglie's presentation at the conference was a summary of a recently published paper.[11]

13.3.2.1 The pilot wave

The pilot wave theory assumes that the material particle and the continuous wave, represented by the solution ψ of the Schrödinger equation, are distinct realities. Writing $\psi = Ae^{i\phi}$ de Broglie postulates that the velocity of the particle is given by

$$\vec{v} = \frac{\hbar}{m}\vec{\nabla}\phi. \tag{13.12}$$

In the published paper, de Broglie presented this idea as provisional, awaiting the further development of the double solution. In his report at the Solvay Conference, he said that a proper physical theory should consist of many waves in three-dimensional space so that the ψ wave in $3N$-dimensional space could only be considered as a "fictional" mathematical tool. Nevertheless, in the conference report de Broglie applied the theory to an N-body system so that ψ, A and ϕ are functions of $q = (x_i, y_i, z_i)$ with i running from 1 to N. Then the velocity (13.12) is the velocity of the point representing the system in the $3N$-dimensional space and the velocity components are

$$v_i = \frac{\hbar}{m_i}\frac{\partial}{\partial x_i}\phi. \tag{13.13}$$

Thus, given the position of the system point at a particular instant of time, its position for all later times is determined by (13.13). If the initial position is not known—and according to the uncertainty principle, it cannot be known exactly—the initial distribution of system points will be taken to be given by Born's expression for the probability of finding the particle in a volume element $d\tau$ in $3N$-dimensional space:

$$P(q)\, d\tau = |\psi|^2\, d\tau. \tag{13.14}$$

Calculating the particle trajectories for two-slit diffraction gives the result shown in Fig. 13.3. While the particle follows only one trajectory and goes through only one of the slits, that trajectory is influenced by the wave function, which in turn is influenced by the presence or absence of the second slit. This shows that each ψ wave represents a class of motions, with the path chosen by each individual particle depending on its initial position. The guiding field produces a bunching of trajectories

[11] De Broglie, L., *La mechanique ondulatoire et la structure atomique de la matiere et du rayonnement*, (Wave mechanics and the atomic structure of matter and radiation), J. de Physique, 8, 225 (1927) received April 1, 1927.

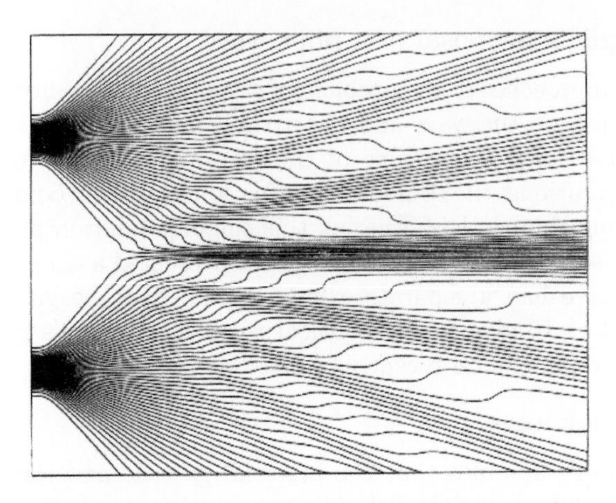

Fig. 13.3 Particle trajectories emanating from two slits at the left. The bunching of trajectories seen on the right is due to the quantum potential $Q = -\frac{\hbar^2}{2m}\frac{1}{A}\left(\nabla^2 A\right)$: the particles are accelerated out of the valleys of the potential and bunch up in regions where the potential plateaus. (Fig. 3 from Philipidis, C., et al, 1979 op. cit., p. 304). (Reproduced with the permission of Prof. B. Hiley.)

which reproduces the fringe pattern of the usual wave theory. This effect is due to the "quantum potential" (see below). This calculation was only done much later (1979).[12]

We can gain some insight into this model by writing the Schrödinger equation (for simplicity we consider a single particle) as

$$i\hbar\frac{\partial}{\partial t}\left(Ae^{iS/\hbar}\right) = -\frac{\hbar^2}{2m}\nabla^2\left(Ae^{iS/\hbar}\right) + V(x)\,Ae^{iS/\hbar} \tag{13.15}$$

Separating the real and imaginary parts, we obtain the two equations

$$0 = \frac{\partial}{\partial t}S + \frac{1}{2m}\left(\vec{\nabla}S\right)^2 + V(x) - \frac{\hbar^2}{2m}\frac{1}{A}\left(\nabla^2 A\right) \tag{13.16}$$

$$\frac{\partial}{\partial t}A = -\frac{1}{m}\left[\left(\vec{\nabla}A\right)\cdot\left(\vec{\nabla}S\right) + \frac{1}{2}A\left(\nabla^2 S\right)\right]. \tag{13.17}$$

From the second equation, we find the equation for the probability density, with $\rho = A^2$:

$$\frac{\partial}{\partial t}\rho = \frac{\partial}{\partial t}A^2 = 2A\left(\frac{\partial}{\partial t}A\right) = -\frac{1}{m}\vec{\nabla}\cdot\left(\rho\vec{\nabla}S\right). \tag{13.18}$$

Eq. (13.16) is the Hamilton-Jacobi equation of classical physics $\left(\vec{p} = \vec{\nabla}S\right)$ with the added term $Q = -\frac{\hbar^2}{2m}\frac{1}{A}\left(\nabla^2 A\right)$, which was called the quantum potential by Bohm (Fig. 13.4).[13]

[12] Philippidis, C., Dewdney, C. and Hiley, B. J., *Quantum interference and the quantum potential*, *Nuovo Cimento B* 52, 15 (1979) received Dec. 27, 1978.

[13] Bohm, D., *A suggested interpretation of the quantum theory in terms of hidden variables, I and II*, Phys. Rev. 85, 166 (1952) received July 5, 1951 and Phys. Rev. 85, 180, (1952) received Jul 5, 1951.

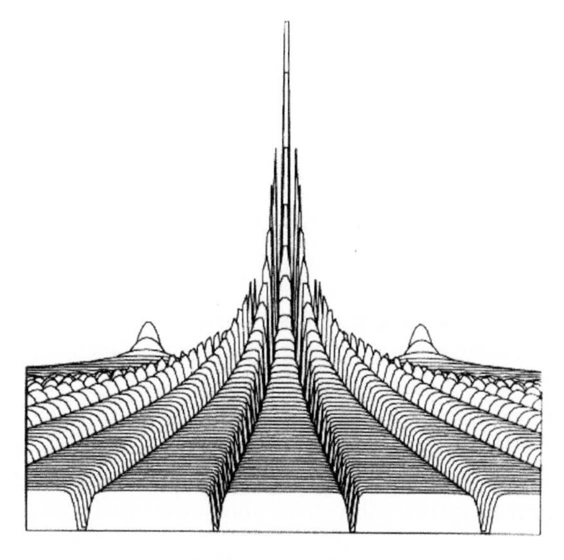

Fig. 13.4 The quantum potential for two Gaussian slits viewed from the detector plane in a two-slit experiment. (Philipidis, C., et al, 1979, op. cit., p. 304). (Reproduced with the permission of Prof. B. Hiley.)

De Broglie thought that this would account for the curved motion of the rays in the case of diffraction at a slit or edge. The quantum potential would act on particles passing close to an edge so that their trajectories would curve and form the diffraction pattern. Calculation of the quantum potential for the case of two-slit interference demonstrates that it does indeed have this effect.[14] Eqs. (13.16) and (13.17) were first found by Madelung[15] (October 1926).

The wave function collapse problem does not occur here as the system point is always in one or the other of the component waves depending on the initial conditions of the instance in question.

Without being able to adjust the initial positions to be different than that required by a given wave function, the results of this theory agree exactly with ordinary wave mechanics, but it constitutes a deterministic theory: probability is only introduced with the initial conditions.

Eq. (13.18) is the equation of conservation of probability with a probability current density

$$\rho\vec{v} = \rho\vec{\nabla}S/m. \tag{13.19}$$

This is identical with our previous result for the current density:

$$\vec{j} = \frac{\hbar}{2im}\left(\psi^*\vec{\nabla}\psi - \psi\vec{\nabla}\psi^*\right) = \rho\vec{\nabla}S/m\,. \tag{13.20}$$

[14] Philipidis, C., et al., op. cit., p. 304.

[15] Madelung, E., *Quantum theory in a hydrodynamic form*, Zeit. Phys. 40, 322, (1926), received Oct. 25, 1926.

Thus de Broglie's hypothesis (13.13) appears as a natural result of the Schrödinger equation. In fact the whole argument is a reversal of the argument that led from the analogy between classical mechanics and geometrical optics to the Schrödinger equation. In this way of looking at things, the particles maintain their character throughout.

While this model represents a tremendous achievement, allowing us to bypass all arguments about indeterminism and offering a direct, intuitive solution to the problem of wave-particle duality, it seems to be something of a *non sequitur*: once we have a solution to the Schrödinger equation for a given problem, we have all the information we need for predicting experimental results. The use of (13.13), while illuminating, provides no additional experimentally verifiable information. Another flaw is that the particle motion has no reciprocal action on the ψ wave. Such a one-way interaction does not appear anywhere else in physics.

After the conference, interest in this model rapidly disappeared due to a counter-argument presented by Pauli. De Broglie himself stopped working on the problem and promoted the mainstream interpretation until 1951, when David Bohm[16] challenged Pauli's argument and reawakened interest in pilot wave models.

13.3.2.2 Pauli's counterexample to the pilot wave

Pauli's argument is based on the fact that if the wave function is a superposition

$$\psi(q) = \sum_n c_n \phi_n(q), \tag{13.21}$$

then, in regions where more than one of the $\phi_n(q)$ are nonzero (overlapping regions), de Broglie's prescription (13.13) would predict rapidly varying velocities due to the interference between the overlapping terms, and this disagrees with experiments in the case of scattering experiments where the initial and final velocities of the interacting particles are seen to be constant. Pauli called attention to the case of a particle moving in two dimensions (x, y) scattering from a rotor located at the origin $(x = y = 0)$. The stationary states of the rotor are represented by wave functions $\Phi_m(\phi) = e^{im\phi}$ and the initial state of the system would be given by

$$\psi_o(x, y, \phi, t) = e^{i(k_x x + k_y y)} e^{im\phi} e^{-i\omega t} \tag{13.22}$$

where ω is the energy of the state divided by $\hbar$. The initial state is thus nonzero for all values of $-\infty < \phi < \infty$. Since the interaction between the particle and the rotor must be periodic in ϕ, and vanishes at some distance from the origin, Pauli pointed out that the system was analogous to diffraction at a grating. The inelastically scattered particles would be equivalent to the diffracted waves, with the different diffraction orders corresponding to the different amounts of energy transferred to the rotor. As the initial state was of infinite extent, the equivalent grating would also be infinitely long. Thus, the different orders would remain overlapping for all times and the problem of the non-classical behavior of the velocity according to (13.13) would remain. In particular, the velocity of the rotor would not be constant, and the model is not

[16]Bohm, D., op. cit., p. 297.

compatible with the requirement that the rotor be in a stationary state before and after the collision.

Pauli felt that this was an inherent problem with the pilot wave model. We now know that this is incorrect.

De Broglie pointed out that in the diffraction problem we can only speak of the beam diffracted in a given direction if the incident wave is limited laterally (collimated), because otherwise all the beams would overlap with themselves and the incident beam. In the case of limited (collimated) beams, the diffracted waves would separate in space and the system point, finding itself in one of the waves, would correspond to a constant velocity. We can demonstrate this argument by writing the wave function as a sum of incident and scattered waves,

$$\psi(x, y, \phi, t) = a e^{i m_o \phi} e^{-i\omega_o t} e^{i\left(k_x^{(0)} x + k_y^{(0)} y - \omega_k^{(0)} t\right)} \tag{13.23}$$

$$+ b \sum_m e^{i m \phi} e^{-i\omega_m t} f_m \left(k_x^{(m)} x + k_y^{(m)} y - \omega_k^{(m)} t\right) \tag{13.24}$$

where the initial state of the rotor (particle) is given by $m_o, \omega_o \left(k_{x,y}^{(0)}, \omega_k^{(0)}\right)$ and $k_{x,y}^{(m)}, \omega_k^{(m)}$ represent the energy and momentum of the particle after exchanging energy with the rotor, which made a transition from the state $e^{i m_o \phi}$ to the state $e^{i m \phi}$. Eventually the packets represented by f_m will become separated in space (if the incident beam is collimated) and the system point, finding itself in one of the packets (depending on its initial location), will be moving with a constant velocity. We have discussed this issue in Section 10.3.1.

Kramers asked, in the case of a single photon reflecting from a mirror, how can we account for the sudden change of momentum of the mirror? De Broglie replied, "No theory gives the answer to Mr. Kramers's question." We now know, and this is an essential point missed by all participants in the early discussions, that in order to treat this problem we have to include the coordinates of the mirror. We take x_m to represent the coordinates of the mirror, which is initially in a state $f(x_m)$. If the photon is initially in the state $e^{i k_x x_p}$ (momentum $\hbar k_x$), the initial state of the system will be

$$\psi_{\text{inc}}(x_o, x_m) = e^{i k_x x_p} f(x_m). \tag{13.25}$$

As a result of the interaction, the system will make a transition to a state

$$\psi_{\text{refl}}(x_o, x_m) = e^{i(-k_x x_p)} f(x_m) e^{i 2 k_x x_m} \tag{13.26}$$

corresponding to the reflected photon moving in the $(-x)$ direction and the mirror recoiling with momentum $\Delta p = 2\hbar k_x$. The same argument applies in the ordinary Schrödinger theory as well.

Although de Broglie felt he had adequately answered Pauli's objections, he abandoned the theory, only taking it up again after Bohm's revival of the idea in 1951–52. The reasons for this were presented in a book he published in 1956[17] in which he

[17]De Broglie, L., *Une tentative d'intepr$\hat{e}$tation, causale er non lineaire de la m$\hat{e}$canique onfulatoire (La theorie de la double solution)*, (Gauthier-Villars, 1956). English translation: de Broglie, L.; Knodel, A..; Miller, J.C., *Non-linear wave mechanics: a causal interpretation* (Elsevier, 1960).

discussed his view of the pilot wave theory and its development, his proposal of the double solution,[18] and the relation between his and Bohm's understanding of the pilot wave theory. The reasons he gave for his rejection of the pilot wave theory, and to his disagreement with Bohm, who saw the wave function as physically real were:

- Since the ψ wave is representative of a probability and conditioned by the knowledge of the user, it was difficult to conceive of it as a physical reality, capable of guiding a particle.
- For many-particle systems, the fact that the ψ wave propagates in the purely abstract $3N$-dimensional space speaks against it being a physical wave.
- The fact that the wave function was thought to collapse as the result of a measurement speaks against it having an independent physical existence.

Another objection, raised by Francis Perrin, was that in the case of scattering, the magnitude of the scattered wave decreases with distance as $(1/r)$, so that its amplitude will become infinitesimally small far away from the scattering center. Although the quantum potential Q does not depend on the magnitude of the wave, it is difficult to see how, if the ψ wave were really a physical agent acting on the particle, the action could be the same no matter how small the wave amplitude.

With the advent of Bohm's work, these objections seemed to have less force and the model was seen by many physicists as a viable, causal alternative to the standard interpretation. At the least, it is a concrete example of a hidden variable theory consistent with quantum mechanics, something that people thought (falsely) that von Neumann[19] (1932) proved was impossible. We discuss von Neumann's "proof" in Chapter 14. There is a significant literature pertaining to the pilot wave and related ideas. See the book by Peter Holland for a detailed discussion.[20]

13.4 Einstein and Bohr: the battle of the century?

13.4.1 Einstein's contribution to the published discussions

In the general discussion at the end of the conference Einstein considered an electron beam diffracting through a slit and then continuing on until it hits a photographic film curved to form a hemisphere centered on the slit O (Fig. 13.5).

Due to diffraction, we have a spherical de Broglie wave beyond the slit which arrives at the screen P. According to Einstein, there are two possible interpretations of the theory.

1. The de Broglie-Schrödinger waves apply only to an ensemble of particles, a cloud of electrons extended in space. It gives no information on the individual processes.

[18]The idea of the double solution was that in addition to the ordinary solution to the Schröedinger equation there was another, singular solution representing the actual position of the particle. De Broglie was never able to work out the solution in a satisfactory manner.

[19] Neumann, J. von, *Mathematische Grundlagen der Quantenmechanik*, (Springer, 1932), English translation, *Mathematical Foundations of Quantum Mechanics*, (Princeton, 1955 and new edition 2018).

[20]Holland, P.R., *The Quantum Theory of Motion: An Account of the de Broglie-Bohm Causal Interpretation of Quantum Mechanics* (Cambridge University Press, 1995).

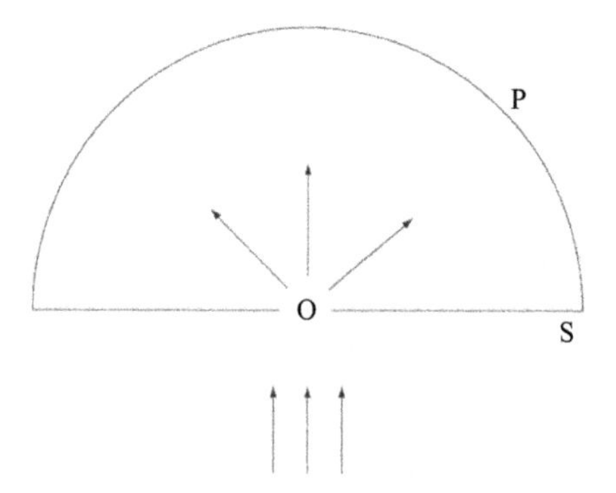

Fig. 13.5 Geometry of the situation considered by Einstein with an electron beam passing through a slit O and hitting a curved photographic plate P.

2. The theory is a complete description of individual processes. Each particle is described by a wave packet of small spatial extension and small angular divergence. This wave packet is diffracted and then arrives at the screen.

According to interpretation 1, $|\psi|^2$ expresses the probability that a particle is found in the given region, say at a point on the screen, while according to 2, it represents the probability that at a given instant the same particle is found at different regions of the screen. Only the interpretation 2 allows for the conservation laws to apply to individual processes. However, there are some objections to 2. If $|\psi|^2$ is envisaged as the probability that a given particle is found at a certain space-time point, then it could happen that the same elementary process could produce an action in two or more places on the screen. Einstein went on to say:

> But the interpretation according to which $|\psi|^2$ expresses the probability that *this* particle is located in a given region, is based on a very special mechanism of action at a distance, which impedes the wave, continually spread in space, from producing an action in *two* regions of the screen.

Thus, if the wave function is valid for the description of a single particle, the wave function must collapse instantaneously after detection of the particle at a point, to reflect the fact that the probability of finding the particle elsewhere has gone to zero. Or as Bohr expressed it 20 years later:

> If the particle is detected at one point on the screen, then it is out of the question of ever observing this electron at another point on the screen, although the ordinary laws of wave propagation offer no room for a correlation between two such events.

According to Einstein this requires that, in addition to the wave, we localize the particle during the propagation. This is the approach taken by de Broglie in his pilot wave model, which Einstein thought was on the right track: "I think that M. de Broglie is right to search in this direction." Otherwise 2 implies, according to Einstein's view, a contradiction with the postulate of relativity.

As another objection to interpretation 2, Einstein referred to the fact that for a multi-particle system (N particles), 2 requires the use of a $3N$-dimensional space. In this representation two configurations of a system which differ only by the exchange of two identical particles correspond to two different points of the space, which does not agree with the new quantum statistics. According to modern usage, we take the sum of the permuted and the original non-permuted wave functions as representing the physical state.

Bohr's reply is not recorded in the published conference proceedings but according to notes found in his files he replied, "I don't understand what precisely is the point that Einstein wants to make." Then he called attention to the fact that every measurement implies an interference with the measured system. The existence of individuals is complementary to the wave picture "we cannot in the description of space and time account for all features."

He then went on to defend the use of $3N$-dimensional space as being essentially a technical device. We will return to this point in Section 13.5.

13.4.2 Einstein and Bohr: off the record discussions

13.4.2.1 The fifth Solvay Conference, 1927

For many observers, these unrecorded discussions were the major event at the conference, eclipsing in importance everything else. The only detailed account was published by Bohr 20 years later in a collective work honoring Einstein.[21]

A few days after the conference Ehrenfest wrote a letter to his colleagues at home in Leiden. He gave a graphic description of the discussions:[22]

> It was delightful for me to be present during the conversations between Bohr and Einstein. Like a game of chess. Einstein all the time with new examples. In a certain sense a sort of Perpetuum Mobile of the second kind to break the UNCERTAINTY RELATION. Bohr from out of philosophical smoke clouds constantly searching for the tools to crush one example after the other. Einstein like a jack-in-the-box, jumping out fresh every morning. Oh, that was priceless. But I am almost without reservation pro Bohr and contra Einstein. His attitude to Bohr is now exactly like the attitude of the defenders of absolute simultaneity towards him.

After commenting on Einstein's single slit example discussed above, Bohr went on to ask if it would not be possible to make a more precise prediction of the point on the screen where the particle will be detected. For instance, if the plate with the slit were allowed to move, it would be possible to determine its recoil momentum and hence we would have more information concerning the direction the particle took after leaving the slit. What we have is a two-body system consisting of the particle and the plate with the slit. In the course of the discussions, interest was focused on an arrangement where a screen with two slits separated by a spacing d, was placed after the first slit (Fig. 13.6).

The momentum transferred to the first slit can indicate which of the two following slits the particle went through if the momentum of the first slit can be determined to an uncertainty $\Delta p < \hbar k\theta$ where $p = \hbar k$ is the momentum of the particle and the angle is

[21] Bohr, N., *Discussion with Einstein on Epistemological Problems in Atomic Physics,* in Schilpp, P.A., ed., Albert Einstein, Philosopher-Scientist, Vol.1 Harper, (1949).

[22] Ehrenfest, P., to students and associates, Nov. 3, 1927, original text in Ruedinger, E., ed., *Niels Bohr Collected Works, Vol. 6, Foundations of Quantum Physics I (1926-1932),* (North-Holland 1985), p.. 4i5, English translation p. 37. quoted in Mehra, J. and Rechenberg, H., *The Historical Development of Quantum Theory, Vol.6, The Completion of Quantum Mechanics (1926-1941), Part 1,* (Springer, 2000), p. 251.

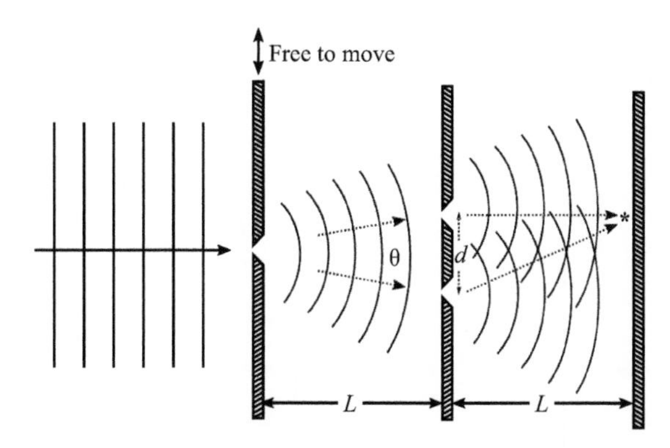

Fig. 13.6 Experimental arrangement discussed by Bohr and Einstein. The plate with the slit is free to move. After Schilpp, *Albert Einstein, Philosopher-Scientist.*

$\theta = d/L$ with d the distance between the two slits. Then the uncertainty in the position of the first slit will be $\Delta x_1 = 2\pi/k\theta$. The angular separation between the interference fringes behind the double slit is given by $\alpha = 2\pi/kd$ so the fringes will be separated by $y = L\alpha = L2\pi/kd = 2\pi/k\theta = \Delta x_1$. Hence the uncertainty in the position of the first slit, which will cause an equal uncertainty in the fringe position, will be enough to destroy the interference pattern. The measurement of the interference pattern and the determination of the particle trajectory are seen to be complementary, and we see how the mutually exclusive experimental conditions (whether the first slit is fixed or free to move) determine which of the complementary phenomena appear.

Bohr went on to describe another experimental arrangement discussed by Einstein. A photon incident on a half-silvered mirror will be recorded on one, and only one, of two photographic plates placed at a great distance in front and behind the mirror. If we replace the plates by mirrors, we can observe interference between the two reflected wave trains returning to the semi-reflecting mirror. In one case the photon chooses one side; in the other case, it travels over both possible trajectories. Recall Dirac's proposal that the choice is made only when interference is impossible and Heisenberg's idea that the choice is made "by the observer." According to de Broglie, the photon is always in one or the other wave packet, but the guiding wave accounts for the interference. As we saw in Chapter 12, the second quantization picture would imply that the photon is absorbed into a wave function (disappears into the "cloud" as it were) only to reappear when it is detected.

As Bohr expressed it, "It is just arguments of this kind that recall the impossibility of subdividing quantum phenomena and reveal the ambiguity of ascribing customary physical attributes to an atomic object."

13.4.3 Quantitative approach to Bohr's argument concerning two-slit interference

Jumping ahead in the historical development, in 1978, two recent PhDs at the University of Texas, Austin, (W.K. Wootters and W.H. Zurek) who had been stimulated by participating in a seminar on the fundamentals of quantum mechanics organized by

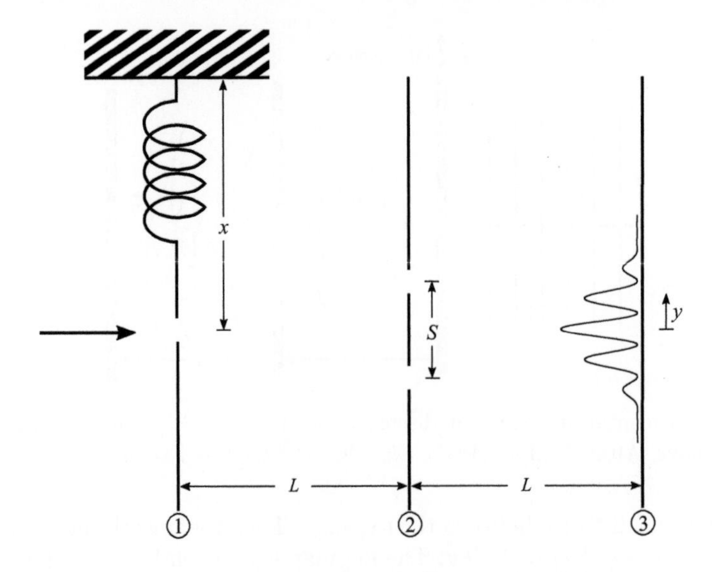

Fig. 13.7 Schematic for the double-slit experiment.

John Wheeler (see Chapter 17 for more details on the background) did a calculation[23] that quantified the relation between knowledge of which slit a particle went through in the two-slit interference experiment and the loss of visibility of the fringes. Following Bohr's idea of treating the plate with the two slits quantum mechanically the authors proposed that one of the plates in the apparatus, (plate 1 in Fig. 13.7 carrying the first slit) was free to move and coupled to a spring so that its motion would be that of a simple harmonic oscillator and one could choose to measure either the position or momentum of the plate.

In the ground state of the oscillator the position of the first slit x is given by the wave function

$$\psi(x) = \frac{1}{\pi^{1/4}a^{1/2}}e^{-\frac{x^2}{2a^2}} \tag{13.27}$$

while the momentum of the slit, p, is described by the wave function ($p = \hbar k$):

$$\phi\left(k\right) = \frac{a^{1/2}}{\pi^{1/4}}e^{-\frac{k^2a^2}{2}}. \tag{13.28}$$

We work in the Fraunhofer limit where plate 3 is far from plate 2 so all angles are small. When the first slit is centered ($x = 0$) the path length difference for rays going through each of the slits and arriving at the screen at a point y, is sy/L and the phase shift between the two rays is $\phi_o = ksy/L$ so that the wave at the point y on plate 3 will be proportional to

$$f(y) = e^{i\Phi}\left(1 + e^{i\phi_o}\right) \tag{13.29}$$

[23]Wootters, W.K. and Zurek, W.H., *Complementarity in the double-slit experiment: Quantum nonseparability and a quantitative statement of Bohr's principle*, Phys. Rev. D **19**, 473 (1979) received July 10, 1978.

and the intensity at that point will be ($k_o = ks/L$)

$$f(y) f^*(y) = 2(1 + \cos\phi_o) = 2(1 + \cos(k_o y)) \tag{13.30}$$

forming an interference pattern as y varies.

If x increases, there will be an additional path difference sx/L so that the interference pattern will be given by

$$I_x(y) \sim (1 + \cos(k_o(y - x))). \tag{13.31}$$

Each position, x, of the slit will contribute an interference pattern $I_x(y)$. So, if we choose to measure x, the observed values will be distributed with a probability $|\psi(x)|^2$ and the interference pattern, averaged over all the possible values of x, will be

$$F(y) = \int dx\, |\psi(x)|^2 I_x(y) \sim 1 + e^{-\left(\frac{k_o a}{2}\right)^2} \cos k_o y. \tag{13.32}$$

The momentum of plate 1 is an indicator of which slit the particle went through. If the momentum of the plate is down (positive) the particle has most likely been deflected up and so we would guess that it will have passed through the upper slit (A) and vice-versa. In order to go through one of the slits the particles would have to be deflected by an angle $\vartheta = \pm s/2L$ so that the momentum transfer will be $\pm k\vartheta = \pm k_o/2$.

If the initial momentum of the plate is k, whose probability is given by $|\phi(k)|^2$, (Eq. 13.28), the momentum of the plate, after a particle has gone through the slits, will be $\kappa = k \pm k_o/2$ and κ will be described by the probability distribution $|\phi(\kappa \mp k_o/2)|^2$. These distributions are shown in the figure.

Each curve represents the conditional probability $P(\kappa|A)$, or $P(\kappa|B)$, where $P(\kappa|A, (B))$ represents the conditional probability that we will observe the screen momentum to be κ if the particle is known to have passed through slit $A(B)$. What we need to know is $P(A,(B)|\kappa)$, the conditional probability that the particle will have passed through slit $A(B)$ if we have observed the slit to have the momentum κ. Bayes'

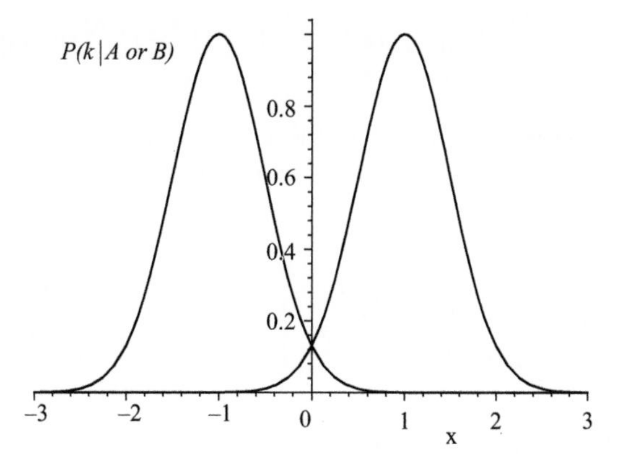

Fig. 13.8 Probability distribution as a function of the slit position x.

theorem makes the connection between the two. If we note that the joint probability of two propositions being true $P(M,N)$ can be written in two ways in terms of the conditional probability:

$$P(M,N) = P(M|N)\,P(N) = P(N|M)\,P(M) \tag{13.33}$$

so

$$P(M|N) = P(N|M)\,\frac{P(M)}{P(N)} \tag{13.34}$$

where $P(M), P(N)$ are called the prior probabilities of M, N. Therefore we can write

$$\frac{P(A|\kappa)}{P(B|\kappa)} = \frac{P(\kappa|A)\,P(A)}{P(\kappa|B)\,P(B)} = \frac{P(\kappa|A)}{P(\kappa|B)} = \frac{e^{-(k-k_o/2)^2/a^22}}{e^{-(k+k_o/2)^2/a^22}}. \tag{13.35}$$

Note that $P(A|\kappa) + P(B|\kappa) = 1$, since we limit ourselves to events producing a detected particle at screen 3, and for these events, a particle must have passed though either slit A or B. The amplitude at the screen 3 is the sum of amplitudes coming through the slits (f_A, f_B real)

$$f_3(y) = f_A e^{i\phi_A} + f_B e^{i\phi_B} \tag{13.36}$$

so that the intensity will be

$$|f_3(y)|^2 = 1 + 2f_A f_B \cos(\phi_A - \phi_B) \tag{13.37}$$

$$= 1 + 2f_A f_B \cos k_o y = 1 + 2P_A^{1/2}P_B^{1/2}\cos k_o y. \tag{13.38}$$

The same interference pattern would be obtained if the widths of the slits were taken in the ratio P_A/P_B. The fringe visibility, V. is defined as the peak to valley ratio of the fringes (13.38), ($\gamma = P_A$):

$$V(\gamma) = \frac{1 + 2P_A^{1/2}P_B^{1/2}}{1 - 2P_A^{1/2}P_B^{1/2}} = \frac{1 + 2\gamma^{1/2}(1-\gamma)^{1/2}}{1 - 2\gamma^{1/2}(1-\gamma)^{1/2}}. \tag{13.39}$$

The plot shows the visibility as a function of $\gamma = P_{A.}$ The interference pattern is surprisingly robust; for $P_A = 0.2, (P_B = 0.8)$ the visibility is still $V(0.2) = 9.0$, i.e., a very respectible interference pattern.

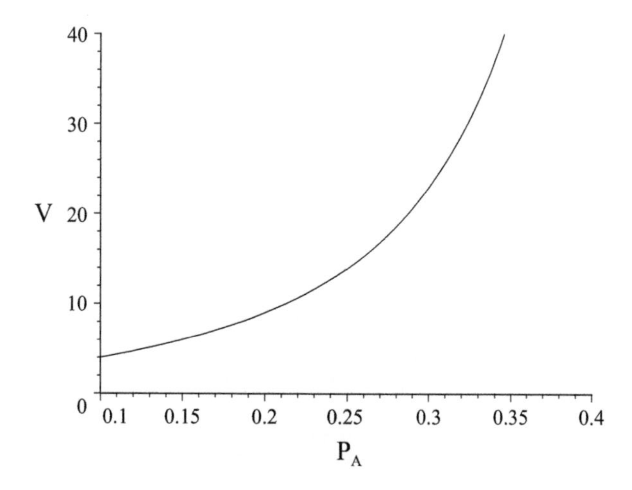

Fig. 13.9 Fringe visibility as a function of $P_A = \gamma$.

13.4.3.1 The sixth Solvay Conference, 1930

The official publication of the proceedings of this conference, which was devoted to magnetism, do not contain anything relevant but, according to Bohr's account in the 1949 volume,[24] there was a "dramatic" discussion behind the scenes between him and Einstein. This has been regarded as Bohr's greatest triumph over Einstein and has been discussed in great detail in the literature. It was hailed as giving a deep insight into the intricacies of quantum mechanics. The discussion involved a box ("photon box"), with reflecting walls, containing radiation and equipped with a shutter that can be opened for a very short time by a clock mechanism. A means of weighing the box is provided so that after the shutter opens for a very short time to release a photon, the box can be weighed, taking as much time as we want and thus determining the energy lost by the box due to the escape of the photon using the relation $E = mc^2$. In this way the energy of the emitted photon and the time of emission can be determined with an accuracy unlimited by the uncertainty principle. A conceptual drawing of the box made by Bohr in what he called a "pseudo-realistic" style is shown in Fig. 13.10.

After a sleepless night, Bohr was able to show that taking account of the uncertainty principle applied to the box, combined with the general relativistic gravitational red shift (later work showed that the red shift argument could be dispensed with), measuring the energy of the box and, by implication, the measurement of the photon energy, would be subject to the uncertainty principle where the relevant time uncertainty is in the time taken to perform the weighing operation, which Bohr visualized as adding weights to the box so that the spring returns to its original length. (Note that this time is not necessarily related to the opening time of the slit.) Thus Bohr claims that the "use of the apparatus as a means of accurately measuring the energy of the photon will prevent us from controlling the moment of its escape." Bohr's reply

[24]Op. cit., p. 310.

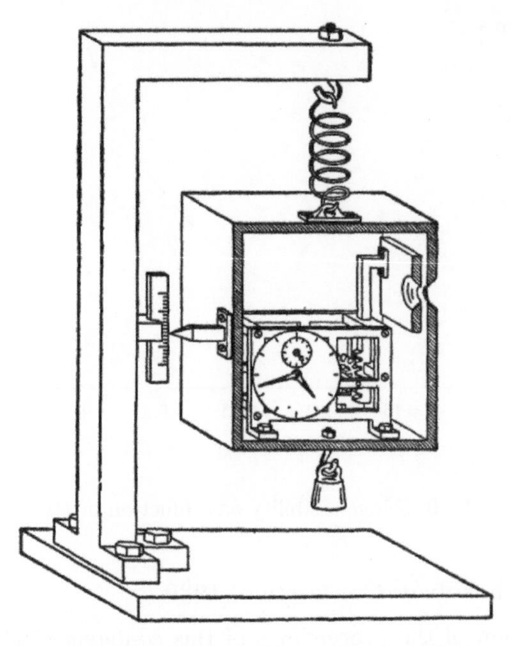

Fig. 13.10 Experimental arrangement discussed by Bohr and Einstein, as described in *Albert Einstein, Philosopher-Scientist*, Note 4. p. 295. (Reproduced with permission, Elsevier Science and Technology Journals, from *Niels Bohr, Collected Works Volume 7*, Kalckar, J. ed., Chapter 'Discussion with Einstein on Epistemological Problems in Atomic Physics,' p. 367. Fig. 8 (Elsevier, 1996).)

if correct, (there is a substantial literature discussing this question[25]) answered the question whether or not the two measurements (energy and time) can be carried out simultaneously to an accuracy higher than allowed by the uncertainty principle. But several researchers have considered that this argument was inessential to Einstein's argument. Bohr himself describes Einstein's argument as

> ...after the preliminary weighing of the box...and the subsequent escape of the photon, one was still left with the choice of either repeating the weighing or opening the box and comparing the reading of the clock with a standard time scale. Consequently, we are...still free to choose whether we want to draw conclusions either about the energy of the photon or about the moment it left the box. Without in any way interfering with the photon between its escape and its later interaction with other suitable measuring instruments, we are, thus, able to make accurate predictions either to the moment of its arrival or the energy that would be liberated during its absorption. Since however, according to the quantum mechanical formalism, the specification of the state of an isolated particle cannot involve both a well-defined connection with the time scale and an accurate fixation of the energy, it might thus appear as if this formalism did not offer the means of an adequate description.

[25]E.g., M. Jammer, *The Philosophy of Quantum Mechanics: The Interpretations of Quantum Mechanics in Historical Perspective* (New York: Wiley, 1974).

Or as Einstein would say, quantum mechanics was incomplete. Bohr then went on to say

In fact we must realize that in the problem in question we are not dealing with a *single* specified experimental arrangement, but are referring to *two* different, mutually exclusive arrangements. In the one the balance together with another piece of apparatus like a spectrometer is used for the study of the energy transfer by a photon; in the other a shutter regulated by a standardized clock together with an apparatus of a similar kind, accurately timed relatively to the clock, is used for the study of the time of propagation of a photon over a given distance. In both cases, as also assumed by Einstein, the observable effects are expected to be in complete conformity with the predictions of the theory.

The problem again emphasizes the necessity of considering the whole experimental arrangement.

Bohr said that a similar paradox occurs for the simpler case of a particle going through a single slit. We can choose to measure the momentum or the position of the slit after the photon has passed through and make predictions as to the photon's position or momentum. Einstein would argue that we can choose whether the photon has a definite position or momentum without exerting any physical influence on it as it is traveling away from the slit at the velocity of light. Thus the exact values of position and momentum must have existed before we made our choice and quantum mechanics is incomplete as it does not allow for this possibility. Bohr's reply is that the subsequent measurement of position and momentum require different experimental arrangements, so the physical situation is different in the two cases.

In these cases we see the germ of the Einstein, Podolski, Rosen argument where, as we will see, the photon box or slit arrangement is replaced by a second particle traveling away from the first particle at a great speed.

Einstein would support this argument, applied to various physical systems, for his entire life. In the volume[26] in which Bohr presented his account of the earlier discussions, Einstein presented his thinking as follows:

Consider a system consisting of two partial systems S_1 and S_2 which are spatially separated at a given time. The complete system is described by a wave function ψ_{12}. According to Einstein, all quantum physicists would agree that if we make a measurement on S_1, the results of the measurement, along with a knowledge of ψ_{12}, will give us a definite wave function ψ_2 of the system S_2. However, the character of ψ_2 depends on *what kind* of measurement we make on S_1. Einstein then states that there is a "real factual" situation of S_2, which is completely independent of what is done with the system S_1. This independence must be held to as a fundamental principle. Now, according to the type of measurement that is made on S_1, we get a very different ψ_2 for S_2. But according to our fundamental principle, the real situation of S_2 is independent of what happens to S_1 so that we can find different types of wave functions for the same real situation of S_2. Faced with this we can conclude that either:

1. The measurement of S_1 (telepathically) changes the real situation of S_2.
2. Things which are spatially separated from each other do not possess real independent states.

[26]See Note 4, p. 310.

Since both alternatives were unacceptable to Einstein, he concluded that quantum mechanics is incomplete and its statistical character is a necessary consequence of this incompleteness. The remarks by both Bohr and Einstein cited above show quite clearly that the issue between them concerned the entanglement between two spatially isolated systems. Einstein's gedanken experiments, e.g., the photon box, were not at all meant as attacks on the uncertainty principle.[27]

It is interesting to note, as remarked by Bohr, that Einstein never questioned the validity of the predictions of quantum mechanics, even though there was no experimental evidence dealing with these issues of non-locality and separability. It was only much later that experimental tests were conceived and eventually carried out. The work of John Bell[28] and others has shown that all local theories, in which the behavior of a system cannot be influenced by distant systems (Einstein's fundamental principle), are in contradiction with experiment. Experimental results are always in agreement with the predictions of quantum mechanics. We discuss this further in Chapter 15.

From the modern point of view, we would say the electromagnetic field emitted from the slit, which is open for time τ, will be a superposition of waves with frequencies described by a distribution $F(\omega)$, which is nonzero in a band $\Delta\omega \approx 1/\tau$ centered on the original frequency ω_o. By conservation of energy, the box will be in a superposition of states with energy $(E_o - \hbar\omega)$, E_o being the energy of the box before the photon was emitted. Measurement of the energy of the box would give a result corresponding to a small range $\delta\omega < \Delta\omega$ of the possible photon frequencies ω consistent with the uncertainty principle applied to the measurement. If the measurement results are centered around a frequency ω', the photon frequency will be found to be in the range $\delta\omega$ around ω'. As a result of the measurement, the wave function of the photon "collapses" into a wave group with a frequency spread of $\delta\omega$. Repeated measurements on an ensemble of similarly prepared systems would give results for $(E_o - \hbar\omega)$ distributed according to $F(\omega)$, so that the results for the ensemble will satisfy the uncertainty principle with the opening time τ. The electromagnetic field would propagate in a vacuum as a pulse of fixed width τ. So Einstein was right in so much as a measurement on the box would significantly alter the wave function of the photon flying away at the velocity of light. Bohr's argument relying on different instruments to measure time or frequency seems irrelevant here as we are only measuring the frequency in every case. The modern resolution of Einstein's dilemma is that the theory is really non-local, separated entities do not possess independent states, and the measurement of S_1 can change the state of S_2. Faced with the choice between Einstein's fundamental principle of the independence of spatially separated systems and the possibility of mutual influence of such systems, nature seems to have chosen the latter. That in itself does not exclude the existence of some variables not recognized by quantum mechanics, i.e., quantum mechanics could still be incomplete, but the possibility of any theory restricted to local variables has been ruled out experimentally in connection with Bell's theorem.

[27]For a more detailed discussion of the development of Einstein's ideas see: Howard, D., 'Nicht sein kann was nicht sein darf' or the prehistory of EPR, 1909-1935, Einstein's early worries about the quantum mechanics of composite systems. In Miller, A.I., ed., *Sixty-two years of uncertainty*, (Plenum Press, 1990), p. 61.

[28]Bell.J.S., *On the Einstein-Podolsky-Rosen paradox*, Physics **1**, 195 (1964); reprinted in Bell, J.S., op cit., p. 299.

13.5 The question of 3*N* dimensions

If we write the Hamiltonian for a system of N particles and apply the quantization rule of replacing the $3N$ momentum components p_i $(1 \leq i \leq N)$ of all the particles by $(\hbar/i)(\partial/\partial q_i)$, where q_i is the position coordinate conjugate to p_i, we obtain a single partial differential equation in $3N$ dimensions. The fact that its solution $\psi(q_i)$ represents a wave in $3N$-dimensional space rather than in ordinary three-dimensional space was a subject of some discussion at the 1927 Solvay Conference. Several participants felt that this was an indication that the wave function had to be considered as a mathematical tool rather than a physical object. We have already seen Einstein's objections above.

Schrödinger interpreted $\rho = \psi^*(q_i)\psi(q_i)$ as a charge density. The system point continuously fills the $3N$-dimensional q_i space, spending a time proportional to $\rho(q_i)$ in each volume element $d\tau$. The real system is a superposition of all possible configurations of the classical system with ρ as a weight function. Each point contributes a charge density in three-dimensional space that is given by, for the nth particle,

$$\rho_n(\vec{r}_n) = \int \psi^*\psi(q_i)\,d\tau' \tag{13.40}$$

where the prime indicates the integration is carried out over all the $3N$ coordinates except those pertaining to the nth particle. These charge clouds must not be thought of as acting on each other or as being acted on by an external field, as these interactions are included in the Schrödinger equation that determines $\psi(q_i)$ in the first place. However, Schrödinger proposed that they could serve as the source of a radiation field. For charge distributions confined to a region small compared to the wavelength, the radiation is determined by the electric dipole moment of the charge distribution, which is calculated as

$$M = \int \vec{M}_{\text{cl}}\psi^*\psi(q_i)\,d\tau' \tag{13.41}$$

with the classical dipole moment $M_{\text{cl}} = \sum_i e_i q_i$. Then, with

$$\psi(q_i, t) = \sum_n c_n \psi_n(q_i)\,e^{-i\omega_n t}, \tag{13.42}$$

the dipole moment will contain terms oscillating with the difference frequencies $(\omega_n - \omega_l)$ with amplitudes $c_n^* c_l$. While this gives a nice physical reason for the Bohr frequency condition, it requires the amplitudes of both the initial and final states to be nonzero, which contradicts the idea according to which a system originally in an initial state with $c_n = 1$, $c_l = 0$ makes a transition to a state with $c_n = 0$, $c_l = 1$. While these ideas are interesting, Dirac's treatment of the radiation process was preferred by most workers in the field.

Pauli spoke up several times in favor of the $3N$-dimensional space. Following Bragg's report on X-ray scattering, he stated that it is necessary to do the calculations in $3N$ dimensions. As an example, the probability of scattering from wave vector $\vec{k}_i$ to $\vec{k}_f$ will be proportional to

$$\int e^{i\left(\vec{k}_i - \vec{k}_f\right)\cdot\vec{r}} \left|\psi\left(q_i\right)\right|^2 d\tau. \tag{13.43}$$

The result obtained by assuming a three-dimensional density cannot be correct.

In the general discussion he went on to argue that the multi-dimensional space is only a technical means of formulating the laws of the interaction between particles, which certainly cannot be described simply in ordinary space and time. He suggested that this could be replaced by another method. As we have seen, Dirac had quantized the characteristic vibrations of a cavity filled with radiation and introduced a function ψ which was a function of the excitation level of each of these vibrations. Similarly Jordan and coworkers used the excitations of ordinary, four-dimensional matter waves as the argument of a state function (second quantization). These functions give, in the language of particles, the probability of the number of particles present for which given kinematical properties take certain values. We have shown that this procedure yields exactly the same results as the Schrödinger theory in multi-dimensional space. In addition, it avoids Einstein's objection concerning the interchange of identical particles in $3N$ dimensions.

Born agreed with the position that second quantization solved the problem. He called attention to the fact that Dirac and Jordan had shown that it is possible to base the theory on vibrations in three-dimensional space if we consider the eigenfunction itself, not as an ordinary number but as one of Dirac's q-numbers (that is, an operator) and quantize its amplitude as a function of time. In this way the number of electrons itself appears as a quantum number, and the electrons appear as discontinuities of the same nature as the stationary states.

13.6 Conclusion

It seems that in some sense both Heisenberg and Langevin were correct. In the light of modern conceptions, there certainly was considerable confusion and some significant disagreements as to what the theory really meant. On the other hand, Heisenberg was correct in that the Bohr, Heisenberg, and Born interpretation (what came to be called the Copenhagen interpretation) attained a dominant position with respect to alternate points of view, although Einstein, Schrödinger, and de Broglie would remain unsatisfied with the theory for the rest of their lives.

14

Von Neumann's mathematical foundations of quantum mechanics: redux

14.1 Introduction

John von Neumann's seminal book *Mathematical Foundations of Quantum Mechanics* (MFQM) was published in 1932.[1] This book was already mentioned in Chapter 11; however, its further discussion appears prudent, as it presents the full statistical theory of quantum mechanics based on the Hilbert space formulation in an axiomatic fashion. MFQM presents the final snapshot of the early development of the theory and establishes much of the descriptive language of the subject that is still used today. In this sense, the book is very much a classic, but more so in the Mark Twain sense that it is a book that "everybody wants to have read and nobody wants to read."

After the Solvay Conferences of 1927 and 1930, the "Copenhagen" interpretation of Bohr, Heisenberg, Born, Jordan, and others was accepted by the majority of working physicists. However, there were a number of physicists who, unable to accept the acausal features of that interpretation, held out the hope that the random elements in the theory could be shown to arise in a way similar to the way statistical fluctuations arise in classical statistical mechanics. In that case, the random elements would represent the behavior of a number of classical variables which cannot be followed in detail. These hypothetical variables are called "hidden variables." As Pauli wrote in a letter to Heisenberg (June 15, 1935),[2]

> Old men like Laue and Einstein are haunted by the idea that quantum mechanics is indeed *correct* but *incomplete.* It is possible to *extend it by propositions that are not contained within it, without changing the propositions that it already contains.* (I call such a theory - *incomplete.* Example the kinetic theory of gases) . . . such an extension to quantum mechanics is not possible without changing its content. [Emphasis in original.]

This last statement is a reference to a theorem proved by von Neumann in his book. As we will see below, this theorem shows that the existence of hidden variables is incompatible with a model based on vectors in Hilbert space with Hermitian

[1]Neumann, J. von, *Mathmatische Grundlagen der Quantenmechank*, (Springer, 1932).
We will refer to pages in the English translation: *Mathematical Foundations of Quantum Mechanics*, trans. R. T. Beyer (Princeton: Princeton University Press, 1955). A new edition was recently published, Nicholas A. Wheeler (ed.) Robert T. Beyer (translator) (Princeton University Press, 2018).

[2]Meyenn, Karl von, *Wolfgang Pauli: Scientific Correspondence with Bohr, Einstein, Heisenberg a.o. Volume II: 1930-1939*, (Springer, 1985), p.402.

The Historical and Physical Foundations of Quantum Mechanics. Robert Golub and Steven K. Lamoreaux, Oxford University Press.
 DOI: 10.1093/oso/9780198822189.003.0014

operators transforming the vectors. The main part of the proof is concerned with showing that dispersion-free ensembles—groups of identical systems which would have the same definite values of all observables—cannot exist if quantum mechanics is valid. In quantum mechanics, observables represented by noncommuting operators cannot simultaneously have such values. That is, in Hilbert space, noncommuting operators cannot have the same eigenstate.

The existence of hidden variables, allowing definite prediction of all physical properties, would imply that the ensembles discussed in quantum mechanics are capable of being broken up into subensembles, each with definite values of the hidden variables and hence of all variables (i.e., dispersion-free ensembles). Von Neumann shows that within a Hilbert space description, any subensembles of a quantum mechanical ensemble have the same statistical properties as the original ensemble and thus are also not dispersion free.

Indeed, von Neumann was first and foremost a mathematician; Hilbert was one of his doctoral examiners. His predilection toward proofs and logical formalism is evident in MFQM and makes its reading difficult, compounded by the fact that much of the material had never before been discussed at length, making the exposition often overcomplicated simply because each idea carries extra baggage in its justification. However, he occasionally strays from mathematical rigor, as on p. 183 of MFQM, where he states:

> In short, everything has been done which one should not do when working in correct mathematical fashion. As a matter of fact, this kind of negligence is present elsewhere in theoretical physics, and the present treatment actually will produce no disastrous consequences in our quantum mechanical applications. Nevertheless it must be understood that we have been careless.

In later years, when von Neumann was in Princeton, Pauli quipped to him, "If physics was proofs, you would be a great physicist." After the publication of MFQM, von Neumann had little input to the ensuing development of quantum mechanics and focused on such "mundane" subjects as game theory, mathematical logic, hydrodynamics, and the invention of digital computers.

The principal goal of MFQM was to bring axiomatic and logical rigor to the subject, without applying the formalism directly to any physical problem. Much of MFQM presents a recasting of the Dirac-Jordan transformation theory into a form that does not rely on the delta function (see pp. 25–27 of MFQM, where von Neumann refers to the delta function as "a fiction" that one might accept; he does not accept it, except for a few special cases). As mentioned in Chapter 11 (see Fig. 11.1), this function was widely viewed with a jaundiced eye before it was placed on its own rigorous mathematical foundation by L. Schwartz[3] (who in 1950 received the Fields Medal for his work on distributions). As such, von Neumann's treatment of this subject is not germane to our exposition. However, von Neumann's introduction of the density matrix, the statistical formulation of the measurement process, and his concept of entropy in quantum mechanics, are all used largely unaltered from his original forms. The first two of these were essential to his discussion of "hidden parameters" and the completeness of the mathematical foundations of quantum mechanics.

[3]Schwartz, L., *Introduction to the Theory of Distributions* (University of Toronto Press, 1952).

A central question at the time of MFQM's publication was whether quantum mechanics was *complete* in the logical sense that it truly captures all of physical reality (as opposed to the functional completeness we discussed in Chapter 11, which addresses whether a set of orthogonal functions over some range can be used to expand any piecewise continuous and differentiable function over that range). The issue of completeness was at the forefront of mathematics regarding the question of whether it was possible to describe a set of axioms and inference rules in the context of symbolic logic that could be used to prove (or disprove) all mathematical assertions, that is, prove that the axioms of arithmetic are consistent. This was the goal of, for example, Bertrand Russell and Alfred Whitehead's *Principia Mathematica*, and it is the second problem on the list of 23 unsolved problems that Hilbert put forward in 1900. Hilbert's sixth problem is whether all of physics can be axiomatized (which means, is there a Theory of Everything?). Hilbert's problems were discussed at his retirement festschrift,[4] where in a side discussion Kurt Gödel presented the first of his incompleteness theorems, with von Neumann solely taking a keen interest (their earlier work was complementary, and later they collaborated). Gödel's theorem can be näively thought of as stating that in any consistent axiomatic logical system, you can prove the things that can be proven, but there will always be some statements that cannot be proven either true or false within the system (an example of such a statement is "This statement is false"). In particular, this theorem means that Hilbert's Problem 2 is resolved in the negative. Axioms themselves can never be proven, which brings into question Hilbert's Problem 6, for physics is an experimental science, and it is impossible to prove that all the axioms are known, as based on existing knowledge that is cast as axioms, making the argument circular.

The logical completeness of quantum mechanics was questioned because the theory is statistical and requires the abandonment of causality in the sense that the outcome of an experiment, after specifying the initial conditions, cannot be exactly predicted. This is strikingly manifest in the noncommutativity of the position and momentum operators for a particle. Based on his mathematical interests, von Neumann's questioning the completeness from a logical and axiomatic standpoint is understandable. The principal question in this regard, which he addressed, is whether there are "hidden parameters" that would eliminate uncertainty or scatter in experimental measurements (translated as 'dispersion' in MFQM) and thereby restore classical causality.

In this chapter, we will review several sections of von Neumann's book that remain pertinent today, starting with his treatment of measurement theory, followed by his invention of the density matrix, both of which have remained largely unaltered, and his "no hidden parameters" proof, which is widely misused and misunderstood. We will conclude with a discussion of von Neumann's entropy, which remains of importance in quantum information theory. These topics will be presented in a different order than in MFQM, in what might be a slightly more logical if not a more pedagogical sequence.

[4]Festschrift David Hilbert Zu Seinem Sechzigsten Geburtstag Am 23. Januar 1922 Gewidmet Von Schülern Und Freunden, (Festschrift for the sixtieth birthday of David Hilbert on Jan 23, 1922 dedicated by his students and friends.), University of Michigan Library, 2006.

There is much lore surrounding MFQM. For example, it is often cited as containing a proof of completeness (in the sense of a complete set of orthogonal functions), but von Neumann states himself that he is not providing a proof in MFQM. Similarly, the "no hidden parameter" proof is a statement that within the framework of a Hilbert space formulation of quantum mechanics, extra hidden parameters are not possible, and their inclusion would require a reformulation of quantum mechanics. This is different from the statement that hidden variables are not possible in a fundamental, absolute sense, which has been an erroneous interpretation that has only grown stronger over the years. In part, the Copenhagen school originally oversold von Neumann's conclusion since it appeared to give a firm theoretical underpinning to the school's interpretation, and later J. S. Bell referred to the proof as wrong, misguided, and silly. We will show that the proof is contextually correct; however, von Neumann's presentation is sufficiently convoluted that it continues to be subject to misinterpretation.

Most interestingly, von Neumann never commented on the general misunderstanding of his "no hidden parameters" proof, which was not limited to the Copenhagen school, before his death in early 1957. One is left with the overall impression that von Neumann was satisfied that the mathematical foundations of quantum mechanics were sound, and after 1932 he paid little attention to the development of the subject, directing his energy to other fields of interest.

14.2 Von Neumann's measurement theory

The measuring process outlined in Chapter VI of MFQM mirrors the currently accepted "postulates" of quantum mechanics that were discussed in Chapter 11 in the context of the transformation theory. Von Neumann divides an experimental system into three regions labeled I, II, and III, as shown in Fig. 14.1.

A state $|\psi\rangle$ is prepared at the left, evolves causally under the action of a Hamiltonian H in Region I, and is then sent to a measurement apparatus (or operator) in Region II, the output of which is seen by an observer in Region III. For the specific case shown, the state $|\Psi'\rangle$ has "collapsed" into State 3 of the measurement appara-

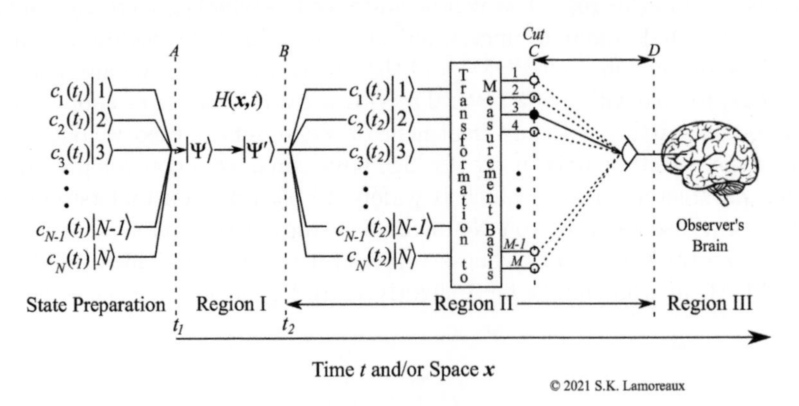

Fig. 14.1 Von Neumann's concept of an experiment; see text for a description. (Brain image credit: ID 34191041 ©Mariia Hlushchenko, Maglyvi Dreamstime.com.)

tus (which implies a specific value of some physical parameter). This collapse is not described by the Schrödinger equation, which describes the causal development of the system, and it is this collapse that introduces randomness into the theory. The time or position where the collapse occurs is referred to as the Heisenberg or von Neumann "*Cut*" and could be anywhere in Region II, between the lines B and D. Certainly, after the line D, which represents the division between the quantum and classical world (which itself is somewhat arbitrary), the propagation of the state is irreversible, but the point of irreversibility is likely earlier, as labeled by the *Cut* C, where the wave function collapses into a single eigenstate of the measurement operator. Experiments with entangled states are consistent with the notion that the collapse occurs at the moment of measurement, which happens somewhere within the measurement apparatus. This phenomenon will be described in depth in a later chapter, and a consequence of this is that quantum mechanics is *nonlocal.*

To recapitulate, Region I is the quantum system that is set up by conscious or natural design and allowed to evolve to some "final state." At the end of this evolution period, the state is sent to Region II where it encounters the measurement device or operator. At some time or space point (labeled as the *Cut* C) within Region II (between the dashed vertical lines B and D), the state "collapses" to give a specific and single output state of the measurement device; this is required because classical devices can only exist in non-overlapping, distinguishable states. The physical properties that are defined by the measurement device are encoded on this state.

Von Neumann's concept of the measurement process was to establish propositions that connect the eigenfunctions of H that are used to construct the state $|\Psi'\rangle$ to eigenstates, which eventually are the observable output states of the measurement device. The eigenstates of H labeled by $|n\rangle$ are not necessarily the same as those of the measurement device labeled by $|m\rangle$, and we can define a map (matrix) that connects, or projects, the N states onto the M states. The total number of states M and N does not need to be equal. Initially, the process of mapping $|\Psi'\rangle$ onto the $|m\rangle$ states is reversible (if $M \geq N$), and we recast the state in the m basis as

$$|\Psi'\rangle \rightarrow |\Psi'_M\rangle = \sum_m^M c_m |m\rangle. \tag{14.1}$$

At some point, labeled as the *Cut* in Fig. 14.1, the measurement output is chosen to be one of the $1...M$ states. This is von Neumann's collapse of the wave function, which we can see is a sort of sifting among the measurement operator eigenstates; each microstate (e.g., each individual particle with its given $|\Psi\rangle$) is destined to take a single path into the physical world and the paths are distributed according to the respective probabilities.

The measurement projection operator onto a specific state $|m\rangle$ is simply

$$\mathbf{P}_m = |m\rangle\langle m| \,, \tag{14.2}$$

which when applied to a state $|n\rangle$ yields

$$\mathbf{P}_m|n\rangle = |m\rangle\langle m|n\rangle. \tag{14.3}$$

For a probability conserving system, with observables represented by Hermitian operators, the measurement states (eigenfunctions of the measurement operator) form a complete set, so that

$$\sum_{m=1}^{M} |m\rangle\langle m| = 1. \tag{14.4}$$

The ultimate result of the measurement process are the probabilities of measurement device outcomes, that is,

$$w_m = \langle \Psi'_M | \mathbf{P}_m | \Psi'_M \rangle = \langle \Psi'_M | m \rangle \langle m | \Psi'_M \rangle = c^*_m c_m, \tag{14.5}$$

which manifests the statistical nature of the theory.

The boundary of Region III is defined as where the output information of the measurement device enters the mind of a conscious observer (MFQM, p. 419). In this region there absolutely can be no further evolution of the state vector; the output of the measurement device is one and only one of its eigenvectors, labeled $1 \ldots M$ in the figure; in the case shown in Fig. 14.1, "3" is the output state. The commitment to this state is (perhaps) final only when the information is perceived; however, Heisenberg and von Neumann showed that the *Cut* could lie anywhere between B and D without changing the result.

Region III always lies outside of the calculation; the collapse has already occurred at the *Cut* and the output of the measurement is being conducted to the consciousness of an external observer in this region. As a concrete example, the measurement device eigenstates might be photons in the directions $1 \ldots M$ that are then absorbed at specific points on the retina of the observer, resulting in specific signals that are eventually perceived by the "ego" of the observer. Indeed, in this picture, the *Cut* can be moved all the way to D. There is no absolute definition of the boundary between Regions II and III; it simply represents where classical physics starts. Within this region an observer's conscious registration of the output of the measurement device occurs, e.g., whether we include the optic nerve in Region III (some conscious processing might occur in this nerve, so perhaps there is reason to include it) is debatable. The central question is whether the existence of an observer can affect the measurement process.

So far, we have considered the case that Regions I+II feed Region III through the measurement device. This can be thought of as the Heisenberg representation of the system, which describes the equation of motion of the expectation values of the system with the wave function itself being of no direct importance.

However, von Neumann considers a second alternative case (see pp. 420–421 of MFQM for a discussion) in which the dynamics can be cast in terms of Region I feeding Regions II+III together. In this case we have the evolved state in Region I directly measured with the observer as part of the measurement device, and the wave function itself, not the expectation values, is of central importance. This can be thought of as the Schrödinger representation. The equivalence of the two representations is proven with much effort in MFQM and we do not reproduce the derivation here as this result was obtained earlier by Schrödinger and is discussed in Section 7.4. Heisenberg later (in 1935)[5] gave a proof that the position of the *Cut* could be arbitrarily moved (within certain limits) and used this to argue against the possibility of hidden variables, as the values of hidden variables would have to be fixed at the moment of the wave function collapse (at the *Cut)* and it is difficult to see how that fixation could happen at a point that can be shifted arbitrarily. In fact, it is possible to experimentally demonstrate that the system remains as a quantum superposition up to the moment of an irreversible measurement, and we will discuss this later in conjunction with entanglement.

14.3 No hidden parameters proof

In MFQM, von Neumann placed quantum mechanics on a firm mathematical footing based on operators and vectors in Hilbert space and showed that it did not need to rely on the mathematically dubious δ-function introduced by Dirac. He also spent considerable effort discussing the possibility that the statistical behavior associated with the quantum states was due to fluctuations in some unknown parameters, whose variation would lead to the random behavior, just as averaging over the positions and velocities of the individual molecules leads to statistical behavior in classical statistical mechanics. These parameters are called "hidden variables" and von Neumann showed that if hidden variables existed, quantum mechanics would be wrong, even for the known physical observables. This has led to much confusion in the literature, and for years von Neumann's demonstration was interpreted as proving that hidden variables could not exist at all.[6]

Von Neumann considers quantum mechanics to be characterized by the relation for the expectation value of a physical variable represented by the Hermitian operator $\mathbf{R}$, in the state represented by the Hilbert space vector $|\phi\rangle$:

$$\langle R\rangle = \langle\phi\,|\mathbf{R}|\,\phi\rangle. \tag{14.6}$$

This is the inner product of the state $|\phi\rangle$ with the state $\mathbf{R}\,|\phi\rangle$.

We expand $|\phi\rangle$ in a complete orthonormal set $|\psi_j\rangle$:

$$|\phi\rangle = \sum_j |\psi_j\rangle\,\langle\psi_j|\,\phi\rangle, \tag{14.7}$$

so that

$$\begin{aligned}
\langle\phi\,|\mathbf{R}|\,\phi\rangle &= \sum_{j,k} \langle\phi|\,\psi_j\rangle\,\langle\psi_j|\,\mathbf{R}\,|\psi_k\rangle\,\langle\psi_k|\,\phi\rangle & (14.8)\\
&= \sum_{j,k} \langle\psi_k|\,\phi\rangle\,\langle\phi|\,\psi_j\rangle\,\langle\psi_j|\,\mathbf{R}\,|\psi_k\rangle & (14.9)\\
&= \sum_{j,k} \langle\psi_k|\,\rho\,|\psi_j\rangle\,\langle\psi_j|\,\mathbf{R}\,|\psi_k\rangle = \sum_{j,k} \rho_{kj}R_{jk} & (14.10)\\
&= \mathrm{Tr}\,(\rho\mathbf{R}) & (14.11)
\end{aligned}$$

[5]Heisenberg, W., '1st eine deterministische Erganzung der Quantenmechanik moglich?' (Is a deterministic extension of quantum mechanics possible?) in Meyenn, 1985, op. cit. p. 409-418.

[6]Our discussion overlaps partly with that of Jeffrey Bub in 'von Neumann's "No Hidden Variables" Proof: A Re-appraisal', *Found. Phys.* **40**, 1333 (2010), arXiv:1006.0499 [quant-ph].

where ρ is the projection operator onto the state $|\phi\rangle$ given by

$$\rho = |\phi\rangle \langle\phi| . \tag{14.12}$$

The operator ρ is called the density matrix or statistical operator and represents all the information available about the quantum state, equivalent to the wave function. It allows the introduction of statistical mixtures of different states and/or particles.

Equation (14.12) holds when the system is in a single quantum state, called a pure state. In the case when the system is described by a statistical ensemble consisting of systems in various states $|\phi_i\rangle$ with probabilities w_i, what is called a mixed state, the density operator is given by

$$\rho = \sum_i w_i |\phi_i\rangle \langle\phi_i|. \tag{14.13}$$

If (14.6) is considered an assumption it can never be disproved, so von Neumann takes one step back and replaces (14.6) by a set of assumptions which can lead to (14.11). Without specifying a definite method for calculating expectation values, he lists a number of properties he would expect to be features of such a method. Among these are

$$\langle \alpha R\rangle = \alpha \langle R\rangle \tag{14.14}$$

where α is a number and

$$\langle R + S + T + ..\rangle = \langle R\rangle + \langle S\rangle + \langle T\rangle + ... \tag{14.15}$$

Von Neumann was well aware of the problem connected with the addition of operators that do not commute. Equation (14.15) is certainly valid in quantum mechanics since it follows from (14.11), but to cover the more general case, von Neumann defined the sum of the operators $R + S + T + ...$ to be that operator which had the expectation value given by the right-hand side of (14.15) (i.e., an implicit definition). Indeed, he considers explicitly the case of the Hamiltonian for the electron in a hydrogen atom, in which

$$E = \langle H\rangle = \left\langle \frac{p^2}{2m} - \frac{e^2}{r} \right\rangle = \left\langle \frac{(p_x^2 + p_y^2 + p_z^2)}{2m} \right\rangle - \left\langle \frac{e^2}{\sqrt{x^2 + y^2 + z^2}} \right\rangle. \tag{14.16}$$

Measurement of the first term on the rhs of (14.16) requires a momentum measurement while that of the second term is a coordinate measurement. The sum is measured in an entirely different way. Each measurement requires an entirely different apparatus. The variables p_i and x_i (and hence functions of them) cannot be individually simultaneously precisely determined due to their noncommutativity. Nonetheless, as in the present case of the hydrogen atom, the sum of $p^2/2m$ and $-1/r$ is defined and has a precise constant value, which is the energy. The role of a hidden variable theory is to allow the determination of the expectation values of p and r, as can be done in classical mechanics, and provide an internal framework that corresponds to classical expectations. In such a theory, quantum mechanical uncertainty would arise simply due to our inability to see these hidden parameters or variables.

Von Neumann then proceeds to derive (14.11) from Eq. (14.15). This means that in any case where (14.15) fails, (14.11) and hence quantum mechanics will also not be valid.

Starting with the representation of an operator as a matrix,

$$\mathbf{R} = \sum_{mn} |m\rangle \langle m| \mathbf{R} |n\rangle \langle n| = \sum_{mn} |m\rangle \langle n| R_{mn}, \tag{14.17}$$

we take the expectation value of both sides (note that this applies for any method of taking expectation values that satisfies (14.14) and (14.15)):

$$R = \langle \mathbf{R} \rangle = \left\langle \sum_{mn} |m\rangle \langle n| R_{mn} \right\rangle = \sum_{mn} \langle |m\rangle \langle n| \rangle R_{mn} \tag{14.18}$$

$$= \sum_{mn} \rho_{nm} R_{mn} = \mathrm{Tr}\,(\rho \mathbf{R}) \tag{14.19}$$

where we defined $\rho_{nm} = \langle |m\rangle \langle n| \rangle$ the expectation value of the operator $|m\rangle \langle n|$.

14.3.1 Implications of hidden variables

If the statistical variations in experimental results were due to averaging over unknown "hidden variables," the ensembles described by quantum states would in fact consist of separate subensembles, each with some exact value of all physical variables. These values would have to be eigenvalues of the corresponding operators in order for the results to agree with observations, which are found to always yield eigenvalues, in agreement with the current version of quantum mechanics. In addition, it would be possible, at least in principle, to separate out these subensembles, so that each of the separate ensembles would have the property that all variables had exact values and there would be no scatter in the measured values of observables. Von Neumann called such states "dispersion-free" states, and the resulting subensembles he called "homogeneous ensembles." A "homogenous ensemble" is an ensemble that can only be divided into subensembles that are identical to the original ensemble. If the statistical behavior associated with the quantum states were due to hidden variables, these quantum states would represent "inhomogeneous ensembles."

14.3.2 "Dispersion-free" states and homogeneous ensembles in quantum mechanics

Starting with Eq. (14.11), von Neumann writes[7] the dispersion (mean square fluctuation) of the variable represented by the operator $\mathbf{R}$ as

$$\sigma^2 = \mathrm{Tr}\left[\rho(R - \mathbf{R})^2\right] = \mathrm{Tr}\left[\rho(\mathbf{R^2} - R^2)\right] = \langle \mathbf{R^2} \rangle - \langle \mathbf{R} \rangle^2. \tag{14.20}$$

This is, in general, nonzero. It is zero if $\mathbf{R}|\phi\rangle = R|\phi\rangle$, which means that $|\phi\rangle$ is an eigenstate of $\mathbf{R}$, and the system is in that eigenstate.

[7]Von Neumann's proof is cast in modern notation by James Albertson in *Von Neumann's Hidden-Parameter Proof*, Am. J Phys. **29**, 478 (1961).

On the other hand, if the state is pure and comprises two non-degenerate eigenstates of the measurement operator $\mathbf{R}$ with eigenvalues r_1 and r_2, $|\phi\rangle = c_1|r_1\rangle + c_2|r_2\rangle$ and if $c_1^*c_1 + c_2^*c_2 = w_1 + w_2 = 1$, then

$$R = \langle\phi|\mathbf{R}|\phi\rangle = (c_1^*\langle r_1| + c_2^*\langle r_2|)\,\mathbf{R}\,(c_1|r_1\rangle + c_2|r_2\rangle) = w_1 r_1 + w_2 r_2 \tag{14.21}$$

$$\langle R^2\rangle = \langle\phi|\mathbf{R^2}|\phi\rangle = (c_1^*\langle r_1| + c_2^*\langle r_2\rangle\,\mathbf{R^2}\,(c_1|r_1\rangle + c_2|r_2\rangle) = w_1 r_1^2 + w_2 r_2^2. \tag{14.22}$$

Using $1 - w_1 = w_2$ and $1 - w_2 = w_1$, we find

$$\begin{aligned}
\sigma^2 = R^2 - [R]^2 &= w_1 r_1^2 + w_2 r_2^2 - (w_1 r_1 + w_2 r_2)^2 \\
&= (w_1 - w_1^2) r_1^2 + (w_2 - w_2^2) r_2^2 - 2 w_1 w_2 r_1 r_2 \\
&= w_1 w_2 r_1^2 + w_2 w_1 r_2^2 - 2 w_1 w_2 r_1 r_2 \\
&= w_1 w_2 (r_1 - r_2)^2,
\end{aligned}$$

which is zero only if c_1 or c_2 is zero or if $r_1 = r_2$, contrary to our assumptions.

We call a state "dispersion free" when

$$\mathrm{Tr}\left[\rho\mathbf{R^2}\right] = (\mathrm{Tr}\,[\rho\mathbf{R}])^2 \tag{14.23}$$

for all Hermitian measurement operators $\mathbf{R}$.

As discussed earlier, we can construct any operator by multiplying the projection operator for each eigenstate of the operator by the eigenvalue for that state. Therefore, let us simply consider $\mathbf{R}$ to be the projection operator onto a state $|\phi\rangle$, so that $\mathbf{R} = \mathbf{P}_\phi = |\phi\rangle\langle\phi|$. Let us further make use of the *idempotence* of the projection operator: when $\mathbf{P}_\phi$ operates on *any* state $|\Psi\rangle$ that is formed from a complete set of eigenfunctions that include $|\phi\rangle$,

$$(\mathbf{P}_\phi)^N\,|\Psi\rangle = \mathbf{P}_\phi|\Psi\rangle = c_\phi|\phi\rangle\,, \tag{14.24}$$

which means that the same result is obtained when $\mathbf{P}_\phi$ is applied multiple times.

With $\mathbf{R} = \mathbf{P}_\phi$ in Eq. (14.23), we obtain

$$\langle\phi|\rho|\phi\rangle = \langle\phi|\rho|\phi\rangle^2 \tag{14.25}$$

for all $|\phi\rangle$. Therefore,

$$\langle\phi|\rho|\phi\rangle = 0 \quad \text{or} \quad \langle\phi|\rho|\phi\rangle = 1. \tag{14.26}$$

This is true for any normalized state $|\phi\rangle$. Let us take $\phi = c_1|\phi_1\rangle + c_2|\phi_2\rangle$ and vary c_1 and c_2 in a continuous manner, so that $|\phi\rangle$ starts at $|\phi_1\rangle$ and ends at $|\phi_2\rangle$ (MFQM p. 320–321). During this process, the relation $\langle\phi|\rho|\phi\rangle = 0$ or 1 must hold over the entire variation, and we therefore conclude that

$$\rho = 0 \quad \text{or} \quad \rho = 1 \tag{14.27}$$

for a dispersion-free ensemble. This means that ρ is a diagonal matrix with elements all 0 or all 1. Furthermore, this implies that $\rho = \rho^2$. The case of all zero diagonal elements is trivial because in this case $R = 0$ always, which is not possible in a realistic system.

The obvious dispersion-free case of ρ being in a pure state of the $\mathbf{R}$ basis is possible for one or even for several measurement operators $\mathbf{R}$, but it cannot simultaneously be true for *all* measurement operators since some of them do not commute. Thus, such a state would not be dispersion free because it would still show dispersion in the measurements of some observables.

In the case of $\rho = 1$ we are faced with a normalization problem, in that

$$\mathrm{Tr}\,(\rho) = N \to \infty \tag{14.28}$$

where N is the number of elements (dimension of the state space). However, this sum should be unity if ρ represents the system average density matrix. This is because the diagonal elements are the probabilities of being in each eigenstate, and

$$\sum_n w_n = 1 = \mathrm{Tr}\,(\rho). \tag{14.29}$$

This would seem to preclude the possibility of constructing a dispersion-free density matrix because it cannot be normalized.

One might expect that extending the density matrix to infinite dimension might be a way to eliminate the intrinsic scatter due to finite measurement resolution. This possibility is discussed in MFQM. If we attempt to calculate σ^2 as before, we need to take the difference between infinite terms, so σ^2 is either undefined or nonzero in the case $\rho = 1$ because the trace diverges.

Thus, we must conclude that if the assumptions (14.14) and (14.15) hold, dispersion-free states do not exist.

Next, von Neumann asks if there is a possibility for the existence of subensembles of the density matrix which each have different but exact dispersion-free expectation values for a set of physical parameters, leading to dispersion of the full ensemble. The idea is that such subensembles could be chosen to produce a dispersion-free value of some hidden parameter, and that the dispersion of the full ensemble is due to the spread in the exact values among the subensembles. These subensembles would be homogenous ensembles. He then proves that any homogenous ensemble is a quantum state and any quantum state is a homogeneous ensemble. This means quantum states cannot be divided into subensembles with different properties.

Consider dividing the density matrix into two parts such that

$$\rho = \rho_1 + \rho_2 \tag{14.30}$$

ρ would constitute a homogenous ensemble if $\rho_1 = c_1\rho$ and $\rho_2 = c_2\rho$. Von Neumann proceeds by assuming we have a ρ with such properties.

Then he writes $\rho_{1,2}$ as the following Hermitian operators:

$$\rho_1 = \frac{\rho|f_0\rangle\langle f_0|\rho}{\langle f_0|\rho|f_0\rangle} \tag{14.31}$$

and

$$\rho_2 = \rho - \frac{\rho|f_0\rangle\langle f_0|\rho}{\langle f_0|\rho|f_0\rangle} \tag{14.32}$$

where $|f_0\rangle$ is a state vector such that $\rho|f_0\rangle \neq 0$. Because $|f_0\rangle$ is arbitrary, with only the constraint that its projection by ρ is nonzero, this decomposition is general. Also, it is obvious that $\rho = \rho_1 + \rho_2$ as required by the definition.

For any state $|\psi\rangle$, the probability that the system is in that state is

$$w_\psi = \langle\psi|\rho|\psi\rangle \geq 0 \tag{14.33}$$

for all $|\psi\rangle$, which follows from the definition of ρ. This means that ρ is a *positive definite operator*. Therefore

$$\langle f|\rho_1|f\rangle = \frac{|\langle f|\rho|f_0\rangle|^2}{\langle f_0|\rho|f_0\rangle} \geq 0, \tag{14.34}$$

which shows that ρ_1 is positive definite. From a theorem proved on p. 101 of MFQM, we know that

$$|\langle n|\mathbf{R}|m\rangle|^2 \leq \langle n|\mathbf{R}|n\rangle\langle m|\mathbf{R}|m\rangle, \tag{14.35}$$

so we immediately see that

$$\langle f|\rho_2|f\rangle = \frac{\langle f|\rho|f\rangle\langle f_0|\rho|f_0\rangle - |\langle f|\rho|f_0\rangle|^2}{\langle f_0|\rho|f_0\rangle} \geq 0. \tag{14.36}$$

Thus, both ρ_1 and ρ_2 are Hermitian (by construction) and positive definite, which also means that the trace is greater than zero.

Now define a state

$$|\phi\rangle = \frac{\rho|f_0\rangle}{K} \tag{14.37}$$

where K is chosen so that $\langle|\phi\rangle = 1$. Since we have assumed ρ represents a homogenous ensemble, $\rho_1 = c_1\rho$. Then

$$\rho|f\rangle = \frac{1}{c_1}\rho_1|f\rangle = \frac{1}{c_1}\frac{\rho|f_0\rangle\langle f_0|\rho|f\rangle}{\langle f_0|\rho|f_0\rangle} = \frac{K}{c_1}|\phi\rangle\frac{\langle f_0|\rho|f\rangle}{\langle f_0|\rho|f_0\rangle} = \frac{K^2/c_1}{\langle f_0|\rho|f_0\rangle}|\phi\rangle\langle\phi|f\rangle \tag{14.38}$$

$$= C|\phi\rangle\langle\phi|f\rangle. \tag{14.39}$$

Thus

$$\rho = C|\phi\rangle\langle\phi| = C\mathbf{P}_\phi\,, \tag{14.40}$$

and we see that if ρ represents a homogeneous ensemble then $\rho = \mathbf{P}_\phi$ and represents a quantum state. Every homogeneous ensemble is a quantum state.

Next we turn around and assume that we have a system in a quantum state $(\rho = \mathbf{P}_\phi)$. Von Neumann then proves that a quantum state is necessarily a homogeneous ensemble. Start with a state $|f\rangle$ orthogonal to $|\phi\rangle$, i.e., $\langle f|\phi\rangle = 0$ so that $\rho|f\rangle = \mathbf{P}_\phi|f\rangle = |\phi\rangle\langle\phi|f\rangle = 0$. With $\rho = \rho_1 + \rho_2$, where $\rho_{1,2}$ are both positive definite, we have the inequality

$$0 \leq \langle f|\rho_1|f\rangle \leq \langle f|\rho_1|f\rangle + \langle f|\rho_2|f\rangle = \langle f|\rho|f\rangle = 0\,, \tag{14.41}$$

so that $\langle f|\rho_1|f\rangle = 0$. By considering Eq. (14.35), replacing $|n\rangle$ by $|g\rangle$, $|m\rangle$ by $|f\rangle$, and $\mathbf{R}$ by ρ_1 we find $\langle g|\rho_1|f\rangle = 0$ for any state $|g\rangle$. Since $|f\rangle$ is any state orthogonal to $|\phi\rangle$ and every $|f\rangle$ is also seen to be orthogonal to $\rho_1|g\rangle$, we must have

$$\rho_1|g\rangle = c_g|\phi\rangle \tag{14.42}$$

with c_g a constant *that can depend on* $|g\rangle$. Taking $|g\rangle = |\phi\rangle$ yields

$$\rho_1|\phi\rangle = c'|\phi\rangle. \tag{14.43}$$

Any state $|h\rangle$ can be written as

$$|h\rangle = |\phi\rangle\langle\phi|h\rangle + |h'\rangle \tag{14.44}$$

where $|h'\rangle$ is orthogonal to $|\phi\rangle$. Then

$$\rho_1|h\rangle = \rho_1|\phi\rangle\langle\phi|h\rangle + \rho_1|h'\rangle \tag{14.45}$$

$$= c'|\phi\rangle\langle\phi|h\rangle = c'\mathbf{P}_\phi|h\rangle \tag{14.46}$$

where $\rho_1|h'\rangle = 0$ because h' is orthogonal to $|\phi\rangle$. Therefore, $\rho_1 = c'\rho$ and, because $\rho_1 + \rho_2 = \rho$, then $\rho_2 = (1 - c')\,\rho$ and any quantum state is a homogeneous ensemble.

If hidden parameters existed, we could pick specific values of those parameters and form subensembles such that a measurement of that subensemble would produce exact values for some parameters or reduce the dispersion of some parameters. Von Neumann's point with this proof was to show that quantum ensembles are not dispersion free simply because they are composed of distinct subensembles, each of which has a different value of some hidden parameter, with the statistical nature of the full ensemble being due to scatter from the different values associated with each subensemble.

To reiterate, there are two reasons for this. First, there are no dispersion- or scatter-free ensembles because for such $\rho = 1$. Second, if we were to postulate that dispersion results from an ensemble representing a collection of subensembles which each have a specific value for some hidden parameter, that postulate would fail because the subensembles would be homogeneous ensembles which themselves are quantum states and hence not dispersion free. As we have seen, the original ensemble, representing a quantum state, is a homogeneous ensemble, precluding its breakup into different subensembles.

It is understood that all these results hold only in the case that quantum mechanics is valid.

14.3.3 No hidden variables "theorem"

Either of these two results would rule out hidden variables coexisting with quantum mechanics. Dispersion-free states are impossible and the quantum states are homogeneous ensembles so that it is impossible, according to quantum mechanics, to break up an ensemble into subensembles with different physical properties. According to von Neumann:

> But this [the existence of hidden variables] is impossible for two reasons: First, because then the homogeneous ensemble in question could be represented as a mixture of two different ensembles, contrary to its definition. Second, because the dispersion-free ensembles, which would have to correspond to the "actual" states (i.e., which consist only of systems in their own "actual" states), do not exist.[8]

[8]MFQM, p. 324

After the reinvention of de Broglie's pilot wave theory by Bohm (1952), John Bell surmised the following on reading Bohm's paper:

But in 1952 I saw the impossible done. It was in papers by David Bohm. Bohm showed explicitly how parameters could indeed be introduced into nonrelativistic wave mechanics, with the help of which the indeterministic description could be transformed into a deterministic one. More importantly, in my opinion, the subjectivity of the orthodox version, the necessary reference to the "observer" could be eliminated.[9]... This [assumption (14.15)] is true for quantum mechanical states, *it is required by von Neumann of the hypothetical dispersion free states also* [emphasis added]. At first sight the required additivity of expectation values seems very reasonable, and it is the non-additivity of allowed values (eigenvalues) which requires explanation. Of course the explanation is well known: A measurement of a sum of noncommuting variables cannot be made by combining trivially the results of separate operations on the two terms—it requires a quite distinct experiment. For example the measurement of σ_x for a magnetic particle might be made with a suitably oriented Stern-Gerlach magnet. The measurement of σ_y would require a different orientation and the measurement of $(\sigma_x + \sigma_y)$ a third and different orientation. But this explanation of the non-additivity of allowed values also established the non-triviality of the additivity of expectation values.

Most importantly for our discussion, Bell goes on to make the following point:

The latter (14.15) is a quite peculiar property of quantum mechanical states, not to be expected *a priori*. There is no reason to demand it individually of the hypothetical dispersion-free states, whose function it is to reproduce the *measurable* properties of quantum mechanics when *averaged over*. [Emphasis added.][10]

Others have come to a similar conclusion. For example, a year after the publication of MFQM the same point concerning (14.15) was made by Grete Hermann, a philosophy student who was defending the philosophical tradition that causality was a necessary constituent for any scientific view of the world. She'd had extensive discussions with Heisenberg and claimed that as (14.15) is valid only in quantum mechanics, where there are no dispersion-free states, von Neumann's argument was circular.[11]

David Mermin[12] characterized the assumption (14.15) as "silly" and quoted Bell in a published interview:

Yet the von Neumann proof, if you actually come to grips with it, falls apart in your hands! There is *nothing* to it. It's not just flawed, it's *silly*! ... When you translate [his assumptions] into terms of physical disposition, they're nonsense ... The proof of von Neumann is not merely false but *foolish*! [13]

Bell's point, quoted above with emphasis, is correct: In dispersion-free states the expectation value of a variable would have to be one of the eigenvalues of the corresponding operator, and so equation (14.15) could not apply. But contrary to Bell's point above, von Neumann does not require (14.15) to be true of dispersion-free states. The whole point of his argument is that the existence of dispersion-free states would

[9]Bell, J.S., *On the impossible pilot wave*, *Found. Phys.* **12**, 989 (1982).

[10]Bell, J. S., *On the Problem of Hidden Variables in Quantum Mechanics*, *Rev. Mod. Phys.* **38**, 447 (1966).

[11]Quoted in Dieks, D., *Von Neumann's Impossibility Proof: Mathematics in the Service of Rhetorics*, *Stud. Hist. Phil. Mod. Phys.* **60**, 136 (2017), arXiv:1801.09305v1 [physics.hist-ph].

[12]Mermin, D., *Hidden variables and the two theorems of John Bell*, *Rev. Mod. Phys.* **65**, 803 (1993).

[13]Bell, J.S., Interview in *Omni*, May 1988, p. 88.

mean that (14.15) *is false* and so (14.11) and (14.19), and hence quantum mechanics, would not be valid if there were dispersion-free states. As von Neumann writes:

> It should be noted that we need not go any further into the mechanism of the "hidden parameters," since we now know that the established results of quantum mechanics can never be rederived with their help. In fact, we have even ascertained that it is impossible that the same physical quantities exist with the same functional connections, if other variables (i.e., "hidden parameters") should exist in addition to the wave functions. Nor would it help if there existed other, as yet undiscovered, physical quantities, in addition to those represented by the operators in quantum mechanics, because the relations assumed by quantum mechanics (i.e., **I.**, **II.**)[14] would have to fail already for the currently known quantities, those that we discussed above. It is therefore not, as is often assumed, a question of a re-interpretation of quantum mechanics—the present system of quantum mechanics would have to be objectively false, in order that another description of the elementary processes than the statistical one be possible.[15]

To reiterate, the logical argument, which has the appearance of circularity (sort of a *reductio ad absurdum*) but is valid, is based on the following three propositions:

A. The sum of expectation values assumption, Eq. (14.15).
B. $\langle \mathbf{R} \rangle = \mathrm{Tr}(\rho \mathbf{R})$, that is, the trace of ρ times an operator gives the expectation value of the operator.
C. Dispersion-free states do NOT exist.

Then von Neumann has shown that A $\Rightarrow$ B $\Rightarrow$ C.

If this conclusion is taken as the argument against hidden variables, then Bell, Hermann, and others are correct because A. does not hold for dispersion-free states, and it is therefore a circular, silly argument.[16]

On the other hand, von Neumann's argument is that A *does* fail for dispersion-free states and this means that B would be false. In that case (the existence of dispersion-free states), quantum mechanical relations would have to fail in situations where they are known to apply.

14.4 Von Neumann entropy

Quantum information theory can trace its origin to MFQM, in which von Neumann introduces the concept of entropy to express the degree of order or disorder of a quantum ensemble. He makes a connection with entropy as defined in classical thermodynamics. As shown in Fig. 14.2, von Neumann's calculation of the entropy is performed on a box K of volume V with N particles in a mixture of states and which form an ideal gas.

The density matrix is the only reasonable way to address the properties of a statistical mixture and this was the motivation for its introduction. As we have seen already,

[14]Relations **I.** and **II.** are found on pp. 313-314 of MFQM and correspond to our Eqs. (14.14) and (14.15). It is clear that von Neumann is implying that he is employing all of the previously used conditions, **A'.**, **B'.**, α), β) defined on pp. 311-312 of MFQM, in addition to **I.** and **II.**, as stated earlier on p. 323.

[15]MFQM, p. 324.

[16]D. Dieks, op. cit. p. 334.

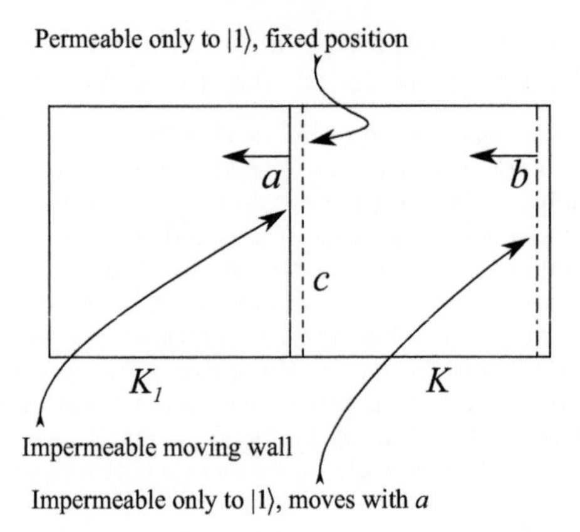

Fig. 14.2 Von Neumann's entropy machine in which two identical boxes K, K_1 of same volume V are brought together, forming a common wall on one side a that is then made movable, and a filter c permeable only to particles in state $|1\rangle$ is introduced at a fixed location at a's initial position. A second filter b that allows all particles to pass except those in state $|1\rangle$ is moved along with a so that the volume between the surfaces a and b is constant. As a, b are moved, all of the particles in state $|1\rangle$, and only those, are swept into K_1. The volume K_1 can be tapped off, and the process repeated, by introducing new collection volumes K_2, K_3, K_4...for each state.

the diagonal elements of the density matrix give the probabilities, w_n, to be in the states $|n\rangle$

$$\rho = \sum_n w_n |n\rangle\langle n| = \sum_n w_n \mathbf{P}_n \tag{14.47}$$

where $\mathbf{P}_n$ is the projection operator onto $|n\rangle$. As shown in Fig. 14.2, we can perform a sort of *gedanken* experiment where particles in a specific state are separated from the initial volume K, in a reversible manner, into unique volumes K_n, each of which contains $N_n = w_n N$ particles corresponding to the states $|n\rangle$. Next, these volumes are each isothermally compressed to give the same density as the original system, that is, to volumes $V_n = w_n V$. The change in entropy for an isothermal compression of the volume for state $|n\rangle$ is the heat Q_n liberated in this process divided by the constant temperature T at which the compression was performed (T can be held constant by placing the system in close contact to a large thermal reservoir). Q_n can be calculated from the ideal gas law, which gives $dQ_n = pdV = nk_BTdV/V$, so that

$$\Delta Q_n = \int dQ_n = N_n k_B T \ln\left(\frac{V_n}{V}\right) = w_n N k_B T \ln\left(\frac{w_n V}{V}\right) = w_n N k_B T \ln w_n. \tag{14.48}$$

Setting $k_B = 1$, the total entropy increase for this process, repeated for each state, is

$$\Delta S = \sum_n \frac{\Delta Q_n}{T} = N \sum_n w_n \ln w_n. \tag{14.49}$$

Finally, the particles in each volume K_n can be subjected to unitary transformations that bring them all to some arbitrarily chosen state $|\phi\rangle$; a unitary transformation does not increase the entropy but can change the energy. The individual volumes can be added together with no increase in entropy because they are now all in the same state with the same density. The sum of these volumes equals the initial volume because $\sum_n w_n = 1$.

If we had started in a pure state, the change in entropy would be zero, so we can assign $S = 0$ to a pure state. We have determined the total entropy change in bringing the mixed state to a pure state, and because the entropy is zero in the final state, in the initial state it was

$$S = -N \sum_n w_n \ln w_n. \tag{14.50}$$

If ρ is diagonal, then $\rho \ln \rho$ can be expanded for each element as $w_n \ln w_n$, where it is understood that the logarithm operation on a matrix is effected as a Taylor series. Therefore, we can rewrite the entropy as

$$S = -N \sum_n w_n \ln w_n = -N \,\mathrm{Tr}(\rho \ln \rho) \tag{14.51}$$

where ρ is normalized by dividing by N so its trace is unity.

The probabilities are all $0 \leq w_n \leq 1$ therefore $w_n \ln w_n \leq 0$ and equals zero only when $w_n = 0$ or 1. This can be obtained from continuity considerations or by the fact that if $w_n = 0$ initially it does not contribute to the calculation of S. If the state is pure, then $\rho = |\phi\rangle\langle\phi|$ and $w_\phi = 1$ and $S = 0$.

The elements of the density matrix can be written in terms of the state amplitudes (numbers) c_n as $\rho_{mn} = c_m c_n^*$ and we see immediately that the diagonal elements of the density matrix are the state probabilities. If we start with a pure state, for which only one of the c_n is nonzero, then $c_n c_n^* = 1$ and the entropy is zero. Let us now investigate the effect of a unitary transformation on a pure state ρ which will generate both diagonal and off-diagonal elements (coherences). It is not obvious that the state entropy will remain zero.

Noting that the natural log in the definition of the von Neumann entropy is to be understood as a power series, a simple term-by-term examination shows that

$$\ln(U\rho U^\dagger) = U[\ln \rho]U^\dagger. \tag{14.52}$$

Let us define a new matrix determined by a unitary transformation operator U with $\rho = U\rho_D U^\dagger$ and ρ_D is diagonal. Then

$$S(\rho) = -N \,\mathrm{Tr}\left(U\rho_D U^\dagger \ln(U\rho_D U^\dagger)\right) = -N \,\mathrm{Tr}\left(U\rho_D \ln(\rho_D) U^\dagger\right) \tag{14.53}$$

$$= -N \,\mathrm{Tr}\left(U^\dagger U\rho_D \ln(\rho_D)\right) = -N\mathrm{Tr}\left(\rho_D \ln(\rho_D)\right), \tag{14.54}$$

where the next-to-last last step made use of the fact that the trace is invariant under a unitary transformation $\mathrm{Tr}\,(ABC) = \mathrm{Tr}\,(CAB)$. Consequently, because the time evolution operator is a unitary operator, the von Neumann entropy of an isolated system evolving under the action of a Hamiltonian is constant in time. Note that this result, that the entropy is constant under a unitary transformation, is general for any density matrix, not only for a pure initial state.

If there are processes that cause decoherences (while not affecting the state probabilities), the entropy of the system must increase. This can be most readily seen if starting from a pure state. After a suitable unitary transformation, off-diagonal density matrix elements will be generated, and therefore there will be more than one diagonal element. Because the diagonal elements are probabilities, they must be positive and their sum must be unity. If we wait long enough, the off-diagonal elements will all decay to zero (see Section 17.1), and then it is trivial to calculate the entropy as $-\sum_n c_n c_n^* \ln c_n c_n^* > 0$. Therefore, per these definitions, the off-diagonal elements or coherences must carry negative entropy.

The close connection between the von Neumann entropy and the Shannon entropy H in information theory is obvious,[17]

$$H = -K \sum_{i=1}^{k} p_i \log(p_i) \tag{14.55}$$

where K is a positive constant that depends on the units of measurement, and p_i is the probability of a communications system producing a discrete value x_i. For a binary string, the usual definition is

$$H(p) = -(p \log_2 p + (1-p)\log_2(1-p)) \tag{14.56}$$

where p is the ratio of ones to zeros, which is the same as the von Neumann entropy for a system of spin-1/2 particles (up to numerical factors, e.g., for $\log_2$ vs. ln). According to one source,[18] Shannon gives credit to von Neumann for the term entropy in the context of information theory in the following quote (from circa 1938):

> My greatest concern was what to call it. I thought of calling it "information", but the word was overly used, so I decided to call it "uncertainty." When I discussed it with John von Neumann, he had a better idea. Von Neumann told me, "You should call it entropy, for two reasons: In the first place your uncertainty function has been used in statistical mechanics under that name, so it already has a name. In the second place, and more important, nobody knows what entropy really is, so in a debate you will always have the advantage."

It is of interest that Shannon in his Mathematical Theory of Communications cites Tolman's book as providing a connection to entropy in statistical mechanics.[19] Here, Tolman recasts Boltzmann's H theorem in the context of phase space volume elements (which implies a quantum treatment and a particle wavefunction), and obtains essentially the same equation for the entropy as the von Neumann entropy, perhaps not unexpectedly. Apparently Shannon did not know about von Neumann's work on the subject, done earlier and published in 1932, while Tolman mentions it cryptically

[17] Shannon, C.E., *Mathematical Theory of Communication*, Bell System Tech. Jour. **27**, pp. 379-423, 623-656 (1948). See Part I, Sec. 6 and Appendix II.

[18] Tribus, M. and McIrvine, E.C., *Energy and information*, Scientific American, 224 (September 1971).

[19] Tolman, R.C. *The Principles of Statistical Mechanics* (Oxford University Press, 1938). See Eq. (47.7), p. 135.

in passing (p. 257). This earlier work would appear be the fundamental motivation behind the naming suggestion von Neumann gave Shannon for his information function, in particular given its essentially identical form as the von Neumann entropy. Of course, all are extensions of the Gibbs entropy, however, Gibbs did not know about quantization so his phase space volumes were labeled as microstates and were uncountably infinite in number; that problem was solved by Planck with the introduction of his eponymous constant.

14.5 Conclusion

The foundational techniques and open issues discussed in MFQM remain largely relevant today. Von Neumann showed that, within the context of the Hilbert space formulation, there is no mathematically possible way to eliminate "dispersion" or scatter from measurements, which is simply a consequence of the measurements yielding eigenvalues of the measurement operator. If the measured state is not in an eigenstate of the measurement operator, the measurement results will follow a statistical distribution. This is a simple consequence of the discreteness of the eigenvalues of the measurement operator. On the other hand, if the measurement operator is constructed on a finer and finer scale, the measurements will be distributed over a larger and larger number of measurement eigenvalues. Such a distribution or scatter means that the classical concept of causality—that specific measurement results are obtained for a system prepared with some specific initial conditions—is lost in quantum mechanics. It is shown conclusively in MFQM that this results simply from having only fixed eigenvalues available for the measurement process, and for a measured state that is not in an eigenstate of the measurement operator, the results will be weighted by the probabilities of being in those eigenstates.

It is further shown in MFQM that it is not possible to divide up an ensemble into subensembles that might produce scatter-free measurement results. Subensembles that might be selected to have the same values for some hidden parameter that would allow the elimination or reduction of dispersion have, within the Hilbert space formulation of the theory, the same statistical properties as the original ensemble. This is proven on very general grounds.

The myths and confusion regarding von Neumann's "no hidden parameters" proof were initially and largely fueled by the Copenhagen school, which used this proof as a basis for establishing the completeness (in the sense that there are no further fundamental principles to be discovered) of the usual form of quantum mechanics. We have seen, and von Neumann admits himself, that the proof only exists for questions within the context of the Hilbert space formulation of the theory. Von Neumann's proof was later maligned, most notably by J. S. Bell, who did not fully appreciate its structure, and who was echoing the Copenhagen school's stance on the universal validity and scope of the proof. Bell stated that the proof is "so wrong as to be silly," but he was mistaken in his assertions, as we have shown in this chapter.

Although von Neumann states directly that the Hilbert space formulation of quantum mechanics does not allow for hidden parameters (hidden variables), he does say that their existence is an open experimental question, but their inclusion would require a substantial reworking of the theory. This would almost certainly result in measurable consequences.

Well before presentation of his "no hidden parameter" proof, von Neumann states (MFQM, p. 210),

> We shall show later (IV.2) that an introduction of hidden parameters is certainly not possible without a basic change in the present theory. For the present, let us re-emphasize only these two things: The (quantum state) Φ has an entirely different appearance and role from the $q_1...q_k$, $p_1...p_k$ in classical mechanics and the time dependence of Φ is causal and not statistical: Φ_{t_0} determines all Φ_t uniquely, as we saw above.
>
> Until a more precise analysis of the statements of quantum mechanics will enable us to test objectively the possibility of the introduction of hidden parameters (which is carried out in the place quoted above), we shall abandon this possible explanation. We therefore adopt the opposite point of view. That is, we admit as a fact that those natural laws which govern the elementary processes (i.e., the laws of quantum mechanics) are of a statistical nature. (The causality of the macroscopic world can in any event be simulated by the leveling action which is manifest whenever many elementary processes operate simultaneously, i.e., by the "law of large numbers.") ... Accordingly, we recognize (14.6) as the most far-reaching pronouncement on elementary processes. [I.e., expectation values add linearly, and from this, the fact that the expectation value of an operator $\mathbf{R}$ is $\text{Tr}\rho\mathbf{R}$ can be derived.]
>
> This concept of quantum mechanics, which accepts its statistical expression as the actual form of the laws of nature, and which abandons the principle of causality, is the so-called statistical interpretation. It is due to M. Born, and is the only consistently enforceable interpretation of quantum mechanics today—i.e., of the sum of our experiences relative to the elementary processes. It is this interpretation to which we shall conform in the following (until we can proceed to a detailed and fundamental discussion of the situation).

If hidden variables existed, one could pick specific subensembles to reduce the dispersion of noncommuting variables. However, von Neumann showed that all subensembles have the same statistical properties within the context of the Hilbert space formulation of quantum mechanics. The essential lesson is that quantum mechanics itself cannot be used to ascertain its own completeness; von Neumann's proof shows that there is internal consistency in the theory, but we cannot go beyond that per Gödel's theorem as we discussed earlier in this chapter.

The open questions discussed in MFQM, for example the location of the "Heisenberg cut" where a quantum system collapses to an eigenstate of the measurement apparatus, remain open today. We are left with the problem that each potential measurement does not have a value (in the sense of knowing in which specific measurement eigenstate the system will end up) before the measurement; we will discuss this later in regard to the EPR paradox. It is sometimes suggested that we not pay attention to the wave function itself but consider only the Heisenberg picture which describes the equations of motion of the observables. This would be reasonable except in modern physics it is possible to perform measurements on single quantum systems and the results of such are eigenvalues of the measurement operator, as indicated in Fig. 14.1. It is only after repeated measurements that the statistical character of a mixed state can be observed.

We can further ask what consequences hidden variables might have on the fundamental theory, in regard to von Neumann's statement: "The present system of quantum mechanics would have to be objectively false in order that another description of the elementary process than the statistical one be possible." For example, deviations from the standard theory would require significant modification of the Pauli exclusion principle to account for extra degrees of freedom in specifying electron states (both spin and momentum), as well as modification of quantum statistical mechanics (e.g., the Sackur-Tetrode equation for the entropy of an ideal gas would need to include extra degrees of freedom to account for the fine scale structure). These issues have not been, to our knowledge, addressed in the literature, with the exception of a letter from Pauli to de Broglie and a published contribution to a conference.[20] Pauli states in no uncertain terms that hidden variables are completely at odds with the spin-statistics theorem and that he could see no way to incorporate them into quantum mechanics without upsetting statistical physics. The breadth and broadness of the complex of phenomena that can be perfectly described in the context of quantum mechanics means that any attempt to modify the theory must be considered in light of collateral deleterious complications.

We suggest that there are two types of hidden variables: one type that suppress dispersion in measurements, which is the type addressed by von Neumann, and a second type introduced to explain nonlocal correlations in entangled systems. The former type might be most evident in statistical ensembles and not directly detectable beyond their modifications of the thermal properties of physical systems, while the latter can be investigated through tests of Bell's inequality with entangled systems, which we will discuss in Chapter 15.

In reading MFQM one is left with the impression that von Neumann wanted to put into writing his thoughts on quantum mechanics as he was leaving the field. The book is sometimes haphazardly written, and one is left wondering what might have happened if he had embraced the δ-function and fully developed the theory of distribution functions. By the time MFQM was written in 1932 the fundamental theory was firmly established; the open questions remained but were less acutely painful than they were to the slightly earlier generation of physicists who were faced with throwing out the comfort of absolute knowledge to be replaced by fundamental uncertainty.

Over the ensuing years, there have been numerous attempts to inject hidden variables into the theory, perhaps most famously by Bohm (Chapter 13). He suggested that his theory was able to violate von Neumann's "no hidden variable" proof because it contained hidden variables associated with both the observed system and the measuring instrument, while von Neumann only considered hidden variables connected to the observed system. In fact, the de Broglie-Bohm theory is only a hidden variable theory in waiting. In a letter to Bohm, Pauli stated

> I do not see any longer the possibility of any logical contradiction as long as your results agree completely with those of the usual wave mechanics and as long as no means is given to measure the values of your hidden parameters both in the measuring apparatus and in

[20] Pauli, W. "Remarques sur le problem des parametres cache dans la mecanique quantiqe et sur la theorie de l'onde pilote," (Remarks on thje hidden paramter problem in quantum mechanics and the theory of the pilot wave), in *Louis de Broglie, Physicien et Penseur* (Paris: Michel, 1953). See also CERN-ARCH-PMC-06-071 at the Pauli Archives, http://cds.cern.ch/record/96311?ln=sv. and Pauli, W. to Fierz, M., Jan. 6, 1952 in Meyenn, K. von, 'Scientific Correspondence with Bohr, Einstein, Heisenberg, a.o. Volume IV, Part I: 1950-1952,' (Springer, 1996).

the observed system. As far as the whole matter stands now, your "extra wave-mechanical predictions" are still a check, which cannot be cashed.[21]

Until we can produce a state where the hidden variables (particle positions) are better known than is allowed by the quantum state, the theory has no consequences. The very act of producing such a state would prove that quantum mechanics was in disagreement with experiment, which is what von Neumann showed would have to be the case if hidden variables were to exist. It is actually not a deterministic theory as claimed, as probability is introduced at the outset, where Bohm has to assume that the initial conditions on the hidden variables are randomly distributed according to $|\psi|^2$. Nevertheless, as Bell stated, the fact that "... the subjectivity of the orthodox version, the necessary reference to the 'observer' could be eliminated"[22] is a very attractive feature of the theory. We will return to these questions in later chapters, and in particular the Einstein-Podolsky-Rosen paradox as it relates to the nonlocality of measurements and the collapse of the wave function.

[21]Letter [1313], W. Pauli to D. Bohm, in Wolfgang Pauli, *Scientific Correspondence with Bohr, Einstein, Heisenberg, a.o.*, Vol. IV, Part I: 1950-1953, ed. Karl von Meyenn (Berlin: Springer Verlag 1996).

[22]J. S. Bell, *On the impossible pilot wave*, *Found. Phys.* **12**, 989 (1982).

15

Einstein and Schrödinger renew the assault on quantum mechanics

15.1 Introduction

After de Broglie's seeming capitulation and Bohr's apparently successful countering of Einstein's arguments at the Solvay conferences of 1927 and 1930, the "Copenhagen interpretation" of Bohr, Heisenberg, and Born was accepted by the majority of working physicists. Interest largely turned away from fundamental questions of the nonrelativistic theory to applications such as the physics of atomic nuclei, electrons in metals, chemical forces, cosmic rays, beta decay, and the problems of extending Dirac's quantum electrodynamics to a full theory. However, Einstein was not satisfied and continued to work on the problem of whether quantum mechanics was indeed complete.

In 1931, with Tolman and Podolsky, he published a short letter in *Physical Review*.[1] The issue was whether the uncertainty principle applied to the past, since certain authors had erroneously suggested that it did not. We consider the case of a box containing a photon gas that has two shutters, allowing the simultaneous release of two photons (Fig. 15.1).

The authors assumed that the momentum of the photon going directly to the observer at O (and hence its energy) could be exactly measured, along with the time of arrival at O. This would allow the determination of the time of opening of the two slits. Assuming the box is weighed before and after the release of the two photons, the

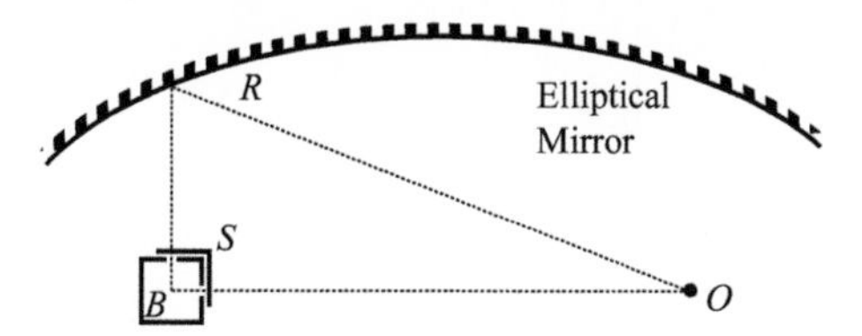

Fig. 15.1 Experiment discussed by Einstein, Tolman, and Podolsky. B is a box that emits one photon through each of the two slits S, which open simultaneously for an instant. Both photons eventually end up at the location of the observer O, one directly, and one after reflecting at R.

[1]Einstein, A., Tolman, R. C. and Podolsky, B., *Knowledge of past and future in quantum mechanics*, Phys. Rev. **37**, 780 (1931).

The Historical and Physical Foundations of Quantum Mechanics. Robert Golub and Steven K. Lamoreaux, Oxford University Press.
 DOI: 10.1093/oso/9780198822189.003.0015

total energy of the photons would be known and hence the energy of the photon which took the much longer path SRO, chosen to be long enough to allow the completion of the second weighing before the arrival of the photon at O. The energy and arrival time of this second photon at O could then be precisely predicted, in violation of the energy-time uncertainty principle. Thus, our assumption that the momentum of the first photon could be exactly measured must be incorrect. Indeed, any measurement of the momentum will change the momentum in a way that leaves an uncertainty in the exact time this change took place, which means that the opening time of the shutters, calculated using the measured arrival time of the first photon, will have a corresponding uncertainty, preventing a precise prediction of the time of arrival of the second photon. This shows that a measurement of the momentum of a particle in the past must also be restricted by the uncertainty principle.

What we see here is Einstein continuing to think about what came to be called the photon box in ways that avoided Bohr's objection that weighing the box will alter the reading of a clock attached to it. Here the argument showed the consistency of quantum mechanics. We know of Einstein's further thinking from letters and from the testimony of colleagues with whom he discussed these issues.

In 1932 Ehrenfest published a paper with the title *Basic questions about quantum mechanics*[2] in which he asked a number of questions that he felt had not been adequately addressed by the physics community. Among these was the fact that (as we have previously discussed) the Schrödinger equation for N particles is written in a $3N$-dimensional space. As a partial differential equation, the Schrödinger equation tells us that the value of the wave function at a single point is influenced by infinitesimally close neighboring points (in $3N$-dimensional space). But if, for simplicity, we take the case of two particles with one degree of freedom each, say (x_1, x_2), then an infinitesimal neighborhood in the multiparticle space can consist of points corresponding to the particles being very far away from each other, say $x_2 \gg x_1$, as seen in Fig. 15.2.

Pauli published a reply[3] to Ehrenfest's questions. He pointed out that the use of $6N$-dimensional phase space is quite common in classical mechanics. However, that applies only to systems of particles; classical wave equations are always applied in

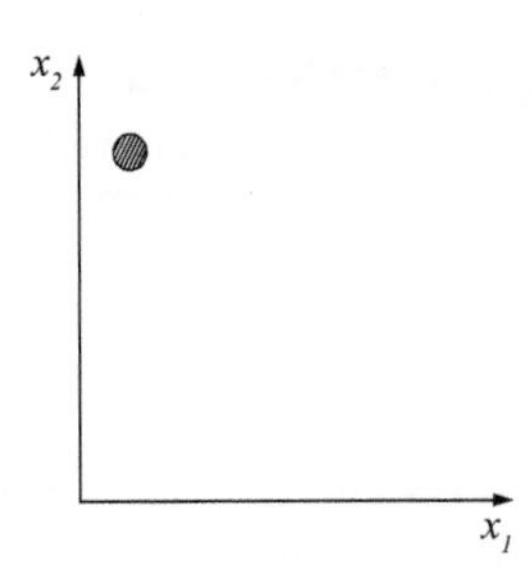

Fig. 15.2 Points corresponding to two widely separated particles are close together in 3N dimensional space, here two particles with one degree of freedom each.

[2]Ehrenfest, P. *Einige die Quantenmechanik betreffende Erkundigungsfragen* [Some Inquiry Questions Concerning Quantum Mechanics], Zeit. Phys. 78. 555, (1932), received 16 Aug., 1932.

[3]Pauli, W. *Basic questions about quantum mechanics*, Zeit. Phys. 80, 573, (1933), received 17 Dec., 1932.

three dimensions, so the question of nonlocality does not arise there. In fact, the $3N$-dimensional space does imply nonlocal correlations.

At one point, to compensate for the perturbing of the clock by the weighing procedure, Einstein considered that after the box was weighed and the photon emitted the experimenter could choose either to reweigh the box—determining the photon's energy—or to read the clock and compare the result with a standard clock so as to compensate for the gravitational shift and predict the time of arrival of the photon after it was reflected by a fixed mirror at a known position. Thus, without interfering in any way with the photon, we can make accurate predictions of either its moment of arrival or its energy. Since the state of the photon is unchanged by the choice of measurement, it would seem that the state of the photon would have to contain information about both quantities. However, according to quantum mechanics, the specification of a state cannot involve an accurate specification of both the energy and its time of arrival at a given point. This would imply that quantum mechanics does not offer an adequate description.[4] Here we see the germ of the argument that would be further developed.

In 1933 Einstein attended a lecture by Rosenfeld on quantum field theory. Rosenfeld reported that in discussions after the lecture Einstein considered the case of two particles approaching each other and interacting for a short time. Now an observer could measure (far from the interaction region) the momentum of one of the particles and would then be able to deduce the momentum of the other particle. If instead the observer measured the position of the particle, he would know the position of the second particle. This is a correct deduction from quantum mechanics, but it seems paradoxical that the state of the second particle can be influenced by a measurement performed on the first at a time when there was no interaction between the particles. We see here that the photon box has been replaced by a second particle, and this is another example of the nonlocality in the theory that Einstein had long objected to, starting with his discussion of the collapse of the wave function at the 1927 Solvay conference.

This argument would be elaborated in detail in a paper published in 1935. In 1945 Einstein wrote to Epstein,[5] explaining that he had arrived at the ideas in the 1935 paper by means of the following thought experiment. To avoid the problems with gravitation, he now considers a photon box mounted on a frictionless horizontal rail. There is either an absorber or a reflector mounted on the end of the rail. A slit in the wall of the box is opened at a precisely known moment, after which the observer could either lock the box to the rail and measure the position of the box and hence predict the time of arrival of the photon at the end of the rail, or he could allow the box to recoil and measure the velocity (momentum) of the box using the Doppler effect with low frequency radiation (so as not to significantly alter the momentum of the box) and thus know the momentum of the photon. Thus either the energy or the time

[4]Jammer, M., *The Philosophy of Quantum Mechanics: The Interpretations of Quantum Mechanics in Historical Perspective* (New York: Wiley, 1974), p. 170.

[5]Einstein to Paul Epstein, 8 November 1945., https://www.christies.com/en/lot/lot-6210432, quoted in Howard, D., '"Nicht Sein Kann was Nicht Sein Darf," or the Prehistory of EPR, 1909-1935: Einstien's Early Worries About the Quantum Mechanics of Composite Systems' in Miller, A.I., ed. *Sixty-Two Years of Uncertainty: Historical, Philosophical, and Physical Inquiries into the Foundations of Quantum Mechanics*, Plenum Press, New York and London, p.61

of arrival of the photon can be predicted exactly, and both properties would have to be assigned to the photon since its properties cannot depend on a procedure carried out by a distant observer. The alternative—that a measurement on the box could physically alter the state of the distant photon—was unacceptable to Einstein since it would involve an action propagating with a velocity greater than c. Again Einstein is homing in on the nonlocality associated with the quantum theory in its present form.

15.2 Einstein attacks quantum theory

This was the headline of an article that appeared in the *New York Times* on May 4, 1935, Fig. 15.3.[6] The article itself is based on an upcoming feature to appear in *Science News Letter* and has a byline "Copyright 1935 by Science Service."[7] The article is a description of a paper by Einstein, Podolsky, and Rosen that was soon to be published in the May 15 issue of *Physical Review.*[8]

The *Times* article quotes Podolsky (who was likely the sole author of the press release), and gives a confused description of the physics being discussed in the paper.

Einstein reacted very strongly to this article, and a note from him appeared a few days later,[9] in which he states:

> Any information upon which the article "Einstein Attacks Quantum Theory" in your issue of May 6 is based was given to you without my authority. It is my invariable practice to discuss scientific matters only in the appropriate forum and I deprecate advance publication of any announcement in regard to such matters in the secular press.

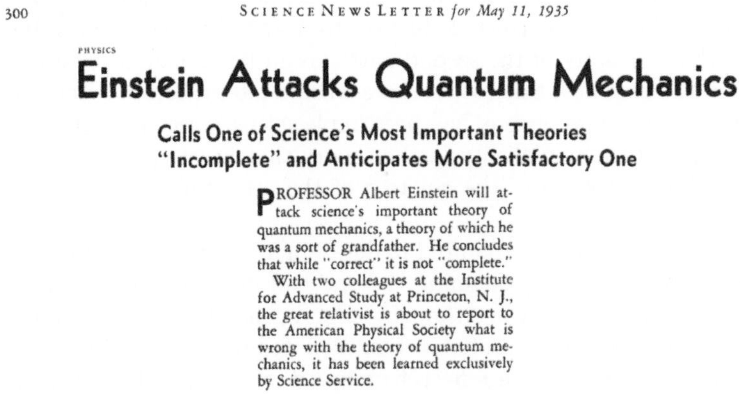

300 SCIENCE NEWS LETTER *for May 11, 1935*

PHYSICS

Einstein Attacks Quantum Mechanics

Calls One of Science's Most Important Theories "Incomplete" and Anticipates More Satisfactory One

PROFESSOR Albert Einstein will attack science's important theory of quantum mechanics, a theory of which he was a sort of grandfather. He concludes that while "correct" it is not "complete."

With two colleagues at the Institute for Advanced Study at Princeton, N. J., the great relativist is about to report to the American Physical Society what is wrong with the theory of quantum mechanics, it has been learned exclusively by Science Service.

Fig. 15.3 The introduction to the announcement of the EPR study (by Polodosky) that appeared on p. 300 of the May 11, 1935 edition of Science News Letter, and which was pre-published by the *New York Times.* ("Fair Use" per communication with Science News, Kathlene Collins, Chief Marketing Officer, Society for Science; www.societyforscience.org and www.sciencenews.org.)

[6] *New York Times*, May 4, 1935, Section BOOKS SOCIAL NEWS-ART-BOOKS, p 11.

[7] Available at https://www.sciencenews.org/wp-content/uploads/2010/11/EPR.pdf, and through Jstor https://www.jstor.org/stable/3911008?origin=JSTOR-pdf.

[8] Einstein, A., Podolsky, B. and Rosen , N., *Can quantum-mechanical description of physical reality be considered complete?*, Phys. Rev. 47, 777, (1935), received 15, May, 1935.

[9] A. Einstein, *New York Times* May 7, 1935 (Section BOOKS ART, page 21).

The stated purpose of the *Physical Review* article is to show that quantum mechanics in its current form is "incomplete." This is defined as follows. First an "element of reality" is defined as a property that can be predicted with certainty without disturbing the system in any way, and then a complete theory is one where every element of reality has a counterpart in the theory. Quantum mechanics is usually considered complete, as the information given by the wave function corresponds exactly to all the information that can be obtained through measurements on the system. Since every measurement produces an unavoidable (and uncontrollable) disturbance of the system, it changes the wave function and thus also the information available for future measurements.

However, it seems that information can be obtained about a system without making a measurement on the system itself. We consider two systems, I and II, which are known to have interacted between the times $t = 0$ and $t = T$, but do not interact at all for times $t > T$. The Schrödinger equation should apply to the state of the combined system for $t > T$. We can expand the eigenstate of the combined system in a series of complete, orthonormal eigenfunctions (depending on the coordinates of system I alone) of an operator $\mathbf{A}_I$ representing an observable A of system I. The eigenfunctions $u_n(x_1)$ satisfy $\mathbf{A}_I u_n(x_1) = a_n u_n(x_1)$, and we can write the wave function of the combined system as

$$\Psi(x_1, x_2) = \sum_n \psi_n(x_2)\, u_n(x_1), \tag{15.1}$$

where the expansion coefficients are functions of x_2 alone.

Now if we measure the quantity A and find the result a_k, we know that the system II must be in the state $\psi_k(x_2)$, following the general rules of quantum mechanics for a measurement. If instead we chose to measure a different observable B_I, we would expand the combined system wave function in the eigenfunctions $v_s(x_1)$ of $\mathbf{B}_I$ $(\mathbf{B}_I v_s(x_1) = b_s v_s(x_1))$:

$$\Psi(x_1, x_2) = \sum_s \phi_s(x_2)\, v_s(x_1). \tag{15.2}$$

If a measurement of B_I were to yield the result b_r, system II would be left in the state $\phi_r(x_2)$. Thus the system II can be left in states with two different wave functions, but because the systems are not interacting, no real change can take place in system II as a result of anything we do to system I. We are left with two different wave functions describing the same physical reality.

We can even choose to measure two noncommuting observables. As an example take

$$\Psi(x_1, x_2) = \int_{-\infty}^{\infty} e^{\frac{i}{\hbar}(x_1 - x_2 + x_o)P}\, dP \tag{15.3}$$

where x_o is a constant. Letting $\mathbf{A} = \mathbf{P}_1 = \frac{\hbar}{i}\frac{\partial}{\partial x_1}$ which has the eigenfunction $u_p(x_1) = \exp\left(\frac{i}{\hbar}Px_1\right)$, we see that after a measurement of P_1 that yields the result P_o, system II will be in the state

$$\psi_P(x_2) = e^{-\frac{i}{\hbar}(x_2 - x_o)P_o}, \tag{15.4}$$

that is, an eigenfunction of $\mathbf{P}_2 = \frac{\hbar}{i}\frac{\partial}{\partial x_2}$ with eigenvalue $-P_o$. We can instead choose B_I to be a coordinate of system I, say x_1, with eigenfunction $v_x(x_1) = \delta(x_1 - x)$

(eigenvalue x). Replacing the sum in (15.2) by an integral and using the orthogonality of the eigenfunctions $v_s(x_1)$, we find in this case

$$\phi_x(x_2) = \int_{-\infty}^{\infty} \Psi(x_1, x_2)\, v_x^*(x_1)\, dx_1 = \int_{-\infty}^{\infty} \left[\int_{-\infty}^{\infty} e^{\frac{i}{\hbar}(x_1 - x_2 + x_o)P} dP\right] \delta(x_1 - x)\, dx_1 \tag{15.5}$$

$$= \int_{-\infty}^{\infty} e^{\frac{i}{\hbar}(x - x_2 + x_o)P} dP = \frac{2\pi}{\hbar}\delta(x - x_2 + x_o). \tag{15.6}$$

It is thus possible for system II to be in an eigenfunction of one or the other noncommuting observable; it can have either a definite position or a definite momentum, depending on what the experimenter decides to measure on system I. Thus, both are elements of reality, but the final wave function of II contains an exact value only for either P_2 or x_2, so the wave function in each case is an incomplete description. Since the systems are not interacting when the measurements are made, anything done to I cannot have any effect on the state of II, so this system must be in the same physical state no matter what the experimenter has done to system I. The same state would then be described by two radically different wave functions. If we wanted to maintain that quantum mechanics was complete, we would have to allow that two noncommuting observables could have exact values at the same time. As this is explicitly forbidden by quantum mechanics, we have to conclude that quantum mechanics is incomplete.

The authors referred to another possibility, namely that their definition of reality was too restrictive. We could postulate that two or more physical quantities can only be simultaneously elements of reality when they can be simultaneously measured and predicted. In this case the quantities P and x in the above example are not simultaneously real, thus making

> the reality of P and x depend on the process of measurement carried out on the first system, which does not disturb the second system in any way. No reasonable definition of reality could be expected to permit this.

We will see that this rejected possibility is very similar to the ideas that Bohr used in his attempt to refute these arguments.

Einstein's comments in the *New York Times*, and later in correspondence with Schrödinger, indicates that he needed additional time to understand the result. Very soon after the publication of the EPR paper, Bohr commented that if one simply follows the established rules of quantum mechanics, the correct answer will be obtained,[10] while Schrödinger embraced the result and showed that there is a new, nonclassical feature of quantum mechanics and invented the term "entanglement" to describe the origin of the correlations between separated states.[11]

Note that the question of the experimental verification or invalidation of the quantum mechanical effects discussed above does not enter into consideration. The argument claims that even if the quantum predictions are correct, the theory cannot be considered as a complete description.

[10]Bohr, N., *Can quantum-mechanical description of physical reality be considered complete?*, Phys. Rev. **48**, 696, (1935), received 13 July, 1935.

[11]Schrödinger, E., *Discussion of probability relations between separated systems*, Proc. Cambridge Philos. Soc. **31**, 555, (1935), received 24 Aug, 1935.

15.2.1 Elaborations and modern representations of the EPR problem

Before continuing the study of the discussions following the EPR paper, is worth looking at several elaborations and modern interpretations of the EPR effect. A simpler system for study was introduced by David Bohm[12] and is based on two electrons combined into a spin singlet state, as

$$\Psi\rangle = \frac{1}{\sqrt{2}}\left[|+\rangle_a|-\rangle_b - |-\rangle_a|+\rangle_b\right], \tag{15.7}$$

where a, b label a particular electron, and $|+\rangle$, $|-\rangle$ refer to the eigenstates of the z component of the spin, s_z with eigenvalues $\pm 1/2$. (See Chapter 19 for a detailed discussion of spin). Bohm introduced this state as a *gedanken* experiment although it can be prepared with some difficulty using modern experimental techniques. Let us assume the two electrons move with the same velocity but in opposite directions. (Note that the exchange symmetry requirements for this state depend on the both the spatial and spin wave functions, and an antisymmetric state can be constructed.) Let us consider two observers called, conventionally, Alice (A) and Bob (B), who can analyze the spin direction with something like a Stern-Gerlach magnet which we could call a polarizer or an analyzer, depending on if it is used to prepare a state, or analyze a state (we will use the term polarizer for both uses; the use as an analyzer will be implied by context, and follows the terminology used in optics). This is because a Stern-Gerlach magnet sends the $+$ and $-$ states in different directions; suitably placed detectors will determine which spin state b was in. Similarly, in optics, we can separate horizontally and vertically photons into two different direction by use of a calcite crystal.

Let us consider a case where A measures particle (electron) a to be in the $+$ state, and does so with probability of 50%. B then will find the spin of b in the $-$ state every time A is found to be in the $+$ state (of course, we do not know which specific electron is a or b; these are labels that refer to the two directions). The question arises: How is this possible, when there is no communication between the detector systems? The two detectors could be separated by a light year, and if the detection is at the same time, B's undetermined state seems to instantly jump to a state of determined spin. Similarly, if A measures particle a to be in the $-$ state, B will measure $+$. It appears that A can instantly signal to B. The choice of A's quantization axis *appears* to determine the quantization axis of B.

This nonlocal correlation might suggest that information must be exchanged somehow, and there is a notion that such correlations could lead to faster-that-light communications, for example. If such were possible, then it would also be possible to send information into the past as shown in the following space-time diagram:

Let us suppose a device capable of sending an instantaneous superluminal signal exists in a laboratory (Fig. 15.4). Such a device could, for example, simultaneously turn on lights separated, say, by 1 m, along the $\pm x_a$ axis. We can set up such a device in any reference frame, and view the x_a axis from another comoving reference frame, as shown. Such a signal propagates parallel to the x_a (or x_b) axis so that t_a (or t_b) is constant in the frame that contains the signaling device. As shown in the figure,

[12]Bohm, D., *Quantum Theory*, Prentice-Hall, (1951), p.614.

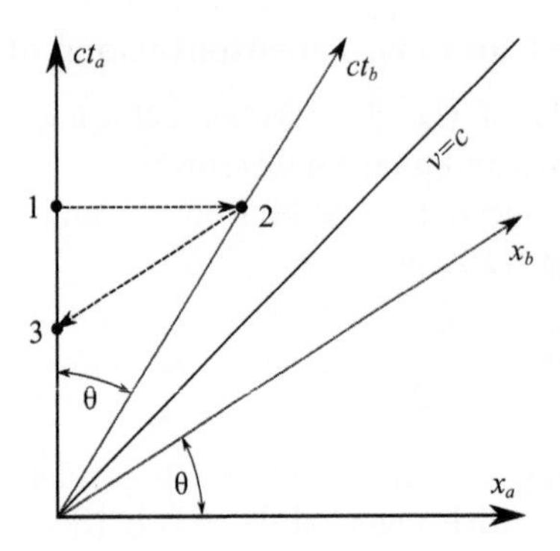

Fig. 15.4 An instantaneous superluminal signal is sent in the a frame parallel to the x_a axis (which could be indicated by instantly and simultaneously turning on light bulbs at intervals along the Arrow 1-2 axis). This signal can be observed in a comoving frame at velocity v, at the end of Arrow 1-2 (parallel to x_a). At this instant, using a similar superluminal device in the comoving frame, a second signal is sent by an observer in the b frame in a direction parallel to $-x_b$, along Arrow 2-3. This signal could be observed in the a frame, at the end of Arrow 2-3, corresponding to a time before the original signal was sent from the a frame.

by signaling between two reference frames, it is possible to send information into the past.

However, we have neglected two important points. First, the measurement process is passive, and not deterministic, in the sense that particular detector settings are presented to the quantum state to be measured, and if the state of the particle happens to be "correct" a detection will occur. A particle is detected at A with a particular setting of A's axis with a probability of 50%. *If* B is set to the same axis, a coincidence *will* occur. Such a detection at A together with the coincidence measurement at B occurs 50% of the time and is expected from the postulates of quantum mechanics—that the results of measurements are distinct eigenvalues of the measurement operator, and in the present case the state has equal probability to be in either measurement eigenstate. If the axes of the A and B detectors are not aligned, then the probability of detecting a coincidence is reduced, and becomes zero when they are perpendicular. The second point is now obvious: We cannot decide if A or B determined the measurement basis, but we can determine if they were parallel or perpendicular. Even if a delay was introduced for the particles traveling toward B so that the detection occurs at A, the ambiguity remains unresolved; we could delay A relative to B and the results of a post detection coincidence study will be the same.

As such, it is not possible to use the polarizers' settings to send superluminal signals between A and B. In principle, we could set up a system where electron pairs are produced at very specific times with specific velocities, so we know when each of a pair reach A or B respectively. If we know A and B are set to the same angle, we know that when A measures + then B will measure −. Let us now assume the controller of the A polarizer, Alice, tried to send signals to Bob who is watching the B polarizer. Alice has a scheme where she rotates her polarizer by 90° if she wants to send a binary bit "0" to Bob, and leaves it at 0° to indicate a binary "1." For the second 0° case, Bob observes + and − electrons with equal probability, and knows what Alice detected. For the first 90° case, Bob still observes + and − electrons with equal probability and there are no clues that Alice has changed her measurement basis. It is not until later when the specific events are compared that we can discover that the respective +, − measurements are highly correlated, uncorrelated, or highly anti-correlated, depending on the polarizers' angle settings.

Such a correlation might appear to have classical analogs, however, those analogs tend to fall short. For example, consider a pair of identical twins. Because the genetics between the two are highly correlated, if we observe that one of the twins has brown eyes, we can immediately conclude with high certainty that the other twin also has brown eyes; or if one has blue eyes, so does the other. This is an example of a hidden variable that was contained in the DNA of the twins, and we could have in principle elucidated the eye colors without doing any observations other than taking and analyzing a saliva sample from one of the twins. Knowledge of the distant twin's eye color does not allow us to transmit any information between the remote locations, and in that regard the analogy works.

In the case when quantum mechanics is valid, we have a choice of the measurement basis, and until the measurement is made, the state of each particle is undetermined, and for the singlet electron case we have discussed, measurements of individual particles in the a or b directions appear as completely undetermined random states. In the classical example of twins, the eye color was predetermined. We could say that the electron polarizations were predetermined, but then the entangled state Eq. (15.7) would be incorrect. Given the fact that the electrons alone appear as unpolarized means that this angle would serve as a type of hidden random variable that would need to be averaged over. We will discuss this in some detail in regard to photons.

For the quantum case, we apparently have no freedom to introduce hidden variables. They are not allowed based on von Neumann's proof that was discussed at length in Chapter 14. In addition, the degrees of freedom introduced by assuming hidden variables would alter quantum statistical mechanics and the exclusion principle. However it remains an experimental question whether hidden variables exist, and even if Eq. (15.7) exists or is correct for separated particles. We will discuss these questions later in this chapter in the context of a specific experimental scenario.

15.3 Reactions to the Einstein Podolsky Rosen (EPR) argument

Needless to say, the arguments presented elicited a strong reaction on all sides, with the discussion continuing to this day. As we will see in Section 15.5, 30 years later John Bell would propose a quantitative test of whether the correlations predicted by EPR-type arguments could be explained by assuming that the particles, after separation, could be assigned all the properties necessary to determine the outcome of all possible measurements. Subsequent measurements showed that this was not possible and that quantum mechanics contains a fundamental nonlocality. Ironically Einstein's work highlighted this nonlocality—the strangest property of quantum mechanics—which he at one stage referred to as "spooky action at a distance."

As this nonlocality, together with the question of wave function collapse, is the main feature of quantum mechanics which is responsible for all the difficulties with interpretation of the theory, it is very instructive to look at the immediate reactions of some of the main workers in the field.

15.3.1 Pauli

On June 15, 1935, Pauli wrote to Heisenberg,[13] calling the publication of EPR a "catastrophe" which was in "danger of producing a confusion of public opinion, namely in America." He suggested that Heisenberg write a rejoinder for publication in *Physical Review* ("with formulas"). He then explained the EPR argument as follows: A system made up of two subsystems (1,2) is in an eigenstate of $(p_1 + p_2)$ and $(x_2 - x_1)$. As these commute, they can have the same eigenstate:

$$\Psi(x_1, x_2) = e^{\frac{i}{\hbar}\frac{(x_1+x_2)P_o}{2}} \delta(x_2 - x_1 - x_o). \tag{15.8}$$

Such a state can be produced by having two particles simultaneously go through two slits positioned in a screen. If the screen is free to recoil, we can measure its momentum before and after the passage of the particles, thus knowing $(p_1 + p_2)$. $(x_1 - x_2)$ is fixed at a distinct moment in time by the distance between the slits.

We can choose to measure either p_2 or x_2 on system 2, leaving system 1 in a different eigenstate for each measurement. Now Einstein says that since measurements on 2 cannot affect system 1, there must be a reality of system 1 for itself, independent of what measurements are made on 2. It is absurd to assume that 1 changes due to measurements on 2—that it is transformed from one state to another—so the quantum mechanical description of 1 is correct but incomplete. A complete description of the state of 1 would contain all the properties that can be predicted with certainty after all possible measurements on 2.

Pauli went on to say that

> completely independent of Einstein we should give more emphasis, in a systematic foundation of quantum mechanics, to the forming and separation of composite systems than has been done up to now. As Einstein correctly understood, this is a very fundamental point in quantum mechanics which has a direct connection to your [Heisenberg's] considerations of the "cut" and the possibility of moving it to an arbitrary position. ...The idea is going around old men like Laue and Einstein that quantum mechanics is indeed correct, but incomplete. ...Perhaps you could, in an authoritative manner make it clear that an extension of quantum mechanics (through new concepts) is not possible without changing its contents.

He then said that these issues were "trivialities for us."

In a letter to Schrödinger a few weeks later (July 9, 1935),[14] Pauli referred to Schrödinger's framing of the hidden variable question in terms of an ensemble represented by a wave function (pure state). The question is whether all members of the ensemble are identical. In the case of hidden variables, the answer would be no: the ensemble members would be distinguished by the different values of the hidden variables. Pauli responded that if we had an eigenstate of momentum **P**, we cannot divide

[13] Meyenn, Karl von, *Wolfgang Pauli: Scientific Correspondence with Bohr, Einstein, Heisenberg a.o. Volume II: 1930-1939*, (Springer, 1985), p.402.

[14] loc. cit., p. 419.

the ensemble up into subensembles with different values of position $\mathbf{Q}$ without contradicting the results of quantum mechanics. In the case of mean values $\langle \mathbf{F} \rangle$ of quantities like $\mathbf{F} = \mathbf{F}_1(\mathbf{P}) + \mathbf{F}_2(\mathbf{Q})$ there would be no difference, but the statistical fluctuations $\left\langle (\mathbf{F} - \langle \mathbf{F} \rangle)^2 \right\rangle$ would be different in the case of (i) a mixture of ensembles each with a different value of Q and fixed $P = P_o$ or (ii) in the usual quantum mechanical total ensemble. Consider functions such as

$$\mathbf{F} = \mathbf{P}^2 + \omega^2 \mathbf{Q}^2 \tag{15.9}$$

$$\mathbf{J} = \mathbf{p}_x \mathbf{y} - \mathbf{p}_y \mathbf{x}. \tag{15.10}$$

They have a continuous distribution according to (i) but discrete values (the eigenvalues of the operators) according to (ii).

So it is not possible—as the conservative old men would like—to declare that the statistical predictions of quantum mechanics are correct and in spite of that to set a hidden causal mechanism behind it! In this sense it seems to me that the system of quantum mechanical laws is logically closed (complete in the sense of the axiomatics)—in contrast to the kinetic theory of gases. On the other hand there are the problems of a relativistic quantum theory (and there I am not such a strong believer in Saint PAM) [referring to P. A. M. Dirac].

15.3.2 Heisenberg

Heisenberg responded in a letter (July 2, 1935) in which he reported that Bohr was planning to publish a response to EPR in *Physical Review*. He gave Pauli's formulation a more physical basis by stating that the EPR experiment was identical to a screen with two slits separated by a distance a. Two photons pass through the screen, one through each slit. The distance between the slits fixes $(x_1 - x_2)$ and measuring the recoil of the screen would fix $(p_{x_1} + p_{x_2})$. The experimenter can choose to fix the screen or let it recoil and measure its momentum, thus controlling the wave function of the particles which are far from the screen. In the following weeks Heisenberg apparently wrote a manuscript which was both a response to Pauli's suggestion and to a paper that Schrödinger was about to publish. He sent a copy to Bohr (August 28, 1935), asking for comment and expressing the wish to submit it for publication in two weeks' time. The manuscript was never published but appears in Pauli's published correspondence as an appendix to the above letter.[15]

The conclusion of EPR was that, as quantum mechanics was incomplete, it was necessary to search for a more complete theory involving additional physical quantities. As a response to this statement, the title of Heisenberg's paper was "Is a deterministic extension of quantum mechanics possible?" As an example, he called attention to a decaying radium nucleus and asked if it could have (unknown until now) properties, in addition to those that fix its stationary state, whose knowledge would allow an exact prediction of the time of its decay.

As our observations are always described by classical quantities, the question arises whether quantum mechanics should only apply to atomic-scale systems and the measuring apparatus should be described by classical mechanics, or whether quantum

[15] W. Pauli, *Scientific Correspondence with Bohr, Einstein, Heisenberg, a.o.*, Vol II: 1930-1939, ed. by K. von Meyenn, A. Hermann and V. F. Weisskopf (Berlin: Springer Verlag, 1985), p. 409 ff.

mechanics should also be applied to the apparatus and only the observation of the apparatus be considered as a classically described process? Where should the cut between the wave function description (superposition of possible states) and the classical description be placed? That is, where does the collapse of the wave function take place? This question had previously been discussed by von Neumann, as we have seen. Within wide limits, the predictions of any experimental result are independent of the location of the cut. However, there are some detectors that cannot be described classically, as is the case with the detection of neutrons by induced radioactivity. In that case, there is a chain of events: neutron capture, followed by decay of the detector nucleus, followed by detection of a decay product in a counter or photographic plate. So the cut cannot be moved arbitrarily in the direction of quantum mechanical validity. The wave function of the quantum system always evolves continuously and deterministically as predicted by the Schrödinger equation. After the cut, the system evolves deterministically according to classical physics. The random element (probabilities) is only introduced at the cut. With every observation there is (an only partially controllable) perturbation of the system.

Now let us assume that a quantum system where quantum mechanics allows the prediction of the probabilities of a set of possible outcomes also possesses unknown properties (hidden variables) whose knowledge would allow exact predictions of the outcomes, independent of how the outcomes are observed. The motion of the system both before and after the cut is deterministic (the Schrödinger equation before the cut and classical physics after the cut), and probabilistic behavior can only occur at the cut. This means that the hidden variables can only act at the cut, which cannot be physically fixed at a definite point in the measurement chain. This free choice in the position of the cut is decisive for quantum mechanics. Consider again the process of α decay. The energy is distributed over a width $\sim \hbar/\tau$ (τ is the average decay time) which is relatively narrow, so that decay α's could be diffracted by a lattice in sharply defined directions (diffraction peaks). Each α is detected at a specific point, and the wave function would collapse at the point of detection. If there were hidden variables that prescribed the direction of the individual α's, we would be able to predict which part of the lattice each α would hit and the diffraction pattern would be washed out. If we were to simply detect the decay α's, the wave function would have to collapse at the time of emission, whereas if the particles are diffracted by a suitable lattice, the collapse can only take place after the diffraction, at the specific point of detection. Thus it is not possible to have a hidden variable theory where the action of the hidden variables is independent of how the observations are made. The argument can be applied to the more accessible case of an atom emitting light, for instance, the sodium D line. Note the similarities to the discussion between Dirac and Heisenberg at the 1927 Solvay conference.

15.3.3 Bohr

According to reports by Rosenfeld, Bohr was shocked by the EPR paper and initiated a period of intense activity with the goal of understanding and rebutting its arguments. This lasted for a few months and on July 13, 1935 (almost exactly four months after EPR was published), the *Physical Review* received a paper[16] containing Bohr's thinking related to the issues raised by EPR.

Bohr began by restating EPR's reality criterion, which had been applied to different variables, as they could each be predicted on the basis of relevant measurements. However, in quantum mechanics it is never possible to give definite values to canonically conjugate variables. Bohr went on to mention that the impossibility of controlling the interaction between an object and the measuring instrument due to the quantum of action leads to the necessity of a

> final renunciation of the classical ideal of causality and a radical revision of our attitude to the problem of physical reality. ...a criterion of reality like that proposed by the named authors (EPR) contains...an essential ambiguity when it is applied to the actual problems with which we are here concerned.

Here Bohr seems to be embracing the limited criterion of reality rejected by EPR, namely that only quantities that are simultaneously predictable are simultaneously real, in spite of EPR's statement that "no reasonable definition of reality would permit this."

Bohr then went on to give the following formal derivation of the EPR effect. We have two systems defined by (q_1, p_1) and (q_2, p_2) which satisfy the commutation relations

$$[\mathbf{q}_1, \mathbf{p}_1] = [\mathbf{q}_2, \mathbf{p}_2] = i\hbar \tag{15.11}$$

with all other commutators equal to zero. We can replace these by new canonical variables (Q_1, P_1), (Q_2, P_2) obtained by a transformation

$$\mathbf{q}_1 = \mathbf{Q}_1 \cos\theta - \mathbf{Q}_2 \sin\theta \qquad \mathbf{p}_1 = \mathbf{P}_1 \cos\theta - \mathbf{P}_2 \sin\theta \tag{15.12}$$

$$\mathbf{q}_2 = \mathbf{Q}_1 \sin\theta + \mathbf{Q}_2 \cos\theta \qquad \mathbf{p}_2 = \mathbf{P}_1 \sin\theta + \mathbf{P}_2 \cos\theta. \tag{15.13}$$

The new coordinates will satisfy commutation relations of the form $[\mathbf{Q}_2, \mathbf{P}_2] = i\hbar$, $[\mathbf{Q}_1, \mathbf{P}_2] = 0$, so that Q_1 and P_2 can have exact values simultaneously. Since

$$\mathbf{Q}_1 = \mathbf{q}_1 \cos\theta + \mathbf{q}_2 \sin\theta \tag{15.14}$$

$$\mathbf{P}_2 = -\mathbf{p}_1 \sin\theta + \mathbf{p}_2 \cos\theta\,, \tag{15.15}$$

we see that with fixed (Q_1, P_2), a measurement of either q_2 or p_2 will allow the prediction of q_1 or p_1, respectively. After explaining the two-slit analogy to the EPR experiment, which corresponds to $\theta = -\pi/4$ in the above equations, Bohr pointed out that even though there is a completely free choice of which quantity (q_2, p_2) to measure by a process which does not directly interfere with system 1, we are really "concerned with a *discrimination between different experimental procedures which allow of the unambiguous use of complementary classical concepts.*" (Italics in original.) If we want to measure the position of one particle, the screen with the two slits has to be rigidly fixed to a support which defines the frame of reference, which means all information on the momentum is lost because an uncontrollable amount of momentum can be absorbed by the support. If, on the other hand, we choose to measure the momentum of

[16] Bohr, N., *Can quantum mechanical description of reality be considered complete*, Phys. Rev., 48, 696 (1935), received 13 July, 1935.

the screen, there will be an inevitable uncontrollable displacement which will make it impossible for us to deduce the position of the screen from the measurement on the particle and we will not be able to predict the position of the second particle. Now we see that the EPR criterion of reality concerns an ambiguity regarding the meaning of the expression "without in any way disturbing a system." Of course there is no mechanical disturbance. But "there is essentially the question of *an influence on the very conditions which define the possible types of predictions regarding the future behavior of the system.*" (Italics in original.) "Since these conditions constitute an inherent element of the description of any phenomenon to which the term 'physical reality' can be properly attached," the EPR argument does not justify the conclusion that quantum mechanics is essentially incomplete.

Bohr's argument seems to be that we must consider the composite system as a whole, and that when we choose to measure either position or momentum, we are in fact changing the whole system so there is nothing in the fact that the wave function of the undisturbed system is changed that can challenge the consistency of the theory. Bohr is not denying that the correlations pointed out by EPR exist, that you can "drive" the second system into different quantum states without interacting with it. He is arguing that this does not mean that quantum mechanics is incomplete.

Both Bohr and EPR agreed that the results being discussed were predicted by quantum mechanics and the question of experimental verification did not arise because the issue ultimately was whether an extension of quantum mechanics was necessary or not.

Today the essential point seems to be that EPR-type effects do indeed exist and that quantum mechanics is a nonlocal theory. We will see how Bell derived a quantitative test which allowed the experimental confirmation of this feature of the theory. Whereas EPR wanted to justify continuing the search for a "more complete" (local, hidden variable) theory, the ultimate result of their work has been to show that, in fact, a local theory can never reproduce the experimentally observed correlations.

15.3.4 Schrödinger

The publication of EPR stimulated Schrödinger to investigate the same problem. In the next few months he published a set of papers[17] in which he analyzed the problem of separated systems in an attempt to highlight the aspects of the theory that he found most disconcerting. He began by showing that an expansion of the form (15.1) where both sets of functions are an orthogonal set is always possible. He then called composite wave functions of this form "entangled" and called the act of collapsing the wave function by making a measurement on one subsystem "disentanglement." Any further measurement on either system will have no effect on the distant system. He then proceeded to further pick apart the properties of the entangled state and showed in another way that the quantum mechanical result precludes the experimental results being determined by pre-ordained "objectively real" quantities.

[17] Schrödinger, E. *Discussion of probability relations between separated systems*, Proc. Camb. Philos. Soc. **31**, 555 (1935), received 8/24/35) and *Probability relations between separated systems*, Proc. Camb. Philos. Soc.; **32**, 446 (1936) received 21 April, 1936

Turning his attention to the composite system discussed above, i.e., two particles going through two slits and hence being in a simultaneous eigenstate (Ψ) of $P = p_1 + p_2$ and $X = x_1 - x_2$, so that

$$P\Psi = p'\Psi \text{ and } X\Psi = x'\Psi, \tag{15.16}$$

he shows that it is possible to measure any observable on system 1 by making measurements on system 2. Taking a Hermitian operator $F(x_1, p_1)$ represented by an analytic function of the operators (x_1, p_1), he shows that its value is given by $F(x_2 + x', p' - p_2)$. This follows from (15.16):

$$F(x_2 + x', p' - p_2)\,\Psi = F(x_1, p_1)\,\Psi\,. \tag{15.17}$$

Suppose system 1 has fixed values

$$x_1 = x_1',\ p_1 = p_1'. \tag{15.18}$$

Then we might think that $F(x_1, p_1) = F(x_1', p_1')$, but this is not possible. Take as an example

$$F(x_1, p_1) = \frac{p_1^2}{b} + bx_1^2 = \frac{p_1^2}{m\omega_o} + m\omega_o x_1^2 = \frac{1}{\omega_o}\left[\frac{p_1^2}{m} + m\omega_o^2 x_1^2\right] = (2n+1)\,\hbar, \tag{15.19}$$

where we put $b = m\omega_o$ for clarity and made use of the fact that measuring the operator in square brackets will give one of its eigenvalues. This means that

$$\left[\frac{p_1^2}{m} + m\omega_o^2 x_1^2\right]\psi_n = (2n+1)\,\hbar\,\omega_o\psi_n\,, \tag{15.20}$$

which cannot be the result of substituting the values (15.18) into $F(x_1, p_1)$ for every b. We are thus faced with the situation where the "one to one correspondence between the answers of the two systems necessarily extends to *all* pairs of observables," but this is accompanied by the bewildering "complete lack of insight into the relationship between the different answers in *one* system."

Probing deeper into the situation, Schrödinger goes on to consider the wave function for the composite system

$$\Psi(q_1, q_2) = \sum_n c_n\, g_n(q_1)\, f_n(q_2) \tag{15.21}$$

where f_n, g_n are systems of orthogonal eigenstates of the Hermitian operators f, g. We will work in the modern notation where (15.21) is written in terms of abstract states as

$$|q_1, q_2\rangle = \sum_n c_n\,|g_n\rangle_1\,|f_n\rangle_2 \tag{15.22}$$

If we were to measure the observable f on system 2 with the result f_i then system 1 would be in the state $|g_i\rangle_1$. Now if we choose to measure another quantity, say h, on system 2 we have to rewrite (15.22) in terms of its eigenstates,

$$|f_n\rangle_2 = \sum_i |h_i\rangle_2 \langle h_i | f_n\rangle_2 \tag{15.23}$$

This is a unitary transformation, so it can be inverted for any unitary matrix $\langle h_i | f_n\rangle_2$,

$$|h_n\rangle_2 = \sum_i |f_i\rangle_2 [\langle h_n | f_i\rangle_2]^{-1} = \sum_i |f_i\rangle_2 \langle f_i | h_n\rangle_2, \tag{15.24}$$

thus providing a set $|h_n\rangle_2$ corresponding to any arbitrary set of parameters $\langle h_i | f_n\rangle_{2i,n}$ (subject to the unitarity constraint).

Then (15.22) becomes

$$|q_1, q_2\rangle = \sum_i \left[\sum_n \frac{c_n \langle h_i | f_n\rangle_2}{\sqrt{w_i}} |g_n\rangle_1 \right] |h_i\rangle_2, \tag{15.25}$$

where we introduced $w_i = \sum_n |c_n|^2 |\langle h_i | f_n\rangle_2|^2$ so that the state in square brackets is normalized. It is the state into which we would drive system 1 if we found system 2 in the state $|h_i\rangle_2$. The probability of this is given by w_i. According to (15.23) we can choose the arbitrary coefficients $\langle h_i | f_n\rangle_2$ by suitably choosing the observable h that we will measure on system 2, which will leave system 1 in the state

$$|\psi\rangle_1 = \sum_n \frac{c_n \langle h_i | f_n\rangle_2}{\sqrt{w_i}} |g_n\rangle_1. \tag{15.26}$$

Since the $|g_n\rangle_1$ are a complete set, we see that we can effectively drive system 1 into any state we wish with a probability w_i solely by acting on the widely separated system 2. Schrödinger described his results as "these conclusions, unavoidable within the present theory, but repugnant to some physicists including the author."

He noted that if the original state was a mixture, that is, the coefficients c_n in (15.22) had randomly fluctuating phases, we would not be able to produce a coherent superposition in (15.26) and the resulting state of system 1 would also be a mixture.

15.3.5 Furry

W. H. Furry, a professor at Harvard University, also published an analysis of the EPR system.[18] He took the argument one step further by considering the case where, in addition to measuring the quantity h on system 2, we also measure the quantity μ on system 1. To handle this case we expand

$$|g_n\rangle_1 = \sum_j |\mu_j\rangle_1 \langle \mu_j | g_n\rangle_1 \tag{15.27}$$

[18]Furry, W.H, *Note on the Quantum-Mechanical Theory of Measurement*, Phys. Rev. 49, 393, (1936) received 12 November, 1935.

and substitute this into (15.25):

$$|q_1, q_2\rangle = \sum_{i,j} \left[\sum_n \frac{c_n \langle h_i | f_n \rangle_2}{\sqrt{w_i}} \langle \mu_j | g_n \rangle_1 \right] |h_i\rangle_2 |\mu_j\rangle_1. \tag{15.28}$$

Then the probability that a measurement of μ on system 1 would give μ_j and a measurement of h on system 2 would give h_i is the square of the amplitude of the corresponding eigenstate:

$$\left| \sum_n \frac{c_n \langle h_i | f_n \rangle_2}{\sqrt{w_i}} \langle \mu_j | g_n \rangle_1 \right|^2. \tag{15.29}$$

Furry compared this result with what would be expected if, after separation of the two systems, they possessed independently real properties. That is, if during their interaction each system made a random transition to a definite state, the state (15.22) would go over to a mixture of states $|g_n\rangle_1 |f_n\rangle_2$ with probabilities $|c_n|^2$. This means that each member of an ensemble would be in a definite state. The states of the two systems would remain correlated. In this case the result is obtained by assigning random phases to the c_n, so that (15.29) would be replaced by

$$\sum_n \left| \frac{c_n \langle h_i | f_n \rangle_2}{\sqrt{w_i}} \langle \mu_j | g_n \rangle_1 \right|^2 = \sum_n \frac{|c_n|^2 |\langle h_i | f_n \rangle_2|^2}{w_i} |\langle \mu_j | g_n \rangle_1|^2. \tag{15.30}$$

According to quantum mechanics, system 1 is in the pure superposition state (15.26) after the measurement on system 2. This is incompatible with the idea that each member of the ensemble is in a definite state and shows that it is untenable to consider that only the knowledge of the state of system 1 is affected by measurements made on system 2.

Furry summed up his results as follows:

> We have seen that the assumption that a system when free from mechanical interference necessarily has independently real properties is contradicted by quantum mechanics...a system and the means used to observe it are to be regarded as related in a more subtle and intimate way than was assumed in classical theory.

In a letter to the editor of *Physical Review* written a few months later (March 2, 1936),[19] Furry wrote in regard to the measurement of position in the EPR experiment: "No matter how far apart the particles are when we try to collect one of them, the relative probabilities of finding it in different places are strongly affected by the 'interference term' in the cross section; it is not really free." Thus he is emphasizing the role that interference plays in the EPR effect.

15.3.6 Schrödinger's cat

Schrödinger continued his analysis of quantum mechanics in another series of papers[20] in which he examined the measurement problem and the EPR effect, graphi-

[19] Furry, W.H., *Remarks on Measurement in Quantum Theory*, Phys. Rev. 49. 476, 1936, received March 2, 1936.

[20] Schrödinger, E., *Die gegenwärtige Situation in der Quantenmechanik* [The Present Status of Quantum Mechanics], Naturwissenschaften **23**(48), 807 (1935); **23**(49), 823 (1935); **23**(50), 844 (1935) [English translation in Wheeler, J. A. and Zurek, W. H. eds., *Quantum Theory and Measurement* (Princeton: Princeton University Press, 1983), p.152].

cally illustrating the measurement problem (wave function collapse) by contemplating the possibility of putting a cat into a superposition state of being dead or alive.

He considered various interpretations and concluded that none of them were adequate. As we have seen, the quantum mechanical prescription is that if we want to predict the results of a measurement of a given quantity A, we expand the state $|\Psi\rangle$ of the system in eigenstates of A ($A|a_n\rangle = a_n|a_n\rangle$):

$$|\Psi\rangle = \sum_n |a_n\rangle \langle a_n |\Psi\rangle. \tag{15.31}$$

Then the average value of A is given by the expectation value

$$\langle\Psi| A |\Psi\rangle = \sum_n |\langle a_n |\Psi\rangle|^2 a_n. \tag{15.32}$$

An individual measurement will give one of the a_n with probability $\mathrm{P}_n = |\langle a_n |\Psi\rangle|^2$, so this is compatible with an ensemble model where each member of the ensemble is in one of the states $|a_n\rangle$. However, the ensemble model fails in cases where interference effects are important, as in the EPR results. We have seen that if each member of an ensemble had a definite value of p_1, q_1, we would predict values of (15.19) in disagreement with the predictions of quantum mechanics. Then the example of the cat shows that the quantum mechanical prediction that each member of the ensemble is in a superposition state leads to absurdities. Schrödinger presented various other arguments in an attempt to find a viable interpretation.

Schrödinger presented another argument concerning the question of whether the probabilistic predictions of quantum mechanics can be seen as applying to an ensemble of systems, each with a definite value of the "smeared out" variable. Consider a harmonic oscillator in an energy eigenstate, say with energy $E_1 = \frac{3}{2}\hbar\omega_o$. Then if we think of this state as representing an ensemble of systems, each with a definite value of the coordinate q, the values of q must have a well-defined maximum given by $V(q_o) = E_1$. But this is not the case: arbitrarily large values of q are allowed by the wave function and have a real physical importance as shown by the theory of α–decay. On the other hand, a decaying nucleus must be in such a smeared out state that neither the time of decay or the direction of the emitted α particle are fixed. We have no problem with that so long as it applies inside the nucleus. The emitted particle will be represented by a spherical wave centered on the nucleus and would totally cover a surrounding photographic detector. But the detector does not show a uniform illumination but short, spatially well-defined pulses, now here, now there. If the decaying nuclei are in a cloud chamber, we see the ion pairs distributed along lines corresponding to trajectories emanating from each decaying nucleus. Schrödinger referred to a paper by Lise Meitner which showed cloud chamber tracks of protons produced by bombarding aluminum with α particles in an (α, p) reaction (Fig. 15.5). No spherical waves are to be seen.

Schrödinger then presented his most famous paradox:

We can also construct completely burlesque cases. Think of a cat locked in a closed box (a "hell machine") which also contains a radioactive source and a counter. The counter is

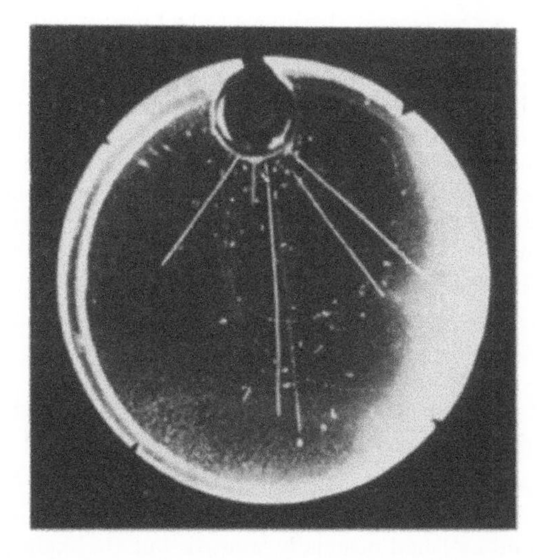

Fig. 15.5 Cloud chamber tracks produced by protons. (Reprinted with permission from Springer Nature Customer Service Centre GmbH: Springer Nature, Science of Nature.) (L. Meitner, *Atomkern und periodisches System der Elemente*, Naturwissenschaften 22, 733 (1934).)

connected to a relay which, in the case the counter is triggered by a decay product, moves a hammer which breaks a glass containing cyanide and kills the cat. The source is adjusted so that in one hour there is a 50% chance that one decay takes place. At the end of one hour the source will be in a superposition state of having one or zero decayed nuclei, and the cat will be in a superposition state of being dead or alive. These cases have the common property that an original indeterminacy (superposition) valid on the atomic level is transformed into an uncertainty on a large scale which is decided (state collapse) by a direct observation. This hinders us from accepting a "washed out" model as a picture of reality. The difference between the two views is like the difference between a photograph that has been blurred because the camera is out of focus or a picture of a cloud.

This is the problem that led to the introduction of the *cut* by von Neumann and Heisenberg.

Concerning the measurement process, Schrödinger wrote:

By every measurement we must assign to ψ a sudden change, which depends on the result of the measurement and is not predictable. This has not the least to do with the regular propagation (Schrödinger equation) between two measurements. This abrupt change is the most interesting point in the whole theory. Exactly this requires a break with naive realism. On these grounds ψ cannot take the place of a model or a "real thing." Not because a "real thing" or a model cannot experience unforeseen abrupt changes, but because from the realistic standpoint, observation is a natural process like all others and cannot produce an interruption in natural processes. ...A variable has in general no definite value before I measure it, so that measurement does *not* mean finding the value that it *had.*

He also gave the following interpretation of entanglement:

The strange theory of measurement, the apparent jumping of the ψ functions and the contradictions of entanglement all result from the simple way in which the calculational machinery of quantum mechanics allows two separated systems to be brought together (theoretically) to form a single system. It seems to be predestined for that. When two systems interact their ψ functions do not interact but they immediately stop existing and a single ψ function, for the composite system, takes their place. At first this is the product of two single functions which, since each function depends on completely different variables than the other, is a function of all these variables, or may be said to "play" in a region of much higher dimensions than the single functions. As soon as the systems begin to work on each other the total function stops being a product and does not decay into factors, which can be attributed to each of the systems, even when the systems are again separated. So we have (until the entanglement is lifted by a real observation) only a single description of both systems in that region of higher dimensions.

15.3.7 Einstein

It seems that Einstein was not too pleased with the published form of the EPR argument. In June 1935 (three months after the publication of EPR) he wrote to Schrödinger:[21]

This [the EPR paper] was written after much discussion by Podolsky for reasons of language. It did not come out as good as I had originally wanted: but the main point was, so to say, slopped up by pedantry. ...We describe a composite system consisting of two subsystems A and B by means of its wave function ψ_{AB}. The description refers to a time in which the interaction has practically stopped working. This wave function can be expanded in the eigenfunction of two observables, or systems of observables, α of system A and β of system B. We can write

$$\psi_{AB} = \sum_{m,n} c_{mn}\psi_m(x_1)\varkappa_n(x_2) \tag{15.33}$$

When we make a measurement of α on A this reduces this to

$$\psi_B = \sum_n c_{mn}\varkappa_n(x_2) \tag{15.34}$$

This is the ψ function of the subsystem B in the case where I have made an α measurement on A.

Now, instead of the eigenfunctions of the observables α and β, I can do the expansion in terms of the eigenfunctions $\underline{\alpha}$ and β, where $\underline{\alpha}$ is a system of other commuting variables.

$$\psi_{AB} = \sum_{\underline{m},n} \underline{c_{mn}}\psi_{\underline{m}}(x_1)\varkappa_n(x_2) \tag{15.35}$$

A measurement of $\underline{\alpha}$ results in

$$\underline{\psi_B} = \sum_n \underline{c_{mn}}\varkappa_n(x_2) \tag{15.36}$$

The important point is that ψ_B and $\underline{\psi_B}$ are completely different. I claim that this difference nullifies the hypothesis that the ψ description can be in a one to one relation with reality (the real state). After the collision the real state of (AB) consists of the real state of A and the real state of B, both states not having anything to do with the other. The real state of B can now not depend on what measurements I undertake on A. But then there are two equally valid ψ_B corresponding to the same state of B, which contradicts the hypothesis of a one to one or complete description of the real state. Note: Whether the ψ_B and $\underline{\psi_B}$ are eigenfunctions of observables in B or $\underline{B}$ is irrelevant to me.

[21]Einstein to Schrödinger, 19, June, 1935, Albert Einstein Archives, Hebrew University of Jerusalem, AEA 22-47. See also Fine, A., *The Shaky Game*, University of Chicago Press, (1986), p.35 ff

Now I want to note that I do not believe we have to be content with an "incomplete" description of the real state, but that we should look for a complete description. (Emphasis in original.)

Thus Einstein presented the EPR argument. In later years he presented the argument many times with various changes.

In March 1936, a year after the appearance of the EPR publication, he published a rather long statement of his views on physics[22] which Jammer[23] called "Einstein's credo." Concerning quantum mechanics, he wrote:

There has hardly ever before been a theory which delivered a key to the understanding and calculation of so many different experimental facts as quantum mechanics. In spite of this I believe that it has the tendency to lead us into error in our search for a united foundation of physics. According to my opinion it is an incomplete representation of the real structure, even if it is the only possible theory that can build on the fundamental concepts of material points and forces (quantum corrections of classical mechanics). The incompleteness of the theory corresponds to the statistical character of the proposed laws.

If the ψ function (apart from special cases) only provides statistical predictions of measurable quantities, this is not only due to the fact that the procedure of measurement introduces unknown, only statistical elements (Bohr), but also because the ψ function does not at all describe the state of a single system. The Schrödinger equation determines the time variation of the system ensemble (with or without external influences).

He then went on to give a succinct statement of the EPR problem:

A system consists of two subsystems A and B which interact for only a limited time. After the interaction we can determine the physical state of A by the most complete possible set of measurements. Quantum mechanics then allows us to determine the ψ function for B from the ψ function of the entire system and the results of the measurements. However the result depends on which variables of A were measured (e.g., coordinates or momenta). Since there can only be one physical state of B after the interaction—which reasonably cannot be thought to depend on the measurements that I perform on the system A, separated from B—it follows that the ψ function is not unambiguously associated with the physical state. This association of several ψ functions to the same physical state shows again that the ψ function cannot be understood as the complete description of a physical state of a single system. The association of a ψ function with a statistical ensemble removes that difficulty.

However, we have seen above that interference effects due to the system being in a superposition state are crucial to the EPR effect, and this is in fact evidence that quantum mechanics does indeed apply to single systems.

Einstein then discussed the fact that a weak perturbation, causing relatively small changes in a wave function and correspondingly small changes in the statistical densities, can cause significant changes in a few of the single systems.

The changes in the single systems remain completely unexplained by such a viewpoint, they are even completely eliminated from discussion by the statistical viewpoint.

I now ask: Does any physicist really believe that we will never have any insight into these significant changes of single systems, their structure and their causal relationships, in spite of the fact that every single process is observable in such wonderful inventions as the Wilson cloud chamber and the Geiger counter. To believe this is indeed logically free of contradiction, but it is so contrary to my scientific instinct that I cannot give up seeking for a complete understanding.

[22]Einstein, A., *Physik und realität*, Journal of the Franklin Institute **221**, 313 (1936); Physics and Reality. English translation p. 349.

[23]Jammer, M., *The Philosophy of Quantum Mechanics.*, (Wiley, 1974).

In 1949 a two-volume collection of essays titled *Albert Einstein, Philosopher-Scientist* was published.[24] The essays discussed various aspects of Einstein's work, and several of them deprecated Einstein's doubts about quantum theory. This is the place where Niels Bohr gave his detailed account of the discussions with Einstein at the Solvay Conferences. The work closed with an essay by Einstein in which he replied to criticisms. In this essay he re-emphasized his feeling that quantum mechanics can only apply to an ensemble of systems. In relation to the EPR work he said the following:

Of the "orthodox" quantum theoreticians whose position I know, Niels Bohr's seems to me to come nearest to doing justice to the problem. Translated into my own way of putting it, he argues as follows:

If the partial systems form a total system which is described by its ψ function, ψ_{AB}, there is no reason why any mutually independent existence (state of reality) should be ascribed to the partial systems A and B viewed separately, *not even if the partial systems are spatially separated from each other at the particular time under consideration.* The assertion that, in this latter case, the real situation of B could not be (directly) influenced by any measurement taken on A is, therefore, within the framework of quantum theory, unfounded and (as the paradox shows) unacceptable.

By this way of looking at the matter it becomes evident that the paradox forces us to relinquish one of the following two assertions:

(1) the description by means of the ψ function is complete.

(2) the real states of spatially separated objects are independent of each other.

On the other hand, it is possible to adhere to (2), if one regards the ψ function as the description of a (statistical) ensemble of systems (and therefore relinquishes (1)). However this view blasts the framework of the "orthodox quantum theory."

15.3.8 Bohm

In 1951 David Bohm published a textbook on quantum mechanics[25] in which he gave a very detailed analysis of the measurement problem according to the consensus (Copenhagen) interpretation. He showed by an explicit calculation that, during the course of a measurement, the interaction between the object and the instrument would cause the phases of the individual terms in the superposition of states to become randomized, so that one could say that each member of the ensemble had a definite value of the observed quantity. However, it was still not possible to say which value would turn up in a particular case.

Bohm also introduced an alternative to the EPR experiment which was to have far-reaching consequences. He considered a molecule consisting of two spin-1/2 atoms in a state of total spin zero. If the molecule is broken up by a process that conserves angular momentum, the result will be the two atoms flying away from each other with zero total angular momentum. Defining the eigenstates of the z component of the spin by

$$\sigma_z(i)\,|\pm\rangle_{z,i} = \pm\,|\pm\rangle_{z,i} \tag{15.37}$$

where $i = 1, 2$ distinguishes the two atoms, the state with zero total angular momentum is given by (see Chapter 19)

$$|\Psi_o\rangle = \frac{1}{\sqrt{2}}\left[|+\rangle_{z,1}\,|-\rangle_{z,2} - |-\rangle_{z,1}\,|+\rangle_{z,2}\right]. \tag{15.38}$$

[24] Schilpp, P.A., *Albert Einstein, Philosopher-Scientist,* (Harper, 1949).

[25] Bohm, D., *Quantum Theory,* (Prentice-Hall, 1951).

Note that each state has a position-dependent part which is not indicated. The states $|\pm\rangle_1$ refer to the particle traveling in a given direction and the states $|\pm\rangle_2$ to the particle traveling in the opposite direction.

This can be rewritten in terms of the eigenstates of another component of the spin, say σ_x. We have

$$|\pm\rangle_x = \frac{1}{\sqrt{2}}\left[|+\rangle_z \pm |-\rangle_z\right] \tag{15.39}$$

so that

$$|\Psi_o\rangle = \frac{1}{\sqrt{2}}\Big[|+\rangle_{x,1}\,|-\rangle_{x,2} - |-\rangle_{x,1}\,|+\rangle_{x,2}\Big]. \tag{15.40}$$

Analogous to the EPR proposal, if we were to measure $\sigma_{z,2}$, the state (15.38) would be forced into a state where $\sigma_{z,1}$ had a definite value opposite to that which was measured for $\sigma_{z,2}$. On the other hand, if we were to measure $\sigma_{x,2}$, the form (15.40) indicates that $\sigma_{x,1}$ would have a definite value. Thus as in EPR the choice of measurement made on atom 2 changes the state of atom 1, but since atom 1 cannot be influenced by anything done to atom 2, we conclude that the same physical state of 1 is associated with two different quantum states. Note Bohm rejects EPR's assumption that "the world can correctly be analyzed into elements of reality, each of which is a counterpart of a precisely defined mathematical quantity appearing in a complete theory." Instead quantum mechanics assumes that physical properties exist "only in an imprecisely defined form. ...only as potentialities, which are more definitely realized in interaction with an appropriate classical system such as a measuring apparatus."

For noncommuting observables,

> neither exists in a given system in a *precisely* defined form, but that both exist together in a roughly defined form, such that the uncertainty principle is not violated. Either variable is potentially capable of becoming better defined at the expense of the degree of definition of the other, in interaction with a suitable measuring apparatus. ...in a very accurate description they cannot be regarded as belonging to the electron (object) alone; for the realization of these potentialities depends just as much on the systems with which it interacts as on the electron (object) itself.

This is Bohm's presentation of the basics of the Copenhagen representation. After publishing his book, which is both a deep and clear exposition of the consensus view of quantum mechanics, Bohm seems to have become rather disillusioned and published his hidden variable interpretation of quantum mechanics in *Physical Review* (received July 5, 1951) shortly thereafter.[26]

15.3.9 Bohm and Aharonov

Bohm (along with Aharonov) returned to the EPR problem a few years later.[27] In this paper they proposed another alternate form of the experiment, one which is not only feasible but had already been carried out by Wu and Shaknov.[28]

[26]Bohm, D., *A Suggested Interpretation of the Quantum Theory in Terms of "Hidden" Variables, Part I*, Phys. Rev. 85, 166, (1952), received 3 July, 1951, and *Part II*, Phys.Rev. 85, 180, (1952), received 5 July, 1951.

[27]Bohm, D. and Aharonov, Y., *Discussion of Experimental Proof for the Paradox of Einstein, Rosen, and Podolsky*, Phys. Rev. **108**, 1070, (1957) received 10 May 1957.

[28]Wu, C.S. and, Shaknov, I., *The Angular Correlation of Scattered Annihilation Radiation*, Phys. Rev. **77**, 136, (1950) received 21 November 1949. This experiment was suggested by Wheeler, J.A., *Polyelectrons*, Ann. N.Y. Acad. Sci. **48**, 219 (1946); see p. 235.

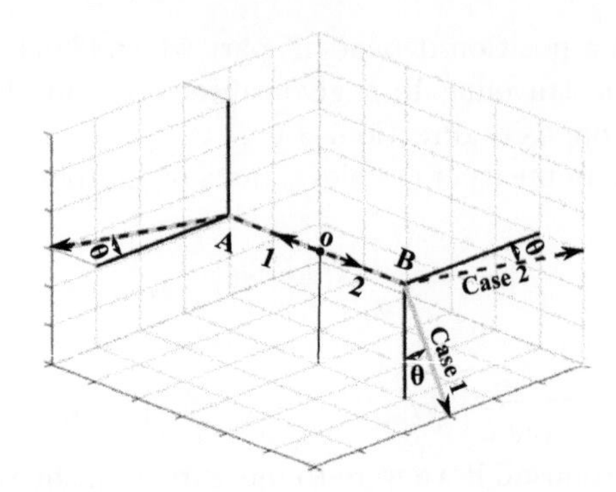

Fig. 15.6 Photon annihilation experiment.

The proposal was to use the two photons emitted when an electron-positron pair annihilates. According to theory, the angular momentum of the annihilating state should be zero (singlet state), and so the total angular momentum of the photons should also be zero. This means that the state of the two photons should be given by

$$|\Psi_{2\gamma}\rangle = \mathbf{a}^{\dagger}_{k1,x}\mathbf{a}^{\dagger}_{k2,y}|0\rangle - \mathbf{a}^{\dagger}_{k1,y}a^{\dagger}_{k2,x}|0\rangle \tag{15.41}$$

where $|0\rangle$ represents the vacuum state and $\mathbf{a}^{\dagger}_{k1,x}$ is the operator that creates a photon with wave vector $\overrightarrow{k}_1$ polarized along the x direction and $\overrightarrow{k}_1 + \overrightarrow{k}_2 = 0$. Then all the usual EPR arguments apply: measurement of the polarization of photon 2 fixes the polarizations state of photon 1 to be perpendicular. Measurement of the polarization along a rotated set of axes:

$$x = x' \cos\alpha + y' \sin\alpha \tag{15.42}$$

$$y = -x' \sin\alpha + y' \cos\alpha \tag{15.43}$$

so that

$$\mathbf{a}^{\dagger}_{k,x} = \mathbf{a}^{\dagger}_{k,x'} \cos\alpha + \mathbf{a}^{\dagger}_{k,y'} \sin\alpha \tag{15.44}$$

$$\mathbf{a}^{\dagger}_{k,y} = -\mathbf{a}^{\dagger}_{k,x'} \sin\alpha + \mathbf{a}^{\dagger}_{k,y'} \cos\alpha \tag{15.45}$$

yields the same form when substituted into (15.41):

$$|\Psi_{2\gamma}\rangle = \frac{1}{\sqrt{2}}\left[a^{\dagger}_{k1,x'}a^{\dagger}_{k2,y'}|0\rangle - a^{\dagger}_{k1,y'}a^{\dagger}_{k2,x'}|0\rangle\right]. \tag{15.46}$$

This means that the polarization states of the two photons will be correlated: they will always represent orthogonal polarizations.

The experiment consisted of observing the scattering of the two annihilation photons on aluminum (Fig. 15.6). The ratio R of the number of scattering events when the two photons scatter in the same plane (Case 2) to the number when they scatter in perpendicular planes (Case 1) is calculated according to quantum mechanics to be 2.00. The Wu and Shaknov experiment obtained $R = 2.04 \pm .08$.

The purpose of the experiment was to confirm the fact that the annihilating state was indeed a singlet, so that (15.41) was a correct description of the photon state.

Bohm and Aharonov applied an argument similar to Furry's. They calculated the ratio R for the situation where each photon was in a definite polarization state, definitely related to the state of the other photon, i.e., that the superposition state (15.41) was replaced by a mixture. In this case the largest possible predicted value of R was 1.5. As we will see, this was the first of a host of similar experiments which attempted an experimental test of the possibility that the observed EPR-type correlations could be explained by each component of the compound system being considered as being in a real definite state, i.e., that the correlations could be produced by local effects. In almost all experiments, the results indicated that this was impossible.

A form similar to (15.41) applies to material particles according to Jordan's second quantization. Remember that the operators $a_k, a_k^\dagger$ represent changes in the amplitudes of functions in three-dimensional space. Thus the nonlocal correlations can, in this way, be described in three dimensions without recourse to the $3N$-dimensional space.

15.4 Summary of historical commentary

We have given some indication as to how Einstein's thinking about the nonlocal effects implied by quantum mechanics developed following the Solvay conferences of 1927 and 1930, culminating in the work of Einstein, Podolsky, and Rosen (1935). While this work presented its argument in the form of "elements of reality" and claimed that the result meant that quantum mechanics was "incomplete", Einstein later expressed some annoyance at the formulation given in the paper and offered a much simpler argument. The mutual influences predicted by quantum mechanics represented a nonlocal influence ("spooky action at a distance," as he called it at one time) of measurements of one subsystem on the physical state of the other, and thus, he argued, quantum mechanics could not be the final form of the theory. He never questioned the validity of the quantum mechanical predictions, feeling that there must be a local theory that could make the same predictions. He was supported in this by Schrödinger, and the two of them worked for the rest of their lives in seeking an improved form of the theory, which they envisaged as a nonlinear field theory where particles would emerge as solutions of the wave equations. They never succeeded in this, although Schrödinger angered Einstein at one time by announcing to the press that he had found a solution, only to retract it shortly thereafter.

Following the publication of the EPR paper, several authors elaborated the quantum mechanical description and suggested more practical experiments that would display the same type of effect. When carried out, these experiments showed that the nonlocal correlations predicted by quantum mechanics were experimental facts.

Starting with the argument he advanced in the 1927 Solvay Conference involving the collapse of the wave function implied by the detection of a particle described by a wave function spread out in space, Einstein had argued that quantum mechanics applied only to ensembles, not to individual systems. He claimed that the EPR correlations were consistent with this, but in fact, as we have seen, they depend on the compound state being a superposition of possible states, which must apply to a single system. It is this superposition, what Schrödinger called an "entangled" state, which results in the fact that the free choice of basis for the measurements of the first subsystem determines the state of the second subsystem, i.e., we can take either (15.41) or (15.46) as the state of the composite system. Several of the authors discussed above have made this point.

It is interesting to note that none of the critical comments on the EPR paper referred to von Neumann's proof concerning hidden variables. Instead several of the authors gave their own, more concise, demonstrations of the inconsistency of hidden variables with quantum mechanics. This is likely in part due to Bell, who claimed that von Neumann's proof of no hidden variables is flawed. As we discussed in Chapter 14, this was a misundertanding by Bell; von Neumann had said that to incorporate hidden variables would require a major modification of quantum mechanics, that they could not be incorporated into the existing theory. This is quite different than saying hidden variables are an absolute impossibility.

15.5 Bell inequalities

Many years after the EPR conjecture, Bell proved a theorem that showed there is a quantitative limit on the degree of correlation between measurement observables that can be produced by a local hidden variable theory. Subsequent experimental measurements exceeded this limit, supporting quantum mechanics without the inclusion of hidden variables.

To summarize the discussion so far, in the absence of hidden variables, the apparent paradox in quantum mechanics is that the results of measurements on one particle of an entangled pair appear to be instantaneously affected by the type of measurement made on the other particle. By introducing the electron singlet state that we discussed earlier, Bohm sharpened the EPR discussion by considering the correlations between particle spins, in which case the effect is easily seen.

Bell further developed these concepts by noting that when unobserved or hidden variables are introduced, calculations must be averaged over the hidden variables when particles are subjected to measurement processes that are indirectly sensitive to those variables. For example, the expectation value of the correlation between the measurements at the two detectors must be modified as

$$E(\alpha,\beta) = \int_\Lambda P(\alpha,\beta,\lambda)p(\lambda)d\lambda \tag{15.47}$$

where α and β are measurement settings at the two separated detectors and $p(\lambda)d\lambda$ is a normalized probability distribution over a variable (or set of variables) λ. For example,

$$P(\alpha,\beta,\lambda) = |\langle\Psi,\lambda|\vec{\sigma}_A\cdot\vec{\sigma}_B|\Psi,\lambda\rangle|$$

is the quantum state correlation for the case of variable polarizer angles α and β for polarizers A and B respectively. This averaging reduces the degree of correlation, and in this way Bell derived his celebrated inequalities, which subsequently led to a series of experimental tests. These tests have established entanglement and nonlocality as undisputed features of the theory.[29]

We will show a specific use of the Bell inequalities in the next section.

15.6 Entangled photons

It is easier to make entangled photon pairs than entangled particle pairs, so many of the experimental studies of quantum entanglement have employed photons. There are many possible correlated photon sources, for example, a cascade emission from an excited atomic state. A three-level atom in a cascade decay can produce two photons at different times, but in an entangled state. An experiment could be arranged to detect the later photon first, which would require sending information backward in time if the apparent nonlocal correlation were due to some sort of unknown communication related to the states or their detection. We will discuss a simpler system.

15.6.1 Generation of entangled photon pairs via parametric down-conversion

With modern quantum electronics it is straightforward to produce a tightly controlled entangled state between two photons.[30] In fact the system described here has been duplicated in undergraduate physics teaching laboratories with great success.

Parametric down-conversion of a 400 nm photon in a BBO (β-barium borate) crystal produces two 800 nm photons with the same polarization (perpendicular to that of the 400 nm beam). The down-converted photons emerge in a cone with half-angle ϕ around the incident beam, with the range of possible ϕ determined principally by the crystal properties. The two photons produced by a single conversion event come out at the same time on opposite sides of the laser beam and propagate in the directions $+\phi$ and $-\phi$. The rate of production depends on the angle θ between the optical axis of the BBO crystal and the incident 400 nm light polarization as $\cos^2\theta$.

If we take two very thin BBO crystal sheets and orient one with the optical axis vertical (V) and the other with the optical axis horizontal (H) and have the incident light polarization at 45° to both, an entangled state can be produced. For linearly polarized photons, horizontal and vertical polarizations form a convenient basis set.

Such a system produces an entangled pair, with each photon moving in a distinct direction labeled a and b, as

$$|\Psi\rangle = \frac{1}{\sqrt{2}}\left[|V\rangle_a|V\rangle_b + |H\rangle_a|H\rangle_b\right]. \tag{15.48}$$

We can also describe these states in a basis rotated (about the propagation direction) by an angle α for the a path, as

$$|V_\alpha\rangle_a = \cos\alpha|V\rangle_a - \sin\alpha|H\rangle_a \tag{15.49}$$

[29]See, for example, Whitaker, M.A.B., *The EPR Paper and Bohr's Response: A Re-Assessment* Found. Phys. **34**, 1305-1339 (2004) for a discussion of the history of these concepts.

[30]Dehlinger, D. and Mitchell, M.W., *Entangled photon apparatus for the undergraduate laboratory,* Am. J. Phys. **70**, 898 (2002), and the accompanying paper, *Entangled photons, nonlocality, and Bell inequalities in the undergraduate laboratory*, Am. J. Phys. **70**, 903 (2002).

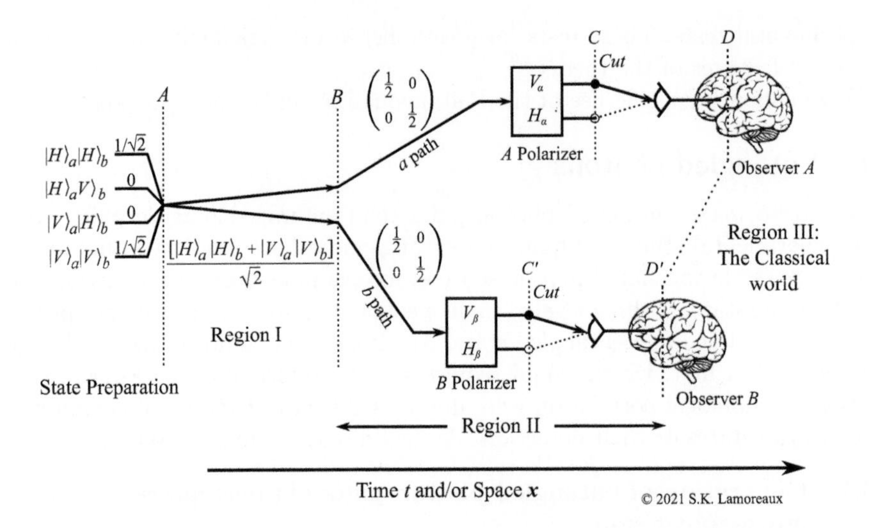

Fig. 15.7 An entangled photon Bell inequality demonstration experiment which includes the possibility of introducing a time delay in one path. At the separation point labeled by the vertical line at B, the particles become spatially separated and are thus distinct and can be independently measured. An attempt to elucidate the quantum state in either path alone results in a completely mixed state, represented by a diagonal density matrix with elements 1/2.(Brain image credit:ID 34191041 ©Mariia Hlushchenko, Maglyvi Dreamstime.com)

$$|H_\alpha\rangle_a = \sin\alpha|V\rangle_a + \cos\alpha|H\rangle_a \tag{15.50}$$

and similarly for the b path using a basis rotated by an angle β. We can use this as the measurement basis when the A and B polarizers are set to angles α and β, as shown in Fig. 15.7.

First, let us determine the probability of detecting a V polarized photon at A in this measurement basis:

$$P_{Va}(\alpha) = \langle\Psi|(\cos\alpha|V\rangle_a - \sin\alpha|H\rangle_a)(\cos\alpha\ {}_a\langle V| - \sin\alpha\ {}_a\langle H|)|\Psi\rangle \tag{15.51}$$

$$= \frac{1}{2}(\sin^2\alpha + \cos^2\alpha) = \frac{1}{2}. \tag{15.52}$$

A similar result can be obtained for B. We conclude that the photons in the A or B directions are unpolarized, and since the only quantum numbers for the measurement basis are H and V, the density matrix is a completely mixed state. Any *particular* photon has some specific polarization (it goes either to the H or V channel), but on average there is no polarization information.

The projection operator for *simultaneously* detecting two photons in the vertical directions at A and B in this basis is

$$P_{VV}(\alpha,\beta) = |V_\alpha\rangle_a|V_\beta\rangle_b\ {}_a\langle V_\alpha|\ {}_b\langle V_\beta| \tag{15.53}$$

$$= (\cos\alpha|V\rangle_a - \sin\alpha|H\rangle_a)(\cos\alpha\ {}_a\langle V| - \sin\alpha\ {}_a\langle H|) \tag{15.54}$$

$$\times\ (\cos\beta|V\rangle_b - \sin\beta|H\rangle_b)(\cos\beta\ {}_b\langle V| - \sin\beta\ {}_b\langle H|). \tag{15.55}$$

The probability to detect a coincidence is the expectation value of $P_{VV}(\alpha,\beta)$, which can be calculated using Eq. (15.48) as

$$\begin{aligned}\langle P_{VV}(\alpha,\beta)\rangle = \langle\Psi|P_{VV}(\alpha,\beta)|\Psi\rangle &= \frac{1}{2}(\cos\alpha\cos\beta+\sin\alpha\sin\beta)^2 \\ &= \frac{1}{2}\cos^2(\alpha-\beta). \end{aligned} \tag{15.56}$$

This result tells us that when the polarizers are set to the same angle, half of the time they will measure a $V_\alpha V_\alpha$ pair. This is the maximum coincidence rate because half of the total number of pairs are in the other (unmeasured) eigenstate, $H_\alpha H_\alpha$. This reflects the EPR paradox: How does one detector know the angle of the other detector's polarizer when they are physically separated? This correlation persists even if we introduce a time delay (path length increase), as shown in the a path of Fig. 15.7 and quickly (before detecting at A) change the angle of the A polarizer after a detection at B. With such timing, for one detector to receive information about the other would require superluminal communication.

Let us now compare this result to what might be expected in a classical system where there is no entanglement.

15.6.2 Photon polarization as a hidden variable

In the experiment described above, two entangled photons were produced by parametric down-conversion. If the polarization of the photons in either of the two propagation directions is measured separately, it appears as a completely unpolarized mixed state. We can describe the system in a semiclassical picture if we assume that the polarization was somehow encoded on the photons when they were produced at the same point in space and time (a local hidden variable). This can be modeled in an *ad hoc* fashion by assuming that the electric fields of the photons were both in the same random direction γ when they were produced, and we can think of this as a sort of hidden variable, one that we cannot directly access. If the polarizers at the detectors are independently set to angles α and β, the probability of detecting a photon at either detector (assuming perfect efficiency) is

$$P_A = \cos^2(\alpha-\gamma); \quad P_B = \cos^2(\beta-\gamma). \tag{15.57}$$

The average probability of detecting an A polarized photon with random initial polarization direction given by γ (which we might think of as playing the role of a hidden variable λ) is

$$\overline{P_A} = \frac{1}{2\pi}\int_0^{2\pi} d\gamma\cos^2(\alpha-\gamma) = \frac{1}{2} \tag{15.58}$$

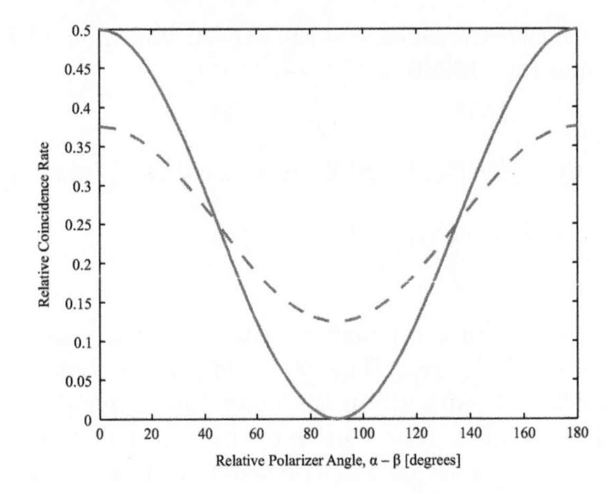

Fig. 15.8 The relative coincident rate between detectors A and B as a function of difference angle, for the quantum (solid) and semiclassical (dashed) cases.

which is the same as $\overline{P_B}$ and is independent of the polarizer angle as expected.

Let us now calculate the probability of detecting simultaneously the photons of a pair as a function of the polarizer angles α and β for the A and B polarizers, respectively. We must average over the random hidden variable γ because that is required to reproduce the apparently random polarizations of the two photons when they are considered by themselves:

$$\overline{\vec{P}_A \cdot \vec{P}_B} = \frac{1}{2\pi}\int_0^{2\pi} d\gamma \cos^2(\alpha-\gamma)\cos^2(\beta-\gamma) = \frac{1}{8} + \frac{1}{4}\cos^2(\alpha-\beta). \tag{15.59}$$

This can be compared to the quantum mechanical result (15.56) that we derived earlier,

$$\langle P_{VV}(\alpha,\beta)\rangle = \frac{1}{2}\cos^2(\alpha-\beta). \tag{15.60}$$

The most important difference between these cases is a reduction in the maximum to minimum angular variation (a reduction in contrast or correlation), together with an offset for the hidden variable case, as shown in Fig. 15.8. This offset term arises because even if the α and β axes are orthogonal, the angle γ will almost always lie somewhere between them (except in the rare instances when it is exactly parallel to either) and thereby allow coincident photons at both detectors. The overall effect of the hidden variable is to reduce the effective contrast, or correlation, between the photon polarizations.

For the quantum case, the detections represent eigenstates of the measurement operator. The detection of one photon ensures that the second photon will not be detected if its polarizer is in an orthogonal state. The high-contrast quantum mechanical effect is spoiled in the presence of hidden variables.

We could also perform the semiclassical calculation by assuming that the initial state is a product of definite polarization states,

$$|\Psi\rangle_{def} = (\cos\gamma|V\rangle_a + \sin\gamma|H\rangle_a)(\cos\gamma|V\rangle_b + \sin\gamma|H\rangle_b), \tag{15.61}$$

determine the expectation value of $P_{VV,def}(\alpha, \beta, \gamma)$, and average over γ. However, the amount of arithmetic is alarming. We will do this calculation later using the density matrix, which provides a great simplification. Spoiler alert: the result will be identical.

15.6.3 Bell inequality for two-particle states

By introducing a sort of *ad hoc* hidden variable γ, we demonstrated that the introduction of a local hidden variable reduces the degree of correlation between the polarizations of photon pairs. In fact, it can be proven that *any* local hidden variable theory will produce a lower correlation than the quantum case. This might seem counterintuitive; however, we saw in the semiclassical calculation that the hidden variable causes coincidences even when the polarizers are orthogonal.

Bell showed that any hidden variable theory will reduce the degree of correlation between measurements on entangled subsystems, and that there is a maximum correlation that *any and all* hidden variable theories cannot exceed, and most importantly, this maximum is below the quantum mechanical correlation.

The derivation proceeds as follows. For any hidden variable theory, the probability distribution of the hidden variable is given by some function $\rho(\lambda)$ where

$$\rho(\lambda) \geq 0 \quad \text{and} \quad \int_\Lambda d\lambda\, \rho(\lambda) = 1 \tag{15.62}$$

where Λ represents the range of the variable or set of variables λ.

Let us take a specific example: the correlations between the A and B detectors for the two-photon state $|\Psi\rangle$ that we have been considering. We know that a measurement at A or B will always yield a count at the H OR V polarizer output, assuming perfect detection efficiency, but never both. (We might only measure one polarization output, as is often done in the case of polarizers that are absorptive along one polarization axis, but when one photon is not detected, the existence of the other is implied in the sense that we know the detection eigenvalue.)

If we detect a V photon in A, we might also detect a V photon in B, with the probability depending on the polarizer angle settings. Similarly, if we detect a V photon in B, then there is a probability to detect a V photon in A. Any single measurement produces a 1 or 0, depending on whether there was or was not a detection. We worked out the ensemble average probabilities as $\langle P_{VV}(\alpha, \beta)\rangle$. However, to determine the correlation function in order to assess the effect of a hidden variable, we need a slightly different form. Consider a *particular* photon pair i. The measurement operation can produce only 0 or 1 for any particular state:

$$\begin{aligned} P_{i,Va}(\alpha) &= 0 \text{ or } 1 \\ &= \frac{1}{2} + \frac{1}{2}a(\alpha) \end{aligned} \tag{15.63}$$

$$\begin{aligned} P_{i,Vb}(\beta) &= 0 \text{ or } 1 \\ &= \frac{1}{2} + \frac{1}{2}b(\beta) \end{aligned} \tag{15.64}$$

$$\begin{aligned} P_{i,VV}(\alpha,\beta) &= 0 \text{ or } 1 \\ &= \frac{1}{2} + \frac{1}{2}\mathrm{ab}(\alpha,\beta) \end{aligned} \tag{15.65}$$

(note that $\mathrm{ab}(\alpha,\beta)$ is not necessarily the product of $a(\alpha)$ and $b(\beta)$). The new functions $a(\alpha)$, $b(\beta)$, and $\mathrm{ab}(\alpha,\beta)$ have values ± 1 which represent the eigenvalues V $(= +1)$ or H $(= -1)$, or for $\mathrm{ab}(\alpha,\beta)$, their products $HH = VV = 1$; $HV = VH = -1$. (Although we might not measure the H photons at the detectors, their existence is implied as discussed earlier.) The values ± 1 are thus eigenvalues of the new forms of the measurement operators. Reinterpreting the projective measurement in these terms provides a means of converting P_{VV} into an ensemble average of the possible polarization eigenvalues. This is what we need to form the correlation function, and there is a simple linear relationship between the two expressions.

This relationship can be derived as follows for the function $\mathrm{ab}(\alpha,\beta)$, related to the quantum correlation function defined below. First, we can take the expectation values of the two sides to obtain

$$\langle P_{VV}(\alpha,\beta)\rangle = \frac{1}{2} + \frac{1}{2}\langle \mathrm{ab}(\alpha,\beta)\rangle. \tag{15.66}$$

Next we can average both sides of the equation over all angles, subtract those averages from the respective sides, and divide by the averages (assuming the average is not zero). We can then define the correlation function as

$$E(\alpha,\beta) = \frac{\langle \mathrm{ab}(\alpha,\beta)\rangle - \overline{\langle \mathrm{ab}(\alpha,\beta)\rangle}}{\overline{\langle \mathrm{ab}(\alpha,\beta)\rangle}} \tag{15.67}$$

where $\overline{\cdots}$ indicates an average over all angles. Doing the same thing to the left-hand side of the relationship results in

$$E(\alpha,\beta) = \frac{\langle P_{VV}(\alpha,\beta)\rangle - \overline{\langle P_{VV}(\alpha,\beta)\rangle}}{\overline{\langle P_{VV}(\alpha,\beta)\rangle}}. \tag{15.68}$$

For the quantum case that we have calculated already, $P_{VV}(\alpha,\beta) = (1/2)\cos^2(\alpha-\beta)$, so that

$$E_q(\alpha,\beta) = \frac{(1/2)\cos^2(\alpha-\beta) - (1/4)}{(1/4)} = \cos(2(\alpha-\beta)). \tag{15.69}$$

This has the desired properties; when $\alpha - \beta = 0°$ it equal to 1, and when $\alpha - \beta = 90°$ it is equal to -1, corresponding to $VH = 1 \times (-1)$.

Let us now introduce a hidden variable λ into the two-particle state. λ, we imagine, carries all the information needed to determine the results of measurements at A and B. We might expect that such a variable really does exist because it seems strange that one photon "knows" about the measurement basis of the other, so perhaps this is all decided beforehand and needs to be included in the wave function (e.g., the case we considered by including γ as the hidden photon polarization direction variable). Note that $\text{ab}(\alpha, \beta, \lambda)$ is $+1$ if both photons are in the same H, V state, and -1 if not. To impose locality (i.e., no noncausal connection between the states when they are at A and B because the setting of one polarizer cannot affect the measurement of the other), we require that measurements at A and B respond independently:

$$\text{ab}(\alpha, \beta, \lambda) = a(\alpha, \lambda)b(\beta, \lambda). \tag{15.70}$$

This imposition is a hypothesis that has testable consequences. The expectation value for the polarization correlation is determined by averaging over the unobserved and unselected hidden variable λ:

$$E(\alpha, \beta) = \int d\rho\, a(\alpha, \lambda)b(\beta, \lambda) \tag{15.71}$$

where $d\rho = \rho(\lambda)d\lambda$ is defined in Eq. (15.62). Of course, there is no averaging for the quantum case because there is no λ.

If $\alpha = \beta$, we know that the polarizations will be perfectly correlated, and therefore

$$a(\alpha, \lambda) = b(\alpha, \lambda), \tag{15.72}$$

and the correlation expectation value is

$$E(\alpha, \beta) = \int d\rho\, a(\alpha, \lambda)a(\beta, \lambda). \tag{15.73}$$

To elucidate a difference between ordinary quantum mechanics and a hidden variable theory, it is necessary to look at the differences between the correlation expectation values for three different directions, α, β, and ϕ,

$$\Delta = E(\alpha, \beta) - E(\alpha, \phi) = \int d\rho\, [a(\alpha, \lambda)a(\beta, \lambda) - a(\alpha, \lambda)a(\phi, \lambda)], \tag{15.74}$$

which can be written in a slightly different form by use of $a(\beta, \lambda)^2 = 1$:

$$\Delta = \int d\rho\, a(\alpha, \lambda)a(\beta, \lambda)\, [1 - a(\beta, \lambda)a(\phi, \lambda)]. \tag{15.75}$$

Taking the absolute value of both sides, and using the fact that $|\int f(x)dx| \leq \int |f(x)|dx$, and $|a(\alpha, \lambda)a(\beta, \lambda)| = 1$, we find

$$|\Delta| \leq \int d\rho\, |a(\alpha, \lambda)a(\beta, \lambda)| \cdot |\,[1 - a(\beta, \lambda)a(\phi, \lambda)]\,| \tag{15.76}$$

$$= \int d\rho\, |(1 - a(\beta, \lambda)a(\phi, \lambda)| \tag{15.77}$$

$$= |1 - E(\beta, \phi)|. \tag{15.78}$$

We have thus arrived at a Bell inequality:

$$|E(\alpha, \beta) - E(\alpha, \phi)| \leq |1 - E(\beta, \phi)|. \tag{15.79}$$

For the quantum case, $E_q(\alpha, \beta) = \cos(2(\alpha - \beta))$,(15.69). Taking $\alpha = 45°$, $\beta = 22.5°$, and $\phi = -22.5°$, we arrive at

$$\sqrt{2} \leq 1, \tag{15.80}$$

and we see that the quantum case does not support the hypothesis that $\mathrm{ab}(\alpha, \beta, \lambda) = a(\alpha, \lambda)b(\beta, \lambda)$ and/or that there is a hidden variable λ. The results of many experiments have yielded a failure of this inequality, and there is universal agreement that quantum mechanics, without including hidden variables, is correct as a fundamental theory. The conclusion is that quantum mechanics is a nonlocal theory and that entanglement is a reality.

We can rederive our previous semiclassical result by similarly averaging over the hidden variable, introduced as γ in Eq. (15.59). Using the same angles as before in the Bell inequality, $\alpha = 45°$, $\beta = 22.5°$, and $\phi = -22.5°$, we find the inequality satisfied as

$$\frac{\sqrt{2}}{2} \leq 1. \tag{15.81}$$

A further investigation shows that there is no combination of angles where the inequality is violated.

If we write for any general correlation

$$E(\alpha, \beta) = \eta \cos(2(\alpha - \beta)) \tag{15.82}$$

then η parameterizes the degree of correlation. A numerical study shows that the Bell inequality is violated for all angles α and β if $\eta > 2/3$.

For the cases considered here, it is clear that quantum mechanics does not satisfy the Bell inequality because the degree of correlation is too high. When the degree of correlation is reduced to $\eta = 2/3$ or below, the Bell inequality is satisfied. It might be tempting to assume that states corresponding to such low values of η are not entangled; however, that is not the case, as we will show. A violation of the Bell inequality requires that there be quantum entanglement, but an entangled state does not necessarily lead to a violation.

15.6.4 Wigner's proof of the Bell inequality

Eugene Wigner[31] has given a very nice direct proof of Bell's theorem. He starts with two spin 1/2 particles in a singlet state (total spin =0):

$$|+\rangle_1 |-\rangle_2 - |-\rangle_1 |+\rangle_2$$

[31] Wigner, E.P., "Interpretation of quantum mechanics", in Wheeler, J.A. and Zurek, W.H., *Quantum Theory and Meaurement*, p.291, (Princeton U.P., 1983).

where $|+\rangle_1$ is the state with the spin of particle 1 in the positive direction and this holds in any basis. If each particle goes through a polarization analyzer the probability of the two spins being positive or negative simultaneously is

$$P_{++} = P_{--} = \frac{1}{2}\sin^2(\theta_{12}/2) \tag{15.83}$$

where θ_{12} is the angle between the axes of the two analyzers. The probability of a different result from the two analyzers is

$$P_{+-} = P_{-+} = \frac{1}{2}\cos^2(\theta_{12}/2). \tag{15.84}$$

These results follow from the quantum mechanics of spin 1/2 (see Chapter 19). Now we assume that the spins will be measured by analyzers with their axes in one of three possible directions, $\vec{e}_1, \vec{e}_2, \vec{e}_3$. If the results are determined by averaging over some hidden variables there will be a definite probability for each possible outcome. Wigner introduces the notation

$$(+-+,--+) \tag{15.85}$$

to indicate the probability that the first particle will have spin up in the directions $\vec{e}_1, \vec{e}_3$ and spin down in the direction $\vec{e}_2$ while the second particle has spin down in the directions $\vec{e}_1, \vec{e}_2$ and spin up in the direction $\vec{e}_3$. In fact this probability is zero due to (15.83), which shows that the probability of the two particles being up (or down) in the same direction is zero. Since the hidden variables are supposed to completely determine the results these quantities are all fully defined and always positive. Taking account of (15.83) we see there are $2^3 = 8$ such possible nonzero probabilities.

Now consider measurements in the $\vec{e}_1, \vec{e}_2$ directions. The probability of both particles giving spin up $(+)$ that is particle 1 giving a result $(+)$ in direction $\vec{e}_1$ and particle 2 giving the result $(+)$ in the $\vec{e}_2$ direction is then

$$(+-+,-+-) + (+--,-++) = P_{++} = \frac{1}{2}\sin^2(\theta_{12}/2). \tag{15.86}$$

If measurements are made in the $\vec{e}_2, \vec{e}_3$ directions we find

$$(++-,--+) + (-+-,+-+) = \frac{1}{2}\sin^2(\theta_{23}/2) \tag{15.87}$$

and for the $\vec{e}_1, \vec{e}_3$ directions:

$$(++-,--+) + (+--,-++) = \frac{1}{2}\sin^2(\theta_{13}/2). \tag{15.88}$$

Then by adding (15.86) and (15.87) we find (using (15.88))

$$\frac{1}{2}\left[\sin^2(\theta_{12}/2) + \sin^2(\theta_{23}/2)\right] = (+-+,-+-) + (+--,-++) + .. \tag{15.89}$$

$$+ (++-,--+) + (+--,-++) \tag{15.90}$$

$$= (+-+,-+-) + (+--,-++) + \frac{1}{2}\sin^2(\theta_{13}/2) \tag{15.91}$$

$$\geq \frac{1}{2}\sin^2(\theta_{13}/2) \tag{15.92}$$

This is violated for $\theta_{12} = \theta_{23} = \pi/3$, $\theta_{13} = 2\pi/3$ which results in $1/4 \geq 3/8$ which is obviously false. Wigner then goes on to state that a hidden variable theory which does not violate Bell's theorem (15.92) requires that the hidden variables specify the state of the measuring apparatus as well as that of the spins. There would have to be correlations between the state of the spins and the directions in which the spins will be measured. The theory discussed above is called a "local" hidden variable theory. All theories of this kind are ruled out by the experimental violation of the Bell inequality.

15.7 Entanglement in the density matrix

15.7.1 Some properties of the density matrix

As we have seen (Chapter 14), von Neumann based his presentation of quantum mechanics on the existence of what is now called the density matrix. The density matrix ρ is an operator that represents the state of the system under study. Its defining property is that the average value of any observable, represented by a Hermitian operator $\mathbf{A}$, is given by

$$\langle A\rangle = \mathrm{Tr}\,(\rho\mathbf{A}). \tag{15.93}$$

This is to be understood as the average of the results of a large number of identical measurements on identically prepared systems, i.e., an *ensemble* of systems.

In the case of a mixture, i.e., when the system is known to be in one of a set of states $|\psi_i\rangle$ with probability P_i, the density matrix is given by

$$\rho = \sum_i P_i\,|\psi_i\rangle\,\langle\psi_i|. \tag{15.94}$$

This is a diagonal matrix. So we can find out whether a given density matrix represents a pure state or a mixture by diagonalizing it. If more than one of the eigenvalues is nonzero, ρ represents a mixture, while the density matrix of a pure state will have only one nonzero eigenvalue, which will be equal to one since the sum of the eigenvalues is $\sum_i P_i = 1$. For a pure state $|\psi_i\rangle$, ρ is a projection operator: $\rho = |\psi_i\rangle\,\langle\psi_i|$. This is seen to be equivalent to the criterion that $\rho^2 = \rho$ for a pure state, a condition which defines projection operators.

Given any state of a system, we can construct a density matrix which allows the calculation of all physical results according to (15.93). On the other hand, it is not always possible to associate a unique state vector with a given density matrix. Weinberg[32] has given the example of a 2×2 density matrix for a spin-1/2 system in the s_z basis

$$\rho_1 = \begin{bmatrix} .675 & .175 \\ .175 & .325 \end{bmatrix} \tag{15.95}$$

which has the eigenvectors $|\psi_1\rangle = \begin{bmatrix} -0.923\,88 \\ -0.382\,68 \end{bmatrix}$ with eigenvalue 0.75, and $|\psi_2\rangle = \begin{bmatrix} 0.382\,68 \\ -0.923\,88 \end{bmatrix}$ with eigenvalue 0.25. Remembering that the state vector for a spin pointing in a direction making an angle θ with the z-axis is given (in the s_z basis) by $|\psi\rangle = \begin{bmatrix} \cos\theta/2 \\ \sin\theta/2 \end{bmatrix}$ we see the density matrix (15.95) represents a state with a 75% probability of making an angle $\theta = 2\cos^{-1}(-.92388) = -\pi/4$ and a 25% probability of an angle $\theta = 3\pi/4$, with the z-axis. Now the point is that the density matrix ρ_1 can also be obtained by taking the mixture (all operators are in the s_z basis)

$$.5\,|+\rangle\langle+| + .15\,|-\rangle\langle-| + .35\,|X\rangle\langle X| \tag{15.96}$$

where $|X\rangle = [|+\rangle + |-\rangle]/\sqrt{2}$ is the eigenstate of s_x with eigenvalue $+1/2$. In matrix form

$$.5\begin{bmatrix} 1 & 0 \\ 0 & 0 \end{bmatrix} + .15\begin{bmatrix} 0 & 0 \\ 0 & 1 \end{bmatrix} + \frac{.35}{2}\begin{bmatrix} 1 & 1 \\ 1 & 1 \end{bmatrix} = \rho_1. \tag{15.97}$$

15.7.2 Density matrix for two two-state particles

In the case of an EPR-like system consisting of two isolated, distant subsystems (a, b) that had previously been in contact, we know that a measurement result on one system appears to instantaneously determine the measurement result of the other system. If the state of the system is given by

$$|\psi\rangle = |+\rangle_a\,|-\rangle_b - |-\rangle_a\,|+\rangle_b \tag{15.98}$$

a measurement result on a will force the wave function of b to be in one state or the other, but the density matrix of b, representing an ensemble of identical measurements on b, will not be changed as the probabilities for any outcome of b remain the same. The change in the wave function can only be observed by taking into account the correlations between the measurements of a and b, which can only be done by exchanging information between the two observers.

Weinberg considered that this instantaneous change of the wave function due to a measurement made an arbitrary distance away cannot be taken seriously, and suggested that quantum mechanics should be formulated exclusively in terms of the density matrix; we will discuss the implications of this later. Let us therefore consider the states of two identical noninteracting two-state particles (two-level systems that

[32]Weinberg, S., *Quantum mechanics without state vectors*, Phys. Rev A **90**, 042102 (2014), received 8, June, 2014 arXiv:1405.3483v1 [quant-ph].

can be represented by spin-1/2 states). Let us label the quantum states as 1, 2 for particles a and b, and general states of each as

$$|a\rangle = a_1|1\rangle_a + a_2|2\rangle_a; \qquad |b\rangle = b_1|1\rangle_b + b_2|2\rangle_b. \tag{15.99}$$

We can form a new system state by taking the product of these two states,

$$|\Psi\rangle = |a\rangle|b\rangle, \tag{15.100}$$

which satisfies the Schrödinger equation for the Hamiltonian $H_a + H_b$. (Including particle interactions such as $H_{ab} = e^2/r_{ab}$, an electrostatic interaction between the particles, complicates matters, so for the ensuing discussion we will not consider such effects.)

Instead of writing the state as two separate systems, we can instead define a new state space based on the outer (tensor) product of the a and b states. We can write the state as a linear combination of the products of single-particle eigenstates, so that

$$|\Psi\rangle = |a\rangle \otimes |b\rangle = \sum_{i,j=1,2} a_i b_j |i\rangle_a |j\rangle_b \tag{15.101}$$

and the Hamiltonians are combined as $H = H_a \otimes \hat{I} + \hat{I} \otimes H_b$. This allows writing states that cannot be factorized into single-particle states as given by Eq. (15.99). Such states are *entangled*, so that the result of a measurement on particle a appears to dictate what result will be obtained for a measurement on b, as we have discussed already.

The density matrix for the tensor product $\rho_a \otimes \rho_b$ is straightforward to work out; letting $a_{mn} = a_m a_n^*$, $b_{mn} = b_m b_n^*$, the matrix is

$$\begin{aligned} \rho_{a\otimes b} &= \begin{pmatrix} a_{11} & a_{12} \\ a_{21} & a_{22} \end{pmatrix} \otimes \begin{pmatrix} b_{11} & b_{12} \\ b_{21} & b_{22} \end{pmatrix} \\ &= \begin{pmatrix} a_{11}\begin{pmatrix} b_{11} & b_{12} \\ b_{21} & b_{22} \end{pmatrix} & a_{12}\begin{pmatrix} b_{11} & b_{12} \\ b_{21} & b_{22} \end{pmatrix} \\ a_{21}\begin{pmatrix} b_{11} & b_{12} \\ b_{21} & b_{22} \end{pmatrix} & a_{22}\begin{pmatrix} b_{11} & b_{12} \\ b_{21} & b_{22} \end{pmatrix} \end{pmatrix} \end{aligned} \tag{15.102}$$

and therefore the density matrix can be written in expanded form as

$$\rho_{a\otimes b} = \begin{pmatrix} a_{11}b_{11} & a_{11}b_{12} & a_{12}b_{11} & a_{12}b_{12} \\ a_{11}b_{21} & a_{11}b_{22} & a_{12}b_{21} & a_{12}b_{22} \\ a_{21}b_{11} & a_{21}b_{12} & a_{22}b_{11} & a_{22}b_{12} \\ a_{21}b_{21} & a_{21}b_{22} & a_{22}b_{21} & a_{22}b_{22} \end{pmatrix}. \tag{15.103}$$

An operator that acts on either a or b alone that we want to operate on the combined density matrix can be formed by

$$\mathbf{A}_{ab} = \mathbf{A}_a \otimes \hat{I} + \hat{I} \otimes \mathbf{A}_b. \tag{15.104}$$

Given two two-state systems, there are four different maximally entangled states that can be constructed:

$$|\Psi_\pm\rangle = \frac{1}{\sqrt{2}}\left(|1\rangle_a|1\rangle_b \pm |2\rangle_a|2\rangle_b\right) \tag{15.105}$$

$$|\Phi_\pm\rangle = \frac{1}{\sqrt{2}}\left(|1\rangle_a|2\rangle_b \pm |2\rangle_a|1\rangle_b\right). \tag{15.106}$$

Note that these states cannot be rewritten as products of two single-particle states, as in Eq. (15.99). This can be taken as a definition of entanglement. Let us take $|\Psi_+\rangle$, which is the entangled photon state we considered before, and find its density matrix:

$$\rho_{ab} = |\Psi_+\rangle\langle\Psi_+| = \begin{pmatrix} \frac{1}{2} & 0 & 0 & \frac{1}{2} \\ 0 & 0 & 0 & 0 \\ 0 & 0 & 0 & 0 \\ \frac{1}{2} & 0 & 0 & \frac{1}{2} \end{pmatrix}. \tag{15.107}$$

It is easy to show that $\rho_{ab}^2 = \rho_{ab}$, and this is true for all of the maximally entangled states, which means they are *pure states.*

Some caution is required here. The density matrix Eq. (15.103), as written, represents a product state. However, because ρ_{ab} does not represent a product state, it cannot be put in this form. We will discuss a test to determine when the density matrix of a two-particle system cannot be factored into two separate (non-entangled) single-particle states.

Before we do that, let us consider the state b alone, without regard to state a. This can be done by taking the trace over a which requires that we return to the operator definition of the density matrix. We can recast Eq. (15.103) with the products of eigenstates as

$$\rho_{ab,kl} = c_{kl}|m\rangle_a|i\rangle_b\ {}_b\langle j|\ {}_a\langle n| \tag{15.108}$$

where $k = 2m+i-2$ and $l = 2n+j-2$, and i, j, m, n are 1 or 2 for the two eigenstates for each particle. The constants satisfy $c_{kl} = c_{lk}^*$ to maintain hermiticity; however, the c_{kl} are not in general determined from single-particle states as were the $a_{mn}b_{ij}$ factors in Eq. (15.103).

The indices for the density matrix run through $k, l = 1, 2, 3, 4$, and of course ρ_{ab} is a 4×4 matrix as expected. Thus the i, j, m, n of the eigenstates are mapped onto the 16 density matrix elements, in the appropriate location in the 4×4 matrix. Therefore

$$\rho_b = \sum_{h=1,2} c_{kl}\ {}_a\langle h|m\rangle_a|i\rangle_b\ {}_b\langle j|\ {}_a\langle n|h\rangle_a \tag{15.109}$$

$$= c_{i,j}|i\rangle_b\ {}_b\langle j| + c_{2+i,2+j}|i\rangle_b\ {}_b\langle j|. \tag{15.110}$$

Perhaps not surprisingly, this is the sum of the upper left and the lower right 2×2 blocks of Eq. (15.103). We can extract these blocks by multiplying ρ_{ab} from the left by a 4×2 matrix and on the right by a 2×4 matrix as

$$\rho_b = \mathrm{Tr}_a\, \rho_{ab} = \begin{pmatrix} 1\,0\,0\,0 \\ 0\,1\,0\,0 \end{pmatrix} \rho_{ab} \begin{pmatrix} 1\,0 \\ 0\,1 \\ 0\,0 \\ 0\,0 \end{pmatrix} + \begin{pmatrix} 0\,0\,1\,0 \\ 0\,0\,0\,1 \end{pmatrix} \rho_{ab} \begin{pmatrix} 0\,0 \\ 0\,0 \\ 1\,0 \\ 0\,1 \end{pmatrix} \tag{15.111}$$

$$= \begin{pmatrix} \frac{1}{2} & 0 \\ 0 & \frac{1}{2} \end{pmatrix}. \tag{15.112}$$

Similarly, the matrices to perform a trace over b can be constructed (but it's not the sum of distinct blocks):

$$\rho_a = \mathrm{Tr}_b\, \rho_{ab} = \begin{pmatrix} 1\,0\,0\,0 \\ 0\,0\,1\,0 \end{pmatrix} \rho_{ab} \begin{pmatrix} 1\,0 \\ 0\,0 \\ 0\,1 \\ 0\,0 \end{pmatrix} + \begin{pmatrix} 0\,1\,0\,0 \\ 0\,0\,0\,1 \end{pmatrix} \rho_{ab} \begin{pmatrix} 0\,0 \\ 1\,0 \\ 0\,0 \\ 0\,1 \end{pmatrix} = \begin{pmatrix} \frac{1}{2} & 0 \\ 0 & \frac{1}{2} \end{pmatrix}. \tag{15.113}$$

The conclusion is that either a or b considered by itself is in a completely mixed state, e.g., the photons are unpolarized as a statistical average—any *particular* photon is detected as an H or V eigenstate but, apparently, randomly.

Let us now attempt to find a signature of entanglement. Referring back to Chapter 14, Section 3, the definition of the von Neumann entropy is

$$S = -\mathrm{Tr}\, \rho \ln \rho \tag{15.114}$$

and thus $S = 0$ for a pure state (ρ_{ab}). However, ρ_a and ρ_b represent completely mixed states and therefore *maximum entropy* density matrices. This can be seen by considering a general form

$$\rho = \begin{pmatrix} p & 0 \\ 0 & 1-p \end{pmatrix} \tag{15.115}$$

for $0 \le p \le 1$, in which case

$$S = -p \ln p - (1-p)\ln(1-p) \tag{15.116}$$

and so S is maximum when $p = 1/2$, e.g., for ρ_a and ρ_b.[33]

Even though the combined state of the system has zero entropy because it is a pure state, the states of the two subsystems are maximum entropy states. This situation is only possible in quantum mechanics: classically the entropy of a composite system is at least as large as the entropy of any of its components. In analogy, a classical house is at least as dirty as its dirtiest room, whereas in a quantum house, every room could be dirty, but when the house is looked at as a whole, it appears perfectly clean![34] The same result is obtained for all four maximally entangled states.

[33]This is identical in form to the Shannon entropy in information theory, except in that case the function is $\log_2$, which results in a multiplicative factor, and p refers either to the number of errors in a bit stream or the ratio of 1s to 0s in a bit stream.

[34]The analogy is due to C. Bennett.

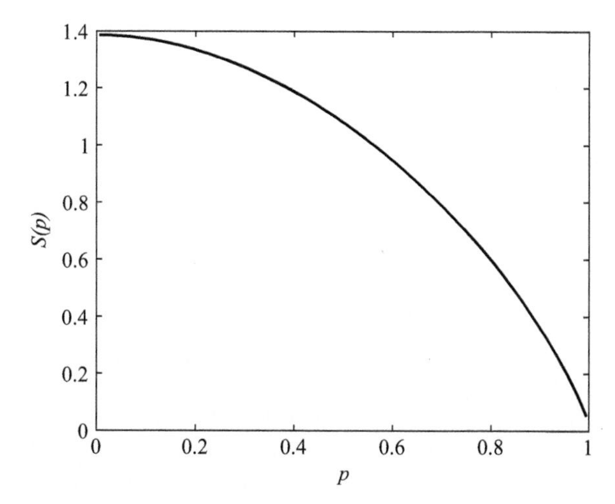

Fig. 15.9 The entropy as a function of p, where $1-p$ is the fraction of fully mixed state combined with a maximally entangled state.

Let us now add some impurity to the pure state. In the two-particle state basis (four total states), the completely mixed state is

$$\rho_{mix} = \begin{pmatrix} \frac{1}{4} & 0 & 0 & 0 \\ 0 & \frac{1}{4} & 0 & 0 \\ 0 & 0 & \frac{1}{4} & 0 \\ 0 & 0 & 0 & \frac{1}{4} \end{pmatrix}. \tag{15.117}$$

The degree of purity is controlled by the parameter $0 \leq p \leq 1$ as

$$\rho_p = (1-p)\rho_{mix} + p\rho_{ab} = \frac{1}{4}(1-p)\hat{I} + p\rho_{ab} = \frac{1}{4}\begin{pmatrix} 1+p & 0 & 0 & 2p \\ 0 & 1-p & & 0 \\ 0 & 0 & 1-p & 0 \\ 2p & 0 & 0 & 1+p \end{pmatrix} \tag{15.118}$$

which preserves the normalization, $\mathrm{Tr}\,\rho_p = 1$. The four eigenvalues ξ_i of this matrix are (three equal) $(1-p)/4$ and (one) $(1+3p)/4$, which sum to one, as expected. The entropy as a function of p, calculated by summing $-\xi_i \log \xi_i$, is shown in Fig. 15.9.

How small can p be while the system can still remain entangled? There is no obvious way to determine this for any general dimensional system, but for low dimensions, the PPT (positive partial transpose) criterion, also called the Peres-Horodecki criterion, can be employed. This criterion represents a necessary condition for a two-particle quantum mechanical system to be separable into a product of two single-particle states, in which case the density matrix is also separable into a product of the two particles' density matrices.[35]

[35]Horodecki, M., Horodecki, P., Horodecki, R., *Separability of mixed states: necessary and sufficient conditions*, Phys. Lett. A **223**, 8 (1996); A. Peres, *Separability criterion for density matrices*, Phys. Rev. Lett. **77**, 1413 (1996).

The criterion for ρ_p to be separable is that all of the eigenvalues of the partially transposed density matrix are positive, which means that it is an allowed density matrix (the eigenvalues are the state populations and must be greater than or equal to zero). This transpose is done by interchanging terms $a_{ij}b_{mn} \to a_{ij}b_{nm}$) (which is equivalent to breaking the 4×4 density matrix into four 2×2 blocks and taking the transposes of the four blocks, as in Eq. (15.103)). In the case of ρ_p, the partially transposed matrix is

$$\mathrm{T}_b\,\rho_p = \frac{1}{4}\begin{pmatrix} 1+p & 0 & 0 & 0 \\ 0 & 1-p & 2p & 0 \\ 0 & 2p & 1-p & 0 \\ 0 & 0 & 0 & 1+p \end{pmatrix}, \tag{15.119}$$

where T_b is the b-submatrix transpose operator. If a system is not entangled, we might expect that varying the submatrices of the density matrix should not result in a disallowed system: for a state expressible as a product as in Eq. (15.103), the partial transpose corresponds to transposing the density matrix of one of the component subsystems, and the transposed matrix still has the same eigenvalues, so it is still an allowed density matrix. This consideration provides some insight into the PPT criterion.

The partially transposed matrix has four eigenvalues (three equal) $(p+1)/4$ and $(1-3p)/4$. The smallest eigenvalue becomes negative when $p > 1/3$, implying for that for $p > 1/3$ the system is entangled. Note that there are no unusual effects in the entropy for $p = 1/3$, as shown in Fig. 15.9, indicating that the von Neumann entropy is not a test for entanglement.

15.7.3 Bell state correlation calculated using the density matrix

Let us now investigate the effects of p on a photon-based Bell inequality experiment. First we will need to write Eq. (15.53) for $P_{VV}(\alpha,\beta)$ as a matrix operator. Note that this operator acts simultaneously on both the a and b states, so we will need to work out its elements in the product state basis. This is straightforward, and defining $c_\alpha = \cos\alpha,\ c_\beta = \cos\beta,\ s_\alpha = \sin\alpha$, and $s_\beta = \sin\beta$, we have (see Eq. 15.53)

$$P_{VV}(\alpha,\beta) = \begin{pmatrix} c_\alpha^2 c_\beta^2 & -c_\alpha^2 c_\beta s_\beta & -c_\alpha s_\alpha c_\beta^2 & c_\alpha s_\alpha c_\beta s_\beta \\ -c_\alpha^2 c_\beta s_\beta & c_\alpha^2 s_\beta^2 & c_\alpha s_\alpha c_\beta s_\beta & -c_\alpha s_\alpha s_\beta^2 \\ -c_\alpha s_\alpha c_\beta^2 & c_\alpha s_\alpha c_\beta s_\beta & s_\alpha^2 c_\beta^2 & -s_\alpha^2 c_\beta s_\beta \\ c_\alpha s_\alpha s_\beta s_\beta & -c_\alpha s_\alpha s_\beta^2 & -s_\alpha^2 c_\beta s_\beta & s_\alpha^2 s_\beta^2 \end{pmatrix}. \tag{15.120}$$

The expectation value for the pure state ρ_{ab} is

$$\langle P_{VV}(\alpha,\beta)\rangle_{ab} = \mathrm{Tr}\left[\rho_{ab} P_{VV}(\alpha,\beta)\right] \tag{15.121}$$

$$= \frac{1}{2}\left[c_\alpha^2 c_\beta^2 + s_\alpha^2 s_\beta^2 + 2c_\alpha s_\alpha c_\beta s_\beta\right] = \frac{1}{2}\left[c_\alpha c_\beta + s_\alpha s_\beta\right]^2 \tag{15.122}$$

$$= \frac{\cos^2(\alpha-\beta)}{2} \tag{15.123}$$

which is the same as our previous result; the density matrix formalism produces the same result that was obtained using the pure quantum state alone.

In the case of the mixed state, there are no corresponding pure states to consider. The density matrix brings a substantial simplification. For the mixed state,

$$\langle P_{VV}(\alpha,\beta)\rangle_{mix} = \mathrm{Tr}\left[\rho_{mix}P_{VV}(\alpha,\beta)\right] \tag{15.124}$$

$$= \frac{1}{4}\left[c_\alpha^2 c_\beta^2 + c_\alpha^2 s_\beta^2 + s_\alpha^2 c_\beta^2 + s_\alpha^2 s_\beta^2\right] \tag{15.125}$$

$$= \frac{1}{4} \tag{15.126}$$

which means that if the photons at a and b are on average unpolarized, and if there is no correlation in the polarization, then there is 50% probability of detecting one of a pair at either, so the overall probability is 25%.

With these results, it is easy to determine that

$$\langle P_{VV}(\alpha,\beta)\rangle_p = (1-p)\langle P_{VV}(\alpha,\beta)\rangle_{mix} + p\langle P_{VV}(\alpha,\beta)\rangle_{ab} = \frac{1-p}{4} + \frac{p\cos^2(\alpha-\beta)}{2} \tag{15.127}$$

and when $p = 1/2$ we get the same result as we did for the simple product state $|\Psi\rangle_{def}$, e.g., with the two photon polarizations fixed at an angle γ when they were produced (see Eq. (15.59)). It is interesting to recall that if $p > 1/3$ the two photons are entangled; however, the simple product state $|\Psi\rangle_{def}$ for which $p = 1/2$ does not violate Bell's inequality. The implication is that the Bell inequality is not a universal test for entanglement which might exist to some degree even if the inequality is satisfied.

We can further explore the degree of entanglement in $|\Psi\rangle_{def}$. Let us determine the density matrix for this case. The two-photon state Eq. (15.61) is

$$|\Psi\rangle_{def} = (\cos\gamma|H\rangle_a + \sin\gamma|V\rangle_a)(\cos\gamma|H\rangle_b + \sin\gamma|V\rangle_b), \tag{15.128}$$

and we avoided taking the matrix elements squared at that time because of the alarming amount of algebra. However, use of the density matrix simplifies the calculations. The density matrix can be readily determined as (let $\cos\gamma = c_\gamma,\ \sin_\gamma = s_\gamma$)

$$\rho_{def}(\gamma) = \begin{pmatrix} c_\gamma^4 & c_\gamma^3 s_\gamma & c_\gamma^3 s_\gamma & c_\gamma^2 s_\gamma^2 \\ c_\gamma^3 s_\gamma & c_\gamma^2 s_\gamma^2 & c_\gamma^2 s_\gamma^2 & c_\gamma s_\gamma^3 \\ c_\gamma^3 s_\gamma & c_\gamma^2 s_\gamma^2 & s_\gamma^2 c_\gamma^2 & c_\gamma s_\gamma^3 \\ c_\gamma^2 s_\gamma^2 & c_\gamma s_\gamma^3 & c_\gamma s_\gamma^3 & s_\gamma^4 \end{pmatrix} \tag{15.129}$$

This will need to be averaged over γ from 0 to 2π. The odd terms average to zero, so the density matrix becomes

$$\rho_{def} = \begin{pmatrix} \frac{3}{8} & 0 & 0 & \frac{1}{8} \\ 0 & \frac{1}{8} & \frac{1}{8} & 0 \\ 0 & \frac{1}{8} & \frac{1}{8} & 0 \\ \frac{1}{8} & 0 & 0 & \frac{3}{8} \end{pmatrix}. \tag{15.130}$$

Note that ρ_{def} satisfies

$$\mathrm{T_b}\,\rho_{def} = \rho_{def}, \tag{15.131}$$

and because we started with an allowed density matrix with non-negative eigenvalues of 0, $\frac{1}{4}$ (twice), and $\frac{1}{2}$, this density matrix is *not* entangled according to the PPT criterion. The correlation can be determined from the average over γ of

$$\begin{aligned}\langle P_{VV}(\alpha,\beta)\rangle &= \langle \mathrm{Tr}\, \rho_{def}(\gamma) P_{VV}(\alpha,\beta)\rangle = \mathrm{Tr}\, \langle \rho_{def}(\gamma)\rangle P_{VV}(\alpha,\beta) && (15.132)\\ &= \mathrm{Tr}\left[\rho_{def} P_{VV}(\alpha,\beta)\right] && (15.133)\\ &= \frac{1}{2} c_\alpha s_\alpha c_\beta s_\beta + \frac{3}{8}(c_\alpha^2 c_\beta^2 + s_\alpha^2 s_\beta^2) + \frac{1}{8}(c_\alpha^2 s_\beta^2 + c_\beta^2 s_\alpha^2) && (15.134)\\ &= \frac{3}{8}(c_\alpha c_\beta + s_\alpha s_\beta)^2 + \frac{1}{8}(c_\alpha s_\beta - c_\beta s_\alpha)^2 && (15.135)\\ &= \frac{3}{8}\cos^2(\alpha-\beta) + \frac{1}{8}\sin^2(\alpha-\beta) = \frac{1}{8} + \frac{1}{4}\cos^2(\alpha-\beta), && (15.136)\end{aligned}$$

which is the same result that we obtained earlier by use of geometrical arguments. Because one of the eigenvalues of ρ_{def} is zero, ρ_{def} might be close to being an entangled system. Writing the density matrix in a different form allows us to determine how much reduction of the completely mixed component is needed to make one of the eigenvalues negative, and thus satisfy the PPT criterion for entanglement. The density matrix can be rewritten by separating out ρ_{mix} as

$$\rho_{def} = \begin{pmatrix} \frac{3}{8} & 0 & 0 & \frac{1}{8} \\ 0 & \frac{1}{8} & \frac{1}{8} & 0 \\ 0 & \frac{1}{8} & \frac{1}{8} & 0 \\ \frac{1}{8} & 0 & 0 & \frac{3}{8} \end{pmatrix} = \frac{1}{2}\begin{pmatrix} \frac{1}{2} & 0 & 0 & \frac{1}{4} \\ 0 & 0 & \frac{1}{4} & 0 \\ 0 & \frac{1}{4} & 0 & 0 \\ \frac{1}{4} & 0 & 0 & \frac{1}{2} \end{pmatrix} + \frac{1}{2}\begin{pmatrix} \frac{1}{4} & 0 & 0 & 0 \\ 0 & \frac{1}{4} & 0 & 0 \\ 0 & 0 & \frac{1}{4} & 0 \\ 0 & 0 & 0 & \frac{1}{4} \end{pmatrix} = \frac{1}{2}\rho_{def,0} + \frac{1}{2}\rho_{mix}. \quad (15.137)$$

where a new matrix $\rho_{def,0}$ is defined.

The effect of changing the relative amount of ρ_{mix} can be investigated by introducing a parameter $0 \le p \le 1$, so that

$$\rho_{def,1-p} = (1-p)\rho_{mix} + p\rho_{def,0} \quad (15.138)$$

which produces $\rho_{def,0}$ when $p = 1$. Noting that $\mathrm{T}_b\, \rho_{def,1-p} = \rho_{def,1-p}$, the PPT criterion can be directly applied. A numerical calculation shows that if $p > 1/2$ one eigenvalue becomes negative and the system is entangled. Slightly surprisingly, that means $\rho_{def,1-p}$ is not an allowed density matrix for $p > 1/2$. However, we can calculate the two-photon correlation as

$$\begin{aligned}\langle P_{VV}(\alpha,\beta)\rangle_{def,1-p} &= p\langle P_{VV}(\alpha,\beta)\rangle_{def,0} + (1-p)\langle P_{VV}(\alpha,\beta)\rangle_{mix}\\ &= \frac{p}{2}\cos^2(\alpha-\beta) + \frac{1-p}{4}, \quad (15.139)\end{aligned}$$

with the understanding that it is physically valid only for $p \le 1/2$, and this is the same result that was obtained for ρ_p.

In closing this discussion, let us recapitulate a few points. Previously, we found that if we write $E(\alpha,\beta) = \eta\cos(2(\alpha-\beta))$, the Bell inequality will be violated if $\eta > 2/3$. In the case of ρ_p, we found that $\eta = p$, and therefore Bell's inequality will be violated only if $p > 2/3$. We also found that ρ_p is entangled for $p > 1/3$, which is a factor of two smaller than required to violate the Bell inequality, but in general depends on the form of the density matrix. In the case of $\rho_{def,1-p}$ we are limited in the amount of the mixed density matrix that can be removed and still have a valid density matrix, and such a state will always satisfy the Bell inequality.

We can therefore conclude that entanglement is a necessary requirement for a violation of Bell's inequality, but satisfying the inequality is NOT proof of a lack of entanglement.

15.7.4 What does the Bell inequality tell us?

Weinberg has suggested that quantum mechanics should be formulated in terms of the density matrix rather than the state vector, which avoids the whole issue of the wave function changing instantaneously when one component of an entangled system is measured. Stating that quantum mechanics should be based solely on the density matrix is really just a restatement of the statistical nature of the theory, and underscores the fact that experimental results are ensemble averages of measurements on individual states. This formulation also precludes the possibility of assigning specific meanings to single measurements that can only be understood in the context of an average.

The point is that we can't form any conclusion about the relative angles between the two polarizers based on a single coincidence measurement.

Part of the problem in discussing the Bell inequalities and similar effects is that there is often language like "setting a detector and making a measurement, thereby forcing a particular result on a distant detector." This is perhaps a misrepresentation of what is going on. Measurement is 100% passive. In the measurement process, a particle is presented with different outcomes at the detection state selection apparatus, and the choice is random, based on the probability to be in any particular state.

By focusing on the state vector, and saying that the state vector of one particle is instantly changed by the measurement of the other particle, an important point is missed. The only thing we know for sure is that if one particle is in a particular eigenstate, and if we are doing a coincidence measurement, then there will be no coincident signal if the detector for the second particle is set to measure an orthogonal eigenstate. Setting the detectors to orthogonal states does not stop detections, only coincident ones. To decide that the detector polarizers are orthogonal requires a long observation, with many single but not coincident detections in each detector. Only then can we surmise that the bases are orthogonal.

Nonetheless, the fact that separated measurements indicate a correlation in the separated states with no apparent additional label in the state is a consequence of quantum mechanics that has no classical analog.

We can probe the nature of the state vector by doing statistical analysis of the correlation of measurement outcomes. If we put in a hidden variable, then we are never sure of what the eigenstate will be compared to the settings on our experiment. Take out the hidden variable, and we know that there are only two projection eigenstates. When a hidden variable is added to explain why measurements are correlated, the correlation is instead reduced, which is perhaps counterintuitive. From the quantum mechanical point of view that only the two measurement eigenstates are possible (but we can set those at will), the classical world might appear as strange.

The density matrix is merely a way to keep the algebra straight while doing ensemble averages; this averaging results in mixed states which are difficult to represent with pure state wave functions. The density matrix really only needs to be introduced at the time of measurement, but if we have an ensemble interacting with a random field, using the density matrix and the Liouville equation and all those tools really makes the calculation simpler.

The correlation between separated particles is really strange and new, and completely nonclassical. According to the postulates of quantum mechanics, the results of measurements are eigenvalues (or projections onto eigenstates) of the measurement operator. If both measurements are the same, the same results are obtained for both particles, always a coincidence, but that happens only half the time for the Bell-type states we've been discussing. So it's 50-50 whether the two particles are in a particular eigenstate, and we can never predict whether we will get a count in one or the other detector. That is the usual quantum mechanical result. We can ask, how could the two particles know to be in a particular state without a hidden variable? The answer is that they don't know, but that it is a 50-50 choice. We could ask the same thing about a one-particle state: How does the particle know when we change measurement basis states? The difference is that the single particle acts locally; in the case of two separated particles, the correlation is nonlocal.

If the measurement eigenbases are different, it then appears that one measurement provides the eigenbasis, and the other will be a projection onto that basis. But we can't say which of the two measurements was the projection, or which particle was in an eigenstate of the measurement basis. Therein lies the rub; the problem is fraught with uncertainty and the notion of one particle determining the state of the other is in fact not a possibility or perhaps not even relevant.

15.8 Conclusion

As physics is an experimental science, the debate on the EPR paradox can only be settled by experimental measurement. The introduction of Bell's inequalities provided a path to clear experimental signatures of the nonlocal properties of quantum mechanics resulting from entanglement. These experiments have established that a state is not in any particular measurement eigenstate until the measurement process occurs, which is one of the fundamental postulates of quantum mechanics. The mystery is that we are free to choose the bases for the measurements of two entangled particles independently and remotely; however, the two particles seem to be aware of each other's bases, in that one or the other appears to determine the basis, with the remaining particle's measurement outcome instantly fixed by its projection on this remotely-determined basis. This is the fundamental notion of nonlocality, and it is an undeniable feature of quantum theory.

In the entangled photons example considered in this chapter, there is a quantifiable difference between the quantum and "hidden variable" (photon polarizations determined before measurements) cases. The experimental evidence for no hidden variables is overwhelming. Indeed, it is interesting that quantum mechanics, with its fundamentally necessary statistical interpretation, actually has a stronger degree of correlation than a semiclassical theory where the polarization information is carried as an internal hidden variable.

We are left to ponder what this means. There are a few points to consider. First, again consider the photon experiment that was discussed above. When the polarizers are set to different angles, we don't know which photon was in the measurement eigenbasis; all we know is that the total coincidence rate is lower than when the polarizers are at the same angle. Only 50% of the photons at each detector are actually detected because the polarization appears as random. The most important point is that the correlation is built upon the detected photons. If the first photon of an entangled pair at polarizer A passes through unattenuated, and if polarizer B is set to the same angle, we know that the second photon will pass through polarizer B unattenuated. Overall the efficiency is 50%. If the A and B polarizers are set to be orthogonal, then if a photon transmits through A, the second coincident photon of the pair will never transmit through B, and vice versa. If the relative angle is set to some $\theta = \alpha - \beta$, then if a photon transmits through A, the probability that the second will transmit through B is $\cos^2 \theta$, or vice versa. We can't say which polarizer determined the eigenbasis, only that the probability of detecting both photons (in coincidence) depends on $\cos^2 \theta$. The problem arises if we try to ascribe this correlation to some internal variable; when we do that, we get a result that does not correspond to observations, as quantified through the use of the Bell theorem.

It is remarkable that EPR unwittingly discovered a new feature of quantum mechanics in their attempts to prove that the theory is incomplete. In a certain sense, it was incomplete, but only in its understanding. Schrödinger recognized that entanglement is a feature of the theory, and eventually the understanding that this can be experimentally tested was developed, initially by Bohm, but principally by Bell. Nowadays we have entire industries built around the well-established notion of entanglement (quantum information), and its experimental verification is irrefutable. In a certain sense Bohr was correct when he said, "Use the mathematics and it will provide the correct result." Of course that leaves us craving an understanding in terms of our classical notions of reality. Of all the quantum phenomena, perhaps nonlocality and entanglement are the most difficult to accept, and Einstein referred to nonlocality as "spooky action at a distance." We know of course that the apparent nonlocal effects are not action at a distance, but a type of correlation between system components that has no classical analog, and that is consistent with the measurement postulates of quantum mechanics. So although quantum mechanics strikes us as weird, at least it is consistent. As pointed out by Gödel, consistency within an axiomatic system is not proof of its completeness.

In attempting to call attention to what seemed a weakness of quantum mechanics, EPR and Schrödinger discovered entanglement (in spite of Pauli's comment that it was all "trivial" for guys like us), which led to the experimental demonstration that local theories could not reproduce the correct predictions of quantum mechanics. Entanglement also turns out to be the foundation of quantum computing (see Chapter 21), and we might expect that the results of repeated measurements as needed in quantum computing can be directly theoretically elucidated through use of the density matrix.

It is important to note, the seldom mentioned fact, that In both quantum mechanics and classical local theories the existence of correlations is related to a conservation law. For example, if we were measuring the correlation of electron spins from an original singlet state, simultaneously measuring two spin-up electrons would violate conservation of angular momentum (as applied in Equ. 15.72). In the quantum case there is a new type of correlation due to the choice of measurement basis as first emphasized by EPR. It is important to note, the seldom mentioned fact, that in both quantum mechanics and classical local theories the existence of correlations is related to a conservation law. For example, if we were measuring the correlation of electron spins from an original singlet state, simultaneously measuring two spin-up electrons would violate conservation of angular momentum (as applied in Equ. 15.72). In the quantum case there is a new type of correlation due to the choice of measurement basis as first emphasized by EPR.

16
Weimar culture and quantum mechanics

16.1 Introduction

During the decade following the introduction of Bohr's atomic model in 1913, research focused on applying the Bohr-Sommerfeld model and Bohr's correspondence principle to problems beyond the hydrogen atom: many-electron atoms and the influence of applied electric (Stark effect) and magnetic (Zeeman effect) fields on atomic spectra. By 1923–24 it was evident that the model was insufficient and a new approach was needed. Heisenberg's "reinterpretation" of kinematical variables (the introduction of matrices) in the summer of 1925 was followed by two years of intensive work during which the modern form of the nonrelativistic theory was established, and most of the interpretational issues were identified and to some extent dealt with. These results were obtained by a relatively small number of mostly young (many were postdocs and even graduate students), mostly German-speaking (Dirac and de Broglie being notable exceptions) scientists. Heisenberg, Pauli, Jordan, Dirac, Wigner, and von Neumann were all born between 1900 and 1903.

On reading the original papers with their often tightly formulated arguments, as well as the large amount of correspondence between the major contributors, we get the impression that they were living in a rather calm environment where they were able to devote their energy to their physics research, unburdened by the day-to-day cares of ordinary life.

Nothing could be further from the truth. Between 1914 and 1918, Europe was consumed by the "Great War", and afterward Germany went through a profound transformation of the whole society, during which citizens were asked to give up all the ideas and beliefs that had been part of their identity, all of the customs, cultural symbols, and rituals that had been the foundation of a society based on the absolute rule of the kaiser. Many were unable to make the transition, and a significant fraction of the population remained hostile to the post-war democratic republic. The post-war government—known as the Weimar Republic because the remarkably democratic constitution had been adopted in the city of Weimar[1]—was subject to extremes of political and economic instability due mainly to the pressures exerted by the victorious powers, who were activated by the belief in German war guilt and the desire to make Germany pay for the immense war damage, suffered mainly by France and Belgium.

[1]Berlin was too dangerous at the time due to political violence in the streets.

The Historical and Physical Foundations of Quantum Mechanics. Robert Golub and Steven K. Lamoreaux, Oxford University Press.
 DOI: 10.1093/oso/9780198822189.003.0016

The Weimar Republic was remarkably successful for a number of years, dealing with economic and political crises and forming the backdrop for rapid cultural and technological changes and the development of many features of what we now see as a modern society, until the pressures associated with the Great Depression that started in 1929 proved too much for the democratic system. The political parties were unable to compromise and political power was taken over by the president (Hindenburg, who had commanded the army in World War I), who ruled by decree, as was foreseen in the constitution. This led to Hitler's appointment as chancellor in 1933, followed by his seizure of power.

The question naturally arises: Did the social and cultural environment (milieu) have an influence on the scientific output of the physicists living in such unstable times? This question has been discussed by many historians of science. While many working physicists are appalled by the notion that the development of their science might have been influenced by anything other than the logical development of the field informed by experimental observation, Erwin Schrödinger for one was more open to the idea of possible cultural influences, as we shall see.

In any case it seems that awareness of these arguments might be helpful for a deeper understanding of quantum theory.

16.2 The Weimar Republic, a brief history

After the German high command agreed to an armistice in the First World War, mutinies and revolution broke out in many places in Germany.[2] Unrest in the streets of Berlin forced the Kaiser to abdicate on November 9 1918, even before the armistice was signed on November 11. Workers' and soldiers' councils (modeled on the Russian soviets) were formed in many regions. A provisional government was formed with Friederich Ebert as president. One of its first acts was to forge an agreement with the military, under General Groener, and the "Freikorps"—groups of armed men, deserters, and returning soldiers—to suppress the various rebellions. The government agreed not to try to reform the power structure of the army in exchange for the army's help in suppressing rebellion. The government supported the maintenance of order and discipline in the army, which gained a new entry to power and prestige. This pact with the army has been called the original sin of the republic and insured right-wing, anti-democratic forces a strong influence in the years to come. Unrest continued even as a national assembly was formed in Weimar in January 1919 to write a constitution, which would be ratified in August 1919. The constitution provided for a strong president, with a seven-year term and the power to dissolve the Reichstag (parliament), appoint the chancellor, and to take charge (rule by presidential decree) in an emergency.

In Munich a Soviet-style republic was formed in April, following the murder of the prime minister of the revolutionary government, Kurt Eisner, and a series of general strikes. The Bavarian government fled Munich but regrouped and returned with the

[2] Storer, C., *A Short History of the Weimar Republic* (London: I. B. Tauris, 2013); Weitz, E. D., *Weimar Germany: Promise and Tragedy*, (New and Expanded Edition) (Princeton: Princeton University Press, 2013); Hiden, J. *The Weimar Republic*, second ed.(Harlow, Essex: Addison Wesley Longman, 1996).

support of the German army, voluntarily formed citizen forces, and the "Freikorps", which suppressed the revolutionary government with much brutality.

The then 17-year-old high school student Werner Heisenberg described the conditions some 50 years later:[3]

Munich was in a state of utter confusion. On the street people were shooting at one another, and no one could tell precisely who the contestants were. Political power fluctuated between persons and institutions few of us could have named. Pillage and robbery (I was burgled myself) caused the term "Soviet Republic" to become a synonym of lawlessness, and when, at long last a new Bavarian government was formed outside of Munich, and sent its troops into the city we were all of us hoping for a speedy return to more orderly conditions.

Through an acquaintance, Heisenberg was recruited by his friend's father, who "took command of a company of volunteers," to act as a guide to the invading army. "Our adventures were over after a few weeks; then the shooting died down and military service became increasingly monotonous." Heisenberg's main duties consisted of writing reports and running errands, such as delivering guns where needed. His last year in high school began in September 1919, some months after these events.

Max Born, writing some 50 years after the event,[4] described conditions in Berlin at the end of 1918. The students at the Berlin University had formed a council modeled after the workers' and soldiers' soviets. They had deposed and locked up the rector and other members of the administration. On account of his left-wing views, Einstein had been asked to intervene, and he in turn asked Born and the psychologist Max Wertheimer to assist him. The student soviet was meeting in the Reichstag (parliament) building. There were large crowds of citizens and soldiers surrounding the building, but Einstein was soon recognized and the trio was escorted into the student council's meeting room. New statutes for the university were being discussed and after they had received a positive vote, the chairman asked Einstein what he thought of them. According to Born, Einstein replied:

The German universities' most valuable institution is academic freedom...the lecturers are in no way told what to teach, and the students are able to choose which lectures to attend...Your new statutes abolish all this and replace it by precise regulations. I would be very sorry if the old freedom were to come to an end.

The student council decided it had no jurisdiction in the matter of the rector's detention and referred the trio to the new government. Conditions in the Reich Chancellor's Palace, the seat of the new government, seemed rather chaotic. As Born describes it, there were

people running about the corridors...more or less shabbily dressed and carrying briefcases—socialist delegates and delegations from the workers' and soldiers' councils. The main hall was full of excited people talking in loud voices. But Einstein was recognized at once, and we had no difficulty in getting through to the newly appointed President Ebert...who said we would appreciate that he was unable to pay attention to minor matters that day when the very existence of the Reich itself was in the balance.

[3]Heisenberg, W., *Physics and Beyond: Encounters and Conversations* (New York: Harper & Row, 1971), p. 7.

[4]*The Born-Einstein Letters, Correspondence between Albert Einstein and Max and Hedwig Born from 1916 to 1955* with commentaries by Max Born, Macmillan, 1971, p. 149-151 and Einstein to Born 7 Sept., 1944, p. 148.

Ebert wrote a note to the responsible minister, "and in no time at all our business had been concluded." Years later, in September 1944, Einstein wrote to Born:

Do you remember, less than 25 years ago that we were together in a tram going to the Reichstag, convinced we could help make them real democrats? How naive we were, even as 40-year-old men. I can only smile when I think about it. We didn't understand how much more is contained in the backbone than in the brain and how much firmer it is.

As can be seen from the timeline (table 16.1), the early years of the republic were plagued with political assassinations, unrest in the streets, and galloping inflation. It was also a time of technological and cultural dynamism.

The economy was under great pressure as a result of the losses of territory and the associated economic activities and the demands for reparations payments by the Allies. Germany lost 13% of its territory and 6 million citizens, 14.6% of its arable land, 74.5% of its iron ore, 68% of its zinc ore, 26% of its coal production, as well as the potash mines, and the textile industry in Alsace. It was forced to give up all merchant ships over 1600 tons, half of those between 1000 and 1600 tons, and a quarter of its fishing fleet, rail locomotives, and rolling stock, plus the property in the lost areas. It had acquired a war debt of 1.5×10^{11} marks and would be required to pay reparations of comparable size (see timeline).

With regard to the Treaty of Versailles and its enforcement, the Allied leaders were under domestic political pressure to impose a harsh peace. Germany was seen as guilty for initiating the war and for the immense destruction which resulted. The new German government was forced to sign a "war guilt clause" recognizing Germany's responsibility, which in turn alienated the German people from the new republic and strengthened the hand of those who longed for the former aristocratic state. On the other hand, the German population believed Germany had fought a defensive war and felt that the new democratic state should be treated leniently. To some extent, all sides were victims of their own wartime propaganda. The signing of the Treaty of Versailles by a left-of-center coalition led to them being branded as traitors and undermined support for the republic.

The pre-war political parties mostly reorganized after the war and a few new ones were formed at the beginning of the Weimar period, notably the Communist and Nazi Parties. Many of those in positions of power under the old regime—judges, prosecutors, civil servants, and military officers—remained in power. The taking over of the old state apparatus practically unchanged was a grave historical error.

From left to right, the political parties of the Weimar period were

Communist Party of Germany (KPD) Founded December 1918.

Independent Social Democrats (USPD) Left wing of the SPD who opposed the party's support of the war. Dissolved in December 1920, with members splitting between KPD and SPD.

Social Democrats (SPD) Party supporting parliamentary democracy and workers' interests, prepared to compromise with parties on its right, fiercely anti-communist—the "socialists who would not socialize."

German Democratic Party (DDP) Bourgeois, progressive, liberal, intellectual, middle-class professionals, including Jews. A small minority of academics and industrialists, non-socialist democrats. Supported by AEG (German General Electric) and Siemens.

German People's Party (DVP) Monarchist, anti-socialist, mouthpiece of industry, favoring the rollback of gains won by the workers in the revolution, management control of enterprises, and an independent middle class. Included the largest number of academics of any party. It cooperated with the republic after 1923 and favored maximizing exports. The party opposed immigration (due to closet anti-Semitism). Its founder, Gustav Stresemann, initially a monarchist who supported imperialistic war aims, became a strong supporter of the republic and served as premier and foreign minister several times. He was a powerful voice for reason and compromise who worked to restore Germany.

Center Party Roman Catholic, a religious rather than a class-based party, with a diverse social base. Supported compromise with the republic but opposed (with DDP) socialist threats to private property. Divided between a liberal wing that supported social reform and a conservative, authoritarian wing. Liberals dominated the party at first but the conservatives gained control after 1929.

German National People's Party (DNVP) Consisted of landed aristocracy, certain business leaders, army officers, and high-level bureaucrats. The party supported high tariffs to keep food prices high and opposed the Treaty of Versailles and all subsequent international treaties. It was a racist party that favored a hierarchical society and a strong military, believed Germany had been betrayed by Jews and socialists, and even supported political assassinations. It was supported by many university professors, about one-third of the academics.

National Socialist German Workers Party (Nazi) (NSDAP) Formed February 1920.

Millions of voters read only the newspapers belonging to their party, hardening attitudes.

The so-called Weimar coalition—SPD, DDP, DVP, and the Center Party—supported the republic. With so many parties, there was always the need to build a coalition, but the coalitions were papering over deep divisions (coalitions without consensus). As time went on, the governing coalitions moved steadily to the right. The timeline gives an impression of the important developments. Because of the requirement for large reparations payments and the loss of so many economic assets, inflation steadily increased. At first inflation helped: rising prices stimulated investment and the expansion of manufacturing. There was a boom in German exports and rapid growth in the new middle class, with the rise of mass consumption and department stores. However, reparations payments became more and more difficult. Following Germany's default on payments in July 1922, the French and Belgians, accusing the Germans of using the inflation as an excuse to delay or avoid reparations payments, lost their patience. In January 1923, they sent troops into the Ruhr—a heavily industrialized region containing a concentration of mines, steel mills, and other heavy industry—to seize what wealth they could. The German government responded by

Table 16.1: Timeline of events in the Weimar Republic.

Time	Political and Economic Developments	Art and Technology	Physics
1918			
Sept	German army high command agrees to armistice and formation of a civilian government		Pauli to Sommerfeld in Munich
Nov	Mutiny in Kiel; revolution; kaiser's abdication; proclamation of republic; armistice signed; Ebner-Groener pact, army supports republic; big business - union pact; workers' and soldiers' soviets	*Dada* in Berlin; Taut-Gropius Arbeitsrat fur Kunst; first volume of Spengler's *Decline of the West* published	
Dec	Founding KPD	November Group of politically engaged artists, writers, and designers *Dada* in Cologne	
1919			
Jan	Spartacus uprising in Berlin; elections for national assembly; murder of Rosa Luxemburg and Karl Liebknecht	Otto Dix founds Dresdner Secession	
Feb	National assembly in Weimar; murder of Eisner, prime minister of Bavaria, in Munich; Ebert elected president	Expansion of telephone network	
March	Left-wing unrest in Berlin, crushed	Founding of Hamburg University	
April	Soviet republic in Munich...	Gropius founds *Bauhaus* in Weimar	Seventeen-year-old Heisenberg serves in army suppressing revolution (until June)
May	...repressed with much brutality	First transatlantic commercial flight	
June	Scheidemann government resigns; Bauer chancellor; Treaty of Versailles signed		
Aug	Constitution ratified by national assembly		
Nov			Max Planck and Fritz Haber receive Nobel Prize; Eddington confirms general relativity
Dec		Krupp steel works transitions from manufacturing arms to making railway engines	
1920			
Jan	Exchange rate 64.8 marks/$...was 8.9 in previous year; Freikorps disbanded		

Time	Political and Economic Developments	Art and Technology	Physics
Feb	Nazi Party (NSDAP) formed; Hitler announces manifesto (anti-Semitic)	*Cabinet of Dr Caligari*, expressionist horror film, is released	
March	Right-wing Kapp Putsch, defeated by general strike of trade unions		Sommerfeld introduces a fourth quantum number
	Communist uprising in Ruhr defeated by troops		
June	Elections: right wing strengthens, "bourgeois" parties form coalition government (Center, DDP, DVP); unrest over increased food prices	International *Dada* Fair in Berlin	
Aug	Disarmament law; arms withdrawn from country		Einstein visits Copenhagen for the first time
Oct		Foundation of Association for Support of German Science	Heisenberg registers at Uni Munich and meets Pauli
Nov	Swastikas banned in schools		
Dec	Split in USPD: left wing joins KPD, remaining return to SPD		
		Klee to *Bauhaus*; Grosz and Herzfeld prosecuted; Beckmann exhibit;	
1921			
Jan	Reparations set at 269 billion marks, payable over 42 years		
Feb	Founding of German of air mail		
March	France occupies part of Ruhr to enforce reparations payments; Communist action		Bohr Institute founded at Copenhagen
April	Miners' strikes to protest French action		Born starts his institute at Gttingen
June	Reparations reduced to 132 billion marks		
July	Hitler chairman of NSDAP; exchange rate 76.7 marks/$		
Aug	Erzberger (Treaty of Versailles negotiator) assassinated; US-German peace treaty		
Sept	AVUS auto race track opens in Berlin		
Oct			Pauli to Born in Gttingen
Nov	SA formed (Nazi Party paramilitary organization)		

Time	Political and Economic Developments	Art and Technology	Physics
	Henry Ford's *The International Jew-The World's Foremost Problem* published in German: huge success, possible source of Hitler's *Mein Kampf*	Otto Dix makes the *Circus* series of etchings; Grosz visits Moscow, disenchanted	Einstein attempt to demonstrate existence of photons refuted by Ehrenfest
1922			
Jan	Rathenau becomes foreign minister; exchange rate 191.8 marks/$		
Feb	Strikes on railways and factories		
March		Film *Nosferatu* is released	
April	Treaty of Rapallo with Soviet Union; economic and military cooperation with Soviet Union		Pauli to Hamburg (big city man); Stern-Gerlach experiment demonstrates "spatial" quantization
May		International Congress of Progressive Artists	
June	Rathenau assassinated; acid attack on Scheidemann (SPD)	Radio transmission of pictures Berlin-USA	"Bohr Festival" at Göttingen; Bohr meets Heisenberg and Pauli
Aug	Large demonstrations of ultra-nationalists in Munich		
Sept	Right of USPD returns to SPD	*Dada* Congress	Einstein cancels talk at Leipzig because of anti-Semitic threats
Oct	Beginning of hyperinflation; Mussolini marches on Rome, seizes power		Heisenberg spends six months with Born; Einstein, director of Kaiser Wilhem Institute for Physics, appoints von Laue as his substitute;
Nov	Cuno chancellor (head of shipping line); Mussolini becomes dictator	Art exhibition sent from USSR to Berlin First *Bauhaus* exhibit	Einstein (1921) and Bohr (1922) receive Nobels
1923			
Jan	French and Belgians occupy entire Ruhr; strikes; passive resistance; French arrest strikers and seize railways and mines; exchange rate 18,000 marks/$; first NSDAP conference in Munich		
April		Freud publishes *Ego and Id*	
May	Miners' strike		Compton effect is published
June			Pauli: anomalous Zeeman effect suggests replacing derivatives by differences
July			Einstein -> Copenhagen; Heisenberg (barely) gets PhD

Time	Political and Economic Developments	Art and Technology	Physics
Aug	4.6 million marks/$; Cuno resigns; Stresemann chancellor; "Great coalition"		
Sept	Passive resistance ended in Ruhr; loaf of bread costs 10.37 million marks		De Broglie matter - wave relation
Oct	Communist uprisings in Hamburg, Saxony, Rhineland; new central bank "Rentenbank"	Radio broadcasting begins; Berlin Templehof airport opened	Heisenberg → Born's assistant; Pauli → Hamburg after a year in Copenhagen; Heisenberg quotes Born as looking for a "discretization" of atomic physics
Nov	SPD quits government; hyperinflation; barter; food riots; Rentenmark (1:10^{12}) replaces mark; Stresemann resigns; inflation contained; Hitler Putsch fails in Munich; students strongly support Hitler		
1924			
Jan	French withdraw from Ruhr	Dix publishes *The War* triptych; Roth Rebellion; *New Objectivity* movement develops among artists; first political radio broadcast	Failure to explain the anomalous Zeeman effect and the spectrum of helium point to the need for a "new" quantum theory involving "discretization" (Born, Heisenberg) and rejecting the idea of definite electron orbits (Heisenberg, Pauli)
Feb	Unemployment 2.6 million		BKS theory: no photons but only statistical energy conservation
March			Heisenberg visits Bohr; Kramers derives dispersion formula using quantum mechanics
April	Hitler sentenced to five years for failed Putsch		
May	Elections swing to communists and nationalists	Opel introduces production line in Germany	
June		George Grosz chairman of Communist Artists Association; opening of Institute for Social Research, Frankfurt	Born derives Kramers's dispersion formula by replacing derivatives with differences
July			Bose derives Planck's law using statistics; Einstein applies Bose statistics to atoms
Sept	Dawes Plan reduces reparations, proposes French leave Ruhr, allows foreign loans to Germany	First advertising on German radio	Heisenberg goes to work with Bohr until May 1925
Oct		First Zeppelin crosses Atlantic	

Time	Political and Economic Developments	Art and Technology	Physics
Nov		Thomas Mann publishes *The Magic Mountain*	De Broglie thesis approved with support of Einstein
Dec	New elections support to moderates and SPD; stabilization begins; Hitler released after serving nine months	First German radio exhibition in Berlin	
		Leica camera, 35 mm film introduced—great technical innovation; new “Deutsches Museum” dedicated to science and technology opens in Munich; first phonograph with an amplifier	
1925			
Jan	DNVP join government		Pauli introduces a double-valued fourth quantum number and the exclusion principle; Kramers and Heisenberg extend dispersion formula to incoherent scattering of light; Einstein relates Bose statistics to de Broglie waves
Feb	NSDAP re-founded after Hitler released from prison, Hitler gains more control; President Ebert dies	There are 200,000 cars and 100,000 motorcycles in Germany	
April	Hindenburg (field marshall in WWI) elected president		Experimental refutation of BKS theory
June		Exhibition on *New Objectivity* defines movement, also called *Magic Realism*	Heisenberg’s first paper on matrix mechanics, written on Helgoland (North Sea) because of hay fever
July	*Mein Kampf, vol.1* published by Hitler; French withdraw from Ruhr		Heisenberg’s first paper (“reformulation” of QM) received, appears September 1925; Elsasser paper showing (at Born’s suggestion) observed electron diffraction is consistent with de Broglie waves received; Born recognizes importance of non-commutation
Sept			Born and Jordan paper on matrix mechanics received
Oct	Locarno Treaty (recognition of Germany’s western borders)	*Bauhaus* closes in Weimar, moves to Dessau	Goudsmit and Uhlenbeck interpret Pauli’s new quantum number as spin

Time	Political and Economic Developments	Art and Technology	Physics
Nov	SS founded as personal bodyguard for Hitler		Pauli solves the hydrogen atom using Heisenberg's matrix method; three-man paper on matrix mechanics (Born, Heisenberg, and Jordan) received; Dirac invents q-numbers, relates commutators to Poisson brackets
Dec	Locarno Treaty signed; IG Farben founded	Josephine Baker performs in Berlin	Schrödinger develops wave equation while on ski vacation (Arosa, Switzerland) with unknown woman; Lanczos field-like formulation of QM with integral equations (equivalence of matrix - integral equations shown 20 years earlier by Hilbert)
		New Objectivity exhibition continues; *Bauhaus* to Dessau; Otto Dix's first show	
1926			
Jan		Lufthansa airline founded	Schrdinger solves wave equation for hydrogen atom (paper received January, published February); Born in USA introduces time-dependent operators with Norbert Wiener; Pauli submits calculation of hydrogen spectrum in crossed E and B fields; Dirac solves hydrogen atom
Feb	2.4 million unemployed		Three-man paper (Born, Heisenberg, Jordan) appears; Fermi found statistics for particles obeying exclusion principle; Schrdinger wave equation for many-electron systems
March			Schrdinger proof of wave mechanics - matrix mechanics equivalence received; Heisenberg and Jordan reproduce spectra for many-electron atoms using spin and relativity
April	Germany and Soviet Union sign treaty agreeing to neutrality		Dirac applies q-numbers to Compton effect
May			Schrdinger paper on perturbation theory received; Heisenberg -> Bohr's assistant at Copenhagen
June			Dirac gets doctorate at Cambridge

Time	Political and Economic Developments	Art and Technology	Physics
July	Second Nazi Party conference; founding of Hitler Youth		Born - probabilistic interpretation of the wave function; Schrödinger—time dependent wave equation—lectures at Munich on his way to Copenhagen; Heisenberg disapproves wave theory, solves He atom using wave mechanics to calculate matrix elements
Aug			Dirac finds relation between symmetry of wave function and Fermi or Bose statistics
Sept	Germany joins League of Nations		Dirac develops transformation theory while at Copenhagen
Oct	Von Seeckt forced out as head of army, German–English industrial conference		Schrödinger visits Copenhagen, no agreement on interpretation of wave and matrix mechanics
Dec	Stresemann receives Nobel Peace Prize; Hitler speaks to industrialists	*New Bauhaus* opens in Dessau	
	Industrial production reaches pre-war levels	Max Beckmann gets Honorary Empire Prize for German Art; film *Berlin: Symphony of a Great City* is released	
1927			
Jan	DNVP back in government; first rally of NSDAP	Film *Metropolis* is released	
Feb			Dirac quantizes electromagnetic field
March	First street battle NSDAP-KPD in Berlin		Davission-Germer paper on electron diffraction received; Heisenberg paper on uncertainty principle conceived in Bohr's absence and submitted after some disagreement with Bohr
April	Ten-hour workday replaces eight-hour workday		
July	Universal unemployment insurance; abortion reform (partial); heavy industry support for NSDAP		
Sept			Como Conference—Bohr presents complementarity principle and what will be known as the Copenhagen interpretation
Oct			Solvay Conference—"completion" of the theory; Jordan and Klein—second quantization for many-body (boson) systems

Time	Political and Economic Developments	Art and Technology	Physics
		All Quiet on the Western Front published; Nazi book *Art and Race* published	
1928			
Jan			Dirac equation manuscript received; Jordan and Wigner—second quantization for fermions
Feb		Electric fast trains	
March		Association of Revolutionary Visual Artists (Asso) is founded	
May	Elections support to SPD from DNVP	Press exhibit, Cologne	
June	"Grand coalition": SPD, Center Party, DDP, and DVP		
July	Congress in Copenhagen demands legalization of homosexuality		
Aug	Germany, France, and USA sign Kellogg-Briand Pact banning war	*The Threepenny Opera* (play by Brecht, music by Weill) opens in Berlin—exposes the hypocrisies of capitalism	
Sept	Unemployment 650,000		
Oct	Hugenberg leader of DNVP (opposes republic)	Company to develop sound films	
Dec	2.8 million unemployed		

Fig. 16.1 Trillion mark note, as discussed by Storer, op. cit. Note 2, p. 392 (Public Domain).

declaring passive resistance, encouraging workers to stay away from the factories and mines in the occupied territory. The resulting loss of production, coupled with government spending on unemployment and related social payments, led to an accelerating inflation. By September 1923, when passive resistance was ended, a loaf of bread cost 10.37 million marks and communist uprisings had broken out in several places.

Due to the inflation, foreign currency had immense purchasing power. Max Born financed a family vacation on the £22 he received as royalties from England.

In order to get control of the situation, a new bank was created to issue a new currency, the "Rentenmark," backed by mortgages on all the land used by industry and agriculture. The inflation was so rapid that they waited until October when the dollar exchange rate was exactly 4.2×10^{12} (the old exchange rate had been 4.2 marks to the dollar) so they could introduce the new mark at a rate of 10^{12} to the old one. Figure 16.1 shows a 10^{12} mark note (in German 1 billion=10^{12}) issued by the German railroad on October 27, 1923.

This stopped the inflation but at the cost of brutal austerity and unemployment in the millions. Twenty-five percent of government employees were fired. There were major reductions in social welfare payments. Employers went on the offensive. By the spring of 1924 there was a 12-hour workday in factories and workers could be fired at will. Large segments of the population lost everything. The middle classes were particularly hard hit, resulting in a brutal social leveling—the "proletarianization" of the middle classes. There was a downward slide in the economic and social status of professors and students. The republic lost the support of the middle class in the inflation and that of the workers in the subsequent stabilization. However, the stabilization resulted in an economic revival financed largely by US capital.

Political power shifted right while the Communists also gained. In July 1924 the Allies agreed to reduced reparations, Germany was allowed to accept foreign loans, and the economy began to stabilize.

At the beginning of November 1923, Hitler attempted to take over the government in Munich by force. The attempt failed in spite of strong support from the university students.

The shocking social and economic effects of the Ruhr occupation led the Allies to reconsider Germany's difficulties, and a new French leader was prepared to increase cooperation with Germany. Reparations were adjusted downward and the French withdrew from the Ruhr in July 1925, followed by the evacuation of the first zone on the left bank of the Rhine in December 1925. In 1926 Germany joined the League of Nations and 1927 saw the Allies reduce their remaining forces in the Rhineland. By 1927 industrial production had reached the pre-war (1913) level, wages were rising, and the workday was shortened.

The economic conditions improved steadily during this time, but a large fraction of the population remained hostile to the government due to nationalist sentiments and to the suffering induced by the inflation. Between 1924 and 1929, 25.5 billion marks flowed into Germany in the form of loans and investments, and the total reparations payments amounted to 22.9 billion marks. The government was unable to ensure that the recovery benefited all sections of society. The net effect of the crisis was the transfer of wealth from the lower middle class to industrialists and financiers. Table 16.1 is an attempt to give an overview of the developments inpolitics, economics, art and technology and physics.

16.3 Weimar culture[5]

The political and economic upheavals and rapid modernization created a heightened sense of anxiety among large segments of the population, and this was reflected in the culture. In 1918 the Dada movement, a group of artists rebelling against art, proclaimed the death of conventional art. The Dada manifesto expressed despair due to the violence of the war and the enormous scale of death and destruction, as well as the hopes that emerged with the subsequent revolution.

While Germany became a springboard for new ideas and technology, many remained alienated and viewed the avant-garde as degenerate and decadent. Urbanization, which was accelerated by the collapse of the monarchy and the development of the new republic, led to an association of the big cities—especially Berlin—with the avant-garde, and a kind of cultural split between Berlin and the rest of the country.

In Berlin daily newspapers devoted large amounts of space to the visible arts and the outstanding accomplishments of science. There was a continuous flow of artistic, scientific, and commercial improvisation.

To people in the countryside, Berliners seemed like visionaries and dreamers, living in a delusion. By some Berlin was seen as the cesspool of Germany, with too many Slavs and East European Jews. On the other hand, many Berliners felt contempt for outsiders, considering them provincial philistines, reactionaries, superstitious and dim-witted peasants, aristocrats, and craven officers.

This dissonance affected the physics community, where many resented and tried to overcome the dominance of the Berlin physicists and their journals. A group of nationalist and reactionary physicists, which began to form at the outset of the war, opposed the generally more progressive Berlin physicists who controlled the German Physical Society (DPG, Deutsche Physikalische Gesellschaft). Among the points of

[5]Gay, P., *Weimar Culture: The Outsider as Insider*, (New York: Norton, 2001); Barron, S., and Eckmann, S., eds., *New Objectivity: Modern German Art in the Weimar Republic, 1919-1933* (Munich: Delmonico Books, 2015); Gale, M.and Wan, K., eds., *Magic Realism: Art in Weimar Germany 1919-33 (London: Tate, 2018);* Pfeiffer, I.,*ed.*, *Splendor and Misery in the Weimar Republic* (Munich: Hirmer, 2017).

contention were the management of the scientific journals and the methods of citation. On several occasions the DPG came close to splitting.

In 1919 the Berliners in the DPG founded *Zeitschrift für Physik* to the consternation of W. Wien, a leader of the conservative group who felt that the new journal would compete with the traditional *Annalen der Physik* edited by him and Planck. It is interesting to note that Wien had wanted to fail Heisenberg for the experimental part of his doctorate examination, and Schrödinger published his main papers in *Annalen der Physik*. However, he also published in *Die Naturwissenschaften*, a journal which emphasized non-mathematical discussion of scientific and philosophical issues and was edited by Arnold Berliner, who was Jewish.

In December 1919, Sommerfeld asked Einstein to intervene against the new journal: "The new journal...troubles me greatly...Wien has already resigned from the society [DPG] because he did not receive any notice of it and sees it as hostile to the *Annalen*."

In 1920 Philipp Lenard joined Wien in promoting new rules for reviewing papers submitted for publication, which strove to increase the number of citations to the German literature. In a letter to Wien in August 1920, Lenard wrote, "everything that stands in high esteem...is to be sent to the devil. A kind of mass suicide of the German spirit whereby the Jews act now as a medium."[6] In 1925 the *Zeitschrift für Physik*, which had become the most important physics journal and had published many foreign authors, published a paper in English, an action which inflamed feelings so much that Lenard resigned from the DPG and Max Wien resigned as its chairman. Politically Lenard supported Hitler in 1924 but did not join the Nazi Party until 1937. Stark joined the Nazi Party in 1930, Jordan in 1933. In 1933 Gustaf Mie became a supporting member of the SS. In 1937 Stark published (in the SS magazine) a violent attack on Sommerfeld and Heisenberg as the "white Jews of science" and the ambassadors of the "Einstein spirit" in Germany. This was related to the struggle over the selection of Sommerfeld's successor.

There was continuous criticism, by members of this group, of both quantum mechanics and relativity.

These tensions within the physics community took place against a backdrop of dramatic changes in the outer society. The war had changed the way that women worked, behaved, and looked. After the war, there were more women than men, and women became more assertive. After losing their jobs at the end of the war, women re-entered the work force in great numbers in the 1920s. There was a sexual revolution where gender roles and conceptions of sex were challenged. Liberal attitudes to sex became more commonplace. There were a million abortions a year. By 1929 there were more than 65 bars for gay men and over 50 for lesbians in Berlin.

Every political movement encouraged communion with nature; various hiking, bicycling, and nudist clubs promoted a healthy lifestyle. Every party depicted its enemies as sick, depraved bodies; Jewish stereotypical images were common.

[6]Our treatment of the strife among physicists is based on Wolff, S. L., "The Establishment of a Network of Reactionary Physicists in the Weimar Republic", in *Weimar Culture and Quantum Mechanics*, ed. by Carson, C.. Kojevnikov, A. and Trischler, H. (London: Imperial College Press; Singapore: World Scientific, 2011), pp. 294-318.

At a Lutheran conference in 1924, speakers told of the necessity of standing against the Jewish threat to the national character, a threat posed by a totally degenerate urban spirit borne by the Jewish race.

Spengler's *Decline of the West* (first volume published in 1918, second volume in 1922), which supported the cultural pessimism of the time, had sold 100,000 copies by 1926. Spengler extolled the virtues of war and a militaristic society, a melding of nationalism and socialism, promoting national unity and struggle as the way out of the crises.

By 1925 Berlin had four million residents, and its population continued to increase by 80,000–100,000 per year. By 1928 it was the third largest city in the world. Slums were cleared; transport was modernized; modern architecture was spreading. In 1926 a radio tower for broadcasting was erected. At the same time, poverty and deprivation increased, as did youth crime rates, as well as prostitution and serial killings targeting prostitutes.

The introduction of mass consumer culture resulted in an improved standard of living for many but also fostered the spread of alienation. Films and the new illustrated magazines established the tyranny of mass taste. At the same time German cities became the home of a new spirit of inquiry and experimentation in the arts and sciences. A new young and well-educated middle class encouraged the growth of consumer culture and urbanization. This was seen by the right as a rise of decadence and a threat to traditional values.

There was an expansion of the welfare state. Women obtained equal voting rights in Germany before they gained the vote in the United States, Great Britain, or France.

Austrian and German scientists were boycotted by the international community for most of the 1920s, but they won more Nobel prizes between 1928 and 1932 than the British and Americans together.

New ideas in philosophy were produced by the "Frankfurt school"—Adorno, Marcuse and others—who believed Weimar was only a waystation to socialism. The philosopher Martin Heidegger's membership in the Nazi Party resulted in his work being the subject of widespread controversy, but he gave respectability to the contemporary German obsession with unreason, naturalism, and non-intuitive thinking. He had a far-reaching influence, and his books, whose meaning was barely decipherable, were devoured.

For Heidegger, urban life was a shallow spectacle; modernity favored seeing over understanding. He lived in a mountain hut for considerable periods. He embedded man in the world, supporting the unity of man and his surroundings. Being has meaning only in the links between subject and observer, things and agents. Man is thrown into the world, lost and afraid; he must learn to face nothingness and death. Reason and intellect are inadequate guides to the secret being—thinking is the mortal enemy of understanding.

His criticism of modernity and his pleas for the individual to recognize his deep embeddedness in a national and racial community were common to all right-wing philosophy of the Weimar period.

On the other hand the architect Erich Mendelsohn felt that the spiritual triumphs associated with the revolution would be the basis for the development of new forms of architecture and sought to resolve the tensions of modern life by grasping modernity.

Fig. 16.2 Einstein Tower, part of the Astrophysical Institute Potsdam, designed by Bauhaus architect Erich Mendelsohn, built 1920–1921. (blickwinkel / Alamy Stock Photo)

He constructed an "Einstein tower" in Potsdam (Fig. 16.2). Designed to house a solar observatory for the study of general relativistic effects on the sun, the building was meant to express the ideas of general relativity by its non-classical curves. It was built between 1919 and 1921 after a fundraising drive by the "Einstein foundation" that had been founded by Erwin Freundlich with the support of Einstein. Active research on the gravitational red shift on the sun was carried out in this observatory under Freundlich's leadership. The tower was damaged in World War II and restored in 1999. It is considered one of the landmarks of expressionist architecture.

In the arts, the revolution had inspired a bold imagining of a harmonious, beautiful future. The expressionist sensibility, ranging from the depths of despair to the heights of joy, reflected the trauma of the war and the initial hope of the revolution. Despair and hope inspired the arts.

The Cabinet of Dr. Caligari, a silent film released in 1920 and now available on the internet, highlighted the expressionist themes of alienation, madness, and the evils of authority. The original intention—to show the brutality and insanity of authority—was negated by the addition of a final scene which gave the authorities the appearance of decency and portrayed rebellion against authority as a delusion. Along with other expressionist films, it probed the psychology of the characters and emphasized emotional complexity and the layered levels of consciousness.

Fig. 16.3 George Grosz, *The Eclipse of the Sun*, oil on canvas. (©1926 Estate of George Grosz / Licensed by VAGA at Artists Rights Society (ARS), NY.)

By 1924 a more sober, practical approach to everyday reality became noticeable with a new naturalism in art ("Neue Sachlichkeit"—new objectivity) that seemed more suited to relative political stability and strong economic development. This represented a repudiation of the exorbitant intoxication of expressionism, replacing it with an enthusiasm for immediate reality, a desire to take things objectively.

The Magic Mountain by Thomas Mann expressed the conflict between the love of life and the love of death, between artistic achievement and physical survival. The novel takes place in a sanatorium for tubercular patients. Tuberculosis was seen as a symbol of the infatuation with death.

The painters George Grosz, Otto Dix, and others adopted a "warts and all" approach, which was also evident in literature and cinema. Grosz's work reflected the loathing that so many Germans felt for their elites (Figs. 16.3 and 16.4). The violence of the war had undermined deference toward authority, while the revolution that led to the Weimar Republic resulted in the feeling that barriers had been broken and all things were possible.

The architecture and designs of furniture and household objects produced by the "Bauhaus" combined beauty and function with the goal of improving everyday life. This was seen as dangerously foreign by some of those on the political right. The Bauhaus had been founded as a school of architecture and design by Walter Gropius, who expressed the hope that modernist aesthetics would transform man and society and lead to a new age of harmony and creativity. Modern architecture would heal the rift between technology and beauty, man and nature, individual and society, and fulfill the revolutionary promise of renewal and rebirth. It would overcome alienation and make mankind whole again. The cure for the ills of modernity was more of the right

Fig. 16.4 George Grosz, *The Pillars of Society*, oil on canvas. (©1926 Estate of George Grosz / Licensed by VAGA at Artists Rights Society (ARS), NY.)

kind of modernity. For others, modernism represented nothing but factory products installed in people's living space, a triumph of mechanism over true spirituality, which degraded man to a material being. The new buildings were attacked as representing "nomadic" (a term sometimes applied to Jews) architecture.

The youth movement, driven by a hunger for wholeness and estranged from the republic, yearning for a romantic past, sought inspiration from the poets and searched for an organic philosophy of life.

Rilke, who died in 1926, was the favorite poet of all the youth movements and became the inspiration of a worldwide fanatical sect. His barely comprehensible poems supported amateurish pseudo-religious needs, discrediting the intellectuality that had dominated the west for millennia. In August 1914 he had invoked the god of war and the suffering he would bring.

The "Wandervogeln" (hiking birds) organizations promoted a pantheistic love of nature and a mystical love of the fatherland. Many were anti-Semitic but others accepted Jews. There were groups associated with the major parties—Communists, Socialists, Nazis. They saw their rambling and singing around campfires as a haven from a Germany they could not respect or understand. The right wing sought the reawakening of a "genuine Germanness" while the left longed for a communally constructed society.

By 1925 the cultural landscape was calmer; the time for radical experimentation was over.

The Weimar Republic established representative democracy, enabling people to live freer lives. It led the world in social legislation and sexual tolerance. It presided over a great flowering of art, literature, science, and scholarship. It produced a generation of probing, searching artists and intellectuals, laid the foundations of theoretical physics, transformed music, philosophy, and sociology—changing the way many people think. A welfare state and an eight-hour working day were introduced; women gained the vote and the freedom to choose their lifestyles. It was one of the most open and progressive societies in Europe. But the social and economic stresses induced by the rapid social, cultural and economic changes eventually enabled Hitler to take power (1933), when the democratic system was unable to agree on a response to the immense problems caused by the Great Depression (1929).

One possible answer as to why this particular moment, this place, proved to be a cradle for so much creativity is that the old society had been proven bankrupt by the war and the revolution had brushed it aside for a while. The future seemed unbounded and open.

16.4 Physics in the Weimar Republic

The main contributors to the development of quantum mechanics were living in Germany during the Weimar period and were exposed to the political, economic, and cultural turbulence described above. We have no problem with the assertion that the work of artists, novelists, and architects is influenced by the general cultural milieu, but to ask the same question concerning the work of physicists raises several problems. To be accepted, a physical theory has to satisfy several requirements; it has to be self-consistent and agree with all experiments. It is often considered as representing an objective truth that anyone can check for themselves. Physicists tend to present new ideas as following from some logical thought pattern in order to make the ideas easier for their colleagues to understand. The development of physics is seen by most physicists as being "internally" determined. That is, new theories follow from the need to explain experiments that cannot be explained by existing theories, without contradicting the results of those experiments that can be so explained. Given this, is there any place in the development of physical theories where the cultural milieu might be able to exert an influence on the content of the theories? There is quite an extensive literature discussing this question.

Before addressing this issue, we give a brief description of the interaction of some physicists with the evolving Weimar Republic.

16.4.1 Physicists in Weimar[7]

The Weimar constitution gave the government a mandate to cultivate science and scholarship, and the attitude of the administration was in general favorable to supporting the sciences and academic life. The Weimar government wanted to demonstrate to the academic community that democracy was not hostile to higher culture. Since the academics, especially scientists and technologists, had supported the imperial regime as the guarantor of their special position in society and feared that the new reforms would destroy higher culture and eliminate their leading social position, the government acted to counteract those feelings with a generous support of science.

German society accepted the academics' new conception of international leadership in science as a substitute for the military and economic power lost in the war. Science was seen as the only element of prestige left to Germany. Thus in 1918 Max Planck told the Prussian Academy of Sciences that the one thing no enemy had taken from Germany is the position of German science in the world. The mission of the academy was to maintain and defend it with every available means. After complete disarmament and severe breakdown of the economy, German science was the only asset left. However, this was countered by an increasing distrust of rationalism, as we will see.

We start with the reaction of some physicists to the turmoil of the early revolutionary days of the republic.

In early December 1918, just after the end of the war, Sommerfeld wrote to Einstein: "I hear...that you believe in the new age and want to cooperate with it—God preserve your beliefs. I find it all unspeakably miserable and stupid. Our enemies, the greatest liars and scoundrels, we the greatest morons. Not God, but money rules the world." Einstein replied a few days later:

> It is true that I have more hopes for this age in spite of the many ugly things that you refer to. I see that the political and economic organization of our planet is progressing. If England and America can agree, there will be no more significant wars. ...If at present the transition period seems pretty hard for us, it is my opinion that it is not completely unearned. I am convinced that the culture loving Germans will soon be as proud as ever—with more grounds than before 1914. I do not believe that the present disorganization will result in lasting damage.

Following this exchange, Einstein went to Zurich to present a series of lectures.

Ewald, Sommerfeld's assistant at the time, who experienced the revolution in Munich, wrote in a letter to Epstein in May 1919: "The reaction is again strong and the contrast between the educated and the workers is greater than ever. The Bourgeoisie is unorganized and powerless. The peace conditions hardly let Germany breathe, with the result that even peacefully inclined men like me see the recourse to weapons as the best way forward." He told of meeting Laue, who was serving as a radio-lieutenant.

[7]Sources for this section are *Albert Einstein - Max Born,* Briefwechsel 1916-1955 (Munich: Nymphenburger, 1969) [English translation: The Born-Einstein Letters (London: Macmillan, 1971)]; Pauli, W., *Scientific Correspondence with Bohr, Einstein, Heisenberg a.o.*, vol. 1, 1919-1929, ed. Hermann, A., Meyenn, K. von and Weisskopf, V. F. (New York: Springer Verlag, 1979);
Mehra, J. and Rechenberg, H., *The Historical Development of Quantum Theory*, vol. 2, *The Discovery of Quantum Mechanics*, 1925 (New York: Springer Verlag, 1982);
Meyenn, K. von ed., *Quantenmechanik und Weimarer Republik* (Braunschweig: Vieweg, 1994);
Hermann, A., ed., *Einstein - Sommerfeld Briefwechsel* (Basel: Schwabe and Co., 1968).

Wilhelm Wien, one of the main representatives of the reaction, circulated a letter among physicists which encouraged resistance against English influence in physics. Some physicists agreed with this but others, including Einstein, Warburg, and many physicists of the younger generation, euphorically greeted the new society and hoped for a new, democratic spirit in scientists. They rejected the nationalism, militarism, and anti-Semitism that was the leading opinion in German academic circles at the time.

Sommerfeld was a fervent nationalist who felt that the accusations of war guilt against Germany were completely unjustified and whose political views had some influence on his two master's students, and Heisenberg. In a letter to his former colleague Epstein, a Russian citizen who had been imprisoned in Germany during the war and went to Caltech after the war, Sommerfeld wrote, "The great lie about the German war guilt will last a few years, not forever."

In January 1921 Born wrote to Epstein that American physics, which had been equal or superior to German physics in experiment, would now take the lead in theory as well, and asked him to encourage US physicists to publish new work in German journals in the German language. "German science is not lying on the floor as much as the German state, and foreigners will not be ashamed to publish in Germany." On the other hand he went on to say, "You have no idea how many valuable men are succumbing to the emergency."

After the war the Allies instituted a boycott of German science, including scientific societies, scientists themselves, and German scientific journals. In 1926 the annual meeting of the Kaiser-Wilhelm-Gesellschaft (forerunner of the Max Planck Institutes) was told that

> we are in a permanent competition with the science of other peoples...after the lost war and our current poverty, also in relation to the next generation of researchers, we have a difficult task ahead. Other countries have overtaken us in many lines of research. We must catch up with them and surpass them. ...Our scientific capital and our special enjoyment of scientific work are some of the great advantages we possess in competition with other nations. We must preserve them.

The blockade of the Allies resulted in the complete break in relations with the French and English physicists. The isolation was broken by close contact with Denmark, Switzerland, and the Netherlands. Niels Bohr promoted a regular exchange between German and Danish physicists. Copenhagen also provided contacts with physicists from other nations, so the Bohr Institute became a central communication hub for knowledge of atomic physics.

Scientific contacts with America normalized relatively quickly. In 1922–23 Sommerfeld visited the United States for a lecture tour with the goal of improving the American opinion of German science. He was followed by several prominent German physicists—Einstein, Born, and Schrödinger—who made similar tours.

In October 1917 Einstein had been named director of the Kaiser-Wilhelm Institute in Berlin with a salary of 5,000 marks. This was raised in October 1922 to 20,000 marks to offset inflation. At this time, von Laue became acting director because of Einstein's extensive travels. Grants were given mainly for carrying out experiments and purchasing equipment, with the emphasis on work relative to quantum theory. Max Born received a grant to support Stern's experiments to measure the velocity distribution of a molecular beam, which were carried out in his institute for theoretical physics in Frankfurt, where the budget available for experiments was 3,000 marks.

An exception to the preference for experiments was the granting of a stipend to Jordan, who received a grant for the years 1924–26.

In June 1919 Einstein gave his views on the political situation in a letter to Born:

[The Versailles] conditions are hard but will never be enforced. They are more to satisfy the enemy's eye than his stomach...The French are motivated by fear. ...The hardships resulting from the errors of the French are alleviated by a slovenliness which never fails. . .Eventually, Germany's dangerousness will go up in smoke, together with the unity of her opponents...may a hard bitten...determinist be allowed to say, with tears in his eyes, that he has lost faith in humanity? The impulsive behavior of contemporary man in political matters is enough to keep one's faith in determinism alive.

I am convinced that in the next few years things will be less hard than in those we have recently lived through.

Born's chair in Frankfurt was financed by a wealthy jeweler named Oppenheim.

In a letter of December 1919, Einstein says how, at a jubilee celebration for the University of Rostock, the representatives of the new government were attacked by "the academic dignitaries while the ex-Grand Duke was given a seemingly endless ovation. No revolution can help against such inborn servility." He mentions the hardship caused by the devaluation of property and continues:

The behavior of the Allies is beginning to appear disgusting even to my standards. It seems that my hopes for the League of Nations will not be realized. [In fact Germany would only be admitted to the League in 1926.] Nevertheless France seems to be suffering severely in spite of the coal imports as can be seen from its recent restrictions on railway passenger traffic. Here all the fixed and movable property is being bought up by foreigners, to the point of us becoming an Anglo-American colony. Just as well that we do not have to sell our brains or make an emergency sacrifice of them to the state.

In December 1919 Einstein wrote (from Berlin) to Ehrenfest, "Here, there is strong anti-Semitism and wild reaction, at least by the educated." In April 1920, in another letter to Ehrenfest, Einstein wrote, "Need and hunger are frightful in the city. The children's death rate is depressing. Nobody knows what to do. The state has sunk to extreme powerlessness while the main powers—the sword, money and extreme socialist gangs—struggle among themselves."

In January 1920 Einstein wrote to Born:

streams of blood will have to flow: the forces of reaction are also growing more violent all the time. ...France is playing a rather sorry role in all this. ...victory is very hard to bear. ...They [the English] and the Americans are sending emergency supplies. But little can be done in the face of this mass suffering.

The peace treaty certainly goes too far. But since its fulfillment is quite impossible it is better that its demands are objectively impossible to fulfill, rather than just intolerable.

In the same letter he continues:

That business about causality causes me a lot of trouble, too. Can the quantum absorption and emission of light ever be understood in the sense of the completely causal requirement or would a statistical residue remain? I must admit that there I lack the courage of my convictions. But I would be very unhappy to renounce *complete* causality. (Emphasis in original.)

Around this time the inflation was so bad that Born held public lectures on general relativity to earn funds to support the Stern-Gerlach experiment on spatial quantization that was being conducted in his institute, and he managed to collect 6,000 marks in entry fees.

In September 1920 there was an important meeting of the Association of German doctors and scientists. Einstein and Born attended together. At this meeting Philipp Lenard made an undisguised anti-Semitic attack on Einstein and relativity which caused Einstein to lose his temper. This was the beginning of the enmity between Lenard, who was later joined by Stark and others, and Einstein. Lenard invented the difference between "German" and "Jewish" physics. This meeting was, according to Born, the first indication of the "great danger of anti-Semitism to German science."

In October 1920 Born wrote: "in Göttingen just recently I saw Runge, reduced to a skeleton and correspondingly changed and embittered."

A Russian student of Born wrote to him describing conditions in Russia at this time:

> I accepted a chair in Moscow as much as a year and a half ago, I dare not return there for risk of starving to death...the harvest had been extremely poor. ...Scientific life has almost ceased to exist here. No journals are being published...for the last three years we have not received any foreign journals. ...As professor I earn approximately 1.5×10^4 rubles per month. With this one can buy approximately one pound of bread per day. Everyone is allowed to buy only 150 g of bread at the controlled price. ...One pound of butter costs approximately 2×10^3 rubles, sugar somewhat more. A pair of boots costs 3-6 $\times 10^4$ rubles, etc. One frequently meets people who spend their entire month's salary in a single day. How this squares with the continuity equation of money is extremely puzzling...Firewood costs about 100 rubles per kilogram! But very few people are able to pay that much.

In January 1921 Einstein wrote to Born:

> You need not be so depressed by the political situation. The huge reparation payments and the threats are only a kind of moral nutrition for the dear public in France, to make the situation appear rosier to them. The more impossible the conditions, the more certain it is that they are not going to be put into practice.

With the support of the government, an "Emergency Society for German Science" (Notgemeinschaft der Deutschen Wissenschaft) was founded in October 1920, which was another source of funds to support research.

In early 1921 Franck moved to Göttingen and Born raised 68,000 marks for him from an industrialist. In a letter to Einstein, Born pointed out that Wien had obtained a million marks to refurbish his institute in Munich. On October 21, 1921, Born wrote to Einstein as head of the Kaiser-Wilhelm Association asking for rapid finalization of a grant for a new X-ray apparatus for Franck that was offered for 100,000 marks. If the purchase was not made by October 31, the price would rise to 150,000 marks. In the same letter Born wrote, "my aversion to the Allies grows because they are so disgustingly hypocritical. The Germans have, indeed, robbed and stolen when they could but they have not blathered on about saving civilization, etc."

Born lost his inheritance when a debtor sent him a note for the entire amount (50,000 marks) that was actually worth about one mark.

In September 1921 Einstein had to refuse an invitation from Sommerfeld to lecture in Munich because of hostility from Nazi students.

In April 1922 Born wrote to Einstein asking if he could find a place (perhaps in Holland) for his (Born's) assistant. Born was supporting him with 2,000 marks per month from a private fund but that was not enough for a family to live on. As a Hungarian Jew he would not get a post in Germany even if he was "habilitiert" (had obtained the postdoctoral degree necessary to teach in a German university). Einstein replied that even a leading theorist like Fokker could only obtain a modest post in a high school.

In August 1922 Born wrote that he and his family were going on vacation to Italy in September, financed by £22 that Born had received for the translation of his book and converted into lira.

In September 1922 Pauli wrote to Bohr asking him to send some Danish krone to pay for his upcoming visit to Copenhagen: "As a result of the recent collapse of the mark I am not able to pay for the trip with my German money." Bohr sent 200 Danish krone. Also around this time Einstein withdrew from presenting a review of relativity to the Association of German Doctors and Scientists because of anti-Semitic threats. Heisenberg, who attended the meeting, told how people had passed out anti-Einstein propaganda claiming that relativity was a speculation blown up by the Jewish press. Heisenberg thought at first that it was the work of a lunatic, but it was in fact a famous experimentalist.

In November 1922 Schrödinger wrote to Pauli that he was lucky to be in Copenhagen and stated that he was unable to work in Germany where the prices fluctuated so strongly because of the currency rates. He also wrote that the cost of mailing a letter was so terribly dear that he would have to reduce the frequency of writing.

In April 1923 Born wrote:

> We cannot form any rational political opinions because, just as in war, the truth is systematically falsified. The madness of the French strengthens German nationalism and weakens the republic. I am thinking a lot about how I can spare my son from taking part in a future revenge war.

Born had gotten to know Henry Goldman, of Goldman Sachs, who gave significant support to Born's institute and greeted Born and his wife as guests in his Fifth Avenue apartment in New York during the 1926 Christmas holidays.

In July 1923 Langevin went to Berlin for a pacifist demonstration. Pauli, who had left Copenhagen for Hamburg, wrote to Bohr that when he had left Copenhagen he was afraid that the political and economic conditions would have hindered the scientific institutes, but he was happy to see that at least in Hamburg that was not at all the case. A few days later Pauli was on holiday at the seaside.

In December 1923 Pauli wrote to Kramers in Copenhagen (Bohr was away at that time) asking him to send 200 krone from the money that Pauli had left behind when he left for Germany so that he could travel to Munich (to hear Heisenberg lecture) and then to Vienna.

A physics conference was planned for September 1923 in Bonn, which had been occupied by French and Belgian troops since January. The Bonn physicists were hoping for a good attendance as a sign of support from the other German physicists, who in turn were not sure if attending would be seen as a sign of support for the occupation.

Pauli again asked Bohr to send him 100 krone of his Danish money as he had some special expenses. "These everlasting money stories are horrible."

16.4.2 Schrödinger's thinking on cultural influences on science

One person who thought about the influence the cultural milieu can exert on the content of physical theories was Erwin Schrödinger. In February 1932, when the Weimar Republic was in a state of crisis, unable to find an adequate solution to the problems caused by the Great Depression, and Hindenburg had just been re-elected president, defeating Hitler in a boisterous campaign, Schrödinger gave a lecture[8] to the Prussian Academy of Sciences entitled "Is natural science determined by the milieu?"

He began by looking for any chinks in the armor of the objectivity surrounding science, where subjectivity might be able to penetrate, and identified the choice, out of the enormous set of possible experiments, of what experiments to actually perform or, more generally, what subjects to study. "Our interests, and their influence on the direction of future work, are a wide open entrance for subjectivity...Of the many possible questions we can pose to nature we choose some as very important and interesting and others less so." Among several examples, he mentioned the diffraction of light by a wire, discovered by Grimaldi in 1650 but ignored for centuries because Newton's particle theory of light focused interest on reflection and refraction, and the values of the elastic constants of solids, at one time only interesting to machine and bridge builders, but then subject to precision measurements when the Einstein-Debye theory of lattice vibrations allowed predictions of their values. He noted that workers in the same field around the world usually agree on what is interesting at a given time, which might indicate some objective, internal scientific determination. However, global travel and the circulation of scientific journals mean that most of these researchers know each other personally, read the same journals, discuss each new development, and build a consensus. As Schödinger pointed out,

> This agreement as to what is interesting is far from being proof that subjectivity is excluded, there is a good portion of fashion involved. While international checks guarantee objectivity regarding the evaluation of a given result, the direction in which results are sought may be subjectively determined. Our culture is a whole—scientists speak about politics, read novels and poems, go to the theatre, travel, play music, look at paintings, sculpture, architecture—so the cultural milieu must play a role in determining what interests us.

Schrödinger then attempted to identify characteristics of modern physics that may be culturally determined. He listed five such characteristics.

1. *What in art is called "pure objectivity."*

In the artistically influenced culture of material things, houses, furniture, and other household objects are designed solely on functional grounds. The idea is that the real functional form will then be seen to be beautiful. Large empty spaces on inner or outer walls or furniture are seen as acceptable. It is ridiculous and ugly to have superfluous decoration and unmotivated clutter.

In physics this shows up in the fact that our physical world picture is limited to observable facts, containing as few auxiliary concepts as possible, and says nothing

[8]Reprinted in K. von Meyenn, ed., *Quantenmechanik und Weimarer Republik* (Braunschweig: Vieweg, 1994).

about things that are not observable. All unnecessary concepts should be removed from the theory, which should be based on a description by means of only observable quantities.

As an example, take an urgent question from the old (Bohr-Sommerfeld) quantum theory and consider the transition of an atom from the very sharp state E_1 to the state E_2 with emission of a photon of frequency $\nu = (E_1 - E_2)/h$. Intermediate energies (between E_1 and E_2) are not allowed. Does the atom jump instantaneously from one state to the other? But the wave train emitted has a large length (0.5 meter), so its emission requires a significant time. Which energy does the atom have during this time, E_1 or E_2? In either case we have to give up instantaneous conservation of energy. In the new quantum theory, it has no meaning to ask what energy the atom has at a given instant of time except when the energy is measured. The measurement will take time (uncertainty principle), as much time as the duration of the wave train if we want enough accuracy. The possible violation of conservation of energy is removed by the new quantum theory. The measurement itself consists of an interaction with the system that delivers the energy necessary to restore the balance. So, to overcome dilemmas such as this one, we remove certain categories from the realm of questions that are meaningful to ask, e.g., the energy that the system possesses independent of measurement or the real instant of occurrence of a quantum jump.

2. *The need for disruption, a preference for freedom and lawlessness.*

In all fields of culture, whether political, social, or religious, or in questions of artistic taste, we see that today, more than in previous times, serious doubts are being raised about whether previously held beliefs and opinions are correct and worthwhile. Authorities are no longer blindly accepted, and everything needs to be justified. What distinguishes the present situation is the universality of this phenomenon, which has affected all parts of our culture, and the fact that the critics of previously held beliefs include many serious, able thinkers.

What is interesting for Schrödinger is that this applies also to the physics of his time, and quantum theory has even put in doubt the "causality dogma." This can never be tested experimentally. As Schrödinger expressed it, "The relation between cause and effect...is not something that we find in nature but it is a feature of our thinking about nature. We have complete freedom to keep it or give it up, ...according to what results in a simpler complete description of nature." In his paper proposing the probabilistic interpretation of the wave function, Max Born also wrote that the rejection of causality was an individual choice. According to him (pre-von Neumann) one could always imagine an underlying causal layer.

3. *Relativity of ideas—invariance theory.*

Here Schrödinger discusses mostly special relativity. Of interest to us is the following:

The question is whether the claims of natural science are invariant with respect to the cultural milieu, or whether they need a culture as a coordinate system and with a significant change of milieu, while they would not become false, their meaning and interest would be significantly altered.

4. *Mass control.*

Modern society has highly developed techniques to handle numerically large ensembles of people whose needs require an individual treatment in a finite time: voters, taxpayers, customers, motor vehicle owners and drivers, for example. The mass production of consumer goods requires dealing not only with large numbers of people but also with large numbers of identical items that need to be kept track of.

In the pre-computer age, the methods available for such tasks included registration, card files, catalogs, billing books, teams of civil servants, etc.

Mass control through proper organization in a labor economy in which efforts and decisions are made one time for all is reflected in the methods of mathematical analysis, which completely determine the face of physics today. Consider a simple problem in hydrodynamics, where the motion of each particle is seen as due to all the other particles, whose motion must also be determined. It is surprising that the human mind is even capable of solving such an enormously complicated problem. Today we set it as a one-hour exam problem because we solve it by means of a single partial differential equation which makes a statement that is valid at every time and every point for every single particle. The trick is to formulate the statement so that it is the same at every position and thus is open to an economical, factory-like treatment. As another example Schrödinger calls attention to vectors and tensors, where a single letter with indices can represent 20, 40, or more quantities satisfying systems of 20 or 40 equations.

5. *Statistics.*

The methods of mass control include the use of statistics, which play an important role in modern physics and astronomy. In physics the need for statistics can seem like limitation, a resignation that we cannot attain detailed knowledge of individual processes. In astronomy the role of statistics is more positive, as the detailed knowledge of a given star is not very meaningful. Which star is redder or whiter is not of interest, but the statistics of large numbers of stars show us the life cycle of stars, the relation between stellar distance and velocity, and much else.

In social statistics we can, to some extent, predict the laws according to which various statistical indicators will shift when the external conditions change. It is a goal of higher cultures to reach the required order and legal behavior without too detailed an interference in the life of individuals by studying the average inclinations of people and their statistical variations so that at least on the average a bearable common life will be secured.

The washing out of detailed knowledge of individual events in favor of a higher knowledge of relationships is a characteristic of both quantum mechanics and public life.

16.4.3 Forman's thesis concerning the influence of Weimar culture on quantum mechanics

As we have seen, during the years 1918–1927 German society under the Weimar Republic underwent a number of severe political and economic crises and experienced a rapid rate of technological, social, and cultural change.

It seems natural to ask if the development of quantum mechanics, which took place largely in this society, was influenced in any way by the cultural environment.

Was the development of quantum mechanics purely a question of the internal, logical development of the field in response to new and more accurate experimental results enabled by technical advances, or was the development somehow influenced by events in the wider society? Was there a reason that quantum mechanics developed at this time in Weimar Germany other than technical reasons like the long-standing tradition of spectroscopic work in Germany? In addition to Schrödinger, this question was also addressed by Paul Forman in his doctoral thesis and a series of following publications.[9] His ideas have resulted in a number of books and conferences dedicated to his proposal. Forman singles out the acausal nature of quantum mechanics and investigates the possible external (cultural) influences on its adoption. As Born pointed out in the first papers suggesting a statistical interpretation of the wave function,[10] the acceptance of acausality was up to the individual. One could always maintain that the observed random behavior was the result of not taking into account the influence of unknown ("hidden") variables whose exact knowledge would allow exact prediction, just as in the case of classical statistical mechanics, where the statistical behavior is understood to be based on averaging the causal motions of a large number of atoms. (This was well before von Neumann's controversial 1932 proof that the existence of hidden variables would mean that quantum mechanics would have to be objectively false; see Chapter 14.)

Thus, opinions can differ over whether events appear random due to lack of knowledge of the basic mechanisms or whether nature really possesses a fundamental acausal property. Both views agree with experiment. It seems that only such differences, allowed from an internal scientific viewpoint, could be influenced by personal or environmentally determined arguments.

The strong rejection of the de Broglie "pilot wave" theory at the 1927 Solvay Conference, based on an erroneous argument by Pauli (Section 13.3.2), is contrary to Born's flexible attitude and could be considered as evidence of cultural preferences at work.

Forman starts his paper with a quote from Gustav Mie's inaugural lecture as professor at the University of Freiburg in January 1925:

> It is interesting to observe that even physics, a discipline rigorously bound to the results of experiment, is led into paths which run perfectly parallel to the paths of the intellectual movements in other areas [of modern life].

[9]P. Forman, *The Environment and Practice of Atomic Physics in Weimar, Germany: A Study in the History of Science* (PhD thesis, U. C. Berkeley, 1967); *Weimar Culture, Causality and Quantum Theory, 1918-1927: Adaptation by German Physicists and Mathematicians to a Hostile Intellectual Environment*, Hist. Stud. Phys. Sci. **3**, 1-114 (1971); *Scientific Internationalism and the Weimar Physicists: The Ideology and Its Manipulation in Germany after World War I*, Isis **64**, 151-180 (1973); *The Financial Support and Political Alignment of Physicists in Weimar Germany*, Minerva **12**, 39-66 (1974); *The Reception of an Acausal Quantum Mechanics in Germany and Britain*, in The Reception of Unconventional Science, ed. S. H. Mauskopf (Boulder: Westview Press, 1979), https://www.researchgate.net/publication/288324999; *Kausalität, Anchaulichkeit und Individualität, oder wie Wesen und Thesen, die der Quantenmechanik zugeschrieben, durch kulturelle Werte vorgeschrieben wurden*, Sonderheft 22: Wissenssoziologie, 393-406 (Wiesbaden: Verlag für Sozialwissenschaften, 1981).

[10]M. Born, *Quantenmechanik der Stoßvorgänge*, Z. Phys. **37**, 863-867 (1926); *Quantenmechanik der Stoßvorgänge*, Z. Phys. **38**, 803-827 (1926).

Forman's argument consists of three points:

1. Weimar culture as a hostile intellectual environment. Rationalism, utilitarianism, and the exact sciences were looked on with disdain by the general (and academic) public (cultural milieu) in the Weimar period. Although science and scientists had enjoyed high prestige under the imperial regime, the connection of science with technology and the war effort contributed to the backlash against it after the defeat. The idea of causality, equated as it was with mechanism and rigorous determinism, became the object of particular opprobrium. Causality symbolized all that was despised in science and was seen as hampering the free expression of the exuberant feelings animating the existential "life philosophy" and limiting the notion of free will. In spite of the deep hostility to science and mathematics, Weimar culture was one of the most creative environments in the entire history of these fields.

2. Adaptation of ideology to the intellectual environment. As a reaction to Point 1, academics were attracted to new romantic, existential ideas (life philosophy). A general tendency rather than a particular philosophical school, "life philosophy" rejected both mechanism and determinism (seen as inseparable from causal explanation), as well as the idea of reason as the primary instrument of knowledge, instead emphasizing intuition and direct, unanalyzed experience. Exact scientists tried to accommodate their thinking to these trends. They repudiated positivism and previous utilitarian justifications for the pursuit of their science. In contrast, scientific contributions to military and industrial technologies were considered as the most positive aspect of science during the war. With the defeat in 1918, this association became a liability.

3. Dispensing with causality: Adaptation of knowledge to the intellectual environment. In order to achieve conciliation between their science and the values of a hostile milieu, many scientists, without persuasive scientific motivation, gave up on causality before a rational quantum mechanics justified it. Moves to reconstruct the foundations of the field can be seen as a reaction to the negative prestige. In order for physics to improve its image, it would have to dispense with causality and rigorous determinism, the most abhorrent features of the existing physical world picture.

16.4.3.1 Part 1: Weimar culture as a hostile intellectual environment

Forman claims that a precise and detailed analysis of public lectures by some of the main figures of quantum physics shows the gradual shift in positions on causality. The synchronism in time of these shifts and the fact that they preceded the formulation of quantum mechanics are indicative of the influence of external factors.

Max Born's wife, Hedwig, who dabbled in poetry and play writing, and had a long philosophical correspondence with Einstein, wrote that the company of "objective" natural scientists inspired in her "the feeling of being cast upon a lunar landscape."

Rudolf Steiner was the founder of a philosophy of life (Anthroposophy) and an organization which today has branches in Germany and 50 other countries, and some of whose products (e.g., Waldorf schools, cosmetics) are internationally known. He got into a dispute with Max von Laue, who believed that Steiner's view represented the natural sciences as "bearing the guilt for the current [1922] world crisis...and the whole of the intellectual and material misery" bound up with it.

Many scientists remarked on the anti-scientific currents in contemporary thought. In a published lecture (February 1923), Max Planck stated that

the belief in miracles of the most various forms—occultism, spiritualism, theosophy...penetrates wide circles in the public, educated and uneducated, more mischievously than ever, despite the stubborn defensive efforts directed against it from the scientific side.

Einstein remarked in a newspaper article (July 1921) that it is "peculiarly ironic that many people believe that in the theory of relativity one may find support for the anti-rationalistic tendency of our days." In an article written in 1927 for a popular monthly magazine, Sommerfeld noted that "in Munich probably more people get their living from astrology than are active in astronomy." He continued:

The belief in a rational world order was shaken by the way the war ended and the peace dictated; consequently one seeks salvation in an irrational world order. But the reason must lie deeper, for astrology, spiritualism and Christian Science are flourishing among our enemies also. We are thus evidently confronted once again with a wave of irrationality and romanticism like that which a hundred years ago spread over Europe as a reaction against the rationalism of the eighteenth century and its tendency to make the solution of the riddle of the universe a little too easy.

This anti-rationalist outlook also penetrated the Prussian Ministry of Education. In 1919 C. H. Becker, the highest civil servant in the ministry, wrote that the problem was "the overvaluing of the purely intellectual in our cultural activity, the exclusive predominance of the rationalistic mode of thought which had to lead and has led to egotism and materialism of the crassest form." Concerning education, he believed that "our entire educational system is too exclusively oriented toward the intellect. We must acquire again reverence for the irrational."

His superior, the Social Democrat Minister of Culture, K. Haenisch, agreed:

the German people, having suffered for decades from the plight of mechanism and materialism...if in our spiritual life not only the intellectual but also the irrational is to receive its due, then the barriers will have to be broken down which presently separate the universities and the people.

These ideas were influential in the formulation of the new reforms for the secondary school curriculum where mathematics and science teaching was reduced in favor of "cultural" subjects. Accepting that the "economic-political, technical and positivistic age...now lies behind us," the ministry refused to justify the curriculum on utilitarian grounds. Felix Klein remarked, "This school reform signifies for our educational system the end of the century of science."

Hugo Dingler, a philosopher of physics, wrote in 1926 that "this new collapse of science in whose midst we stand...consists in the collapse of the belief in the certainty of the experimental principle"— the possibility of establishing the truth of a theory by its agreement with experiment.

Spengler's *Decline of the West* appeared in 1918. It would eventually sell 100,000 copies and had an enormous influence. Einstein remarked in a letter to Born (January 1920):

Spengler has not spared me either. Sometimes one agrees with his suggestions in the evening, and then smiles about them next morning. One can see that the whole of his monomania had its origin in school teacher mathematics...These things are amusing; if someone should say the exact opposite tomorrow with sufficient spirit, it is amusing once more, but the devil only knows what the *truth* is.

Spengler reviewed the history of many extinct civilizations, using the *Encyclopedia Britannica* as his main source, and purported to extract a model of the life cycle of a civilization. Physics and mathematics played a prominent role in his treatment of western civilization. He thought that science and mathematics were strongly culture dependent. To him the causality principle was an artificial construction erected as a defense against the more fundamental, irrational notion of *destiny*, which is an "indescribable inner certainty."

The principle of causality is late, unusual and only for the energetic intellect of higher cultures, a secure, somewhat artificial possession. Out of it speaks fear of the world. Into it the intellect banishes the demonical in the form of a continually valid necessity, which rigid and soul-destroying is spread over the physical world picture.

The natural scientist, the thinker in systems, whose entire mental existence is founded on the principle of causality, is a "late" manifestation of the hatred of the powers of destiny, of the incomprehensible.

As proof of the dominance of "destiny" over "causality" Spengler cited the radioactive decay of a uranium atom, struck by destiny while its neighbors remain untouched.

Spengler felt that there was no immanent, invariant criterion of knowledge. The science of a period depended totally on its attitude to life. Science was, moreover, a relatively short-lived phenomenon appearing in the late stage of the life of a civilization.

Before us there stands a last spiritual crisis that will involve all Europe and America. ...The tyranny of reason—of which we are not conscious, for the present generation is its apex—is in every culture an epoch between man and old man and no more...the history of higher cultures shows that "science" is a transitory spectacle, belonging only to the autumn and winter of their life cycle and that...a few centuries suffice for the complete exhaustion of its possibilities. ...the last stage of western science...ever smaller, narrower and more unfruitful researchers...in physics and in chemistry, in biology as in mathematics, the great masters are dead, and we are experiencing today the decrescendo of the stragglers who arrange, collect, conclude...

Western European physics has reached the end of its possibilities...doubt that has arisen about...the unchallenged foundations of physical theory, about the meaning of conservation of energy, the concepts of mass, space, absolute time...What deep and utterly unconscious skepticism lies...in the rapidly increasing use of...statistical methods, which aim only at the probability of the results and forego in advance the absolute exactitude of the laws of nature, as understood in hopeful earlier generations.

These ideas were an integral part of an analysis of western culture which helped shape the thoughts and inclinations of the educated middle classes in Weimar Germany.

16.4.3.2 Part 2: Physicists' adaptation of their ideology

In this section Forman discusses the response of scientists at the level of ideology—their justification of scientific activity, their intellectual stance, their spirit, and their confidence in the future of science. We will concentrate on examples before 1927, by which time quantum mechanics had more or less its modern form.

Wilhelm Wien, one of the most prominent German physicists of the period, expressed himself as follows on the motivation for doing physics. In June 1914 he stated that physics and chemistry "have created the solid foundations upon which the pillars of our industry are erected." In May 1918 he spoke of the "mutual support and stimulation" which physics and technology have and continue to offer each other. His position was quite different in February 1920, when he presented two possible points of view concerning "the significance of physical research." The first was that physics aimed to achieve "human domination over the recalcitrant forces of nature." The second view, and the one preferred by Wien, was that the pursuit of physics is "free of all striving towards a goal." Physical research is nothing but the expression of "the pure human instinct for inquiry...it arises solely from an inner need of the human spirit."

In November 1925 he went so far as to say that the the goal of science is *culture*. "The significance of scientific achievement can ultimately only be measured by the effect which it has upon the intellectual life; the results of research are worthless if they are not taken up into the culture."

In the introduction to Volume 4 of the *Handbuch der Physik* (1929), Hans Reichenbach wrote that doing physics "is a need, that it grows up out of the human being just like the wish to live, or to play, or to form a community with others."

These and similar comments by other physicists were made in lectures to general academic audiences. While it is impossible to say if they reflected the core beliefs of the speakers, they at least indicate a desire to appear in agreement with the prevalent ideas of the time.

Max Born, whose key role in the introduction of acausality has already been noted, wrote in the introduction to his book *Einstein's Theory of Relativity* (1920) that the scientist concentrates on objective properties of the world in order to reach agreement with the maximum number of other people, that only this type of knowledge can be agreed on with other, foreign egos. However, "the best that can be found in this way are not the experiences of the soul, not the sensual impressions, imaginings, feelings, but abstract concepts of the simplest type, numbers, logical forms, the working tools of the exact sciences." The scientist experiences the "growth of the surely known regions of the world, and when he feels this, the pain of the loneliness of the soul vanishes and a bridge is built to kindred spirits."

Forman points out that the academic chemists never felt "at all bashful about discussing technical applications and justifying their science through them."

16.4.3.3 Part 3: "Dispensing with causality": adaptation of knowledge to the intellectual environment

Franz Exner, professor of physics in Vienna and mentor to Schrödinger, was an early opponent of causality in physics. This was undoubtedly based on the success of statistical mechanics. In a lecture given in association with his appointment as rector of the University of Vienna (1908), he said that "all physical processes are the results of random events" and argued that all physical laws were the result of the law of large numbers. The lack of laws in the social sciences was due, in his view, to the relatively small number of chance events underlying the phenomenon being studied. In a series of lectures published in 1919, he stated that no laws of physics are exact, and causality is not valid. All macroscopic laws are statistical in character, and the apparent regularity arises out of random motions. Physical laws are a creation of man, not nature.

According to Exner, "We have no right to postulate causality even if it is necessary to understand nature. Nature does not care if we understand or not."

Of course, since Exner held these sentiments long before the outbreak of war, they are not relevant to Forman's discussion of the post-war cultural milieu; in fact, Exner can serve as a counterexample to Forman's thesis that the unique properties of Weimar culture strongly influenced physicists to give up causality.

It is interesting to note the similarity of Exner's views concerning causality to those that were expressed much later (1932) by von Neumann in his book *Mathematical Foundations of Quantum Mechanics*:

The place of causality in today's physics can be described as follows. In macroscopic physics there is no experience which supports it, and there cannot be any because the seemingly causal order of the world (i.e., objects visible to the naked eye) has certainly no other cause than the "law of large numbers"—completely independent of whether the laws underlying the elementary processes are causal or not. That macroscopically identical objects behave identically has little to do with causality: they are absolutely not identical, since those coordinates that determine the state of the atoms never agree, and the macroscopic viewpoint averages over these coordinates. ...First in the elementary processes in atoms can the question of causality really be tested, but here, according to our current state of knowledge, everything speaks against it, since the only available formal theory that halfway agrees with experiment, the Quantum Mechanics, is in strong contradiction with it.

Another physicist who was an early opponent of causality was Hermann Weyl. In his paper (1920) on *The Relation of the Causal to the Statistical Approach in Physics*,[11] Weyl asks:

Are statistics merely a shortcut to certain consequences of causal laws, or do they imply that no rigorous causal interconnection governs the world, and that instead "chance" is to be recognized alongside law as an independent power restricting the validity of the law? The physicists are today entirely of the first opinion.

He, however, already held a different view. As early as 1918–19 Weyl asserted that "at the basis of statistics lies an independent principle which is not to be reduced to causality." In a more detailed description of his views, he wrote:

Thus the rigid pressure of natural causality relaxes, and there remains, without any prejudice to the validity of natural laws, room for autonomous decisions, causally absolutely independent of one another, whose locus I consider to be the elementary quanta of matter. These "decisions" are what is actually real in the world.

Weyl felt that causal laws (because they were symmetric under time reversal) were incompatible with "our most fundamental experience, the unidirectionality of time."

Although Weyl did change his opinion on causality, he did so very early in the Weimar period, probably toward the end of 1918. Weyl's change of mind was based on his adoption of an existentialist philosophy.

Forman next shows that three physicists "converted" to acausality in the summer and fall of 1921: Schottky in June, von Mises in September, and Nernst in October.

Schottky based his questioning of causality in June 1921 on the statistical nature of the interactions of atoms with radiation. Since the laws governing these interactions

[11]Weyl, H., *Das Verhältnis der kausalen zur statistischen Betrachtungsweise in der Physik.* (The Relation of the Causal to the Statistical Approach in Physics) Schweizerische Medizinische Wochenschrift, 50:737–741. GA II, 113–122, (1920).

were not known, he considered that it is no longer possible to "conceive of the course of events like a continually and uniformly flowing stream." The elementary acts of absorption and emission were "without direct cause and without direct effect." Due to the statistical, acausal nature of atomic events, "the law of causality itself, with its complete conditioning of the coming phenomena by the present and past phenomena, appears...to be placed in doubt."

In February 1920 von Mises regarded a physical explanation as being synonymous with a causal explanation: "We see now...how a new and simply enormous field of phenomena, the multiplicity of the chemical elements, is drawn into the realm of causal explanation." By September 1921, his view on causality had changed completely. "Every electrical, every thermal, every optical process is a statistical phenomenon and as such fundamentally incompatible with the concept of causality." He no longer believed in the possibility of causal explanation even for the phenomena of classical mechanics:

> Can the temporal course of every motion of a...mass be unambiguously determined by specifying the initial state and assuming some appropriate force law to be acting?...it is highly improbable that this goal of classical mechanics could ever be attained, and that other...considerations are destined to relieve or to supplement the rigid causal structure of the classical theory...within the purely empirical mechanics there are phenomena of motion and equilibrium which will forever escape an explanation on the basis of the differential equations of mechanics.

In a lecture given in October 1921, Nernst argued against causality and exact physical laws because he saw them as equivalent to rigorous determinism and, as such, eliminating the possibility of free will. "Can philosophy and natural science really assert with certainty that...every human action is the unambiguous result of the circumstances prevailing at the moment? If absolutely rigorous laws of nature controlled the course of all events, one would in fact scarcely be able to escape from this conclusion." The "conception of the principle of causality as an absolutely rigorous law of nature laced the mind in Spanish boots, and it is therefore at present the obligation of research in natural science to loosen these fetters sufficiently so that the free stride of philosophical thought is no longer hindered."

Nernst went on to propose that if there were exact laws of nature, the fluctuations of the ether would disrupt the motions of atoms, and since it is not possible to control these fluctuations, it is impossible to work with identically prepared systems, a kind of causal explanation of the failure of causality. Rather than the validity of causality, the main question for Nernst becomes whether natural processes are comprehensible or whether the human mind is incapable of following these processes. Nernst favors the latter, stating that "only statistical mean values of the course of events are accessible to our scientific knowledge." He thus connects the rejection of causality with the repudiation of the ability of the rational mind to understand nature.

Forman argues that the coincidence in time of these conversions, coupled with the fact that there were no specific developments in physics at this time indicating the need for an acausal theory, indicates that these conversions are examples of the "capitulation to those intellectual currents in the German academic world" which he discussed previously.

However, there were a number of experimental observations showing statistical behavior at the atomic level, namely the radioactive decay of atoms and Einstein's use of the A and B probability coefficients in describing the emission and absorption of radiation by atoms.

Perhaps one of the most interesting cases of rejection of causality pointed out by Forman is that of Schrödinger who, influenced by Exner, launched a strong attack on causality in his inaugural lecture as a professor at the University of Zurich (December 1922).

> In the past four or five decades physical research has demonstrated perfectly clearly that for at least the overwhelming majority of phenomena, the regularity and invariance of whose development has led to the postulate of general causality, the common root of the observed rigorous lawfulness is—*chance.*

Since all macroscopic laws were statistical, he considered it a strange dualism to propose a layer of underlying causal laws at the microscopic level. "In the world of visible phenomena we have clear intelligibility [i.e., statistical laws], but behind this a dark, eternally incomprehensible commandment, an enigmatic *'must'.*" The solution to the problems of atomic physics, he believed, would require "liberation from the rooted prejudice of absolute causality."

Ironically, four years later, having undergone a complete reversal, Schrödinger saw his attempt to formulate a continuous, causal theory taken into the service of a statistical theory by Max Born, and remained strongly opposed to the Copenhagen interpretation, with its acausal quantum jumps, after the establishment of the theory in its final form.

Einstein and Planck remained staunch defenders of causality. In June 1922 Planck told a meeting of the Prussian Academy that

> when the quantum hypothesis shall have been developed sufficiently so that one can properly speak of a quantum theory, that will be the proper moment to consider its consequences for our scientific-causal thought. ...precisely at the present time not inconsiderable dangers to the sure advance of scientific work have arisen from various sides.

This is in agreement with Forman's suggestion that the abandonment of causality was mainly due to influences external to physics itself.

As a reminder of the turbulence of that time, in late 1921 Einstein had accepted an invitation from Planck to speak in defense of causality at the next meeting of the Association of German Natural Scientists and Medical Doctors the following summer, but he later cancelled his plans after the assassination of Rathenau (June 24, 1922).

According to Forman it was only in 1924 that physicists became convinced of the inadequacy of the current theory of the interaction of atoms with radiation, and the theory proposed by Bohr, Kramers, and Slater (BKS) relied heavily on probability. The original idea, due to Slater, was that light quanta emitted from atoms would follow the Poynting vector of a virtual radiation field. Bohr convinced Slater that the same result would be obtained with a somewhat simpler model according to which the virtual field emitted by one atom determined the transition probability of another atom (absorber). Slater agreed, even though this meant that conservation of energy and momentum did not hold for individual processes, but only on average. This could easily be tested by measurements on the Compton effect, and the theory was rapidly disproved experimentally after receiving a widespread positive reception. Pauli for one never accepted these ideas: "I consider it very good luck that the idea of BKS was...so quickly disproved by experiment." Otherwise, "the proposal of BKS would have perhaps been an obstacle to progress in theoretical physics for a long time."

Einstein, had also staunchly opposed the BKS theory. In April 1924 he responded to it with a strong statement of his position on causality:

I do not want to be forced to abandon strict causality without defending it in new stronger ways than has so far been done. The idea that an electron, exposed to radiation would freely choose the moment and direction in which to jump is unacceptable. If so, I would rather be a shoemaker or an employee in a gambling house than a physicist.

Forman claims to have shown that there was a will among physicists to believe that causality was invalid *before* the invention of an acausal wave mechanics. With the introduction of matrix and then wave mechanics, physicists realized that the rejection of causality could be put on a firmer basis. While, as we have seen above, Wien adapted his stated justification for doing physics to perceived public preferences, he remained a strong defender of causality throughout and strove to defend physics itself from the influence of the cultural milieu. He considered Sommerfeld's approach to the theory of atomic spectra, i.e., the introduction of quantum numbers to select allowed classical orbits, as a kind of "number mysticism" which he hoped "would be supplanted by the cool logic of physical thought." "A physics in which mysticism governs, ...relinquishes the ground from which it draws its strength." He suggested that those who doubted that "insight into the causal interconnections of natural processes will continue to be possible" are suffering from mental exhaustion and thus inclined to the pessimistic ideas of Spengler's *Decline of the West.*

Wien felt that his defense of causality was strengthened by Schrödinger's wave mechanics, which was seen by Schrödinger, after he had converted back to causality, as a causal space-time description of atomic processes. (It is interesting to note that Schrödinger published his papers in *Annalen der Physik*, which was favored by the more conservative physicists outside of Berlin, while the Berlin physicists, who tended to be more progressive, and more Jewish, published mainly in *Zeitschrift für Physik.*)

But this idea of wave mechanics as a causal theory was short lived. As we have seen, in June 1926 Max Born proposed the probabilistic interpretation of the wave function by showing that it was the obvious way to describe scattering phenomena, so that by March 1927 Heisenberg could write: "Because all experiments are subject to the laws of quantum mechanics...quantum mechanics establishes definitively the fact that the law of causality is not valid."

Schrödinger, meanwhile, had converted back to causality in the fall of 1925. In August 1926 he explained his change of views in a letter to Wien:

Today I no longer want to agree with Bohr that a single event like, e.g., the interaction of an electron with an atom, is absolutely random, i.e., perfectly indeterminate. I no longer believe that this assumption, that I supported so strongly four years ago, brings any advantages. I reject Bohr's idea that a space-time description is impossible. Physics is not only atomic research, science is not only physics and life is not only science. The purpose of atomic physics is to fit the results of our experience in this field into the rest of our usual thought. This usual thought, at least that which is relevant to the external world, moves in space and time.

On the other hand, Max Born was enthusiastic with regard to these new developments. In a Berlin newspaper he wrote that

> such a conception of nature [i.e., governed by causality] is deterministic and mechanistic. There is no place in it for freedom of any sort, whether of the will or of a higher power. And it is that which makes this view so highly valued by all "good rationalists." But happily physics has discovered new laws which give it an entirely different character.

We have seen above how Forman tried to establish that cultural pressures during the chaotic years of the Weimar Republic led several prominent physicists to give up causality, at least in public, in an attempt to recover some of the prestige that scientists had lost as a result of Germany's defeat in the war. He claims that this renunciation of causality occurred well before it was required by the internal development of the field, but this is only partially true since phenomena such as radioactive decay and the emission and absorption of photons were already seen as statistical prior to 1918 (Einstein introduced his A and B coefficients in 1916). Then, when the internal developments led to the introduction of probabilistic behavior as a fundamental property of quantum mechanics, a large group of physicists were only too happy to accept this viewpoint as a way to reconcile their professional lives with the prevailing cultural attitudes.

Among those who resisted an acausal interpretation were Einstein, Schrödinger, Planck, and Wien. As an example, we cite Einstein in a 1932 interview with James G. Murphy:

> MURPHY: ...it is now the fashion in physical science to attribute something like free will even to the routine processes of inorganic nature.
>
> EINSTEIN: That nonsense is not merely nonsense. It is objectionable nonsense. ...Quantum physics has presented us with very complex processes and to meet them we must enlarge and refine our concept of causality.
>
> MURPHY: You'll have a hard job of it, because you will be going out of fashion...scientists live in the world just like other people. ...They cannot escape the influence of the milieu in which they live. And that milieu at the present time is characterized largely by the struggle to get rid of the causal chain in which the world has entangled itself.

Forman points out that the physicists who were readiest to repudiate causality had either progressive political views or a close interest in and contact with contemporary literature. Many progressives bowed to the zeitgeist, almost repudiating the basis of scientific work. They rejected utilitarianism and adopted the popular existentialist "life philosophy." They promoted the satisfaction of innate psychological needs as the justification for doing physics.

Except for Einstein, the defenders of causality were conservative politically and/or interested in classical literature; some were outright reactionaries. They refused to bend to the anti-rationalist milieu and insisted on a deterministic description of nature and the capacity of the human mind to understand it.

All this indicated to Forman that his model "cannot be the whole truth;" in order to account for its applicability to some physicists and inapplicability to others, we need to account for personal and intellectual propensities as well as external pressures from the cultural milieu.

16.4.4 Reactions to Forman's thesis

Forman's arguments have prompted a wide range of responses in the form of scholarly papers, books, and conferences. We can only call attention to a limited selection of these.

Hans Radder[12] examines the role of Kramers and asks whether he might be a counterexample to Forman's thesis. Kramers was Dutch and had lived in Denmark since 1916, so he was not directly exposed to the crises of the Weimar Republic. This raises the question of whether the "Weimar cultural sphere" extended outside of Germany. Kramers was a strong opponent of causality and supported the BKS theory, where emission and absorption of radiation in distant atoms are not causally connected. He felt that science describes the formal aspects but not the essence of phenomena. His views of the world did not change suddenly as in the examples of conversion cited by Forman, but were the result of long-standing religious convictions. Radder leaves open the question of whether the Weimar cultural pessimism and repulsion of causal materialism penetrated to Denmark, but quotes the rector of the University of Copenhagen on the occasion of the dedication of Bohr's institute for theoretical physics (1921)—"it is a delight that such an institution can be erected in this age which has such a bad repute for materialism"—as an indication that it might have.

Stephen G. Brush,[13] writing in 1980, takes a longer-term perspective regarding the influence of culture on the development of science. He sees the Weimar era as part of a long cyclic history and believes that changes in scientific theories correlate with fluctuations in cultural styles, but that does not mean that the cultural milieu is responsible for the scientific changes. Rather he considers that both science and culture undergo oscillatory fluctuations. When a cultural or scientific trend goes too far, practitioners tend to oppose the trend, and it starts to swing back in the other direction. So the culture-science nexus behaves to some extent like a set of coupled pendulums: if they are correlated at one time, they will tend to remain correlated. Even if the coupling is weakened, the correlations will continue due to past interactions. He points out that his model differs from Forman's in that he ascribes a larger role to internal factors than Forman does. He finds Forman's evidence useful as indicating the influence of a "romantic cultural background on quantum mechanics."

As major categories of his analysis, between which the oscillations take place, Brush chooses Realism and Romanticism. Realism in science is characterized by materialism and the belief in the existence of an objective real world that is independent of our observation of it. In the culture, Realism is characterized by greater public support for science and technology, politically liberal or socialist attitudes, and the tendency to favor increased rights for women and the lower classes. In science, Romanticism is characterized by empiricism and instrumentalism, which considers that the goal of science is to provide the most economical description of observations and ridicules the attempt to discover the nature of unobservable quantities. In the larger culture, Romanticism is characterized by idealism, holism, and a tendency to favor arts and

[12] H. Radder, *Kramers and Forman's Thesis*, Hist Sci. **21**(2), 165 (1983).

[13] S. G. Brush, *The Chimerical Cat: Philosophy of Quantum Mechanics in Historical Perspective*, Soc. Stud. Sci. **10**, 393 (November 1980).

regard science with a hostile view. It is also associated with politically conservative governments.

Schrödinger's wave mechanics was conceived in a Realist spirit but was, with the probabilistic interpretation of the wave function, complementarity, and the uncertainty principle, taken over into supporting the Romantic worldview.

The cycles that Brush identifies are
1800 → Romanticism, 1850 → Realism,
1890 → Neo-Romanticism, 1905 → Neo-Realism,
1918 → Romanticism III, 1945 → Realism III,
1968 → Romanticism IV

Brush gives a description of each of these phases and their influence on scientific attitudes and developments. The Neo-Realism phase that began in 1905 was characterized by Einstein's treatment of Brownian motion, special relativity, the discovery of photons, and an emphasis on mathematical structure over mechanism. World War I broke the pattern and Romanticism III (1918–1945) emerged prematurely. As we have seen, this was accompanied by an opposition to rational, causal explanation and a favorable attitude to empiricism, indeterminism, and holism. We have seen how Schrödinger acknowledged a close correlation between scientific theories and the fashion of the times. At that time the cultural climate was unfavorable to his own attempt at a Realist interpretation of quantum mechanics, but there were also strong internal reasons for the rejection of his ideas (spreading of the wave packets; Chapter 10).

Among the founders of quantum mechanics, Planck, Einstein, de Broglie, and Schrödinger remained Realists, but their philosophical views carried little weight against the trend of Romanticism III. The founders who embodied Romantic characteristics were Bohr, Heisenberg, Born, Pauli, and von Neumann. On the Romantic side, Bohr was the leading spokesman with his ideas of complementarity and the "Copenhagen interpretation." Heisenberg was the most forceful exponent of the "instrumentalist" claim that theories must deal only with relations between observables. He felt that "speculation about whether there is a true universe behind the statistical one is without value and meaningless." He rejected the materialist belief in an objective real world whose smallest parts exist independently of observation. Von Neumann's proof that hidden variable theories cannot reproduce the observable consequences of quantum mechanics—as long as quantum mechanics agrees with experiment, the theory must remain fundamentally indeterministic—was accepted by physicists until the 1950s as a protection of the Copenhagen interpretation. Bell's 1964 analysis of a presumed error in von Neumann's proof (Chapter 14), which was a reaction to Bohm's demonstration of a hidden variable (pilot wave) theory that did not contradict quantum mechanics, was taken by some physicists as an encouragement to work on hidden variables. This led to Bell's proposal of a test that would have ominous consequences for so called "local" Realism.

The Realism III phase, which coincided with a shift in scientific leadership to the victorious United States, was accompanied by considerable public interest in science and technology, substantial government funding, application of physics to other fields, such as plate tectonics, cosmology, and space science, the discovery of the transistor, and the development of digital computers. The discovery of the double helix structure of DNA was partly inspired by Schrödinger's book *What is Life?*[14] rather than by more holistic viewpoints. Most physicists paid lip service to Copenhagen but largely ignored issues of interpretation. Realism III ended with cutbacks in research funding and the loss of momentum of the Realist trends, which had failed to overthrow the Copenhagen interpretation. The experimental violation of the Bell inequality, conceived as a Realist attempt to disprove the Copenhagen interpretation, required either the total abandonment of a Realist philosophy or a revision of the concepts of space-time, as we saw in Chapter 15. This strongly "internal" influence coincided with the introduction of Romanticism IV, which was accompanied by emphasis on holistic systems, born again religions, general emphasis on spiritual realms, "the Age of Aquarius," student demonstrations and organizations for changing society, and anti-scientific movements. Wheeler (1977) extended the notion that it was observations which created reality, to the entire universe, which only comes into existence as a result of the presence of observers, viewing the universe as a self-excited circuit in which the universe gives rise to the observer and the observer gives a meaningful existence to the universe:[15] "the universe, through some mysterious coupling of future with past, required the future observer to empower past genesis."

In this way, Brush fits the development and interpretation of quantum mechanics into a broad scheme of the relationship of scientific and cultural evolution over the past two centuries. He points out that this cannot account for new theories and discoveries, but "it may suggest why the community reacts favorably or unfavorably to them."

Forman himself was working during a period of backlash against science and rationality (Romanticism IV), and some of the attitudes he observed were remarkably similar to those expressed during the Weimar period (Romanticism III). Forman called attention to remarks by Lewis Branscom, head of the then National Bureau of Standards (1971) and W. D. McElroy, director of the NSF (1970). Branscom, writing in 1971, observed that "astrology is booming, there are three professional astrologers in this country for every astronomer."

McElroy noted the same cultural trend, but he, in contrast, thought the Romantic worldview had something positive to contribute to the scientific one:

> In my view, the science community generally should consider more carefully...the new romanticism, emphasizing man as an emotional and feeling creature as well as a reasoning one. A healthy dose of this view may counterbalance some of the extreme emphasis on rational thinking I suspect is endemic within the science community.

They, and Forman, were writing in the aftermath of the turmoil connected with the "events" of 1968, which was also a period of widespread attacks on rationality and was followed shortly by another revolution in physics, the establishment of the Standard Model.

In the volume *Weimar Culture and Quantum Mechanics*[16] copies of Forman's main

[14]Schrödinger, E., *What is Life? & Mind and Matter*, (Cambridge University Press, 1967).

[15]J. Wheeler, "Genesis and Observership", in *Proceedings of the Fifth International Congress of Logic, Methodology and Philosophy of Science*, Part 2, ed. by R. E. Butts and J. Hintikka (Boston: Reidel, 1977).

[16]C. Carson, A. Kojevnikov and H. Trischler, eds., *Weimar Culture and Quantum Mechanics, Selected Papers by Paul Forman and Contemporary Perspectives on the Forman Thesis* (London: Imperial College Press; Singapore: World Scientific, 2011).

papers are republished along with the papers presented at a conference, "The Cultural Alchemy of the Exact Sciences: Revisiting the Forman Thesis," held at the University of British Columbia in March 2007. In the first chapter, "The Forman Thesis: 40 Years After," the editors give an overview of the intellectual debate that followed Forman's publications. They point out that Forman's argument "that the cultural values prevalent at a given place and time could influence the results of discipline-bound research, i.e., the very content of scientific knowledge," which "at the time of its introduction created uproar as it explicitly contradicted generally accepted and cherished beliefs about science," has "become commonly used in cultural studies of science. ...and has arguably been the most influential article ever published in the historical studies of science."

A. Kojevnikov calls attention to a letter that Ehrenfest, a professor in Leiden, wrote to Bohr in June 1919.[17]

> It is remarkable that precisely here, in the circles of men having much to do with technology, production, industry, patents, etc., opinions develop so uniformly about perspectives of culture. Overall there is building up an uncannily intensive reaction *against rationalism*...If I am not entirely mistaken, in the next 5-10 years we will see the following happening at the institutes of higher learning (including technical!). Professors raised as relatively *rational* and disciplined individuals will despairingly and uncomprehendingly face the complaints and demands of a relatively *"mystical"* student body. At the same time, scientifically less clear but personally warmer teachers will gain the main influence over students. (Emphasis in original.)

Ehrenfest's observations seem to offer some confirmation of Forman's thesis that following the war there was a general revulsion against rationalism, not only in the general public but also among engineers and exact scientists. The implications of the letter go beyond Forman in identifying the interaction with students as an important point of contact between scientists and the cultural milieu. It also adds additional evidence that the cultural trends discussed by Forman penetrated beyond the borders of Weimar Germany.

In the same volume Olival Freire Jr[18] considers that the existence of Bohm's theory serves to corroborate Forman's claims of a cultural bias being responsible for the opposition to causality. The rejection of de Broglie's earlier formulation of the pilot wave theory on the basis of Pauli's incorrect argument, which, as we have pointed out, had been correctly refuted by de Broglie at the Solvay Conference of 1927, is taken as strong evidence of the Forman hypothesis at work.

16.5 Conclusion

In this chapter, which obviously goes beyond physics, we have attempted to give an overview of life and culture in the Weimar Republic in the years leading up to the creation of quantum mechanics, with emphasis on the observations of prominent physicists. We have seen how this has served as a case study for historians of science

[17] Kojevnikov, A., "Philosophical Rhetoric in Early Quantum Mechanics 1925-27: High Principles, Cultural Values and Professional Anxieties", ibid., p. 320.

[18] Freire, O. Jr., "Causality in Physics and in the History of Physics: A Comparison of Bohm's and Forman's Papers", ibid., p. 397.

to explore the question of the relation between internal and external influences on the development of scientific theories. One additional point is that while there were several hundred physicists working on atomic physics in Germany, the debates concerning the interpretation and philosophical significance of quantum mechanics were confined to a very small number. However, it was the large "silent majority" which effectively decided the outcome by the way they chose to present their published results and the works that they cited.

17

Further development of the interpretation of quantum theory

17.1 Introduction

In Chapter 13 we showed that immediately following the establishment of quantum mechanics as a theory capable of making accurate predictions as to the outcomes of a wide range of experiments, the question of what the theory really meant began to trouble workers in the field. We have seen how those who had made the most significant contributions to the development of quantum mechanics had different ideas as to how to interpret the wave function and other aspects of the theory. This was most apparent at the fifth Solvay Conference, which took place in October 1927, two and a half years after Heisenberg's original paper had introduced matrices as the basic elements of the theory.

One of the main points of discussion at the Solvay Conference was the meaning of the fact that states of a system (often represented by wave functions) can be expressed as superpositions of other states. Thus a general state $|\psi\rangle$ of a system can be written as a sum of eigenstates of a Hermitian operator, say, the Hamiltonian H, as

$$|\psi\rangle = \sum_n c_n |\phi_n\rangle , \tag{17.1}$$

where $H|\phi_n\rangle = E_n |\phi_n\rangle$ and the $c_n = \langle\phi_n |\psi\rangle$ are arbitrary complex numbers satisfying $\sum_n |c_n|^2 = 1$ so that $|\psi\rangle$ is normalized ($\langle\psi |\psi\rangle = 1$). It was clear from the beginning that the phase relations between the c_n were of crucial importance, as we discussed in Section 13.2. The average energy in the state $|\psi\rangle$ is

$$\langle E\rangle = \langle\psi| H |\psi\rangle = \sum_n |c_n|^2 E_n, \tag{17.2}$$

which led to the interpretation of $|c_n|^2$ as the probability of being in the state $|\phi_n\rangle$ given that the system was in the state $|\psi\rangle$, that is, the probability of obtaining the value E_n if we measured the energy in the state $|\psi\rangle$.

In Chapter 14 we described how von Neumann introduced a new physical process that took the system directly from the initial state $|\psi\rangle$ to the state $|\phi_n\rangle$ corresponding to the value of energy that was measured. Between these measurement events (jumps), the system was seen as undergoing unitary evolution described by the Schrödinger equation, according to which superpositions evolve into other superpositions. Such

The Historical and Physical Foundations of Quantum Mechanics. Robert Golub and Steven K. Lamoreaux, Oxford University Press.
 DOI: 10.1093/oso/9780198822189.003.0017

unitary evolution can never produce irreversible dynamics; in order to explain how $|\psi\rangle \to |\phi_n\rangle$, an additional process ("collapse" of the wave function in the Copenhagen interpretation) is necessary.

The line spectra emitted by atoms are taken as indicating that atoms are always in one definite energy state, as first postulated by Bohr, and that they make transitions between these states when they emit or absorb light. However, we have seen in Section 12.3 that, according to the theory, an atom in a given energy eigenstate will evolve, due to the interaction with the electromagnetic field, into a state of the type (17.1). But the light emitted always has a definite frequency, so that the atom must be in a definite energy state after the emission, thus requiring the "collapse" of the wave function as postulated by von Neumann.

The theory as presented by von Neumann is in agreement with all known experiments, so it has been considered as "the" quantum theory by a majority of working physicists since von Neumann's publication. However, there has always been a number of physicists who felt (and feel) uneasy with this viewpoint. For example, no statement is made, in this standard theory, concerning the meaning of the state $|\psi\rangle$ (17.1). Does this mean that the system in actually in one of the states $|\phi_n\rangle$ but we do not know which one? Is the state $|\psi\rangle$ and its associated probabilities $|c_n|^2$ merely an expression of our ignorance as to the true state? If so, what do we make of the fact that in certain circumstances we can observe interference effects between the various terms in Eq. (17.1)? Is that telling us that the system is really in several states at the same time?

One group of physicists feels that the element of probability introduced into the theory is the manifestation of a deeper layer of reality in the same way that the thermal fluctuations of the pressure in a gas are due to the motions of the individual atoms in the gas. That is, they think that there are "hidden variables," unknown to us at the moment, whose knowledge would allow us to predict the exact outcome of every measurement of e.g., energy. We have shown in Section 14.2 how von Neumann proved that hidden variables are incompatible with the present form of quantum mechanics. The de Broglie-Bohm theory (Section 13.3.2), which contains (unobservable) hidden variables but makes exactly the same predictions as quantum mechanics, shifts the origin of probabilities to the unknown initial conditions on these hidden variables.

Another, growing group of physicists feels quite strongly that the theory is incomplete if it cannot provide a description of the collapse of the wave function and that the cause of the collapse should be found in the theory. However, unitary evolution as described by the Schrödinger equation can strictly speaking never result in a collapse. There is a body of work that proposes adding terms to the Schrödinger equation to bring about the collapse, i.e., to drive the state (17.1) into one of its component states (see Section 17.7). Others try to push the bounds of unitarity by trying to derive the collapse, or something close to it, based on the Schrödinger equation alone in open systems, which do not undergo a unitary evolution even though the total universe does.

There is a classical analoge to the non-unitary collapse problem, namely the question of how irreversible behavior can arise when all the classical equations of motion—Newton's equations and Maxwell's equations—are symmetric under time reversal. The appearance of irreversibility is a difficult issue with an enormous literature, and it can to some extent be considered an unsolved problem. One possible solution is to note that in systems with large numbers of constituents, the time needed to return arbitrarily close to a given initial state is immensely long (Poincaré time).

We have met something reminiscent of collapse in the quantum theory in Section 12.3.3 where we discussed the Wigner-Weisskopf theory of light emission from an atom: an atom in an excited state decays into a lower energy state, while the probability of returning to the original state is extremely small because the final state consists of a superposition of photon states with a continuous energy distribution. In these cases the treatment is approximate, so some argue that, e.g., even if the interference induced by the superposition (17.1) is extremely small (and experimentally undetectable), the collapse has not really occurred.

Another group argues that the collapse never occurs: all branches of the superposition remain active, but we as observers are associated with just one of them, so that all our observations are consistent with a collapse. This idea, first introduced by Everett (1957) and called by him the "relative states" approach, is called by others the "many worlds" or "many universes" theory. These ideas have a growing number of supporters. In fact it appears that in some sense physics is splitting into factions. We have "Bohrians," "Bohmians," and "Everettians" as well as followers of "decoherence" and many others. All of these ideas have in common the fact that they do not predict any behavior inconsistent with the standard von Neumann theory and so are not empirically testable. For this reason many physicists choose to ignore these questions and work on the applications of quantum mechanics to an enormous variety of physical systems.

The key question is the relation of (17.1) to macroscopic bodies, which are never observed in a superposition. In a letter to Born (January 1, 1954)[1], Einstein the problem as follows:

Let ψ_1 and ψ_2 be two solutions of the same Schrödinger equation. Then $\psi_1 + \psi_2$ is also a solution with the same ability to describe a possible real state. If the system is a macro-system and ψ_1 and ψ_2 are narrow with respect to the macro-coordinates then in most cases this [*that* $\psi_1 + \psi_2$ *describe a possible real state of the system*] is no longer the case.

Schrödinger expressed the absurdity of applying quantum mechanics to macroscopic objects by introducing the famous thought experiment of a living cat confined in a closed box with a decaying atom with a half-life of one hour and an apparatus that would release poison in the event that a decay took place. After one hour the atom would be in a superposition state of decayed and undecayed and the cat...would be in a superposition of dead and alive. Since macroscopic objects are never in superposition states, this thought experiment brings up the crucial problem of where the transition from the quantum to the classical realm occurs or, in other words, what constitutes the "measurement" that causes the wave function to collapse into a single component of the superposition.

In his discussion with Einstein at the Solvay Conference (Section 13.4), Bohr applied the uncertainty principle to macroscopic objects in order to establish the consistency of the theory. On the other hand, a main tenet of the Copenhagen interpretation

[1] *The Born-Einstein Letters Correspondence between Albert Einstein and Max and Hedwig Born from 1916 to 1955 with commentaries by Max Born*, (Macmillan, 1971), p. 213.

(Section 10.5) is that the macro-world must be described classically in order for us to be able to communicate the results of experiments, and the transition from quantum to classical behavior, i.e., the collapse of the superposition takes place at the famous Heisenberg-von Neumann cut. If the uncertainty principle applies to classical objects, it seems that the only reasonable place to apply the cut is at the moment of observation by a conscious observer. This conclusion has been drawn by Wigner, among others, and in a well-known work by London and Bauer, which we discuss in Section 17.3.

Another strong departure of quantum mechanics from classical physics is entanglement: the fact that two systems which interact at some time and then separate are correlated in such a way that measurements on one of the systems can influence the results of subsequent measurements on the second system. This phenomenon was brought to the attention of the physics community by the work of Einstein, Podolsky, and Rosen, which we discussed in Chapter 15. The authors never suggested that such effects did not exist; rather, they argued that the existence of such effects meant that the quantum theory was "incomplete." In fact, these effects have been shown to exist and now find applications in quantum computing and communication.

In this chapter we will trace the development of ideas on the interpretation of quantum theory in the years following the EPR work. Out of necessity we will have to focus on a limited number of authors.

17.2 Schrödinger

In Section 15.3.4 we discussed a series of papers written by Schrödinger after the work of EPR, in which he gave a detailed discussion of entanglement. In addition to these technical arguments, he published a long article which appeared as three papers in *Naturwissenschaften* (*Sciences*), a German journal similar to *Scientific American* without the illustrations, which published (and publishes) mostly non-mathematical articles on a broad range of scientific topics. In this article[2] he addressed the interpretation of superposition and entanglement and discussed what appeared to be inconsistencies in the theory.

Schrödinger begins by pointing out that one possible interpretation of the statistical uncertainties introduced by quantum mechanics is that the wave function applies to an ensemble of identically prepared systems, each member of the ensemble being in a completely determined (but unknown to us) state. This interpretation turns out to be untenable, and he considers a harmonic oscillator in the first excited state $\left(E = \frac{3}{2}\hbar\omega\right)$ as a counterexample. In this state, the coordinate is smeared out over a significant range. If this were due to different members of the ensemble having different coordinates, the coordinates would have to be limited to those for which $V(x) \leq \frac{3}{2}\hbar\omega$, i.e., tunneling would be prohibited. If we were to ascribe to the system at every instant an exact (but unknown) state in which all variables are determined but not known to us, there is no possible assumption concerning these values which does not contradict some quantum mechanical prediction.

[2]Schrödinger , E., *The present situation in quantum mechanics*(in German), Naturwissenschaften **23**, 807; 823; 844 (1935).

An interesting question is how this situation (tunneling into classically forbidden regions) is handled in the de Broglie-Bohm model. In this model the quantum potential can lower the total effective potential allowing the particle to enter classically forbidden regions.[3]

The wave function ψ gives a clear expression of the smearing of all variables at every time. The laws of motion are as clear and deterministic as classical laws. Problems only arise when the uncertainty reaches large, visible objects where the concept of smearing is certainly false. An atom emits light in a spherical wave but when the wave reaches a screen, the screen reacts at a single point: this point at one time, a different point at another time, etc. It is here that Schrödinger introduces the thought experiment with the cat.

Schrödinger notes that there is an important distinction within conventional quantum theory between the quantum object and the instrument used to observe it. Every measurement suspends the law of the continuous time variation of ψ and replaces it by changes which follow no known law (collapse) and are dictated by the result of the measurement. But measurement is a natural process and so cannot be subject to special laws; it cannot violate the regular motion of natural systems. The fact that the ψ function breaks up during a measurement indicates that the ψ function cannot be a reflection of objective reality. Moreover, the discrete change in ψ depends on the value obtained in the measurement, which is not predictable.

If, instead, the wave function represents the maximum possible knowledge of the system, then the collapse represents a loss of knowledge. Since knowledge cannot be lost, the disconcerting implication is that the object itself must also change by discontinuous jumps in unpredictable ways.

Schrödinger argues that we must try to view the interaction between the object and the instrument objectively. The instrument and object form a system with a ψ for the total system. As we will see in more detail later, the wave functions for the partial systems (object and instrument) are correlated even if the parts are isolated from each other. Measurements on one partial system can influence the expectations for the other part. An essential point is that the two separated partial systems must have formed one system sometime in the past, i.e., they must have had a significant interaction and then separated. After the measurement, which can run automatically with no observer, the wave function for the complete system does not contain a specific value for the position of the instrument pointer. The ψ function of the object has not made a jump, nor has it evolved according to continuous laws for the object alone. It has, however, propagated according to the continuous laws of the total wave function and is entangled with the ψ function of the instrument. The ψ function of the object has split into a conditional disjunction of experimental outcomes. The object can only be removed from this superposition when a living subject obtains knowledge of the result of the measurement. It is only then that the process can be considered as a measurement, even though we really prepared everything objectively. Looked at in this way, the discontinuous jump seems like a mental act since the object is completely

[3]Dewdney, C., and Hiley, B.J., *A quantum potential description of one-dimensional time-dependent scattering from square barriers and square wells*, Found. Phys. 12, 27, (1982), received 26 January 1981, and Holland, P.R., *Quantum Theory of Motion*, (Cambridge UP, 1982), p. 198-203.

separated from the instrument at this time. But this is incorrect. The ψ function of the object was lost in the entanglement. What does not exist cannot be changed. The observation has no physical influence on the object. The object's individual ψ was destroyed by the entanglement and resurrected by the observation. After this linking of the measurement problem with entanglement, Schrödinger goes on to give a general discussion of two entangled systems similar to that of EPR.

The issues raised in Schrödinger's exposition of standard quantum theory remain problematic today, after 85 years of development. While the Copenhagen interpretation suffices to give agreement with every experiment so far performed, it seems unsatisfactory to many researchers who continue to seek alternate interpretations.

In the following we will show how a selection of other physicists have attempted to resolve these issues.

17.3 London and Bauer

In 1939, Fritz London and Edmund Bauer took up the measurement problem, starting with the following opening salvo:

> Physicists are to some extent sleepwalkers, who try to avoid such issues and are accustomed to concentrate on concrete problems. But it is exactly these questions of principle which nevertheless interest non-physicists and all who wish to understand what modern physics says about the analysis of the act of observation itself.
>
> Although these problems have already been the subject of deep discussions [see especially von Neumann, 1932, *and Chapter 14 of the present book*], there does not yet exist a treatment both concise and simple. This gap we have tried to fill.[4]

The authors state that their work shows how ψ represents information acquired by the observer and that each new measurement modifies it. There are two stages to a measurement: the coupling of the quantum system being measured to an instrument and observation of the instrument by a conscious observer. The observation results in a new wave function, i.e., the wave function collapses and the cut occurs when a conscious observer intervenes.

They then give an intensive review of quantum mechanics with emphasis on the transformation theory. The same state $|\psi\rangle$ can be expanded in different basis sets, say $|u_k\rangle$ or $|v_j\rangle$, as

$$|\psi\rangle = \sum_k |u_k\rangle \langle u_k |\psi\rangle = \sum_j |v_j\rangle \langle v_j |\psi\rangle, \tag{17.3}$$

where the amplitudes are linearly related as

$$\langle u_k |\psi\rangle = \sum_j \langle u_k |v_j\rangle \langle v_j |\psi\rangle = \sum_j S_{kj} \langle v_j |\psi\rangle \tag{17.4}$$

for any $|\psi\rangle$, and it is easy to see that S is unitary, $S \cdot S^\dagger = \sum_j \langle u_k |v_j\rangle \langle v_j |u_{k'}\rangle = \delta_{kk'}$.

[4] F. London and E. Bauer, "Theory of observation in quantum mechanics", originally published in French as "La Theorie de l'Observation en Mecanique Quantique", (Paris: Hermann et Cie, 1939). English translation in *Quantum Theory and Measurement*, ed. Wheeler, J. A. and Zurek, W.H. (Princeton: Princeton University Press, 1983).

London and Bauer characterize quantum probabilities as a bit strange. They claim that the wave function ψ has an objective character like the wave field in optics and represents the maximum possible information about the object. It represents potential probabilities which come into play only by an actual measurement.

For definiteness let the states $|u_k\rangle$ represent the energy eigenstates. Then in a pure state as given by Eq. (17.3), the probability of finding a particular energy E_k is given by the square of the amplitude, $|\langle u_k |\psi\rangle|^2$ and $\langle E\rangle = \sum_k E_k |\langle u_k |\psi\rangle|^2$. This is obviously an arithmetic average. However, the average in the pure state $|\psi\rangle$ of another arbitrarily chosen physical quantity F is of the form (using (17.3))

$$\langle\psi| F |\psi\rangle = \sum_{k,l} \langle\psi |u_l\rangle \langle u_l| F |u_k\rangle \langle u_k |\psi\rangle. \tag{17.5}$$

Unless F happens to be diagonal in the energy basis, the expression in (17.5) is unlike any classical arithmetic average.

In addition to this quantum uncertainty in a pure state, there could be an uncertainty in the actual state, described by ordinary probabilities. This is called a mixed state, a mixture of several different pure states. These would give different results. For a mixed state

$$\langle F\rangle = \sum_k \langle u_k| F |u_k\rangle |\langle u_k |\psi\rangle|^2. \tag{17.6}$$

The amplitudes $\langle u_k |\psi\rangle$ are complex numbers and can be written as $\sqrt{p_k}\, e^{i\alpha_k}$. Then the expectation value of F (17.5) in the pure state is

$$\langle\psi| F |\psi\rangle = \sum_{k,l} \sqrt{p_k p_l}\, e^{i(\alpha_k-\alpha_l)} \langle u_l| F |u_k\rangle. \tag{17.7}$$

If the phases were random, averaging over the phases would give

$$\langle\psi| F |\psi\rangle = \sum_{k,l} p_k \langle u_k| F |u_k\rangle, \tag{17.8}$$

the same as (17.6).

London and Bauer consider pure and mixed states to be manifestations of different types of probability. In pure states the probabilities are only "potential," while in mixed states different pure states occur randomly with "ordinary" probabilities. Following von Neumann, they introduce the density or statistical operator. In a state which is a mixture of pure states $|n\rangle$, the expectation value of an observable G is given by

$$\langle G\rangle = \sum_n p_n \langle n| G |n\rangle. \tag{17.9}$$

Expanding $|n\rangle$ in a complete set $|k\rangle$ (i.e., choosing a coordinate basis in the Hilbert space), $|n\rangle = \sum_k |k\rangle \langle k |n\rangle$ we find

$$\langle G\rangle = \sum_{n,j,k} p_n \langle n | j\rangle \langle j| G | k\rangle \langle k | n\rangle \tag{17.10}$$

$$= \sum_{j,k} \langle j| G | k\rangle \sum_n p_n \langle k | n\rangle \langle n | j\rangle = \sum_{j,k} \langle j| G | k\rangle \langle k| \rho | j\rangle \tag{17.11}$$

$$= \mathrm{Tr}\,[G\rho] \tag{17.12}$$

with the statistical (density) matrix given by

$$\langle k| \rho | j\rangle = \sum_n p_n \langle k | n\rangle \langle n | j\rangle. \tag{17.13}$$

We can write a density operator as

$$\widehat{\rho} = \sum_n p_n | n\rangle \langle n| . \tag{17.14}$$

For a pure state all the p_n are zero except for one of the elements if the pure state is a member of the coordinate basis. From (17.14) we have for the trace of ρ

$$\mathrm{Tr}\,[\rho] = \sum_{n,k} p_n \langle k | n\rangle \langle n | k\rangle = \sum_{n,k} p_n |\langle k | n\rangle|^2 = \sum_n p_n = 1. \tag{17.15}$$

For a pure state (17.14) becomes $\widehat{\rho} = | n\rangle \langle n|$ and we see that

$$(\widehat{\rho})^2 = | n\rangle \langle n| \, n\rangle \langle n| = | n\rangle \langle n| = \widehat{\rho}. \tag{17.16}$$

For a mixture we find

$$(\widehat{\rho})^2 = \left[\sum_n p_n | n\rangle \langle n|\right] \left[\sum_m p_m | m\rangle \langle m|\right] \tag{17.17}$$

$$= \sum_{n,m} p_m p_n | n\rangle \langle n| \, m\rangle \langle m| = \sum_n p_n^2 | n\rangle \langle n|. \tag{17.18}$$

As $\widehat{\rho}$ is Hermitian, it can always be diagonalized and written in the form (17.14) with $\sum_n p_n = 1$, so that $p_n - p_n^2 \geq 0$.

In the case of a mixed state $\widehat{\rho}$ describes an ensemble of identical systems, each in a different pure state. It is similar to the distribution function of classical statistical mechanics, so we should expect that $\widehat{\rho}$ will play a role in quantum statistical mechanics.

Classically the entropy is given by (k_B is Boltzmann's constant)

$$S = -k_B \sum_i [n_i \ln n_i - N \ln N] \tag{17.19}$$

with n_i the number of systems in the state i and N the total number of systems in the ensemble. Taking the occupation probability of a state to be $p_i = n_i/N$, we find

$$S = -k_B N \sum_i p_i \ln p_i\,. \tag{17.20}$$

If we evaluate $\widehat{\rho}$ in the basis in which $\widehat{\rho}$ is diagonal (remember that the trace is invariant to transformation of the basis), this is equal to

$$S = -k_B N \mathrm{Tr}\left(\widehat{\rho} \ln \widehat{\rho}\right), \tag{17.21}$$

which is the von Neumann definition of entropy. For a pure state the diagonal elements of $\widehat{\rho}$ are either 0 or 1, so that $S = 0$ for a pure state, and for a mixed state the entropy is always positive. Just as in classical statistical mechanics, we seek the density operator that maximizes S (17.21) subject to the constraints $\mathrm{Tr}\left(\widehat{\rho}\right) = 1$ and $N\mathrm{Tr}\left(\widehat{\rho}H\right) = E$ with E the total energy of the system and H the Hamiltonian. The solution is

$$\widehat{\rho} = \frac{e^{-\beta H}}{Z\left(\beta\right)} \tag{17.22}$$

with $Z\left(\beta\right) = \mathrm{Tr}\left(e^{-\beta H}\right)$ and $\beta = 1/k_B T$. Substituting into (17.21), we find

$$S = k_B N \mathrm{Tr}\left(\frac{e^{-\beta H}}{Z\left(\beta\right)}\left(\beta H + \ln Z\left(\beta\right)\right)\right) = -k_B N\left(\beta \frac{\partial \ln Z}{\partial \beta} - \ln Z\right), \tag{17.23}$$

where the average energy is given by

$$E = N\mathrm{Tr}\left(\widehat{\rho}H\right) = N\mathrm{Tr}\left(\frac{e^{-\beta H}}{Z\left(\beta\right)}H\right) = -N\frac{\partial \ln Z}{\partial \beta}, \tag{17.24}$$

and the Helmholtz free energy is

$$F = E - TS = -N\left(\frac{\partial \ln Z}{\partial \beta} - k_B T\left(\beta \frac{\partial \ln Z}{\partial \beta} - \ln Z\right) = -Nk_B T \ln Z\right). \tag{17.25}$$

17.3.1 Composite systems

Following this review of the quantum theory (we have only presented the highlights), London and Bauer turn to "the aspect of quantum mechanics...which contains the very essence of the theory, the feature responsible for the appearance of probabilities." As shown by von Neumann, a pure state will always remain pure as it undergoes the unitary evolution described by the Schrödinger equation. The state ψ will be known for all future times if it is known for one time. Now we consider a system composed of two subsystems, each in a pure state, that interact for some time and then separate. Let the coordinates of the first system be denoted by x and those of the second system by y.

Initially both systems are in pure states that can be expanded in an appropriate basis as

$$\psi_I\left(x\right) = \sum_k a_k u_k\left(x\right) \tag{17.26}$$

$$\psi_{II}\left(y\right) = \sum_j b_j v_j\left(y\right), \tag{17.27}$$

and the complete system can be described by the product of the two functions,

$$\Psi(x,y) = \psi_I(x)\,\psi_{II}(y) = \sum_{k,j} a_k b_j u_k(x)\,v_j(y). \tag{17.28}$$

With the Hamiltonian

$$H = H_I(x, p_x) + H_{II}(y, p_y), \tag{17.29}$$

the Schrödinger equation separates into two independent equations, but in the presence of an interaction $H_i(x, y, p_x, p_y)$, this is no longer the case. If $u_k(x)$, $v_j(y)$ each represent a complete basis in their domains, then the product $u_k(x)\,v_j(y)$ forms a complete set in the domain (x, y), so that the composite wave function can always be written as

$$\Psi(x,y) = \sum_{k,j} A_{kj} u_k(x)\,v_j(y) \tag{17.30}$$

where A_{kj} cannot in general be written as a product $a_k b_j$. This state of the entire system remains a pure state. The basis for the entire system is denoted by two indices (kj), and the density matrix for the whole system can be written as

$$\rho_{kj,lm} = A_{kj}A^*_{lm} = \langle kj|\,\Psi\rangle\,\langle\Psi|\,lm\rangle, \tag{17.31}$$

which represents a pure state. Now we focus on system I. Let $F(x)$ be a dynamical quantity in system I with matrix elements $\langle lm|\,F\,|kj\rangle = F_{lk}\delta_{mj}$. Then the expectation value of F in the pure composite state $|\Psi\rangle$ is

$$\langle\Psi|\,F\,|\Psi\rangle = \sum_{kjlm} \langle\Psi|\,lm\rangle\,\langle lm|\,F\,|kj\rangle\,\langle kj|\,\Psi\rangle \tag{17.32}$$

$$= \sum_{kjlm} \langle\Psi|\,lm\rangle\,F_{lk}\delta_{mj}\,\langle kj|\,\Psi\rangle \tag{17.33}$$

$$= \sum_{kl}\left[\sum_m \langle km|\,\Psi\rangle\,\langle\Psi|\,lm\rangle\right] F_{lk} = \sum_{kl} \rho^I_{kl} F_{lk}, \tag{17.34}$$

where

$$\rho^I_{kl} = \sum_m \langle km|\,\Psi\rangle\,\langle\Psi|\,lm\rangle \tag{17.35}$$

is the trace of the density matrix of the composite system with respect to the states of system II. The crucial feature of the matrix in (17.35) is that it can be interpreted as a density matrix of a mixed state of system I. To see this, it is helpful to introduce the pure states (of system I), which are normalized by construction:

$$\left|\Psi_I^{(m)}\right\rangle = \frac{\sum_k |k\rangle\,\langle km|\,\Psi\rangle}{\left[\sum_k |\langle km|\,\Psi\rangle|^2\right]^{1/2}}. \tag{17.36}$$

Observe that, from (17.36), $\langle km|\,\Psi\rangle = c_m\langle k|\,\Psi_I^{(m)}\rangle$, where $c_m = \left[\sum_l |\langle lm|\,\Psi\rangle|^2\right]^{1/2}$. Substitution of this into (17.35) yields

$$\rho_{kl}^{I} = \sum_{m} c_m^2 \langle k|\Psi_I^{(m)}\rangle\langle\Psi_I^{(m)}|l\rangle, \tag{17.37}$$

By invoking completeness of the basis $|km\rangle$, the sum of the positive coefficients $\sum_m p_m = \sum_m c_m^2$ can be shown to be unity. Thus, the p_m can be identified as probabilities, meaning that ρ_{kl}^{I} in (17.37) is, indeed, a density matrix of a mixed state. Similar arguments apply to subsystem II, so that it will also be in a mixed state $\left(\rho_{mn}^{II}\right)$ while the total system remains in a pure state.

Substituting (17.36) into (17.37) we find

$$\rho_{kl}^{I} = \sum_{m} p_m \left\{ \frac{\sum_{k'} \langle k\,|k'\rangle\,\langle k'm|\,\Psi\rangle}{\left[\sum_k |\langle km|\,\Psi\rangle|^2\right]^{1/2}} \right\} \left\{ \frac{\sum_{k''} \langle \Psi|\,k''m\rangle}{\left[\sum_k |\langle km|\,\Psi\rangle|^2\right]^{1/2}} \langle k''|\,l\rangle \right\} \tag{17.38}$$

$$= \sum_{m} p_m \frac{\langle km|\,\Psi\rangle\,\langle\Psi|\,lm\rangle}{\sum_k |\langle km|\,\Psi\rangle|^2}. \tag{17.39}$$

In the case where the amplitudes $\langle km|\,\Psi\rangle$ factor into a term for each subsystem,

$$\langle km|\,\Psi\rangle = \langle k|\,u\,(x)\rangle\,\langle m|\,v\,(y)\rangle\,, \tag{17.40}$$

we see that

$$p_m = \sum_{k} |\langle k|\,u\,(x)\rangle|^2\,|\langle m|\,v\,(y)\rangle|^2 = |\langle m|\,v\,(y)\rangle|^2, \tag{17.41}$$

where we have used the completeness of the basis $\{|k\rangle\}$ and the fact that the states $|u\,(x)\rangle$ are normalized. Similarly,

$$\left|\Psi_I^{(m)}\right\rangle = \frac{\sum_k |k\rangle\,\langle k|\,u\,(x)\rangle\,\langle m|\,v\,(y)\rangle}{\left[\sum_k |\langle k|\,u\,(x)\rangle\,\langle m|\,v\,(y)\rangle|^2\right]^{1/2}} = \sum_{k} |k\rangle\,\langle k|\,u\,(x)\rangle = |u\,(x)\rangle. \tag{17.42}$$

Using (17.41) and (17.42) in (17.37), with $p_m = c_m^2$,

$$\rho_{kl}^{I} = \sum_{m} p_m\,\langle k|\,\Psi_I^{(m)}\right\rangle \left\langle \Psi_I^{(m)}\right|\,l\rangle = \sum_{m} p_m\,\langle k|\,u\,(x)\rangle\,\langle u\,(x)|\,l\rangle, \tag{17.43}$$

that is, ρ_{kl}^{I} is the density matrix of a pure state.

When the amplitudes factor as in (17.40), both subsystems are seen to be in pure states, but as the interaction transforms the amplitudes so that they can no longer be expressed as a product, the individual subsystems transform into mixtures. This transformation of a subsystem cannot be represented by a unitary transformation within the Hilbert space of that subsystem. The entropy of the pure state of the composite system is zero while the individual subsystems, being mixtures, have positive entropy. This paradox is resolved because $\Psi\,(x, y)$ contains relations (correlations) between the subsystems, which do not appear when we regard each subsystem separately.

This is a truly remarkable situation. Mixtures appear out of a pure state when we choose to take notice of only part of a complex system. The division of a complex system into separate subsystems produces a non-unitary change in the density matrix of each subsystem because in regarding each subsystem separately we are neglecting the correlations between the two subsystems. This mechanism plays an important role in several of the following interpretations.

17.3.2 Measurement as an objective process

London and Bauer then apply these ideas to the measurement process. System I (coordinates x) is taken to be the object under study and system II (coordinates y) is the measuring instrument. Let $F(x)$ be the system I observable we wish to measure and $|u_k(x)\rangle$ its eigenfunctions with eigenvalues f_k $(F|u_k(x)\rangle = f_k|u_k(x)\rangle)$, and let $G(y)$ represent an observable of system II such as a pointer position (that the measurement interaction will correlate with the values f_k), whose eigenfunctions are $|v_\alpha(y)\rangle$, $(G|v_\alpha(y)\rangle = g_\alpha|v_\alpha(y)\rangle)$. Before the measurement, the complete system is described by the state

$$|\Psi(x,y)\rangle = |v_o(y)\rangle \sum_k |u_k(x)\rangle \langle u_k(x)|\, \psi_o(x)\rangle, \tag{17.44}$$

where $|\psi_o(x)\rangle$ is the state of the object that is being measured and $|v_o(y)\rangle$ is the state of the pointer before the measurement. After the measurement interaction, the complete system is described by the wave function

$$|\Psi'(x,y)\rangle = \sum_k |u_k(x)\rangle \left|v_{\alpha(k)}(y)\right\rangle A(k), \tag{17.45}$$

where the measurement has correlated each value of k with a unique $\alpha = \alpha(k)$. As shown above, this state will be a mixture for each subsystem taken on its own, i.e., the instrument will be found to be in a mixture state so that it will never be seen to be in a superposition, as long as the amplitudes $\left\langle u_k(x)\, v_{\alpha(k)}(y)\right|\Psi'(x,y)\rangle = A(k)$ cannot be factored (as in Eq. 17.40) into a factor depending on the object and another depending on the instrument. Then the probability of finding the result $\alpha(k)$ for the pointer position (and f_k for the value of F) is the square of the amplitude, $\left|\left\langle u_k(x)\, v_{\alpha(k)}(y)\right|\Psi'(x,y)\rangle\right|^2$.

While we have, in this way, eliminated any phase coherence between the components of the original state of the object, this does not constitute a measurement. An observation will choose one of the possible results f_k, following which the wave function of the object will be a pure state corresponding to the observed f_k. According to London and Bauer, it is the consciousness of the observer which changes the mixture state of the system and the observer into a pure state. This then is von Neumann's non-unitary "collapse" process, initiated by a conscious observer. We can extend our system to include the observer, whose states are denoted by $|w(z)\rangle$, so that after the measurement (we denote the correlated values of the quantum numbers for each subsystem by k), the complete system is described by the wave function

$$|\Psi'(x,y,z)\rangle = \sum_k a_k |u_k(x)\rangle \left|v_{\alpha(k)}(y)\right\rangle |w_k(z)\rangle. \tag{17.46}$$

This would be the state as objectively seen from outside; we do not know what state (k) the system is in. But to the observer only the object and instrument are part of the external world. The observer knows by introspection which state they are in and is capable of cutting the chain of superpositions by declaring, "I am in the state $|w_k(z)\rangle$; I see $G = g_k$." Only a conscious observer can separate themselves from the state $|\Psi'(x, y, z)\rangle$, and as a result of their observation set up a new objectivity by attributing a new state ($|u_k(x)\rangle$) to the object. We will see that this idea serves as the basis of the Everett "relative state-many worlds" interpretation.

These arguments are essentially those that had been presented by von Neumann in his classic book,[5] but given in a more concise form and with a simpler notation by London and Bauer. We have used the Dirac notation, which is clearer and more in line with modern usage.

Many physicists through the years have had difficulty accepting these ideas about the role of consciousness as in some way solving the measurement problem, and the search for a definitive alternative continues. John Bell's attitude is shown by the following statement:[6]

> What exactly qualifies some systems to play this role [*measuring instrument*]? Was the world wave function waiting to jump for thousands of millions of years until a single-celled living creature appeared? Or did it have to wait a little longer for some more highly qualified measurer—with a PhD? If the theory is to apply to anything but idealized laboratory operations, are we not obliged to admit that more or less "measurement-like" processes are going on more or less all the time more or less everywhere? Is there ever then a moment when there is no jumping and the Schrödinger equation applies?

We describe some of the attempts to find an alternative interpretation in the remainder of this chapter.

17.4 David Bohm

As time went on, the interest in these questions of interpretation seemed to wane among working physicists. David Bohm, a young physicist who had been a student of Oppenheimer's at Berkeley, decided, while lecturing on quantum mechanics at Princeton, to write a book in which the physical and philosophical implications of quantum mechanics would be emphasized without diminishing the role of mathematics. The book supported the then dominant Copenhagen interpretation. It was published in February 1951.[7] Evidently after writing the book Bohm became somewhat disillusioned with the Copenhagen interpretation, and on 5 July 1951, two manuscripts were received by the Physical Review[8] in which Bohm proposed a new "interpretation of the theory based on hidden variables" which was almost identical with the pilot wave theory that had been proposed by de Broglie at the Solvay Conference in 1927 (Chapter 13), thus completely repudiating the position he had taken in his textbook.

[5]Neumann, J. von, *Mathematische Grundlagen der Quanten Mechanik* (Berlin: Springer, 1932). English translation: *Mathematical Foundations of Quantum Mechanics* (Princeton: Princeton University Press, 1955).

[6]Bell, J. S., "Quantum Mechanics for Cosmologists," in *Quantum Gravity: An Oxford Symposium*, ed. C. Isham, R. Penrose and D. Sciama (Oxford: Oxford University Press, 1981); reprinted in J. S. Bell, *Speakable and Unspeakable in Quantum Mechanics* (Cambridge: Cambridge University Press, 1987).

[7]Bohm, D., *Quantum Theory* (Englewood Cliffs, NJ: Prentice-Hall, 1951).

[8]Bohm, D., *A Suggested Interpretation of the Quantum Theory in Terms of Hidden Variables I and II*, Phys. Rev. **85**, 166; 180 (1952), received July 5, 1951.

This had been an eventful time for Bohm.[9] On 4 December 1950, he had been arrested for contempt of Congress based on his refusal to give names to an executive committee of the House Committee on Un-American Activities on 23 May 1950, at the "Hearings Regarding Communist Infiltration of Radiation Laboratory and Atomic Bomb Project at the University of California, Berkeley, California." He was released on bail shortly after his arrest, and Princeton suspended him immediately after. Although he was acquitted at a trial on 31 May 1951, his contract with Princeton was allowed to terminate in June 1951, and in October he left the United States for Sao Paulo, Brazil, where he had obtained a position with references from Einstein and Oppenheimer. We have already discussed the pilot wave approach in connection with the Solvay Conference. Here we examine Bohm's attempt at a (later abandoned) defense of the Copenhagen interpretation in his book cited above.

In his textbook Bohm had offered strong support to the Copenhagen interpretation, pointing out the distinction between the physics of quantum and classical systems and the necessity of classical systems for the study of microscopic systems. He concluded the book with the statement:

...quantum theory has actually evolved in such a way that it implies the need for a new concept or [*sic*] the relation between large scale and small scale properties of a given system. In this chapter we have discussed two aspects of this new concept:

1. Quantum theory presupposes a classical level and the correctness of classical concepts in describing this level.
2. The classically definite aspects of large-scale systems cannot be deduced from the quantum mechanical relationships of assumed small scale elements. Instead, classical definiteness and quantum potentialities complement each other in providing a complete description of the system as a whole.

Indeed, he suggested that "the successful extension of quantum theory to the domain of nuclear dimensions may perhaps introduce more explicitly the idea that the nature of what can exist at the nuclear level depends to some extent on the macroscopic environment."

Within the Copenhagen interpretation, "the definition of small scale properties of a system is possible only as the result of interaction with large scale systems undergoing irreversible processes."

"In line with the above suggestion," Bohm proposed that "irreversible processes taking place in the large scale environment may also have to appear explicitly in the fundamental equations describing phenomena at the nuclear level."

These ideas have been questioned by later workers, who, as we shall see, feel that the classical behavior of macroscopic systems should emerge from quantum mechanics itself without any extra hypotheses.

A major concern of Bohm's book is the meaning of the phase in a superposition. For a wave function $\psi(x) = R(x)\, e^{i\phi(x)}$ with $R(x)$, $\phi(x)$ real, the phase $\phi(x)$ at different points controls the momentum distribution (momentum $p = \hbar k$):

[9]Peat, F. D., *Infinite Potential: The Life and Times of David Bohm* (New York: Basic Books, 1997).

$$\Phi(k) = \int \psi(x) e^{-ikx} dx, \tag{17.47}$$

and the phase relations between $\Phi(k)$ for different values of k control the shape of $\psi(x)$. To get a narrow packet in x we need a coherent superposition of a wide range of k values. If we start with a $\psi(x)$ that is a broad wave packet and measure the position with a resolution Δx much smaller than the initial width of the packet, we will wind up with a wave function broken up into independent packets, each of size Δx and with no definite phase relations between them, so that the momentum distribution will be significantly broadened compared to the initial momentum distribution. The fact that a measurement results in randomizing the phases of the elements of a superposition is a key point in Bohm's interpretation. To show that, he considers a quantum system S for which we wish to measure a given property and a (classical) apparatus A. The measurement is mediated by an interaction between the two, so that the total Hamiltonian is given by

$$H = H_S(x) + H_A(y) + H_{int}(x, y), \tag{17.48}$$

where (x, y) are the coordinates of the (system, apparatus) respectively. He introduces the idea of an impulsive measurement, in which $H_{int}(x, y)$ is very large for a short time. During this time H_S, H_A are negligible, and the Schrödinger equation is

$$i\hbar \frac{\partial \psi}{\partial t} = H_{int}\psi. \tag{17.49}$$

The quantity to be measured is designated as M (eigenstates $|m\rangle$, eigenfunctions $\langle x|m\rangle$, eigenvalues m) and we take $H_{int} = H_{int}(M, y)$. In this case H_{int} is diagonal in the eigenstates $|m\rangle$ and M will not change during the interaction. We expand the state of the complete system in the eigenstates of M as

$$|\Psi(x, y)\rangle = \sum_m |m(x)\rangle \langle m(x)|\Psi(x, y, t)\rangle = \sum_m |m(x)\rangle a_m(y, t), \tag{17.50}$$

which on substitution in (17.49) yields

$$i\hbar \sum_m |m(x)\rangle \frac{\partial}{\partial t} a_m(y, t) = H_{int}(M, y) \sum_m |m(x)\rangle a_m(y, t) \tag{17.51}$$

$$i\hbar \frac{\partial}{\partial t} a_m(y, t) = H_{int}(m, y) a_m(y, t). \tag{17.52}$$

So for each eigenstate $|m(x)\rangle$ the apparatus will evolve differently during the measurement and the needle pointer (y) will be correlated with the eigenvalue (m). This will only constitute a measurement if the values of y are sufficiently separated for different values of m.

These ideas were then applied to the concrete example of the Stern-Gerlach experiment. This was briefly mentioned in Section 4.2.2. The experiment consists of sending a beam of atoms or molecules through an inhomogeneous magnetic field. For a spin-1/2 atom, the interaction of the atom with the field is $H_{int} = -\vec{\mu} \cdot \vec{B}(z) = -\mu\vec{\sigma} \cdot \vec{B}(z)$

where μ is the magnitude of the magnetic moment and $\vec{\sigma}$ is the vector of Pauli matrices. We can describe the situation classically. There will be a force on the atom, $\vec{F} = \vec{\nabla} H_{int} = \mu \frac{\partial B_z}{\partial z}$ (neglecting other components), so that the momentum in the z direction (perpendicular to the beam direction) will change by

$$\Delta p_z = \mu_z \frac{\partial B_z}{\partial z} \tau = \mu_z \frac{\partial B_z}{\partial z} \frac{L}{v} \tag{17.53}$$

with $\tau = L/v$ the length of time the atoms, with velocity v, spend in the gradient field which extends over a distance L. Neglecting the deflection of the beam inside the magnet, the atoms will have a velocity in the z direction of $\Delta p_z/m$ ($m =$ mass) after leaving the magnet. At a time t after leaving the magnet, the atoms will be displaced in z by

$$z = \frac{\mu_z}{m} \frac{\partial B_z}{\partial z} \frac{L}{v} t. \tag{17.54}$$

Atoms with different z components of magnetic moment will differ in position by

$$\Delta z = \frac{\Delta \mu_z}{m} \frac{\partial B_z}{\partial z} \frac{L}{v} t, \tag{17.55}$$

and if this is larger than the initial width of the beam, differences in the directions of travel will be detectable. The result of the Stern-Gerlach experiment, astounding at the time, was that only discrete values of the μ_z component were found (Fig. 4.4). The situation becomes more interesting if we look at it from the standpoint of quantum theory.

Let the eigenstates of σ_z be given by $|\pm\rangle$ ($\sigma_z |\pm\rangle = \pm |\pm\rangle$), and take the state of the beam atoms as they enter the magnet to be

$$|\psi_o\rangle = f_o(z)(c_+ |+\rangle + c_- |-\rangle). \tag{17.56}$$

As in Eq. (17.50), we write the state for later times as

$$|\psi(t)\rangle = f_+(z,t)|+\rangle + f_-(z,t)|-\rangle, \tag{17.57}$$

where $f_\pm(z,0) = f_o(z) c_\pm$. With the interaction Hamiltonian $H_{int} = \mu B' z \sigma_z$ where ($B' = \partial B_z/\partial z$), the Schrödinger equation is

$$i\hbar \left(\frac{\partial f_+(z,t)}{\partial t} |+\rangle + \frac{\partial f_-(z,t)}{\partial t} |-\rangle \right) = \mu B' z (f_+(z,t)|+\rangle - f_-(z,t)|-\rangle). \tag{17.58}$$

Decoupling the terms, we have

$$i\hbar \frac{\partial f_+(z,t)}{\partial t} = \mu B' z f_+(z,t) \tag{17.59}$$

$$i\hbar \frac{\partial f_-(z,t)}{\partial t} = -\mu B' z f_-(z,t). \tag{17.60}$$

Solutions satisfying the initial conditions are

$$f_\pm(z,t) = f_o(z) c_\pm e^{\mp i\mu B' z t/\hbar}. \tag{17.61}$$

We are assuming an impulsive measurement where H_{int} dominates the other terms in the Hamiltonian (17.48) while the atoms are in the magnet. Treating the motion along the beam classically, we can take the time in the magnet as $\tau = L/v$. At the exit of the magnet, the state (17.57) will be determined by $f_\pm(z,\tau)$. To study the motion after leaving the magnet, we write

$$f_o(z) = \int g(k)\, e^{ikz} dk \tag{17.62}$$

and find from (17.57) and (17.61)

$$|\psi(z,\tau)\rangle = \int g(k)\, dk \left[c_+ e^{i(k-\mu B'\tau/\hbar)z} |+\rangle + c_- e^{i(k+\mu B'\tau/\hbar)z} |-\rangle\right]. \tag{17.63}$$

From this point on, the atoms move as free particles with the Hamiltonian $H_A = -(\hbar^2/2m)\,\partial^2/\partial z^2$, so that the state develops as

$$|\psi(z,t)\rangle = \int g(k)\, dk \left[c_+ e^{i[(k-\mu B'\tau/\hbar)z - \omega_+(t-\tau)]} |+\rangle + c_- e^{i[(k+\mu B'\tau/\hbar)z - \omega_-(t-\tau)]} |-\rangle\right], \tag{17.64}$$

which is seen to be a solution of the Schrödinger equation with the Hamiltonian H_A if

$$\omega_\pm = (\hbar/2m)\,(k \mp \mu B'\tau/\hbar)^2. \tag{17.65}$$

After leaving the magnet at time τ, the packets corresponding to the two spin states are seen to be moving with values of the z component of momentum which have been altered by $\Delta p = \pm\mu B'\tau$. In order for the two packets to eventually separate, the momentum difference between them must be larger than the initial spread of momentum, $\delta p_o \geq \hbar/\delta z_o$, where δz_o is the original width of the beam:

$$\Delta p = \mu B'\tau \gg \delta p_o \geq \hbar/\delta z_o. \tag{17.66}$$

The phase of each of the two wave packets must therefore vary by

$$\delta\phi = \frac{\mu B'\tau}{\hbar}\delta z_o \gg 1. \tag{17.67}$$

Taking typical numbers: $L = 10$ cm (length of magnet), $v = 10^4$ cm/sec (velocity of atoms), and $B' = 10^4$ gauss/cm, we find $\tau = L/v = 10^{-3}$ sec, which gives $\frac{\mu B'}{\hbar} = 1.4 \cdot 10^6 \cdot 2\pi \cdot 10^4$ rad/sec. Thus with δz_o in cm,

$$\delta\phi \approx \pm 10^8 \delta z_o. \tag{17.68}$$

The state of the system is a superposition of the two spin states, each multiplied by a wave packet. Each of the packets is moving with a group velocity (see 17.65)

$$v_\pm = \frac{\partial\omega_\pm}{\partial k} = (\hbar/m)\,(k_o \mp \mu B'\tau/\hbar), \tag{17.69}$$

where we have assumed $g(k)$ is a narrow peak centered at $k = k_o$. So the wave packets of the two spin states are separating in time and after a time such that $(\mu B'\tau/m)\,t \gg \delta z$, where δz is the width of the wave packet, $f_o(z)$, at the entrance of the magnet, the wave packets will no longer overlap. If the initial spin state was an eigenstate of σ_x so that $c_+ = c_- = 1/\sqrt{2}$, the beam would split into two separate beams of equal intensity corresponding to $\sigma_z = \pm 1$. Measuring which beam an atom is in is equivalent to measuring the spin. The position of the atom thus serves as the classical "pointer variable" indicating the value of the spin.

In order to make a measurement of the spin, the wave packets have to be separated by a macroscopic distance larger than their width. This means that the phase of the packets will be spread out by a huge amount (17.68), so that the relative phase of the two packets will be completely uncorrelated, thus demonstrating the destruction of the possibility of interference by a measurement. Then the average of any spin function cannot be influenced by interference terms, and after the "measurement" process (after the apparatus has functioned) but before the actual observation, it will have the same value as if the atom occupied one of the spin states with probability $|f_\pm|^2$. We effectively have a statistical ensemble of separate states. Interference will also be prevented by the fact that after the interaction the packets no longer overlap, but this in itself does not preclude interference if the beams were to be brought back together in the future. However, this does not eliminate the need for a collapse-like event.

The observer will then discover which state the system is actually in and will replace the statistical ensemble by a single state corresponding to the actually observed spin. "The sudden replacement of the statistical ensemble of wave functions by a single wave function represents absolutely no change in the state of the spin, but is analogous to the sudden changes in classical probability functions which accompany an improvement of the observer's information."[10]

If we leave out the conscious observer, we can imagine that the z coordinate of the atom is measured objectively by an apparatus without any human intervention. The entire system—spin, z coordinate of the atom, and apparatus to measure z—will be in another pure state after the measurement, which is described by a unitary evolution of the combined system. We have seen above that the phase relations between the different parts of the wave function corresponding to different values of σ_z are so complicated that no further process can show any interference. The system, functioning without any human interference, goes over into a state where the physical results are the same as if the system were in a statistical ensemble of states. A human observer would put the system into a genuine statistical ensemble of states. This process would destroy the phase relations but this will make no significant difference in the behavior of the system because the interference effects were already negligible. This is a completely objective description of measurement not involving human observers.

If we would attempt to bring the packets back together in order to demonstrate interference, the relative phases would depend very sensitively on the fine details of the apparatus. The slightest error would result in large changes in the relative phases. Anticipating future work on "decoherence" caused by the environment (see below), Bohm points out that any apparatus suitable for the last stage must be macroscopic in nature and thus have a large number of degrees of freedom coupled together, and each degree of freedom would then introduce further large and complicated phase shifts

[10]Bohm, D., op. cit. p. 604.

in the combined wave function of the whole system. As an example, friction in the axle of a meter needle will excite complicated thermal motions of the atoms in the shaft and bearings, introducing new complicated phase shifts into the composite wave function. In quantum theory measured quantities can only take on definite values as the result of a coupling to a macroscopic, classically described system undergoing irreversible processes. The irreversibility of the "collapse" as formulated by von Neumann is an intrinsic part of the theory.

With its demonstration that a system's properties depend partly on the apparatus which we choose to measure those properties, quantum theory calls into question the fundamental assumption that the universe can be regarded as made up of distinct and separate parts interacting according to causal laws. In quantum theory the intrinsic properties of a part of the universe can only be defined in connection with its interaction with some other part; different interactions imply different intrinsic properties. The universe is basically a single, indivisible unit.

However, as creatures with finite abilities, we are forced to deal with abstracted parts of the universe. As London and Bauer have shown, a pure state of a composite system looks like a mixture if we single out one part of the system and ignore the rest. This division of the universe into parts is crucial if we want to talk about a quantum system or a measuring apparatus or an "environment" capable of injecting some irreversibility into the system.

Both Bohm and London and Bauer have attempted to understand the meaning of a superposition $\psi = \sum_n a_n \psi_n$. Both agree that an essential feature is the phase relations between the a_n, and note, as did Born and Heisenberg in 1927, that a measurement will randomize these phases. According to London and Bauer, this comes about by neglecting the correlations between the quantum system under study and the measuring apparatus, while Bohm shows that the interaction between the system and the apparatus leads to uncontrollable phase shifts. In Bohm's case each individual member of an ensemble will have a definite phase but these will vary so rapidly between ensemble members that they are effectively randomized. In either case the result is that the superposition with random phases is indistinguishable from the situation where the system is in one of the states ψ_n in the superposition with probability $|a_n|^2$.

17.5 Hugh Everett III and the world's second most important PhD thesis(?)

Hugh Everett III arrived at Princeton as a beginning graduate student in the fall of 1953. It must have been a very exciting place for a beginning physicist. Einstein, Wigner, Oppenheimer, and von Neumann were all present. David Bohm had been expelled less than two years earlier. In his first year Everett was registered in the mathematics department, and he held an NSF fellowship to work on game theory, a very strong interest of his.[11] In the fall of 1954 Bohr visited the Institute of Advanced Study, and Everett and some other graduate students had the opportunity of

[11] Biographical information is from Byrne, P., *The Many Worlds of Hugh Everett III: Multiple Universes, Mutual Assured Destruction, and the Meltdown of a Nuclear Family* (New York: Oxford University Press, 2010); Byrne, P., 'Everett and Wheeler, the Untold Story', in *Many Worlds? Everett, Quantum Theory and Reality*, ed. Saunders, S., Barret J., , Kent, A. and Wallace, D., (Oxford: Oxford

discussing the problems of quantum theory with him, Aage Petersen (Bohr's assistant), and each other. At this time Everett switched to the physics department (his initial adviser was Frank Shoemaker, a well-known accelerator builder, but he soon changed to John A. Wheeler, a prominent theorist who had been Feynman's PhD adviser) and started working on his PhD thesis concerning the interpretation of quantum mechanics.

It was a time of great upheaval in the United States. The Korean War had ended in July 1953 in a stalemate with the People's Liberation Army, and Senator Joseph McCarthy was leading a purge of "Communists" in government, science, and culture, culminating in the "Army-McCarthy" hearings in the summer of 1954. It was a time of great technological and cultural change. The transistor, invented in 1947, was gradually finding more and more applications. The Soviet Union launched *Sputnik*, the first man-made satellite, in 1957, and the rapid spread of television was upending American national culture and politics.

A commentator we have already met summed up the situation:

> Within the country: concentration of tremendous financial power in the hands of the military; militarization of the youth; close supervision of the loyalty of the citizens, in particular of the civil servants, by a police force growing more conspicuous every day. Intimidation of people of independent political thinking. Subtle indoctrination of the public by radio, press and schools. Growing restriction of the range of public information under the pressure of military secrecy.
>
> The armaments race between the U.S.A. and the U.S.S.R....assumes hysterical character.[12]

The first successful test of an atomic bomb by the Soviet Union in 1949 spurred efforts to develop even more powerful thermonuclear weapons. Due to their overwhelming destructive power, development of these H-bombs raised serious concerns from many physicists, even those who had previously worked on the development of fission bombs. John von Neumann, on the other hand, was a strong proponent of building the H-bomb:

> With the Russians it is not a question of whether but when. If you say why not bomb them tomorrow I say why not today? If you say today at 5 o'clock, I say why not one o'clock?
>
> ...I think in particular that the US-USSR conflict will probably lead to an armed "total" collision, and that a maximum rate of armament is therefore imperative.[13]

Robert Oppenheimer, who had directed the Los Alamos atomic bomb project and who as director of the Princeton Institute of Advanced Study opposed the construction of the H-bomb, had his security clearance taken away in June 1954, after a long, controversial hearing before the Atomic Energy Commission.

University Press, 2010), pp. 521-541; and Shikhovtsev, E. B., "Biographical Sketch of Hugh Everett III," ed. Ford, K. W. (2003), https://space.mit.edu/home/tegmark/everett/everettbio.pdf

[12]Einstein, A., "National Security', contribution to Mrs. Eleanor Roosevelt's television program concerning the implications of the H-bomb" (February 13, 1950), in A. Einstein, *Ideas and Opinions*, ed. Bargmann, S., (New York: Bonanza Books, 1954), quoted by Peter Byrne OUP (2010), op. cit., p. 453.

[13]Poundstone, W., *Prisoner's Dilemma: John von Neumann, Game Theory, and the Puzzle of the Bomb* (New York: Anchor Books, 1992), pp. 143, 260, a biography of von Neumann quoted by Byrne (2010) op. cit., p. 453.

So it seems Princeton people were heavily involved in the political currents of the time. It is striking that two of the three main challenges to the standard interpretation of quantum theory came out of this milieu and the third, "decoherence," was strongly promoted by W. H. Zurek, who as a graduate student at the University of Texas in Austin came into contact with Wheeler after he moved there.

In an interview recorded in 1974 between Charles Misner, who had been a graduate student with Everett, and himself, Hugh claimed that he had been steered to thinking about fundamental issues of quantum mechanics by listening to discussions between Misner and Aage Petersen (Bohr's assistant). When Misner reminded Everett that Wheeler, in connection with some solutions he had found to the equations of General Relativity, "was preaching this idea that you ought to just look at the equations, and if they were the fundamentals of physics, why you followed their conclusions and give them a serious hearing."[14] Everett agreed that this was the case.

Everett handed a completed thesis to John Wheeler, his supervisor, in January 1956. The thesis contained, as we will see, a vigorous criticism of Bohr and von Neumann. Wheeler, who had worked closely with Bohr (in 1939 they had written a seminal paper on fission, highlighting the importance of isotopes and predicting Pu^{239} would undergo fission), was reluctant to be involved with anything critical of him. But he seemingly admired Everett's work, partly because it opened the way to applying quantum mechanics to the universe as a whole, something that was felt necessary for the construction of a quantum theory of gravitation. Wheeler put a hold on Everett's thesis until he had a chance to discuss it personally with Bohr, which he was able to do during a visit to Copenhagen in May 1956. To Wheeler's disappointment, Bohr was not willing to say anything positive about Everett's work. In April 1956, Everett joined the Pentagon, where his father had a high-level position, and began doing operations research for the Weapons Systems Evaluation Group. He studied the most efficient targeting strategies for nuclear weapons. His work had an influence on the planning for nuclear war, summarized by MAD (mutually assured destruction), and played a role in the disarmament negotiations which contributed to an eventual reduction of tensions. Around January 1957, Wheeler and Everett reworked the thesis, with the goal of making it less offensive to Bohr, into a much shorter work which was accepted by Princeton in April 1957 and published in *Reviews of Modern Physics* along with a paper by Wheeler offering extremely strong support, comparing Everett's innovation to that of Copernicus.[15] These papers were published as part of the proceedings of a conference held in Chapel Hill, North Carolina, that had been attended by Wheeler but not by Everett. The editor of the proceedings, Bryce DeWitt, included Everett's short thesis because of Wheeler's support. Wheeler offered Everett an instructor's position and in the following years tried several times without success to get an academic position for Everett, who he felt had great talent for fundamental research, but Everett spent the remainder of his career in operations research, being involved in starting several companies and eventually undertaking civil projects like trying to ferret out

[14]Quoted by Byrne 2010 op. cit., p. 453.

[15]H. Everett III, *Relative state formulation of quantum mechanics*, Rev. Mod. Phys. **29**, 454 (1957) and J. A. Wheeler, *Assessment of Everett's "Relative state" formulation of quantum theory*, Rev. Mod. Phys. **29**, 463 (1957).

patterns of discrimination in government and private organizations by statistical analysis.[16] He also invented an extension to the Lagrange multiplier method which played an important role in simplifying complex optimization problems.

Everett's thesis was ignored for about 10 years until Bryce DeWitt had the original long version of the thesis published.[17]

Everett's interpretation of quantum mechanics was a direct response to Bohm's declaration:[18]

> If the quantum theory is to be able to provide a complete description of everything that can happen in the world, however, it should also be able to describe the process of observation itself in terms of the wave functions of the observing apparatus and those of the system under observation. Furthermore, in principle, it ought to be able to describe the human investigator as he looks at the observing apparatus and learns what the results of the experiment are, this time in terms of the wave functions of the various atoms that make up the investigator, as well as those of the observing apparatus and the system under observation. In other words, the quantum theory could not be regarded as a complete logical system unless it contained within it a prescription in principle for how all of these problems, were to be dealt with.

The thesis began with a strong critique of the then current interpretation. With a nod to Einstein's famous elevator, and in anticipation of Wigner's friend[19], Everett imagines an observer A, capable of making a measurement of some quantity and recording the result, sealed in a room with a system S, completely isolated, floating in outer space. There is also an external observer, B. The composite system (C=A plus S) is the object system for observer B, who considers that his wave function $\Psi_B(C)$ is a complete description of C. This wave function will contain non zero amplitudes over several values of the recorded result, so that, as far as B is concerned, the recorded values have no objective existence although A would have observed a definite recorded value (collapsed wave function). Either A is wrong in assuming the collapse with its probabilistic elements or else B's wave function is an inadequate description of the situation. In this case there must be an alternate description for systems containing observers (or recording apparatus) and we need some criterion for designating observers. This of course was the Bohrian view, namely that the world consists of classical objects capable of being altered semi-permanently (recordings) by interactions with quantum objects, these interactions being limited by the "uncontrollable exchange of the quantum of action" which imposes a random, discontinuous behavior. However, in his famous debate with Einstein,[20] Bohr saved the day for his interpretation by applying the uncertainty principle to macroscopic objects. This latter point had been noted by Everett.

To Everett, the discussion of the two observers showed that the standard view of quantum mechanics is untenable in a situation with more than one observer. In the

[16]Byrne, P. (2010), op. cit., p. 453.

[17]DeWitt, B. S., and Graham, N., eds., *The Many Worlds Interpretation of Quantum Mechanics* (Princeton: Princeton University Press, (1973).

[18]Bohm, D., *Quantum Thoery,* (Prentide-Hall, 1951), p.582

[19]Wigner, E.P., "Remarks on the mind-body question", in Good, I.J., ed. *The Scientist Speculates,* (Heinemann, 1961), p. 282, reprinted in Wigenr, E.P., *Symmetreis and Reflections,* (Indiana UP, 1967).

[20]Bohr, N., "Discussion with Einstein on Epistemological Problems in Atomic Physics" in *Albert Einstein, Philosopher-Scientist,* ed. Schilpp, P.A. (New York: Harper, 1959), p. 220.

discussion at the end of his thesis, Everett sharpened his criticism of the "Copenhagen interpretation" developed by Bohr, in which

phenomena can only be understood by the use of different, mutually exclusive (i.e., "complementary") models in different situations. All statements about microscopic phenomena are regarded as meaningless unless accompanied by a complete description (classical) of an experimental arrangement.

While undoubtedly safe from contradiction, due to its extreme conservatism, it is perhaps overcautious. We do not believe that the primary purpose of theoretical physics is to construct "safe" theories at severe cost in the applicability of their concepts, which is a sterile occupation, but to make useful models which serve for a time and are replaced as they are outworn.

Another objectionable feature of this position is its strong reliance upon the classical level from the outset, which precludes any possibility of explaining this level on the basis of an underlying quantum theory. (The deduction of classical phenomena from quantum theory is impossible [*in the Bohrian view*] simply because no meaningful statements can be made without pre-existing classical apparatus to serve as a reference frame.) This interpretation suffers from the dualism of adhering to a "reality" concept (i.e., the possibility of objective description) on the classical level but renouncing the same in the quantum domain.

These comments, of course, did not appear in the version of the thesis that was originally published.

After reviewing several possible solutions to the two-observer conundrum, Everett produces the solution that he will develop in the body of the thesis:

Alternative 5: To assume the universal validity of the quantum description, by the complete abandonment of Process 1 [*collapse*]. The general validity of pure wave mechanics, without any statistical assertions, is assumed for all physical systems, including observers and measuring apparatus. Observation processes are to be described completely by the state function of the composite system which includes the observer and his object-system, and which at all times obeys the wave equation (Process 2).

Processes 1 and 2 refer to von Neumann's characterization, where Process 2 is the unitary evolution of a quantum state determined by the Schrödinger equation, and Process 1 is the famous collapse leading to a definite but randomly chosen eigenstate of the measured observable, and taking place at the notorious "cut"; that is, Process 1 is the discontinuous change of a state $|\psi\rangle$ brought on by the observation of a quantity with eigenfunctions $|\phi_1\rangle$, $|\phi_2\rangle$,... into just one of the eigenfunctions $|\phi_i\rangle$ with probability $|\langle\phi_i\,|\psi\rangle|^2$.

The title of Everett's thesis changed as the thesis itself evolved.[21] The original (long) thesis was titled "Wave Mechanics Without Probability" and was published as "The Theory of the Universal Wave Function" in 1973.[22] The short thesis accepted by Princeton was called "On the Foundations of Quantum Mechanics" and the paper in *Reviews of Modern Physics*, probably due to Wheeler's urging,[23] "Relative State Formulation of Quantum Mechanics."[24]

[21]Osnaghi, S.. Freitas,F. and Freire. O. Jr., *The origin of the Everettian heresy*, Stud. Hist. Philos. Mod. Phys. **40**, 97 (2005).

[22]DeWitt and Graham (1973), op. cit.

[23]Wheeler, J. A. and Ford, K. , *Geons, Black Holes and Quantum Foam: A Life in Physics* (New York: Norton, 1998), p. 270.

[24]Everett H., II, *Rev. Mod. Phys.* **29**, 454 (1957), known as the "short thesis".

17.5.1 "Relative" states

If a system consists of two or more subsystems, there is, in general, no well-defined state for a given subsystem independent of the states of the other subsystems. For any state of a given subsystem, there is a unique correlated state of the remainder of the system that Everett called the "relative" state, but which is more appropriately called the "correlated" or "related" state. Again the reference to relativity is clear. Everett often refers to a well-defined state as an "absolute" state.

The connection was made explicit in a letter[25] of 1957 from DeWitt to Everett and Wheeler: "The conventional interpretation of the formalism of quantum mechanics in terms of an 'external' observer seems to me similar to Lorentz's original version (and interpretation) of relativity theory, in which the Lorentz-Fitzgerald contraction was introduced ad hoc. Everett's removal of the 'external' observer may be viewed as analogous to Einstein's denial of the existence of any privileged inertial frame."

Consider a system S consisting of two subsystems S_1 and S_2. Let $|\phi_i\rangle_1$ and $|\eta_j\rangle_2$ be complete sets of orthonormal states for S_1 and S_2 respectively. Then a general state of the whole system (S) can be written

$$|\Psi\rangle_s = \sum_{i,j} a_{ij} |\phi_i\rangle_1 |\eta_j\rangle_2, \tag{17.70}$$

which is a definite pure state for S. However, S_1 and S_2 do not possess pure states, but if we specify that S_1 is in a specific state $|\phi_k\rangle_1$ then we see that S_2 will be in the pure state

$$|\psi_2\rangle^r_{S_1k} = \sum_j a_{kj} |\eta_j\rangle_2, \tag{17.71}$$

where the notation designates the state of S_2 "relative" to S_1 being in the state $|\phi_k\rangle_1$. Then we can write the state of the composite system as

$$|\Psi\rangle_s = \sum_k |\phi_k\rangle_1 |\psi_2\rangle^r_{S_1k}. \tag{17.72}$$

As Everett notes, "subsystem states are generally correlated with one another. One can arbitrarily choose a state for one subsystem, and be led to their relative state for the remainder. Thus we are faced with a fundamental *relativity of states*...It is meaningless to ask the absolute state of a subsystem—one can only ask the state relative to a given state of the remainder of the system."

17.5.2 Measurement

A measurement consists of a measuring apparatus interacting with an object system. The result is that the state of the measurement apparatus no longer has any independent meaning. "It can be defined only *relative* to the state of the object system. In other words there exists only a correlation between the states of the two systems." But in real life we always observe physical objects to be in definite states. Is it possible to describe the fact that instrument pointers are always found to be in one position

[25]Quoted in S. Osnaghi, F. Frettas and O. Freire, Jr, op. cit. 457.

at a time, with a pure wave mechanical theory based only on unitary evolution? In order to do this Everett introduces what is perhaps his most important innovation, namely a model observer O, considered as a robotic recording device with memory. He represents the state of the observer as

$$\left|\Psi^O[A, B, ...C]\right\rangle, \tag{17.73}$$

which means that the observer has recorded the events $A, B, ...C$, in that order, [....] representing memory entries irrelevant to the case of interest.

Everett discusses measurements starting with the initial state of an observer and an object system S,

$$\left|\Psi^{S+O}\right\rangle = |\phi_k\rangle_S \left|\Psi^O[.....]\right\rangle, \tag{17.74}$$

so that S is in a particular eigenstate, $|\phi_k\rangle_S$, of the variable being measured. An interaction will act as a measurement if it transforms this state into

$$\left|\Psi^{(S+O)}\right\rangle = |\phi_k\rangle_S \left|\Psi^O[...\alpha_k..]\right\rangle, \tag{17.75}$$

i.e., it does not change the state of the object system if the system is in an eigenstate of the measured quantity. The record α_k signifies that the observer is aware that S is in the eigenstate $|\phi_k\rangle_S$. It is crucial that the state of the object is unchanged. Only then can the measurement be repeatable.

In the general case where S is not in an eigenstate, we have

$$\left|\Psi^{(S+O)}\right\rangle = \sum_i a_i |\phi_i\rangle_S \left|\Psi^O[...\alpha_i..]\right\rangle \tag{17.76}$$

where $a_i = \langle\phi_i \left|\psi^S\right\rangle$ and $\left|\psi^S\right\rangle$ is the initial state of S. This is a superposition of states, each with the observer's memory holding a definite observed value (α_i) and the system S being in the corresponding eigenstate $|\phi_i\rangle_S$. This formalism can be applied to repeated measurements on the same system or on an ensemble of identically prepared systems by the same or different observers.

In the state (17.76) neither the system nor the observer are in a definite state but they have become correlated. In each element of the superposition, the system S is in a definite state $|\phi_i\rangle_S$, and the observer is in a state corresponding to having definitely observed the fact that S is in $|\phi_i\rangle_S$. Repeating the measurement of the same quantity on the state in (17.76), we find the result

$$\left|\Psi'^{(S+O)}\right\rangle = \sum_i a_i |\phi_i\rangle_S \left|\Psi^O[...\alpha_i, \alpha_i..]\right\rangle, \tag{17.77}$$

reflecting the fact that, for each separate term, the observer has obtained the same result two times in succession, i.e., the measurements are repeatable. A measurement by a second observer would result in a similar superposition, where in each term the second observer will have a definite result which always agrees with the result of the first observer. So all observations will have the element of permanency (in as much as repeated measurements of the same quantity by the same observer will agree) and seem to be objective (because different observers will agree with each other as to the observed results).

We now consider the case of successive measurements by a single observer on an ensemble of identical systems $(S_1, S_2, ...)$ prepared in the same state, which can be expressed in the basis of the eigenstates of the measured observable as

$$\left|\Psi^{S_1}\right\rangle = \left|\Psi^{S_2}\right\rangle = ... = \sum_i a_i \left|\phi_i\right\rangle. \tag{17.78}$$

The initial state of the complete system is

$$\left|\Psi^{S_1}\right\rangle \left|\Psi^{S_2}\right\rangle ... \left|\Psi^O[.......]\right\rangle, \tag{17.79}$$

and a measurement on the first system, S_1, will result in the state

$$\left|\Psi_1\right\rangle = \sum_i a_i \left|\phi_i\right\rangle_{S_1} \left|\Psi^{S_2}\right\rangle ... \left|\Psi^O[....\alpha_i^1...]\right\rangle. \tag{17.80}$$

Likewise a subsequent measurement of the same quantity, by the same observer, on S_2 will yield

$$\left|\Psi_{12}\right\rangle = \sum_{i,j} a_i a_j \left|\phi_i\right\rangle_{S_1} \left|\phi_j\right\rangle_{S_2} \left|\Psi^{S_3}\right\rangle ... \left|\Psi^O[....\alpha_i^1, \alpha_j^2...]\right\rangle, \tag{17.81}$$

and so on for successive measurements on additional members of the ensemble. After N measurements, the state will be

$$\left|\Psi_{1..N}\right\rangle = \sum_{i,j,k} a_i a_j ... a_k \left|\phi_i\right\rangle_{S_1} \left|\phi_j\right\rangle_{S_2} \cdots \left|\phi_k\right\rangle_{S_N} \left|\Psi^{S_{N+1}}\right\rangle .... \left|\Psi^O[....\alpha_i^1, \alpha_j^2...\alpha_k^N....]\right\rangle. \tag{17.82}$$

The system will then be in a superposition of states, each of which represents the observer having a definite sequence of results in their memory. If a measurement is repeated on a system measured earlier, the same value will be obtained as was observed previously, for each term in the superposition. In every case, the observer sees the system jumping into eigenstates with the observed eigenvalues, and each measured system remains in the observed state. So an observer, described by the superposition, sees the results of the von Neumann quantum mechanical collapse. This shows the brilliance of Everett's idea: the way the observed statistical behavior emerges from a unitary deterministic development of the total wave function. Everett's interpretation of this is that there is only one physical system representing the observer but no single unique state. The observer's state branches at each observation, and all branches exist simultaneously in the superposition. See Fig. 17.1. The observer sees only one branch as actually existing at any time, thus agreeing with the conventional approach which postulates a "collapse" of the superposition into a single term. But, according to Everett, this transition is unnecessary: all branches are equally real, and there is no need to suppose that all but one of the branches are destroyed. Each element of the

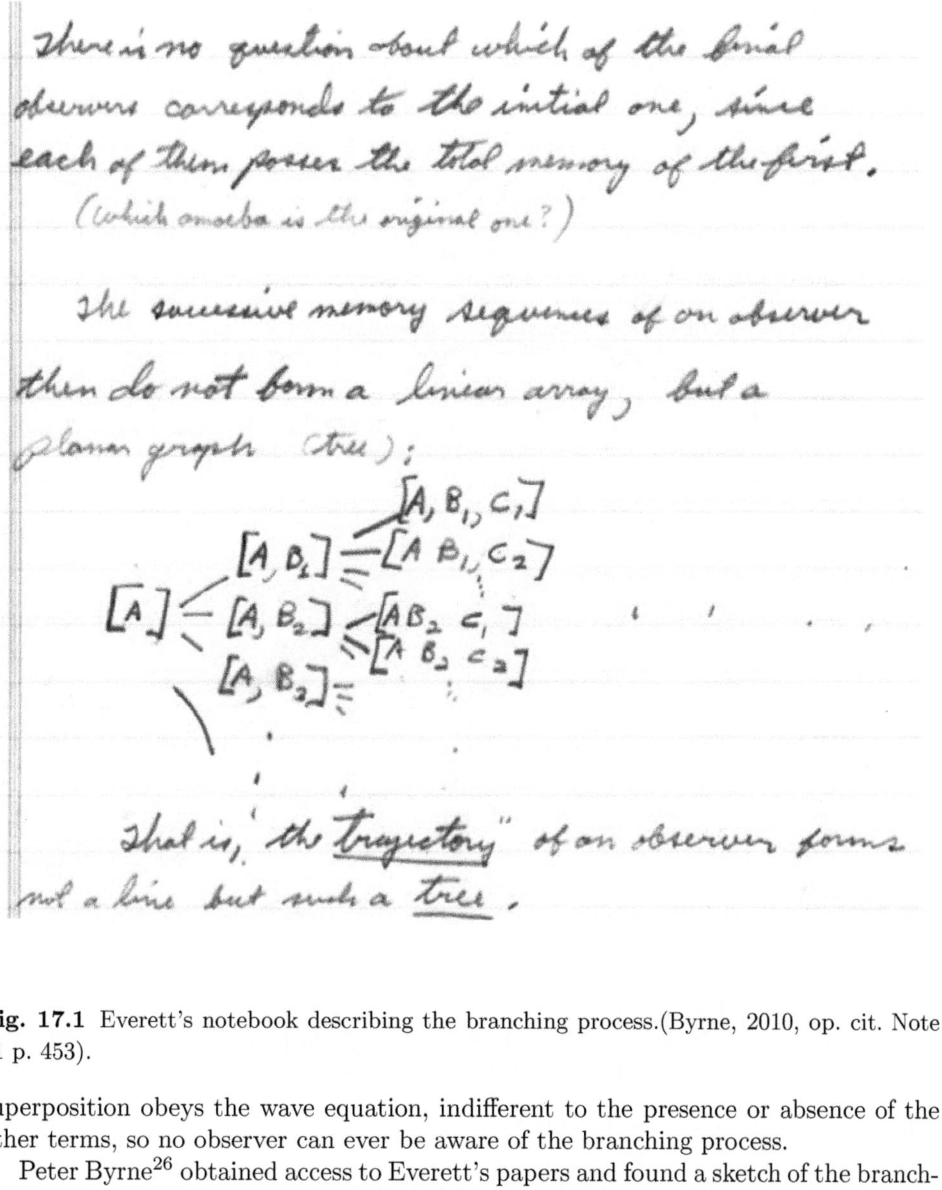

There is no question about which of the final observers corresponds to the initial one, since each of them posses the total memory of the first.
(which amoeba is the original one?)

The successive memory sequences of an observer then do not form a linear array, but a planar graph (tree):

[A] [A, B1] [A, B2] [A, B3] [A, B1, C1] [A B1, C2] [A B2, C1] [A B2, C2]

That is, "the trajectory" of an observer forms not a line but such a tree.

Fig. 17.1 Everett's notebook describing the branching process.(Byrne, 2010, op. cit. Note 11 p. 453).

superposition obeys the wave equation, indifferent to the presence or absence of the other terms, so no observer can ever be aware of the branching process.

Peter Byrne[26] obtained access to Everett's papers and found a sketch of the branching process for the memory state of an observer (Fig. 17.1).

Arguments that the world picture presented by this theory is contradicted by experience, because we are unaware of any branching process, are like the criticism of the Copernican theory that the mobility of the earth as a real physical fact is incompatible with the common-sense interpretation of nature because we feel no such motion. In both cases the argument fails when it is shown that the theory itself predicts that our experience will be what it in

[26]Byrne, P. (2010), op. cit., Note 11 p. 453.

fact is. (In the Copernican case the addition of Newtonian physics was required to be able to show that the earth's inhabitants would be unaware of any motion of the earth.)[27]

According to Everett, the wave function of the universe continuously evolves as indicated, driven by the deterministic, unitary operation of the Schrödinger equation.

17.5.3 The Born rule

Up to this point the theory is incomplete. We require a quantitative measure of the likelihood of a given value α_n appearing in the memory sequences contained in the elements of the superposition (17.82). That is, we need to assign a weight to each element of a superposition of orthonormal states, $\sum_i a_i |\phi_i\rangle$. This measure should be a function of the a_i. In order to have unambiguous values for the a_i, we take the states as normalized, $\langle\phi_i |\phi_i\rangle = 1$. As we cannot fix the overall phase of the states, we have to take the measure m to be a function of the magnitude of a_i: $m = m(|a_i|)$. The measure should satisfy an additivity requirement. If we have an element of a superposition $\alpha |\phi'\rangle$ which then branches into another superposition due to the deterministic unitary time development,

$$\alpha |\phi'\rangle \Rightarrow U_t \alpha |\phi'\rangle = \sum_{i=1}^{n} a_i |\phi_i\rangle, \tag{17.83}$$

where U_t is the unitary time development operator $\left(U_t^\dagger U_t = 1\right)$, the measure assigned to $|\phi'\rangle$ should be the sum of the measures applied to the individual $|\phi_i\rangle$:

$$m(\alpha) = \sum_{i=1}^{n} m(a_i). \tag{17.84}$$

Since $|\phi'\rangle$ and the $|\phi_i\rangle$ are normalized, we see that

$$|\alpha|^2 = \sum_{i=1}^{n} |a_i|^2 \tag{17.85}$$

and

$$m(\alpha) = m\left(\left[\sum_{i=1}^{n} |a_i|^2\right]^{1/2}\right) = \sum_{i=1}^{n} m(|a_i|) = \sum_{i=1}^{n} m\left(\left[|a_i|^2\right]^{1/2}\right). \tag{17.86}$$

Defining $g(x) = m(\sqrt{x})$, we then have

$$g\left(\sum_{i=1}^{n} |a_i|^2\right) = \sum_{i=1}^{n} g\left(|a_i|^2\right) \tag{17.87}$$

which requires $g(x)$ to be linear, $g(x) = cx = m(\sqrt{x})$, with c a constant. Thus $m(|a_i|) = c|a_i|^2$. The constant c can be fixed by a normalization of the total measure; setting this equal to unity gives $c = 1$.

[27]H. Everett III, *Relative state formulation of quantum mechanics*, Rev. Mod Phys **29**, 454 (1957). (This is known as the "short thesis".)

Using this result, we see that the measure (weight) assigned to the $(i, j...k)^{th}$ element of the superposition (17.82), the observer state with the given memory configuration, will be

$$\mathcal{M}(i, j...k) = |a_i a_j ... a_k|^2 = |a_i|^2 |a_j|^2 .... |a_k|^2, \tag{17.88}$$

that is, the product of the weights for the individual components of the memory sequence. As measure theory is equivalent mathematically to probability theory, we can interpret the weights as probabilities; thus the probability of the observer state in any element of the superposition (17.82) will be just as predicted by standard quantum mechanics using the Born rule.

The statistical assertions of the collapse model will appear valid to the observer. They are not independent hypotheses but follow from pure wave mechanics that starts completely free of statistical postulates. The pure unitary development (Process 2) without probability leads to the probability concepts of the usual formulation.

17.5.4 Splitting universes?

This was Everett's achievement. The burning question is, "While the theory predicts exactly what a given observer experiences, what about the other branches of the superposition?" Everett was adamant that all branches of the superposition were equally "real," and that since each branch represented a solution of the Schrödinger equation independent of the existence of the other branches, there was no way for one branch to have any influence on another branch. In a long footnote in the long thesis, he described the branching process as follows:

Whereas before the observation we had a single observer state afterwards there were a number of different states for the observer, all occurring in a superposition. Each of these separate states is a state for an observer, so that we can speak of the different observers described by the different states. On the other hand, the same physical system is involved, and from this viewpoint it is the same observer, which is in different states for different elements of the superposition (i.e., has had different experiences in the separate elements of the superposition). In this situation we shall use the singular when we wish to emphasize that a single physical system is involved, and the plural when we wish to emphasize the different experiences for the separate elements of the superposition, (e.g., "The observer performs an observation of the quantity A, after which each of the observers of the resulting superposition has perceived an eigenvalue.")

Everett addressed the question of the existence of the other branches in a letter to DeWitt in 1957:

From the view point of the theory, all elements of a superposition (all "branches") are "actual", none any more "real" than another. It is completely unnecessary to suppose that after an observation somehow one element of the final superposition is selected to be awarded with a mysterious quality called "reality" and the others condemned to oblivion. We can be more charitable and allow the others to coexist—they won't cause any trouble anyway because all the separate elements of the superposition ("branches") individually obey the wave equation with complete indifference to the presence or absence ("actuality" or not) of the other elements.[28]

[28]Quoted in S. Osnaghi et al., op. cit., p. 457.

Here we see Everett struggling with the meaning of a superposition, the same problem that has been with us since the beginning of quantum mechanics. It seems that his approach to the remaining branches was to essentially ignore them. He considered the question of the existence of the other branches to be meaningless. Since all branches develop independently according to the Schrödinger equation, there can be no mutual influence, so that we, in our particular branch, can never become aware of the other branches.

It was DeWitt, who became a sort of Boswell to Everett, who introduced the "many worlds" interpretation, an approach which Everett never embraced, and emphasized the ever-growing splitting of the universe. As DeWitt explained,

> This universe is constantly splitting into a stupendous number of branches, all resulting from the measurement-like interactions between its myriads of components. Moreover, every quantum transition taking place on every star, in every galaxy, in every remote corner of the universe is splitting our local world on earth into myriads of copies of itself.

Several commentators consider this an entirely new interpretation. Instead of ignoring the non-pertinent branches of the wave function, DeWitt embraces them as a dramatically new physical fact.[29]

17.5.5 The EPR paradox in Everett's notation

The physical system consists of two identical quantum systems $S_{1,2}$ and two observers $O_{1,2}$. The systems $S_{1,2}$ have interacted in the past but are now separated. There are two observables (A, B) with eigenfunctions $|\alpha_i\rangle$, $|\beta_i\rangle$ respectively. In the long thesis Everett considers the case where O_1 measures A in S_1 and O_2 measures B in S_2. The initial state is given as

$$\left|\psi_o^{S_1+S_2}\right\rangle = \sum_i a_i \left|\alpha_i\right\rangle_{S_1} \left|\alpha_i\right\rangle_{S_2} \Psi^{O_1}[...]\,\Psi^{O_2}[...]\,. \tag{17.89}$$

After O_1 measures A in S_1, the state is

$$\left|\psi_1^{S_1+S_2}\right\rangle = \sum_i a_i \left|\alpha_i\right\rangle_{S_1} \left|\alpha_i\right\rangle_{S_2} \Psi^{O_1}[..\alpha_i..]\,\Psi^{O_2}[...]\,. \tag{17.90}$$

The measurement of B by O_2 in S_2 results in the state

$$\left|\psi_2^{S_1+S_2}\right\rangle = \sum_{i,j} a_i \left|\alpha_i\right\rangle_{S_1} \left(\langle\beta_j\left|\alpha_i\right\rangle_{S_2}\right) \left|\beta_j\right\rangle_{S_2} \Psi^{O_1}[..\alpha_i..]\,\Psi^{O_2}[..\beta_j..]\,. \tag{17.91}$$

We see that the intervention of O_2 does not influence the observations of O_1, since O_1 remains correlated with the state $\left|\alpha_i\right\rangle_{S_1}$ and any subsequent measurements by O_1 will proceed without any change due to observations made by O_2.

On the other hand, if O_2 chooses to measure A in S_2, the resulting state will be

$$\left|\psi_1^{S_1+S_2}\right\rangle = \sum_i a_i \left|\alpha_i\right\rangle_{S_1} \left|\alpha_i\right\rangle_{S_2} \Psi^{O_1}[..\alpha_i..]\,\Psi^{O_2}[..\alpha_i..] \tag{17.92}$$

and the results recorded by O_2 will always be correlated with those of S_1.

[29] B. S. DeWitt, *Quantum mechanics and reality*, Physics Today **23**(9), 30 (1970).

17.5.6 The appearance of "collapse" and of a classical world

Each observer in the superposition (17.76) will experience a "collapse" of the wave function that appears to him as an irreversible process, even though the complete wave function, including the observer, is changing reversibly. The irreversibility is a subjective phenomenon, reflecting the fact that an act of observation transforms the observer into a superposition of states, with each element describing an observer cut off from the other branches of the wave function. As soon as the observation is completed, the composite state is split into a superposition of states, each with a different object state and an observer with knowledge of that state. Only the sum total of observer states contains the complete information about the original object state. There is no possibility of communication between observers described by the separate states, and any single observer can only possess knowledge of the relative state (relative to their state) of the object, which is all that matters to them.

In the case of a macroscopic body in a general state, an observer will not see the body smeared out in a superposition of possible positions or orientations; they will not be aware that the state of the object does not correspond to a definite position and momentum. The interaction with the system is so strong that the observer will be correlated with the system. As a result of the observation interaction, the complete system—observer and object—will be in a superposition, each element of which describes the observer as having recorded the object as having a definite position and momentum. The object will appear to a single observer as obeying the laws of classical mechanics.

Subatomic degrees of freedom do not interact as strongly with the observer. If we consider a free electron and proton in a box, the wave functions of each will be essentially uniform inside the box. Eventually they will come together and form a hydrogen atom by emitting radiation. The wave function for each particle will remain uniform over the box, but now they will be correlated: the conditional probability for the electron to be at a certain position, given the condition that the proton is at a specific point, is no longer uniform but is given by the ground state solution of the hydrogen atom in the relative coordinate. The general state is a superposition of hydrogen atoms with uniformly distributed centers of mass.

17.5.7 Frequently asked questions

17.5.7.1 Many worlds?

Everett's ideas did not seem to attract any significant attention in the decade after their publication. They were too far away from currently accepted viewpoints and just seemed crazy to some readers. It was Bryce DeWitt who had been instrumental in publishing Everett's thesis in 1957, and who called the community's attention to them by publishing the paper "Quantum mechanics and reality" in *Physics Today*, where he emphasized the picture of branching universes (quoted earlier, op. cit. p. 464). Of course, this is the major problem for non-Everettians.

Everett himself never wrote anything about this, and historians who study the question of whether Everett agreed with the idea of branching universes seem to be split. David Deutsch met Everett in Austin, Texas, in May 1977, when Everett attended a seminar there at the invitation of John Wheeler, who had moved there from Princeton. Deutsch, who later went on to become one of the founders of quantum computing, was much impressed by Everett's theory and later claimed that the extraordinary speed of the quantum computer factoring of a very large number comes from the calculation being carried out simultaneously in all the parallel universes. For him the possible speed up of quantum computing was proof of the existence of many universes. (Steane disputes this contention: the fact that a computation done in a certain way requires an enormous amount of resources does not mean that an alternate way of doing the computation necessarily uses the same amount of resources.[30]) Deutsch is quoted as saying that "Everett did not prefer the term 'relative states,' being on the contrary extremely enthusiastic about 'many universes' and being very stalwart, as well as subtle in its defense."[31] On the other hand, Everett's attitude may have been closer to that of Levy-Leblond, who was repelled by the idea of many worlds, which he held to be a "left-over of classical conceptions."

> The coexisting branches here, as the unique surviving one in the Copenhagen point of view, can only be related to "worlds" described by classical physics. ...To me the deep meaning of Everett's ideas is not the coexistence of many worlds, but on the contrary, the existence of a single *quantum* one...this question [*which branch are we on*] only makes sense from a classical point of view, once more. It becomes entirely irrelevant as soon as one commits oneself to a consistent quantum view.[32]

Levy-Leblond had sent a copy of his paper to Everett asking his opinion on this question. Everett replied

> ...the "Many-Worlds Interpretation etc." This, of course, was not my title as I was pleased to have the paper published in any form anyone chose to do it in! I, in effect, had washed my hands of the whole affair in 1956.[33]

Thus the question as to the meaning of the "+" sign in the quantum superposition remains. As Levy-Leblond put it, for the standard interpretation the "+" sign means a classical "or," which then requires an additional postulate (collapse) to make a selection. For the fans of "Many Worlds" the "+" sign represents a classical "and." Levy-Leblond favors a "consistent quantum view" as in the above quote.

So the possibility of applying Everett's theory without many worlds remains, although the majority of Everettians seem to be enthusiastic about this feature, which has had a certain degree of public notice. Fig. 17.2 is a cover of the July 5, 2007, issue of *Nature* in honor of the fiftieth anniversary of Everett's theory.

17.5.7.2 Probability measure and the Born rule

In order to arrive at his goal of showing that the Schrödinger equation alone is sufficient for a compete theory without any additional assumptions concerning its interpretation, Everett felt that he had to show that the probabilistic statements of quantum mechanics (the Born rule) followed directly from the equation. We have seen above

[30]Steane, A.M., *A Quantum Computer Needs Only One Universe* Studies in History and Philosophy of Modern Physics 34B: 469-478, (2003).

[31]Shikhovtsev, E. B., op. cit., p. 17. (note 9).

[32]Levy-Leblond, J.-M., *Towards a proper quantum theory,* in Quantum Mechanics a Half Century Later, eds. J. L. Lopez and M. Paty (Dordrecht: Reidel, 1977).

[33]Byrne, P., op. cit., p. 453.

Fig. 17.2 Cover of *Nature* on July 5, 2007.(Reprinted by permission from Springer Nature Customer Service Centre GmbH: Nature/Springer/Palgrave), *Nature*, Cover, **448** (2007). ©www.nature.com)

how Everett arrived at the Born rule, but this proof was often criticized, even by Everettians. Thus DeWitt's publication of Everett's long thesis was accompanied by a strong attack on the Born rule derivation by N. Graham, a student of DeWitt's.[34] Everett never seemed to take these criticisms seriously, ascribing them to a lack of understanding of his proof. He seems to have always felt that his critics did not understand his argument, which was based on an analogy with classical physics. Just as the phase space volume associated with a trajectory is the only measure that allows probability to be conserved in classical physics, in quantum mechanics, where we are concerned with the "trajectories" of the observer, the only measure allowing a conservation law is the one provided by the Born rule:

To have a requirement analogous to the "conservation of probability" in the classical case, we demand that the measure assigned to a trajectory at one time shall equal the sum of the measures of its separate branches at a later time. This is precisely the additivity requirement which we imposed and which leads uniquely to the choice of square-amplitude measure. Our procedure is therefore quite as justified as that of classical statistical mechanics.[35]

One point that comes to mind is the question of the normalization that leads to Eq. (17.85). Is the normalization $\langle \phi_i | \phi_i \rangle = 1$ determined uniquely, or does it introduce an element of circularity into the argument? In any case, the place of the Born rule in the

[34] Dewitt, B. and Graham, N. (1973), op. cit., 457.

[35] DeWitt, B., and Graham, N. (1973), op. cit., p. 73.

Everettian universe continues to be controversial. Alastair Rae,[36] for one, claims to have proved that the Born rule is incompatible with the Everett model and requires an additional assumption, thus negating Everett's claim that everything follows directly from the Schrödinger equation without additional assumptions; however, Rae seems to ignore the fact that different branches of the wave function can have different intensities.

Due to the limitations of this book, we are unable to follow these arguments any further.

17.5.7.3 "Splitting" basis

The question of the choice of basis, the set of eigenfunctions in which the wave function splits, i.e., the basis $|\phi_i\rangle$ in Eq. (17.76), was raised many times, but again, Everett did not think it was a problem. David Deutsch asked Everett about this when they met in May 1977, and Everett replied that it was determined by the structure of the system, i.e., by the Hamiltonian. This question has been studied in great detail in connection with decoherence, which we discuss in the next section. We will see there how the basis selected is the set of eigenfunctions of an observable that commutes with the measurement Hamiltonian.

17.6 Decoherence

17.6.1 Introduction

Recapitulating, the measurement of a quantity represented by a Hermitian operator Φ on a system that is in the state $|\psi\rangle$ is described by expanding the state $|\psi\rangle$ in an orthonormal $(\langle\phi_m|\phi_n\rangle = \delta_{mn})$, complete $(\sum_n |\phi_n\rangle\langle\phi_n| = 1)$ set of eigenfunctions of Φ $(\Phi|\phi_n\rangle = \phi_n|\phi_n\rangle)$ as

$$|\psi\rangle = \sum_n c_n |\phi_n\rangle \tag{17.93}$$

$$c_n = \langle\phi_n|\psi\rangle, \tag{17.94}$$

where the c_n are complex numbers: $c_n = r_n e^{i\theta_n}$ with r_n and θ_n real numbers. The possible results of a measurement are the eigenvalues ϕ_n, with the probability of obtaining a given eigenvalue equal to $|c_n|^2 = r_n^2$. The physical meaning of the phases θ_n can be seen by calculating the expectation value of another observable, F, in the state $|\psi\rangle$:

$$\langle\psi|F|\psi\rangle = \sum_{n,m} c_m^* c_n \langle\phi_m|F|\phi_n\rangle = \sum_{n,m} \rho_{nm}\langle\phi_m|F|\phi_n\rangle = \mathrm{Tr}(\rho F) \tag{17.95}$$

where

$$\rho_{nm} = c_m^* c_n = r_m r_n e^{i(\theta_n - \theta_m)} \tag{17.96}$$

is the density matrix, and $F_{mn} = \langle\phi_m|F|\phi_n\rangle$ is the matrix element of F in the $|\phi_n\rangle$ basis. The terms with $n \neq m$ (off-diagonal terms in ρ) represent interference terms.

[36] Rae, A., *Everett and the Born rule*, Stud. Hist. Philos. Mod. Phys. **40**, 243 (2009).

If the phase differences are large and fluctuate between different systems in the ensemble, we see that the density matrix becomes $\rho_{nm} = r_n^2 \delta_{mn}$ and the expectation value (17.95) becomes

$$\langle\psi| F |\psi\rangle = \sum_n r_n^2 \langle\phi_n| F |\phi_n\rangle, \tag{17.97}$$

which is indistinguishable from the result that we would find for an ensemble of systems that are each in a definite state with probability r_n^2. This might then be considered as a solution to the measurement problem as each of the systems is in a definite state. Bell[37] called this the replacement of "and" (the superposition of states with random phases) by "or" (the assertion that each system in an ensemble is definitely in one state or another) and rejected it as unacceptable: "The idea that elimination of coherence, in one way or another, implies the replacement of 'and' by 'or', is a very common one among solvers of the 'measurement problem'. It has always puzzled me." We have seen that the same point had been made by Levy-Leblond.

As we saw in Section 9.127, Bohm showed that a measurement leading to distinguishable states of a macroscopic observable would lead to such large values of the phase shifts θ_n that the variations between one member of an ensemble and another would effectively average the phase shifts to zero, but he stopped short of saying that this solved the measurement problem. In more recent times several physicists have developed the idea of the cancellation of the interference terms because of large fluctuations of the phases brought about by interaction with a complex "environment". Some physicists feel that the smearing of the phases described by Bohm, and the resultant vanishing of the off-diagonal terms in the density matrix, is not sufficient to account for a true measurement and to supply a solution to the Everett basis problem. Each member of the ensemble exists in a pure state with a definite phase, so that even though the state may seem to have lost its coherence and appears as if it were a mixed state, there is always the possibility (at least theoretically) of reversing this dephasing.[38] An example cited is the spin-echo effect in nuclear magnetic resonance, which occurs in a system of spins that are fixed at different locations in different magnetic fields $B(\vec{r})$ and undergo Larmor precession for a time long enough that they become completely out of phase and depolarized. However, if the magnetic field is reversed, the precessions will be reversed, and after a time the system polarization will be restored. What is needed is an interaction with an environment that is so complex that the phases can never be restored. In fact, the unitary time evolution predicted by the Schrödinger equation can never turn a pure state into a mixture. This is the same problem of irreversible behavior from reversible fundamental laws that occurs in classical mechanics and which we have met in the Wigner-Weisskopf discussion of the decay of an excited atom.

Putting this aside, environment-induced decoherence offers a possible solution to the preferred basis problem in Everett's scheme. These ideas are summarized under the rubric "decoherence". However, decoherence by itself does not solve Bell's "and/or"

[37] Bell, J. S., *Against measurement*, Physics World **3**(8), 33 (1990).

[38] Zurek, W.H., *Decoherence, einselection, and the quantum origins of the classical*, Rev. Mod. Phys. **75**, 715 (2003), Section III.C.

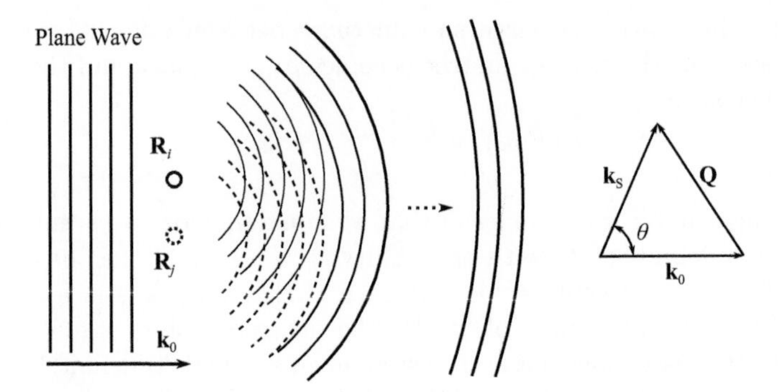

Fig. 17.3 A plane wave of environment particles incident on two fixed scattering centers. a) A pane wave ... b) Vector diagram of $\vec{Q} = \vec{k}_o - \vec{k}_s$.

quandary; for this we need Everett's, or de Broglie-Bohm's, method of selecting a particular branch of the "universal wave function."

17.6.2 Decoherence and the appearance of classical objects

Another result of the interaction of a quantum system with an environment is a quantum mechanical model for the appearance of a classical world. This was first proposed by Joos and Zeh.[39] We will discuss a very simplified model which shows the physical basis of the idea. Consider a massive particle with center of mass coordinate $\vec{R}$ in a state described by a wave function $\psi(\vec{R}) = \langle \vec{R}|\psi\rangle$. The particle is immersed in a background field (the "environment") which may be thermally excited photons in a cavity, the cosmic background radiation, gas molecules in the atmosphere or in a laboratory vacuum, etc. We represent this environmental radiation as a plane wave $\sim e^{i\vec{k}\cdot\vec{r}}$.

The initial state of the system plus environment can be written as

$$\Psi_{in}(\vec{r}, \vec{R}) = e^{i\vec{k}\cdot\vec{r}}\psi_o(\vec{R}) = e^{i\vec{k}\cdot\vec{r}}\langle \vec{R}|\psi_o\rangle. \tag{17.98}$$

We assume the radiation is scattered elastically (no change of energy) from the particle located at $\vec{R}_i$ into a spherical wave (this is called s-wave scattering). This spherical wave will be given by $f_o e^{ik|\vec{r}-\vec{R}_i|}/|\vec{r} - \vec{R}_i|$. ($f_o$ is the scattering amplitude with dimensions of length; for s-wave scattering, the total scattering cross section σ is $\sigma = 4\pi\,|f_o|^2$.) This spherical wave has to be multiplied by the incident wave evaluated at $\vec{R}_i$, so that the state of the environment after scattering will be

$$\Phi_s(\vec{r}) = \frac{f_o e^{i\vec{k}\cdot\vec{R}_i} e^{ik|\vec{r}-\vec{R}_i|}}{|\vec{r} - \vec{R}_i|}. \tag{17.99}$$

For $r \gg R_i$ we can write

$$k|\vec{r} - \vec{R}_i| \approx k\left(r^2 - 2\vec{r}\cdot\vec{R}_i\right)^{1/2} \approx k\left(r - \hat{r}\cdot\vec{R}_i\right) = kr - \vec{k}_s\cdot\vec{R}_i \tag{17.100}$$

[39] Joos, E. and Zeh, H. D., *The emergence of classical properties through interaction with the environment*, Z. Phys. B-Cond. Mat. **59**, 223 (1985).

where $\vec{k_s} = k\hat{r}$ is the wave vector of the scattered wave ($\hat{r}$ is the unit vector in the direction of $\vec{r}$). Thus

$$\Phi_s(\vec{r}) \approx f_o e^{i\vec{k}\cdot\vec{R}_i} e^{-i\vec{k}_s\cdot\vec{R}_i} \frac{e^{ikr}}{r} = f_o e^{i\vec{Q}\cdot\vec{R}_i} \frac{e^{ikr}}{r} \tag{17.101}$$

where $\vec{Q} = \vec{k} - \vec{k}_s$ is the change in wave vector due to the scattering. This is a spherical wave with a phase shift given by $\vec{Q}\cdot\vec{R}_i$. Since we are concerned with the position of the massive particle, we expand the initial state (17.98) in eigenstates of the operator $\vec{\mathbf{R}}$ for the center of mass position $\left(\vec{\mathbf{R}}|\vec{R}_i\rangle = \vec{R}_i|\vec{R}_i\rangle\right)$, as in Eq. (17.93) (we use the summation sign to indicate the relevant integration):

$$|\Psi_{in}\rangle = e^{i\vec{k}\cdot\vec{r}} \sum_i |\vec{R}_i\rangle\langle\vec{R}_i|\psi_o\rangle = e^{i\vec{k}\cdot\vec{r}} \sum_i |\vec{R}_i\rangle\psi_o(\vec{R}_i). \tag{17.102}$$

Using (17.101), the wave function after the scattering will be

$$|\psi_s\rangle = f_o \frac{e^{ikr}}{r} \sum_i e^{i\vec{Q}\cdot\vec{R}_i} \left|\vec{R}_i\right\rangle \psi_o\left(\vec{R}_i\right). \tag{17.103}$$

Then the density matrix is given by (following (17.96))

$$\left\langle\vec{R}_j\right|\rho\left|\vec{R}_i\right\rangle = f_o f_o^* e^{i\vec{Q}\cdot(\vec{R}_j-\vec{R}_i)} \psi_o\left(\vec{R}_j\right)\psi_o^*\left(\vec{R}_i\right). \tag{17.104}$$

This is the result of a single scattering at a single value of $\vec{Q}$. Expanding the exponential, we have

$$\begin{aligned} e^{i\vec{Q}\cdot(\vec{R}_j-\vec{R}_i)} &\approx 1 + i\vec{Q}\cdot\left(\vec{R}_j - \vec{R}_i\right) - \left(\frac{1}{2}\right)\left[\vec{Q}\cdot\left(\vec{R}_j - \vec{R}_i\right)\right]^2 \\ &\approx 1 - \frac{1}{6}\left\langle Q^2\right\rangle\left|\vec{R}_j - \vec{R}_i\right|^2 \approx e^{-\frac{1}{6}\langle Q^2\rangle|\vec{R}_j-\vec{R}_i|^2} \end{aligned} \tag{17.105}$$

where the second step follows from the fact that $\langle\vec{Q}\rangle = 0$, where $\langle\vec{Q}\rangle$ is the average value of Q, and the isotropy of the environment. With a density n of environment particles with an average velocity v, we will have $nv\sigma t$ collisions in time t, and the phase factor (17.105) will become time-dependent undergoing a random walk:

$$e^{i\vec{Q}\cdot(\vec{R}_j-\vec{R}_i)} \approx \exp\left[-\frac{1}{12}k_o^2 n\sigma v\left|\vec{R}_j - \vec{R}_i\right|^2 t\right] = \exp\left[-\frac{t}{\tau_{decoh}}\right] \tag{17.106}$$

$$\tau_{decoh} = \frac{12\lambda^2}{(2\pi)^2\left|\vec{R}_j - \vec{R}_i\right|^2}\tau_o \tag{17.107}$$

where we used $\langle Q^2\rangle = k_o^2\left\langle\sin^2(\theta/2)\right\rangle = k_o^2/2$ and $\tau_o = (n\sigma v)^{-1}$ is the usual relaxation time. This agrees, apart from a numerical factor, with the result of Joos and

Zeh.[40] The further apart the interfering points are, the faster quantum interference effects will decay, i.e., the decay rate of the off-diagonal elements of the density matrix (in the position representation) grows with the distance from the diagonal. For long enough times, only elements very close to the diagonal will be nonzero. In this way the object approaches the behavior of a classical object, but only if we ignore the "and/or" problem. The state of the object remains a mixture of being located in different positions. Eq. (17.107) shows that the decoherence time can be much shorter than the relaxation time, which is usually taken to characterize the interaction with a thermal bath (environment) and is the decay time of the diagonal elements of the density matrix in the position basis (analogous to T_1 in nmr), while τ_{decoh} is the decay time of the off-diagonal elements (analogous to T_2 in nmr).

Following Joos and Zeh, we consider a laboratory vacuum of molecules with a mass of 30 amu at a density of $n = 10^6/\mathrm{cm}^3$ and with an average speed $v = 5 \times 10^4\,\mathrm{cm/sec}$, which gives an average $k \sim 10^9\,\mathrm{cm}^{-1}$. If the size of the object is a cm, the cross section will be $\sigma \sim \pi a^2$, and the time for a particle to be localized to its own size would be obtained by setting $\left|\vec{R}_j - \vec{R}_i\right|^2 \sim a^2$. In this case the decay rate (17.106) would be

$$\frac{1}{\tau} \sim 2.6 \times 10^{28} a^4\,\mathrm{sec}^{-1}. \tag{17.108}$$

For a dust particle of $10\,\mu$ we would have $1/\tau \sim 5 \times 10^{16}\,\mathrm{sec}^{-1}$, for a large molecule of 10^3 Å we find $1/\tau \sim 5 \times 10^8\,\mathrm{sec}^{-1}$, and a molecule of size 10 Å would yield $1/\tau \sim 5\,\mathrm{sec}^{-1}$, so it seems that experimental investigations are feasible. Arndt et al.[41] have performed such an experiment using C^{60} (Fullerene) molecules in an interferometer with $a = 7$ Å. The destruction of coherence as a function of environment pressure is seen in Fig. 17.4, which is a plot of fringe visibility vs. pressure.

The parameters in this experiment $\left(\langle Q^2 \rangle \left|\vec{R}_j - \vec{R}_i\right|^2 \gg 1\right)$ were out of the range of validity of our rough estimate (17.106) but the results are in agreement with a more exact calculation (black curve in the figure). There is a large body of literature on the role of decoherence in the quantum to classical transition, including several books.[42]

17.6.3 Decoherence and the measurement problem

Wojciech H. Zurek was a graduate student working on a thesis in statistical mechanics at the University of Texas, Austin, when John Wheeler transferred there from Princeton.[43] Zurek met Wheeler when he took Wheeler's course on classical electrodynamics and then participated in a course on quantum measurements also given by Wheeler. Zurek became Wheeler's postdoc when he finished his PhD and assisted with

[40] More detailed calculations are in Hornberger, K. and Sipe, J. E., *Collisional decoherence reexamined*, Phys. Rev. A **68**, 012105 (2003).

[41] Arndt, M., Hornberger, K. and Zeilinger, A. *Probing the limits of the quantum world*, Physics World **18**(3), 35 (2005).

[42] Joos, E., Zeh, H.D. et al., *Decoherence and the Appearance of a Classical World in Quantum Theory, 2nd ed.* (Berlin, Heidelberg: Springer, 1996); Schlosshauer, *M., Decoherence and the Quantum to Classical Transition* (Berlin, Heidelberg: Springer, 2007).

[43] Biographical information concerning Zurek is from an e-mail from W. H. Zurek to the authors and D. Kaiser, *How the Hippies Saved Physics: Science, Counterculture and the Quantum Revival* (New York: Norton, 2012).

teaching the quantum measurements course. This led to the publication of an anthology of papers on the measurement problem in quantum mechanics.[44] Together with William Wootters, another graduate student, Zurek published (1978) what had been a term paper for the quantum measurement course, subjecting Bohr's argument concerning the two-slit experiment to mathematical analysis. We discussed that paper in Chapter 13.

Wheeler was interested in the problem of general relativity and its relation to quantum mechanics, including the possible existence of gravitational waves. This led to discussions of gravitational wave detectors, which at the time were conceived of as "Weber bars":[45] large, freely hanging bars of aluminum at very low temperature that were expected to vibrate in resonance with any incident gravitational waves. Quantum mechanics entered in attempts to reduce the noise. Extensive conversations with Wheeler, Wigner (who spent some time in Austin), and others convinced Zurek that, contrary to received wisdom, there were significant open questions in both Bohr's and Wigner's interpretations of quantum mechanics.

At that time the Copenhagen interpretation was accepted by the vast majority of working physicists (as it still is), who looked on questioning of this dogma as something bordering on heresy. This attitude was summarized as "shut up and calculate," as quantum mechanics found success in describing more and more phenomena, and the so-called Standard Model of particle physics had just been established in more or less its present form. According to Kaiser (op. cit.), Wheeler's support and encouragement offered some protection against this attitude. After receiving his PhD (1979), Zurek submitted his first paper on decoherence in April 1981[46] and went on to publish, with

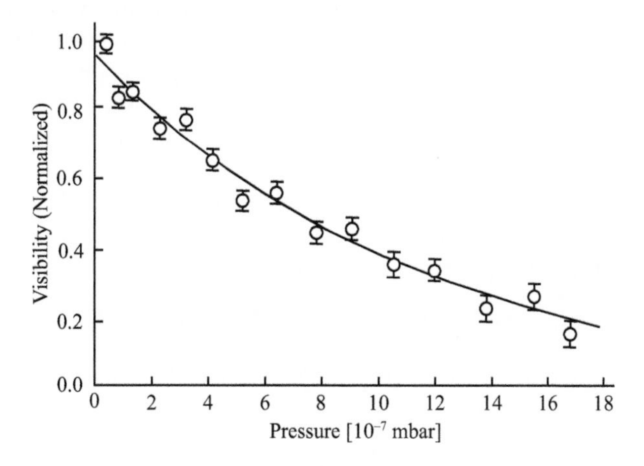

Fig. 17.4 Decay of fringe visibility (coherence) with increasing environment gas pressure, as measured by Arndt et al., op. cit. p.472.

[44]Wheeler, J. A. and Zurek, W.H.., eds., *Quantum Theory and Measurement* (Princeton: Princeton University Press, 1983).

[45]Weber, J., *General Relativity and Gravitational Waves*, Interscience Tracts on Physics and Astronomy, Number 10 (New York: Interscience, 1961), repr. Mineola, NY: Dover, 2004.

[46]Zurek, W. H., *Pointer basis of quantum apparatus: Into what mixture does the wave packet collapse?*, Phys. Rev. D **24**, 1516 (1981).

Wootters, their famous no-cloning theorem,[47] which we discuss below. The paper dealing with the pointer basis was the first in a long line of papers by Zurek investigating the problems of measurement theory and the related question of the possibility of developing all of quantum mechanics from a purely unitary evolution by making use of open systems interacting with an "environment."

Von Neumann had considered that having a second apparatus A' to measure the apparatus A would not help the measurement problem because the resultant state of the whole system (observed system S, apparatus A, second apparatus A') would still be a pure state: $|\psi\rangle = \sum_s b_s |s\rangle |A_s\rangle |A'_s\rangle$, with $|s\rangle$ the eigenstates of a system observable, $|A\rangle, |A'\rangle$ the states of the two apparatus, and H_{AS} ($H_{AA'}$) the Hamiltonian for the interaction between A and S (A and A'). The unitary evolution would then yield

$$|A'_o\rangle |A_o\rangle |\psi\rangle .. \overrightarrow{H_{AS}} |A'_o\rangle \sum_s b_s |A_s\rangle |s\rangle \tag{17.109}$$

$$.. \overrightarrow{H_{AA'}} \sum_s b_s |A'_s\rangle |A_s\rangle |s\rangle \tag{17.110}$$

where the arrow represents unitary evolution under the indicated interaction. This chain could be continued indefinitely, thus requiring the von Neumann-Heisenberg cut and the collapse hypothesis to break the chain. As we have seen, others thought that the chain continued until reaching a human consciousness. Zurek argues that if A' is replaced by an environment whose states are not measured, requiring our observations on A to be averaged over the possible states $|\mathcal{E}\rangle$ of the environment (i.e., taking the trace of the density matrix with respect to $|\mathcal{E}\rangle$), the composite system-apparatus will be in a mixture of states $|A_s\rangle |s\rangle$, thus eliminating any interference effects. Note that the environment can be part of the apparatus, for example, degrees of freedom other than the pointer basis such as phonons in the solid material of the apparatus. This mixture state can be considered as representing a situation where the system is really in one of these states, but we don't know which. The view that this can be considered as a solution of the measurement problem (Bell's "and/or" formulation) seems to be discredited now as discussed above.

Consider our composite system in the initial state

$$|\Phi (t = 0)\rangle = |\psi\rangle |A_o\rangle |\mathcal{E}_o (t = 0)\rangle \tag{17.111}$$

with the system in a pure state $|\psi (0)\rangle = \sum_n c_n |n\rangle$. A measurement of S by the apparatus A at $t = t_1$ will result in the state

$$|\Phi (t_1)\rangle = \sum_n c_n |n\rangle |A_n\rangle |\mathcal{E}_o (t_1)\rangle, \tag{17.112}$$

and an interaction between A and the environment at $t = t_2$ will result in a state

$$|\Phi (t > t_2)\rangle = \sum_n c_n |n\rangle |A_n\rangle |\mathcal{E}_n (t)\rangle, \tag{17.113}$$

[47] Wootters, W. K. and Zurek, W.H., *A single quantum cannot be cloned*, Nature **299**, 802 (October 1982).

which is a pure state. At time t_1 the observable being measured is not defined as the state $|\Phi(t_1)\rangle$ can be expanded in another orthonormal basis of the system S. Calculating the density matrix for the state $|\Phi(t > t_2)\rangle$ and tracing over the environment states, we find

$$\mathrm{Tr}_{\mathcal{E}}\left[|\Phi(t > t_2)\rangle\langle\Phi(t > t_2)|\right] = \sum_n |c_n|^2 |n\rangle\langle n| \otimes |A_n\rangle\langle A_n| \tag{17.114}$$

$$+ \sum_{n \neq m} c_n c_m^* |n\rangle\langle m| \otimes |A_n\rangle\langle A_m| \langle\mathcal{E}_m|\mathcal{E}_n\rangle. \tag{17.115}$$

As $\langle\mathcal{E}_m|\mathcal{E}_n\rangle$ decreases, the state of the system-apparatus approaches a mixture, but this is only a part of what happens. The apparatus-environment interaction selects a "pointer basis" for the apparatus: the basis of eigenstates of the apparatus-environment interaction Hamiltonian.[48]

The argument can be clarified by applying it to a simple model, a two-state system (a spin object) and a two-state atom serving as an apparatus. We denote the basis for the spin as $|\sigma_z, \pm\rangle$, indicating the state with $\sigma_z = \pm 1$. For the atom we use the isotopic spin states $|\tau_z, \pm\rangle$, which indicate the ground and excited state. We can also use superpositions of these states, e.g., $|\sigma_x, +\rangle = (|\sigma_z, +\rangle + |\sigma_z, -\rangle)/\sqrt{2}$. We take the interaction as $H_{int} = -g\,\tau_y \sigma_z$ and the initial state as (dropping the $\otimes$ signs)

$$|\phi_1\rangle = (a|\sigma_z, +\rangle + b|\sigma_z, -\rangle)|\tau_x, +\rangle. \tag{17.116}$$

For the states $|\sigma_z, \pm\rangle$ the evolution operator is

$$U = e^{-iH_{int}t} = e^{\pm ig\tau_y t}, \tag{17.117}$$

which will rotate the original τ_x eigenstate toward the $\pm z$ directions. Think of it as a magnetic field along the y direction with the τ spin precessing around the field. Taking $2gt = \pi/2$ (a $\pi/2$ flip), we obtain the final state

$$|\phi_f\rangle = a|\sigma_z, +\rangle|\tau_z, +\rangle + b|\sigma_z, -\rangle|\tau_z, -\rangle, \tag{17.118}$$

which is a pure state with a definite correlation between the system and apparatus. But this can be brought back to the initial state by further application of the interaction. This cannot be considered a completed measurement as we can rewrite the state in terms of the basis $|\tau_x, \pm\rangle$. In this basis we have

$$|\phi_f\rangle = (a|\sigma_z, +\rangle + b|\sigma_z, -\rangle)|\tau_x, +\rangle + (a|\sigma_z, +\rangle - b|\sigma_z, -\rangle)|\tau_x, -\rangle. \tag{17.119}$$

Now the states $|\tau_x, \pm\rangle$ are correlated with the spin states $a|\sigma_z, +\rangle \pm b|\sigma_z, -\rangle$. In this case there is no definite observable that has been measured by the atom-apparatus. To address this issue, we introduce an environment consisting of N two-level atoms, which we describe by isotopic spin states $|I_z^k, \pm\rangle$ with k running from 1 to N. We consider the basis states as being degenerate and assume that there is no interaction between

[48]This is true when the interaction with the environment dominates over the self-Hamiltonian of the system, as we are implicitly assuming.

the environment atoms. The interaction between the apparatus and environment is taken to be

$$H^{A\mathcal{E}} = \sum_k H_k^{A\mathcal{E}} \tag{17.120}$$

$$H_k^{A\mathcal{E}} = g_k \tau_z I_z^k \prod_{j \neq k} 1_j \tag{17.121}$$

where 1_j is the identity operator in the isotopic spin basis of environment atom j. We assume that the system and the apparatus have interacted to produce the state (17.118), and we say that the interaction $H^{A\mathcal{E}}$ starts at $t = 0$, when the state of the environment atoms is such that the state of the composite system $SA\mathcal{E}$ is given by

$$|\Phi(0)\rangle = |\phi_f\rangle \prod_k \left(a_k \left|I_z^k, +\right\rangle + b_k \left|I_z^k, -\right\rangle\right). \tag{17.122}$$

The unitary evolution operator $U(t)$ is

$$U(t) = e^{-iH^{A\mathcal{E}}t} = \prod_k e^{ig_k \tau_z I_z^k t}. \tag{17.123}$$

This has the effect that for atom (apparatus) eigenstates $|\tau_z, \pm\rangle$, the isospin $\vec{I}$ precesses around the I_z axis and the state $|\Phi(0)\rangle$ evolves into

$$|\Phi(t)\rangle = a|\sigma_z, +\rangle |\tau_z, +\rangle \prod_k \left[a_k \left|I_z^k, +\right\rangle e^{ig_k t} + b_k \left|I_z^k, -\right\rangle e^{-ig_k t}\right] + \tag{17.124}$$

$$+ b|\sigma_z, -\rangle |\tau_z, -\rangle \prod_k \left[a_k \left|I_z^k, +\right\rangle e^{-ig_k t} + b_k \left|I_z^k, -\right\rangle e^{ig_k t}\right]$$

$$\equiv a|\sigma_z, +\rangle |\tau_z, +\rangle |\mathcal{E}, +\rangle + b|\sigma_z, -\rangle |\tau_z, -\rangle |\mathcal{E}, -\rangle \tag{17.125}$$

thus defining $|\mathcal{E}, \pm\rangle$. We see that $\langle\mathcal{E}, \pm|\mathcal{E}, \pm\rangle = 1$ and

$$z(t) = \langle\mathcal{E}, +|\mathcal{E}, -\rangle = \langle\mathcal{E}, -|\mathcal{E}, +\rangle^* = \prod_k \left[|a_k|^2 e^{-i2g_k t} + |b_k|^2 e^{i2g_k t}\right] \tag{17.126}$$

$$= \prod_k \left[\cos 2g_k t - i\left(|a_k|^2 - |b_k|^2\right) \sin 2g_k t\right] \tag{17.127}$$

Using equation (17.124), the density matrix for the state $|\Phi(t)\rangle$ is

$$\rho_{SA\mathcal{E}} = |\Phi(t)\rangle \langle\Phi(t)| \tag{17.128}$$

$$= [a|\sigma_z, +\rangle |\tau_z, +\rangle |\mathcal{E}, +\rangle + b|\sigma_z, -\rangle |\tau_z, -\rangle |\mathcal{E}, -\rangle] \otimes \tag{17.129}$$

$$[a^* \langle\sigma_z, +| \langle\tau_z, +| \langle\mathcal{E}, +| + b^* \langle\sigma_z, -| \langle\tau_z, -| \langle\mathcal{E}, -|], \tag{17.130}$$

and tracing over the environment basis states $\left|I_z^k, \pm\right\rangle$ gives the effective density matrix for the system and apparatus. Noting that $\mathrm{Tr}_{\mathcal{E}} |\mathcal{E}, \pm\rangle \langle\mathcal{E}, \pm| = 1$ and $\mathrm{Tr}_{\mathcal{E}} |\mathcal{E}, -\rangle \langle\mathcal{E}, +| = \{\mathrm{Tr}_{\mathcal{E}} |\mathcal{E}, +\rangle \langle\mathcal{E}, -|\}^* = z$, we find

$$\begin{aligned}\rho_{SA} = {} & |a|^2 |\sigma_z, +\rangle |\tau_z, +\rangle \langle\sigma_z, +| \langle\tau_z, +| + \\ & + |b|^2 |\sigma_z, -\rangle |\tau_z, -\rangle \langle\sigma_z, -| \langle\tau_z, -| + \\ & + ab^* |\sigma_z, +\rangle |\tau_z, +\rangle \langle\sigma_z, -| \langle\tau_z, -| z^* + \\ & + a^*b |\sigma_z, -\rangle |\tau_z, -\rangle \langle\sigma_z, +| \langle\tau_z, +| z.\end{aligned} \tag{17.131}$$

According to (17.127) the time average $\langle z(t)\rangle = 0$, and the average of the magnitude squared is

$$\left\langle |z(t)|^2 \right\rangle = \frac{1}{2^N} \prod_{k=1}^{N} \left[1 + \left(|a_k|^2 - |b_k|^2\right)^2\right] \tag{17.132}$$

which is very small for large N. Thus, unless most of the environment atoms are in eigenstates of the interaction Hamiltonian (a_k or $b_k \to 0$), the magnitude $|z(t)|$ will become much less than its initial value of unity as time goes on and the off-diagonal terms will become very small, so that the state ρ_{SA} (Eq. (17.131)) will evolve into a mixture. The information contained in the initial state of the spin, $a|\sigma_z, +\rangle + b|\sigma_z, -\rangle$, which represents the spin pointing in some arbitrary direction given by

$$a = \cos\theta/2 e^{-i\phi/2} \tag{17.133}$$

$$b = \sin\theta/2 e^{i\phi/2} \tag{17.134}$$

with θ, ϕ the spherical polar angles of the spin direction, has been lost to the environment. The spin is left in a mixture of states whose basis, the "pointer basis," is that basis which couples to the environment basis (through the apparatus), i.e., the basis of states which are the eigenstates of the apparatus-environment interaction. However, as long as N is finite, $|z(t)|$ will return to its initial value infinitely many times, but the times between these recurrences can be arbitrarily large. We have met a similar situation in our discussion of the Wigner-Weisskopf model.

The situation is analogous to the Poincaré recurrences in classical statistical mechanics, and (in both cases) this is the result of the impossibility of obtaining truly irreversible behavior from a reversible theory, unless we wish to impose irreversibility by fiat, as in the Copenhagen model or other "collapse" models discussed below.

The apparent forcing of the state into a mixture by the interaction with a complex environment is considered as being equivalent to selecting a classical state. The preferred states that are selected in this way not only diagonalize the density matrix but are also unaffected by the environment, so that the density matrix remains diagonal while the interaction with the environment continues.

Critics (such as Ruth Kastner[49]) claim that the division of the universe into subsystems already implies classical behavior. But Zurek had already embraced this,[50] stating that we need to consider the universe as divided into separate subsystems in order to discuss semiclassical relations. Without the division of the universe into systems, we cannot assign local wave functions; in fact, it is hard to see how we can do

[49]Kastner, R., *Classical selection and quantum Darwinism*, Physics Today **68** (May 2015), p.8.

[50]Zurek, W. H., *Preferred sets of states, predictability, classicality and environment induced decoherence*, in Physical Origins of Time Asymmetry, ed. J. J. Halliwell, J. Perez-Mercader and W.H. Zurek (Cambridge: Cambridge University Press, 1994).

physics at all. To establish the correspondence between the formalism and observations, we require a relation between the universal state vector and the states of memory (records) of special systems (as detailed by Everett).

The measurement problem cannot even be posed without the division into observed system and apparatus. In the absence of this division, any closed system evolves deterministically (unitary evolution). This evolution takes a composite object from an initial state where each component has definite properties into an entangled state where none of the components has a state of its own. Critics say decoherence does not result in a unique state corresponding directly to our experience, but following Everett, decoherence theory assumes that observers are a part of the universe. The observer is demoted from an all-powerful external being to a subsystem of the universe. A fundamental collapse is unnecessary; the correlation between the memory of the observer and records of past observations in essential.

Only states that can define observers and their knowledge for prolonged periods, much longer than the timescale for information processing by the observer's nervous system, will correspond to a perception. The stable records in memory will then correlate with a specific branch of the universe. This is the "existential" interpretation of quantum mechanics (see below): decoherence defines stable alternative outcomes—states of the observer that can exist in spite of interactions with the environment—which helps define a set of stable options for the state of the system, which are represented by only a small fraction of the available Hilbert space. Every time the system, or the memory of the apparatus, is forced into a superposition, it will decohere on a very short timescale when the options are macroscopically distinguishable. Each alternative becomes a fact to an observer who has recorded that result, resulting in an apparent almost instantaneous reduction of the wave packet.

An essential point is that after a measurement entangles the state, this entangled state can be written in different bases, so that neither the system nor the apparatus can be considered to be in a definite state. As another example[51] of this, consider a system with states $|s_i\rangle$ interacting with an apparatus with states $|A_k\rangle$ so that a measurement takes place:

$$\sum_i \alpha_i |s_i\rangle |A_o\rangle \rightarrow \sum_i \alpha_i |s_i\rangle |A_i\rangle. \tag{17.135}$$

This can be considered as a measurement of the state of the system by the apparatus, but this state is seen to suffer from basis ambiguity, as we can write the state $|s_i\rangle$ in terms of the eigenstates of the conjugate basis,

$$|s_l\rangle = \frac{1}{\sqrt{N}} \sum_{k=0}^{N-1} e^{-i2\pi kl/N} |r_k\rangle \tag{17.136}$$

where N is the dimension of the Hilbert space. Then we can write the final state (17.135) as

[51] Zurek, W.H., *Decoherence, einselection, and the quantum origins of the classical*, Rev. Mod. Phys. **75**, 715 (2003).

$$\sum_l \alpha_l |s_l\rangle |A_l\rangle = \frac{1}{\sqrt{N}} \sum_{k=0}^{N-1} |r_k\rangle \sum_l \alpha_l e^{-i2\pi kl/N} |A_l\rangle \tag{17.137}$$

$$= \frac{1}{\sqrt{N}} \sum_{k=0}^{N-1} |r_k\rangle |B_k\rangle \tag{17.138}$$

with

$$|B_k\rangle = \sum_l \alpha_l e^{-i2\pi kl/N} |A_l\rangle. \tag{17.139}$$

Once we have such a correlated state (sometimes called "premeasurement"), we introduce the interaction with an environment, and ignoring the uncontrollable and unmeasured distribution of the environment states, we average over all possible states of the environment. This averaging is performed by taking the trace of the composite system density matrix over the environment variables. If the environment basis is orthogonal, the result is a diagonal density matrix of the remaining system plus apparatus (reduced density matrix of these subsystems). Recall the argument by London and Bauer showing that taking the trace of the density matrix over one subsystem of a compound system leads to the remaining subsystem being in a mixed state.

The present model, including the build-up of correlations with the environment, solves two problems together: the apparent decay of information during the collapse and the choice of observable. Thus decoherence is useful for both major interpretations of quantum mechanics: it defines the branches in the Everett interpretation and specifies the "cut" in the Copenhagen interpretation. The information is diffusing from the system into the environment. This only occurs with an open system where information is passed on and spreads away from where it originated. The total state of the complete system remains a pure state.

Zurek states that von Neumann and others feel that the destruction of information is necessary to explain the "collapse". But according to Shannon's definition the "information" associated with the original state is the information that is gained when one learns the exact state of the system so the collapse is a kind of cashing in rather than destruction of the information. Von Neumann and others, e.g., Wigner, London and Bauer, also proposed that the collapse first occurred when the measurement chain reached human consciousness, compatible with this interpretation of information.

Information is a measure of the width of a probability distribution; the wider the distribution the more information we gain if we find an exact result. The larger the number of possible outcomes the more information is delivered by determining a specific outcome.

According to (17.107), the decoherence time is

$$\tau_{decoh} = \frac{12\lambda^2}{(2\pi)^2 \left|\vec{R}_j - \vec{R}_i\right|^2} \tau_o, \tag{17.140}$$

so that only localized states $\left(\left|\vec{R}_j - \vec{R}_i\right| \ll \lambda\right)$ will survive the interaction with the environment. The special role of position comes from the fact that the interaction potential between the universe's subsystems (particles, atoms, etc.) is diagonal in position. If an observable commutes with the interaction, a superposition of its eigenstates will be unstable in the presence of the interaction with the environment, and the eigenstates in the superposition will decohere.

When the interaction is periodic in position, i.e., diagonal in momentum, the preferred states will be momentum eigenstates (delocalized in position), as in electrons in a lattice.

In 1991 Zurek published the paper "Decoherence and the transition from quantum to classical"[52] in *Physics Today*, where he presented his ideas at that time. Sometime later[53] the journal published a series of letters critical of Zurek's ideas along with a reply by Zurek. We give a brief review of that discussion as it highlights many of the important issues.

James Anderson referred to Bohm's argument concerning the huge phase shifts occurring between different components of a superposition when a measurement is carried out. He called this decoherence without assumptions. In order for a measurement to take place, the apparatus has to provide a macroscopic distinction between microstates. The interaction with such an apparatus leads to huge phase shifts and the initial coherent superposition becomes a mixed state.

Zurek replied to this that the single quantum correlation, even with a macroscopic but isolated apparatus as considered by Bohm, does not provide enough of a foundation to build the correspondence between classical and quantum reality. This allows only Everett-like pairings of an arbitrary state of the system (including nonclassical superpositions of localized states of an apparatus) with the relative state of the observer. Zurek points out that in his book,[54] Bohm concluded that in spite of his previous argument concerning the large phases, quantum theory presupposes a classical level. It does not deduce classical concepts as a limiting case of quantum concepts. (Recall that shortly after finishing the book Bohm published his revision of the de Broglie pilot wave theory.)

In a later work,[55] Zurek gave the argument in more detail. To illustrate the contrast with Bohm, he considered the state of a system and apparatus entangled by a premeasurement:

$$|\psi_{SA}\rangle = \sum_j a_j \,|s_j\rangle\,|A_j\rangle, \tag{17.141}$$

in which each state picks up a phase shift through the act of amplification that produces macroscopically distinguishable states, as shown by Bohm:

$$|\psi_{SA}\rangle \to \sum_j a_j e^{i\phi_j}\,|s_j\rangle\,|A_j\rangle. \tag{17.142}$$

However, this remains a pure state even though the density matrix,

[52]Zurek, W.H., *Decoherence and the transition from quantum to classical*, Physics Today **44**, 36 (October 1991).

[53]Anderson, J. L., et al., *Negotiating the tricky border between the quantum and the classical*, Physics Today **46**, 13 (April 1993).

[54]Bohm, D., *Quantum Theory* (Englewood Cliffs, NJ: Prentice-Hall, 1951).

[55]Zurek, W.H., *Rev. Mod. Phys.*, op. cit.

$$|\psi_{SA}\rangle\langle\psi_{SA}| = \sum_j |a_j|^2 |s_j\rangle\langle s_j| \otimes |A_j\rangle\langle A_j| + \sum_{j\neq k} a_j a_k^* e^{i(\phi_j-\phi_k)} |s_j\rangle\langle s_k| \otimes |A_j\rangle\langle A_k|, \tag{17.143}$$

may approach a diagonal form as the large variations in the phases between different states will tend to cause the off-diagonal terms to become very small. However, each member of the ensemble is still in a pure state and the phases could, in principle, be undone as in the NMR spin-echo. It is necessary to introduce an environment to produce a true mixed state of the system-apparatus complex. We will see that proponents of the direct collapse model use similar arguments against the decoherence idea.

Only environment-induced decoherence can result in classical states which persist, but whose superpositions are extremely unstable. The environment forces decoherence only when the apparatus is forced into a superposition of states which differ in their effect on the environment. The establishment of the correlation between the system and apparatus occurs once, but the monitoring by the environment is a continuous, unending sequence of correlation-inducing interactions.

G. C. Ghirardi, R. Grassi and P. Pearle point out that the split into system, detector, and environment is arbitrary with no guiding principles as to how to carry it out. They restate the fact that a pure state always evolves into a pure state, so that the mixture state can only be an approximation, and recommend their own approach[56] which involves modifying the Schrödinger equation so that superpositions of macroscopic states rapidly evolve into one or the other of those states, as observed experimentally (see next section).

N. Gisin calls Zurek's treatment a "pragmatic pseudo-philosophy" since Zurek bases his argument on the fact that the difference between a mixture and the exact state can never be detected after several decoherence times.

Albert and Feinberg consider that although states of a macro-system become very quickly entangled with states of the environment, this does *not* mean that superpositions of macroscopically different states can be regarded as the system being in one or another of the component states, but us not knowing which really exists. In quantum mechanics, the world has definite physical properties when in a superposition state that it would not have in the case of being in one of the component states. They state that the practical measurability of these properties is beside the point. Linear quantum mechanics implies the reality of these properties.

In a similar vein, Peter Holland, a strong supporter of the modern de Broglie-Bohm theory,[57] supports the view that the quantum mechanical analysis does not solve the measurement problem. Arranging for ρ to decohere only results in a mixed state in which the component pure states all coexist, not in a unique outcome for a measurement. A treatment based on ψ alone can never yield classical mechanics from quantum mechanics. The de Broglie-Bohm theory assigns a precise position q to a particle in an arbitrary quantum state. The wave function and the set of particle

[56] Ghirardi, G.C. and Rimini, A., "Old and new ideas in the theory of quantum measurement," in *Sixty-two Years of Uncertainty*, ed. A. Miller (New York: Plenum, 1990).

[57] Holland, P., *The Quantum Theory of Motion: An Account of the de Broglie-Bohm Causal Interpretation of Quantum Mechanics* (Cambridge: Cambridge University Press, 1982).

positions of a system at any given time completely determine the future evolution of the system, thus leading to a unique measurement outcome for every initial condition.

V. Ambegaokar accuses Zurek of standard quantum mechanical thinking which, according to him, goes as follows:

a) When in doubt enlarge the system to include an environment.

b) Calculate according to instructions written in Göttingen and published in Copenhagen.

c) Average over unobserved variables to produce a non-unitary, dissipative time evolution in the system of interest.

Zurek began his reply by quoting the letter from Einstein to Born that we mentioned in the introduction to this chapter: "Assume that ψ_1, ψ_2 are solutions of the Schrödinger equation for a macroscopic system and they are both narrow with respect to the macroscopic coordinates. Then $\psi = \psi_1 + \psi_2$ is no longer a solution (because superpositions of macroscopically different states do not exist). Narrowness with respect to macroscopic coordinates is not only independent of the principles of quantum mechanics, but, moreover, incompatible with them."

The openness of macroscopic objects can account for the emergence of classical behavior from quantum theory. For open quantum systems, superposition is not valid. Zurek explained his "existential" interpretation of quantum mechanics: Accept the Everett relative state explanation of wave function collapse, with the set of possible outcomes determined by environment-induced superselection rules whose consequence is decoherence which delineates the classical-quantum border. Classical states—those that are stable in spite of interaction with the environment—are characterized by persistence of properties and predictable evolution; in fact, predictability can be used as a criterion for selecting classical states from the Hilbert space of the system.[58]

17.6.4 Beyond decoherence: quantum Darwinism and the existential interpretation of quantum mechanics

In a recent paper,[59] Zurek summarized his basic idea as follows: Accept the relative state explanation of collapse. Observers see the state of the rest of the universe relative to their own state or the state of their records. In this way quantum theory can be universally valid. This does not mean the acceptance of the "many worlds" model. A small number of degrees of freedom can have objective reality; the rest of the universe is needed to keep redundant records giving the observed states objective reality.

Von Neumann had based quantum mechanics on Hilbert space and the postulate defining expectation values of an observable as the diagonal matrix element, in the current state of the system, of the operator corresponding to the observable. We have seen that this definition leads immediately to the density matrix and the Born rule.

Decoherence and einselection (environment-induced selection) use the reduced density matrix which is obtained by tracing out the environment degrees of freedom, which

[58] Zurek, W.H., "Decoherence and the transition from quantum to classical -REVISITED," arXiv:quant-ph/0306072 [quant-ph] (June 2003).

[59] Zurek, W.H., "Quantum theory of the classical: quantum jumps, Born's rule and objective classical reality via quantum Darwinism," arXiv:1807.02092 [quant-ph] (July 2018).

in turn is based on the probability interpretation and Born's rule. For this reason, decoherence is not a good starting point for fundamental considerations of the origin of classical behavior. In order to carry out Everett's main idea, that everything should follow from a unitarily evolving state, it is necessary to derive the Born rule without introducing any additional assumptions. *(We might ask if the requirement of unitarity is based on the conservation of probability. Remember the goal is to avoid ad hoc introductions of the probability concept.)*

To clarify the point, Zurek[60] introduced five axioms that he claimed were necessary to produce quantum mechanics as we know it. These were:

1. States of a system are described by vectors in Hilbert space.

2. States evolve under unitary transformations described by the Schrödinger equation.

3a. Every observable corresponds to a Hermitian operator.

3b. The only possible result of a measurement is an eigenvalue of the operator corresponding to the observable.

4. After measurement the system is in the eigenstate of the measured observable that corresponds to the observed eigenvalue.

5. If system is in a state $|\psi\rangle$, the probability of a measurement of an observable O yielding an eigenvalue O_i is given by $P_i = |\langle O_i | \psi\rangle|^2$ where $|O_i\rangle$ is a (normalized) eigenstate of the operator O with the eigenvalue O_i. Everett bypassed 4. by stating that the observer perceives the state of the universe reflected in their records.

As we have seen above, the basic problem with this is that the state of a system after measurement can be written in many different basis sets. There has to be something that picks out the set of preferred states and gives them objective classical properties like long time endurance.

Zurek then introduced what he called axiom 0, embracing the point that this was essential to his program: A composite system is described by a vector in the tensor product of the individual Hilbert spaces. The universe consists of systems. In the absence of systems, the measurement problem disappears. Instead you would have a deterministic evolution of an indivisible universe.

Zurek's program then is to derive everything from axioms 0, 1, and 2. That would fulfill Everett's idea that everything follows from a unitary evolving state.

To start with he wishes to prove axiom 4: Results of measurement are orthogonal eigenstates of the operator representing the measured quantity. There is just one result (collapse).

We start by again considering an apparatus (states $|A_i\rangle$) and a system (states $|s_i\rangle$). It is necessary that after a measurement the states $|s_i\rangle$ can be distinguished by the corresponding state of the apparatus. If an initial state $|s_i\rangle |A_o\rangle$ is transformed by a measurement to the state $|s_i\rangle |A_i\rangle$ then the measurement will bring about the transformation

$$|s_1\rangle |A_o\rangle \to U_m |s_1\rangle |A_o\rangle = |s_1\rangle |A_1\rangle \tag{17.144}$$

$$|s_2\rangle |A_o\rangle \to U_m |s_2\rangle |A_o\rangle = |s_2\rangle |A_2\rangle \tag{17.145}$$

[60] Zurek, W.H., Rev. Mod. Phys. op cit.

where U_m is the unitary evolution operator representing the effect of the interaction which performs the measurement. Then taking the dot product between these quantities, we find

$$\langle s_1| \langle A_o| U_m^\dagger U_m |s_2\rangle |A_o\rangle = \langle s_1| \langle A_1| |s_2\rangle |A_2\rangle \tag{17.146}$$
$$\langle s_1| s_2\rangle (1 - \langle A_1| A_2\rangle) = 0. \tag{17.147}$$

If $\langle A_1| A_2\rangle = 1$, we do not have a measurement as the apparatus cannot distinguish the states of the system. In order to have a measurement we require $\langle s_1| s_2\rangle = 0$, so the possible results of a measurement must be orthogonal states. Orthogonality is essential for them to imprint even a minute difference on the states of any other system while retaining their identity. In this way we show that observables are represented by Hermitian operators rather than assuming Hermiticity and deriving orthogonality of the eigenstates. This argument is similar to that of the famous no-cloning theorem, first proved by Wootters and Zurek in 1982.[61] In that proof, they supposed there was a unitary transformation that could perfectly copy an arbitrary state $|s_k\rangle$: $U|s_k\rangle = |s_k\rangle |s_k\rangle$. Taking the inner product of two such cloned states, $U|s_k\rangle, U|s_j\rangle$, we find $\langle s_k| U^\dagger U |s_j\rangle$-$\langle s_k| s_j\rangle = \langle s_k| s_j\rangle \langle s_k| s_j\rangle$ or

$$\langle s_k| s_j\rangle (1 - \langle s_k| s_j\rangle) = 0, \tag{17.148}$$

which has solutions $\langle s_k| s_j\rangle = 0, 1$.[62]

17.6.4.1 Born's rule

Continuing to follow Everett's vision of deducing everything from the Schrödinger equation, i.e., from postulates 0, 1, and 2, Zurek studies the consequences of einselection and the establishment of the pointer basis without introducing the reduced density matrix. In order to do this, he introduces a new derivation of Born's rule, deeming Everett's derivation (even with the improvements suggested by others) inadequate.

Again the argument is based on the existence of an environment. We start with the introduction of the concept of "envariance" (environment-assisted invariance). Consider a system (states $|s_j\rangle$) in a pure state entangled with an environment (states $|\varepsilon_j\rangle$)

$$|\psi_{S\varepsilon}\rangle = \sum_k a_k |s_k\rangle |\varepsilon_k\rangle \tag{17.149}$$

and apply two unitary transformations in succession,

$$U_S = u_s \otimes 1_\varepsilon \tag{17.150}$$
$$U_\varepsilon = 1_s \otimes u_\varepsilon \tag{17.151}$$

where the first operates only on the system and the second only on the environment. Now assume that

[61] Wootters, W. K. and Zurek, W.H., *A single quantum cannot be cloned*, Nature **299**, 802 (1982).

[62] Although this proof shows that creating a device to clone an arbitrary quantum state is impossible, the solution $\langle s_k| s_j\rangle = 0$ means that it is possible to construct a device that can clone a particular set of orthogonal states.

$$|\eta_{S\varepsilon}\rangle = U_S\,|\psi_{S\varepsilon}\rangle \tag{17.152}$$

and

$$U_\varepsilon\,|\eta_{S\varepsilon}\rangle = |\psi_{S\varepsilon}\rangle, \tag{17.153}$$

that is, we have made a transformation on the system alone which can be undone by a transformation of the environment alone. If this is so, it must mean that the local state of S is unaffected by U_S because it is certainly unaffected by U_ε. As an example consider a change of phase of the terms in (17.149) brought about by the operator

$$u_s^{(\phi)} = \sum_j e^{i\phi_j}\,|s_j\rangle\,\langle s_j|\,. \tag{17.154}$$

This can be undone by the operator

$$u_\varepsilon^{(\phi)} = \sum_j e^{-i\phi_j}\,|\varepsilon_j\rangle\,\langle\varepsilon_j|, \tag{17.155}$$

showing the irrelevance of the phases for the local states. We can see this in the standard way by considering the expectation value of an operator operating only on S in the state (17.149) modified by phase shifts:

$$\langle\psi_{S\varepsilon}|\,F_{(S)}\,|\psi_{S\varepsilon}\rangle = \sum_{k,k'} a_k a_{k'}^* e^{i(\phi_k-\phi_{k'})}\,\langle s_{k'}|\,F_{(S)}\,|s_k\rangle\,\langle\varepsilon_{k'}\,|\varepsilon_k\rangle \tag{17.156}$$

$$= \sum_k |a_k|^2\,\langle s_k|\,F_{(S)}\,|s_k\rangle \tag{17.157}$$

where the second equality follows because of the presence of the environment (orthogonality of $|\varepsilon_k\rangle$). This means measurements on S alone cannot distinguish states that started as superpositions with different phases for the a_k; the interaction with the environment results in a decoherence. This is not because the phases are randomized by the interaction with the environment, as would be the case in Bohm's argument, but because they have been delocalized and have lost their significance for S alone. The phases are now global properties of the composite system; they no longer belong to S. We do not need the concept of the density matrix in this argument, only the idea of the entanglement-assisted invariance (envariance) under phase shifts of the pointer state coefficients. The operator $u_s^{(\phi)} \otimes 1_\varepsilon$ acting on a pure state of S that has not interacted with the environment can have real physical effects; measurements can reveal the phase changes. However, in an entangled state $|\psi_{S\varepsilon}\rangle$, $u_s^{(\phi)}$ has no physical effects on S since it can be undone by $1_S \otimes u_\varepsilon^{(\phi)}$, that is, an action on a faraway decoupled system. In the same way, when operating on the state (17.149) with a_k given by $e^{-i\phi_k}$, the swap operator on the system states,

$$U_S\,(k \Leftrightarrow l) = |s_k\rangle\,\langle s_l| + |s_l\rangle\,\langle s_k|, \tag{17.158}$$

acting only on the system can be undone by the operator

$$U_\varepsilon\,(k \Leftrightarrow l) = e^{-i(\phi_k-\phi_l)}\,|\varepsilon_k\rangle\,\langle\varepsilon_l| + e^{i(\phi_k-\phi_l)}\,|\varepsilon_l\rangle\,\langle\varepsilon_k| \tag{17.159}$$

acting only on the environment. The first swap on S alters the global state, but when the coefficients of the superposition differ only in phase, the swap can be undone without any action on S by a counterswap operating only on ε. The global state of $S\varepsilon$ is restored, so the state of S is restored. But S cannot be affected by a counterswap acting only on ε, so we conclude that the state of S was not affected by the original swap. The envariance of the total state under swaps implies the invariance of the local state. After the first swap on S the probabilities of the swapped states must be the same as that of their new environmental partner states. But the probabilities of the partner states in ε are not altered by the swap on S. Thus swapping the states of S exchanges their probabilities, but also leaves them the same, which means the probabilities must have been equal to start with.

To see this in more detail, we start with a pure entangled state between a system $|s_k\rangle$ and apparatus $|A_k\rangle$,

$$|s_1\rangle |A_1\rangle + |s_2\rangle |A_2\rangle, \tag{17.160}$$

and apply a unitary swap to S, $U_S = |s_1\rangle \langle s_2| + |s_2\rangle \langle s_1|$, which produces the state

$$|s_1\rangle |A_1\rangle + |s_2\rangle |A_2\rangle \to |s_2\rangle |A_1\rangle + |s_1\rangle |A_2\rangle. \tag{17.161}$$

After the swap the probability $P(|A_1\rangle)$ of the state $|A_1\rangle$ is equal to the probability $P(|s_2\rangle)$ of $|s_2\rangle$ and $P(|A_2\rangle) = P(|s_1\rangle)$. Now a similar swap on A will restore the original pre-swap state without touching S in any way. The local state of S is restored even though it cannot have been affected by the swap on A. So $P(|s_1\rangle)$ and $P(|s_2\rangle)$ have been exchanged but unchanged; therefore they must be equal, $P(|s_1\rangle) = P(|s_2\rangle) = 1/2$.

If there are N terms in the sum (17.160), all with equal coefficients, we will have $P_k = 1/N$. The elimination of the phases is crucial here: swaps in isolated pure states will, in general, change the phase.

For a state where the components do not all have the same coefficients, say,

$$|\Phi_{SA}\rangle = \alpha |s_1\rangle |A_1\rangle + \beta |s_2\rangle |A_2\rangle \tag{17.162}$$

swapping S results in

$$\alpha |s_2\rangle |A_1\rangle + \beta |s_1\rangle |A_2\rangle \tag{17.163}$$

while swapping A will produce

$$\alpha |s_2\rangle |A_2\rangle + \beta |s_1\rangle |A_1\rangle \tag{17.164}$$

which differs from the original pre-swap state. This state is therefore not envariant and $P(|s_1\rangle) \neq P(|s_2\rangle)$. To proceed we introduce a fine graining, that is, we expand the apparatus states, which we take as normalized, in the basis $|a_k\rangle$ which is the preferred pointer basis einselected by the environment $|\varepsilon_k\rangle$:

$$|A_1\rangle = \frac{1}{\sqrt{m_1}} \sum_{k=1}^{m_1} |a_k\rangle \tag{17.165}$$

$$|A_2\rangle = \frac{1}{\sqrt{m_2}} \sum_{k=m_1+1}^{m_1+m_2=M} |a_k\rangle \,. \tag{17.166}$$

Then

$$|\Phi_{SA}\rangle = \alpha\frac{|s_1\rangle}{\sqrt{m_1}}\sum_{k=1}^{m_1}|a_k\rangle + \beta\frac{|s_2\rangle}{\sqrt{m_2}}\sum_{k=m_1+1}^{m_1+m_2=M}|a_k\rangle. \tag{17.167}$$

To as good an approximation as desired; $m_1, m_2, M = m_1 + m_2$ (all integers) can be selected so that

$$\alpha = \frac{\sqrt{m_1}}{\sqrt{M}} \tag{17.168}$$

$$\beta = \frac{\sqrt{m_2}}{\sqrt{M}}. \tag{17.169}$$

Then all terms will have equal coefficients. Since $m_{1,2}$ are integers, this can only be an approximation in most cases, but the approximation can be improved by taking larger values of $m_{1,2}$. Now we suppose that A is entangled with the environment by a suitable interaction:

$$|\Phi_{SA}\rangle \to |\Phi_{SA\epsilon}\rangle = \frac{|s_1\rangle}{\sqrt{M}}\sum_{k=1}^{m_1}|a_k\rangle\,|\varepsilon_k\rangle + \frac{|s_2\rangle}{\sqrt{M}}\sum_{k=m_1+1}^{m_1+m_2=M}|a_k\rangle\,|\varepsilon_k\rangle\,. \tag{17.170}$$

We see that swaps of any of the terms $|s_1\rangle\,|a_k\rangle$ with $|s_2\rangle\,|a_k\rangle$ can be undone by corresponding swaps of the $|\varepsilon_k\rangle$, so we conclude that the probabilities of all the terms must be equal: $P(|a_k\rangle) = 1/M$, and it follows that the probability

$$P(|s_1\rangle) = \sum_{k=1}^{m_1}\frac{1}{M} = \frac{m_1}{M} = |\alpha|^2 \tag{17.171}$$

$$P(|s_2\rangle) = \sum_{m_1+1}^{m_1+m_2=M}\frac{1}{M} = \frac{m_2}{M} = |\beta|^2. \tag{17.172}$$

Note that the entanglement necessary for this argument is allowed by postulate 0: composite systems reside in Hilbert spaces that are tensor products of the component Hilbert spaces. Also notice that it is the requirement that the states be normalized, i.e., that $\langle\psi\,|\psi\rangle$ be finite (Eqs. 17.165 and 17.166), that leads to the appearance of the square of the amplitudes exactly as in the case of Everett's derivation of the Born rule. Does this requirement of normalization somehow implicitly assume a probability interpretation?

17.6.4.2 The appearance of a classical world

Pointer states are (approximately) eigenstates of the apparatus-environment interaction Hamiltonian, so they are stable despite the continuing interaction with the environment. A superposition of such states quickly loses phase coherence and decoheres. The individual pointer states are stable, faithful records which remain correlated with the state of the system in spite of decoherence. Einselection (environment-induced selection) and decoherence are two complementary views of the same process. Decoherence is the destruction of coherence between preferred states while einselection is the consequence. It is the exclusion of all but a small set of pointer states from a larger Hilbert space of the apparatus, and this is what defines the set of possible outcomes of a measurement. Einselected states are stable in spite of interacting with the environment.

This leads to what has been called quantum Darwinism: the quantum states most resilient to decoherence are most fit to survive monitoring by the environment.

Decoherence should not be confused with relaxation that can also be induced by the environment. Relaxation results from the environment perturbing the system whereas decoherence and einselection result from the system perturbing the environment; decoherence requires an entanglement of the system with the environment. The fact that a quantum universe appears classical when seen from within makes sense only in the context of a universe divided up into systems and must be phrased in terms of correlations between systems. No systems—no collapse. Our classical experience does not apply to the universe as seen from outside. The universe is a collection of open systems. The interpretation problem does not arise without interacting systems.

There may be a pure state of the universe including what we consider the environment, but we as observers along with our apparatus are forced to perceive the universe the way we do. We are part of the universe observing from the inside. For us the interaction with the environment specifies what exists.

Decoherence and einselection define branches of the universal state and complete the Everett interpretation. They also justify the Copenhagen interpretation by drawing the border between quantum and classical states.

Following Everett, an observer described by a specific einselected state, including the configuration of the memory bits, will perceive only that state. Collapse results from the einselection and the one-to-one correspondence between the state of the memory and the physical outcome. These distinct memory states of the observer cannot be superposed as enforced by decoherence and einselection. Distinct memory states label inhabitants of different branches of the universe wave function. There is no room for collapse in purely unitary models.

The unitary Schrödinger evolution leads to premeasurement, an entangled state of the system and apparatus. The superposition can always be rewritten in terms of another basis. The preferred basis arises from the introduction of the environment and is the basis that retains correlation between the apparatus and system in spite of decoherence.

Quantum states are often said to describe the knowledge of an observer about the results of past measurements that prepared the system. The key argument against the objective existence of a quantum state is that it is impossible to determine the state of an isolated system without knowing the observables that were used to prepare it. Measurements of different observables than were used in the preparation will change the state. However, continuous monitoring of the einselected (environment selected) state leads to an objective existence of the state. We consider a state to objectively exist if multiple observers can independently determine the state without changing it.

Relative objective existence is a consequence of the existence of many records of the same set of system states. All observers agree on the branch that they find themselves in. Redundancy of environmental records is a measure of the Darwinian fitness of states which persist the longest under environmental decoherence. In a system monitored by the environment, the (einselected) state of the system coincides with what is known to be. As an example of the uniqueness of interaction with the environment, consider experiments with trapped ions which simulated decoherence by applying classical noise. The experiment is repeated many times with different instances of the applied noise, and the results are averaged over many noise realizations. The results show the same behavior as would be obtained if the ions had been interacting with an environment. However, in each individual measurement the state of the system is still pure; the simulated decoherence could be reversed if the time dependence of the noise was recorded and used to generate the relevant unitary transformation of the system.

In general, the continued action of the system self-Hamiltonian will cause the system to continuously evolve so that there are no exact pointer states. However, approximate pointer states still exist and they can be found by studying the evolution of the system, as they will be the states that are the most stable, much more stable than superpositions of these favored states.

The decoherence idea is now seen as not being fundamental enough as it was initially based on the Born rule, which is an unnatural assumption for a fundamental theory of the classical realm. In this sense decoherence did not go far enough and it is necessary to derive the Born rule from postulates 0, 1, and 2, as has been done using the notion of "envariance."

Everett's insight, the realization that the relative states settle the problem of collapse under unitary evolution, is crucial to this program. However this does not force us to accept the "many worlds interpretation."

Before there can be a collapse, a set of preferred states (one of which will be realized) must be chosen. Zurek states that "there is nothing in the writings of Everett that would hint at a criterion for such preferred states and nothing to indicate that he was aware of this problem."[63] While this seems to be true, David Deutsch has reported a conversation he had with Everett on the occasion of Everett's visit to Austin in May 1977 which we discussed above. Deutsch had asked Everett "what defines the Hilbert space basis with respect to which one defines universes." Everett replied it was the structure of the system, i.e., the Hamiltonian. Everett did not seem to think this was an important issue.

According to Zurek, senses were developed to maximize survival. When nothing can be gained from prediction there is no evolutionary reason to develop the capability of prediction. Only classical states have robust predictable consequences.

17.7 (Spontaneous) Direct wave function collapse

According to von Neumann's description of the standard (Copenhagen) interpretation of quantum mechanics, the measurement of an observable A on a system whose state

[63] Zurek, W.H., "Relative states and the environment: einselection, envariance, quantum Darwinism and the existential interpretation," arXiv:0707.283v1 [quant-phy] (July 2007).

was a superposition of the eigenstates of A would result in the collapse of the state into one of the eigenstates. The particular resulting eigenstate could not be predicted for each individual measurement, but the probability distribution of the resulting eigenstates was given by the theory. A major feature of the theory is that in the case of a number of interacting systems (von Neumann chain), the location of the collapse (von Neumann-Heisenberg cut) was irrelevant as it had no effect on the final result. This was found unsatisfactory by many people as it appeared to invoke a vague and unphysical process. We have seen in this chapter how various physicists have attempted to go beyond von Neumann and give a deeper insight into the physical processes involved. Zurek and Zeh tried to account for the collapse by introducing an "environment," a system akin to a thermal bath, containing many degrees of freedom, interacting with the system under study via a Hermitian Hamiltonian.

In 1986, a few years after the introduction of the decoherence idea, Ghirardi, Rimini, and Weber[64] proposed a model in which the wave function simply collapses spontaneously. In order to not disagree with observations in the atomic domain, the collapse had to be limited to large, practically macroscopic systems. The idea was to search for a modification of the dynamics of macroscopic objects that would suppress the linear combination of states corresponding to the same macroscopic object being located in well-separated spatial regions. Thus the dynamical equations would induce transitions from pure states to statistical mixtures. Note that the authors consider the transition to a statistical mixture—the vanishing of the off-diagonal elements of the density matrix—to be a satisfactory resolution of the problems, in contrast to Zurek, who invokes Everett's model. This transition to a mixture is accomplished by introducing a stochastic term corresponding to a localization process into the dynamical equations. This is equivalent to an approximate position measurement being carried out on the system at random times. The basis problem is solved by fiat, so to speak: the localization takes place in coordinate space. A possible physical origin of this effect is deliberately ignored. The authors worked with the equation of motion for the density matrix. A year later Bell[65] showed how the GRW process plays out on the wave function. We start with this exposition.

In this picture the wave function $\psi(t, r_1 ... r_N)$ evolves according to the Schrödinger equation except that it makes jumps to new states at random times, with the probability per unit time given by N/τ, where N is the number of degrees of freedom and τ is a new natural constant. (In a macroscopic system N is thought to be the number of atoms, but we might ask why it does not include the electrons, protons, and neutrons, perhaps also the quarks and gluons.) The jump is to a new state given by

$$\psi' = j(\vec{x} - \vec{r}_i) \frac{\psi(t, \vec{r}_1 ... \vec{r}_N)}{R_n(\vec{x})} \tag{17.173}$$

where $\vec{r}_i$ is one of the (randomly chosen) degrees of freedom, $R_n(\vec{x})$ is a normalization factor given by

[64] Ghirardi, G.C., Rimini, A. and Weber, T., *Unified dynamics for microscopic and macroscopic systems*, Phys. Rev. D, 34, 470, (1986), received Dec. 17, 1985.

[65] Bell, J. S., 'Are there quantum jumps?' in *Schrödinger: Centenary Celebration of a Polymath*, ed. Kilmister, C. W., (Cambridge: Cambridge University Press, 1987), reprinted in Bell, J. S., *Speakable and Unspeakable in Quantum Mechanics* (Cambridge: Cambridge University Press, 1987).

$$|R_n(\vec{x})|^2 = \int d^3r_1...d^3r_N \, |j(\vec{x}-\vec{r}_i)\,\psi(t,\vec{r}_1...\vec{r}_N)|^2, \tag{17.174}$$

and the "jump factor" $j(\vec{x})$ is normalized so that

$$\int j(\vec{x})\, d^3x = 1. \tag{17.175}$$

The point $\vec{x}$ will be chosen to be the center of a collapse with probability

$$|R_n(\vec{x})|^2 \, d^3x\,. \tag{17.176}$$

GRW take

$$j(\vec{x}) = Ke^{-\left(\frac{x^2}{2a^2}\right)} \tag{17.177}$$

with a another new constant of nature. With $\tau = 10^{15}$ sec and $a = 10^{-5}$ cm, microscopic systems would not be affected but if $N \gtrsim 10^{15}$ the rate would become significant.

Turning now to a typical measurement situation where we have a small system represented by a wave function $\phi(s_1...s_L)$ interacting with a large system $\chi(r_1...r_M)$ where M is large and L is not large. Assume that after a measurement interaction the system is in a state

$$\overline{\psi} = \phi_1(s_1...s_L)\,\chi_1(r_1...r_M) + \phi_2(s_1...s_L)\,\chi_2(r_1...r_M)\,, \tag{17.178}$$

where $\chi_{1,2}(r_1...r_M)$ are pointer states that are separated by a macroscopic distance. Then multiplication by $j(\vec{x}-\vec{r}_i)$ will reduce one or the other (or both) terms to zero. One of the terms will disappear.

We can apply similar reasoning to the EPR situation with $\phi_{1,2}$ representing up and down spin states for the particle on the left and $\chi_{1,2}$ up and down states of the particle on the right. If the counter on the left is closer to the source, the left particle will couple with a macroscopic system first and the GRW jumps would reduce the state of that particle, thus forcing the particle on the right into a definite state. According to Bell, the crux of the problem here is not the correlations between the spins of the two particles but rather that before the first encounter with a detector, there is nothing but the wave function, which is entirely neutral between possibilities. The first measurement makes the decision for both particles so it cannot be revealing a decision already taken (e.g., in the source). The EPR conclusion that quantum mechanics is incomplete, implying the existence of hidden variables and the possibility that these could restore local causality, is difficult to maintain in the case of the imperfect correlations predicted by quantum mechanics for misaligned polarizers, thus summarizing Bell's theorem.

With this introduction we turn to the treatment of GRW. They rewrite the equation of motion for the density matrix by adding a new (non-Hamiltonian) term as follows

$$\frac{d\rho}{dt} = -\frac{i}{\hbar}[H,\rho(t)] - \lambda[\rho(t) - T(\rho(t))], \tag{17.179}$$

where the operator T is defined by

$$\langle q'|\, T(\rho)\, |q''\rangle = e^{-\frac{\alpha}{4}(q'-q'')^2} \langle q'|\, \rho\, |q''\rangle, \tag{17.180}$$

so that we can write (17.179) as

$$\frac{d\rho}{dt} = -\frac{i}{\hbar}[H, \rho(t)] - \lambda\left[1 - e^{-\frac{\alpha}{4}(q'-q'')^2}\right]\rho(t). \tag{17.181}$$

In the limit $\lambda \to \infty$, $\lambda\alpha = \gamma$ =constant $\ll 1$, we have

$$\frac{d\rho}{dt} = -\frac{i}{\hbar}[H, \rho(t)] - \frac{\gamma}{4}\left[(q' - q'')^2\right]\rho(t) \tag{17.182}$$

We note that the last term is equal to the matrix element of the double commutator

$$\frac{\gamma}{4}\langle q'|\, [q, [q, \rho]]\, |q''\rangle \tag{17.183}$$

which represents a general diffusion type motion. For a system interacting with a thermal reservoir in the high temperature limit, Caldeira and Leggett[66] obtain this term with $\gamma/2$ replaced by $\eta k_B T/\hbar^2$ with η the friction constant. Zurek[67] has also applied this term to the decoherence problem, where it determines the timescale of the decoherence process. We see that this time is on the order of

$$\tau_D \approx \frac{4}{\gamma(q' - q'')^2}, \tag{17.184}$$

decreasing for large spatial separations.

Thus spontaneous localization and decoherence seem to be mathematically equivalent. This point was made by Joos,[68] who as one of the originators of the decoherence idea stated that a realistic application of quantum mechanics already yields an irreversible non-unitary dynamics for macroscopic objects, so that there is little motivation for the introduction of a new fundamental dynamics. Naturally GRW replied (in the same issue of *Physical Review D*[69]). They state that while "decoherence" theory escapes the strict application of the N-body Schrödinger equation to a local system by itself by considering an unavoidable coupling with the environment, the "collapse" theory sticks to the local system, accepting a stochastic modification of the dynamical behavior of the elementary constituents. The result is that the wave function disentangles when macroscopic levels are reached, thus allowing the interpretation of the wave function as an objective property of a local system.

According to the decoherence model, there is a pure evolution of the entire system according to the Schrödinger equation, and then tracing out the environment degrees

[66]See Caldeira, A. D. and Leggett, A. J., *Path integral approach to quantum Brownian motion*, Physica A **121**, 587 (1983), received Aug. 25, 1982, revised Jan. 4, 1983.

[67]Zurek, W.H., *Rev. Mod. Phys.* op cit.

[68]E. Joos, Comment on *Unified dynamics for microscopic and macroscopic systems*, Phys. Rev. D **36**, 3285, (1987).

[69]Ghirardi, G. C., Rimini, A. and Weber, T., *Disentanglement of quantum wave functions: answer to 'Comment on "Unified dynamics for microscopic and macroscopic systems"'* Phys. Rev. D, 36, 3287 (1987) received Feb. 6, 1987.

of freedom results in the local system having its long-distance correlations suppressed. The entanglement of the local quantum state is transferred to the environment, which is then neglected; the coherence initially present in the system cannot be detected without inspection of the environment. However, the neglect of the environment cannot be justified in principle and no unitary treatment can explain why only one of the wave function components is experienced.

In the GRW collapse process there is a strong disentanglement: the local system has a definite wave function at all times, and the basic evolution of the wave function contains a stochastic element.

Benatti et al. have shown how these ideas apply to the measurement problem in more detail.[70]

Philip Pearle[71] has proposed a slightly altered model. Whereas according to GRW the state vector changes instantaneously into a partially localized state as a result of random events, in Pearle's model the transition occurs smoothly, so that if we are in a canonical state of the measurement problem where the wave function is a superposition of two packets with widths small compared to their separation, the dynamics will result in one of the packets (randomly selected) growing and the other decaying. The density matrices are identical at long times for the two models, but for short times ($t < 1/\lambda$) there should be a distinction: the changes build up smoothly over several time constants in Pearle's system, while the GRW system exhibits sudden jumps. The reader should ask if the distinction is observable.

As for every other topic considered in this book, there is an immense literature concerned with collapse models. See for example a recent book edited by Shan Gao.[72]

17.8 Second quantization and particle-wave duality

It is striking that none of the proposed interpretations and extensions to nonrelativistic quantum mechanics take the second quantized theory as their starting point, even though this was considered by Jordan and many others as the most refined version of the theory.

In Sections 12.3.4 and 12.3.5 we showed how the quantization of the Maxwell electromagnetic field and the Schrödinger wave field leads to the results that the probability amplitude for an electron or a photon to propagate through an optical system is the same as that predicted by classical wave optics in each case. The initial state where the particle is released at $\vec{r}_S$ is given by

$$|\psi\,(\vec{r}_S, t_S)\rangle = \sum_k u_k^*\,(\vec{r}_S)\, e^{\frac{i}{\hbar}\varepsilon_k t_S} a_k^\dagger\,|0\rangle, \tag{17.185}$$

that is, the wave field is excited into oscillations of frequency $\varepsilon_k/\hbar$ with $\varepsilon_k = \hbar^2 k^2/2m$ in the case of a free electron.

[70]Benatti, F., Ghirardi, G. C.. Rimini, A. and Weber, T., *Quantum mechanics with spontaneous localization and the quantum theory of measurement*, Nuovo Cimento B **100**, 27 (1987).

[71]Pearle, P., *Combining stochastic dynamical state vector reduction with spontaneous localization*, Phys. Rev. A **39**, 2277 (1989).

[72]Gao, S., ed., *Collapse of the Wave Function: Models, Ontology, Origin and Implications* (Cambridge: Cambridge Univerity Press, 2018).

The operators $a_k^\dagger$ acting on the vacuum state leave the system in a superposition of waves, each satisfying the boundary conditions for the physical system. These waves propagate through the system. The amplitude for the particle to be absorbed at the detector position $\vec{r}_D$ is given by Eq. (12.309):

$$M(\vec{r}_D, \vec{r}_S) = \sum_{k,j} \langle 0| a_j u_j(\vec{r}_D) u_k^*(\vec{r}_S) e^{-\frac{i}{\hbar}(\varepsilon_j t_D - \varepsilon_k t_S)} a_k^\dagger |0\rangle \tag{17.186}$$

$$= \sum_k u_k(\vec{r}_D) u_k^*(\vec{r}_S) e^{-\frac{i}{\hbar}\varepsilon_k(t_D - t_S)}. \tag{17.187}$$

In the case of empty space, the eigenfunctions $u_k(\vec{r})$ are given by plane waves

$$u_k(\vec{r}) = \frac{1}{(2\pi)^{3/2}} e^{i\vec{k}\cdot\vec{r}}. \tag{17.188}$$

Replacing the sum by an integral over three-dimensional k space, we find

$$\int d^3k\, e^{i\vec{k}\cdot(\vec{r}_D - \vec{r}_S)} e^{-\frac{i}{\hbar}\varepsilon_k(t_D - t_S)} = \left(\frac{m}{2\pi i\hbar(t_D - t_S)}\right)^{3/2} e^{\frac{i}{2}\frac{m(r_D - r_S)^2}{\hbar(t_D - t_S)}}, \tag{17.189}$$

which yields the Green's function for the nonrelativistic Schrödinger equation as is clear from (12.315).

The mathematics is telling us that the particle plays no role in the propagation from $\vec{r}_S$ to $\vec{r}_D$; in the two-slit experiment the electron does not go through any slit. Between "emission" and "absorption" there is no "electron" in the system, only excited oscillations of the wave field, which because of the quantized transfer of energy to the detector appear as a particle. The electron, or photon, effectively disappears into the "cloud." The discreteness only appears when we measure the state of the detector atom and find it either in the excited or the unexcited state. In fact, the observation of discrete particles is seen to be due to the fact that matter waves exist only in discrete quantum states. According to Jordan, "the non-commutation of the wave amplitudes in three-dimensional space is responsible for the existence of particles and the validity of the exclusion principle." In fact, this viewpoint is very similar to that originally introduced by Einstein in his treatment of an ideal gas.

This view does not add anything to the discussion of superposition and entanglement. An EPR-like state based on two entangled spins can be written as

$$|\psi_{EPR}\rangle = \left[a^\dagger(k)_{(s+)} a^\dagger(-k)_{(s-)} + a^\dagger(k)_{(s-)} a^\dagger(-k)_{(s+)}\right] |0\rangle \tag{17.190}$$

where $a^\dagger(\pm k)_{(s\pm)}|0\rangle$ is a state with momentum $\pm\hbar\vec{k}$ and spin $\pm\hbar/2$. Projection onto a state, say $a^\dagger(k)_{(s-)}|0\rangle$ (moving to the right with $s = -1/2$) results in the state $a^\dagger(-k)_{(s+)}|0\rangle$ and so forth, as in the usual discussion of the EPR effect.

17.9 Conclusion

John Bell called attention to the similarity between the Everett model and the de Broglie-Bohm model. In both cases the wave function evolves deterministically, continuously branching without collapse. In the de Brogie-Bohm model it is the particle (or the system point in an N-particle system) surfing on the wave that singles out a specific branch (i.e., depending on the initial conditions). In Everett's model, this selection is made by the records (memory) of the individual observers.

The Everett model seems to attract more attention as time goes by. As the many-worlds idea is discussed more widely, it comes to seem less and less outlandish. The clarity of the model, combined with the fact that it does what the standard interpretation cannot do, i.e., allow a wave function to represent the entire universe without the need for an external, classical observer, and thus hopefully opens the way for a quantum theory of gravity, means that it will continue to attract interest.

Decoherence refers to the ideas, first proposed by Zeh and Zurek, that ordinary (Hermitian) Hamiltonian interaction with a thermal bath-like environment will cause the off-diagonal elements of the density matrix to become negligibly small. In its early form, this was deemed an adequate solution of the measurement problem, but in later work Zurek turned to Everett's model to account for the appearance of a single state from the mixture implied by a diagonal density matrix.

Based on the fact that a unitary evolution can never reduce a superposition into a pure mixture, GRW suggested that a new, non-unitary process reduces the wave function. The process works at the same rate for all degrees of freedom and would thus be very fast for macroscopic objects while leaving microscopic objects essentially unaffected. Under reasonable assumptions, the density matrix evolution of GRW is the same as that obtained from unitary decoherence. GRW make no attempt to go beyond the production of a diagonal density matrix, i.e., Bell's "and/or" question is not addressed.

Kochen and Specker have given a mathematically more sophisticated proof of Bell's theorem by introducing the idea of "contextuality," i.e., that in quantum systems the results of a measurement depend not only on the quantity measured and the objective properties of the system being measured but also on the "context" of the measurement, that is, the results of other measurements that may be made simultaneously on the system (i.e., with the same experimental arrangement).[73]

Second quantization, in addition to removing the need to consider wave functions in more than three dimensions, throws a unique light on the question of wave-particle duality, namely that the discreteness associated with particles is a result of the quantization of the amplitudes of the wave fields but does not help with the measurement problem. We calculate the probability distribution of, detecting a particle at a given point. Nevertheless this is the modern, mature form of the non-relativistic theory and should be the starting point of any attempt at interpretation.

The goal of the various interpretations of quantum mechanics is an explanation of how the mathematical theory relates to reality. Given the apparent statistical nature of the theory, different interpretations attempt to address such questions as whether the theory is deterministic or random, which elements of quantum mechanics represent reality, and the nature of measurement (collapse of the wave function), to name a few. Quantum mechanics has been subjected to extremely precise experimental tests, both implicitly and explicitly, over an enormous range of energies and for differ-

[73]Kochen, S., and Specker, E. P., *The problem of hidden variables in quantum mechanics*, J. Math. Mech. **17**, 59 (1967).

ent interactions—using one of the earliest interpretations of quantum mechanics, the Copenhagen interpretation. This interpretation remains the simplest and has survived all tests. On the other hand, the Copenhagen interpretation is almost a restatement of the "postulates" of quantum mechanics (one must bear in mind that quantum mechanics is not an axiomatic mathematical theory, but built upon a series of experimentally tested hypotheses—any of which might ultimately prove false—nonetheless we have postulates to form a basis of a theory in an operational and practical sense) and it therefore appears as a contentless recipe—and is the basis of the "shut up and calculate" school.[74]

One might suspect that the uneasiness associated with purely probabilistic interpretations of the theory represents the fear of loss of human determinism and loss of control of the physical world, the loss of dominion over the microscopic world, and these concerns are more psychical than physical.

David Mermin once quipped, "New interpretations appear every year. None ever disappear."[75] This is due to an apparent lack of testable hypotheses for essentially all interpretations expanded from the Copenhagen interpretation. One might recall the laments of Ernst Mach regarding haphazard hypotheses and frivolous models as a basis of physical understanding. We might lament that Ramsey's paper discussed in Chapter 2 has been cited about 30 times in total, compared to 78,000 results for a Google Scholar search of Everett's "Many Worlds". In a nutshell, all interpretations to date have not produced a single cleanly testable hypothesis, so collected together they are about as useful as a stamp collection.

Nonetheless, there is merit for attempts toward a deeper understanding. Schrödinger 's studies of minimum uncertainty states as a proxy for classical particles resulted in the notion of coherent states that are central to modern quantum information. It was in this context that the squeezed state operator presently used in forefront quantum electronics was formulated.

We have endeavored to give an idea of the major trends in the interpretation of nonrelativistic quantum mechanics. Despite the fact that many groups are working in this field and interest in it has increased dramatically, it by no means forms a mainstream topic of current research. In relativistic quantum field theory (QFT) the questions raised here are mostly ignored. One major concern with QFT is that in spite of the successes of the renormalization program and the theory's ability to yield numerical results of amazing accuracy, the presence of infinities seems ugly to some people.[76] In fact this problem is ignored by most physicists, as are the concerns of this chapter. Present day questions about the proper way to do lattice calculations

[74] Mermin, N. David, *What's Wrong with this Pillow?* Physics Today 42, 9 (1989).

[75] Mermin, N. David (2012-07-01). *Commentary: Quantum mechanics: Fixing the shifty split.* Physics Today. 65 (7): 8-10.

[76] Similarly to Einstein, de Broglie and Schrödinger, who disavowed the later forms of the theories they helped establish, Dirac expressed his unhappiness with the infinities in quantum electrodynamics, which he founded: "They use what they call renormalization techniques which involves handling infinite quantities and this is not really a mathematical or logical process...just a set of working rules rather than a correct mathematical theory and I don't like this whole development at all." Paul Dirac interview at Göttingen, 1982, available on YouTube and elsewhere on the web, e.g., https://av.tib.eu/media/11186.

in quantum chromodynamics (QCD) and the status of string theory, to name just a few, seem to have left the considerations of this chapter far behind. In some sense physicists concerned with questions of interpretation are fighting a straw man, a not fully matured version of the theory, that could be considered as superseded Although some of the issues remain in the modern version of the theory it might be productive to reframe the discussion in these terms.

For the vast majority of working physicists, the Copenhagen interpretation or something like it is sufficient to allow them to "shut up and calculate." Many others feel that none of the interpretations is adequate, that we are missing some important ingredient of the overall picture.

Part II

Applications of Quantum Mechanics

18
Operator techniques and the algebraic solutions of problems

18.1 Introduction

Fundamentally, quantum mechanics is a linear operator theory. This allows great freedom in the setting up of problems and in the seeking of solutions, and usually makes perturbative expansions and their summation straightforward. Recursion relationships between the eigenvalues in solutions to a Hamiltonian allow a direct summation of matrix elements, as often required in perturbation theory, without directly solving for the wavefunction.

This inspires us to find solutions to other problems by using *algebraic and operator techniques* and thereby avoiding the handling of mathematical functions that are a throwback to the nineteenth century. Operator methods provide a powerful and effective means to avoid complicated coordinate representations of the wavefunctions.

These methods will include sum rules that are a cornerstone of many subfields of physics; for example their utility in quantum chromodyanamics (QCD) is well-known. Such sum rules were important in the early days of quantum mechanics in testing its applicability to the calculation of atomic structure, and more recently, in QCD sum rules that were used to determine the masses of the light and heavy quarks that are otherwise not directly measurable.

The abstract and compact formalism of the Dirac notation often facilitates the solution to quantum mechanics problems by algebraic techniques. By this, we mean that the wavefunction doesn't necessarily need to be explicitly known in order to calculate the energy spectrum of a Hamiltonian, or to calculate the matrix elements of operators. Technically, the algebraic techniques we will discuss are not purely algebraic as occasionally, using elementary calculus, the Hamiltonian can be recast into a form (e.g., the harmonic oscillator) for which solutions are already known. The use of contour integration to evaluate integrals and sums is also a tool of great importance. For the most part, these techniques are possible solely because quantum mechanics is a linear operator theory and because the solutions to the Schrödinger equation are analytic.

In this chapter we will give a number of explicit examples of operator and algebraic techniques. These examples will illustrate the utility of the symmetry of quantum systems, and allow the determination and classification of degenerate multiplets and manifolds. The dynamical properties of systems can be determined without resolving the eigenvalues or eigenfunctions in a coordinate representation.

The Historical and Physical Foundations of Quantum Mechanics. Robert Golub and Steven K. Lamoreaux, Oxford University Press.

Often, the *completeness* of a set of eigenfunctions over a given range allows them to be used to expand a different set of eigenfunctions defined over the same range. The quantum harmonic oscillator provides sevaral examples of complementarity and completeness, meaning that the states used for a single harmonically bound particle can be applied to other systems. For example, the hydrogen atom can be transformed into a harmonic oscillator by an appropriate coordinate transformation, which will be demonstrated in this chapter. The fields of *functional analysis* and *spectral analysis* address transformations of this type.[1]

18.2 Uncertainty relationships via operator techniques

Consider two physical observables given by Hermitian operators **A** and **B** that do not commute with each other

$$[\mathbf{A},\mathbf{B}] = \hbar\mathbf{F} \neq 0 \tag{18.1}$$

where **F** is another operator that need not commute with **A** and **B**, or is a constant (a number f).

In the classical limit, $\hbar \to 0$, which forces the operators to commute. In the next level (quasi-classical) approximation, the operator **F** can be replaced by a constant, f.

Let us consider some general state $|\psi\rangle$ that is given by a superposition of the **A** and **B** eigenstates,

$$|\psi\rangle = \sum_{i,j} c_{i,j}|a_i b_j\rangle. \tag{18.2}$$

The dispersion in the expectation values of **A** and **B** can be calculated by considering the mean square deviation (dispersion) about the expectation values

$$a = \langle\psi|\mathbf{A}|\psi\rangle; \quad b = \langle\psi|\mathbf{B}|\psi\rangle. \tag{18.3}$$

The expectation values a and b are the sum of the eigenvalues weighted by the state probabilities.

The mean square deviation (dispersion) or variance $(\Delta a)^2$ and $(\Delta b)^2$ can be determined from

$$(\Delta a)^2 = \langle\psi|(\mathbf{A}-a)^2|\psi\rangle \equiv \langle\psi|\mathbf{C}^2|\psi\rangle \tag{18.4}$$

$$(\Delta b)^2 = \langle\psi|(\mathbf{B}-b)^2|\psi\rangle \equiv \langle\psi|\mathbf{D}^2|\psi\rangle \tag{18.5}$$

where $\mathbf{C} = \mathbf{A} - a$ and $\mathbf{D} = \mathbf{B} - b$. Let us introduce a real constant λ and calculate the norm of $(\mathbf{C} + i\lambda\mathbf{D})|\psi\rangle$, using the fact that **C** and **D** are also hermitian:

$$\|\,(\mathbf{C}+i\lambda\mathbf{D})\,|\psi\rangle\|^2 = \langle\psi|\mathbf{C}^2|\psi\rangle + \lambda^2\langle\psi|\mathbf{D}^2|\psi\rangle + i\lambda\langle\psi|\mathbf{CD}|\psi\rangle - i\lambda\langle\psi|\mathbf{DC}|\psi\rangle \geq 0. \tag{18.6}$$

Using the definitions that we introduced,

$$(\Delta a)^2 + \lambda^2(\Delta b)^2 + i\lambda\langle\psi|[\mathbf{C},\mathbf{D}]|\psi\rangle \geq 0 \tag{18.7}$$

[1]Philip M. Morse and Herman Feshbach, *Methods of Theoretical Physics, Vol 1.*(McGraw-Hill, New York-London, 1953). A general treatment of eigenfunctions by a factorization method is given on pp. 788-789.

Because a and b are (real) constants, $[\mathbf{C}, \mathbf{D}] = [\mathbf{A}, \mathbf{B}]$, so

$$(\Delta a)^2 + \lambda^2(\Delta b)^2 + i\lambda\langle\psi|[\mathbf{A}, \mathbf{B}]|\psi\rangle \geq 0. \tag{18.8}$$

This can be minimized by taking the derivative with respect to λ and finding the value that gives 0,

$$2\lambda(\Delta b)^2 + i\langle\psi|[\mathbf{A}, \mathbf{B}]|\psi\rangle = 0 \tag{18.9}$$

which has the solution

$$i\lambda = \frac{\langle\psi|[\mathbf{A}, \mathbf{B}]|\psi\rangle}{2(\Delta b)^2}. \tag{18.10}$$

The minimum value of the inequality is obtained by substituting $i\lambda$ with this result, yielding

$$(\Delta a)^2 + \frac{1}{4}\frac{(\langle\psi|[\mathbf{A}, \mathbf{B}]|\psi\rangle)^2}{(\Delta b)^2} \geq 0. \tag{18.11}$$

This can be rearranged to yield the uncertainty relationship for non-commuting operators,

$$(\Delta a)(\Delta b) \geq \frac{i}{2}|\langle\psi|[\mathbf{A}, \mathbf{B}]|\psi\rangle| \equiv \frac{i}{2}\langle[\mathbf{A}, \mathbf{B}]\rangle = \frac{i}{2}\langle\mathbf{F}\rangle, \tag{18.12}$$

which relates the product of the dispersions to the expectation value of the commutator. It is understood that

$$\Delta a = \sqrt{\Delta a^2} = \sqrt{\langle(\mathbf{A} - a)^2\rangle} = \sqrt{\langle\mathbf{A}^2 - a^2\rangle} \tag{18.13}$$

and similarly for Δb, and this uncertainty might be better referred to as the dispersion. For the eigenstates themselves, there is no uncertainty, but for non-commuting observables, this represents the spread in measurements of each observable as, necessarily independently, obtained.

Of course this derivation fails if $|\psi\rangle$ is in an eigenstate of $\mathbf{A}$ or $\mathbf{B}$, in which case the corresponding variance is zero, meaning that the other variance is at its maximum (which is infinite for a infinite spectrum of eigenvalues). Of course, this analysis also fails if $[\mathbf{A}, \mathbf{B}] = 0$, in which case simultaneous eigenvalues are possible.

18.2.1 Heisenberg's uncertainty principle

The commutator between the position $\mathbf{x}$ and momentum $\mathbf{p}$ is $[\mathbf{p}, \mathbf{x}] = -i\hbar$ which is exact and immediately yields the most famous uncertainty principle,

$$(\Delta p)(\Delta x) \geq \frac{i}{2}\langle[\mathbf{p}, \mathbf{x}]\rangle = \frac{\hbar}{2}. \tag{18.14}$$

Even though there is an uncertainty in p, the total momentum is always conserved. The uncertainty is in the distribution of the momentum between the experimental apparatus and the particle of interest.

18.2.1.1 Time-energy uncertainty relationship

It is tempting to apply the foregoing to the "second" Heisenberg uncertainty relationship between time and energy

$$(\Delta t)(\Delta E) \geq \frac{i}{2}\langle [t, E] \rangle = \frac{\hbar}{2}. \tag{18.15}$$

Unfortunately, time is not an explicit operator in quantum mechanics (especially non-relativistic), but a parameter. In this respect, the time-energy relationship is a quasi-classical case. We mention this as an aside because most elementary quantum mechanics books present this as a derivation.

The time-energy uncertainty relationship was the subject of one of the Bohr-Einstein debates that are described in Chapter 13. Einstein proposed that both the time of emission and energy of a particle that leaves a box could be determined by weighing the box (using a spring suspension in a gravitational field), which changes when the particle leaves, and also determines the particle's energy. Bohr countered by showing that the fluctuations in the position of the box together with the red shift due to general relativity causes the time to be uncertain, and the time-energy uncertainty relationship is always satisfied although this was not the main point of Einstein's argument, (Section 13.4).

A simple mathematical origin of the time-energy relationship is the finite frequency spread required to describe a signal of finite duration. This can be seen as follows. Consider an atom in an excited state that emits energy $E = \hbar\omega$, over a finite time T. The Fourier transform of the emitted light amplitude is

$$\int_{-T/2}^{T/2} e^{-i\omega_0 t} e^{i\omega t} dt = \frac{2\sin((\omega_0 - \omega)T/2)}{\omega_0 - \omega} \tag{18.16}$$

which is the familiar sinc function. The width of this function along the frequency axis is roughly $\Delta\omega = \omega_0 - \omega = 1/T$, and it can be seen that in general $(\Delta\omega)T \sim 1$.

Again, E is conserved, and the uncertainty represents the range of energies to which a system will respond. For example, when an atom absorbs a photon it receives a momentum kick. If the absorbed photon is of lower energy than the transition frequency, when the atom re-radiates, the photon will have, on average, a higher energy than the absorbed photon. This extra energy comes from the center-of-mass motion, and results in the slowing down or cooling of the atom.

18.3 Pictures

The term *representation* has several meanings in quantum mechanics. For example, we have the operator representation of physical observables, the $|x\rangle$ and $|p\rangle$ representations of the wavefunction, matrix representation of operators, and others. These different representations assist, or even make possible, the solution to various problems and calculations. For example, in scattering theory, the $|p\rangle$ representation is useful.

In this section, we will consider different representations to describe the time evolution of the wavefunction. These different representations are often referred to as different *pictures*.

The Schrödinger picture that we know so well ascribes the time dependence to the states while the *Heisenberg picture* has the time dependence in the operators.

The two pictures represent a basis change with respect to time evolution of a state, corresponding to active and passive transformations. It is worth noting that after transforming to the Heisenberg picture, that the Hamiltonian H is not necessarily diagonal, and possibly $[H(t), H(t')] \neq 0$.

Other pictures fall between these two extremes, in particular the *interaction picture* (or *Dirac picture*) ascribes part of the time dependence to the states and part to the operators or Hamiltonian itself. The specific form of this picture depends on the situation.

The reason for introducing these different pictures is that they can simplify the solution of problems. We will use them in the derivation of spin dressing in a later chapter. First, we will elaborate the mathematics of these pictures.

18.3.1 Schrödinger picture

In the Schrödinger picture, the state of a closed quantum system evolves with time according to the Schrödinger equation as

$$i\hbar\frac{\partial}{\partial t}|\psi(t)\rangle = H|\psi(t)\rangle \tag{18.17}$$

and if $|\psi(0)\rangle$ is a superposition of eigenfunctions of a time independent Hamiltonian H, $*H|E_n\rangle = E_n|E_n\rangle*$,

$$|\psi(t)\rangle = \sum_n c_n e^{-i\omega_n t}|E_n\rangle \tag{18.18}$$

and the expectation value of a measurement (hermitian operator) is

$$\langle A\rangle = \langle\psi(t)|\mathbf{A}|\psi(t)\rangle \tag{18.19}$$

and if $[\mathbf{A}, H] \neq 0$ then $\langle A\rangle$ will in general oscillate even if $\mathbf{A}$ has no explicit time dependence. This analysis is, however, completely general and applies even if $\mathbf{A}(t)$ is explicitly time dependent because is does not enter into the Schrödigner equation, but only applied when a measurement is performed—we can change our measurement/observation operators at will and that has no bearing on the evolution of the closed quantum system.

We can incorporate time dependence in another way, by defining an operator of infinitesimal time translation dt as

$$|\psi(t+dt)\rangle = \left(1 - \frac{i}{\hbar}H dt\right)|\psi(t)\rangle. \tag{18.20}$$

(This is an application of the Stone-von Neumann theorem.) Let us take a time interval 0 to t and divide it into N short time intervals $dt = t/N$, and propagate $|\psi(0)\rangle$ in terms of repeated application of the infinitesimal time translation,

$$|\psi(t)\rangle = \left(1 - \frac{i}{\hbar}H\frac{t}{N}\right)^N|\psi(0)\rangle = e^{-iHt/\hbar}\psi(0)\rangle \tag{18.21}$$

where the last step assumes that H is time independent. If H is time dependent, with $[H(t), H(t')] = 0$ for all $t,\ t'$, then

$$|\psi(t)\rangle = e^{-i\int_0^t H(t)dt/\hbar}|\psi(0)\rangle. \tag{18.22}$$

A time evolution (or displacement) operator can be defined as

$$U(t, t_0) = e^{-i\int_{t_0}^t H(t)dt/\hbar} \tag{18.23}$$

in cases where $[H(t), H(t')] = 0$. The time evolution is then simply

$$|\psi(t)\rangle = U(t, t_0)|\psi(t_0)\rangle \tag{18.24}$$

where $U(t, t_0)$ is a unitary operator because $H = H^\dagger$ is hermitian. Specifically,

$$U^\dagger(t, t_0)U(t, t_0) = e^{i\int_{t_0}^t H(t)dt/\hbar}e^{-i\int_{t_0}^t H(t)dt/\hbar} = 1. \tag{18.25}$$

We can obviously conclude that

$$U(t', t_0) = U(t', t)U(t, t_0). \tag{18.26}$$

The Schrödinger picture is most useful when dealing with a time-independent Hamiltonian, in which case the Schrödinger equation for a particle of mass m in coordinate representation is (as shown in an earlier chapter)

$$i\hbar\frac{\partial}{\partial t}\psi(x, t) = \left[-\frac{\hbar^2}{2m}\frac{\partial^2}{\partial x^2} + V(x, t)\right]\psi(x, t), \tag{18.27}$$

and for ψ an eigenstate of H,

$$E_n\psi_n(x) = \left[-\frac{\hbar^2}{2m}\frac{\partial^2}{\partial x^2} + V(x, t)\right]\psi_n(x), \tag{18.28}$$

and the time dependence of $\psi_n(x)\rangle$ is fully contained in a factor $e^{-iE_nt/\hbar}$. This equation determines the spatial (coordinate) eigenfunctions of H as

$$H|n\rangle = E_n|n\rangle; \quad \psi_n(x) = \langle x|n\rangle. \tag{18.29}$$

18.3.2 Heisenberg picture

The Heisenberg picture is most readily developed by use of the time displacement operator. We can write (taking $U(t, 0) = U(t)$),

$$\langle A(t)\rangle = \langle\psi(0)|U^\dagger(t)\mathbf{A}U(t)|\psi(0)\rangle \tag{18.30}$$

where we have used

$$\langle\psi(t)| = \langle\psi(0)|U^\dagger(t) \quad |\psi(t)\rangle = U(t)|\psi(0)\rangle. \tag{18.31}$$

In general, $\mathbf{A}$ can be time dependent. The time derivative of $\langle A(t)\rangle$ is

$$\langle \frac{\mathrm{d}}{\mathrm{d}t} A(t) \rangle = \langle \psi(0)| \left(\frac{i}{\hbar} U^\dagger(t) H \mathbf{A} U(t) + U^\dagger(t) \left(\frac{\partial \mathbf{A}}{\partial t} \right) U(t) - \frac{i}{\hbar} U^\dagger(t) \mathbf{A}(H) U(t) \right) |\psi(0)\rangle$$
$$= \langle \psi(0)| \Bigg(\frac{i}{\hbar} \left[U^\dagger(t) H U(t) \right] \left[U^\dagger(t) \mathbf{A} U(t) \right]$$
$$+ U^\dagger(t) \left(\frac{\partial \mathbf{A}}{\partial t} \right) U(t) - \frac{i}{\hbar} \left[U^\dagger(t) \mathbf{A} U(t) \right] \left[U^\dagger(t) H U(t) \right] \Bigg) |\psi(0)\rangle. \tag{18.32}$$

Defining the Heisenberg operators

$$U^\dagger(t) \mathbf{A_s} U(t) = \mathbf{A}_h; \quad \text{and} \quad U^\dagger(t) H U(t) = H_h \tag{18.33}$$

the time derivative can be expanded as

$$\langle \frac{\mathrm{d}}{\mathrm{d}t} A(t) \rangle = \langle \psi(0)| \left(\frac{i}{\hbar} \left(H_h(t) \mathbf{A}_h(t) - \mathbf{A}_h(t) H_h(t) \right) + U^\dagger(t) \left(\frac{\partial \mathbf{A}_s}{\partial t} \right) U(t) \right) |\psi(0)\rangle. \tag{18.34}$$

This can be rewritten as

$$\frac{\mathrm{d}\mathbf{A}_h(t)}{\mathrm{d}t} = \frac{i}{\hbar} [H_h, \mathbf{A}_h] + \left(\frac{\partial \mathbf{A}_s}{\partial t} \right)_h \tag{18.35}$$

where the subscript s means the operator in the original Schrödinger picture, and subscript h the Heisenberg picture. Note that if H is time-independent or if $[H(t), H(t')] = 0$ then $H_h = H \equiv H_s$. This is often referred to as the Heisenberg equation.

18.3.2.1 Time dependence of commutators in the Heisenberg picture

In the Heisenberg picture, commutation relations can have a different form if their implicit time dependence is included. Let use consider the harmonic oscillator,

$$H = \frac{1}{2} k \mathbf{x}^2 + \frac{1}{2m} \mathbf{p}^2 \tag{18.36}$$

and the time derivatives of the position and momentum operators are

$$\begin{aligned} \frac{d}{dt} \mathbf{x} &= \frac{i}{\hbar} [H, \mathbf{x}] = \frac{\mathbf{p}}{m} \\ \frac{d}{dt} \mathbf{p} &= \frac{i}{\hbar} [H, \mathbf{p}] = -m\omega^2 \mathbf{x}. \end{aligned} \tag{18.37}$$

The mutual equations of motion for the two operators are

$$\begin{aligned} \dot{\mathbf{p}} &= -m\omega^2 \mathbf{x} \\ \dot{\mathbf{x}} &= \frac{\mathbf{p}}{m} \end{aligned} \tag{18.38}$$

which do not have *explicit* time dependence, but we see that values at t, given values at t_0, are

$$\begin{aligned}\mathbf{x}(t) &= \mathbf{x}(t_0)\cos\omega t + \mathbf{p}(t_0)\omega m \sin\omega t \\ \mathbf{p}(t) &= \mathbf{p}(t_0)\cos\omega t - m\omega^2\mathbf{x}(t_0)\sin\omega t.\end{aligned} \tag{18.39}$$

The generalized commutation relations can be directly calculated,

$$[\mathbf{x}(t_0), \mathbf{x}(t)] = \frac{i\hbar}{\omega}\sin\omega(t - t_0) \tag{18.40}$$

$$[\mathbf{p}(t_0), \mathbf{p}(t)] = i\hbar\omega\sin\omega(t - t_0) \tag{18.41}$$

$$[\mathbf{x}(t_0), \mathbf{p}(t)] = i\hbar\cos\omega(t - t_0) \tag{18.42}$$

and when $t = t_0$ the usual relationships are obtained.

18.3.2.2 Ehrenfest's theorem

Ehrenfest's theorem was first elucidated in the context of the Schrödinger equation and picture, however, it is much more simply obtained in the Heisenberg picture. We will recast its derivation, already presented in Section 12.4, in the Heisenberg picture to illustrate the advantages of choosing one picture over another.

This theorem provides relationships between the time derivatives of the expectation values of the position and momentum operators $\mathbf{x}$ and $\mathbf{p}$ to the expectation value of the force $F = -V'(x)$ on a particle of mass m moving in a scalar potential $V(x)$,

$$\frac{d}{dt}\langle\mathbf{p}\rangle = -\frac{\partial}{\partial x}\langle V(\mathbf{x}, t)\rangle, \quad m\frac{d}{dt}\langle\mathbf{x}\rangle = \langle\mathbf{p}\rangle. \tag{18.43}$$

The second relation can be derived directly from the Heisenberg picture, with

$$H = \frac{\mathbf{p}^2}{2m} + V(\mathbf{x}, t) \tag{18.44}$$

and

$$\frac{d\mathbf{x}_h}{dt} = \frac{i}{\hbar}[H_h, \mathbf{x}_h] + \left(\frac{\partial\mathbf{x}_s}{\partial t}\right)_h. \tag{18.45}$$

Working in the Heisenberg picture (drop h subscript), because $\mathbf{x}$ commutes with $V(x)$ and x is time-independent,

$$\begin{aligned}\frac{d}{dt}\langle\mathbf{x}\rangle &= \frac{i}{\hbar}\left\langle\left[\frac{\mathbf{p}^2}{2m}, \mathbf{x}\right]\right\rangle \\ &= \frac{1}{i\hbar 2m}\langle i\hbar 2\mathbf{p}\rangle \\ &= \frac{1}{m}\langle\mathbf{p}\rangle.\end{aligned} \tag{18.46}$$

Similarly, the first relationship can be verified by noting that $\mathbf{p}$ commutes with $\mathbf{p}^2$ and $\mathbf{p}$ does not depend explicitly on time, leaving the following terms in the Heisenberg equation:

$$\frac{d}{dt}\langle \mathbf{p} \rangle = \frac{i}{\hbar} \langle [\mathbf{p}, V(\mathbf{x}, t)] \rangle . \tag{18.47}$$

The expectation value of the commutator can be calculated as

$$\begin{aligned}
\langle \psi(t) | \frac{i}{\hbar} [\mathbf{p}, V(\mathbf{x}, t)] | \psi(t) \rangle &= \int dx\, \psi^*(x,t) \left(V(x,t) \frac{d}{dx} \psi(x,t) - \frac{d}{dx} \left(V(x,t) \psi(x,t) \right) \right) \\
&= - \int dx\ \psi^*(x,t) \left(\frac{d}{dx} V(x,t) \right) \psi(x,t) \\
&= - \left\langle \frac{d}{dx} V(x,t) \right\rangle = \langle F(x,t) \rangle .
\end{aligned}$$

Therefore, we arrive at the result,

$$\frac{d}{dt}\langle p \rangle = \langle F(x,t) \rangle . \tag{18.48}$$

It is in the context of the Heisenberg equation that the quantum equations of motion are analogous to the classical equations of motion. According to Dirac's prescription, the commutator is equivalent to the Poisson bracket multiplied by $i\hbar$.

18.3.3 Interaction picture

The interaction picture is very useful when the Hamiltonian has a component that is easy to solve H_0, and one that is harder to analyze and that is perhaps a perturbation, H_1. We explicitly indicate that H is in the Schrödinger picture, as indicated by the subscripts s,

$$H_s = H_{0s} + H_{1s} . \tag{18.49}$$

Either or both components can be time dependent, and it is easiest to consider the cases where both $[H_{0s}(t), H_{0s}(t')] = 0$ and $[H_{1s}(t), H_{1s}(t')] = 0$. Let us assume that H_{0s} has solutions that are simple to obtain, so we can try to eliminate it from the problem which will allow H_{1s} to be directly studied. By defining a unitary operator,

$$U_I(t) = e^{i \int^t dt H_{0s}(t)/\hbar} \tag{18.50}$$

the states of the time-dependent Schrödinger equation can be transformed to, with initial conditions $|\psi_I(0)\rangle = |\psi_s(0)\rangle$

$$|\psi_I(t)\rangle = U_I(t)|\psi_s(t)\rangle = U_I(t) e^{-i \int^t dt (H_{0s}(t) + H_{1s}(t))/\hbar} |\psi_s(0)\rangle = e^{-i \int^t dt H_{1s}(t)/\hbar} |\psi_s(0)\rangle \tag{18.51}$$

and $|\psi_I(t)\rangle = |\psi_s(0)\rangle$ is constant in the absence of the interaction H_{1s}.

This means the transformed set of states evolve only under the action of H_{1s} which also needs to be transformed as

$$H_{1I}(t) = U_I(t) H_{1s} U_I^\dagger(t) \tag{18.52}$$

and therefore H_{1I} becomes time dependent, or receives an additional time dependence.

The Hamiltonian H_{0s} is not modified by the transformation,

$$H_{0I} = U_I(t)H_{0s}U_I^\dagger(t) = U_I(t)U_I^\dagger(t)H_{0s} = H_{0s} \tag{18.53}$$

because $U_I(t)$ is a function of H_{0s} and therefore the two operators commute.

The time-dependent Schrödinger equation in the interaction picture can be elucidated by replacing $|\psi_s(t)\rangle \to U_I^\dagger(t)|\psi_I(t)\rangle$, in which case

$$i\hbar\frac{d}{dt}\left(U_I^\dagger(t)|\psi_I(t)\rangle\right) = (H_{0s} + H_{1s})\left(U_I^\dagger(t)|\psi_I(t)\rangle\right). \tag{18.54}$$

Expanding the left-hand side,

$$i\hbar\frac{d}{dt}\left(U_I^\dagger(t)|\psi_I(t)\rangle\right) = H_{0s}U_I^\dagger|\psi_I(t)\rangle + i\hbar U_I^\dagger(t)\frac{d}{dt}|\psi_I(t)\rangle, \tag{18.55}$$

and multiplying this and the right-hand side by $U_I(t)$ yields

$$U_I(t)H_{0s}U_I^\dagger|\psi_I(t)\rangle + i\hbar U_I(t)U_I^\dagger(t)\frac{d}{dt}|\psi_I(t)\rangle = U_I(t)(H_{0s} + H_{1s})U_I^\dagger(t)|\psi_I(t)\rangle \tag{18.56}$$

and the H_{0s} terms cancel. Using $U_I(t)U_I^\dagger(t) = 1$ results in

$$i\hbar\frac{d}{dt}|\psi_I(t)\rangle = H_{1I}(t)|\psi_I(t)\rangle \tag{18.57}$$

as expected. This result shows that we can recast the problem in terms of the basis states of H_{1I}.

Let us take the Hamiltonian without H_{1s} as H_{0s}. The time translation operator is then

$$U(t) = e^{-i\int^t dt H_{0s}(t)\hbar} = U_I^\dagger(t). \tag{18.58}$$

With $U(t)$, other operators (observables) are transformed exactly as in the case of the Heisenberg picture,

$$\mathbf{A}(t)_h = \mathbf{A}_I(t) = U(t)^\dagger \mathbf{A}_s U(t) \tag{18.59}$$

where $\mathbf{A}_s$ is usually constant, however, we can allow for an explicit the time dependence as $\mathbf{A}_s(t)$ as before. The essential motivation of the interaction picture is that the effect of H_{0s} is shunted onto operators with H_{1I} controlling the time evolution of the state vector that would be otherwise constant; when $H_{1I} = H_{1s} = 0$, the interaction picture coincides with the Heisenberg picture.

As an example, when the effects of an oscillating field on a spin-1/2 system are studied, transforming into a rotating coordinate system can be considered as transformation to an interaction picture. In this case, we have a magnetic field rotating at frequency ω in the $x-y$ plane, with Hamiltonian $H_{0s}(t) = \hbar\gamma B_1(\sigma_x \cos\omega t + \sigma_y \sin\omega t)$, that was transformed to a static field. Although $[H_{0s}(t), H_{0s}(t')] \neq 0$, the unitary transformation operator $U(t)$ is a rotation operator around $\hat{z}$, and the static $\hat{z}$ magnetic field in the rotating frame becomes $B_z' = B_z - \gamma/\omega$. The problem is thereby reduced to the spin evolution in the static fields $B_z'\hat{z}$ and $B_1\hat{x}$.

The density matrix can also be transformed to the interaction picture, as, in the absence of H_{1h} as

$$\begin{aligned}\rho_I(t) &= \sum_{mn} c_m(t)c_n^*(t)|\psi_{m,I}(t)\rangle\langle\psi_{n,I}(t)| \\ &= \sum_{mn} c_m(t)c_n^*(t)U(t)|\psi_{m,s}(t)\rangle\langle\psi_{n,s}(t)|U^\dagger(t) \\ &= U(t)\rho_s(t)U^\dagger(t) = \rho_s(0)\end{aligned} \tag{18.60}$$

where the subscripts n, I and n, s indicate that state n has been transformed to the I or s pictures, respectively. The evolution of $\rho_I(t)$ when H_{1s} is applied will then be

$$\frac{\partial}{\partial t}\left(|\psi_I(t)\rangle\langle\psi_I(t)|\right) = \left(\frac{\partial}{\partial t}|\psi_I(t)\rangle\right)\langle\psi_I(t)| + |\psi_I(t)\rangle\left(\frac{\partial}{\partial t}\langle\psi_I(t)|\right) \tag{18.61}$$

$$= \frac{H_{1I}}{i\hbar}\left(|\psi_I(t)\rangle\langle\psi_I(t)|\right) + \left(|\psi_I(t)\rangle\langle\psi_I(t)|\right)\frac{H_{1I}}{-i\hbar} \tag{18.62}$$

$$= \frac{1}{i\hbar}[H_{1I}, \rho_I(t)] \tag{18.63}$$

with observables given by

$$\langle \mathbf{A}_I(t)\rangle = \mathbf{Tr}\big(\rho_I(t)\mathbf{A}_I(t)\big). \tag{18.64}$$

In cases where $[H_{0s}(t), H_{0s}(t')] \neq 0$ the interaction picture is still useful, however, the unitary operator $U(t)$ becomes a time-ordered operator that can be expressed as the Dyson series.

18.3.4 Comparison of pictures

The picture (or representation) to address a problem is chosen based on the one that provides the greatest simplification. In some situations, e.g., the derivation of spin dressing, the problem is vastly simplified (if it is not otherwise impossible), by use of the interaction picture.

For determining the coordinate wavefunctions and eigenvalues of a static system, for example, the H atom, the use of the coordinate representation of the Schrödinger equation is usually the most practical approach.

The Heisenberg picture is useful in that it can convert a quantum system into a classical analog, an example of which is the use of a precessing classical magnetic moment used to describe the evolution of the observables of a spin-1/2 system. It is also used in the operator method to solve the harmonic oscillator.

Perhaps the most interesting application of the Heisenberg picture is that it suggests a powerful and very general means to implement transformations on operators. For example, if we have an operator that depends on the polarization state of an optical field (light), we can rotate (transform) that operator to another coordinate system by simply rotating the light polarization; the fundamental structure of an operator does not depend on a specific coordinate system. Usually it is more difficult to transform an entire operator than it is to transform the operator's internal parameters. We will

Table 18.1 A comparison of pictures (or representations).

	Picture, with $U(t) = e^{-i\int^t H_{0s}(t)dt/\hbar}$		
	Heisenberg	Interaction	Schrödinger
Hamiltonian	H_{0s}	$H_{0s} + H_{1s}$ $H_{1I} = U(t)H_{1s}U^\dagger(t)$	H_{0s}
State (ket)	$\|\psi_s(0)\rangle$	$i\hbar\frac{\partial}{\partial t}\|\psi_I(t)\rangle = H_{1I}\|\psi_I(t)\rangle$ $\|\psi_I(0)\rangle = \|\psi_s(0)\rangle$	$\|\psi_s(t)\rangle = U(t)\|\psi_s(0)\rangle$
Observable	$U(t)^\dagger \mathbf{A}_s U(t)$	$U(t)^\dagger \mathbf{A}_s U(t)$	$\mathbf{A}_s$ (constant)
Dens. Matrix	$\rho_h(t) = \rho_s(0)$	$i\hbar\frac{\partial}{\partial t}\rho_I(t) = [H_{1I}, \rho_I(t)]$	$\rho_s(t) = U(t)\rho_s(0)U^\dagger(t)$

see an example of this later in this chapter when the effects of a beam splitter on a coherent state are studied.

Transformations in the Schrödinger picture are usually referred to as *active* because they are applied directly to the wavefunction. In the Heisenberg picture, they are *passive* as they are applied to the operators. In the interaction picture, transformations can be either active or passive, depending on whether they are applied to the wavefunction or to the operators under study.

The purpose of the interaction picture is to move the oscillating phases of the otherwise constant state vector due to H_{0s} onto operators, leaving H_{1I} to alone determine the time evolution of the state vector. This picture is most convenient when considering the effect of a small interaction term H_{1s} that is added to the Hamiltonian of a solved system H_{0s}, and in particular if $U(t)$ can be easily calculated (for example, if $[H_{0s}(t), H_{0s}(t')] = 0$, and a further simplification results if $[H_{1s}(t), H_{1s}(t')] = 0$). The interaction picture simplifies time-dependent perturbation theory for H_{1I}, an example of which is its use in the derivation of Fermi's golden rule.

Table 18.1 shows a direct comparison of the three pictures considered here. It should be noted that the interaction pictures can come in various forms and can have different names. For example, the transformation to a rotating frame to make a rotating field appear as static can be considered a transformation to the interaction picture.

18.4 Ladder operators

Earlier we have introduced the harmonic oscillator ladder operators which also form the creation and annihilation operators in the second quantization and field theory approaches. The notion of a ladder operator is, however, generic and can be used in other situations. Consider two operators $\mathbf{F}$ and $\mathbf{G}$ that satisfy a commutations relationship

$$[\mathbf{F}, \mathbf{G}] = c\mathbf{G} \tag{18.65}$$

where c is a number. Let us consider the eigenfunctions $\mathbf{F}|f_1\rangle = f_i|f_i\rangle$ and assume they form a complete set. Let $\mathbf{G}$ operate on a state $|f_i\rangle$, and determine the states that are formed in the $\mathbf{F}$ basis:

$$\begin{aligned}\mathbf{FG}|f_i\rangle &== (\mathbf{GF} + \mathbf{FG} - \mathbf{GF})|f_i\rangle \\ &= (\mathbf{GF} + [\mathbf{F}, \mathbf{G}]|f_i\rangle \\ &= (\mathbf{GF} + c\mathbf{G})|f_i\rangle \\ &= (f_i + c)\mathbf{G}|f_i\rangle\end{aligned} \tag{18.66}$$

and therefore $\mathbf{G}|f_i\rangle$ is an eigenfunction of $\mathbf{F}$ with eigenvalue $c + f_i$.

If $\mathbf{F} = F^\dagger$ and therefore a hermitian operator, then the eigenvalues of $\mathbf{F}$ are real so c must then also be real.

We assumed that $|f_i\rangle$ are a complete set with eigenvalues $\{f_i\}$, so the eigenvalues of $\mathbf{G}$ as a set are $\{f_i + c\} \subseteq f_i$. The effect of $\mathbf{G}$ is to change the eigenvalue by c. If $c > 0$ then $\mathbf{G}$ is a raising operator (increases f_i). If $c < 0$ it is a lowering operator.

With this definition, for $c > 0$, $\mathbf{G}^\dagger$ is a lowering operator, as can be seen by taking the adjoint of both sides of Eq. (18.65)

$$\begin{aligned}c\mathbf{G}^\dagger &= [\mathbf{F}, \mathbf{G}]^\dagger \\ &= [\mathbf{G}^\dagger, \mathbf{F}^\dagger] = [\mathbf{G}^\dagger, \mathbf{F}] = -[\mathbf{F}, \mathbf{G}^\dagger] \\ \therefore [\mathbf{F}, \mathbf{G}^\dagger] &= -c\mathbf{G}^\dagger.\end{aligned} \tag{18.67}$$

If $c > 0$, then $\mathbf{G}^\dagger$ is a lowering operator, and a raising operator for $c < 0$.

18.5 Harmonic oscillator

We have already considered the harmonic oscillator in some detail; we revisit it here in the context of algebraic methods. This approach was originally developed by Dirac.

Taking the standard form of the system, let the one-dimensional harmonic oscillator comprise a mass m bound by a spring with force constant k so $\omega = \sqrt{k/m}$. It is convenient to define $k = m\omega^2$ so the only parameters are mass and frequency. Define two operators

$$\begin{aligned}\mathbf{a} &= \sqrt{\frac{m\omega}{2\hbar}}\left(\mathbf{x} + \frac{i}{m\omega}\mathbf{p}\right) \\ &= \frac{1}{\sqrt{2}}(\mathbf{X} + i\mathbf{P}) \\ \mathbf{a}^\dagger &= \sqrt{\frac{m\omega}{2\hbar}}\left(\mathbf{x} - \frac{i}{m\omega}\mathbf{p}\right) \\ &= \frac{1}{\sqrt{2}}(\mathbf{X} - i\mathbf{P})\end{aligned} \tag{18.68}$$

where we have introduced two non-dimensional operators $\mathbf{X}$ and $\mathbf{P}$ that will be useful later. These operators, up to a multiplicative constant, can be thought of as the square root and its complex conjugate of the classical Hamiltonian

$$H = \frac{\mathbf{p}^2}{2m} + \frac{m\omega^2}{2}\mathbf{x}^2 = \frac{m\omega^2}{2}\left[\frac{\mathbf{p}^2}{(m\omega)^2} + \mathbf{x}^2\right] \tag{18.69}$$

$$= \frac{m\omega^2}{2}\left[\left(x + i\frac{\mathbf{p}}{m\omega}\right)\left(x - i\frac{\mathbf{p}}{m\omega}\right) - \frac{i}{m\omega}[\mathbf{p}, \mathbf{x}]\right]. \tag{18.70}$$

Classically, the last term on the right-hand side is zero, but the commutator $[\mathbf{x}, \mathbf{p}] = i\hbar$ in quantum mechanics. Therefore, replacing $\mathbf{x} \pm i\mathbf{p}$ with the appropriately scaled $\mathbf{a}, \mathbf{a}^\dagger$,

$$H = \hbar\omega\left(\mathbf{a}\mathbf{a}^\dagger - \frac{1}{2}\right) = \hbar\omega\left(\mathbf{a}^\dagger\mathbf{a} + \frac{1}{2}\right) \tag{18.71}$$

where the two forms result from taking $\pm$ vs. $\mp$ in the preceding equation. They are equal in their effect on a state.

Let us define $\mathbf{N} = \mathbf{a}^\dagger\mathbf{a}$, which means the Hamiltonian can be written in the form

$$H = \hbar\omega\left(\mathbf{N} + \frac{1}{2}\right) \tag{18.72}$$

and the energy eigenvalues of an energy state $|n\rangle$ are $E_n = \hbar\omega(n + 1/2)$. We show later that N has non-negative integer eigenvalues.

A few commutation relationships will be useful. First, noting $[\mathbf{x}, \mathbf{x}] = 0$, $[\mathbf{p}, \mathbf{p}] = 0]$,

$$[\mathbf{a}, \mathbf{a}^\dagger] = \frac{m\omega}{2\hbar}\left[\left(x + i\frac{\mathbf{p}}{m\omega}\right), \left(x - i\frac{\mathbf{p}}{m\omega}\right)\right] \tag{18.73}$$

$$= \frac{m\omega}{2\hbar}\left(\left[x, -i\frac{\mathbf{p}}{m\omega}\right] + \left[i\frac{\mathbf{p}}{m\omega}, \mathbf{x}\right]\right) \tag{18.74}$$

$$= 1. \tag{18.75}$$

Next, consider

$$\begin{aligned} &= [\mathbf{a}^\dagger\mathbf{a}, \mathbf{a}^\dagger] = \mathbf{a}^\dagger\mathbf{a}\mathbf{a}^\dagger - \mathbf{a}^\dagger\mathbf{a}^\dagger\mathbf{a} = \mathbf{a}^\dagger(\mathbf{a}\mathbf{a}^\dagger - \mathbf{a}^\dagger\mathbf{a}) \\ &= \mathbf{a}^\dagger[\mathbf{a}, \mathbf{a}^\dagger] = \mathbf{a}^\dagger \end{aligned} \tag{18.76}$$

and we can immediately see that $\mathbf{a}^\dagger$ has the form of a raising operator ($c = 1$). This means that $\mathbf{a}^\dagger|n\rangle$ are the eigenvalues of the operator $\mathbf{N}$ with eigenvalue

$$\mathbf{N}\mathbf{a}^\dagger|n\rangle = (n + 1)\mathbf{a}^\dagger|n\rangle \tag{18.77}$$

and

$$\mathbf{N}\mathbf{a}|n\rangle = (n - 1)\mathbf{a}|n\rangle. \tag{18.78}$$

Consider the eigenvalue

$$n = \langle n|\mathbf{N}|n\rangle = \langle n|\mathbf{a}^\dagger\mathbf{a}|n\rangle = (\mathbf{a}|n\rangle)^\dagger(\mathbf{a}|n\rangle) \geq 0 \tag{18.79}$$

because the norm has to be ≥ 0. Therefore, the smallest eigenvalue that is possible is $n = 0$.

The unity normalized states that result from the raising or lowering operators can be easily calculated as follows. Consider first the lowering operator,

$$\langle n-1|n-1\rangle \propto \langle n|\mathbf{a}^{\dagger}\mathbf{a}|n\rangle = \langle n|\mathbf{N}|n\rangle = n\langle n|n\rangle \tag{18.80}$$

and we see that $\mathbf{a}|0\rangle = 0$, which will be useful later. Assuming $\langle n|n\rangle = 1$, the normalized states generated by the lowering operator are

$$|n-1\rangle = \frac{\mathbf{a}}{\sqrt{n}}|n\rangle. \tag{18.81}$$

Similarly, for the raising operator,

$$\langle n+1|n+1\rangle \propto \langle n|\mathbf{a}\mathbf{a}^{\dagger}|n\rangle = \langle n|1+\mathbf{a}^{\dagger}\mathbf{a}|n\rangle = (1+n)\langle n|n\rangle \tag{18.82}$$

and determines the normalized states due to the raising operator as

$$|n+1\rangle = \frac{\mathbf{a}^{\dagger}}{\sqrt{n+1}}|n\rangle. \tag{18.83}$$

As a final check of our formalism, let us write

$$\mathbf{x} = \sqrt{\frac{\hbar}{2m\omega}}(\mathbf{a}^{\dagger}+\mathbf{a}) \tag{18.84}$$

$$\mathbf{p} = i\sqrt{\frac{\hbar m\omega}{2}}(\mathbf{a}^{\dagger}-\mathbf{a}) \tag{18.85}$$

and substituting these into the harmonic oscillator Hamiltonian

$$H = \left(\frac{m\omega^2}{2}\right)\left(\frac{\hbar}{2m\omega}\right)(\mathbf{a}^{\dagger}+\mathbf{a})^2 - \left(\frac{1}{2m}\right)\left(\frac{\hbar m\omega}{2}\right)(\mathbf{a}^{\dagger}-\mathbf{a}) = \frac{\hbar\omega}{2}(\mathbf{a}\mathbf{a}^{\dagger}+\mathbf{a}^{\dagger}\mathbf{a}). \tag{18.86}$$

Using $[\mathbf{a},\mathbf{a}^{\dagger}] = 1$ to get $\mathbf{a}\mathbf{a}^{\dagger} = 1+\mathbf{a}^{\dagger}\mathbf{a}$, and $\mathbf{a}^{\dagger}\mathbf{a} = \mathbf{N}$,

$$H = \hbar\omega\left(\mathbf{N}+\frac{1}{2}\right) \tag{18.87}$$

so $E_n = (n+1/2)\hbar\omega$ as expected.

18.5.1 Generation of the harmonic oscillator wavefunctions

Noting that

$$|n\rangle = \frac{\mathbf{a}^{\dagger}}{\sqrt{n}}|n-1\rangle = \frac{\mathbf{a}^{\dagger}}{\sqrt{n}}\frac{\mathbf{a}^{\dagger}}{\sqrt{n-1}}|n-2\rangle = ... = \frac{(\mathbf{a}^{\dagger})^n}{\sqrt{n!}}|0\rangle \tag{18.88}$$

indicates that all the eigenstates can be constructed from the ground state $|0\rangle$.

The ground state wavefunction can be constructed by noting

$$\mathbf{a}|0\rangle = 0 \tag{18.89}$$

and projecting this into the coordinate representation,

$$\langle x|a|x\rangle\langle x|0\rangle = \sqrt{\frac{m\omega}{2\hbar}}\left(x + \frac{\hbar}{m\omega}\frac{d}{dx}\right)\psi_0(x) \tag{18.90}$$

where $\langle \mathbf{x}|\mathbf{p}|\mathbf{x}\rangle = -i\hbar d/dx$ in the coordinate representation of the momentum operator. The wavefuntion is determined by the differential equation

$$\left(x + \frac{\hbar}{\omega m}\frac{d}{dx}\right)\psi_0(x) = 0 \tag{18.91}$$

which has solution

$$\psi_0(x) = Ae^{-m\omega x^2/2\hbar} \tag{18.92}$$

$$A = \left(\frac{m\omega}{\pi\hbar}\right)^{1/4}. \tag{18.93}$$

If we introduce

$$\beta = \sqrt{\frac{\hbar}{m\omega}} \tag{18.94}$$

then

$$\psi_0(x) = \left(\frac{1}{\beta\sqrt{\pi}}\right)^{1/2} e^{-x^2/2\beta^2}, \tag{18.95}$$

Higher n states can be constructed by use of the raising operator together with its normalization. The first state is given by

$$\langle \mathbf{x}|a^\dagger|\mathbf{x}\rangle\langle \mathbf{x}|0\rangle = \psi_1(x) \tag{18.96}$$

where

$$\langle \mathbf{x}|a^\dagger|\mathbf{x}\rangle = \frac{1}{\sqrt{2}\,\beta}\left(x - \beta^2\frac{d}{dx}\right) \tag{18.97}$$

We can directly write

$$\psi_1(x) = \frac{1}{\sqrt{2}\,\beta}\left(x - \beta^2\frac{d}{dx}\right)\left(\frac{1}{\beta\sqrt{\pi}}\right)^{1/2} e^{-x^2/2\beta^2} = \left(\frac{1}{2\beta\sqrt{\pi}}\right)^{1/2} 2\,\frac{x}{\beta}\,e^{-x^2/2\beta^2}. \tag{18.98}$$

This can be repeated ad infinitum to generate any $\psi_n(x)$, and attention must be paid to the $1/\sqrt{n+1}$ normalization required for states generated by $\mathbf{a}^\dagger$.

18.6 Coherent states

18.6.1 Schrödinger and coherent states

As we have seen in Section 10.1 Schrödinger studied the motion of a wave packet, a superposition of one-dimensional harmonic oscillator eigenstates, corresponding to a narrow band of large quantum numbers, to demonstrate his idea that the wave function, $\psi(x.t)$ could replace particles, i.e., that the dynamics of wave packets would reproduce all the phenomenon of particle dynamics. His idea appeared in a short paper in Die Naturwissenschaften.[2]

We can write the normalized eigenfunctions of the linear harmonic oscillator (coordinate x) as (Section 7.3.1)

$$\psi_n(x.t) = \sqrt{\frac{\beta}{2^n n! \pi^{1/2}}} H_n(z) e^{-z^2/2} e^{-i\omega_n t} \tag{18.99}$$

where $z = \beta x$, $\beta = \sqrt{m\omega_o/\hbar}$, ω_o is the frequency of the oscillator, $\omega_n = (n + 1/2)\omega_o$ and $H_n(z)$ are the Hermite polynomials.

Schrödinger wanted to show that a wave packet constructed to be narrow in space would undergo oscillations with the frequency ω_o and remain narrow, thus being capable of representing a particle.

As is the case for many sets of orthogonal polynomials, the Hermite polynomials can be obtained from a generating function

$$g(s, z) = \sum_{n=0} \frac{s^n}{n!} H_n(z) \tag{18.100}$$

by evaluating $d^n g/ds^n|_{s=0} = H_n(z)$. $g(s, z)$ can be shown to be given by

$$g(s, z) = e^{-s^2+2sz} = e^{z^2-(s-z)^2} \tag{18.101}$$

Schrödinger chose a wave packet that could be written in closed form by using (18.100) as a sum rule, that is he took a wave packet

$$\psi(x.t) = \sum_n \frac{\alpha^n}{\sqrt{2^n n!}} \psi_n(x.t) \tag{18.102}$$

$$= \sqrt{\frac{\beta}{\pi^{1/2}}} e^{-i\omega_o t/2} e^{-\frac{z^2}{2}} \left\{ \sum_n \left(\frac{\alpha e^{-i\omega_o t}}{2} \right)^n \frac{1}{n!} H_n(z) \right\} \tag{18.103}$$

where α is a number much greater than 1. For large α the coefficient $(\alpha/2)^n/n!$ has a very narrow peak near $n = \alpha/2$. Applying (18.100, 18.101) to the term in curly brackets we obtain

[2]Schrödinger, E., *Der stetige Übergang von der Mikro- zur Makromechanik* [The continuous transition from micro- to macro-mechanics], Naturwiss. **14**, 644, (1926) published 9 July, 1926.

$$\psi(x.t) = \sqrt{\frac{\beta}{\pi^{1/2}}} e^{-i\omega_o t/2} e^{-\frac{z^2}{2}} \left\{ \exp\left[-\left(\frac{\alpha e^{-i\omega_o t}}{2} \right)^2 + \alpha e^{-i\omega_o t} z \right] \right\} \tag{18.104}$$

$$= \sqrt{\frac{\beta}{\pi^{1/2}}} e^{-i\omega_o t/2} e^{-\frac{z^2}{2}} \left\{ \exp\left[-\frac{\alpha^2 e^{-i2\omega_o t}}{4} + \alpha e^{-i\omega_o t} z \right] \right\} \tag{18.105}$$

so that

$$|\psi(x.t)|^2 = \frac{\beta}{\pi^{1/2}} \exp\left[-z^2 - \frac{\alpha^2}{2} \cos 2\omega_o t + 2\alpha z \cos \omega_o t \right] \tag{18.106}$$

$$= \frac{\beta}{\pi^{1/2}} \exp\left[-z^2 - \frac{\alpha^2}{2} \left(2\cos^2 \omega_o t - 1 \right) + 2\alpha z \cos \omega_o t \right] \tag{18.107}$$

$$= \frac{\beta e^{\alpha^2/2}}{\pi^{1/2}} \exp\left[-\left(z - \alpha \cos \omega_o t \right)^2 \right] \tag{18.108}$$

The result is seen to be a very large amplitude and very narrow Gaussian function whose peak is oscillating with a frequency ω_o and an amplitude

$$x_o = \alpha \sqrt{\frac{\hbar}{m\omega_o}} \tag{18.109}$$

simulating a classical particle. Schrödinger asserted that the lack of any expansion of the packet was due to the discrete nature of the energy levels so that similar behavior would be observed with a wave packet of Rydberg wave functions for the hydrogen atom thus representing an electron in a classical orbit. This is wrong. The reason the wave packet for the harmonic oscillator does not expand is that the energy levels are equally spaced so that any superposition would be a periodic function. At the time of writing this paper Schrödinger was engaged in writing his series of papers on wave mechanics and showing the equivalence with Heisenberg's quantum mechanics, all of which appeared between January and June of 1926.

We can write the wave function chosen by Schrödinger in the Dirac notation as

$$|\psi\rangle = \sum_n \frac{\alpha^n}{\sqrt{2^n n!}} |n\rangle \tag{18.110}$$

which when normalized and applied to the energy eigenstates of the electromagnetic field are just the coherent states (see Eq. 18.135).

18.6.2 Coherent states: operator techniques

It is difficult—in fact, impossible—to prepare monochromatic states of an electromagnetic wave with a specific excitation (number of photons) n. This is due to the fact that a state with a specific $|n\rangle$ has neither energy nor momentum uncertainty and therefore must be of infinite extent in space and time. The generation of a single photon-on-demand is an outstanding open problem in quantum information. What *can* be produced are minimum uncertainty, or *coherent* states, as discussed in the last section and originally proposed by Schrödinger.

The harmonic oscillator $|n\rangle$ states are called Fock states, spread over the distance range between the classical turning points in the harmonic oscillator potential and extend into the classically forbidden region. We have seen that it is possible to construct a well-defined wave packet that bounces back and forth between the turning points in the h.o. potential, and thereby mimic a vibrating classical particle.

Such states can be created in many ways, in addition to Schrödinger's approach. We can take the ground state $|0\rangle$ as a minimum uncertainty state, and displace it some spatial distance x_0 in the potential by shifting the position of the potential minimum by "suddenly" applying a constant force, and expanding the displaced wavefunction (which initially remains unchanged except to replace $x \to x + x_0$ and is an application of the sudden approximation) in terms of the new eigenfunctions.

Alternatively, we can follow Schrödinger, but instead use operators and attempt to find a minimum uncertainty superposition of states. Such states, labeled as $|\alpha\rangle$, will satisfy the minimum uncertainty relationship dictated by the commutator between $\mathbf{X}$ and $\mathbf{P}$, the non-dimensional operators introduced earlier,

$$\mathbf{P} = \sqrt{\frac{1}{m\hbar\omega}}\mathbf{p}; \quad \mathbf{X} = \sqrt{\frac{m\omega}{\hbar}}\mathbf{x}. \tag{18.111}$$

The non-dimensional operators satisfy

$$[\mathbf{X}, \mathbf{P}] = \frac{1}{\hbar}[\mathbf{x}, \mathbf{p}] = i \tag{18.112}$$

and can be expressed in terms of the raising and lowering operators:

$$\mathbf{X} = \frac{1}{\sqrt{2}}(\mathbf{a} + \mathbf{a}^\dagger) \tag{18.113}$$

$$\mathbf{P} = -\frac{i}{\sqrt{2}}(\mathbf{a} - \mathbf{a}^\dagger) \tag{18.114}$$

Because the ground state $|0\rangle$ is a Gaussian, we know that it is an minimum uncertainty state, which can be demonstrated as follows. Using the formalism leading to Eq. (18.13), the dispersion of the ground state can be evaluated (noting that $\langle X\rangle = \langle P\rangle = 0$ for a potential centered at $x = 0$):

$$\langle X^2\rangle = \langle 0|\mathbf{X^2}|0\rangle = \frac{1}{2}\langle 0|(\mathbf{a} + \mathbf{a}^\dagger)^2|0\rangle = \frac{1}{2}\langle 0|\mathbf{a}\mathbf{a}^\dagger|0\rangle = \frac{1}{2}\langle 0|2\mathbf{N} + 1|0\rangle = \frac{1}{2} \tag{18.115}$$

where we use the relation $\mathbf{a}\mathbf{a}^\dagger + \mathbf{a}^\dagger\mathbf{a} = 2\mathbf{N} + 1$; where $\mathbf{N}|n\rangle = n|n\rangle$ as defined before, and that the matrix elements of $\mathbf{a}^2$, $(\mathbf{a}^\dagger)^2$ are zero for a given $|n\rangle$ state. Similarly,

$$\langle P^2\rangle = \langle 0|\mathbf{P^2}|0\rangle = -\frac{1}{2}\langle 0|(\mathbf{a} - \mathbf{a}^\dagger)^2|0\rangle = \frac{1}{2}\langle 0|\mathbf{a}\mathbf{a}^\dagger|0\rangle = \frac{1}{2}\langle 0|2\mathbf{N} + 1|0\rangle = \frac{1}{2}. \tag{18.116}$$

Therefore, for the $|0\rangle$ state, we have

$$\langle(\Delta X)^2\rangle_0\langle(\Delta P)^2\rangle_0 = (\langle X^2\rangle_0 - \langle X\rangle_0^2)(\langle P^2\rangle_0 - \langle P\rangle_0^2) = \frac{1}{4} \tag{18.117}$$

which is the minimum physically allowed.

Let us next calculate the dispersion for any state $|n\rangle$. The previous calculation can be generalized by noting

$$\langle n|(\mathbf{a}+\mathbf{a}^\dagger)^2|n\rangle = 2n+1 \tag{18.118}$$

so that

$$\langle(\Delta X)^2\rangle_n\langle(\Delta P)^2\rangle_n = (\langle X^2\rangle_n - \langle X\rangle_n^2)(\langle P^2\rangle_n - \langle P\rangle_n^2) = \frac{(2n+1)^2}{4} \tag{18.119}$$

and is not a minimum uncertainty state.

An important fact regarding the state $|0\rangle$ is that it appears to be an eigenstate (of a non-hermitian operator),

$$\langle 0|\mathbf{a}^\dagger\mathbf{a}|0\rangle = 0 \to \mathbf{a}|0\rangle = 0|0\rangle \tag{18.120}$$

which suggests that we try to find other eigenstates of $\mathbf{a}$ such that

$$\mathbf{a}|\alpha\rangle = \alpha|\alpha\rangle \quad \text{and} \quad \langle\alpha|\mathbf{a}^\dagger\mathbf{a}|\alpha\rangle = |\alpha|^2. \tag{18.121}$$

Let us check the dispersion of the state $|\alpha\rangle$ by calculating the expectation values,

$$\langle X\rangle = \frac{1}{\sqrt{2}}\langle\alpha|(\mathbf{a}+\mathbf{a}^\dagger)|\alpha\rangle = \frac{1}{\sqrt{2}}(\alpha+\alpha^*) \tag{18.122}$$

$$\langle P\rangle = \frac{i}{\sqrt{2}}\langle\alpha|(\mathbf{a}-\mathbf{a}^\dagger)|\alpha\rangle = \frac{i}{\sqrt{2}}(\alpha-\alpha^*) \tag{18.123}$$

$$\langle X^2\rangle = \frac{1}{2}\langle\alpha|(\mathbf{a}+\mathbf{a}^\dagger)^2|\alpha\rangle = \frac{1}{2}(\alpha+\alpha^*)^2 + \frac{1}{2} \tag{18.124}$$

$$\langle P^2\rangle = -\frac{1}{2}\langle\alpha|(\mathbf{a}-\mathbf{a}^\dagger)^2|\alpha\rangle = -\frac{1}{2}(\alpha-\alpha^*)^2 + \frac{1}{2} \tag{18.125}$$

using $\mathbf{a}\mathbf{a}^\dagger = 1 + \mathbf{a}^\dagger\mathbf{a}$. We find:

$$\langle X^2\rangle_\alpha - \langle X\rangle_\alpha^2 = \frac{1}{2} \quad \langle P^2\rangle_\alpha - \langle P\rangle_\alpha^2 = \frac{1}{2}. \tag{18.126}$$

and therefore

$$\langle(\Delta X)^2\rangle_\alpha\langle(\Delta P)^2\rangle_\alpha = \frac{1}{4}. \tag{18.127}$$

Thus we have shown that $|\alpha\rangle$ is a minimum uncertainty state.

The state $|\alpha\rangle$ can be expanded in Fock states as

$$|\alpha\rangle = \sum_{n=0}^{\infty} c_n|n\rangle \tag{18.128}$$

with eigenvalue

$$\mathbf{a}|\alpha\rangle = \alpha|\alpha\rangle = \sum_{n=0}^{\infty} c_n\sqrt{n}|n-1\rangle = \sum_{n=0}^{\infty} \alpha c_n|n\rangle. \tag{18.129}$$

Therefore,

$$\sqrt{n}c_n = \alpha c_{n-1}. \tag{18.130}$$

If we take $c_0 = 1$, then

$$c_n = \frac{\alpha}{\sqrt{n}}c_{n-1} == \frac{\alpha^n}{\sqrt{n!}}. \tag{18.131}$$

We can thereby determine $|\alpha\rangle$ up to a normalization constant C,

$$|\alpha\rangle = C\sum_{n=0}^{\infty}\frac{\alpha^n}{\sqrt{n!}}|n\rangle \tag{18.132}$$

and C is determined by

$$1 = \langle\alpha|\alpha\rangle = C^2\sum_{n=0}^{\infty}\frac{(\alpha^*\alpha)^n}{n!} = C^2 e^{\alpha^*\alpha} \tag{18.133}$$

and we used the fact that $\langle n|n'\rangle = \delta_{n'n}$, so that

$$C = e^{-|\alpha|^2/2}. \tag{18.134}$$

The coherent state is therefore

$$|\alpha\rangle = e^{-|\alpha|^2/2}\sum_{n=0}^{\infty}\frac{\alpha^n}{\sqrt{n!}}|n\rangle. \tag{18.135}$$

The probability to find the system in the nth Fock state $|n\rangle$ (when it is in the state $|\alpha\rangle$) is

$$P(n) = |\langle n|\alpha\rangle|^2 = \frac{|\alpha|^{2n}}{n!}e^{-|\alpha|^2}. \tag{18.136}$$

The average $n \equiv \bar{n}$ is

$$\bar{n} = \sum_{n=0}^{\infty} nP(n) \tag{18.137}$$

which can be easily evaluated by noting (taking $|\alpha| = \alpha$, a positive real number),

$$\frac{d}{d\alpha}\sum_{n=0}^{\infty}P(n) = \frac{d}{d\alpha}1 = 0 = \frac{d}{d\alpha}\sum_{n=0}^{\infty}\frac{|\alpha|^{2n}}{n!}e^{-|\alpha|^2} = \sum_{n=0}^{\infty}\left(\frac{2n}{\alpha} - 2\alpha\right)P(n) = \frac{2}{\alpha}\bar{n} - 2\alpha \tag{18.138}$$

and therefore we can identify

$$\bar{n} = \alpha^2 \tag{18.139}$$

and we thereby obtain the Poisson distribution,

$$P(n, \bar{n}) = \frac{\bar{n}^n}{n!}e^{-\bar{n}}. \tag{18.140}$$

For example, this is also the probability distribution of the number of photons in a pulse of light with $\bar{n}$ photons on average, when the photons are generated randomly, for example, from a thermal source, or a highly attenuated laser beam.

Sometimes, as in quantum cryptography, we'd like a state with $n = \bar{n} = 1$ but that is in general not possible due to the uncertainty relationship between $\mathbf{P}$ and $\mathbf{X}$. The best that can be done is choose $\bar{n} \sim 1$, however, this can lead to an insecurity in quantum cryptography because of the finite probability of finding $n > 1$ in a pulse. This can be accounted for by post processing privacy amplification in most instances, which reduces the amount of secure information by accounting for the probability of two or more photons on average.

The time dependence of the coherent state is, noting that $\mathbf{a}^\dagger|n\rangle = \sqrt{n+1}|n+1\rangle$,

$$|\alpha, t\rangle = e^{-\alpha^2/2} \sum_{n=0}^{\infty} \frac{e^{-i(n+1/2)\omega t}\alpha^n}{\sqrt{n!}} |n\rangle \tag{18.141}$$

$$= e^{-\alpha^2/2} e^{-i\omega t/2} \sum_{n=0}^{\infty} \frac{\left(e^{-i\omega t}\alpha\right)^n \left(\mathbf{a}^\dagger\right)^n}{n!} |0\rangle \tag{18.142}$$

$$= \exp\left(-\alpha^2/2 - i\omega t/2 + \alpha e^{-i\omega t}\mathbf{a}^\dagger\right) |0\rangle . \tag{18.143}$$

The first and third terms in the exponent, operating on $|0\rangle$, produce a coherent state with oscillating α, with

$$\alpha(t) = e^{-i\omega t}\alpha \tag{18.144}$$

while the second term is a phase factor. Therefore, the full time dependence of the coherent state is

$$|\alpha, t\rangle = e^{-i\omega t/2}|\alpha(t)\rangle. \tag{18.145}$$

Therefore, the expectation values of $\langle X\rangle$ and $\langle P\rangle$ are time dependent as

$$\langle X(t)\rangle = \frac{\alpha(t) + \alpha^*(t)}{\sqrt{2}} \quad \langle P(t)\rangle = i\frac{\alpha(t) - \alpha^*(t)}{\sqrt{2}}. \tag{18.146}$$

Noting that

$$\alpha(t) + \alpha^*(t) = 2\alpha\cos\omega t; \quad i(\alpha(t) - \alpha^*(t)) = -2\alpha\sin\omega t \tag{18.147}$$

we see that $\langle X(t)\rangle$ and $\langle P(t)\rangle$ describe a circle of radius $\sqrt{2}\alpha$ in the X, P plane, so the total energy is

$$E = \frac{1}{2}\hbar\omega(\langle X(t)\rangle^2 + \langle P(t)\rangle^2) = \hbar\omega\alpha^2 \tag{18.148}$$

and

$$\langle X(t)\rangle = -\frac{d}{dt}\langle P(t)\rangle; \quad \langle P(t)\rangle = \frac{d}{dt}\langle X(t)\rangle \tag{18.149}$$

corresponding to the classical equations of motion, as expected by Ehrenfest's theorem.

18.6.3 Coherent state generation by direct calculation of the displacement operator

Another way to generate a coherent state is to use a sudden displacement, which can be done by applying the generator of finite translations to the harmonic oscillator ground state. Using our non-dimensional operators, where α is a (real) number, we define an (Weyl) operator

$$e^{-i\alpha \mathbf{P}} = e^{\alpha(\mathbf{a}^\dagger - \mathbf{a})} = D(\alpha) \tag{18.150}$$

and we anticipate that this operator applied to the $n = 0$ state will produce a coherent state,

$$D(\alpha)|0\rangle = |\alpha\rangle. \tag{18.151}$$

Because $[\mathbf{a}, \mathbf{a}^\dagger] = 1 \neq 0$, we need to be careful in the expansion of the exponential. First consider a function

$$f(\alpha) = e^{\alpha \mathbf{a}^\dagger} e^{-\alpha \mathbf{a}}. \tag{18.152}$$

The derivative with respect to α can be calculated, paying attention to the ordering of operators, and using $\mathbf{a}^\dagger e^{\alpha \mathbf{a}^\dagger} = e^{\alpha \mathbf{a}^\dagger} \mathbf{a}^\dagger$,

$$\frac{d\, f(\alpha)}{d\alpha} = \mathbf{a}^\dagger e^{\alpha \mathbf{a}^\dagger} e^{-\alpha \mathbf{a}} - e^{\alpha \mathbf{a}^\dagger} e^{-\alpha \mathbf{a}} \mathbf{a} = f(\alpha)(e^{\alpha \mathbf{a}} \mathbf{a}^\dagger e^{-\alpha \mathbf{a}} - \mathbf{a}). \tag{18.153}$$

Expanding

$$\mathbf{a}^\dagger e^{-\alpha \mathbf{a}} = \mathbf{a}^\dagger \left(1 - \alpha \mathbf{a} + \frac{\alpha^2}{2!} \mathbf{a}^2 - \frac{\alpha^3}{3!} \mathbf{a}^3 + ...\right) \tag{18.154}$$

and noting that (by use of $[\mathbf{a}^\dagger, \mathbf{a}] = -1$)

$$[\mathbf{a}^\dagger, \mathbf{a}^n] = -n {\mathbf{a}^\dagger}^{n-1}; \quad \mathbf{a}^\dagger \mathbf{a}^n = \mathbf{a}^n \mathbf{a}^\dagger - n \mathbf{a}^{n-1} \tag{18.155}$$

we arrive at a special case of the Baker-Cambell-Hausdorff formula,

$$\begin{aligned} \mathbf{a}^\dagger e^{\alpha \mathbf{a}} &= \sum_{n=0}^{\infty} \left(\frac{(-\alpha)^n \mathbf{a}^n}{n!} \mathbf{a}^\dagger - \frac{n(-\alpha)^n \mathbf{a}^{n-1}}{n!} \right) \\ &= \left(e^{-\alpha \mathbf{a}} \mathbf{a}^\dagger + \alpha \sum_{n=1}^{\infty} \frac{(-\alpha)^{n-1} \mathbf{a}^{n-1}}{(n-1)!} \right) \\ &= e^{-\alpha \mathbf{a}} \mathbf{a}^\dagger + \alpha e^{-\alpha \mathbf{a}}. \end{aligned} \tag{18.156}$$

Therefore,

$$\frac{d\, f(\alpha)}{d\alpha} = f(\alpha) \left(\mathbf{a}^\dagger - \mathbf{a} + \alpha\right) \tag{18.157}$$

which has solution

$$f(\alpha) = e^{\alpha \mathbf{a}^\dagger} e^{-\alpha \mathbf{a}} = e^{\alpha \mathbf{a}^\dagger - \alpha \mathbf{a} + \alpha^2/2} = e^{\alpha \mathbf{a}^\dagger - \alpha \mathbf{a} + [\alpha \mathbf{a}, \alpha \mathbf{a}^\dagger]/2} \tag{18.158}$$

and is true for any two functions (in this case, $\mathbf{a}$ and $\mathbf{a}^\dagger$) that commute with their commutator. This allows us to immediately write

$$D(\alpha) = e^{\alpha(\mathbf{a}^\dagger - \mathbf{a})} = e^{-\alpha^2/2} e^{\alpha \mathbf{a}^\dagger} e^{-\alpha \mathbf{a}} \tag{18.159}$$

where the ordering of the operators is important.

Let us apply this to the $n = 0$ state, and the first operator gives

$$e^{-\alpha \mathbf{a}}|0\rangle = (1 - \alpha \mathbf{a} + \frac{\alpha^2 \mathbf{a}^2}{2} + ...)|0\rangle = |0\rangle \tag{18.160}$$

because the lowering operator $\mathbf{a}|0>= 0$. The coherent state is therefore expanded as

$$D(\alpha)|0\rangle = e^{-\alpha^2/2} e^{\alpha \mathbf{a}^\dagger} = e^{-\alpha^2/2} \sum_{n=0}^{\infty} \frac{\alpha^n \mathbf{a}^{\dagger n}}{n!} |0\rangle. \tag{18.161}$$

Recalling that

$$|n\rangle = \frac{\mathbf{a}^{\dagger n}}{\sqrt{n!}} |0\rangle, \tag{18.162}$$

the expansion of α in terms of $|n\rangle$ is

$$|\alpha\rangle = e^{-\alpha^2/2} \sum_{n=0}^{\infty} \frac{\alpha^n}{\sqrt{n!}} |n\rangle \tag{18.163}$$

which is the same result obtained earlier.

Note that it is possible to extend this to α being a complex variable, in which case

$$D(z) = e^{z\mathbf{a}^\dagger - z^* \mathbf{a}} = e^{z\mathbf{a}^\dagger} e^{z^* \mathbf{a}} e^{-|z|^2/2}. \tag{18.164}$$

We will discuss the meaning of a complex displacement when squeezed states are presented.

18.6.4 Determination of the coherent states by shifting X in the ground state

Let us work in the non-dimensional X representation, with

$$\langle X | n = 0 \rangle = \frac{1}{\sqrt[4]{\pi}} e^{-X^2/2} = \psi_0(X). \tag{18.165}$$

Next consider

$$\psi_0(X + X_0) = \frac{1}{\sqrt[4]{\pi}} e^{-(X+X_0)^2/2} \tag{18.166}$$

and determine the overlap with the harmonic oscillator states in the original coordinate system, as

$$\langle \psi_0(X + X_0) | \psi_n(X) \rangle = \int_{-\infty}^{\infty} dX \frac{1}{\sqrt[4]{\pi}} e^{-(X+X_0)^2/2} \psi_n(X). \tag{18.167}$$

We determine the wavefunctions in coordinate space by use of the $\mathbf{a}^\dagger$ raising operator written in coordinate space,

$$\psi_{n+1}(X) = \frac{\langle X | \mathbf{a}^\dagger | X \rangle}{\sqrt{n+1}} \psi_n(X) = \frac{(X - \frac{d}{dX})}{\sqrt{2(n+1)}} \psi_n(X) \tag{18.168}$$

which can be put into simpler form by noting that

$$\begin{aligned}\frac{d}{dx}e^{-X^2/2}\psi_n(x) &= e^{-X^2/2}\left(-X+\frac{d}{dX}\right)\psi_n(X)\\ &= (-1)(\sqrt{2})e^{-X^2/2}\mathbf{a}^\dagger\psi_n(x)\\ &= (-1)(\sqrt{2})e^{-X^2/2}\sqrt{n+1}\psi_{n+1}(X).\end{aligned} \tag{18.169}$$

Applying this recursively leads to

$$\psi_n(X) = (-1)^n e^{X^2/2}\frac{\frac{d^n}{dx^n}e^{-X^2/2}\psi_0(X)}{\sqrt{2\ n!}}. \tag{18.170}$$

The overlap amplitude is therefore

$$\langle\psi_0(X+X_0)|\psi_n(X)\rangle = \frac{(-1)^n}{\sqrt{2^n\pi n!}}\int_{-\infty}^{\infty} dX e^{-XX_0-X_0^2/2}\frac{d^n}{dX^n}e^{-X^2} \tag{18.171}$$

integrating by parts n times,

$$e^{-X_0^2/2}\int_{-\infty}^{\infty} dX e^{-XX_0}\frac{d^n}{dX^n}e^{-X^2} = X_0^n e^{-X_0^2/4}\int_{-\infty}^{\infty} dX e^{-(X-X_0/2)^2} \tag{18.172}$$

$$= \sqrt{\pi}X_0^n e^{-X_0^2/4} \tag{18.173}$$

and the amplitude is

$$\langle\psi_0(X+X_0)|\psi_n(X)\rangle = \frac{X_0^n}{\sqrt{2^n n!}}e^{-X_0^2/4}. \tag{18.174}$$

The probability of being in state n is the square of the amplitude,

$$P_(n, X_0) = \frac{1}{n!}\frac{1}{2^n}X_0^{2n}e^{-X_0^2/2}. \tag{18.175}$$

The average of n is

$$\bar{n} = \frac{X_0^2}{2} \tag{18.176}$$

and leads to the Poisson distribution,

$$P(n,\bar{n}) = \frac{\bar{n}^n e^{-\bar{n}}}{n!} \tag{18.177}$$

as we found with other techniques.

18.6.5 Properties of coherent states

18.6.5.1 Non-orthogonality

Let us consider two coherent states $|w\rangle$ and $|z\rangle$ with w, z complex. The inner product is

$$\langle w|z\rangle = \sum_n \langle w|n\rangle\langle n|z\rangle = e^{-(|w|^2+|z|^2)/2}\sum_n \frac{(w^*z)^n}{n!} = e^{-(|w|^2+|z|^2)/2+w^*z}. \quad (18.178)$$

Alternatively, we can find

$$\langle z|w\rangle = e^{-(|w|^2+|z|^2)/2+wz^*} \quad (18.179)$$

implying that

$$|\langle w|z\rangle|^2 = \langle w|z\rangle\langle z|w\rangle = e^{-(|w|+|z|^2)+w^*z+wz^*} = e^{-|w+z|^2} \neq \delta(w-z), \quad (18.180)$$

which is to say, the states are not orthogonal. As we will see in the next section, the specification of minimum uncertainty does not specify a unique state. Later we will explore the types of states that can be constructed.

18.6.5.2 Completeness

Starting from the expansion of a general $|z\rangle$ state,

$$|z\rangle = e^{-|z|^2/2}\sum_{n=0}^{\infty}\frac{z^n}{\sqrt{n!}}|n\rangle \quad (18.181)$$

the outer product is

$$|z\rangle\langle z| = e^{-|z|^2}\sum_{m=0}^{\infty}\frac{(z^*)^m}{\sqrt{m!}}\sum_{n=0}^{\infty}\frac{z^n}{\sqrt{n!}}|m\rangle\langle n|, \quad (18.182)$$

which has elements in the harmonic oscillator energy basis of

$$\sum_z \langle m|z\rangle\langle z|n\rangle = \sum_z e^{-|z|^2}\frac{(z^*)^m z^n}{\sqrt{n!m!}}. \quad (18.183)$$

Sum $z = re^{i\phi}$, so that $\sum_z \to \int_0^\infty \int_0^\pi rdrd\phi$, we find

$$\sum_z \langle m|z\rangle\langle z|n\rangle = \int_0^\infty \int_0^\pi \frac{rdrd\phi}{\sqrt{n!m!}} e^{-r^2} r^{(n+m)} e^{i\phi[n-m]} \quad (18.184)$$

$$= 0 \quad \text{for } n \neq m. \quad (18.185)$$

For $n = m$ we find

$$\sum_z \langle m|z\rangle\langle z|n\rangle = 2\pi \int_0^\infty rdre^{-r^2}\frac{r^{2n}}{n!} = \frac{2\pi}{n!}\frac{n!}{2} = \pi \quad (18.186)$$

and therefore the identity operator can be taken as

$$I = \int\int \frac{r\,dr\,d\phi}{\pi} |z\rangle\langle z| \tag{18.187}$$

where the integral is over the entire z plane and we have shown that

$$\frac{1}{\pi}\sum_{z} \langle m|z\rangle\langle z|n\rangle = \delta_{mn}. \tag{18.188}$$

Because the harmonic oscillator states form a complete set, we can conclude that the $|z\rangle$ states span the entire space. Finally, note that

$$\langle w|I|z\rangle \neq 0 \quad \text{for } w \neq z \tag{18.189}$$

as we learned already.

In the sense that the states $|\alpha\rangle$ are complete but not orthonormal, they are referred to as overcomplete. As we will see in the next section, requiring a minimum uncertainty does not uniquely determine a specific coherent state.

18.6.6 Squeezed states

Since the time that Schrödinger and others investigated minimum uncertainty states, a vast literature on variants of such states has been produced. This work has been largely academic until fairly recently. In modern quantum optics and quantum electronics, not to mention their roles in quantum information and computing, such states are existentially critical. With the advent of non-linear parametric components for optical frequencies, and Josephson Parametric Amplifiers (JPAs) for microwave frequencies, new reduced noise measurement techniques are possible, and the development of such is an area of active interest.

The discussion begins with a recognition that $\mathbf{X}$ and $\mathbf{P}$, our non-dimensional position and momentum operators for the simple harmonic oscillator, can be mapped directly to the two phase components of an electromagnetic wave or field.

There are three basic types of squeezing: squeezing the vacuum, possibly to ameliorate zero-point fluctuation noise; amplitude squeezing to reduce amplitude fluctuations; phase squeezing to allow better definition of the phase of the wave. At optical frequencies, thermal photons are essentially non-existent. 300 K, a typical laboratory temperature, corresponds to photons of about 10μm wavelength, much greater than the 0.5μm we normally associate with optical frequencies. For microwaves in the 10 GHz region, the system temperature needs to be below 0.5 K to suppress thermal photons, which requires specialized equipment but is straightforward. In any case, low temperatures are needed for superconducting components to minimize losses, and the usual amplification techniques based on Josephson junction arrays require low temperatures.

18.6.6.1 The standard quantum limit (SQL)

We have shown that the minimum uncertainty state for the harmonic oscillator, which maps directly to a monochromatic electromagnetic field, is an eigenstate of the annihilation operator $\mathbf{a}$. Phase-preserving linear amplification of an electromagnetic field (as a traveling wave) by either a (LASER) gain medium or electronic amplifier, actually increases the uncertainty of the state (e.g., it adds noise), as we will show.[3]

Linear amplifiers are subject to quantum fluctuations, leading to a what is referred to as the standard quantum limit (SQL) in their performance. This limitation can be parameterized as an effective minimum temperature, $T_{SQL} = \hbar\omega/k_B$ for any linear amplifier. The origin of the SQL can be understood as follows. Consider a well defined mode of an electromagnetic field. As we have seen, quantum field theory leads to the introduction of operators to describe the field. In analogy with the harmonic oscillator we introduce conjugate variables p_0 and q_0 as operators that determine the fields; as always the operators commute as $[p_0, q_0] = \hbar/i$. Let us now apply a linear amplification which increases each operator a factor G. It appears that the commutator of the field so produced $[Gp_0, Gq_0] = G^2\hbar/i$, which is incorrect. We can restore the correct commutation for the amplified final field by introducing the operators p_f, q_f as

$$[p_f, q_f] = \hbar/i = [Gp_0, Gq_0] + [p_g, q_g] = \frac{G^2\hbar}{2i} + [p_g, q_g] \tag{18.190}$$

where $[p_g, q_g]$ is introduced to restore the commutation relation. We see immediately that

$$[p_g, q_g] = \frac{(1-G^2)\hbar}{2i}. \tag{18.191}$$

By use of the generalized Heisenberg uncertainty relation, the bounds on the fluctuations on p_g and q_g are determined by the value of their commutator,

$$\langle|\Delta p_g|^2\rangle\langle|\Delta q_g|^2\rangle \geq \frac{(G^2-1)^2\hbar^2}{4} \tag{18.192}$$

(the zero-point is a minimum uncertainty state, so the equality holds). The first term on the right-hand side of Eq. (18.190) produces a final field energy of $G^2\hbar\omega/2$, while the second term results in field energy of $(G^2-1)\hbar\omega/2$. Referring the final field energy to the initial field energy before amplification,

$$\frac{1}{G^2}\left[\frac{G^2\hbar\omega}{2} + \frac{(G^2-1)\hbar\omega}{2}\right] = \frac{\hbar\omega}{2} + \frac{(G^2-1)\hbar\omega}{2G^2} \approx 2 \times \frac{\hbar\omega}{2} \tag{18.193}$$

in the limit of large G. This result shows final state fluctuations that imply twice the zero-point energy for the initial state; this is the SQL. A principal implication is that the final state after amplification is, as might be expected, not a minimum uncertainty, or coherent, state.

In contrast to linear receivers, single-photon detectors such as photomultipier tubes—which provide a strongly coupled measurement of a quantum state—can in principle be arbitrarily noiseless when operated at sufficiently low temperatures to reduce thermal photons. However, even at very low temperature, there is still noise

[3]Carlton M. Caves, *Quantum limits on noise in linear amplifiers*, Phys. Rev. D **26**, 1817 (1982); see also H.A. Haus and J. A. Mullen, *Quantum Noise in Linear Amplifiers*, Phys. Rev. **128**, 2407 (1962) and A.A. Clerk et al., *Introduction to quantum noise, measurement, and amplification*, Rev. Mod. Phys. **82**, 1155 (2010).

associated with measuring sources, for example, for a finite number of photons in a wavepacket (coherent state), there are amplitude fluctuations that introduce noise (Poisson statistics). It might be possible to reduce this noise in certain situations.

For example, if only one phase component of the electromagnetic field is measured, we could imagine reducing ΔX and increasing ΔP while maintaining the fundamental minimum $\Delta X \Delta P \leq 1/2$, using our non-dimensional operators. Or the phase noise could be reduced at the cost of amplitude noise. Such process are referred to as squeezing.

For linear amplifiers, the vacuum state can be altered in a similar way to allow a signal of a particular phase to be amplified without adding zero-point noise; this is phase dependent amplification for which the effects of zero-point noise can be reduced, and thereby improve the sensitivity of a receiving system. This type of squeezing has been used in the Advanced LIGO gravitational wave detector,[4] and in the HAYSTAC dark matter experiment.[5] For Advanced LIGO, the noise at the interference minimum was reduced, which together with other improvements increased the gravitation detection rate by an order of magnitude. In HAYSTAC, the goal was to detect radiofrequency photons generated by the conversion of axions, a putative particle that was introduced as a solution to the strong-CP problem, that might comprise the dark matter halo of our galaxy. The conversion is affected by a strong magnetic field that permeates a resonant cavity, and an improved signal to noise has been obtained by reflecting a squeezed vacuum state from the cavity.

18.6.6.2 The squeezing operator

The notion of squeezing is easily understood by generalizing the lowering operator

$$\mathbf{a} = \frac{1}{\sqrt{2}} [\mathbf{X} + i\mathbf{P}], \tag{18.194}$$

the eigenfunctions of which

$$\alpha|\alpha\rangle = \mathbf{a}|\alpha\rangle \tag{18.195}$$

are coherent states. This operator can be generalized as

$$\mathbf{a}_\lambda = \frac{1}{\sqrt{2}} \left(\lambda \mathbf{X} + \frac{i}{\lambda} \mathbf{P} \right) \tag{18.196}$$

and similarly for the raising operator,

$$\mathbf{a}^\dagger{}_\lambda = \frac{1}{\sqrt{2}} \left(\lambda \mathbf{X} - \frac{i}{\lambda} \mathbf{P} \right) \tag{18.197}$$

where λ is a real positive number. These generalized operators provide a relative scaling between $\mathbf{X}$ and $\mathbf{P}$ such that $\mathbf{X}_\lambda \mathbf{P}_\lambda = XP$. Therefore,the commutator of the two elements of these operators still satisfy

[4]https://www.ligo.caltech.edu/news/ligo20201028

[5]K.M. Backes et al., *A quantum enhanced search for dark matter axions*, Nature **590**, 238 (2021).

$$\left[\lambda\mathbf{X}, \frac{1}{\lambda}\mathbf{P}\right] = [\mathbf{X}, \mathbf{P}] = i \tag{18.198}$$

and will therefore have minimum uncertainty eigenstates in the same manner as the case of coherent states $\mathbf{a}|\alpha\rangle = \alpha|\alpha\rangle$ that we have already studied, as will be shown. This results from the commutator being a number and therefore commuting with $\mathbf{X}$ and $\mathbf{P}$; however, there is a consequence in that $\Delta P \neq \Delta X$ for eigenstates of the $\mathbf{a}_\lambda$ operator. The parameter λ can be complex, which corresponds to a relative scaling between X and P together with a rotation of the minimum uncertainty state in the $\mathbf{X}$, $\mathbf{P}$ plane if we take those as the components of a two-dimensional vector (phasor).

The ubiquitous and nearly universal form of the squeezing operator was developed in 1970 by D. Stoler.[6] We are seeking solutions to the eigenvalue problem

$$\mathbf{a}_\lambda|s\rangle = s|s\rangle \tag{18.199}$$

where the eigenvalue s labels a squeezed coherent state.

Let us begin by writing $\mathbf{a}_\lambda$ in terms of the usual lowering and raising operators,

$$\mathbf{a}_\lambda = (\lambda\mathbf{X} + \frac{i}{\lambda}\mathbf{P}) = \frac{1+\lambda^2}{2\lambda}\mathbf{a} + \frac{\lambda^2-1}{2\lambda}\mathbf{a}^\dagger. \tag{18.200}$$

If we take $\lambda = e^r$ where r is a real positive number (this can be generalized to a complex number), then

$$\mathbf{a}_\lambda = \mathbf{a}\cosh(r) + \mathbf{a}^\dagger\sinh(r). \tag{18.201}$$

The problem is to find solutions to the equation

$$\mathbf{a}_\lambda|s\rangle = (\mathbf{a}\cosh r + \mathbf{a}^\dagger\sinh r)|s\rangle = s|s\rangle \tag{18.202}$$

where s is the eigenvalue of the generalized lowering operator. This problem can be solved if we can find some unitary operator $\mathbf{S}$ such that

$$\mathbf{S}^\dagger(\mathbf{a}\cosh r + \mathbf{a}^\dagger\sinh r)\mathbf{S} = \mathbf{a} \tag{18.203}$$

in which case the squeezed state is given by

$$|s\rangle = \mathbf{S}|\alpha\rangle \tag{18.204}$$

because then we would have

$$\mathbf{a}_\lambda\mathbf{S}|\alpha\rangle = s\mathbf{S}|\alpha\rangle \tag{18.205}$$

$$\mathbf{S}^\dagger\mathbf{a}_\lambda\mathbf{S}|\alpha\rangle = \mathbf{a}|\alpha\rangle = \alpha|\alpha\rangle = s|\alpha\rangle \tag{18.206}$$

Let us consider the unitary operator

$$\mathbf{S}(r) \equiv \mathbf{S}_r = e^{r(\mathbf{aa}-\mathbf{a}^\dagger\mathbf{a}^\dagger)/2} \tag{18.207}$$

[6]David Stoler, *Equivalence Classes of Minimum Uncertainty Packets*, Phys. Rev. D **1**, 3217 (1970).

where r is real positive number. Furthermore,

$$\mathbf{S}_r^\dagger = \mathbf{S}_r^{-1} = \mathbf{S}_{-r}. \tag{18.208}$$

The effect of $\mathbf{S_r}$ on $\mathbf{a}$ can be evaluated using Hadamard's lemma,[7]

$$e^{\mathbf{A}}\mathbf{B}e^{-\mathbf{A}} = \mathbf{B} + [\mathbf{A},\mathbf{B}] + \frac{1}{2!}[\mathbf{A},[\mathbf{A},\mathbf{B}]] + ... \tag{18.209}$$

for which $\mathbf{A} = r(\mathbf{aa} - \mathbf{a}^\dagger\mathbf{a}^\dagger)/2$ and $\mathbf{B} = \mathbf{a}$. and we can determine the following relationships by use of $[\mathbf{a},\mathbf{a}^\dagger] = 1$,

$$[\mathbf{A},\mathbf{B}] = -\frac{r}{2}[\mathbf{a}^\dagger\mathbf{a}^\dagger, \mathbf{a}] = r\mathbf{a}^\dagger \tag{18.210}$$

$$[\mathbf{A},[\mathbf{A},\mathbf{B}]] = \frac{r^2}{2}[\mathbf{aa}, a^\dagger] = r^2\mathbf{a} \tag{18.211}$$

$$[\mathbf{A},[\mathbf{A},[\mathbf{A},\mathbf{B}]]] = -\frac{r^3}{2}[\mathbf{a}^\dagger\mathbf{a}^\dagger, \mathbf{a}] = r^3\mathbf{a}^\dagger \tag{18.212}$$

and the pattern is easily seen,

$$e^{\mathbf{A}}\mathbf{B}e^{-\mathbf{A}} = \mathbf{a}\sum_{n=0,2,4...}\frac{r^n}{n!} + \mathbf{a}^\dagger\sum_{n=1,3,5,...}\frac{r^n}{n!} = \mathbf{a}\cosh r + \mathbf{a}^\dagger\sinh r, \tag{18.213}$$

and $\mathbf{S}_r$ is indeed the transform that will convert a coherent state to a squeezed state,

$$|s\rangle = |r,\alpha\rangle = \mathbf{S}_r|\alpha\rangle \tag{18.214}$$

with

$$\mathbf{S}_r = e^{r\left(\mathbf{aa}-\mathbf{a}^\dagger\mathbf{a}^\dagger\right)/2}. \tag{18.215}$$

The products $\mathbf{aa}$ and $\mathbf{a}^\dagger\mathbf{a}^\dagger$ suggest that squeezing requires two simultaneous interactions and thus results from a non-linear process.

A squeezed vacuum state can be displaced

$$|\alpha,r\rangle = D(\alpha)\mathbf{S}_r|0\rangle. \tag{18.216}$$

which is the general form of a squeezed coherent state for a quantum harmonic oscillator. The effects of different diplacements and squeezing operations will be discussed shortly.

[7]Messiah, A., *Quantum Mechanics* (Wiley, New York, 1976), p. 339, problem 4.

Let us calculate the uncertainties in X and P for the squeezed state. We can do this by determining the operators $\mathbf{X}$ and $\mathbf{P}$ in terms of $\mathbf{a}_\lambda$ and $\mathbf{a}_\lambda^\dagger$:

$$\mathbf{X} = \frac{\mathbf{a}_\lambda + \mathbf{a}_\lambda{}^\dagger}{\lambda\sqrt{2}}; \quad \mathbf{P} = -i\lambda\frac{\mathbf{a}_\lambda - \mathbf{a}_\lambda^\dagger}{\sqrt{2}}. \tag{18.217}$$

First, we need the commutator

$$[\mathbf{a}_\lambda, \mathbf{a}_\lambda^\dagger] = \frac{1}{2}[\lambda\mathbf{X} + \frac{i}{\lambda}\mathbf{P}, \lambda\mathbf{X} - \frac{i}{\lambda}\mathbf{P}] = -i(\mathbf{XP} - \mathbf{PX}) = 1. \tag{18.218}$$

The expectation values of $\mathbf{X}$ and $\mathbf{P}$ and their squares are (dropping the subscript λ)

$$\begin{aligned}
\langle X\rangle =& \frac{1}{\lambda\sqrt{2}}\langle s|(\mathbf{a} + \mathbf{a}^\dagger)|s\rangle = \frac{s + s^*}{\lambda\sqrt{2}} \\
\langle X^2\rangle =& \frac{1}{2\lambda^2}\langle s|(\mathbf{a} + \mathbf{a}^\dagger)^2|s\rangle \\
=& \frac{1}{2\lambda^2}\langle s|(\mathbf{aa} + \mathbf{a}^\dagger\mathbf{a}^\dagger + \mathbf{aa}^\dagger + \mathbf{a}^\dagger\mathbf{a})|s\rangle \\
=& \frac{1}{2\lambda^2}\langle s|(\mathbf{aa} + \mathbf{a}^\dagger\mathbf{a}^\dagger + 1 + 2\mathbf{a}^\dagger\mathbf{a})|s\rangle \\
=& \frac{(s^*)^2 + s^2 + 2ss^* + 1}{2\lambda^2} \\
(\Delta X)^2 =& \langle X^2\rangle - \langle X\rangle^2 = \frac{1}{2\lambda^2} = \frac{e^{-2r}}{2}
\end{aligned} \tag{18.219}$$

and similarly,

$$\begin{aligned}
\langle P\rangle =& \frac{-i\lambda}{\sqrt{2}}\langle s|(\mathbf{a} - \mathbf{a}^\dagger)|s\rangle = -i\lambda\frac{s - s^*}{\sqrt{2}} \\
\langle P^2\rangle =& -\frac{\lambda^2}{2}\langle s|(\mathbf{a} - \mathbf{a}^\dagger)^2|s\rangle \\
=& -\frac{\lambda^2}{2}\langle s|(\mathbf{aa} + \mathbf{a}^\dagger\mathbf{a}^\dagger - \mathbf{aa}^\dagger - \mathbf{a}^\dagger\mathbf{a})|s\rangle \\
=& -\frac{1}{2\lambda^2}\langle s|(\mathbf{aa} + \mathbf{a}^\dagger\mathbf{a}^\dagger - 1 - 2\mathbf{a}^\dagger\mathbf{a})|s\rangle \\
=& -\lambda^2\frac{(s^*)^2 + s^2 - 2ss^* - 1}{2} \\
(\Delta P)^2 =& \langle P^2\rangle - \langle P\rangle^2 = \frac{\lambda^2}{2} = \frac{e^{2r}}{2}
\end{aligned} \tag{18.220}$$

and therefore the squeezed state satisfies the Heisenberg uncertainty principle,

$$(\Delta X)^2(\Delta P)^2 = \frac{1}{4}, \tag{18.221}$$

with uncertainty reduced in one quadrature and increased in the other. In the case of complex r, we can think of the squeezing as producing a linear combination of X and P coherent states to produce two new orthogonal states, X' and P' rotated relative to X, P. X and P can be taken together as a phasor in the complex plane, as shown in Fig. 18.1, with X' and P' rotated by ϕ. In a similar manner, a complex α results in a displacement of the vacuum state at an angle relative to the X axis.

We therefore see that the minimum uncertainty state of the harmonic oscillator is not unique, and this is the fundamental reason why coherent states are overcomplete.

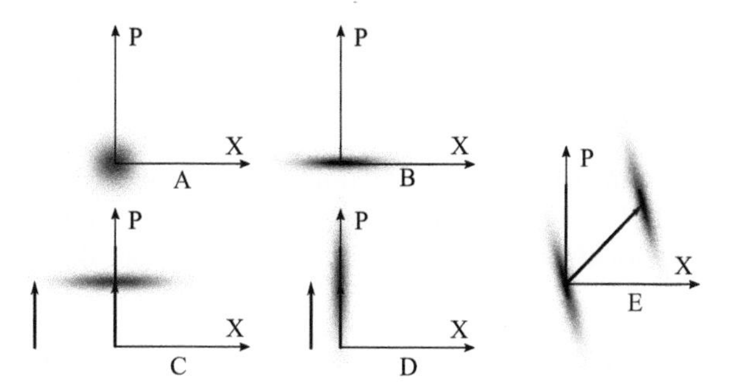

Fig. 18.1 A. Vacuum state, B. squeezed vacuum, C. displaced squeezed vacuum, amplitude squeezing, D. displaced vacuum, phase squeezing, E. mixed phase-amplitude squeezing, which is shown in detail as squeezing of the ground state along one axis oblique to X and P, and then displacement in another oblique direction.

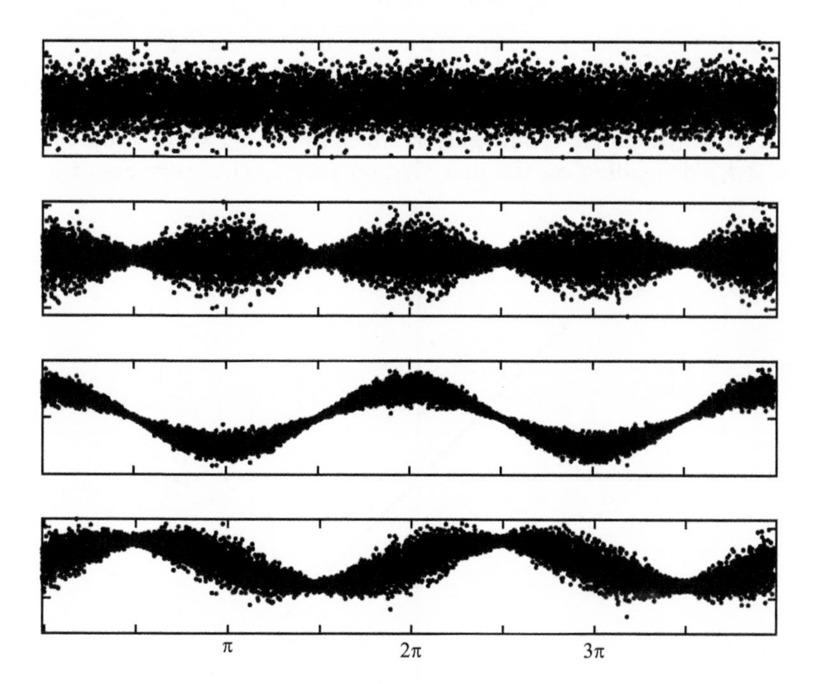

Fig. 18.2 Time-dependent signal for various squeezing options. Top: Zero-point noise for the field vacuum state. 2nd: Phase-squeezed vacuum. 3rd: Displaced phase-squeezed vacuum. Bottom: Displaced amplitude squeezed vacuum.

18.6.7 The effect of a beam splitter on Fock states

A beam splitter divides a light beam into two parts, as shown in the figure. For the current discussion, let us assume the division is equal (a 50%-50% beam splitter). There are four ports, and we consider two as input ports and two as output ports.

Referring to the Fig. 18.3, a is an input port that divides the light between ports c (transmission) and d (reflection). Light entering b is divided between c (reflection) and d (transmission).

Let us assume the light polarizations are the same for the two incident light beams. By energy conservation, we expect the total input power (total field squared) to be unchanged after the beam splitter. Given transmission t and reflection r amplitudes as real numbers, for all directions we will take $|t|^2 = |r|^2 = 1/2$ for simplicity. We need to consider the phases for both t and r, both a and b. Let us take input electric fields as real

$$|E|_a^2 + |E|_b^2 = |E_c|^2 + |E_d|^2 = |E_a t e^{i\phi_{ta}} + E_b r e^{i\phi_{rb}}|^2 + |E_a r e^{i\phi_{ra}} + E_b t e^{i\phi_{tb}}|^2. \quad (18.222)$$

Expanding and simplifying, we get

$$E_a E_b rt \left[e^{i\phi_{ta}} e^{-i\phi_{rb}} + e^{-i\phi_{ta}} e^{i\phi_{rb}} + e^{-i\phi_{ra}} e^{i\phi_{tb}} + e^{i\phi_{ra}} t e^{-i\phi_{tb}} \right] = 0 \quad (18.223)$$

and we arrive at a condition on the phases of

$$\cos(\phi_{ta} - \phi_{rb}) + \cos(\phi_{tb} - \phi_{ra}) = 0 \quad (18.224)$$

which ensures unitarity. Because $\cos(\theta \pm \pi) = -\cos\theta = -\cos(-\theta)$,

$$\phi_{ta} - \phi_{rb} = \pm(\phi_{tb} - \phi_{ra} \pm \pi). \quad (18.225)$$

If we have a truly symmetric beam splitter, we can take $\phi_{ta} = \phi_{tb} = 0$, and then can take $\phi_{ra} = \pm\phi_{rb}$ depending on the first sign on the rhs. In every case we will have

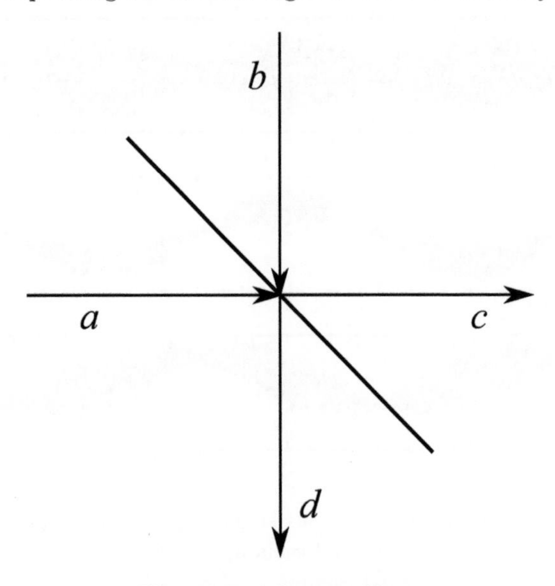

Fig. 18.3 A beam splitter.

$\phi_{ra}, \phi_{rb} = \pm\pi/2$. We can represent the effect of a beam splitter by a matrix, and in this case (taking the + sign)

$$\frac{1}{\sqrt{2}}\begin{pmatrix} 1 & i \\ i & 1 \end{pmatrix}\begin{pmatrix} E_a \\ E_b \end{pmatrix} = \frac{1}{\sqrt{2}}\begin{pmatrix} E_a + iE_b \\ E_b + iE_a \end{pmatrix} = \begin{pmatrix} E_c \\ E_d \end{pmatrix}, \tag{18.226}$$

i.e., a $\pi/2$ phase shift between the transmitted and reflected beam for a symmetric beam splitter.

The effect of a beam splitter on a Fock state can be most easily discussed in the Heisenberg picture ($\mathbf{a}, \mathbf{a}^\dagger$ are functions of time), where we let the beam splitter act on the operators $\mathbf{a}^\dagger{}_a$ and $\mathbf{a}^\dagger{}_b$ that create the input state:

$$|n\rangle_a|m\rangle_b|0\rangle_c|0\rangle_d = \frac{(\mathbf{a}^\dagger{}_a)^n}{\sqrt{n!}}\frac{(\mathbf{a}^\dagger{}_b)^m}{\sqrt{m!}}|0\rangle_a|0\rangle_b|0\rangle_c|0\rangle_d \tag{18.227}$$

where we define $\ell = m+n$. In the Heisenberg picture, as the beam propagates through the beam splitter, the state of the photons is assumed invariant while the operators are transformed as,

$$\frac{1}{\sqrt{2}}\begin{pmatrix} 1 & i \\ i & 1 \end{pmatrix}\begin{pmatrix} \mathbf{a}^\dagger{}_a \\ \mathbf{a}^\dagger{}_b \end{pmatrix} \Longrightarrow \frac{1}{\sqrt{2}}\begin{pmatrix} \mathbf{a}^\dagger{}_a + i\mathbf{a}^\dagger{}_b \\ \mathbf{a}^\dagger{}_b + i\mathbf{a}^\dagger{}_a \end{pmatrix} = \begin{pmatrix} \mathbf{a}^\dagger{}_c \\ \mathbf{a}^\dagger{}_d \end{pmatrix}. \tag{18.228}$$

Therefore, operators that would be applied before the beam splitter can be rewritten in terms of operators at the exit of the beam splitter,[8]

$$\mathbf{a}^\dagger{}_a \Longrightarrow \frac{\mathbf{a}^\dagger{}_c - i\mathbf{a}^\dagger{}_d}{\sqrt{2}}; \;\; \mathbf{a}^\dagger{}_b \Longrightarrow \frac{-i\mathbf{a}^\dagger{}_c + \mathbf{a}^\dagger{}_d}{\sqrt{2}}. \tag{18.229}$$

These operators acting on the vacuum states produce the states at the exit:

$$|0\rangle_a|0\rangle_b|\gamma\rangle_c|\delta\rangle_d = -\frac{1}{\sqrt{n!}}\left(\frac{\mathbf{a}^\dagger{}_c - i\mathbf{a}^\dagger{}_d}{\sqrt{2}}\right)^n \frac{1}{\sqrt{m!}}\left(\frac{-i\mathbf{a}^\dagger{}_c + \mathbf{a}^\dagger{}_d}{\sqrt{2}}\right)^m |0\rangle_a|0\rangle_b|0\rangle_c|0\rangle_d. \tag{18.230}$$

The output state is labeled as γ and δ because it is clear that the output cannot be a single Fock state due to the binomial distribution of the raising operators when their sum is taken to some power; terms of the form

$$\binom{\ell}{j}(\mathbf{a}^\dagger_\mathbf{c})^j(\mathbf{a}^\dagger_\mathbf{d})^k \tag{18.231}$$

where $j+k = m+n = \ell$ indicate that the number of photons is evenly divided between the two output arms. It is easiest to see in the case $n = 0$ and $m \to \infty$ that the outputs c and d will be equal coherent states $|m/2\rangle$ in the case of a 50% beam splitter. After performing the exponentiation and taking the product, the sum can be simplified by noting that $[\mathbf{a}^\dagger_\mathbf{a}, \mathbf{a}^\dagger_\mathbf{b}] = 0 = [\mathbf{a}^\dagger_\mathbf{c}, \mathbf{a}^\dagger_\mathbf{d}]$ because the operators are defined to operate on distinct arms.

[8]Christopher Gerry and Peter Knight, *Introductory Quantum Optics* (Cambridge University Press, 2004), Sec. 6.2.

In the case $n = m = 1$, we have for the state at the exit

$$\left(\frac{\mathbf{a}^\dagger{}_c - i\mathbf{a}^\dagger{}_d}{\sqrt{2}}\right)\left(\frac{-i\mathbf{a}^\dagger{}_c + \mathbf{a}^\dagger{}_d}{\sqrt{2}}\right)|0\rangle_a|0\rangle_b|0\rangle_c|0\rangle_d = \frac{-i}{2}((\mathbf{a}^\dagger{}_c)^2 + (\mathbf{a}^\dagger{}_d)^2)|0\rangle_a|0\rangle_b|0\rangle_c|0\rangle_d$$
$$= \frac{-i}{2}|0\rangle_a|0\rangle_b(\sqrt{2}|2\rangle_c|0\rangle_d + \sqrt{2}|0\rangle_c|2\rangle_d) = \frac{-i}{\sqrt{2}}|0\rangle_a|0\rangle_b(|2\rangle_c|0\rangle_d + |0\rangle_c|2\rangle_d). \tag{18.232}$$

Thus, two photons simultaneously incident on a beam splitter will emerge both together from port c OR both together from port d. The is the HOM effect (Hong-Ou-Mandel) and is a proof of second quantization of the photon.[9]

The same analysis can be applied to coherent states and to squeezed states. In these cases, either the displacement operator $D_a(\alpha)$ or the squeeze operator can be written in terms of the transformed creation and annihilation operators, and it can be shown that the properties of the initial states are modified when passing through the beam splitter in the expected manner. For example, for a coherent state with real α,

$$D_a(\alpha) = e^{\alpha(\mathbf{a}^\dagger{}_a - \mathbf{a}_a)} \Longrightarrow e^{\alpha(\mathbf{a}^\dagger{}_c - \mathbf{a}_c - i\mathbf{a}^\dagger{}_d - i\mathbf{a}_d)/\sqrt{2}} = e^{(\alpha/\sqrt{2})(\mathbf{a}^\dagger{}_c - \mathbf{a}_c)}e^{(-i\alpha/\sqrt{2})(\mathbf{a}^\dagger{}_d + \mathbf{a}_d)} \tag{18.233}$$

because the c and d operators commute. The second factor on the right had side represents a displacement operator for X instead of P, and this reflects the fact that the beam splitter introduces a $\pi/2$ phase shift between the reflected and transmitted states. The Baker-Campbell-Hausdorff fomula can be applied here, and it is easily seen that we have a coherent state identical, up to a phase factor, with that generated by the P displacement operator. The average number of photons in a state $D_a(\alpha)|0\rangle_a$ is $\bar{n} = |\alpha|^2$ so we will see an incoming coherent state become a product of two coherent states, with

$$\left|\frac{\alpha}{\sqrt{2}}\right|^2 = \frac{|\alpha|^2}{2} = \frac{\bar{n}}{2} \tag{18.234}$$

and therefore one-half the initial number of photons will be in each arm. We could do a similar calculation for $D_b(\beta)$ for a coherent state simultaneously incident on the b port, and a similar separation would independently occur due to the fact that the a and b states are orthogonal, with subsequent interference effects between the c and d depending on the relative phases.

As a final note, beam splitters are used in optical interferometers where photons are presented a choice over the path they will take, after which the paths are combined and interference due to possible phase shifts is observed. For this interference to occur, the overlap of the final paths must be very accurate, and the apparatus must be extremely stable. However, the requirements are not so severe as one might expect. Because the paths are determined by the principle of least action, small perturbations only affect the phase in second order. This is a theorem due to Stodolsky and Chiu,[10] who pointed out that the phase difference for paths that almost meet classically in space is not

[9] Hong, C. K., Ou, Z. Y. and Mandel, L. *Measurement of subpicosecond time intervals between two photons by interference*, Phys. Rev. Lett. **59** 2044 (1987), received 10 July 1987.

[10] Chiu, C. and Stodolsky, Leo, *Theorem in matter-wave interferometry*, Phys. Rev. D 22, 1337 (1980), received 7 September, 1979.

affected by first-order variations of the beam. Likewise, when some external influence, such as a potential, is applied the first-order effect arises only from a direct phase change, for example a change in velocity along one path, and not from changes in the path geometry.

18.7 Two-dimensional harmonic oscillator

The Hamiltonian for a particle of mass m bound in a two-dimensional harmonic oscillator potential in the $x - y$ plane but free to move in the z direction is

$$H = \frac{\mathbf{p}_x^2}{2m} + \frac{\mathbf{p}_y^2}{2m} + \frac{\mathbf{p}_z^2}{2m} + \frac{\omega^2}{2m}(\mathbf{x}^2 + \mathbf{y}^2) = H_x + H_y + H_z. \tag{18.235}$$

This solution to $H|\psi\rangle = E|\psi\rangle$ for the energy eigenfunctions is easily solved by separation of variables, as

$$H|\psi_x\rangle|\psi_y\rangle|\psi_z\rangle = \left[\frac{p_z^2}{2m} + \hbar\omega(n_x + \frac{1}{2}) + \hbar\omega(n_y + \frac{1}{2})\right]|\psi_x\rangle|\psi_y\rangle|\psi_z\rangle \tag{18.236}$$

and we see a simple product of one-dimensional harmonic oscillator states for x and y, and a free particle state for z, completely solve this problem. We can neglect the z part of the solution, e.g., work in the center-of-mass frame. The energy levels are then

$$E_{n_x n_y} = \hbar\omega(n_x + \frac{1}{2}) + \hbar\omega(n_y + \frac{1}{2}) = \hbar\omega(n_x + n_y + 1) = \hbar\omega(n + 1). \tag{18.237}$$

These states, for large $n > 1$, are degenerate, i.e., many have the same energy and are indistinguishable based on this eigenvalue. The degree of degeneracy is determined by the number of possible ways two non-negative integers can be summed to give a fixed value, that is, if we choose n, how many possible ways are there to add $n_x + n_y = n$? The degree of degeneracy is $n + 1$.

18.7.1 Angular momentum states of the 2-d harmonic oscillator

Classically, we know that the solution to the two-dimensional harmonic oscillator can have circular motion. Let us calculate the z component of the angular momentum of a given n_x, n_y state.

$$\mathbf{L}_x = \mathbf{x}\mathbf{p}_y - \mathbf{y}\mathbf{p}_x. \tag{18.238}$$

Now $\mathbf{x}$ and $\mathbf{p}_x$ can be expressed in terms of $\mathbf{a}_x$ and $\mathbf{a}^\dagger{}_x$,

$$\mathbf{x} = \frac{1}{\sqrt{2}\beta}(\mathbf{a}_x + \mathbf{a}^\dagger{}_x); \quad \mathbf{p}_x = \frac{i\hbar\beta}{\sqrt{2}}(\mathbf{a}^\dagger{}_x - \mathbf{a}_x) \tag{18.239}$$

and similarly for the y operators, where $\beta = \sqrt{m\omega/\hbar}$. Noting that $[a_x, a_y] = 0$,

$$\mathbf{L}_z = i\hbar(\mathbf{a}_x\mathbf{a}^\dagger{}_y - \mathbf{a}^\dagger{}_x\mathbf{a}_y). \tag{18.240}$$

The Hamiltonian can be expressed in a similar fashion,

$$H = \hbar\omega(\mathbf{a}^\dagger{}_x\mathbf{a}_x + \mathbf{a}^\dagger{}_y\mathbf{a}_y + 1) \tag{18.241}$$

and therefore

$$\frac{1}{i\hbar^2\omega}[\mathbf{L}_z, H] = [\mathbf{a}_x\mathbf{a}^\dagger{}_y, \mathbf{a}^\dagger{}_x\mathbf{a}_x + \mathbf{a}^\dagger{}_y\mathbf{a}_y] - [\mathbf{a}^\dagger{}_x\mathbf{a}_x + \mathbf{a}^\dagger{}_y\mathbf{a}_y, \mathbf{a}^\dagger{}_x\mathbf{a}_y]. \tag{18.242}$$

Recalling that $[\mathbf{a}_x, \mathbf{a}^\dagger{}_x] = 1$ and $[\mathbf{a}_y, \mathbf{a}^\dagger{}_y] = 1$,

$$[\mathbf{a}_x\mathbf{a}^\dagger{}_y, \mathbf{a}^\dagger{}_x\mathbf{a}_x + \mathbf{a}^\dagger{}_y\mathbf{a}_y] = \mathbf{a}_x\mathbf{a}^\dagger{}_y - \mathbf{a}_x\mathbf{a}^\dagger{}_y = 0 \tag{18.243}$$

$$[\mathbf{a}^\dagger{}_x\mathbf{a}_y, \mathbf{a}^\dagger{}_x\mathbf{a}_x + \mathbf{a}^\dagger{}_y\mathbf{a}_y] = -\mathbf{a}^\dagger{}_x\mathbf{a}_y + \mathbf{a}^\dagger{}_x\mathbf{a}_y = 0 \tag{18.244}$$

and therefore

$$[H, \mathbf{L}_z] = 0 \tag{18.245}$$

and thus we should be able to find a set of eigenfunctions common to both H and $\mathbf{L}_z$.

Let us introduce two new operators,

$$\mathbf{a}_\pm = \frac{1}{\sqrt{2}}(\mathbf{a}_x \mp i\mathbf{a}_y), \tag{18.246}$$

which commute as $[\mathbf{a}_\pm, \mathbf{a}^\dagger{}_\pm] = 1$, analogous to the linear harmonic oscillator raising and lowering operators. Expanding the number operator,

$$\mathbf{a}^\dagger{}_\pm\mathbf{a}_\pm = \frac{1}{2}(\mathbf{a}^\dagger{}_x\mathbf{a}_x + \mathbf{a}^\dagger{}_y\mathbf{a}_y \mp i\mathbf{a}^\dagger{}_x\mathbf{a}_y \pm i\mathbf{a}_x\mathbf{a}^\dagger{}_y) \tag{18.247}$$

the Hamiltonian can obviously be expanded as

$$\hbar\omega(\mathbf{a}^\dagger{}_+\mathbf{a}_+ + \mathbf{a}^\dagger{}_-\mathbf{a}_-) = \hbar\omega(\mathbf{a}^\dagger{}_x\mathbf{a}_x + \mathbf{a}^\dagger{}_y\mathbf{a}_y) \tag{18.248}$$

and the angular momentum as

$$\mathbf{L}_x = \hbar(\mathbf{a}^\dagger{}_+\mathbf{a}_+ - \mathbf{a}^\dagger{}_-\mathbf{a}_-) = i\hbar(\mathbf{a}_x\mathbf{a}^\dagger{}_y - \mathbf{a}^\dagger{}_x\mathbf{a}_y) \tag{18.249}$$

as derived above.

Therefore, the operators that determine the number of right and left "circular quanta" are

$$n_+ = \mathbf{a}^\dagger{}_+\mathbf{a}_+; \quad n_- = \mathbf{a}^\dagger{}_-\mathbf{a}_- \tag{18.250}$$

so the Hamiltonian can be written as

$$H = \hbar\omega(n_+ + n_- + 1) = \hbar\omega(n+1) \tag{18.251}$$

and the angular momentum as

$$\mathbf{L}_z = \hbar(n_+ - n_-) = \hbar M \tag{18.252}$$

where we choose M as the (z component) of the angular momentum to distinguish it from m, the mass.

We thus see that the operator $\mathbf{a}^\dagger_+$ increases M by one, whereas $\mathbf{a}^\dagger{}_-$ decreases M by one. Thus these operators are circular quanta creation and annihilation operators.

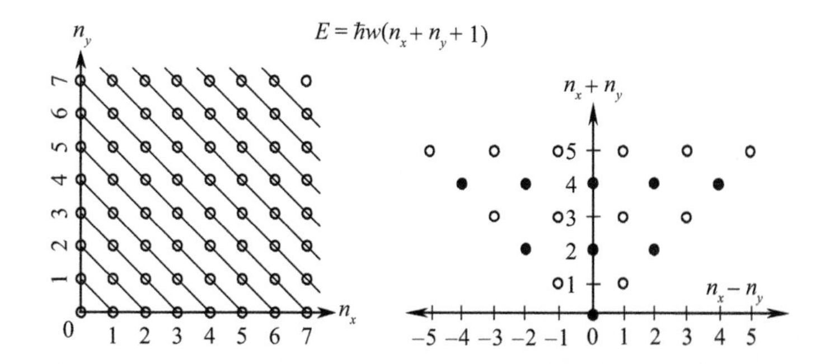

Fig. 18.4 Energy levels of the 2D harmonic oscillator. The lines on the left plot indicated constant energy, on the right the parity (filled dots, even parity; open dots, odd parity) as a function of energy is shown.

The eigenvalues for a constant $n = n_x + n_y = n_+ + n_-$ are shown at the left in Fig. 18.4. In the n_x, n_y basis, lines of constant $n = n_x + n_y$ are shown. Perpendicular to those lines are lines of constant M, and these can be plotted as shown on the right and thereby demonstrate an interesting symmetry in the problem. Changing n_x to $n_x - 1$ and n_y to $n_y + 1$, which keeps n constant, changes M by 2. The right plot shows how M varies with constant n, and indicates that the otherwise ambiguous states have a distinct label when M is included.

18.7.2 2-dimensional harmonic oscillator wave functions

It is straightforward to construct states with a particular n_x and n_y.

$$\langle x, y \,| n_x, n_y\rangle = \frac{1}{\sqrt{n_x! n_y!}} (\mathbf{a}^\dagger{}_x)^{n_x} (\mathbf{a}^\dagger{}_y)^{n_y} \left[\sqrt[4]{\frac{\beta}{\pi}} e^{-\beta^2 x^2} \sqrt[4]{\frac{\beta}{\pi}} e^{-\beta^2 y^2} \right], \tag{18.253}$$

and in the coordinate bases,

$$\mathbf{a}^\dagger{}_x = \frac{1}{\sqrt{2}} \left(\beta x - \frac{1}{\beta} \frac{\partial}{\partial x} \right) \tag{18.254}$$

$$\mathbf{a}^\dagger{}_y = \frac{1}{\sqrt{2}} \left(\beta y - \frac{1}{\beta} \frac{\partial}{\partial y} \right). \tag{18.255}$$

Alternatively, we could generate the wavefunctions in a cylindrical coordinate system by use of the substitution

$$x = \rho \cos\phi; \quad y = \rho \sin\phi; \quad \rho \geq 0; \quad 0 \leq \phi \leq 2\pi \tag{18.256}$$

and then determine the coordinate representation of $\mathbf{a}^\dagger{}_+$ and $\mathbf{a}^\dagger{}_-$. In the exponent of the ground state, this substitution allows the exponents in the x and y factors to be added as $x^2 + y^2 = \rho^2$. Then, $\mathbf{a}^\dagger{}_+$ and $\mathbf{a}^\dagger{}_-$ act directly on this state to give a specific n and M. Specifically, the coordinate representation of these operators is,

$$\mathbf{a}^\dagger{}_+ = \frac{e^{i\phi}}{2}\left[\beta\rho - \frac{1}{\beta}\frac{\partial}{\partial\rho} - \frac{i}{\beta\rho}\frac{\partial}{\partial\phi}\right] \tag{18.257}$$

$$\mathbf{a}^\dagger{}_- = \frac{e^{-i\phi}}{2}\left[\beta\rho - \frac{1}{\beta}\frac{\partial}{\partial\rho} + \frac{i}{\beta\rho}\frac{\partial}{\partial\phi}\right]. \tag{18.258}$$

Each application of $\mathbf{a}^\dagger{}_x\mathbf{a}^\dagger{}_y$ adds a node to the radial wavefunction, but increases the energy by $2\hbar\omega$ because a vibration quantum is added to each oscillator. If there are an unequal number of applications of $\mathbf{a}^\dagger{}_x$ and $\mathbf{a}^\dagger{}_y$ to the ground state, the energy is given by

$$E = \hbar\omega(2\min(n_x, n_y) + |M| + 1). \tag{18.259}$$

We have argued that the number of nodes in the azimuthally symmetric radial part of the wavefunction is simply the $\min(n_x, n_y) \equiv n_r$, and further applications of $|n_x - n_y| = |M|$ generates azimuthal variations, hence angular momentum. We can therefore conclude that the energy levels of the 2d harmonic oscillator are

$$E_{n_r,M} = \hbar\omega(2n_r + |M| + 1) = \hbar\omega(n_x + n_y + 1) = \hbar\omega(n_+ + n_- + 1) \tag{18.260}$$

where $M = n_+ - n_-$. We see that the eigenstates all have unique identifiers and the energy is only doubly degenerate as it depends on $|M|$.

This result can also be obtained by solving the 2d oscillator Schrödinger equation, however, that is not in the spirit of the algebraic and operator methods outlined in this chapter. It is considerable work, and after several variable substitutions, a recursion relation is obtained that diverges except for energies equal to the energy eigenvalues as given above.

18.8 2D harmonic oscillator solution to the H atom

As an example of the application of functional analysis to the solution of a quantum mechanics problem in terms of another known solution, let us consider the hydrogen atom, which can be transformed to a two-dimensional harmonic oscillator.[11] (There are similar methods that do not start from the Schrödinger equation, but use only operators to determine the eigenvalues of the the H atom Hamiltonian—see Section 18.12 for Pauli's matrix mechanics solution.)

The spherical-coordinate system radial equation for the Hydrogen atom can be written as

$$\left[-\frac{\hbar^2}{2m}\frac{1}{r^2}\frac{d}{dr}r^2\frac{d}{dr} + \frac{l(l+1)\hbar^2}{2mr^2} - \frac{e^2}{r} - E_H\right]R_{nl}(r) = 0. \tag{18.261}$$

where n is the number of radial nodes, and l is the angular momentum. The following coordinate transformation (due to Schwinger) will recast this equation into that of the two-dimensional harmonic oscillator:

[11]This method is due to J. Schwinger, and appears only as unpublished lecture notes.

$$r = \frac{\lambda\rho^2}{2}; \quad R_{nl}(r) = \frac{\chi(\rho)}{\rho} \tag{18.262}$$

where the constant $\lambda > 0$. In this coordinate system, the derivative becomes,

$$\frac{d}{dr} = \frac{1}{\lambda\rho}\frac{d}{d\rho}. \tag{18.263}$$

With these substitutions, after multiplying Eq. (18.261) by $-\frac{2m}{\hbar^2}r^2$, the radial equation becomes

$$\left[\frac{1}{\lambda\rho}\frac{d}{d\rho}\frac{\lambda\rho^3}{4}\frac{d}{d\rho} - l(l+1) + \frac{2me^2}{\hbar^2}\frac{\lambda\rho^2}{2} + \frac{2mE_H}{\hbar^2}\frac{\lambda^2\rho^4}{4}\right]\frac{\chi(\rho)}{\rho} = 0 \tag{18.264}$$

which can be simplified to

$$\left[\frac{d^2}{d\rho^2} + \frac{1}{\rho}\frac{d}{d\rho} - \frac{(2l+1)^2}{\rho^2} + \frac{2mE_H\lambda^2\rho^2}{\hbar^2} + \frac{4me^2\lambda}{\hbar^2}\right]\chi(\rho) = 0 \tag{18.265}$$

or, using prime notation for the one-dimensional derivative,

$$\chi''(\rho) + \frac{1}{\rho}\chi'(\rho) - \left(\frac{(2l+1)^2}{\rho^2} - \frac{2mE_H\lambda^2\rho^2}{\hbar^2} - \frac{4me^2\lambda}{\hbar^2}\right)\chi(\rho) = 0. \tag{18.266}$$

This can be compared to the radial equation for the two-dimensional harmonic oscillator obtained by writing the Schrödinger equation in polar (cylindrical) coordinates with radial coordinate ρ, where the angular factor has been accounted for by separation of variables (this procedure is outlined in most mathematical physics books and we do not reproduce it here):

$$R''_{n_r}(\rho) + \frac{1}{\rho}R'_{n_r}(\rho) - \left(\frac{M^2}{\rho^2} + \frac{m^2\omega^2}{\hbar^2}\rho^2 - \frac{2mE_{ho}}{\hbar^2}\right)R_{n_r}(\rho) = 0 \tag{18.267}$$

and we use the label n_r for the harmonic oscillator radial quantum number because we used n in R_{nl} for the hydrogen atom. Through a term-by-term comparison we can deduce the following relations, assuming $E_H \leq 0$ (bound states):

$$\omega = \sqrt{\frac{2\lambda^2|E_H|}{m}}; \quad |M| = 2l+1; \quad E_{ho} = 2e^2\lambda. \tag{18.268}$$

In the last section, we showed that the two-dimensional harmonic oscillator energy levels are

$$E_{ho} = (2n_r + |M| + 1)\hbar\omega = 2e^2\lambda \tag{18.269}$$

where the integers $n_r \geq 0$ and $|M| \leq 2n_r$. We have identified $|M| = 2l+1$ from the term-by-term comparison, therefore we can take

$$(2n_r + 2l + 1 + 1) = 2(n_r + l + 1) = 2(n+1) \tag{18.270}$$

because $n_r + l + 1 = 1, 2, 3, ... = n + 1$ is an integer, and $n = 0, 1, 2....$ So

$$2e^2\lambda = 2(n+1)\hbar\omega = 2(n+1)\hbar\sqrt{\frac{2\lambda^2|E|}{m}} \tag{18.271}$$

and therefore, noting that $E_H < 0$,

$$E_H = -\frac{me^4}{2(n+1)^2\hbar^2} \tag{18.272}$$

which is the well-known result.

The radial wavefunctions are obtained directly from the 2D harmonic oscillator wavefunctions by dividing by ρ and then reversing the substitution by $\rho^2 = 2r/\lambda$. The hydrogen ground state $n = 0$ is obtained from the $|M| = 1$ 2d harmonic oscillator state as (recall that $|M|$ corresponds to $2l + 1 = 1$ and we have $l = 0$ for the ground state).

$$\frac{\chi(\rho)}{\rho} = \frac{1}{\rho}\frac{\beta^2}{\sqrt{pi}}\rho e^{-\beta^2\rho^2/2}. \tag{18.273}$$

Using $\beta^2 = m\omega/\hbar$, $\omega = \sqrt{2\lambda^2|E|/m}$,

$$\begin{aligned} R_{00}(r) &= \sqrt{\frac{2m\lambda^2|E|}{\pi\hbar^2}}e^{-}(\sqrt{2|E|m/\hbar^2})\ r \\ &= \sqrt{\frac{\lambda^2m^2e^4}{\pi\hbar 4}}e^{-}(me^2/\hbar^2)\ r \\ &= \sqrt{\frac{\lambda^2}{\pi a_0}}e^{-}r/a_0 \end{aligned} \tag{18.274}$$

where a_0 is the Bohr radius, and note that the normalization factor is correct if we take $\lambda = 1/a_0$.

18.9 Sum rules and summation techniques

Very often, a sum over states is required in perturbation theory, in the establishment of completeness, the calculation of the total energy in a system (the partition function), and more. Sometimes, the matrix elements of an operator or perturbating Hamiltonian can be derived without actually knowing the wavefunction, as in the case of the Harmonic oscillator.

Sum rule identities can provide a large amount of information in a compact form, in a way that can be directly investigated by experimental measurements. Such measurements can provide a test of the assumed form of fundamental interactions or a test of the perturbative techniques that were employed in deriving the sum rule. Historically, they were important in the early days of quantum mechanics as a test of its application to atomic physics. More recently, QCD sum rules were used to determine the masses of the light and heavy quarks that are otherwise not directly measurable. There have been thousands of papers written on QCD sum rules, some of which are the most highly cited papers in particle physics.[12]

[12] A pedagogical review is presented my Belloni, M. and Robinett, R.W., *Quantum mechanical sum rules for two model systems*, Amer. Jour. Phys. **76**, 798 (2008).

18.9.1 General sum rules

Most sum rules make use of the fact that the quantum states that describe a system form a complete set. Let us take a case where we have energy eigenstates $H|n\rangle = E_n|n\rangle$. Then for some arbitrary Hermitian operator $\mathbf{A}$,

$$\begin{aligned}\sum_k |\langle n|\mathbf{A}|k\rangle|^2 &= \sum_k \langle n|\mathbf{A}|k\rangle\langle k|\mathbf{A}|n\rangle \\ &= \langle n|\mathbf{A}\left(\sum_k |k\rangle\langle k|\right)\mathbf{A}|n\rangle \\ &= \langle n|\mathbf{A}^2|n\rangle\end{aligned} \tag{18.275}$$

where the sum is over all $|k\rangle$ states, both discrete and continuum in which case the sum becomes an integral, and we used the fact that, for a complete set,

$$\sum_k |k\rangle\langle k| = 1. \tag{18.276}$$

Taking $\mathbf{A} = \mathbf{x}$ the dipole matrix element $x-$closure sum rule is obtained,

$$\sum_k |\langle n|\mathbf{x}|k\rangle|^2 = |\langle n|\mathbf{x}^2|n\rangle| \tag{18.277}$$

Next, we will derive the Thomas-Reiche-Kuhn (TRK) sum rule, from the commutation relationship

$$[\mathbf{p}, \mathbf{x}] = \frac{\hbar}{i}, \tag{18.278}$$

implying that

$$\frac{\hbar}{i} = \langle n|\left(\mathbf{px} - \mathbf{xp}\right)|n\rangle = \sum_k \left(\langle n|\mathbf{p}|k\rangle\langle k|\mathbf{x}|n\rangle - \langle n|\mathbf{x}|k\rangle\langle k|\mathbf{p}|n\rangle\right). \tag{18.279}$$

If the Hamiltonian $H = \mathbf{p}^2/2m + V(x)$, then $[\mathbf{x}, V(x] = 0$ and

$$[H, \mathbf{x}] = \frac{1}{2m}[\mathbf{p}^2, \mathbf{x}] = \frac{\hbar}{mi}\mathbf{p} \tag{18.280}$$

and we can replace $\mathbf{p}$ in terms of this commutator, so that

$$\frac{\hbar}{i} = \sum_k \left(\langle n|\frac{mi}{\hbar}[H, \mathbf{x}]|k\rangle\langle k|\mathbf{x}|n\rangle - \langle n|\mathbf{x}|k\rangle\langle k|\frac{mi}{\hbar}[H, \mathbf{x}]|n\rangle\right) \tag{18.281}$$

and operating with H and collecting terms we find

$$\frac{\hbar^2}{m} = \sum_k \left(\langle n|\left(E_n\mathbf{x} - \mathbf{x}E_k\right)|k\rangle\langle k|\mathbf{x}|n\rangle - \langle n|\left(\mathbf{x}|k\rangle\langle k|E_k\mathbf{x} - \mathbf{x}E_n\right)|n\rangle\right) \tag{18.282}$$

and therefore

$$\frac{\hbar^2}{2m} = \sum_k (E_k - E_n)|\langle n|x|k\rangle|^2. \tag{18.283}$$

The TRK sum rule can be generalized for any function $F(x)$ as follows.[13]

$$\begin{aligned}\sum_k (E_k - E_n)|\langle n|F(x)|k\rangle|^2 &= \sum_n (E_k - E_n)\langle n|F(x)|k\rangle\langle n|F(x)|k\rangle^* \\ &= \sum_n (E_k - E_n)\langle n|F(x)|k\rangle\langle k|F^\dagger(x)|n\rangle \\ &= \sum_k \langle n|[F,H]|k\rangle\langle k|F^\dagger(x)|n\rangle \\ &= \langle n|[F(x),H]F^\dagger(x)|n\rangle.\end{aligned} \tag{18.284}$$

Similarly, noting that the matrix elements in the energy basis are symmetric, $\langle n|F(x)|k\rangle = \langle k|F(x)|n\rangle$, and combining these two results,

$$\sum_k (E_k - E_n)|\langle n|F(x)|k\rangle|^2 = \frac{1}{2}\langle n|[F^\dagger(x),[H,F]|n\rangle. \tag{18.285}$$

Because

$$[H,F(x)] = \frac{1}{2m}[\mathbf{p}^2,F(x)] = -\frac{i\hbar}{m}\frac{dF(x)}{dx}\mathbf{p} - \frac{\hbar^2}{2m}\frac{d^2F(x)}{dx^2}, \tag{18.286}$$

the double commutator is

$$[F^\dagger(x),[H,F(x)]] = -\frac{i\hbar}{m}\left[F^\dagger(x), \frac{dF(x)}{dx}\mathbf{p}\right]. \tag{18.287}$$

Therefore, the sum rule for the operator $F(x)$ is

$$\sum_n (E_n - E_k)|\langle n|F(x)|k\rangle|^2 = \frac{\hbar^2}{2m}\left\langle n\left|\frac{dF(x)}{dx}\frac{dF^\dagger(x)}{dx}\right|n\right\rangle \tag{18.288}$$

and if $F^\dagger(x) = F(x)$, then

$$\sum_n (E_n - E_k)|\langle n|F(x)|k\rangle|^2 = \frac{\hbar^2}{2m}\left\langle n\left|\left(\frac{dF(x)}{dx}\right)^2\right|n\right\rangle. \tag{18.289}$$

With this, the TRK sum rule is easily obtained by taking $F(x) = x$.

Taking $F(x) = x^2$ produces the so-called monopole sum rule,

$$\sum_k (E_k - E_n)|\langle n|\mathbf{x}^2|k\rangle|^2 = \frac{2\hbar^2}{m}\langle n|\mathbf{x}^2|n\rangle \tag{18.290}$$

which is used in collective excitations.

[13] Wang, S., *Generalization of the Thomas-Reiche-Kuhn and the Bethe sum rules*, Phys. Rev. A **60**, 262 (1999).

The Bethe sum rule, used in X-ray scattering, and neutron scattering (it is the so-called first moment sum rule in neutron scattering), can also be easily obtained, taking $F(x) = e^{iqx}$,

$$\sum_k (E_k - E_n)|\langle n|e^{iqx}|k\rangle|^2 = \frac{\hbar^2}{2m}\langle n|q^2|n\rangle = \frac{\hbar^2 q^2}{2m}. \tag{18.291}$$

18.9.2 The Casimir force in one dimension

When quantization is applied to unbound electromagnetic modes, free space becomes imbued with zero-point energy which appears as an infinite energy. Thus, the reality of zero-point energy was rejected by many, particularly Pauli, who calculated that if the free space electromagnetic mode energy is summed to a modest limit corresponding to the Compton wavelength of the electron, the mass of empty space in a volume within the moon's orbit would result in a black hole. Clearly this in not observed. Nonetheless, zero-point energy appears as an essential feature of quantum mechanics, and its effect on the universe not fully understood. Its effects are often handled in an ad hoc fashion by use of theoretical techniques, including renormalization and cutoff functions.

Casimir (1948) was the first to propose a way to directly measure the physical effects of zero-point energy.[14] The Casimir force occurs between two closely spaced plane parallel mirrors; the modification of the electromagnetic field zero-point energy between the plates, and its variation as a function of the mirrors' separation, leads to an attractive force between the mirrors, as illustrated in Fig. 18.5.

Let us consider a simpler problem: Vibrations (waves) on a string of variable length a held taut with constant tension between two movable supports, with wave propagation constrained to the $\hat{z}$ direction defined as along the axis between the supports. We will consider only one vibration (polarization) direction.

From continuum mechanics as outlined in Chapter 12, the classical vibrational modes have eigenvalues

$$\omega = ck = 2\pi c/\lambda = \frac{n\pi c}{a} \tag{18.292}$$

where n is the number of half-wavelengths that fit into the cavity, c is the phase velocity (speed of light for electromagnetic waves) and a is the string length or the mirror separation. In second quantization, each eigenmode of a vibrating string or an electromagnetic field in a cavity is excited at discrete energy levels through application

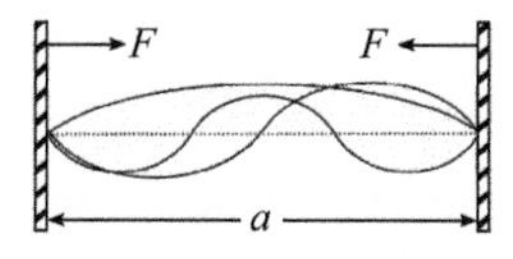

Fig. 18.5 Modification of the electromagnetic modes between two mirrors causes a separation dependence of the zero-point energy, leading to an attractive force between the mirrors.

[14]Casimir, H.B.G., *On the Attraction Between Two Perfectly Conducting Plates*, Proc. Kon. Ned. Acad. Wetenschap **51**, 793 (1948).

of creation and annihilation operators. The energy spectrum for each mode is therefore that of a harmonic oscillator,

$$E_n = \frac{\hbar n\pi c}{a}\left(N_n + \frac{1}{2}\right), \tag{18.293}$$

where N_n is the excitation level of the nth mode. In equilibrium at sufficiently low temperature, $N_n = 0$, however, the zero-point energy of each mode remains.

The zero-temperature force between the mirrors can be calculated as follows, in the one-dimensional case. The total energy is given by the sum of all the ground state modes,

$$E = \sum_{n=1}^{\infty} E_n = \sum_{n=1}^{\infty} \frac{\pi\hbar c n}{2a} = \frac{\pi\hbar c}{2a}\sum_{n=1}^{\infty} n \tag{18.294}$$

and we can use the Ramanujan sum,[15] (which can be justified in this case, as it represents an analytic continuation of the Riemann zeta function, $\zeta(1) = -1/12$.)

$$1 + 2 + 3 + ... = -\frac{1}{12} \tag{18.295}$$

to determine the energy, and thereby the force through the principle of virtual work,

$$E = -\frac{\pi\hbar c}{24a}; \quad F = -\frac{\partial E}{\partial a} = -\frac{\pi\hbar c}{24}\frac{1}{a^2}. \tag{18.296}$$

The force between the mirrors is *attractive*.

Perhaps the reader will balk at the use of the Ramanujan summation. We can instead use complex integration techniques to find the eigenvalues and their sum. We can treat this as an eigenvalue problem that consists of a function and boundary conditions that determines the eigenvalues of k that satisfy the equation

$$\sin kx = 0 = \left(e^{-i|k|x} - e^{i|k|x}\right) = e^{-2i|k|x} - 1 \tag{18.297}$$

at $x = 0$ and $x = a$. In terms of the frequency, from the wave equation, $k^2 = \omega^2/c^2$, and we take k to be a function of ω as

$$k = k(\omega) = \sqrt[+]{\omega^2/c^2} \tag{18.298}$$

and define a function

$$f(\omega) = e^{-2i\omega a/c} - 1. \tag{18.299}$$

The zeros of this function determine the eigenvalues.

We next consider ω to be a complex variable, in which case the eigenvalues (zeros of $f(\omega)$) of interest are those on the real axis at $nc\pi/a$. The number of zeros N_z minus

[15] "=" in this context means "behaves as if." See, e.g., Candelpergher, B., *Ramanujan summation of divergent series*, Lectures Notes in Mathematics 2185, (Springer, 2017). DOI 10.1007/978-3-319-63630-6.

the number of poles N_p in a given region bounded by the contour C is, for any function, is

$$\frac{1}{2\pi i}\oint_C \frac{f'(\omega)}{f(\omega)}d\omega = N_z - N_p, \tag{18.300}$$

which can be easily seen by applying contour integration to a sum of simple poles and zeros. The $f(\omega)$ considered here has only zeros.

This is the so-call *argument theorem* and is valid for any function that is analytic except for poles on and inside a simple closed curve C (a meromorphic function), such as $f(\omega)$ in the present case. $f(\omega)$ has no poles in the right half plane, and the argument theorem can be modified, by multiplying $f(\omega)$ by ω, to give the sum of all the eigenvalues in a region,

$$\frac{1}{2\pi i}\oint_C \omega\frac{f'(\omega)}{f(\omega)}d\omega = \left[\sum_n \omega_n\right]_{f(\omega_n)=0}. \tag{18.301}$$

There are no poles for $f(\omega)$ so there is no sum over poles.

The path C is taken as semicircle of radius R in the right-half plane and a line from iR to $-iR$, and then taking $R \to \infty$, to encircle the eigenvalues along the $+\omega$ axis. For ω sufficiently large, there are no contained modes for mirrors or a string made of realistic materials, so there is no contribution from the semicircle. This gives the sum of ω_n along the positive real axis. Introducing $i\xi = \omega$, and $f(\imath\xi) = g(\xi)$, and then reversing the integration limits, the zero-point energy as a function of a is therefore

$$\begin{aligned} E(a) &= \frac{\hbar}{2}\frac{1}{2\pi i}\int_{\infty}^{-\infty} i\xi\frac{\partial f(i\xi)}{\partial i\xi}\frac{1}{f(i\xi)}d(i\xi) \\ &= -\frac{\hbar}{2}\frac{1}{2\pi}\int_{-\infty}^{\infty} \xi\,\frac{\partial g(\xi)}{\partial \xi}\frac{1}{g(\xi)}\,d\xi \\ &= -\frac{\hbar}{2}\frac{1}{2\pi}\int_{-\infty}^{\infty} \xi\,\frac{\partial \log g(\xi)}{\partial \xi}\,d\xi \\ &= \frac{\hbar}{2}\frac{1}{2\pi}\int_{-\infty}^{\infty} \log g(\xi)\;d\xi \end{aligned} \tag{18.302}$$

where the last step is an integration by parts, and

$$g(\xi) = e^{2|\xi|a/c} - 1. \tag{18.303}$$

By the principle of virtual work, the force is

$$\begin{aligned} F(a) &= -\frac{\partial E(a)}{\partial a} = -\frac{\hbar}{2}\frac{1}{2\pi}\int_{-\infty}^{\infty}\frac{1}{g(|\xi|)}\frac{2|\xi|}{c}e^{2|\xi|a/c}d\xi \\ &= -\frac{\hbar}{\pi c}\int_0^{\infty}\xi\frac{1}{g(\xi)}(g(\xi)+1)\;d\xi \\ &= -\frac{\hbar}{\pi c}\int_0^{\infty}\left[\xi + \frac{\xi}{g(\xi)}\right]\;d\xi. \end{aligned} \tag{18.304}$$

The first term in the integral does not depend on a and is therefore a constant "background" force, which happens to be infinite. This term can be ignored, e.g., if we look at the difference in force for two different values of a, it does not contribute. This infinite term represents a general problem frequently encountered when considering zero-point energy.

We thus arrive at

$$F(a) = -\frac{\hbar}{\pi c}\int_0^\infty \frac{\xi}{e^{2\xi a/c}-1}d\xi = -\frac{\hbar c}{4\pi}\frac{1}{a^2}\int_0^\infty \frac{x}{e^x-1}\,dx = -\frac{\pi\hbar c}{24}\frac{1}{a^2} \tag{18.305}$$

which is the same as obtained previously by use of the Ramanujan sum.

This is one of the few ways to directly observe the effects of zero-point energy. The above calculation can be generalized to three dimensions, and in the electromagnetic case the force is[16]

$$\frac{F(a)}{A} = -\frac{\pi^2\hbar c}{240a^4} \tag{18.306}$$

where A is the area of the mirrors, assuming they are of extent much greater than a.

The Casimir force was measured with good precision only after the late 1990s, 50 years after its prediction, although the general existence of a short-range attractive force had been observed many times before.[17] The force becomes "large" compared to thermal fluctuation forces at distances where a is much less than the thermal wavelength,

$$\lambda = \frac{\pi\hbar c}{k_B T} \approx 20\ \mu\text{m}. \tag{18.307}$$

Thus, the force becomes "large" at distances smaller than 1 μm, meaning it is larger than the thermal force, and coincidentally also larger than typical background electrostatic effects due to patch potentials on, e.g., gold-coated surfaces.

As tempting as it is to assign zero-point energy to the electromagnetic modes, it is also possible that the zero-point mode excitation is due to the oscillators in the mirrors themselves. It should also be noted that this is not the retarded van der Waals forces between the molecular dipoles in the two walls; in the first place, the retardation distance for metal plates is infinite (pole at zero in the electrical permittivity). In the second place, the van der Waals forces between molecules is not additive; the introduction of a third molecule alters the interaction between the first two. The Casimir force results from the imposed boundary conditions.

The fact that the amount of energy contained in free space diverges when zero-point energy is assigned to it suggests that this assignment might be only for computational convenience.

18.9.3 The Dalgarno-Lewis summation method

Consider a Hamiltonian H_0 and its complete set of eigenstates $|n\rangle$ with energies E_n. Let us add an perturbing Hamiltonian H'. The first-order (in H') energy change to a state $|n\rangle$ is

$$E_n^{(1)} = \langle n|H'|n\rangle \tag{18.308}$$

[16]See, e.g., Milonni, P.W., *The Quantum Vacuum* (Academic Press, San Diego, CA, 1994), Chapter 7.

[17]The first modern high-accuracy measurement is reported in Lamoreaux, S.K., *Demonstration of the Casimir Force in the 0.6 to 6 μm Range*, Phy. Rev. Lett. **78**, 5 (1997).

which is a single integral and therefore simple enough. In second order, the energy change is

$$E_n^{(2)} = \sum_{k \neq n} \frac{\langle n|H'|k\rangle\langle k|H'|n\rangle}{E_n - E_k}, \tag{18.309}$$

which is obtained from the first-order correction to the wavefunction,

$$|n^{(1)}\rangle = \sum_{k \neq n} |k\rangle \frac{\langle k|H'|n\rangle}{E_n - E_k}. \tag{18.310}$$

The second order correction to the energy, which uses the first-order correction to the wavefunction, is often very complicated, with its sum over a possibly infinite number of states, compared to the first-order correction. Unless there is a closed-form recursion relation among the matrix elements, the sum cannot in general be evaluated without making some approximations.

The matrix solution directly solves the Schrödinger equation, order by order, where the superscript (e) means the exact solution,

$$\begin{aligned} E_n^{(e)} &= E_n + E_n^{(1)} + E_n^{(2)} + \dots. \\ |n^{(e)}\rangle &= |n\rangle + |n^{(1)}\rangle + |n^{(2)}\rangle + \dots \end{aligned} \tag{18.311}$$

with $H_0|n\rangle = E_n|n\rangle$. The order-by-order Schrödinger equation is

$$\begin{aligned} (E_n - H_0)|n\rangle &= 0 \\ (E_n - H_0)|n^{(1)}\rangle &= (H' - E_n^{(1)})|n\rangle \\ E_n^{(1)} &= \langle n|H'|n\rangle \\ E_n^{(2)} &= \langle n|(H' - E_n^{(1)}|n^{(1)}\rangle \end{aligned} \tag{18.312}$$

and we can obtain the second-order solution to the energy by doing a single integral if we can solve the second equation in the group above for $|n^{(1)}\rangle$. Of course, we already have a solution for it, Eq. (18.310), but it contains an infinite sum. This inspires us to solve the above equations directly by other means for $|n^{(1)}\rangle$.[18]

Let us define a hermitian operator $\mathbf{F}_n$, which is unique for each $|n\rangle$ such that

$$[\mathbf{F}_n, H_0]\ |n\rangle = (H' - E_n^{(1)})|n\rangle. \tag{18.313}$$

Then, taking the inner product with $\langle k|$ on both sides of the equation, noting that $\langle k|n\rangle = \delta_{kn}$,

$$\begin{aligned} \langle k|\ [\mathbf{F}_n, H_0]\ |n\rangle &= \langle k|(H' - E_n^{(1)})|n\rangle \\ \langle k|\mathbf{F}_n H_0 - H_0 \mathbf{F}_n]|n\rangle &= \langle k|H'|n\rangle - E_n^{(1)}\delta_k n \\ (E_n - E_k)\langle k|\mathbf{F}_n|n\rangle + E_n^{(1)}\delta_k n &= \langle k|H'|n\rangle. \end{aligned} \tag{18.314}$$

[18]Dalgarno, A., and Lewis, J.T., *The exact calculation of long-range forces between atoms by perturbation theory*, Proc. Roy. Soc. **A233**, 70 (1956).

This can be substituted into the sum for the second order energy correction,

$$\begin{aligned} E_n^{(2)} &= \sum_{k\neq n} \frac{\left[(E_n - E_k)\langle k|\mathbf{F}_n|n\rangle + E_n^{(1)}\delta_{kn}\right]}{E_n - E_k} \\ &= \sum_{k\neq n} \langle n|H'|k\rangle\langle k|\mathbf{F}_n|n\rangle \\ &= \sum_{\text{all k}} \langle n|H'|k\rangle\langle k|\mathbf{F}_n|n\rangle - \langle n|H'|n\rangle\langle \mathbf{F}_n|n\rangle \\ &= \langle n|H'\mathbf{F}_n|n\rangle - E_n^{(1)}\langle n|\mathbf{F}_n|n\rangle, \end{aligned} \tag{18.315}$$

and the unperturbed energy E_n is eliminated from the second-order correction to the energy, and we have eliminated the sum over states, at the cost of finding a function $\mathbf{F}_n$ that satisfies the equation

$$[\mathbf{F}_n, H_0]|n\rangle = (H' - E_n^{(1)})|n\rangle. \tag{18.316}$$

This function can be also used to obtain the first-order correction to the wavefuntion,

$$|n^{(1)}\rangle = (\mathbf{F}_n - \langle n|\mathbf{F}_n|n\rangle)|n\rangle \tag{18.317}$$

that might be needed to match boundary conditions.

All of this might seem like considerable work for little or no gain. However, this process can be used to address otherwise intractable problems, and can be extended to higher orders.

We will demonstrate the power of this technique by solving a problem where $\mathbf{F}_n = F_n(\mathbf{r}) = F_n$, that is, it is a function of spatial coordinates only.[19]

Let us take

$$H_0 = -\frac{\hbar^2}{2m}\nabla^2 + V(\mathbf{r}). \tag{18.318}$$

It is straightforward to work out the commutation

$$[F_n, H_0]\psi(\mathbf{r}) = \frac{\hbar^2}{2m}\left(\psi_n(\mathbf{r})\nabla^2 F + 2\vec{\nabla}F\cdot\vec{\nabla}\psi_n(\mathbf{r})\right) \tag{18.319}$$

because F_n commutes with $V(\mathbf{r})$.

If we can solve the inhomogeneous differential equation for F_n, the second order correction can be directly obtain by a single integral. The equation for $F_n(\mathbf{r}) \equiv F_n$, $\psi_n(\mathbf{r}) \equiv \psi_n$ is

$$\psi_n\nabla^2 F_n + 2(\vec{\nabla}F_n)\cdot(\vec{\nabla}\psi_n) = \frac{2m}{\hbar^2}(H' - E_n^{(1)})\psi_n. \tag{18.320}$$

Taking a specific case, let us calculate the polarizability moments of the hydrogen atom in the ground state. The Hamiltonian and wavefunction are

$$H_0 = -\frac{\hbar^2}{2m}\nabla^2 - \frac{e^2}{r};\quad \psi_0(r) = \frac{1}{\sqrt{\pi a^3}}e^{-r/a} \tag{18.321}$$

with $a = \hbar^2/me^2$.

[19] Charles Schwartz, *Calculations in Schrödinger Perturbation Theory*, Ann. of Phys. **2**, 156 (1959).

The perturbing Hamiltonian for a 2^l-pole electrostatic potential is

$$H' = -e\xi P_l(\cos\theta)r^l \tag{18.322}$$

where P_l is the lth Legendre polynomial ($l > 0$) and ξ is the strength of the field (gradient). For the spherically symmetric ground state, $E_0^{(1)} = 0$, so we need to calculate only $E_0^{(2)}$. Let us take the case $l = 1$ for a static and homogeneous electric field applied to the H atom (the Stark effect).

To calculate all of the matrix elements for every state of the H atom, which in principle includes positive energy (unbound) states, appears as impossible. Even if only a numerical solution exists to Eq. (18.320), we are still far ahead in many instances. However, in this case, finding F_0 for ψ_0 of the hydrogen atom is straightforward.

For the H atom ground state, $E_0^{(1)} = 0$, and

$$\vec{\nabla}\frac{1}{\sqrt{\pi a^3}}e^{-r/a} = -\frac{\hat{r}}{a}\psi_0(r) \tag{18.323}$$

so the differential equation for F_0 is, from Eq. (18.320),

$$\nabla^2 F_0 - \frac{2}{a}\frac{\partial F_0}{\partial r} = -\frac{2me\xi}{\hbar^2}P_1(\cos(\theta))r. \tag{18.324}$$

The Laplacian in spherical coordinates is

$$\nabla^2 = \frac{1}{r^2}\frac{\partial}{\partial r}\left(r^2\frac{\partial}{\partial r}\right) + \frac{1}{r^2\sin\theta}\frac{\partial}{\partial\theta}\left(\sin\theta\frac{\partial}{\partial\theta}\right) + \frac{1}{r^2\sin^2\theta}\frac{\partial^2}{\partial\varphi^2}, \tag{18.325}$$

and there is no φ dependence in our current problem so the term $\partial^2/\partial\phi^2$ is zero. The Legendre functions are eigenfunctions of this equation, so

$$\frac{1}{r^2\sin\theta}\frac{\partial}{\partial\theta}\left(\sin\theta\frac{\partial}{\partial\theta}\right)P_{l'}(\cos\theta) = \frac{l'(l'+1)P_{l'}(\cos\theta)}{r^2}. \tag{18.326}$$

We can therefore expect that $F_0(r,\theta)$ is a function of the form

$$F_0 = f(r/a)P_l'(\cos\theta), \tag{18.327}$$

and for this to match the rhs of Eq. (18.324), $l' = l = 1$.

The differential equation for $f(r)$ is

$$\frac{1}{r^2}\frac{d}{dr}r^2\frac{d}{dr}f(r) - \frac{2}{a}\frac{d}{dr}f(r) - \frac{l(l+1)}{r^2}f(r) = -\frac{2me\xi}{\hbar^2}r^l \tag{18.328}$$

which can be simplified to, setting $l = 1$,

$$f''(r) + 2\left(\frac{1}{r} - \frac{1}{a}\right)f'(r) - \frac{2}{r^2}f(r) = -\frac{2me\xi}{\hbar^2}r. \tag{18.329}$$

Let us first try to solve the inhomogeneous part by taking

$$f_i(r) = \frac{me\xi}{\hbar^2}\frac{ar^2}{2} \tag{18.330}$$

which almost solves the differential equation but leaves a constant term proportional to $2a$. If we add a second function of the form a^2r, the constant term is canceled. Thus, the solution is

$$F_0(r,\theta) = \frac{me\xi}{\hbar^2}P_1(\cos\theta)\left(a^2r + \frac{ar^2}{2}\right). \tag{18.331}$$

The elementary integral, using $\langle \mathbf{r}|0\rangle = \psi_0(r)$,

$$E_0^{(2)} = \langle 0|H'F_0(r,\theta)|0\rangle = -\xi^2\frac{9a^3}{4} \tag{18.332}$$

and, for $l = 1$, the dipole polarizability is

$$\alpha = -\frac{2}{\xi^2}E_0^{(2)} = \frac{9}{2}a^3 \tag{18.333}$$

which is the known result.

This technique works with any system, including multielectron atoms, and has found important applications in atomic physics calculations.

18.9.4 Reflection from a periodic potential: forbidden bands

A one-dimensional plane wave e^{ikx} propagating in the $+x$ direction in a solid encounters regularly spaced short-range potentials. For a given spacing and potential magnitude, at low energies, only certain ranges of wave momentum $\hbar k$ can propagate in the solid.

A model for a one-dimensional periodic potential in a solid is due to Kronig and Penney, who assumed a potential of the form of a narrow regularly spaced rectangular potential spike which we take here as $\delta-$functions[20]

$$H = \frac{\mathbf{p}^2}{2m} + \sum_{n=-\infty}^{\infty} -a\delta(x - nd) \tag{18.334}$$

where a is the potential amplitude of the δ-function and $d > 0$ is their regular spacing along the x axis. The δ-function is a stand-in for any narrow width function that falls to zero rapidly.

A usual analysis of this system is to find a wavefunction $\psi(x) = e^{ikx}u(x)$ such that $u(x + d) = u(x)$ (Bloch's theorem).[21] We will obtain the solution to this problem is a simpler manner.

[20] Kronig, R. de L.; Kronig, R. de L and Penney, W.G., (3 February 1931). *Quantum Mechanics of Electrons in Crystal Lattices*, Proceedings of the Royal Society A: Mathematical, Physical and Engineering Sciences. The Royal Society. 130 (814): 499-513. doi:10.1098/rspa.1931.0019. ISSN 1364-5021.

[21] Bloch, Felix, *Über die Quantenmechanik der Elektronen in Kristallgittern* [About the Quantum Mechanics of the Electrons in Crystal Lattices], Zeit. für Phys, (in German). **52**, 555-600 (1929). doi:10.1007/bf01339455. ISSN 1434-6001.

Let us consider a potential $V(x)$ starting at $x = 0$ and in the right-half plane only, so

$$H = \frac{\mathbf{p}^2}{2m} + \sum_{n=0}^{\infty} -a\delta(x - nd). \tag{18.335}$$

If a particle with momentum k described by a plane wave propagating from the $-x$ direction encounters this potential, it will be partially reflected and partially transmitted.

$$\psi_t(x) = Te^{ikx} \quad (\text{for } x > 0); \; \psi_r(x) = Re^{-ikx} \quad (\text{for } x \leq 0) \tag{18.336}$$

where T and R result from the linear array of δ functions. If the transmission falls to zero, it means that the plane wave cannot propagate in the potential, and $|R|^2 \to 1$.

We can determine the reflection and transmission coefficients for the array by first considering a single δ function at $x = 0$. If we have a plane wave propagating from the $-x$ direction toward the right, the wave will be partially transmitted and partially reflected. Let us write for the incident and reflected waves $\psi(x)_-$ for $x < 0$ and the transmitted wave $\psi(x)_+$ for $x > 0$:

$$\psi(x)_- = Ae^{ikx} + Be^{-ikx} \quad \psi(x)_+ = Ce^{ikx}; \quad k = \frac{\sqrt{2mE}}{\hbar} \tag{18.337}$$

where E is the kinetic energy of the particle. Continuity of the wavefunction at $x = 0$ requires that

$$A + B = C. \tag{18.338}$$

We can derive a second constraint from the Schrödinger equation, as

$$-\frac{\hbar^2}{2m}\frac{d^2}{dx^2}\psi_\pm(x) = E\psi_\pm(x) - V(x)\psi_\pm(x), \tag{18.339}$$

where we take $\psi_+(x)$ for $x > 0$ and $\psi_-(x)$ for $x < 0$. Let us integrate this equation over an infinitesimally small interval $[-\epsilon, \epsilon]$ around $x = 0$. In this limit, the term with E does not contribute, so

$$-\frac{\hbar^2}{2m}\int_{-\epsilon}^{\epsilon}\frac{d^2}{dx^2}\psi_\pm(x)dx = -\int_{-\epsilon}^{\epsilon} -a\delta(x)\psi_\pm(x). \tag{18.340}$$

The integrals can be easily calculated, and in the limit $\epsilon \to 0$

$$\frac{d}{dx}\psi_+(x)\bigg|_0 - \frac{d}{dx}\psi_-(x)\bigg|_0 = -\frac{2ma}{\hbar^2}\psi(0)_\pm = -\frac{2ma}{\hbar^2}C \tag{18.341}$$

$$= ik\left(C - (A - B)\right) \tag{18.342}$$

and we already have $A + B = C$,

$$ikA - ikB - ikC = \frac{2ma}{\hbar^2}\,C. \tag{18.343}$$

Let us define $\beta = ma/\hbar^2 k$. Then we have two sets of equations to determine $B/A = t$ and $C/A = r$,

$$A + B = C; \qquad A - B - C = -2i\beta(C) \tag{18.344}$$

which can be easily solved to yield

$$B = i\beta C \quad B = \frac{i\beta A}{1 - i\beta} \tag{18.345}$$

and therefore the reflection and transmission coefficients for a single δ–function ($a < 0$, attractive potential) are

$$r = \frac{B}{A} = \frac{i\beta}{1 - i\beta}; \quad t = \frac{C}{A} = \frac{1}{1 - i\beta}; \quad \beta = \frac{ma}{\hbar^2 k}. \tag{18.346}$$

There is a trick to solve self-similar problems, in this case, we can add another δ–function, at $x = -d$, immediately to the left of the linear array of δ–functions that begin at $x = 0$. We then can calculate the reflection R' and transmission T' coefficients when this new potential is added. If we move the observation point of the reflected wave to the left by $-d$ (to eliminate the phase shift due to the shift of the comb location), the self-similar solution dictates that $R' = R$; it should be the same after the addition potential because we are adding a single additional one to an infinite number. The allowed k that will propagate given a and d are those k such that a self-consistent solution $R' = R$ exists, such that $|R|^2 \leq 1$.

We could try to solve this as a boundary value problem, however, we don't know the effective k vector in the right half plane, and we need that to match the boundary conditions.

When a wave propagating from the $-x$ direction reflects from the additional δ-function and from $V(x)$, there will be an interference between the reflected waves, resulting in a net reflection toward the left, as illustrated in Fig. 18.6.

$$\begin{aligned} R' &= r + t^2 R e^{2ikd} + t^2 R^2 r e^{4ikd} + t^2 R^3 r^2 e^{6ikd} + \ldots \\ &= r + \frac{t^2}{r} \sum_{j=1}^{\infty} (Rr)^j e^{2ijdk}. \end{aligned} \tag{18.347}$$

(As a check of the solution, at $t \to 1$ and $r \to 0$,

$$R' = Re^{2ikd} \tag{18.348}$$

which simply shows a phase shift because we have moved a distance d from $x = 0$ and the reflected wave is moving in the $-x$ direction.) Because all the factors are less than one in magnitude, we can use

$$\sum_{n=1}^{\infty} z^n = \frac{z}{1 - z} \tag{18.349}$$

to compute the sum which is valid for $|z| < 1$.

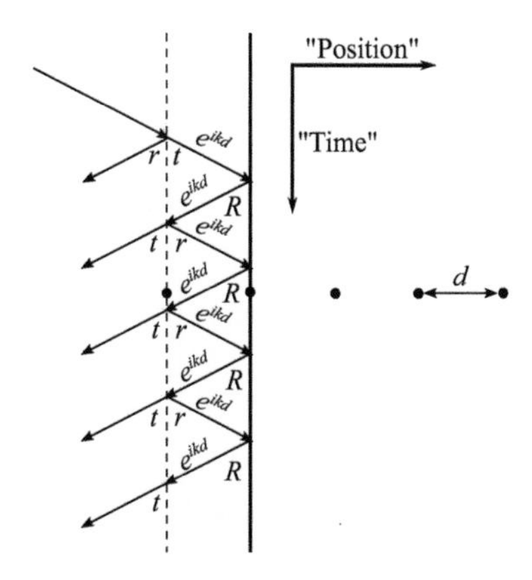

Fig. 18.6 Reflections between the original set of δ functions and the additional one at $-d$ are added up as shown. We can think of a wave bounding back and forth, acquiring a phase change and reduced amplitude with each reflection. Introducing the pretend "time" axis allows the individual reflections to be separated; otherwise they would lie on top of each other and therefore be difficult to visualize in the one-dimensional system.

Self-consistency requires that the additional δ-function added to the left of $V(x)$ should not change R, therefore

$$R' = R = r + \frac{t^2}{r}\frac{Rre^{2ikd}}{1 - Rre^{2ikd}}. \tag{18.350}$$

Then

$$R^2 + \left(\frac{t^2}{r} - r - \frac{e^{-2ikd}}{r}\right) R + e^{-2ikd} = 0, \tag{18.351}$$

and defining

$$A = 1;\ \ B = \frac{t^2}{r} - r - \frac{e^{-2ikd}}{r};\ \ C = e^{-2ikd} \tag{18.352}$$

provides the solution via the quadratic formula as

$$R = \frac{-B \pm \sqrt{B^2 - 4AC}}{2}. \tag{18.353}$$

The allowed k are those that produce $|R|^2 \leq 1$, given t, r, and d (t, r are determined by Eq. (18.346)). This is most efficiently done via numerical calculation because the amount of algebra is alarming. Above we use 'the δ-function, however, r and t can be calculated for any very short-range potential.

The magnitude of the effective wave vector Q in the material (that is, for $x > 0$) can be determined by the reflection coefficient from the potential step,

$$\mathrm{R} = \left|\frac{k - Q}{k + Q}\right|^2. \tag{18.354}$$

The solution obtained here *appears* as significantly different than that obtained via Bloch's theorem. In some sense, determining the allowed k by reflection might better

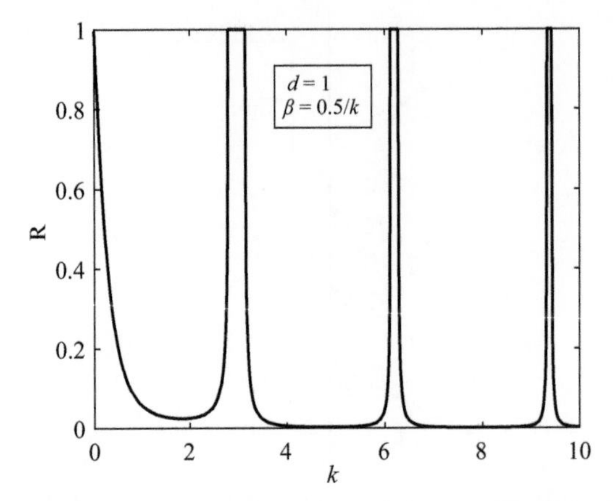

Fig. 18.7 The reflection coefficient for R calculated using the reflection technique compared to the Bloch theorem analysis; the two methods produce indistinguishable results.

mimic reality in some situations. For example, if we want to describe electrons moving from a metal into a semiconductor, the potential step would seem to be a better model and more useful model than determining the wavefunctions in an infinite region.

However, we can compare these results for the allowed k and R to those determined by the exact form of the Bloch theorem solution. The allowed k, for given d and a, can be found from[22]

$$\cos(qd) = \cos(kd) - \beta \sin(kd) \tag{18.355}$$

where the allowed k are those values that result in $-1 \leq \cos(qd) \leq 1$, in which case

$$\pm qd = \arccos\left(\cos(kd) - \beta \sin(kd)\right). \tag{18.356}$$

After q is determined, the back-reflected wave propagating in the lattice for a given forward propagating wave is[23]

$$\mathrm{R}_b = \left| \frac{1 - e^{i(\pm q - k)d}}{1 - e^{i(\pm q + k)d}} \right|^2 . \tag{18.357}$$

Figure 18.7 is a plot of R and R_b on top of each other as a function of k, taking the allowed $\pm$ solutions for both, for $d = 1$ and $\beta = 0.5/k$. When $\mathrm{R} = 1$, the wave is totally reflected, e.g., it does not propagate in the lattice, so these can be taken as disallowed k (band gaps). The two plots are identical, the difference is at the level of round off errors (parts in 10^{10} differences) in the numerical computations for each.[24]

[22]See, e.g., Griffiths, D. and Schroeter, Darrell F., *Introduction to Quantum Mechanics, 3rd Ed.* (Cambridge Univ. Press, 2018); Sec. 5.3.2.

[23]Writing the Bloch wavefunction as $\psi(x) = Ae^{ikx} + Be^{-ikx}$ allows $R_b = A/B$ to be readily determined.

[24]The technique to use wave reflection to solve boundary value problems has been studied extensively, see e.g., Ignatovich, V.K., *An algebraic approach to the propagation of waves and particles in layered media*, Physica B: Condensed Matter, **175**, (1991), pp. 33-38, ISSN 0921-4526, https://doi.org/10.1016/0921-4526(91)90685-8.

18.9.5 Particle on a ring subjected to a δ-function potential

Let us consider a particle of mass m bound to freely move in one dimension along the circumference of a ring of radius R. The quantized energies are determined from

$$\frac{p^2}{2m}\psi(r,\phi,\theta) = \frac{p_\phi^2}{2m}\psi(\phi) = -\frac{\hbar^2}{2mR^2}\frac{\partial^2}{\partial\phi^2}\psi(\phi) = E\psi(\phi) \tag{18.358}$$

and there is no motion allowed along $r = R$ or the polar angle $\theta = 0$. The normalized solution is

$$\psi(\phi) = \frac{e^{i\alpha\phi}}{\sqrt{2\pi}}; \quad \alpha^2 = \frac{2mER^2}{\hbar^2}. \tag{18.359}$$

The solutions must be periodic in $\phi = 0, 2\pi$ which means the energy can only take on discrete values as[25]

$$2\pi\alpha_n = 2n\pi; \quad E_n = \frac{n^2\hbar^2}{2mR^2} = \beta n^2 \tag{18.360}$$

for $n = 0, \pm 1, \pm 2, \pm 3 ...$, and the wavefunctions are

$$\psi_n(\phi) = \frac{e^{in\phi}}{\sqrt{2\pi}}. \tag{18.361}$$

However, we point out that only the probability density (wavefunction squared) needs to be single valued, so in principle the wavefunction could be periodic as $n/2$ instead of n. We will show later that in three dimensions this is not allowed. It is a debatable point in this problem, however, a wavefunction with $n/2$ periodicity can have discontinuities in its derivative, or in the wavefunction itself, and some states will not be orthogonal to the $n = 0$ state. Also, the parity operator (coordinate inversion) commutes with the Hamiltonian, so we expect the states to have definite parity. Only integer values (and zero) of n have definite parity.

It is interesting that there is no zero-point energy for the $n = 0$ state. In principle, the states on the ring are unbound (there is no boundary condition forcing $\psi(\phi) = 0$). However, there is certainly zero-point energy in the $\hat{r}$ and $\hat{z}$ directions and we are ignoring those motions (by assuming infinite binding energy to the ring). Note that the states are degenerate in $\pm n$.

The angular momentum of the states is

$$L_z = mvr = m\sqrt{\frac{2E_n}{m}}R = n\hbar \tag{18.362}$$

and we could have determined the energy levels by angular momentum quantization. The $n = 0$ state is a constant amplitude and has no angular momentum because the $\mathbf{p}_\phi$ momentum operator gives zero.

[25]The particle on a ring was apparently first considered by W. Pauli in assessing the validity of the Born approximation in scattering theory, see *Wolfgang Pauli Scientific Correspondence with Bohr, Einstein, Heisenberg, A.O., Vol. I: 1919-1929,* A. Hermann, K.v. Meyenn, and V.F. Weisskopf, eds. (Springer-Verlag, New York, 1979). Letter [143], Pauli to Heisenberg, 19 Oct 1926, p. 340.

Let us now add a potential at $\phi = 0$ of the form

$$V(\phi) = \beta\delta(\phi). \tag{18.363}$$

The first-order perturbation correction to the energy is

$$E_n^{(1)} = \int d\phi \; \psi_n(\phi)\psi_n^*(\psi)\beta\delta(\phi) = \beta. \tag{18.364}$$

To determine the second-order corrections to the energy, we need matrix elements of the form

$$V_{nn'} = \int d\phi\psi_n^*(\phi)\psi_{n'}(\phi)\beta\delta(\phi) = \beta. \tag{18.365}$$

The second-order energy shift is

$$E_n^{(2)} = \sum_{n \neq n'} \frac{|\langle n'|V|n\rangle|^2}{E_n - E_{n'}}. \tag{18.366}$$

We can immediately see a problem because for $n \neq 0$, the sum included $E_n = E_{-n}$. We can attempt to deal with this by finding a transformation such that

$$\langle n|V|n'\rangle = 0 \tag{18.367}$$

for all $n = -n'$. The usual means to address this problem is to diagonalize the perturbation matrix $V_{nn'}$. Let us find the eigenvalues of $V_{nn'}$ using the basis states Eq. (18.361):

$$\det(V - \lambda I) = \frac{\beta}{2\pi}\begin{pmatrix} 1-\lambda & 1 \\ 1 & 1-\lambda \end{pmatrix} = 0 \tag{18.368}$$

and it is easy to find that the two eigenvalues of the matrix are $\lambda = 0, 2$ and the corresponding eigenvectors are

$$\frac{1}{\sqrt{2}}\begin{pmatrix} 1 \\ 1 \end{pmatrix}; \; \frac{1}{\sqrt{2}}\begin{pmatrix} 1 \\ -1 \end{pmatrix} \tag{18.369}$$

and in this basis, the degeneracy is lifted. In fact, these two states differ in energy by β/π with the $(1, -1)$ state not shifted. Let us investigate the structure of these states. Taking the prescribed linear combination of states, we see

$$\psi_{n\pm} = \frac{1}{\sqrt{2}}\left(\frac{e^{in\phi} \pm e^{-in\phi}}{\sqrt{2\pi}}\right) \to \frac{\sin\phi}{\sqrt{\pi}} \text{ for } -; \; \frac{\cos\phi}{\sqrt{\pi}} \text{ for } +. \tag{18.370}$$

We can immediately see that

$$\begin{aligned} \langle n_-|V|n_-\rangle &= 0 \\ \langle n_+|V|n_+\rangle &= \int_0^{2\pi} d\phi\beta\delta(\phi)\frac{\cos^2\phi}{\pi} = \frac{\beta}{\pi} \end{aligned} \tag{18.371}$$

because $\sin(0) = 0$. The states had a hidden symmetry, in that the perturbing and original Hamiltonians are invariant under the operation

$$
\begin{aligned}
\mathbf{P}\delta(\phi) &= \delta(-\phi) = \delta(\phi) \\
\mathbf{P}|n\rangle &= |-n\rangle \\
\mathbf{P}|n_+\rangle &= |n_+\rangle \\
\mathbf{P}|n_-\rangle &= -|n_-\rangle
\end{aligned}
\tag{18.372}
$$

so $[H, \mathbf{P}] = 0$ and $\mathbf{P}$ is a constant of the motion, however, the original states are not eigenfunctions of $\mathbf{P}$. (Note that $\mathbf{P}$ is a reflection across the $(r, \phi = 0)$ polar coordinate axis; in even-dimensional space, and our problem is two dimensional, the parity operator is equivalent to a rotation.) We could have sought linear combinations that have a definite $\mathbf{P}$ symmetry without diagonalizing the perturbation matrix.

Thus, the states $|n_-\rangle$ are not shifted at all by the perturbation. The second order correction to the + states is

$$
E^{(2)}_{n_+} = \sum_{n \neq n'=1}^{\infty} \frac{|\langle n'_+|V|n_+\rangle|^2}{E_n - E_{n'}} = \frac{\beta^2}{\alpha\pi^2}\left(\sum_{n=1,\neq n'}^{\infty} \frac{1}{n^2 - n'^2} + \frac{1}{2n^2}\right)
\tag{18.373}
$$

where the last term is the contribution from the $n = 0$ state. For the $n = 1$ level, the second order shift is

$$
E^{(2)}_{n_+} = \frac{\beta^2}{\alpha\pi^2}\left(\sum_{n'=2}^{\infty} \frac{1}{1 - n'^2} + \frac{1}{2}\right) = \frac{\beta^2}{\alpha\pi^2}\left(-\frac{3}{4} + \frac{1}{2}\right) = -\frac{\beta^2}{4\pi^2\alpha}
\tag{18.374}
$$

and these are a good example of a summation over all states in second-order perturbation theory.

18.10 Benzene molecule

The particle on a ring is a useful model for the benzene molecule, which has six carbon (C) atoms in a ring, and each C atom provides a delocalized electron that is approximately free to move around the ring. This is roughly the Kekulé picture, proposed in 1865, in which the ring contains alternating double and single bonds, as shown in Fig. 1.3. Later, the picture was modified where positions of the bonds are allowed to oscillate. This was needed to explain the lack of some isomers of bromobenzene, dibromobenzene, and tribromobenzene.

In our ring model, two electrons can exist in the ground state, provided they are in a spin-singlet state, and if we take symmetric spatial wavefunctions of the excited state, the $n = \pm 1$ states can each hold two singlet-spin state electrons, these all together form an approximation to the molecular ground state.

We can do slightly better. The electrons are smeared out around the ring, however, the C atoms are electrostatically screened point charges. Therefore, electrons will be attracted to the C atoms with potential α, and if localized at a particular C atom, also attracted to the two adjacent C atoms but with lower potential energy, β. The electrons are sufficiently smeared out that we can ignore their mutual interaction in this simple model.

A model Hamiltonian for the six identical locations on the ring is

$$H = \begin{pmatrix} \alpha & \beta & 0 & 0 & 0 & \beta \\ \beta & \alpha & \beta & 0 & 0 & 0 \\ 0 & \beta & \alpha & \beta & 0 & 0 \\ 0 & 0 & \beta & \alpha & \beta & 0 \\ 0 & 0 & 0 & \beta & \alpha & \beta \\ \beta & 0 & 0 & 0 & \beta & \alpha \end{pmatrix} \tag{18.375}$$

where α and β are less than zero for an attractive potential. This matrix can be rewritten in terms of discrete translation operators which move the C atoms by one position along the ring,

$$H = \beta\mathbf{T} + \beta\mathbf{T}^{\dagger} + \alpha\mathbf{I} \tag{18.376}$$

with the matrix

$$T = \begin{pmatrix} 0 & 1 & 0 & 0 & 0 & 0 \\ 0 & 0 & 1 & 0 & 0 & 0 \\ 0 & 0 & 0 & 1 & 0 & 0 \\ 0 & 0 & 0 & 0 & 1 & 0 \\ 0 & 0 & 0 & 0 & 0 & 1 \\ 1 & 0 & 0 & 0 & 0 & 0 \end{pmatrix} \tag{18.377}$$

and $\mathbf{T}^{\dagger} = \mathbf{T}^{-1}$. This matrix moves each atom by one position around the ring,

$$\mathbf{T}\begin{pmatrix} a \\ b \\ c \\ d \\ e \\ f \end{pmatrix} = \begin{pmatrix} b \\ c \\ d \\ e \\ f \\ a \end{pmatrix}; \quad \mathbf{T}^{\dagger}\begin{pmatrix} a \\ b \\ c \\ d \\ e \\ f \end{pmatrix} = \begin{pmatrix} f \\ a \\ b \\ c \\ d \\ e \end{pmatrix}, \tag{18.378}$$

so $\mathbf{T}^6 = 1$. If we have an eigenvector then $\mathbf{T}v_\lambda = \lambda v_\lambda$, and together with the knowledge that $\mathbf{T}^6 = 1$,

$$\mathbf{T}^6 v_\lambda = \lambda^6 v_\lambda. \tag{18.379}$$

The eigenvalues of the 6×6 translation operator are the sixth roots of $\lambda_n = \sqrt[6]{1}$. Once the eigenvalues are known, it is straightforward to find elements of the eigenvectors from $\mathbf{T}v_n = \lambda_n v_n$, $n = 1...6$,

$$\lambda_n = e^{2\pi i n/6}; \quad v_{n,j} = e^{2\pi i j n/6}. \tag{18.380}$$

The eigenvalues of the Hamiltonian are

$$E_n = \alpha + \beta e^{2\pi i n/6} + \beta^{-2\pi i n/6} = \alpha + 2\beta\cos(n\pi/3) \tag{18.381}$$

because the eigenvalues of $\mathbf{T}^{\dagger} = \lambda_n^*$. The lowest energy state has $n = 3$ while the highest energy state has $n = 6$. This reflects the patterns in the eigenvectors,

$$v_3 = \begin{pmatrix} 1 \\ -1 \\ 1 \\ -1 \\ 1 \\ -1 \end{pmatrix}; \quad v_6 = \begin{pmatrix} 1 \\ 1 \\ 1 \\ 1 \\ 1 \\ 1 \end{pmatrix} \tag{18.382}$$

and for v_3 it is easy to see that the β contribution to the energy is zero (minimum) while for v_6 it is 2α, the maximum. It should be noted that the eigenvectors are not normalized to unity but to six (electrons), the total energy is $6E_n$, because $\langle n|H|n\rangle = E_n\langle n|n\rangle = 6E_n$.

It should be noted that this model system is not meant to be a realistic representation of the Benzene molecule, but is a toy system used here to illustrate the effectiveness of operator techniques in the solution of problems.

18.11 Angular momentum: an operator approach

Angular momentum falls among the most fundamental physical properties and theoretical principles of quantum mechanics. Its conservation is a cornerstone of both classical and quantum physics, and the first signs of quantization in the hydrogen atom was explained as due to the quantization of orbital angular momentum, by which the energy levels can be determined with good accuracy.

Orbitial angular momentum, due to the motion of particles bound in a potential well, is quantized in units of $\hbar$, with $h = 2\pi\hbar$ Planck's constant introduced, as we have seen, in his derivation of the black body spectrum. In Appendix C, we discussed the universal character of this constant.

Intrinsic angular momentum, for example, the "spin" of the electron, is quantized in units of $\hbar/2$. Orbital and intrinsic angular momentum apparently have quite different origins; in the case of the electron, the origin of intrinsic spin remains a mystery. As such, spin is a property of the electron like mass or charge, except it has two values. This falls in contrast to orbital angular momentum which can take any (quantized) value and can be fully understood, other than its quantization, classically.

Orbital angular momentum states arise naturally in the solution of the H atom by the separation of variables. The coordinate representation of these states are the spherical harmonics which were originally discovered and studied in the mid-nineteenth century because they arise when the method of separation of variables is used to solve Laplace's equation in a spherical-coordinate system.

We will introduce Clebsch-Gordan coefficients, which were also first introduced in the mid-nineteenth century for combining spherical harmonics. It is both fortuitous and not too surprising that mathematics born in classical physics can be directly applied to quantum mechanics. The mathematical apparatus associated with the spherical harmonics, which includes various addition and recursion relationships, applies directly to quantum mechanics due to the similarity between the Schrödinger equation and Laplace's equation as a consequence of the appearance of ∇^2 in both equations.

18.11.1 Generator of spatial rotations

We will use the tradition of non-bold letters to specify angular momentum operators, except for the total angular momentum $\mathbf{L}$, and use units of $\hbar = 1$ for this operator and its eigenvalues.

The center-of-mass (external) wavefunction of any system is described by both its linear momentum states and angular momentum. For an isolated system, both forms of momentum are conserved, and because $\mathbf{L}\cdot\mathbf{p} = \mathbf{p}\cdot\mathbf{L} = 0$, they commute with each other and with the Hamiltonian for free motion. Even if the system is bound by a complicated Hamiltonian that is not rotationally invariant, the aggregate system isolated in space has its angular momentum conserved, and is in an angular momentum quantum state. The total angular momentum is the sum of every element of the system.

If we have a Hamiltonian H that is spherically symmetric (e.g., is rotationally invariant) describing the internal interactions of a system, we can seek an operator that commutes with H, and thus serve as an additional internally conserved quantity. Let us define an infinitesimal rotation operation as

$$x' = x - \epsilon y; \quad y' = \epsilon x + y \tag{18.383}$$

where ϵ is an infinitesimal. If we have a function $f(x, y)$, when rotated it becomes

$$\begin{aligned} f(x', y') = f(x - \epsilon y, y + \epsilon x) &= f(x, y) - \epsilon y \frac{\partial f(x, y)}{\partial x} + \epsilon x \frac{\partial f(x, y)}{\partial y} \\ &= f(x, y) + \epsilon \left(x\frac{\partial}{\partial y} - y\frac{\partial}{\partial x} \right) f(x, y) \end{aligned} \tag{18.384}$$

where we are neglecting higher order terms. We can recognize the quantity in parentheses as an analog of classical angular momentum, because

$$x\frac{\partial}{\partial y} - y\frac{\partial}{\partial x} = \mathbf{r} \times \left(\hat{x}\frac{\partial}{\partial x} + \hat{y}\frac{\partial}{\partial y} \right) = \frac{i}{\hbar} \left[\mathbf{r} \times \mathbf{p} \right]_z \tag{18.385}$$

and we singled out the z component of classical angular momentum. The foregoing discussion applies directly to the other components. Let us define the rotation operator in terms of the angular momentum operator, and set $\hbar = 1$, so $L_z/\hbar \to L_z$:

$$d_z(\epsilon) = 1 + i\epsilon L_z. \tag{18.386}$$

Then for a large angle θ, if we take many steps of $\epsilon = \theta/N$, operator for a finite angle rotation about $\hat{z}$ is

$$R_z(\theta) = \left(1 + i\frac{\theta}{N} L_z \right)^N = e^{i\theta L_z} \tag{18.387}$$

and we can also define rotation operators for the x and y axes, based on L_x and L_y.

18.11.2 Commutation relationships

It is straightforward to show that these operators, being based on the $\mathbf{x}_i$ and $\mathbf{p}_j$, have the following commutation relationships among themselves,

$$[L_x, L_y] = iL_z; \ [L_y, L_z] = iL_x; [L_z, L_x] = iL_y \tag{18.388}$$

and in general

$$[L_i, L_j] = i\epsilon_{ijk}L_k \tag{18.389}$$

in terms of the ϵ antisymmetric Levi-Civita symbol. The total angular momentum squared commutes with each component,

$$[\mathbf{L}^2, L_i] = 0 \tag{18.390}$$

and the proof for a specific case L_x is

$$\begin{aligned}[\mathbf{L}^2, L_x] &= [L_x^2 + L_y^2 + L_z^2, L_x] = [L_y^2 + L_z^2, L_x] \\ &= L_y[L_y, L_x] + [L_y, L_x]L_y + L_z[L_z, L_x] + [L_z, L_x]L_z \\ &= L_y(-iL_z) + (-iL_z)L_y + L_z(iL_y) + (iL_y)L_z = 0.\end{aligned} \tag{18.391}$$

These commutation relations can be used to define angular momentum and determine all of its properties.

18.11.3 Spatial representations: a prelude to operator techniques

There are a few properties of angular momentum that are most easily seen from the spatial representations of angular momentum states and operators. First, the quantization of angular momentum is a consequence of the requirement that the wavefunction be single valued. The azimuthal part of the Schrödinger equation in a spherical-coordinate system is

$$\frac{\mathrm{d}^2\Phi}{\mathrm{d}\phi^2} + m^2\Phi = 0 \tag{18.392}$$

which has the solution

$$\Phi_{\pm m}(\phi) = Ce^{\mp im\phi} \tag{18.393}$$

where C is a normalizing constant and the single-value requirement restricts m to be an integer (or perhaps a half-integer as it is the probability that must be single valued, not the wavefunction. We will return to this question.). The form of Φ does not depend on the total angular momentum, however, m is the z axis projection of the angular momentum, and therefore has a maximum value for a given $\mathbf{L}$ as we learned from the solution to the hydrogen atom.

Because L_z and L^2 are taken as the eigenvalues that specify an angular momentum state, the expectation values of L_x and L_y should be zero for such states. By use of operator techniques, we can show this simply and directly and circumvent fiddling with the coordinate angular momentum wavefuntions. The proof proceeds as follows: Consider any case where we have two hermitian operators $\mathbf{A}$ and $\mathbf{B}$ that do not commute,

$$[\mathbf{A}, \mathbf{B}] = \mathbf{C}. \tag{18.394}$$

We can calculate the matrix elements for eigenstates of the $\mathbf{A}$ operator, where $\mathbf{A}|a_i\rangle = a_i|a_i\rangle$. Then

$$\langle[\mathbf{A}, \mathbf{B}]\rangle = \langle\mathbf{C}\rangle\,. \tag{18.395}$$

The expectation value for the state $|a_i\rangle$ is

$$\begin{aligned}\langle a_i|[\mathbf{A},\mathbf{B}]|a_i\rangle =&\langle a_i|\mathbf{AB}|a_i\rangle - \langle a_i|\mathbf{BA}|a_i\rangle\\ =&a_i\langle a_i|\mathbf{B}|a_i\rangle - a_i\langle a_i|\mathbf{B}|a_i\rangle\\ =&0\\ =&\langle a_i|\mathbf{C}|a_i\rangle.\end{aligned}$$

This can be directly applied to the angular momentum case where $\mathbf{A} = L_z$, $\mathbf{B} = L_x$ and $\mathbf{C} = L_y$, and also with x and y interchanged. In both cases, the expectation values for $\mathbf{C} = L_x$ or L_y are zero. Therefore, the expectation values of $\langle L_x\rangle$ and $\langle L_y\rangle$ are both zero for an eigenstate of L.

The foregoing provides a vivid illustration of the power and utility of algebraic operator techniques.

18.11.4 Angular momentum operator techniques

Let us enumerate the angular momentum facts that we have at hand:

1. For a rotationally invariant Hamiltonian H, the angular momentum operator $\mathbf{L}$ commutes with the Hamiltonian.
2. The components of $\mathbf{L}$ do not commute with each other, $[L_i, L_j] = \epsilon_{ijk}L_k$, however, $[\mathbf{L}^2, L_i] = 0$. We can therefore say that there are two independent eigenvalues, one for $\mathbf{L}^2$ and another for one component, which we choose to be L_z.
3. L_x, L_y, L_z and $\mathbf{L}^2$ are hermitian operators that represent physical properties of a system (are observables). Therefore, the eigenstates of $\mathbf{L}^2$ and L_z each form a complete set of commuting observables. The eigenvalues for L_z in the case of quantized orbital angular momentum are integers m, but we do not yet know the eigenvalues of $\mathbf{L}^2$. We will temporarily label the states as $|\Lambda, m\rangle$.
4. $\langle\mathbf{L}^2\rangle = \langle L_x^2\rangle + \langle L_y^2\rangle + m^2$, so that there is a maximum $m \le \sqrt{\langle\mathbf{L}^2\rangle}$.
5. We expect $\langle L_x\rangle = \langle L_y\rangle = 0$ when the system is in an eigenstate of L_z.

It would be useful to know the matrix elements of L_x and L_y between the eigenstates of L_z. Let us first consider the matrix elements of the commutator of L_x and L_y with L_z:

$$\begin{aligned}i\langle\Lambda, m'|L_x|\Lambda, m\rangle &= \langle\Lambda, m'|\ [L_y, L_z]\ |\Lambda, m\rangle\\ &= \langle\Lambda, m'|L_yL_z|\Lambda, m\rangle - \langle\Lambda, m'|L_zL_y|\Lambda, m\rangle\\ &= m\langle\Lambda, m'|L_y|\Lambda, m\rangle - m'\langle\Lambda, m'|L_y|\Lambda, m\rangle\end{aligned} \tag{18.396}$$

and therefore

$$\langle\Lambda, m'|L_x|\Lambda, m\rangle = i(m' - m)\langle\Lambda, m'|L_y|\Lambda, m\rangle. \tag{18.397}$$

A similar calculation shows that

$$i\langle\Lambda, m'|L_y|\Lambda, m\rangle = (m' - m)\langle\Lambda, m'|L_x|\Lambda, m\rangle. \tag{18.398}$$

These equations can be subtracted or added, yielding

$$\begin{aligned}\langle\Lambda,m'|\,(L_x\pm iL_y)\,|\Lambda,m\rangle &= (m'-m)\langle\Lambda,m'|\,(iL_y\pm L_x)\,|\Lambda,m\rangle\\ &= \pm(m'-m)\langle\Lambda,m'|\,(L_x\pm iL_y)\,|\Lambda,m\rangle.\end{aligned} \tag{18.399}$$

By introducing new operators $L_\pm = L_x \pm iL_y$, we obtain

$$\langle\Lambda,m'|L_\pm|\Lambda,m\rangle = \pm(m'-m)\langle\Lambda,m'|L_\pm|\Lambda,m\rangle. \tag{18.400}$$

Therefore,

$$\begin{aligned}0 &= (m'-m)\langle\Lambda,m'|L_\pm|\Lambda,m\rangle \mp \langle\Lambda,m'|L_\pm|\Lambda,m\rangle\\ &= (m'-m\mp 1)\langle\Lambda,m'|L_\pm|\Lambda,m\rangle.\end{aligned} \tag{18.401}$$

To have non-trivial matrix elements, this imposes $(m'-m\mp 1)=0$ or $m'\pm 1=m$, and therefore $L_\pm$ are angular momentum ladder operators.

The commutation relations for the ladder operators can easily be determined as

$$[L_z, L_\pm] = \pm L_\pm\ ; \qquad [L_+, L_-] = 2L_z. \tag{18.402}$$

It is straightforward to show that

$$\begin{aligned}L^2 &= \frac{L_+L_- + L_-L_+}{2} + L_z^2\\ &= L_z^2 + L_z + L_-L_+\\ &= L_z^2 - L_z + L_+L_-.\end{aligned} \tag{18.403}$$

18.11.4.1 Eigenvalues of $\mathbf{L}^2$ and L_z

The expansion of $\mathbf{L}^2$ in terms of L_z, L_+, and L_- can be used to determine the eigenvalues of $\mathbf{L}^2$ because $m^2 \le \mathbf{L}^2$, implying $L_+|m_{max}\rangle = 0$. Therefore,

$$\mathbf{L}^2|\Lambda,m_{max}\rangle + (L_z^2 + L_z - L_-L_+)|\Lambda,m_{max}\rangle = (m_{max}^2 + m_{max})|\Lambda,m_{max}\rangle \tag{18.404}$$

and for $|\Lambda,m_{min}\rangle$,

$$\mathbf{L}^2|\Lambda,m_{min}\rangle = (L_z^2 - L_z - L_+L_-)|\Lambda,m_{min}\rangle = (m_{min}^2 - m_{min})|\Lambda,m_{min}\rangle. \tag{18.405}$$

The relative magnitudes of m_{max} and m_{min} can be determined as follows. If the coordinate system is partially inverted, for example $x\to -x$, or if there is a rotation by π with the rotation axis in the $x-y$ plane, then $L_z \to -L_z$. Either operation is unitary and self-adjoint, with $U^\dagger U = U^2 = 1$, and

$$UL_zU^\dagger = -L_z. \tag{18.406}$$

U commutes with H (which was assumed rotationally invariant), but anticommutes with L_z. Consequently, the eigenstates of H have the same spectrum of eigenvalues in the inverted system. Because H and L_z commute, their mutual spectra of simultaneous eigenvalues must remain invariant, so L_z and $-L_z$ have the same set of eigenvalues.

We have shown that there are maximum and minimum values m_{min} and m_{max}, and thereby we can deduce that $m_{min} = -m_{max}$. Furthermore,

$$\Lambda = (m_{max}^2 + m_{max}) = l(l+1) = \mathbf{L}^2 \tag{18.407}$$

where we have set $l = m_{max}$, and $l(l+1)$ is independent of m so can identify l as the independent eigenvalue Λ that determines magnitude of the angular moment of the state.

Given that the difference in m for neighboring states is one (i.e., the effect of L_+ and L_- is to change m by ± 1), it will take N steps to raise a state from $m_{min} = -m_{max}$ to m_{max},

$$N = m_{max} - m_{min} = 2m_{max} \tag{18.408}$$

and therefore m_{max} can be either integer or half-integer. For the integer case, there is a spectrum of states with m from $-l, -l+1, ..., -1, 0, 1, ..., l-2, l-1, l$, which is consistent with our previous knowledge that the solution to the azimuthal Schrödinger equation are $e^{im\phi}$ where m is an integer with $-m_{max} \leq m \leq m_{max}$.

Alternatively, the spectrum for the half-integer case is $-l, -l+1, ..., -1/2, 1/2, ..., l-1, /$, with the absence of an $m = 0$ state.

For both cases, there are $2l+1$ states for a given l. Feynman suggests that[26] $\mathbf{L}^2 = l(l+1)$ can be understood by assuming that the sum of squares of all three angular momentum components is a constant, and that on average

$$\langle L^2 \rangle = \langle L_x^2 + L_y^2 + L_z^2 \rangle = 3\langle L_z^2 \rangle = \frac{3\sum_{m=-l}^{l} m^2}{2l+1}. \tag{18.409}$$

The sum can be evaluated as

$$\sum_{m=-l}^{l} m^2 = \frac{l(l+1)(2l+1)}{3}, \tag{18.410}$$

and therefore

$$\langle L^2 \rangle = l(l+1). \tag{18.411}$$

18.11.4.2 No half-integer l?

Because we are forced to deduce that $2m$ is an integer, it is also possible that $|m| = 1/2$ is the smallest possible magnitude, and l is half-integer. We might think this is not allowed because the azimuthal wavefunction $e^{-m\phi} = e^{-in\phi}e^{-i\phi/2}$ where $m = n + 1/2$ is not single valued, but requires two full rotations to return to its initial state. This is the usual reason for requiring that m be an integer in electromagnetism and other classical systems that are solved by the separation of variables. However, from the postulates of quantum mechanics, it is the *probability* that must be single valued, so

[26] Feynman, R. P., Leighton, R. B. and Sands, M., *The Feynman Lectures on Physics* (Addison-Wesley, Reading, MA, 1965), Vol. II, p. 34-11. This simple formula is not widely known, perhaps because it appears in Vol. II, and not in connection with angular momentum algebra in Vol. III. See Milonni, P.W., *Why $l(l+1$ instead of l^2?*, American Journal of Physics **58**, 1012 (1990); doi: 10.1119/1.16284.

this argument falls short. Because the electron's spin is 1/2, we need to accept the possibility of non-single-valuedness of the wavefunction in quantum mechanics.

In fact, unlike the case of orbital angular momentum which has well know wavefunctions (spherical harmonics), it is impossible to construct a coordinate or *spatial* wavefunction for half-integer spin. This can be demonstrated as follows.[27] We can construct a state that satisfies

$$L_+|l,l\rangle = 0\ ;\quad L_z|l,l\rangle = l|l,l\rangle \tag{18.412}$$

by considering these to be differential equations that determines the spatial wavefunction. Specifically,

$$L_+ = L_x + iLy = -i\left(y\frac{\partial}{\partial z} - z\frac{\partial}{\partial y}\right) + \left(z\frac{\partial}{\partial x} - x\frac{\partial}{\partial z}\right) \tag{18.413}$$

and

$$L_z = -i\left(x\frac{\partial}{\partial y} - y\frac{\partial}{\partial x}\right). \tag{18.414}$$

It can be readily verified that the unique solution to Eq. (18.412) is

$$\langle\mathbf{r}|l,l\rangle = c_l(-x-iy)^l = c_l u^l \tag{18.415}$$

where c_l is a normalization factor, and $u = -x - iy$ (we will also use $v = x - iy$). We can ignore the normalization factor.

The lowering operator (not normalized) in this coordinate system becomes

$$L_- = \left(2z\frac{\partial}{\partial u} + v\frac{\partial}{\partial z}\right) \tag{18.416}$$

and

$$L_-^2 = 4z^2\frac{\partial^2}{\partial u^2} + 2vz\frac{\partial^2}{\partial u\partial z} + 2v\frac{\partial}{\partial u} + 2vz\frac{\partial^2}{\partial z\partial u} + v^2\frac{\partial^2}{\partial z^2}. \tag{18.417}$$

Because u^l does not depend on z, we can set $z = 0$ for the ensuing discussion, and therefore

$$L_-^2 = 2v\frac{\partial}{\partial u}. \tag{18.418}$$

If the lowering operator is applied to u^l enough times, the result should be zero. Consider the case where the number of states $2n$ is even. The two lowest $|m|$ states are therefore $m = \pm\frac{1}{2}$, and to get n positive m states separated by $\Delta m = 1$, the starting function is $u^{n-\frac{1}{2}}$. Applying the lowering operator $2n$ times should give zero if this a valid state,

$$\begin{aligned} L_-^{2n}u^{n-1/2} &= \left(2v\frac{\partial}{\partial u}\right)^n u^{n-1/2} = \frac{(2n-1)(2n-3)...(3)(1)}{2^n}2^n v^n u^{-1/2} \\ &= (2n-1)!!v^n u^{-1/2} \neq 0. \end{aligned} \tag{18.419}$$

We can conclude that the starting function $f_{n-\frac{1}{2}}(u) = u^{n-\frac{1}{2}}$, which is a unique solution to the differential equation $L_+f_j(u) = 0$ and to $L_zf(u) = (n-\frac{1}{2})u$, is not a valid

[27]Biedenharn, L.C. and Louck, J.D. , *Angular Momentum in Quantum Physics* (Addison-Wesley, Reading, MA, 1981). Sec. 6.5.

angular momentum state because the ladder operator does not give zero at the required minimum $L_z = -m_{max}$. We can further conclude that because the unique half-integer initial state is invalid, that half-integer states are not possible in general for orbital angular momentum (coordinate) wavefunctions.

Of course the electron has spin angular momentum of 1/2, however, this is an *intrinsic* spin, which is a property like charge and mass, except it comes in two values. There is no known spatial wavefunction for the intrinsic spin. However, the operator methods we have developed are directly applicable to spin-1/2, even though there are no coordinate representations of these operators or wavefunctions.

18.11.4.3 Matrix elements of the angular momentum raising and lowering operators

We can determine the matrix elements and therefore the *selection rules* of the raising and lowering operators by recalling Eq.(18.403), where it is shown

$$\begin{aligned} L_+L_- &= \mathbf{L}^2 - L_z^2 + L_z \\ L_-L_+ &= \mathbf{L}^2 - L_z^2 - L_z. \end{aligned} \tag{18.420}$$

The expectation value in a given L, m state for either operator is

$$\begin{aligned} \langle l, m|L_\mp L_\pm|l, m\rangle &= \langle l, m|\mathbf{L}^2 - L_z^2 \mp L_z|l, m\rangle \\ &= \left(l(l+1) - m(m \pm 1)\right). \end{aligned} \tag{18.421}$$

Therefore,

$$L_\pm|l, m\rangle = \sqrt{l(l+l) - m(m \pm 1)}|l, m \pm 1\rangle. \tag{18.422}$$

18.11.5 Addition of angular momentum

The need to add angular momentum comes up frequently, for example, the combining of electron spin with orbital angular momentum in an atom (addition for states of a single particle), or when adding the angular momenta of several particles.

In analogy to the addition of classical angular momentum, given l_1 and l_2 new angular momentum states can be created, labeled by j. Classically, the total angular momentum depends on the relative orientation of the vectors $\mathbf{L_1}$ and $\mathbf{L_2}$. In quantum mechanics, the angular momentum is subject to spatial quantization, with the total j eigenvalue given along a specific direction. The values of j range from $j = l_1 + l_2$ to $j = |l_1 - l_2|$, corresponding to the two vectors being parallel, adding, or anti-parallel, subtracting, and at angles corresponding to steps of j by 1 in between.

When two angular momenta are added, the total angular momentum state can be expressed as a sum of products of the original states. Our task is to determine the new $|j, m\rangle$ states from an appropriate combination of $|l_1, m_1\rangle|l_2, m_2\rangle$ product states, for which

$$\begin{aligned} J_z|j, m\rangle = &\quad (L_{z1} + L_{z2}) \sum_{m_1, m_2} c_{m_1, m_2}|l_1, m_1\rangle|l_2, m_2\rangle \\ = &\quad \sum_{m_1, m_2} c_{m_1, m_2}(m_1 + m_2)|l_1, m_1\rangle|l_2, m_2\rangle \end{aligned} \tag{18.423}$$

$m = m_1 + m_2$ is the eigenvalue. Conservation of angular momentum supplies an overarching constraint, which implies an inviolable requirement that

$$m = m_1 + m_2 \tag{18.424}$$

with l_1, l_2 and j providing specifications for the maximum and minimum m values. That is to say, the values of j can range between

$$j_{max} = l_1 + l_2,\ l_1 + l_2 - 1,\ l_1 + l_2 - 2,\ ....|l_1 - l_2| = j_{min} \tag{18.425}$$

with m ranging from $-j$ to j for each j. This problem is already done for us in the guise of *Clebsch-Gordan (CG) coefficients*. These were introduced in the nineteenth century for adding spherical harmonics, and they can be looked up in a table or calculated by Mathematica, for example. The point is that there is very little *physics* at stake, however, the *techniques* are of central importance in the general theory of quantum mechanics. Which states can be coupled defines *selection rules* for the mutual action of a Hamiltonian, for example, in the decay of an excited state (selection rules are manifest in the context of the Wigner-Eckart theorem, to be described later).

When two angular momenta are added, the tensor product of the state spaces is being formed. The number of elements in the product space is

$$(2l_1 + 1)(2l_2 + 1) = 4l_1 l_2 + 2l_1 + 2l_2 + 1. \tag{18.426}$$

This can be compared with the total number of j states, based on $l_1 - l_2 \leq j \leq l_1 + l_2$ where we have assumed that $l_1 \geq l_2$. The total number of different j states is

$$\sum_{k=-l_2}^{l_2} (2(l_1 + k) + 1) = 2l_1(2l_2 + 1) + 2l_2 + 1 = 4l_1 l_2 + 2l_1 + 2l_2 + 1 \tag{18.427}$$

and is equal, as expected. This adds justification to our earlier assertion on the range of possible j which was based on a classical argument.

By use of the CG coefficients, a state is simply constructed by

$$|j, m\rangle = \sum_{m=m_1+m_2} \langle l_1, m_1, l_2, m_2|j, m\rangle |l_1, m_1\rangle |l_2, m_2\rangle \tag{18.428}$$

and the CG coefficients $\langle l_1, m_1, l_2, m_2|j, m\rangle$ will be zero if the constraints, $l_1 + l_2 \geq j \geq |l_1 - l_2|$, as already given above, are not met. It can sometimes be helpful to label the $|j, m\rangle$ state that comes from l_1 and l_2 as $|l_1, l_2, j, m\rangle$ because there are multiple ways to arrive at the state j, however, we will not use those additional labels in our ensuing discussions.

Similarly, a state can be deconstructed as

$$|l_1, m_1\rangle |l_2, m_2\rangle = \sum_j \sum_{m=m_1+m_2} \langle j, m|l_1, m_1, l_2, m_2\rangle |j, m\rangle \tag{18.429}$$

where is is understood that j and the CG coefficients are those associated with the basis set comprising the tensor product $l_1 \otimes l_2$, and that

$$\langle l_1, m_1, l_2, m_2|j, m\rangle = \langle j, m|l_1, m_1, l_2, m_2\rangle \tag{18.430}$$

because the CG coefficients, as derived using $L_\pm$ operators, are real.

What if we were on a desert island, without access to the Library of Congress from which we could check out a book with tables of CG coefficients, and needed to add two angular momenta? We can do this by using the $L_{\pm -}$ operators. Recall that

$$L_{\pm}|l,m\rangle = \sqrt{l(l+1) - m(m \pm 1)}|l, m \pm 1\rangle. \tag{18.431}$$

First, form the maximum j and m state, from given l_1 and l_2 states,

$$|l_1 + l_2, l_1 + l_2\rangle = |l_1, l_1\rangle|l_2, l_2\rangle \tag{18.432}$$

and apply the operator

$$J_- = L_{1-} + L_{2-} \tag{18.433}$$

where it is understood that L_{1-} only operates on the $|l_1, m_1\rangle$ states, and similarly for L_{2-}. This operator is applied to each side, with $j = l_1 + l_2$, and $m = l_1 + l_2$.

$$\begin{aligned} J_-|j,j\rangle &= \sqrt{j(j+1) - j(j-1)}|j, j-1\rangle = \sqrt{2j}|j, j-1\rangle \\ &= (L_{1-} + L_{2-})|l_1, l_1\rangle|l_2, l_2\rangle \\ &= \sqrt{2l_1}|l_1, l_1 - 1\rangle|l_2, l_2\rangle + \sqrt{2l_2}|l_1, l_1\rangle|l_2, l_2 - 1\rangle \end{aligned} \tag{18.434}$$

The state is therefore

$$|j, j-1\rangle = \sqrt{\frac{l_1}{j}}|l_1, l_1 - 1\rangle|l_2, l_2\rangle + \sqrt{\frac{l_2}{j}}|l_1, l_1\rangle|l_2, l_2 - 1\rangle. \tag{18.435}$$

We shouldn't need to check the normalization, but as this is our first time,

$$\frac{l_1}{j} + \frac{l_2}{j} = \frac{l_1 + l_2}{l_1 + l_2} = 1, \tag{18.436}$$

as expected.

This can be repeated $2(l_1 + l_2)$ times, until the minimum $m = -l_1 - l_2$ is attained, and the full spectrum of states $|j,j\rangle, |j, j-1\rangle ... |j, -j\rangle$ can be constructed.

The states $|j-1, m\rangle$ can be generated by starting with the largest $m = j - 1$ possible for $j - 1$. This is the state that is orthogonal to $|j, j-1\rangle$, and is constructed as follows

$$|j', j-1\rangle = \sqrt{\frac{l_2}{j}}|l_1, l_1 - 1\rangle|l_2, l_2\rangle - \sqrt{\frac{l_1}{j}}|l_1, l_1\rangle|l_2, l_2 - 1\rangle \tag{18.437}$$

but we don't necessarily know the value of j'; we might assume it is $j - 1$. Let us determine whether $|j-1, j-l\rangle$ is an eigenfunction of $\mathbf{J}^2$, and if so, what is its eigenvalue? This can be determined by operating on the state with

$$\begin{aligned} \mathbf{J}^2 &= (\mathbf{L}_1 + \mathbf{L}_2)^2 = \mathbf{L_1}^2 + \mathbf{L_2}^2 + 2\mathbf{L_1} \cdot \mathbf{L_2} \\ &= \mathbf{L}_1^2 + \mathbf{L}_2^2 + 2L_{z1}L_{z2} + 2L_{x1}L_{x2} + 2L_{y1}L_{y2} \\ &= \mathbf{L}_1^2 + \mathbf{L}_2^2 + 2L_{z1}L_{z2} + L_{1+}L_{2-} + L_{1-}L_{2+} \end{aligned} \tag{18.438}$$

where we have used the fact that L_1 and L_2 commute, and $L_\pm = L_x \pm iL_y$. Noting that $L_+|l,l\rangle = 0$ for any l, and that $L_{1+}L_{2-}|l_1,l_1-1\rangle|l_2,l_2\rangle = 2\sqrt{l_1 l_2}\,|l_1,l_1\rangle|l_2,l_2-1\rangle$ (for $L_{1-}L_{2+}$, interchange l_1 and l_2)

$$
\begin{aligned}
\mathbf{J}^2|j',j-1\rangle &= (\mathbf{L}_1+\mathbf{L}_2)^2\left(\sqrt{\frac{l_2}{j}}|l_1,l_1-1\rangle|l_2,l_2\rangle - \sqrt{\frac{l_1}{j}}|l_1,l_1\rangle|l_2,l_2-1\rangle\right) \\
&= ((l_1+1)l_1 + (l_2+1)l_2 + 2l_1l_2 - 2(l_1+l_2)) \\
&\quad\times\left(\sqrt{\frac{l_2}{j}}|l_1,l_1-1\rangle|l_2,l_2\rangle - \sqrt{\frac{l_1}{j}}|l_1-1,l_2\rangle|l_2,l_2\rangle\right) \\
&= ((l_1+l_2-1)(l_1+l_2))\,|j',j-1\rangle \\
&= j'(j'+1)|j',j-1\rangle \quad \text{if } j' = l_1+l_2-1 = j-1
\end{aligned}
\tag{18.439}
$$

which is the answer that was anticipated. The process of lowering m can be be used on $|j-1,m-1\rangle$ to generate its spectrum of m states. Similarly, the state orthogonal to $|j-1,m-2\rangle$ is $|j-2,m-2\rangle$ and the process continues through $j = |j_1 - j_2|$.

The number of possible CG coefficients increase rapidly with increasing l, approximately as $4l_1l_2$. Fortunately, the CG coefficients have a known functional form, and can be found in tables or generated using Mathematica and similar programs. The CG coefficients are also tabulated as the Wigner $3j$ symbols.

Three angular momenta can be added, but the algebra gets messy quickly. There are tables of coefficients for this, too, in which case the $6j$ symbols are used; these represent a sum over three $3j$ symbols ($3+3=6$), because for three states, three different pairwise sums are possible. For four angular momentum, there are $9j$ symbols representing a sum over six $3j$ symbols (for four different angular momentum states, six different pairwise sums are possible $3+6=9$.).

The angular momenta addition coefficients can be calculated in closed form using group theory techniques or by using Schwinger's method that will be discussed later in this chapter.

18.11.6 Orthogonality of Clebsch-Gordan coefficients

Consider two different states of the form

$$|j,m\rangle = \sum_{m_1,m_2} \langle j_1,m_1,j_2,m_2|j,m\rangle|j_1,m_1\rangle|j_2,m_2\rangle, \tag{18.440}$$

and

$$|j',m'\rangle = \sum_{m_1',m_2'} \langle j_1',m_1',j_2',m_2'|j',m'\rangle|j_1',m_1'\rangle|j_2',m_2'\rangle. \tag{18.441}$$

Let us take their inner product,

$$\begin{aligned}
\langle j', m'|j, m\rangle &= \delta_{jj'}\delta_{mm'} \\
&= \sum_{\substack{m_1,m_2\\ m'_1,m'_2}} \langle j_1, m_1, j_2, m_2|j, m\rangle\langle j'_1, m'_2, j'_2, m'_2|j', m'\rangle\langle j'_1, m'_1|\langle j'_2, m'_2|j_1, m_1\rangle|j_2, m_2\rangle \\
&= \sum_{\substack{m_1,m_2\\ m'_1,m'_2}} \langle j_1, m_1, j_2, m_2|j, m\rangle\langle j'_1, m'_2, j'_2, m'_2|j', m'\rangle\delta_{j_1,j'_1}\delta_{j_2j'_2}\delta_{m_1m'_1}\delta_{m_2m'_2} \\
&= \sum_{m_1,m_2} \langle j_1, m_1, j_2, m_2|j, m\rangle\langle j_1, m_1, j_2, m_2|j', m'\rangle = \delta_{jj'}\delta_{mm'}.
\end{aligned} \tag{18.442}$$

This is possible because the CJ coefficients are real numbers.

Because the states $|j, m\rangle$ form a complete orthonormal set, combining j_1 and j_2,

$$1 = \sum_{m=-j}^{j} |j, m\rangle\langle j, m| =$$

$$\sum_{j=|j_1-j_2|}^{j_1+j_2} \sum_{\substack{m_1+m_2\\=m=-j}}^{j} \langle j_1, m_1, j_2, m_2|j, m\rangle\langle j'_1, m'_1, j'_2, m'_2|j, m\rangle|j_1m_1\rangle|j_2m_2\rangle|\langle j'_1m'_1\langle j'_2m'_2|$$

Taking the inner product

$$\langle j'_1m'_1|\langle j'_2m'_2|1|j_1m_1\rangle|j_2m_2\rangle =$$

$$= \sum_{j=|j_1-j_2|}^{j_1+j_2} \sum_{\substack{m_1+m_2\\=m=-j}}^{j} \delta_{j_1j'_1}\delta_{j_2j'_2}\delta_{m_1m'_1}\delta_{m_2m'_2}\langle j_1, m_1, j_2, m_2|j, m\rangle\langle j'_1, m'_1, j'_2, m'_2|j, m\rangle$$

Therefore,

$$\sum_{j=|j_1-j_2|}^{j_1+j_2} \sum_{\substack{m_1+m_2\\=m=-j}}^{j} \langle j_1, m_1, j_2, m_2|j, m\rangle\langle j_1, m'_1, j_2, m'_2|j, m\rangle = \delta_{m'_1m_1}\delta_{m_2m'_2}. \tag{18.443}$$

18.11.7 Matrix representation of angular momentum states and operators

First, a note on notation:

- **L**, l, and similar represent orbital angular momentum states.
- **S**, $\dot{s}$, and similar represent intrinsic spin-1/2 angular momentum states that have no spatial representation.
- **J**, j, and similar are combinations of two or more spatial and/or intrinsic angular momenta.

The matrix representation of operators will allow us to combine intrinsic and orbital angular momentum. In a coordinate representation, the intrinsic components of the angular momentum will remain as a matrix (ket) whereas the spatial components will be functions of coordinates.

With the foregoing operator algebra, we can determine vectors and square matrices that can represents states and operators. The vector has dimension $2j+1$, the matrices dimension $(2j+1) \times (2j+1)$. Let us first investigate $s = 1/2$, for which the operator matrices are 2×2. Second, we will investigate $l = 1$, for which the operator matrices are 3×3.

For $s = 1/2$, the state vector is a two element vector that requires two complex numbers to be specified, implying four real numbers, with unity normalization reducing the total independent parameters to three real numbers. We can write (a, b) for the complex amplitudes of the $m = 1/2, \ -1/2$ states. The s_z operator is chosen to be diagonal,

$$\langle m'|s_z|m\rangle = \begin{pmatrix} 1/2 & 0 \\ 0 & -1/2 \end{pmatrix} \tag{18.444}$$

and the matrices for s_x and s_y can be determined as

$$\begin{aligned} \langle m'|s_x|m\rangle &= \langle m'| \left(\frac{s_+ + s_-}{2} \right) |m\rangle \\ &= \frac{1}{2} \left(\sqrt{s(s+1) - m(m+1)}\delta_{m+1,m'} + \sqrt{s(s+1) - m(m-1)}\delta_{m-1,m'} \right), \end{aligned}$$

Collecting into an array,

$$s_x = \begin{pmatrix} 0 & 1/2 \\ 1/2 & 0 \end{pmatrix} \tag{18.445}$$

and in a similar fashion,

$$s_y = \begin{pmatrix} 0 & -i/2 \\ i/2 & 0 \end{pmatrix}. \tag{18.446}$$

Note that these matrices are hermitian.

With these matrices explicitly written out, we can see that

$$\mathbf{S}^2 = s_x^2 + s_y^2 + s_z^2 = \frac{3}{4} \begin{pmatrix} 1 & 0 \\ 0 & 1 \end{pmatrix} = s(s+1)I = \frac{1}{2} \left(\frac{1}{2} + 1 \right) I = \frac{3}{4} I \tag{18.447}$$

which gives the the correct eigenvalue (3/4) and it becomes obvious why everything commutes with $\mathbf{S}^2$. The raising and lowering operators are

$$s_+ = s_x + i s_y = \begin{pmatrix} 0 & 1 \\ 0 & 0 \end{pmatrix} \; ; \quad s_- = s_x - i s_y = \begin{pmatrix} 0 & 0 \\ 1 & 0 \end{pmatrix}. \tag{18.448}$$

Note that $s_+^2 = s_-^2 = 0$ because two steps will take any state outside of its allowed range of m.

Following the same development for $l = 1$, The state vector is three dimensional as (a, b, c) for the amplitudes of the $m = 1, 0, 2$ states. The L_z operator is chosen to be diagonal,

$$\langle m'|L_z|m\rangle = \begin{pmatrix} 1 & 0 & 0 \\ 0 & 0 & 0 \\ 0 & 0 & -1 \end{pmatrix} \tag{18.449}$$

and the matrices for L_x and L_y can be determined

$$\langle m'|L_x|m\rangle = \langle m'|\left(\frac{L_+ + L_-}{2}\right)|m\rangle$$
$$= \frac{1}{2}\left(\sqrt{l(l+1) - m(m+1)}\delta_{m+1,m'} + \sqrt{l(l+1) - m(m-1)}\delta_{m-1,m'}\right) \tag{18.450}$$

and collected into an array,

$$L_x = \frac{1}{\sqrt{2}}\begin{pmatrix} 0\,1\,0 \\ 1\,0\,1 \\ 0\,1\,0 \end{pmatrix} \tag{18.451}$$

and in a similar fashion,

$$L_y = \frac{1}{\sqrt{2}}\begin{pmatrix} 0 & -i & 0 \\ i & 0 & -i \\ 0 & i & 0 \end{pmatrix}. \tag{18.452}$$

Note that these matrices are hermitian.

With these matrices explicitly written out, we can see that

$$\mathbf{L}^2 = 2\begin{pmatrix} 1\,0\,0 \\ 0\,1\,0 \\ 0\,0\,1 \end{pmatrix} = 1 \times (1+1)I \tag{18.453}$$

which is the correct eigenvalue, and, again, it becomes obvious why everything commutes with $\mathbf{L}^2$. The raising and lowering operators are

$$L_+ = L_x + iL_y = \sqrt{2}\begin{pmatrix} 0\,1\,0 \\ 0\,0\,1 \\ 0\,0\,0 \end{pmatrix} \;;\quad L_- = L_x - iL_y = \sqrt{2}\begin{pmatrix} 0\,0\,0 \\ 1\,0\,0 \\ 0\,1\,0 \end{pmatrix}. \tag{18.454}$$

Note that $L_+^3 = L_-^3 = 0$ because three steps will take any state outside of its allowed range of m.

18.11.7.1 Rotations

With the matrix representation of angular momentum, it is straightforward to work out the rotation operators. We showed that angular momentum is the generator of infinitesimal rotations, and that the infinitesimals could be added up to give a finite rotation angle. For $l = 1$, with a rotation along the z axis which is easy to work out, (with I the identity matrix),

$$D_z(\theta) = e^{-i\theta L_z} = I + \sum_{n=1}^{\infty}\begin{pmatrix} \frac{(-i\theta)^n}{n!} & 0 & 0 \\ 0 & 0 & 0 \\ 0 & 0 & \frac{(i\theta)^n}{n!} \end{pmatrix} = \begin{pmatrix} e^{-i\theta} & 0 & 0 \\ 0 & 1 & 0 \\ 0 & 0 & e^{i\theta} \end{pmatrix}, \tag{18.455}$$

where traditionally D or d is used as the label of the rotation operator, and $\mathcal{D}(\mathbf{R})$ is used to represent a generic rotation of an object. For a general angular momentum state j, the rotation matrices will be of dimension $(2j+1) \times (2j+1)$.

For a general rotation, we need three rotation directions. In classical mechanics, the Euler angles are used, and those are useful in quantum mechanics also. A general rotation is

$$\mathcal{D}(\mathbf{R}) = \mathcal{D}(\gamma, \beta, \alpha) = D_z(\gamma) D_y(\beta) D_z(\alpha). \tag{18.456}$$

The dimensions of D_y and D_z depend on j, so it is therefore useful to specify $D^{(J)}(\mathbf{R}) \equiv D^{(J)}$ with $\mathbf{R}$ understood as being present. Considering rotations one by one, we will write $D^{(J)} = d_z^{(j)}(\gamma) d_y^{(j)}(\beta) d_z^{(j)}(\alpha) \equiv d_z^{(j)} d_y^{(j)} d_z^{(j)}$ where the angles are understood as being present, and will be included only if there is some ambiguity.

However, we must be aware that the $D^{(J)}$ operates on quantum mechanical state vectors, not coordinate vectors, as can be seen from the rotation of a vector about the z axis by θ for a classical system, in which case

$$\mathcal{D}_{z,classical}(\theta) = \begin{pmatrix} \cos\theta & -\sin\theta & 0 \\ \sin\theta & \cos\theta & 0 \\ 0 & 0 & 1 \end{pmatrix}. \tag{18.457}$$

This looks nothing like the quantum state rotation operator for $L = 1$. Of course, this comparison only makes sense for $l = 1$ because the matrices are 3×3 for both the classical and quantum rotation matrices. But how could two operations that we expect to be the same in a classical sense appear as so very different? The similarity can be elucidated as follows. An eigenstate of L_x can be written in the $|l, m_l\rangle$ basis as

$$|1, 1\rangle_x = \begin{pmatrix} \frac{1}{2} \\ \frac{1}{\sqrt{2}} \\ \frac{1}{2} \end{pmatrix} \tag{18.458}$$

and operate on with $d_z^{(1)}(\theta)$,

$$|1, \theta\rangle_x = d_z^{(1)}(\theta) \begin{pmatrix} \frac{1}{2} \\ \frac{1}{\sqrt{2}} \\ \frac{1}{2} \end{pmatrix} = \begin{pmatrix} \frac{e^{-i\theta}}{2} \\ \frac{1}{\sqrt{2}} \\ \frac{e^{i\theta}}{2} \end{pmatrix}. \tag{18.459}$$

Let us now calculate the expectation values of L_x and L_y. In the L_z basis,

$$L_x = \frac{1}{\sqrt{2}} \begin{pmatrix} 0 & 1 & 0 \\ 1 & 0 & 1 \\ 0 & 1 & 0 \end{pmatrix} \tag{18.460}$$

and

$$L_y = \frac{1}{\sqrt{2}} \begin{pmatrix} 0 & -i & 0 \\ i & 0 & -i \\ 0 & i & 0 \end{pmatrix}. \tag{18.461}$$

The expectation values are

$$\langle L_x \rangle = {}_x\langle 1, \theta | L_x | 1, \theta \rangle_x = \cos\theta \tag{18.462}$$

and

$$\langle L_y \rangle = {}_x\langle 1, \theta | L_y | 1, \theta \rangle_x = \sin\theta. \tag{18.463}$$

Now let us compare this result to a classical unit (angular momentum) vector pointing in the x direction,

$$\begin{pmatrix} x \\ y \\ z \end{pmatrix} = \begin{pmatrix} 1 \\ 0 \\ 0 \end{pmatrix} \tag{18.464}$$

that is rotated about the z axis by use of the classical rotation matrix,

$$R_{z,classical} \begin{pmatrix} 1 \\ 0 \\ 0 \end{pmatrix} = \begin{pmatrix} \cos\theta \\ \sin\theta \\ 0 \end{pmatrix}. \tag{18.465}$$

The classical rotation gives the same x and y components as in the quantum case, implying that the classical rotation is intrinsic with a rotation of the quantum system. In fact, all vectors will transform in this manner, not only angular momentum.

18.11.8 General rotation matrices

So far we have only determined D_z, but also need $D_y(\beta)$ in order to perform a general rotation. The matrix multiplication is slightly more challenging (again, I is the identity matrix).

For spin-1/2, the rotation operator (in the $j = 1/2$ basis) is

$$D^{(\frac{1}{2})}(\mathbf{R}) = e^{-is_z\gamma} e^{-is_y\beta} e^{-is_z\alpha}. \tag{18.466}$$

The s_z factors are easy to determine, however, $d_y^{(\frac{1}{2})}$ requires some care.

$$d_y^{(\frac{1}{2})} = e^{-is_y\beta} = 1 - is_y\beta + \frac{(-is_y\beta)^2}{2!} + \frac{(-is_y\beta)^3}{3!} + \frac{(-is_y\beta)^4}{4!} + ... \tag{18.467}$$

where 1 on the rhs is understood to be the 2×2 identity matrix I. Noting that $s_y^2 = I/4$,

$$\begin{aligned} d_y^{(\frac{1}{2})}(\beta) &= is_y \sum_{n=0}^{\infty} \frac{(-i\beta/2)^{2n+1}}{(2n+1)!} + I \sum_{n=0}^{\infty} \frac{(-i\beta/2)^{2n}}{(2n)!} \\ &= I\cos(-\beta/2) - is_y \sin(\beta/2) \\ &= \begin{pmatrix} \cos(\beta/2) & -\sin(\beta/2) \\ \sin(\beta/2) & \cos\beta/2 \end{pmatrix}. \end{aligned} \tag{18.468}$$

For $l = 1$, considerably more work is required,

$$
\begin{aligned}
d_y^{(1)}(\beta) = e^{-i\theta L_y} &= I + \frac{-i\theta}{\sqrt{2}} \begin{pmatrix} 0 & -1 & 0 \\ 1 & 0 & -1 \\ 0 & 1 & 0 \end{pmatrix} + \frac{1}{2!}\left(\frac{-i\theta}{\sqrt{2}}\right)^2 \begin{pmatrix} -1 & 0 & 1 \\ 0 & -2 & 0 \\ 1 & 0 & -1 \end{pmatrix} + \\
&+ \frac{1}{3!}\left(\frac{-i\theta}{\sqrt{2}}\right)^3 \begin{pmatrix} 0 & 2 & 0 \\ -2 & 0 & -2 \\ & -2 & 0 \end{pmatrix} + \ldots \\
&= I - iL_y \sum_{n=0}^{\infty} \frac{(-1)^n \theta^{2n+1}}{(2n+1)!} + L_y^2 \sum_{n=1}^{\infty} \frac{(-1)^n \theta^{2n}}{(2n)!} \\
&= I - iL_y \sin\theta + \frac{1}{2}(\cos\theta - 1) L_y^2 \\
&= \begin{pmatrix} \frac{1+\cos\theta}{2} & -\frac{\sin\theta}{\sqrt{2}} & \frac{1-\cos\theta}{2} \\ \frac{\sin\theta}{\sqrt{2}} & \cos\theta & -\frac{\sin\theta}{\sqrt{2}} \\ -\frac{1-\cos\theta}{2} & \frac{\sin\theta}{\sqrt{2}} & \frac{1+\cos\theta}{2} \end{pmatrix}.
\end{aligned}
\tag{18.469}
$$

The y rotation matrix calculations become quite cumbersome for large j. However, there are tables and some programs such as Mathematica and Maple that can calculate them. There are algebraic and other methods to determine these matrices in closed form, for example, by use of the Schwinger spin-1/2 angular momentum harmonic oscillator representation that we will discuss in the next section.

18.11.9 The Wigner-Eckart theorem

We will now attempt a heuristic derivation of the Wigner-Eckart therm, and what this derivation lacks in rigor will be compensated for by the illuminating simplicity of the argument. Let us first specialize to $L = 1$, the case of a vector which is the most important.

Very often, an external potential as might be applied to a quantum system in some particular state can be represented as a function of spatial coordinates. When this is the case, the potential can always be written in a form

$$
V(\mathbf{r}) = \sum_{LM} f_L(r) Y_{LM}(\phi, \theta) \tag{18.470}
$$

because the spherical harmonics Y_{LM} form a complete set. The radial function $f_L(r)$ depends only on L because of separation of variables in the Laplacian.

The spherical harmonics for $L = 1$ are given by:

$$
\begin{aligned}
Y_{1,0} &= N_1 \cos\theta \\
Y_{1,\pm 1} &= N_1 \sin\theta e^{\pm i\phi}
\end{aligned}
$$

where the normalizing factor $N_1 = \frac{1}{2}\sqrt{\frac{3}{2\pi}}$.

Thus a vector of magnitude A, where A can be a function of r, in a direction given by the spherical angles (θ, ϕ) has the components

$$A_x = \frac{A}{2N_1}(Y_{1,1} + Y_{1,-1}) = A \sin\theta \cos\phi$$

$$A_y = \frac{A}{2iN_1}(Y_{1,1} - Y_{1,-1}) = A \sin\theta \sin\phi$$

$$A_z = \frac{A}{N_1} Y_{1,0} = A \cos\theta$$

so that calculating the matrix element of any vector operator between angular momentum eigenstates boils down to calculating matrix elements of $Y_{1,M}$:

$$\langle L_3 M_3 | Y_{1,M_2} | L_1 M_1 \rangle = C_{L_1,1,L_3} \int Y_{L_3 M_3}(\Omega) Y_{1,M_2}(\Omega) Y_{L_1 M_1}(\Omega) \, d\Omega \tag{18.471}$$

where $C_{L_1,1,L_3}$ is a factor that depends on the chosen normalization conventions and depends only on the magnitudes of the Ls (no M dependence). Higher order tensor operators are represented by $L > 1$ so we will generalize the argument to calculate the matrix element of Y_{L_2,M_2}.

Then we write $Y_{L_2,M_2}(\Omega), Y_{L_1 M_1}$ as $\langle \Omega | L_2, M_2 \rangle, \langle \Omega | L_1 M_1 \rangle$ and use Eq. (18.429) to write

$$|L_2, M_2\rangle |L_1, M_1\rangle = \sum_{L, M = M_1 + M_2} |L, M\rangle \langle L, M | L_1, M_1, L_2, M_2 \rangle .$$

The matrix element (18.471) is then

$$\begin{aligned} \langle L_3 M_3 | Y_{L_2,M_2} | L_1 M_1 \rangle &= \sum_{L, M = M_1 + M_2} \langle L_3, M_3 | L, M \rangle \langle L, M | L_1, M_1, L_2, M_2 \rangle \\ &= \langle L_3, M_3 | L_1, M_1, L_2, M_2 \rangle \end{aligned} \tag{18.472}$$

using the completeness of the states $|L, M\rangle$. Thus the matrix element of any vector will be proportional to a Clebsch-Gordon coefficient and, since that applies to the angular momentum vector itself, to the matrix elements of the angular momentum of the system. This imposes *selection rules* so that the matrix element is only nonzero if the trio $(L_1, M_{1,}), (L_2, M_2), (L_3 M_3)$ is such that $L_1 + L_2 \geq L_3 \geq |L_1 - L_2|$ and $M_3 = M_1 + M_2$.

This is one of the most useful theorems that is widely used in quantum mechanics, especially in atomic and nuclear physics. However, this theorem really is a feature of the spherical harmonics, in particular, resulting from their completeness and orthogonality, which is as much a classical as a quantum concept.

This formalism also works in the case where $J = L + S$, i.e., when there is also intrinsic spin in the system. In this case the reduced matrix element is a combination of orbital angular momentum and intrinsic spin matrix elements which do not have a spatial integral but nonetheless follow the rules of addition of angular momentum.

There are different forms of the Wigner-Eckart theorem, and the above derivation can be easily put into any convenient form which usually involves a rescaling of the reduced matrix element (the product of all the factors that are independent of the Ms) and including the normalization factors for the spherical harmonics.

18.11.10 Angular momentum from harmonic oscillator states

Although such states are not necessary physical, any angular momentum state l can be constructed from $2l$ spin-1/2 states, and l can be half-integer. The states are unphysical in that it is generally not possible to have the correct exchange symmetry. However, the angular momentum behavior of the state does not depend on the symmeterization, so we can use these artificial states to develop the algebra of angular momentum. This was first used by Marjoranna, and later Schwinger developed the algebra of angular momentum using spin-1/2 creation and annihilation operators.[28]

From $\mathbf{L} = \mathbf{r} \times \mathbf{p}$ we have $L_z = xp_y - yp_x$. Introducing the following new operators,

$$\begin{aligned} q_1 &= \frac{x + p_y}{\sqrt{2}} \\ q_2 &= \frac{x - p_y}{\sqrt{2}} \\ p_1 &= \frac{p_x - y}{\sqrt{2}} \\ p_2 &= \frac{p_x + y}{\sqrt{2}} \end{aligned} \tag{18.473}$$

which have the following commutation relations:

$$[q_1, q_2] = [p_1, p_2] = 0 \tag{18.474}$$

$$[q_j, p_k] = i\delta_{j,k}. \tag{18.475}$$

Therefore, q_1, p_1 and q_2, p_2 appear in a formal manner as position and momentum operators of two different systems. In terms of these operators, define

$$L_z = \frac{1}{2}(p_1^2 + q_1^2) - \frac{1}{2}(p_2^2 + q_2^2), \tag{18.476}$$

and L_z is the difference between two independent harmonic oscillator Hamiltonians, each having unit mass and unit angular frequency,

$$\begin{aligned} L_z &= H_1 - H_2 \\ H_1 &= \frac{1}{2}(p_1^2 + q_1^2) \\ H_2 &= \frac{1}{2}(p_2^2 + q_2^2). \end{aligned} \tag{18.477}$$

The "energy" spectrum of the harmonic oscillator Hamiltonians H_1 and H_2 are $n_1{+}1/2$ and $n_2{+}1/2$, respectively, with n_1 and n_2 positive integers. Because $[H_1, H_2] = 0$, the spectrum of L_z is

$$(n_1 + 1/2) - (n_2 + 1/2) = n_1 - n_2. \tag{18.478}$$

[28] Schwinger, J., *On Angular Momentum*, DOE Report NYO-3071, January, 1952. https://www.osti.gov/servlets/purl/4389568.

This is the difference of two integer numbers, so it is an integer. However, the states that we will be creating and annihilating are spin-1/2, with n_1 corresponding to $|1/2, 1/2\rangle$ states and n_2 corresponding to $|1/2, -1/2\rangle$ states, so

$$L_z = \frac{(n_1 - n_2)}{2}. \tag{18.479}$$

We can use the harmonic oscillator algebra we developed earlier, noting

$$L_z = \frac{1}{2}(n_1 - n_2) = \frac{1}{2}(a_1^\dagger a_1 - a_2^\dagger a_2), \tag{18.480}$$

and we can define ladder operators that increase or decrease L_z by 1,

$$L_+ = a_1^\dagger a_2; \quad L_- = a_2^\dagger a_1. \tag{18.481}$$

These operators have the usual commutation relationships we investigated earlier, but the twist here is we have two sets of operators that are completely independent of each other,

$$\begin{aligned} [L_+, L_-] &= 2L_z \\ [L_z, L_\pm] &= \pm L_\pm. \end{aligned} \tag{18.482}$$

The total $N = n_1 + n_2$ is

$$N = a_1^\dagger a_1 + a_2^\dagger a_2 \tag{18.483}$$

and

$$\mathbf{L}^2 = J_z^2 + \frac{L_+L_- + L_-L_+}{2} = \frac{1}{2}N\,(N+1). \tag{18.484}$$

The n states can be built from the vacuum state by applying the creation operators. Specifically, we can build a state

$$|n_1, n_2\rangle = \frac{(a_1^\dagger)^{n_1}(a_2^\dagger)^{n_2}}{\sqrt{n_1! n_2!}}|0, 0\rangle = |l, m\rangle \tag{18.485}$$

where $l = n_1 + n_2$ and $m = n_1 - n_2$.

The highest state for a given l is

$$|l, l\rangle = \frac{(a_1^\dagger)^{2l}}{(2l)!}|0, 0\rangle \tag{18.486}$$

and it contains $2l$ excitations because we are combining spin 1/2's to get a particular l.

Because the oscillators are completely uncoupled, $[a_1, a_2] = [a_1^\dagger, a_2] = 0$, etc., e.g., any commutator between the 1- and 2-labeled operators is zero. Therefore, we will be unsuccessful if we try to increase m by one by use of L_+ on the state $|l, l\rangle$. This is because L_+ commutes with $(a_1^\dagger)^{2l}$ and acts directly on $|0, 0\rangle$; the annihilation operator producing $a_2|00\rangle = 0$.

This formalism can be used to calculate the Clebsch-Gordan coefficients in closed form. In fact, it is possible to form raising and lowering operators that apply to both l

and m of a $|l, m\rangle$ state. Some say that the l raising operator is the "true" one, however, it must be recalled that m is the observable angular momentum that determines the physical properties of the state.

Earlier, we made use of the fact that an operator could be transformed before applying the state generation operator (Heisenberg picture, applied to the beam splitter), and that often saves considerable effort.[29] By rotating by an angle β the $a_1^\dagger$ and $a_2^\dagger$ operators, and then using them to calculate a given angular momentum state, the expansion of y-axis rotation matrix $D_y(\beta)$ in terms of $d^l_{mm'}(\beta)$ can be directly determined. The rotation of the a_1 and a_2 operators can be performed in direct analogy to rotation of a spin-1/2 particle around the y axis.

$$\begin{aligned} a_1^\dagger &\rightarrow a_1^\dagger \cos(\beta/2) + a_2^\dagger \sin(\beta/2) = a_{\beta 1}^\dagger \\ a_2^\dagger &\rightarrow a_2^\dagger \cos(\beta/2) - a_1^\dagger \sin(\beta/2) = a_{\beta 2}^\dagger . \end{aligned} \tag{18.487}$$

(Note that in general the $|00\rangle$ state would also need to be transformed, however, it is rotationally invariant.)[30]

When we generate a state

$$|l, m\rangle_\beta = \frac{(a_{\beta 1}^\dagger)^{n_1} (a_{\beta 2}^\dagger)^{n_2}}{\sqrt{n_1! n_2!}} |0, 0\rangle, \tag{18.488}$$

where $l = n_1 + n_2$ and $m = n_1 - n_2$, we can use the binomial theorem to expand

$$\begin{aligned} (a_{\beta 1}^\dagger)^{n_1} =& (a_1^\dagger \cos(\beta/2) + a_2^\dagger \sin(\beta/2))^{n_1} \\ =& \sum_k \frac{n_1! (a_1^\dagger \cos(\beta/2))^{(n_1 - k)} (a_2^\dagger \sin(\beta/2)^k}{(n_1 - k)! k!} \\ (a_{\beta 2}^\dagger)^{n_2} =& (-a_1^\dagger \sin(\beta/2) + a_2^\dagger \cos(\beta/2))^{n_2} \\ =& \sum_{k'} \frac{n_2! (-a_1^\dagger \sin(\beta/2))^{(n_2 - k')} (a_2^\dagger \cos(\beta/2))^{k'}}{(n_2 - k')! k'!} \end{aligned} \tag{18.489}$$

and $|l, m\rangle_\beta$ becomes a double sum over k and k'. This double sum can be compared to

$$\begin{aligned} |l, m\rangle_\beta = D_y(\beta)|l, m\rangle =& \sum_m' |l, m'\rangle d^{(l)}_{m'm}(\beta) |l, m'\rangle \\ =& \sum_m' d^{(l)}_{m'm}(\beta) \frac{(a_1^\dagger)^{n_1} (a_2^\dagger)^{n_2}}{\sqrt{n_1! n_2!}} |0, 0\rangle \end{aligned} \tag{18.490}$$

and collecting and equating terms with the same $(a_1^\dagger)^{n_1} (a_2^\dagger)^{n_2}$ allows the $d^{(l)}_{m'm}(\beta)$ to be directly determined.

[29] Sakurai, J.J. and Napolitano, Jim, *Modern Quantum Mechanics, 2nd ed.* (Addison-Wesley, San Francis, 2011). Sec. 3.9.

[30] Schwinger, J., op. cit. p. 579; Sakurai, J.J. and Napolitano, J., op. cit., p. 581

18.11.11 Angular momentum states of molecules

The angular momentum formalism that we have developed can be applied to any quantized object. In particular, the center-of-mass angular momentum state is of central importance in molecular physics. There are three cases, a spherically symmetric molecule, a symmetric top, and a completely asymmetric molecule. We will not include the effects of vibrational states, and assume the atomic constituents of the molecule are rigidly fixed (a rigid rotator).

18.11.11.1 Spherical molecule

An example of a spherically symmetric molecule is methane. For a spherical system, the moment of inertia is the same for all axes $I_x = I_y = I_z = I$, the Hamiltonian is

$$H = \frac{1}{2I}\mathbf{L}^2 = \frac{l(l+1)}{2I} \tag{18.491}$$

and the states are completely described by the $|l, m\rangle$ states with E independent of m.

18.11.11.2 Symmetric top

An example of this is the ammonia molecule, with three hydrogens at the base of a pyramid with a nitrogen at the apex. In this case, two axes have the same moment, $I_x = I_y$, so

$$H = \frac{1}{2I_x}(L_x^2 + L_y^2) + \frac{1}{2I_z}L_z^2 = \frac{1}{2I_x}(L_x^2 + L_y^2 + L_z^2) + \left(\frac{1}{2I_z} - \frac{1}{2I_x}\right)L_z^2 \tag{18.492}$$

which can be rewritten as

$$H = \frac{1}{2I_x}\mathbf{L}^2 + \left(\frac{1}{2I_z} - \frac{1}{2I_x}\right)L_z^2. \tag{18.493}$$

The $|l, m\rangle$ states are simultaneous eigenvalues of l and m, therefore

$$E_{lm} = \frac{l(l+1)}{2I_z} + \left(\frac{1}{2I_z} - \frac{1}{2I_x}\right)m^2 \tag{18.494}$$

and the eigenstate is approximately given by an $|l, m\rangle$ state. In fact the eigenstate is altered because the form of the Hamiltonian is based on the symmetry axes fixed with the molecule, not the laboratory frame in which the $|l, m\rangle$ states were derived.[31]

18.11.11.3 Asymmetric top

In this case, all three moments of inertia are different. There are no general forms for the energy spectrum, although it is possible to collect the states into symmetry classes that provide information regarding transitions between states.

For very large molecules (DNA), the angular momentum suggests a way to see when a system begins to behave classically. Of course, a free molecule floating in

[31] Landau, L.D. and Lifshitz, E.M., *Quantum Mechanics, Non-Relativistic Theory, 3rd Ed.* (Pergamon Press, Oxford, 1977). pp. 410-419.

space, unperturbed, will be in a state of definite $|j, m\rangle$. However, the total j and m comes from a coherent superposition of the l, s, j of every electron and nucleus in the system. In fact, all of these particles in any single molecule are entangled. However, probing such entanglement is nearly impossible because any attempt to measure one molecular subunit will perturb the entire molecule, unlike when separated photons are measured in a Bell inequalilty experiment. However, it might be possible to explore coherences and entanglement by simultaneously probing opposite sides of a large diameter molecule, or opposite ends of a long molecule.

18.12 Algebraic derivation of the hydrogen spectrum

In 1926, Pauli used Heisenberg's matrix mechanics to give the first quantum mechanical derivation of the energy levels and degeneracies of the hydrogen atom.[32] This was a remarkable feat, however, from a foregoing analysis (Sec. 18.8) where the hydrogen atom Hamiltonian was recast as that of a two-dimensional harmonic oscillator, it is not surprising that this was possible. Apparently, the only problem that can be directly solved by matrix mechanics is the harmonic oscillator. So any problem that can be recast as a harmonic oscillator most likely can be solved in the context of matrix mechanics; the problem lies in determining the appropriate operators that can be used to construct the solution.

Pauli was able to find an algebraic solution because for the hydrogen atom there are *three* conserved quantities: the energy E (eigenvalues of H), the angular momentum $\mathbf{L}$, and a third construct that is known in classical mechanics (for a $1/r$ central potential) as the *Laplace-Runge-Lenz vector.*[33]

18.12.1 Classical treatment of the central potential

In classical mechanics, for a central potential,

$$\mathbf{F}(\mathbf{r}) = f(r)\hat{\mathbf{r}} = f(r)\frac{\mathbf{r}}{r} = \dot{\mathbf{p}}. \tag{18.495}$$

The cross product with $\mathbf{L}$ is, noting that $\mathbf{L} = \mathbf{r} \times \mathbf{p} = m\, \mathbf{r} \times \dot{\mathbf{r}}$,

$$\dot{\mathbf{p}} \times \mathbf{L} = \frac{mf(r)}{r}\left[\mathbf{r} \times (\mathbf{r} \times \dot{\mathbf{r}})\right] = \frac{mf(r)}{r}\left[\mathbf{r}(\mathbf{r} \cdot \dot{\mathbf{r}}) - r^2\dot{\mathbf{r}}\right], \tag{18.496}$$

where the $BAC-CAB$ rule was used to expand the triple cross product. Next, because $\mathbf{L}$ is constant, we can rewrite (using $\mathbf{r} \cdot \dot{\mathbf{r}} = r\dot{r}$)

$$\dot{\mathbf{p}} \times \mathbf{L} = \frac{d}{dt}(\mathbf{p} \times \mathbf{L}) = -mf(r)r^2\left[\frac{\dot{\mathbf{r}}}{r} - \frac{\mathbf{r}\dot{r}}{r^2}\right], \tag{18.497}$$

[32] Pauli, W., *Über das Wasserstoffspektrum vom Standpunkt der neuen Quantenmechanik* [On the hydrogen spectrum from the standpoint of the new quantum mechanics], Z. Phys. **36**, 336 (1926). English translation in B. L. Van der Waerden, ed., *Sources of Quantum Mechanics* (Dover, New York, 1968).

[33] See, e.g., Goldstein, H., *Classical Mechanics* (Addison-Wesley, Reading, MA, 1980), p. 102ff. Laplace was first to derive the components of $\mathcal{A}$ in his treatise on celestial mechanics; it was put into vector form by Gibbs.

which can be condensed to

$$\frac{d}{dt}(\mathbf{p}\times\mathbf{L}) = -mf(r)r^2\frac{d}{dt}\left[\frac{\mathbf{r}}{r}\right]. \tag{18.498}$$

In the special case that $f(r) = -k/r^2$ (i.e., the form of the Coulomb or gravitational force) this equation can be solved,

$$\frac{d}{dt}\left[\mathbf{p}\times\mathbf{L} - mk\frac{\mathbf{r}}{r}\right] = 0 \tag{18.499}$$

implying that

$$\mathbf{p}\times\mathbf{L} - mk\frac{\mathbf{r}}{r} = \mathbf{A} \quad \text{a constant} \tag{18.500}$$

which is a constant vector that lies in the plane of the classical orbit (because $\mathbf{A}$ is perpendicular to $\mathbf{L}$, $\mathbf{A}\cdot\mathbf{L} = \mathbf{0}$)

18.12.2 Application to the hydrogen atom or single-electron ion

For a single-electron atom (hydrogen) or ion, the Hamiltonian is

$$H = \frac{\mathbf{p}^2}{2m} - \frac{Ze^2}{r} \tag{18.501}$$

and $\mathbf{A}$ can be imported to the realm of quantum mechanics as (dividing by m for convenience, and using $k = Ze^2$)

$$\mathbf{A} = -\frac{Ze^2\mathbf{r}}{r} + \frac{1}{2m}\left(\mathbf{p}\times\mathbf{L} - \mathbf{L}\times\mathbf{p}\right), \tag{18.502}$$

where $\mathbf{A}$ was made hermitian by the usual procedure of taking the average between $\mathbf{A}$ and its adjoint $\mathbf{A}^\dagger$. It is easy to see that $\mathbf{A}$ is hermitian, as expected, because $(\mathbf{p}\times\mathbf{L})^\dagger = -\mathbf{L}\times\mathbf{p}$ as we enforced this.

In the ensuing discussion, we will loosely follow Weinberg.[34]

Again, for convenience, we will work in units where $\hbar = 1$, $Ze^2 = 1$ and $m = 1$. Hence, the Hamiltonian is

$$H = \frac{1}{2}\mathbf{p}^2 - \frac{1}{r}, \tag{18.503}$$

and similarly,

$$\mathbf{A} = -\frac{\mathbf{r}}{r} + \frac{1}{2}\left(\mathbf{p}\times\mathbf{L} - \mathbf{L}\times\mathbf{p}\right) \tag{18.504}$$

with $\mathbf{p} = -i\nabla$.

[34] Weinberg, S., *Lectures on Quantum Mechanics* (Cambridge University Press, 2012), Sec. 48.

18.12.3 Review of vector operations and angular momentum operators

In the following, summation notation will be used, meaning repeated indices are to be summed over, and ϵ_{ijk} is the anti-symmetric Levi-Civita symbol.

Relationship 1
The quantum version of the $\mathbf{A}\cdot(\mathbf{B}\times\mathbf{C})$ rule maintains the order of the vector operators $\mathbf{A}$, $\mathbf{B}$, and $\mathbf{C}$ as follows:

$$\mathbf{A}\cdot(\mathbf{B}\times\mathbf{C}) = A_i(\epsilon_{ijk}B_jC_k) = (\epsilon_{kij}A_iB_j)C_k) = (\mathbf{A}\times\mathbf{B})\cdot\mathbf{C} \tag{18.505}$$

where the invariance with cyclic permutations of $\epsilon_{ijk} = \epsilon_{jki} = \epsilon_{kij}$ was used.

Relationship 2
The commutator between $\mathbf{L}$ and any vector operator $\mathbf{u}$, component by component, is

$$[L_i, u_j] = i\epsilon_{ijk}u_k. \tag{18.506}$$

If $i = j$, the commutator is zero.

Relationship 3
For any vector operator $\mathbf{B}$, the ith component of its cross product with $\mathbf{L}$ is

$$(\mathbf{B}\times\mathbf{L})_i = \epsilon_{jki}B_jL_k = \epsilon_{jki}(L_kB_j - i\epsilon_{kjl}B_l). \tag{18.507}$$

It is straighforward to determine $\epsilon_{jki}\epsilon_{kjl} = -2\delta_{il}$, and therefore

$$(\mathbf{B}\times\mathbf{L}) = -\mathbf{L}\times\mathbf{B} + 2i\mathbf{B}. \tag{18.508}$$

As an aside, this relation allows us to immediately conclude that

$$\mathbf{L}\times\mathbf{L} = i\mathbf{L}. \tag{18.509}$$

This result might appear as surprising because classically the cross product between a vector and itself is zero. In quantum mechanics, the components of $\mathbf{L}$ do not commute, leading to the nonzero self-cross-product.

Relationships 4
Using Relationship 1,

$$\begin{aligned}\mathbf{r}\cdot\mathbf{L} &= \mathbf{r}\cdot(\mathbf{r}\times\mathbf{p}) = (\mathbf{r}\times\mathbf{r})\cdot\mathbf{p} = 0 = \mathbf{L}\cdot\mathbf{r}\\ \mathbf{L}\cdot\mathbf{p} &= (\mathbf{r}\times\mathbf{p})\cdot\mathbf{p} = \mathbf{r}\cdot(\mathbf{p}\times\mathbf{p}) = 0 = \mathbf{p}\cdot\mathbf{L}.\end{aligned} \tag{18.510}$$

Relationships 5
Using Relationships 1, 3,

$$\begin{aligned}&\mathbf{p}\cdot(\mathbf{p}\times\mathbf{L}) = (\mathbf{p}\times\mathbf{p})\cdot\mathbf{L} = 0\\ &\mathbf{p}\cdot(\mathbf{L}\times\mathbf{p}) = \mathbf{p}\cdot(-\mathbf{p}\times\mathbf{L} + 2i\mathbf{p}) = 2i\mathbf{p}\cdot\mathbf{p} = 2ip^2.\end{aligned} \tag{18.511}$$

Relationships 6

Using $\epsilon_{ijk}\epsilon_{lmn} = \delta_{il}\delta jm - \delta_{im}\delta_{jl}$

$$\begin{aligned}(\mathbf{p} \times \mathbf{L}) \cdot (\mathbf{p} \times \mathbf{L}) &= \epsilon_i jk\epsilon_m nkp_i L_j p_m L_n \\ &= p_i L_j p_i L_j - p_i L_j p_j L_i = p_i(p_i L_j - i\epsilon_i jkp_k)L_j - p_i L_j p_j L_I \\ &= p^2 L^2 + i(\mathbf{p} \times \mathbf{p}) \cdot \mathbf{L} - p_i(\mathbf{L} \cdot \mathbf{p})L_i = p^2 L^2\end{aligned} \tag{18.512}$$

where the fact that $\mathbf{p} \times \mathbf{p} = 0$ and $\mathbf{L} \cdot \mathbf{p} = 0$ were used.

Similarly,

$$(\mathbf{L} \times \mathbf{p}) \cdot (\mathbf{p} \times \mathbf{L}) = (-\mathbf{p} \times \mathbf{L} + 2i\mathbf{p}) \cdot (\mathbf{p} \times \mathbf{L}) = -p^2 L^2 \tag{18.513}$$

and

$$(\mathbf{L} \times \mathbf{p}) \cdot (\mathbf{L} \times \mathbf{p}) = (\mathbf{L} \times \mathbf{p}) \cdot (-\mathbf{p} \times \mathbf{L} + 2i\mathbf{p}) = p^2 L^2 \tag{18.514}$$

using (18.513) and Relation 3. Then

$$(\mathbf{p} \times \mathbf{L}) \cdot (\mathbf{L} \times \mathbf{p}) = (-\mathbf{L} \times \mathbf{p} + 2i\mathbf{p}) \cdot (\mathbf{L} \times \mathbf{p}) \tag{18.515}$$

$$= -p^2 L^2 - 4ip^2 \tag{18.516}$$

using (18.514) and Relation 5.

Relationships 7

Any function of $r = |\mathbf{r}|$, $f(r)$ will commute with $\mathbf{L}$ and L^2 because both of these operators represent combinations of derivatives with respect to angle. That is to say, $f(r)$ is spherically symmetric and invariant under rotations.

We will encounter terms of the form

$$(\mathbf{L} \times \mathbf{p}) \cdot \mathbf{r} = \mathbf{L} \cdot (\mathbf{p} \times \mathbf{r}) = \mathbf{L} \cdot (-\mathbf{L}) = -L^2, \tag{18.517}$$

and because L^2 commutes with any $f(r)$,

$$(\mathbf{L} \times \mathbf{p}) \cdot \frac{\mathbf{r}}{r} = -\frac{L^2}{r}. \tag{18.518}$$

Similarly,

$$(\mathbf{p} \times \mathbf{L}) \cdot \frac{\mathbf{r}}{r} = \frac{L^2}{r} + 2i\mathbf{p} \cdot \frac{\mathbf{r}}{r}. \tag{18.519}$$

The last term can be expanded,

$$\begin{aligned}i\mathbf{p} \cdot \frac{\mathbf{r}}{r} &= \nabla \cdot \frac{\mathbf{r}}{r} \\ &= (3 + \mathbf{r} \cdot \nabla)\frac{1}{r} \\ &= \frac{3}{r} - \frac{\mathbf{r} \cdot \mathbf{r}}{r^3} + \frac{1}{r}\mathbf{r} \cdot \nabla = \frac{2}{r} + \frac{i}{r}\mathbf{r} \cdot \mathbf{p}\end{aligned} \tag{18.520}$$

where we have used $\mathbf{r}\cdot\mathbf{r} = r^2$. Therefore,

$$(\mathbf{p}\times\mathbf{L})\cdot\frac{\mathbf{r}}{r} = \frac{L^2}{r} + \frac{4}{r} + \frac{2i}{r}\mathbf{r}\cdot p \tag{18.521}$$

and

$$\frac{\mathbf{r}}{r}\cdot(\mathbf{p}\times\mathbf{L}) = \frac{L^2}{r} \tag{18.522}$$

and finally

$$\frac{\mathbf{r}}{r}\cdot(\mathbf{L}\times\mathbf{p}) = -\frac{L^2}{r} + \frac{2i}{r}\mathbf{r}\cdot\mathbf{p}. \tag{18.523}$$

Relationships 8

For calculating the commutator $[H, \mathbf{A}]$, and $[A_i, A_j]$ some additional relationships will be helpful:

$$(\mathbf{r}\times\mathbf{L})_i = \epsilon_{ijk}\epsilon_{mnk}r_j r_m p_n = r_i(\mathbf{r}\cdot\mathbf{p}) - r^2 p_i \tag{18.524}$$

and therefore

$$\mathbf{r}\times\mathbf{L} = \mathbf{r}(\mathbf{r}\cdot\mathbf{p}) - r^2\mathbf{p}. \tag{18.525}$$

In a similar fashion,

$$(\mathbf{L}\times\mathbf{r})_i = -\epsilon_{ijk}\epsilon_{lmn}p_l r_m r_k = p_i r_k r_k - p_k r_i r_k = p_i r^2 - (\mathbf{p}\cdot\mathbf{x})x_i \tag{18.526}$$

and therefore,

$$\mathbf{L}\times\mathbf{r} = -(\mathbf{p}\cdot\mathbf{r})\mathbf{r} + \mathbf{p}r^2. \tag{18.527}$$

We will also need

$$\left[p_i, \frac{1}{r}\right] = -i\frac{\partial}{\partial r_i}\frac{1}{r} + i\frac{1}{r}\frac{\partial}{\partial r_i} = \frac{ir_i}{r^3} \tag{18.528}$$

which holds for each component, so

$$\left[\mathbf{p}, \frac{1}{r}\right] = \frac{i\mathbf{r}}{r^3}. \tag{18.529}$$

18.12.4 Determination of the properties of A and finding $\mathbf{A}^2$

First, given that $\mathbf{A}$ is a vector operator, it should satisfy Relationship 2 ,

$$[L_i, A_j] = \epsilon_{ijk}A_k; \quad [L_i, A_i] = 0 = \mathbf{L}\cdot\mathbf{A} - \mathbf{A}\cdot\mathbf{L}. \tag{18.530}$$

Let us next consider $\mathbf{L}\cdot\mathbf{A}$. We have three terms, using Relationships 2, 3, 4, and 5,

$$\begin{aligned}\mathbf{L}\cdot(\mathbf{L}\times\mathbf{p}) &= (\mathbf{L}\times\mathbf{L})\cdot\mathbf{p} = i\mathbf{L}\cdot\mathbf{p} = 0\\ \mathbf{L}\cdot(\mathbf{p}\times\mathbf{L}) &= \mathbf{L}\cdot(-\mathbf{L}\times\mathbf{p} + 2i\mathbf{p})\cdot\mathbf{p} = 0\\ \mathbf{L}\cdot\frac{\mathbf{r}}{r} &= \frac{1}{r}\mathbf{L}\cdot\mathbf{r} = 0\end{aligned} \tag{18.531}$$

and we can conclude $\mathbf{L}\cdot\mathbf{A} = 0 = \mathbf{A}\cdot\mathbf{L}$.

The squared magnitude of $\mathbf{A}$ is $\mathbf{A}^2 = \mathbf{A}\cdot\mathbf{A}$ can be worked out by use of Relationships 6 and 7. There are nine terms,

$$
\begin{aligned}
\frac{1}{4}(\mathbf{p}\times\mathbf{L})\cdot(\mathbf{p}\times\mathbf{L}) &= \frac{p^2L^2}{4} \\
-\frac{1}{4}(\mathbf{L}\times\mathbf{p})\cdot(\mathbf{p}\times\mathbf{L}) &= \frac{p^2L^2}{4} \\
-\frac{1}{4}(\mathbf{p}\times\mathbf{L})\cdot(\mathbf{L}\times\mathbf{p}) &= \frac{p^2L^2}{4} + p^2 \\
\frac{1}{4}(\mathbf{L}\times\mathbf{p})\cdot(\mathbf{L}\times\mathbf{p}) &= \frac{p^2L^2}{2r} \\
\frac{\mathbf{r}}{2r}\cdot(\mathbf{L}\times\mathbf{p}) &= -\frac{L^2}{2r} + \frac{2i}{r}\mathbf{r}\cdot\mathbf{p} \\
-\frac{\mathbf{r}}{2r}\cdot(\mathbf{p}\times\mathbf{L}) &= -\frac{L^2}{2r} \\
(\mathbf{L}\times\mathbf{p})\cdot\frac{\mathbf{r}}{2r} &= -\frac{L^2}{2r} \\
-(\mathbf{p}\times\mathbf{L})\cdot\frac{\mathbf{r}}{2r} &= -\frac{L^2}{2r} - \frac{2}{r} - \frac{i}{r}\mathbf{r}\cdot\mathbf{p} \\
\frac{\mathbf{r}\cdot\mathbf{r}}{r^2} &= 1
\end{aligned}
\tag{18.532}
$$

and adding up the rhs of these equations yields

$$
\mathbf{A}^2 = p^2L^2 + p^2 - 2\frac{L^2}{r} - \frac{2}{r} + 1 = 1 + 2H(L^2+1), \tag{18.533}
$$

which when Ze^2, m, and $\hbar$ are included, becomes

$$
\mathbf{A}^2 = Z^2e^4 + \frac{2H}{m}\left(\mathbf{L}^2 + \hbar^2\right). \tag{18.534}
$$

Next, we can ask if $[\mathbf{A}, H] = 0$, and we do so, in principle, component by component. Note that $[\mathbf{r}/r, r] = 0$ and $[p^2, \mathbf{p}] = 0$. Therefore, we need to determine two commutators. First,

$$
\begin{aligned}
[\mathbf{p}\times\mathbf{L} - \mathbf{L}\times\mathbf{p}, H] &= [\mathbf{p}, H]\times\mathbf{L} - \mathbf{L}\times[\mathbf{p}, H] \\
&= \left[\mathbf{p}, -\frac{1}{r}\right]\times\mathbf{L} - \mathbf{L}\times\left[\mathbf{p}, -\frac{1}{r}\right] \\
&= \mathbf{L}\times\frac{i\mathbf{r}}{r^3} - \frac{i\mathbf{r}}{r^3}\times\mathbf{L}
\end{aligned}
\tag{18.535}
$$

because $[H, \mathbf{L}] = 0$. We then used Relation 8 to determine $[\mathbf{p}, 1/r]$. The second commutator to evaluate is

$$
\left[\frac{\mathbf{r}}{r}, p^2\right] = \left[\frac{\mathbf{r}}{r}, \mathbf{p}\right]\cdot\mathbf{p} + \mathbf{p}\cdot\left[\frac{\mathbf{r}}{r}, \mathbf{p}\right]. \tag{18.536}
$$

The intermediate commutator is

$$\left[\frac{r_i}{r}, p_j\right] = r_i\left[\frac{1}{r}, p_j\right] + \frac{1}{r}[r_i, p_j] = i\left(-\frac{r_i r_j}{r^3} + \frac{\delta_{ij}}{r}\right). \tag{18.537}$$

Substituting in this result leads to

$$\left[\frac{r_i}{r}, p_j\right] p_j + p_j\left[\frac{r_i}{r}, p_j\right] = i\left(-\frac{r_i r_j}{r^3} + \frac{\delta_{ij}}{r}\right) p_j + ip_j\left(-\frac{r_i r_j}{r^3} + \frac{\delta_{ij}}{r}\right)$$

therefore,

$$\begin{aligned}\left[-\frac{\mathbf{r}}{r}, p^2\right] &= i(\mathbf{p}\cdot\mathbf{r})\frac{\mathbf{r}}{r^3} + i\frac{\mathbf{r}}{r^3}(\mathbf{r}\cdot\mathbf{p}) - i\frac{\mathbf{p}}{r} - i\frac{\mathbf{p}}{r} \\ &= -i\frac{\mathbf{L}\times\mathbf{r}}{r^3} + i\frac{\mathbf{r}\times\mathbf{L}}{r^3}.\end{aligned} \tag{18.538}$$

Both Eqs. (18.535) and (18.538) need an overall factor of two, but that is irrelevant because the sum of these two equations is zero. Therefore, $[\mathbf{A}, H] = 0$. This follows from the fact that classically A was shown to be a constant of the motion and is thus a constant in quantum mechanics.

The energy levels will be determined (18.534) if we can find the eigenvalues of $\mathbf{A}^2$.

With some laborious calculations that follow a similar course, we can derive a further relationship that is needed to determine the eigenvalues of $\mathbf{A}$:

$$[A_i, A_j] = -2i\sum_k \epsilon_{ijk} H L_k. \tag{18.539}$$

This is slightly more complicated than showing $[\mathbf{A}, H] = 0$ so we will omit the derivation which can be effected using the relationships we developed earlier.

18.12.5 Hydrogen energy spectrum and level degeneracies

If the eigenvalues of H are negative (bound states) we can rescale $\mathbf{A}$ and maintain its hermiticity,

$$\mathbf{A}' = \sqrt{\frac{1}{-2H}}\mathbf{A} \tag{18.540}$$

and we then have

$$[A'_i, A'_j] = i\sum_k \epsilon_{ijk} L_k; \quad [L_i, A'_j] = i\epsilon_{ijk} A'_k \tag{18.541}$$

from (18.534, 18.530). Let us add all of the permutations of A'_i, A'_j, L_i, L_j taken in (i, j) ordered pairs:

$$[A'_i, A'_j] + [L_i, L_j] + [A'_i, L'_j] + [L'_i, A'_j] = i\sum_k \epsilon_{ijk} 2(L_k + A'_k) = [(L_i + A'_i), (L_j + A'_j)]. \tag{18.542}$$

This suggests defining operators

$$\mathbf{A}_\pm = \frac{1}{2}[\mathbf{L} \pm \mathbf{A}'] = \frac{1}{2}\left[\mathbf{L} \pm \sqrt{\frac{1}{-2H}}\mathbf{A}\right] \tag{18.543}$$

which have the commutation relations

$$[A_{\pm\ i}, A_{\pm\ k}] = i \sum_k \epsilon_{ijk} A_{\pm\ k}; \quad [A_{\pm\ i}, A_{\mp\ k}] = 0. \tag{18.544}$$

The independent operators $\mathbf{A}_\pm$ are exactly analogous to angular momentum; since the properties of the angular momentum operators come from the commutation relations, we know that

$$\mathbf{A}_\pm^2 = a_\pm(a_\pm + 1) \tag{18.545}$$

and the values of $a_\pm$ are either (independent between $\pm$) non-negative integers (0,1,2 ...), or half-integers (1/2,3/2,5/2,...), and there is no means to choose between either set, so both are valid. Furthermore, because $\mathbf{L} \cdot \mathbf{A} = 0 = \mathbf{A} \cdot \mathbf{L}$,

$$\mathbf{A}_\pm^2 = \frac{1}{4}\left[\mathbf{L}^2 + \frac{1}{-2H}\mathbf{A}^2\right] \tag{18.546}$$

and we see that $\mathbf{A}_+ = \mathbf{A}_- = \mathcal{A}$ to disambiguate this operator from $\mathbf{A}$. Therefore, taking eigenvalues of $\mathcal{A}$ as a,

$$\begin{aligned} \hbar^2 a(a+1) &= \frac{1}{4}\left[\mathbf{L}^2 + \frac{1}{-2E}\mathbf{R}^2\right] \\ &= \frac{1}{4}\left[\mathbf{L}^2 + \frac{1}{-2E} - (\mathbf{L}^2 + 1)\right] \\ &= \frac{1}{-8E} - \frac{1}{4} \end{aligned} \tag{18.547}$$

and the energy is

$$E = -\frac{1}{2(2a+1)^2} = -\frac{1}{2n^2} = -\frac{Z^2 e^4 m}{2\hbar^2 n^2} \tag{18.548}$$

where we have taken the principal quantum number as $n = 2a + 1$, and in the last step restored the units for m, $\hbar$, and Ze^2.

The degeneracy for each n level can be determined by the allowed eigenvalues a_3 of $A_{\pm 3}$ in analogy to angular momentum, which are $a_3 \in \{a - a, -(a-1), ..., a-1, a\}$ and are $2a + 1$ in number for each of the $\pm$ operators. Because the $A_{\pm\ 3}$ eignevalues are independent, there are $(2a + 1)^2$ total possible states for a given $n = 2a + 1$.

The hydrogen atom can be solved by other methods, for example by a transformation that recasts the Hamiltonian to that of a four-dimensional harmonic oscillator.[35] It should be noted that most of these methods do not produce positive energy solutions; for example, they do not address the scattering of an electron from a proton due to the Coulomb interaction. This is because the harmonic oscillator potential is increasing with distance, so the particle is always confined to a finite region compared to the case of a $1/r$ potential.

[35] Cornish,F.H.J., *The hydrogen atom and the four-dimensional harmonic oscillator*, J. Phys. A: Math. Gen. **17**, 323 (1984).

18.13 The WKB approximation: boundary conditions by complex analysis

A good example of the power of algebraic and complex analysis techniques is the derivation of the WKB approximation. In the usual derivation, the *connection condition* between the classically allowed and forbidden region wavefunctions is derived from the exact solution of the wavefunction for a linear potential. However, it is possible to determine the matching condition with exponentials and complex analysis.[36]

Let us consider the one-dimensional Schrödinger equation,

$$\frac{\hbar^2}{2m}\frac{d^2}{dx^2}\psi(x) + (E - U(x))\psi(x) = 0. \tag{18.549}$$

By introducing the substitution

$$\psi = e^{(i/\hbar)\chi(x)} \tag{18.550}$$

the differential equation for $\chi \equiv \chi(x)$ is

$$\frac{1}{2m}\left(\frac{d\chi}{dx}\right)^2 - \frac{i\hbar}{2m}\frac{d^2\chi}{dx^2} = E - U(x) \tag{18.551}$$

and we can seek a solution in powers of $\hbar$ which will allow a quasi-classical description of a quantum system. The expansion is

$$\chi = \chi_0 + \left(\frac{\hbar}{i}\right)\chi_1 + \left(\frac{\hbar}{i}\right)^2\chi_2 + \ldots \tag{18.552}$$

In the first approximation, $\chi = \chi_0$, and the differential equation for χ_0 is, omitting the term with $\hbar$ in Eq. (18.551), where the prime indicates a derivative with respect to x,

$$(\chi')^2 = 2m(E - U(x)) \tag{18.553}$$

and therefore

$$\chi_0 = \pm\int\sqrt{2m(E - U(x))}\,dx. \tag{18.554}$$

The integrand is the classical momentum as a function of position x; there is no difference between the quantum and classical momentum as defined in terms of the expectation value. The phase of the solution to this order is therefore the time-independent part of the classical action

$$\chi_0 = \pm\int p dx = S; p = \sqrt{2m(E - U(x)}, \tag{18.555}$$

and the wavefunction at this approximation is

$$\psi = e^{(\hbar/i)\chi_0} = e^{iS/\hbar}, \tag{18.556}$$

which is the expected form of the wavefunction as the classical limit is approached.

[36] Landau, L.D. and Lifshitz, E.M., op. cit., p. 582, Sec. 47.

The next term in the expansion, using Eq. (18.551) with

$$\chi = \chi_0 + \left(\frac{\hbar}{i}\right)\chi_1 \tag{18.557}$$

results in

$$\chi_0'\chi_1' + \frac{1}{2}\chi_0'' = 0 \tag{18.558}$$

implying that

$$\chi_1' = -\frac{1}{2}\frac{\chi_0''}{\chi_0'} = -\frac{p'}{2p}. \tag{18.559}$$

The solution is easily obtained by integrating, and

$$\chi_1 = -\frac{1}{2}\log p + C = \log(1/\sqrt{p}) + C \tag{18.560}$$

where C is an integration constant. Substituting the result for χ_0 and χ_1 into Eq. (18.552) determines the solution to first order in $\hbar$ as

$$\psi = \frac{A_1}{\sqrt{p}}e^{(i/\hbar)\int p\,dx} + \frac{A_2}{\sqrt{p}}e^{(-i/\hbar)\int p\,dx} \tag{18.561}$$

where we have included both $\pm p$ solutions and the integration constants have been absorbed into A_1 and A_2.

The $1/\sqrt{p}$ factors have a classical interpretation, which is that $1/p \sim 1/v$ and represents the time that a particle spends at a given x. Near the limits of motion where $E \approx U(x)$ the particle moves slowly, hence there is a high probability to find it in those regions.

Classically, the motion into a region where $E < U(x)$ is forbidden and this defines the region of movement of a particle. In quantum mechanics, the particle can tunnel into the forbidden region. Nothing in our analysis so far prevents this possibility, and we can expand our solution to include the forbidden region, noting that in that region $p(x)$ is purely imaginary so that the exponentials are real. The wavefunction can therefore be written as

$$\psi = \frac{B_1}{\sqrt{|p|}}e^{(1/\hbar)\int |p|\,dx} + \frac{B_2}{\sqrt{p}}e^{(-1/\hbar)\int |p|\,dx}. \tag{18.562}$$

Generally, the wavefunction is damped when moving into a forbidden region so only one or the other solutions (the exponentially decaying one) is kept.

The point $x = a$ where $U(a) = E$ is a classical turning point. To determine the coefficients A_1, A_2, B_1, B_2, we must have $\psi \to 0$ for $x \to \infty$ and we must connect the wavefunction on one side of a (the forbidden region) to the wavefunction in the allowed region. Let us take $a > 0$ and the allowed region to be $x \leq a$. Assume $U(x)$ is continuous and smoothly varying (can be Taylor expanded at $x = a$), so that

$$2m(E - U(x)) = \alpha(x - a); \alpha = -\frac{dU}{dx}|_{x=a} < 0. \tag{18.563}$$

For $x > a$, the wavefunction in the region of $x \sim a$ is

$$\psi(x) = \frac{B}{\sqrt[4]{|\alpha|}} \frac{1}{(x-a)^{1/4}} \exp\left[\frac{-1}{\hbar}\int_a^x \sqrt{|\alpha|(x-a)}dx\right]. \tag{18.564}$$

Taking the wavefunction phase to be zero for $x > a$, in the region $x < a$

$$\begin{aligned}\psi(x) = &\frac{A_1}{\sqrt{|\alpha|}} \frac{1}{(a-x)^1/4} \exp\left[\frac{i}{\hbar}\int_a^x \sqrt{|\alpha|(a-x)}dx\right] \\ &+ \frac{A_2}{\sqrt{|\alpha|}} \frac{1}{(a-x)^1/4} \exp\left[\frac{-i}{\hbar}\int_a^x \sqrt{|\alpha|(a-x)}dx\right],\end{aligned} \tag{18.565}$$

with

$$\pm\frac{i}{\hbar}\int_a^x p(x)\, dx = \pm\frac{i}{\hbar}\int_a^x \sqrt{|\alpha|(a-x)}\, dx = \mp\frac{i}{\hbar}\frac{2}{3}\sqrt{|\alpha|}\,(a-x)^{3/2} \tag{18.566}$$

in the allowed region, $\mp$ corresponding, respectively, to the A_1 and A_2 wavefunction terms.

Connecting the exponentially damped wavefunction to the oscillating functions in the allowed region is complicated by the $1/\sqrt{p}$ factors because they diverge at the turning points. However, we can make x a complex variable and take an integration path that avoids $p = 0$. The WKB wavefunctions are analytic on the complex plane, and therefore an integration path along x can be moved away from the real axis.

In particular, if we start with the damped wavefunction in the forbidden region at $a + x$, the integration to $a - x$ in the allowed region can be taken around a semicircle above $x = a$ as

$$x - a \to z = \rho e^{i\phi} \tag{18.567}$$

where ρ is constant and ϕ varies from 0 to π. Therefore,

$$-\frac{1}{\hbar}\int \sqrt{|\alpha| z}\, dz = -\frac{1}{\hbar}\frac{2}{3}\sqrt{|\alpha|}\, z^{3/2}|_0^{\pi} = -\frac{1}{\hbar}\frac{2}{3}\rho^{3/2}(\cos(3\phi/2) + i\sin(3\phi/2)) \tag{18.568}$$

which starts as real at $\phi = 0$ to pure imaginary at $\phi = \pi$, with value

$$-\frac{1}{\hbar}\frac{2}{3}\sqrt{|\alpha|}\,\rho^{3/2}(\cos(3\pi/2) + i\sin(3\pi/2)) = \frac{i}{\hbar}\frac{2}{3}\sqrt{|\alpha|} = \frac{i}{\hbar}\frac{2}{3}\sqrt{|\alpha|}\,(a-x)^{3/2} \tag{18.569}$$

which corresponds to the A_2 term. Similarly, if we take a path under a, the wavefunction connects to the A_1 term.

The $1/p$ factor is not in the integral, so we only need consider its values at the two endpoints:

$$(x-a)^{1/4} \to (a-x)^{1/4}e^{-i\pi/4} \tag{18.570}$$

and we can immediately conclude that

$$A_2 = Be^{-i\pi/4}. \tag{18.571}$$

A similar semicircular path below a connects to the A_1 term as

$$A_1 = Be^{i\pi/4}. \tag{18.572}$$

Thus, the exponentially decaying wavefunction in the forbidden region connects to the sum of the two terms in the allowed region,

$$\psi(x) = \frac{2B}{\sqrt{p}} \cos\left[\frac{1}{\hbar}\int_a^x p(x)\, dx + \frac{\pi}{4}\right], \tag{18.573}$$

and the leading factor of two can be eliminated if we set $B \to B/2$.

Although the wavefunction thus calculated is only approximate, the integrals for various potentials, in particular the Coulomb potential, can often be exactly calculated. The exponential damped WKB wavefunction can be used to extend energy dependent (resonant) nuclear cross sections that require tunneling (for example, p capture by ^{12}C which is part of the stellar CNO cycle) to other energies far from the resonance. The temperature dependence of stellar nuclear reaction rates can be thereby reliably elucidated.

18.13.1 Bohr-Sommerfeld quantization condition

Let us now consider a particle bound by potentials on both sides, with the classical turning points at a to the right and b to the left. The previous analysis can be repeated for the case where the forbidden region is to the left,

$$\psi_b(x) = \frac{B'}{\sqrt{p}} \cos\left[\frac{1}{\hbar}\int_b^x p(x)\, dx - \frac{\pi}{4}\right]. \tag{18.574}$$

We can equate this to the wavefunction for the turning point to the right as

$$\psi_a(x) = \frac{B}{\sqrt{p}} \cos\left[-\frac{1}{\hbar}\int_x^a p(x)\, dx - \frac{\pi}{4}\right] \tag{18.575}$$

We seek constraints on E given $U(x)$ so that these two functions are equal for all x between the turning points. Obviously, $p = p(x)$ is the same for both, and we therefore require

$$B\cos\left[-\frac{i}{\hbar}\int_x^a p(x)\, dx - \frac{\pi}{4}\right] = B'\cos\left[\frac{i}{\hbar}\int_b^x p(x)\, dx - \frac{\pi}{4}\right] \tag{18.576}$$

which implies that

$$\begin{gathered} \frac{1}{\hbar}\int_x^a p(x)\, dx - \frac{\pi}{4} = \pm\left[\frac{1}{\hbar}\int_b^x p(x)\, dx - \frac{\pi}{4}\right] + n\pi \\ B = (-1)^n B' \end{gathered} \tag{18.577}$$

where n is an integer, and we can eliminate the x dependence if we choose the $-$ solution, in which case

$$\left[\frac{1}{\hbar}\int_x^a p(x)\, dx - \frac{\pi}{4}\right] + \left[\frac{1}{\hbar}\int_b^x p(x)\, dx - \frac{\pi}{4}\right] = \frac{1}{\hbar}\int_b^a p(x)\, dx - \frac{\pi}{2} = -n\pi. \tag{18.578}$$

We can cast this in terms of the classical action,

$$\oint p(x)\,dx = 2\int_b^a |p(x)\,dx| = 2\pi\hbar\left(n - \frac{1}{2}\right) \qquad (18.579)$$

where the integral is taken over a full cycle of the particle motion. This determines the allowed energies as

$$\frac{1}{2\pi\hbar}\oint \sqrt{2m(E - U(x)}\,dx = n - \frac{1}{2} \qquad (18.580)$$

and is the Bohr-Sommefield quantization condition in the "old" quantum theory.

Note that if the turning points are "hard walls," the quantization condition is

$$\frac{1}{2\pi\hbar}\oint \sqrt{2m(E - U(x)}\,dx = n \qquad (18.581)$$

because the wavefuntion goes to zero at the turning points, and is analogous to a vibrating string; the condition for this to be true at both turning points requires a wavefunction phase change of $n\pi$ between the turning points. The hard wall case is in the opposite limit from before, that is, the wavefunction changes drastically over a distance less than a wavelength, so the WKB approximation is invalid.

The WKB approximation has its most significant use in nuclear astrophysics where the thermal energies of neutrons, protons, and the low energy nuclear resonant interactions of interest are generally well below the nuclear Coulomb barrier. The WKB approximation provides a simple way to ascribe a temperature dependence to nuclear cross sections and, hence, reaction rates. As an example, the CNO cycle, which occurs in stars about twice the mass of the sun, is an important reaction in nucleosynthesis. The first step in this cycle is the absorption of a proton by a ^{12}C nucleus, however, in order for this to take place, the Coulomb repulsion between the ^{12}C and p needs to be overcome (the ^{12}C nucleus has charge $6e$), requiring an energy

$$E_c = \frac{6e^2}{4\pi\epsilon_0 R_n} = 2.16 \text{ MeV}$$

where R_n is the nuclear charge radius. This energy is much higher than the resonance energy of about 0.480 MeV. The WKB approximation can be used to estimate the tunneling probability, which requires integrating the $1/r$ potential from infinity to R_n.

$$P = e^{-W}$$

$$W \approx 5 \approx \frac{4Z_1Z_2}{\hbar v}\frac{e^2}{4\pi\epsilon_0}\left[\frac{\pi}{2} - \arcsin\left(\frac{E}{E_c}\right)^{1/2} - \left(\frac{E}{E_c}\right)^{1/2}\left(1 - \frac{E}{E_c}\right)^{1/2}\right]$$

Taking E as the resonance energy (0.480 MeV), the probability of a proton with that energy to penetrate the Coulomb barrier is $P = 6.5 \times 10^{-3}$ (or $1/P \approx 153$).

The temperature dependence comes in because the number of protons that tunnel and react is determined by an integral of the energy dependent cross section times the tunneling probability times a Boltzmann factor which gives the proton flux at that energy. Given that 0.480 MeV corresponds to a temperature of 5.6×10^9 K, compared to 0.014×10^9 K for the interior of the sun, we might expect this reaction will be very slow in the sun. The integral for the reaction rate shows that for the CNO cycle, with its several steps, the net rate depends on the temperature to the 17th power, and becomes dominant for stars of mass greater than 1.4 times the mass of the sun. Otherwise $p - p$ reactions are the dominant energy source. On the other hand, for a red giant, the internal temperature is 0.2×10^9 K and the CNO reaction can proceed relatively rapidly.

19

Spin-1/2 and two-level systems

The existence of spin-1/2 in physical systems took a relatively long time to be realized in the early days of the development of quantum mechanics, despite clear experimental signatures that were missed at least three times. The story of spin-1/2 begins with the studies by de Haas and Einstein, and by Bartlett, to measure the angular momentum change associated with either magnetizing or rotating a ferromagnetic material, respectively. As background, the nonexistence of magnetic monopoles suggests that magnetism is due to circulation of electric charges, as Amperian current loops in materials. J. Larmor (1894) proposed his famous theorem in regard to this, to wit, that the effects of a homogeneous magnetic field acting on a system of particles can be described by a homogeneous rotating coordinate system with the rotation axis along the magnetic field direction, with rotation frequency determined by the system properties. This theorem subsequently became of great interest with Zeeman's discovery of his eponymous effect, the splitting of spectral lines by a magnetic field, in 1896. This splitting is also indicative of spatial, or angular momentum, quantization, although the direct physical observation of such in the Stern-Gerlach experiment (see Fig. 4.4) did not occur until 1922.

19.1 Larmor's theorem

If we have a system with a single charged particle of mass m and charge q moving at velocity $\mathbf{v}$ subjected to a central force $\mathbf{F}_c(\mathbf{r})$, and then apply an external magnetic field, the total force is

$$\mathbf{F} = \mathbf{F}_c(\mathbf{r}) + q\mathbf{v} \times \mathbf{B}\,. \tag{19.1}$$

The magnetic component of the Lorentz force is of identical form to the Coriolis force that arises when transforming to a rotating coordinate system,

$$\mathbf{F}_{\text{cor}} = 2m\mathbf{v} \times \boldsymbol{\Omega}\,, \tag{19.2}$$

where $\boldsymbol{\Omega}$ is a vector that specifies the rotation axis and with length given by the rotational angular frequency ω, and $\mathbf{v}$ is the velocity relative to the rotating coordinate system. Let us take the case where $\mathbf{B}$ is constant and along the $\hat{z}$ axis, and further assume that the rotation frequency is constant. In the limit that $|\omega r| \ll |v|$, the centripetal force $\omega^2 r$ can be neglected, and we see immediately that the effect of a magnetic field B along the $\hat{z}$ axis is identical to a transformation to a rotating coordinate system with rotation axis along $\mathbf{B}$ and rotational frequency

$$\omega = \frac{qB}{2m}\,. \tag{19.3}$$

The Historical and Physical Foundations of Quantum Mechanics. Robert Golub and Steven K. Lamoreaux, Oxford University Press.
 DOI: 10.1093/oso/9780198822189.003.0019

Note that the same result will be obtained if we consider many particles of equal mass and charge interacting with each other's central fields, and bound by a central field. This result is only true for weak magnetic fields, but that is the case in which we will be primarily interested. This is Larmor's theorem, derived by Joseph Larmor (the Lucasian Professor at Cambridge from 1903 to 1922, succeeded by Dirac) and was the subject of a series of studies by him published between 1894 and 1897. It is noteworthy that Larmor's publications appeared before the discovery of the electron by J. J. Thomson in 1897 (as discussed in Chapter 1, the electron had already been postulated (in 1874) and named (in 1891) by George Johnstone Stoney). Larmor's analyses, representing a major deviation from the Maxwellian concepts of electrodynamics, were largely accepted after the electron's discovery. Larmor's work suggests that it was he who brought the electron concept to physics. In particular Larmor was one of the first physicists to postulate that the electron and matter are distinct from the luminiferous ether, abandoning the concept that they are the results of vortices or similar artifacts of the ether, and he was among the first to separate matter and void as elements of a fundamental theory.

We can add clarity and meaning to Larmor's theorem if we consider an electron (charge $-|e| = -e$, mass m) orbiting a nucleus at radius r, with velocity v, which forms a prototypical Amperian current loop in the $x - y$ plane, with $\mathbf{L} = mvr\hat{z}$. The current in this loop is

$$I = -\frac{ev}{2\pi r}$$

which has a magnetic moment directed along $\hat{z}$,

$$\mu = \pi r^2 I = -\frac{evr}{2} = -\frac{e}{2m}L$$

where we have used $L = mvr$. Because both μ and $\mathbf{L}$ are the same vectors up to a multiplicative constant, we can write

$$\mu = -\frac{e}{2m}\mathbf{L} \equiv -g\left(\frac{e}{2m}\right)\mathbf{L} = -g\mu_B\mathbf{L}\,. \tag{19.4}$$

The g-factor introduced here (in this case, $g = 1$), which is the ratio of a particle's magnetic moment to $\mu_B L$, will be seen to be very useful.

If L is in units of $\hbar$, then we can define the Bohr magneton as $\mu_B = e\hbar/2m$. When this system is placed in a magnetic field $\mathbf{B}$ it will experience a torque

$$\tau = \mu \times \mathbf{B}\,,$$

resulting in a precession of the angular momentum vector $\mathbf{L}$ as

$$\frac{d\left(\hbar\mathbf{L}\right)}{dt} = \tau = \mu \times \mathbf{B} = -g\left(\frac{e\hbar}{2m}\right)\mathbf{L} \times \mathbf{B} \tag{19.5}$$

around the magnetic field at a frequency (note $\hbar$ can be divided out of both sides of this equation)

$$\omega = -g\left(\frac{e}{2m}\right)B, \tag{19.6}$$

where the sign determines the precession direction. We thus recover the Larmor frequency. If we transform into a frame rotating at this frequency with rotation axis along $\mathbf{B}$, then $\mathbf{L}$ will appear as a constant vector.

Note that the Larmor frequency is half of that for the cyclotron motion of a charged particle moving perpendicular to a magnetic field B,

$$\omega_{\text{cyclotron}} = \frac{qB}{m},$$

which is independent of the velocity in the nonrelativistic limit.

It was in the context of Larmor's theorem that the de Haas and Einstein[1], and Bartlett, experiments of 1915 had to be interpreted. The electron charge had been measured by Millikan in 1913, so the electron properties were well known, and the expectation of these experiments was that g should be 1. The results of these experiments, which were difficult and subject to systematic effects, tended to be adjusted until they gave the "correct" value. This history is elaborated upon by Peter Galison.[2] As it turns out, these experiments were done using iron alloys, where the magnetization is nearly all due to electron spins for which $g = 2$, in contrast to the classically expected $g = 1$. Subsequent experiments over the next few years generally converged to $g = 2$, but there was no basic theoretical understanding of this result. Fig. 19.1

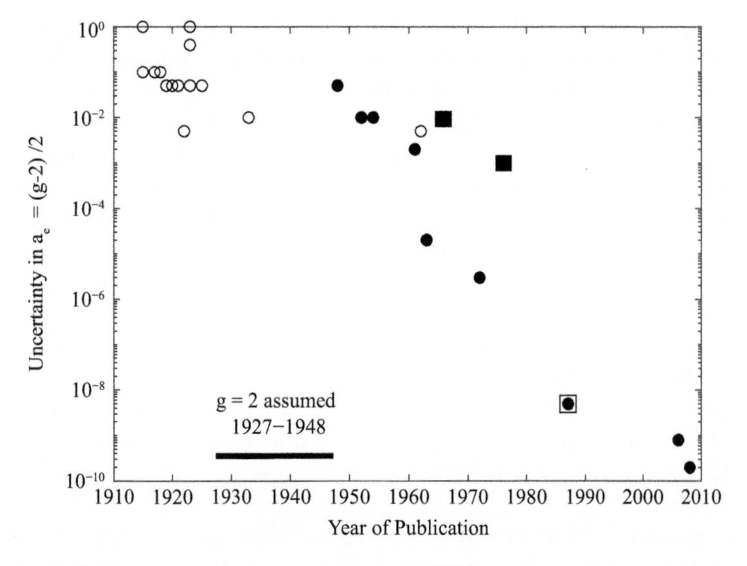

Fig. 19.1 Uncertainty in electron g–factor anomaly $a_e = (g-2)/2$ as a function of year. Open circles are results from measurements of ferromagnetic materials (Galison, op. cit., 599). The filled circles are results from radiofrequency spectroscopy techniques (Commins, op. cit., 600). Squares indicate measurements of the positron and the 1988 square and circle are measurements of both the positron and electron using Dehmelt's technique, which was refined and extended for later measurements.

[1] Einstein, A., *Experimental detection of Amperian molecular currents*, Naturwiss. 3, 237 1915, and Einstein, A., and de Haas, W. J., *Experimental proof of the existence of Ampere's molecular currents*, Royal Netherlands Academy of Arts and Sciences, Proceedings 18, 696 (1915). (https://dwc.knaw.nl/DL/publications/PU00012546.pdf).

[2] Galison, P., *How Experiments End* (Chicago: The University of Chicago Press, 1987), pp. 27-74.

shows the experimental results as a function of time. Thus, the classical model of the orbiting electron failed. After the Dirac equation was discovered, the value of $g = 2$ exactly was assumed, until 1948 when measurement of the ground state $g-$ factor in Gallium and Sodium showed a significant deviation from 2.000... [3] Also of note is that the effective electron spin in ferromagnetics continued to be studied in the 1960s, with the result that $g < 2$ by about 1%, meaning that spin angular momentum accounts for 99% of the observed magnetic moment.

The Stern-Gerlach experiment of 1922 (discussed in Chapter 4) that served as an experimental proof of space (angular momentum) quantization was missing the $L_z = 0$ (i.e., $m = 0$) peak, assuming that the ground state angular momentum of the silver(Ag) atom was due to an orbiting electron with $L = 1$. Nobody caught on that something deep might be going on, partly because the energy spacing between the $m = 1$ and $m = -1$ peaks for $L = 1,\ g = 1$ is the same as for $L = 1/2,\ g = 2$. This issue was further suppressed by comments from Bohr suggesting the $m = 0$ peak should not be visible.

19.1.1 Spin-1/2

These earlier experimental results were finally understood in the context of the anomalous Zeeman effect which occurs in atoms with odd atomic number Z (hydrogen, for example). In such cases, the number of Zeeman sublevels of a state with angular momentum L is an even number rather than the odd number expected from $2L + 1$ when L is an integer. This cannot be explained within the normal Zeeman theory and was attributed to an additional internal rotation of the electron. In 1925 Uhlenbeck and Goudsmit postulated the existence of a new property of the electron, an intrinsic angular momentum of $\hbar/2$, to explain the anomalous Zeeman effect (Kramers had suggested this earlier, but his suggestion was largely ignored). This new intrinsic property was subsequently termed spin by Pauli, perhaps from the notion that it is the extra rotation of the electron which was introduced previously to provide a heuristic explanation of the anomalous Zeeman effect. However, the image of the electron as a spinning sphere is not valid: if it were spinning fast enough to generate an angular momentum of $\hbar/2$, the surface velocity would be much larger than c (assuming some reasonable size for the electron; as far as we know experimentally, it is actually a point particle).

This new property of the electron is *intrinsic*, akin to mass and charge, and as such is a fundamental and fixed property. This is an entirely nonclassical angular momentum that does not arise from a moment of inertia, for example. According to the Uhlenbeck and Goudsmit proposal, the spin of a particle should behave like an angular momentum and, as before, should have an associated magnetic moment

$$\hat{\mu}_s = -g\mu_{\mathrm{B}}\mathbf{s} \tag{19.7}$$

where the minus sign reflects the negative charge of the electron, $\hat{\mathbf{s}}$ is the spin angular momentum operator in units of $\hbar$, g is the constant that was introduced earlier to

[3] Commins, E.D., *Electron Spin and Its History*, Ann. Rev. of Nuc. and Part. Sci.,**62**, 133-157 (2012).

produce the best fit with experiment, and $\mu_B = |e|\hbar/2m$ is the Bohr magneton. The interaction energy of the magnetic moment with a magnetic field is equal to $-\hat{\mu}_s \cdot \mathbf{B}$. It was found that good fits to experimental atomic spectroscopic (fine structure) data were obtained when $g = 2$, which means that the spin gyromagnetic ratio is twice as large as the "classical" orbital gyromagnetic ratio $g = 1$, in spectroscopic cases where the orbital angular L momentum is zero. The Uhlenbeck-Goudsmit electron spin conjecture eliminated this confusion, with one sticking point that will be discussed shortly.

It should be further noted that the spin-orbit interaction results from the magnetic moment of the electron interacting with the magnetic field that arises when the electron moves in the electric field of the nucleus, as suggested by Einstein.

In earlier chapters, we introduced the concept of spin for a particle, in particular in relation to the Stern-Gerlach experiment, Fermi-Dirac statistics, and the exclusion principle. Spin-1/2, the simplest nonzero angular momentum state, can be used as a surrogate for any general two-state or two-level system. As has been discussed already, the eigenvalues of the z component of angular momentum operator $\mathbf{L}_z$ are $m_L = L,\ L-1, ...,\ -L+1,\ -L$, so for a state with spin-1/2, there are only two levels.

Before launching our study of two-level systems, we must again note that it is indeed possible to fully describe the state of a particle in terms of its internal coordinates and properties, separately from the center of mass motion. In particular, the Schrödinger equation is invariant under the Galilean (nonrelativistic) transformation between a moving frame and a frame where the constituents of a multiparticle system have zero net momentum (center of mass frame). Otherwise we would have to contend with the infinite number of momentum states that are available to a quantum system. It is shown in Appendix B that we can indeed ignore center of mass motion (in the case of spatially homogeneous external fields). Because the systems we are considering are nonrelativistic, the time and space coordinates are not varied when transforming between frames, and only Galilean transformations are required.

19.1.2 The spin-orbit interaction and Thomas precession

The Uhlenbeck and Goudsmit proposal of electron spin appeared to be perfect except for one sticking point. For a single atomic valence electron, both the spin magnetic moment and the orbital magnetic moment $g\mu_B\mathbf{L}$ contribute to the net atomic magnetic moment that interacts with an externally applied magnetic field, and in addition the orbital and spin magnetic moments interact with each other. However, the spin magnetic moment appears to be different depending on L and that casts doubt on the electron spin hypothesis. The difference in the apparent gyromagnetic ratio of the electron between $L = 0$ and $L = 1$ states was explained by L. Thomas.[4] It is a relativistic effect that can be understood in several ways.

In the moving frame of an electron that is orbiting a nucleus, the electron experiences a motional magnetic field

$$\mathbf{B} = -\frac{\mathbf{v}}{c^2} \times \mathbf{E}\,, \tag{19.8}$$

[4]Thomas, L. H., *The motion of the spinning electron*, Nature 117, 514, (1926), received 20, February, 1926. See also, Phil. Mag. **3**, 1 (1927).

where $\mathbf{v}$ is the velocity of the electron in its circular orbit and $\mathbf{E}$ is the electric field on the electron as measured in the nucleus (rest) frame. For a single electron orbiting a nucleus, or an electron outside of a closed shell,

$$-e\mathbf{E} = -\nabla V(r) = -\left(\frac{\mathbf{r}}{r}\frac{dV}{dr}\right) \tag{19.9}$$

where $V(r)$ is the potential energy of the electron and e is the magnitude of the electron charge; this direction of electric field is attractive between the electron and nucleus. The magnetic field experienced by the electron is then

$$\mathbf{B} = -\frac{\mathbf{v}}{c^2} \times \frac{1}{e}\left(\frac{\mathbf{r}}{r}\frac{dV}{dr}\right) = -\left(\frac{1}{ec^2}\right)\left(\frac{1}{r}\frac{dV}{dr}\right)\mathbf{v}\times\mathbf{r} = \left(\frac{1}{mec^2}\right)\left(\frac{1}{r}\frac{dV}{dr}\right)\mathbf{L} \tag{19.10}$$

where we have used the fact that $\mathbf{L} = m\mathbf{r}\times\mathbf{v}$. An electron's intrinsic magnetic moment will interact with this field and precess in a counterclockwise direction around $\mathbf{L}$ (which is the same direction as the electron motion). Taking

$$H_{so} = g\mu_B\left(\frac{1}{mec^2}\right)\left(\frac{1}{r}\frac{dV}{dr}\right)\mathbf{L}\cdot\mathbf{s}. \tag{19.11}$$

Regarding the interaction of the electron with a magnetic field (either the field internal to the atom or external fields), the value $g = 2$ is only directly observed in the case that the orbital angular momentum, $L = 0$, and for the total angular momentum $J = L + S$, the orbital part L has $g = 1$ and S has $g = 2$. In the spin-orbit interaction energy we have to take $g = 1$ to agree with experiment. Thomas provided a relativistic explanation of the differing electron g factors, which was originally dismissed because v/c in an atom is small, hence the effect is small. What was forgotten is that the spin-orbit interaction itself is a small relativistic correction of the same magnitude. Thomas' explanation can be cast in a simpler, if not heuristic, form as follows. (This discussion applies in the limit of a small magnetic field, so that the spin direction does not appreciably change during an orbital period, which is valid for the present case.)

In our determination of the spin-orbit interaction, we have neglected an important relativistic effect. Because there is no torque on the electron spin in the absence of a magnetic field, its orientation remains fixed in its rest frame.[5] Let us turn off the motional magnetic field and determine the electron spin direction relative to the laboratory frame as it undergoes circular motion. Again, the electron is in a circular orbit of radius r moving with velocity v, so the orbital frequency is $\omega = v/r$ and the orbital period is $T = 2\pi/\omega = 2\pi r/v$.

Let us divide the circle into N (with $N \gg 1$) infinitesimal line segments (approximate the circle by an N-sided polygon) and define a measuring rod length as $\ell = 2\pi r/N$. The line segments are tangent to the circle and hence always parallel to the electron velocity. When viewed from the electron's frame using this measuring rod, the line segments will appear as shortened by the Lorentz contraction factor,

$$\ell' = \ell/\gamma = \ell\sqrt{1 - v^2/c^2} \approx \ell\left(1 - \frac{v^2}{2c^2}\right). \tag{19.12}$$

[5] Möller, C., *The Theory of Relativity* (Oxford: Oxford University Press, 1955) pp. 53-56; p. 125.

Thus, in a moving frame, the ratio of the circle's circumference to its radius (which, being perpendicular to the velocity, does not suffer a Lorentz contraction) is[6]

$$N\ell'/r = N(\ell/r\gamma) = N(2\pi r/N/r\gamma) = 2\pi/\gamma \approx 2\pi(1 - v^2/2c^2) < 2\pi. \tag{19.13}$$

This suggests that the polygon angles are not the same when viewed from the electron frame. Referring to Fig. 19.2, these angles are increased due to the Lorentz contraction of ℓ. After one full trip around the polygon, the sum of all the $d\phi$ in the lab frame is 2π, by definition. When viewed from the electron frame, the angle changes are $d\phi' = d\phi\gamma$, so there is an additional rotation of the lab frame by

$$\Delta\phi = Nd\phi' - 2\pi = 2\pi\gamma - 2\pi \approx 2\pi\frac{v^2}{2c^2}. \tag{19.14}$$

Relative to the laboratory frame (e.g., the direction of the last line segment), the electron frame coordinates rotate by $-\Delta\phi$ with every orbit cycle. We might have anticipated this result from Eq. (19.13).

This angle accumulates with each orbit cycle, leading to a precession of the electron frame coordinates (in which the electron spin direction is constant, fixed) relative to the lab frame, which can be described by the rotation vector

$$\boldsymbol{\Omega}_t = -\frac{\Delta\phi}{T}\hat{z} = -\omega\left(\frac{v^2}{2c^2}\right)\hat{z} = -\frac{\omega^3 r^2}{2c^2}\hat{z}\,, \tag{19.15}$$

which describes a *clockwise* rotation. Note that this precession occurs in the absence of the motional or other magnetic field. This is the *Thomas precession correction* which

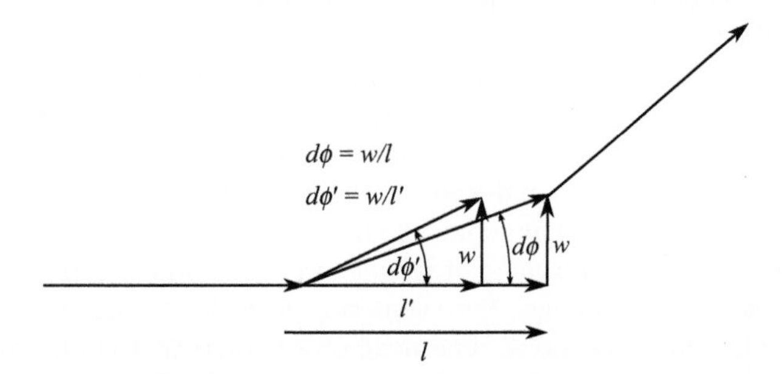

Fig. 19.2 Motion in a circular orbit can be approximated by taking a step of length $\ell = 2\pi r/N$ and then a perpendicular step of length w, around an N-sided polygon, where N is large. We define a measuring rod length in the laboratory frame as $\ell = 2\pi r/N$. In the lab frame, the angular change in direction per step of a vector transported along the circle is $d\phi = \ell/r = 2\pi/N = w/\ell$, and after N steps the angle is 2π. In the moving frame, $\ell \to \ell' = \ell/\gamma$ while $w' = w$ because that translation is perpendicular to the velocity. Thus, the angle change per step of the laboratory frame as viewed from the electron frame is larger by a factor γ, and after N steps the total angle is larger than 2π. This picture was originally suggested by E. M. Purcell.

[6]Landau, L. D. and Lifshitz, E. M., *The Classical Theory of Fields* (Cambridge, MA: Addison-Wesley, 1951), pp. 281-282 (§10-11).

is a precession of the electron rest frame in which the electron spin is fixed, and it is a relativistic effect due to the transformation to the moving electron frame. Therefore, to properly determine the net Larmor frequency of the electron as viewed from the laboratory frame, we must add this precession to the motional (spin-orbit) field Larmor precession.

For circular motion, the centripetal force magnitude is $m\omega^2 r = dV/dr$ and $\mathbf{L}/m\omega r^2 = \hat{z}$, so the Thomas precession vector can be rewritten as

$$\boldsymbol{\Omega}_t = -\frac{1}{2m^2c^2}\left(\frac{1}{r}\frac{dV}{dr}\right)\mathbf{L}. \tag{19.16}$$

To assess the net effect of the Thomas precession rate, it is helpful to convert it into an equivalent magnetic field by the electron magnetic moment $ge/2m = e/m$. (Note that L is NOT in units of $\hbar$ in the present discussion.) Furthermore, the Thomas precession *correction* is in the *clockwise* direction, and therefore opposite to the spin-orbit precession calculated above. This implies a change in the effective magnetic field seen by the electron, when viewed from the laboratory frame, of

$$\mathbf{B}_t = -\left(\frac{1}{2mec^2}\right)\left(\frac{1}{r}\frac{dV}{dr}\right)\mathbf{L}, \tag{19.17}$$

which is exactly half of the motional magnetic field, Eq. (19.10), and of opposite sign. Therefore, the electron spin precesses at half of the frequency expected due to the spin-orbit interaction alone in the case of an atom with angular momentum $L \geq 1$, and this solves the anomalous Zeeman effect problem.

The essential point is that the Thomas precession would occur even in the absence of the motional magnetic field, and to determine the net effect of a field introduced in the electron frame as viewed from the laboratory frame, the Thomas precession must be added. Because it is half the motional magnetic field precession and of opposite sign, we see the spin-orbit effect reduced by a factor of two as required for $g = 2$ to be correct when the electron angular momentum $L \geq 1$.

The last issue for the acceptance of the electron spin and its associated magnetic moment was removed by Thomas. Note that this effect is of order v^2/c^2, for both the spin-orbit and Thomas precessions. The next order correction is due to time dilation in the rotating frame but that is much smaller than the Thomas precession effect. (Time dilation reduces the Larmor frequency by a factor $1/\gamma$). As an aside, around 1897 Larmor had pointed out that time would be dilated for a moving electron.

19.1.3 Other corrections to the electron g-factor

Dirac later showed that spin arises very naturally in a relativistic formulation of the quantum theory of electrons. The Dirac equation not only incorporates spin with $g = 2$, but also leads to negative energy states (positrons) and naturally embodies Thomas precession. The only further significant corrections to the Dirac equation are due to vacuum fluctuations of quantized fields, the first of which is the Uehling correction caused by polarization of the virtual positron/electron cloud around the electron, which causes a slight momentum dependence to the electron charge. Another correction is a slight deviation of the g-factor from two also due to quantized field vacuum fluctuation effects, which has been experimentally measured as

$$\frac{g-2}{2} = a_e = 0.00115965218073(28) \approx \frac{\alpha}{2\pi} = 0.0011614 \quad \text{(first order correction)}$$

where $\alpha \approx 1/137.04$ is the fine structure constant. The number in parentheses is the uncertainty in the last two digits, which represents a part per trillion uncertainty in g, making this the most precisely measured fundamental physical constant. The first order correction was first calculated by J. Schwinger in 1948.

The experimental result is based on the refinement of a measurement technique invented by H. Dehmelt where the orbital (cyclotron) motion of an electron confined in a Penning trap is compared to its spin flip frequency in the same magnetic field. In the absence of vacuum fluctuations, $g = 2$ exactly, and deviation from that causes a detectable anomalous precession. These experiments have been performed by simultaneously measuring the frequency of the cyclotron motion, which can be detected electrically, and detecting the spin flip frequency by driving the system with a near resonant oscillating electric field. Originally, a small magnetic "bottle" (quadratically varying magnetic field) along the homogeneous field was used, which causes a shift in the Penning trap oscillation frequency when a spin flip occurs, thereby providing a "continuous Stern-Gerlach" measurement of the electron spin direction.[7]

The experimental result agrees with theory to within the respective uncertainties, which are at the parts per trillion level (the theory has about three times higher uncertainty), making quantum electrodynamics the most precisely tested fundamental scientific theory.

19.2 Pauli matrices

In matrix mechanics, operators in two-level systems can be represented by 2×2 matrices, which must be Hermitian. The Pauli matrices, which are the same as the spin-1/2 angular momentum matrices up to a factor of two ($2\mathbf{s} = \sigma$ when $\mathbf{s}$ is given in units of $\hbar$), provide a symbolic representation of a general operator for a two-level state as

$$\sigma_x = 2s_x = \begin{pmatrix} 0 & 1 \\ 1 & 0 \end{pmatrix}, \quad \sigma_y = 2s_y = \begin{pmatrix} 0 & -i \\ i & 0 \end{pmatrix}, \quad \sigma_z = 2s_z = \begin{pmatrix} 1 & 0 \\ 0 & -1 \end{pmatrix}. \tag{19.18}$$

The eigenvalues of any of these matrices are ± 1, and the z basis states are

$$\chi_+ = \begin{pmatrix} 1 \\ 0 \end{pmatrix}, \quad \chi_- = \begin{pmatrix} 0 \\ 1 \end{pmatrix}. \tag{19.19}$$

The units of angular momentum are $\hbar/2$ in this representation.

As can easily be verified, these matrices satisfy the commutation laws of angular momentum discussed in Chapter 18,

$$[\hat{\sigma}_x, \hat{\sigma}_y] = 2i\sigma_z \tag{19.20}$$

[7] Dehmelt, H. G., "Experiments with an Isolated Subatomic Particle at Rest" (Nobel Lecture, December 8, 1989).

with cyclic permutations. Similarly, these matrices anticommute as

$$\{\hat{\sigma}_a, \hat{\sigma}_b\} = 2\delta_{ab}\hat{I} \tag{19.21}$$

where I is the identity matrix.

19.2.1 Matrix algebra of the spherical basis Pauli matrices

It is sometimes convenient to work in a new basis formed by taking linear combinations of the Pauli matrices. The basis that corresponds to $|L, m_L\rangle$ states is referred to as the Pauli or spherical basis, and it is a spherical tensor representation. The linear combinations for this basis are

$$2\sigma_\pm = \sigma_x \pm i\sigma_y \tag{19.22}$$

$$\sigma_\pm \sigma_z = \mp\sigma_\pm; \quad \sigma_z\sigma_\pm = \pm\sigma_\pm \tag{19.23}$$

$$\sigma_\pm\sigma_\mp = \frac{1}{2} \pm \frac{1}{2}\sigma_z \tag{19.24}$$

$$\sigma_z\sigma_z = 1; \quad \sigma_\pm\sigma_\pm = 0. \tag{19.25}$$

The commutation relationships are

$$[\sigma_+, \sigma_-] = \sigma_z; \quad [\sigma_\pm, \sigma_z] = \mp 2\sigma_\pm. \tag{19.26}$$

19.3 Vector representation of spin and spinor rotation symmetry

Taking a vector operator $\mathbf{s} = \sigma/2$, the time dependence of this vector is determined by the Heisenberg equation of motion,

$$\frac{d\mathbf{s}}{dt} = \frac{i}{\hbar}[H, \mathbf{s}]. \tag{19.27}$$

When H is the interaction of a magnetic moment $\mu = \hbar\gamma\mathbf{s}$ (which we now take to be a general magnetic moment, not necessarily an electron) with a magnetic field, and γ is the gyromagnetic ratio,

$$H = -\mu \cdot \mathbf{B} = -\hbar\gamma(B_x s_x + B_y s_y + B_z s_z). \tag{19.28}$$

It is straightforward to show that the commutator of H with $\mathbf{s}$ is equal to the cross product $\gamma\mathbf{s} \times \mathbf{B}$, so the equation of motion becomes

$$\frac{d\mathbf{s}}{dt} = \gamma\mathbf{s} \times \mathbf{B} = \tau \tag{19.29}$$

where τ is the torque acting on $\mathbf{s}$; this is the classical equation of a magnetic moment in a magnetic field. It should be noted here that we are considering the equation of motion of an operator in the Heisenberg representation. We can use it to describe the average results for a statistical ensemble relying on a large number of (noninteracting) particles or measurements. In this case the ensemble polarization often decays due to decoherence for s_x and s_y at a rate $1/T_2$, which is called the transverse relaxation, and s_z relaxes at a rate $1/T_1$ toward the thermal or other equilibrium state, and this is called the longitudinal relaxation. Relaxation will be explained later in this chapter.

We thus immediately see the value in thinking of a spin-1/2 system in terms of a vector. This model will serve us well. In terms of the spin wave function or *spinor*, let us take a general state (with $\mathbf{s}$ in units of $\hbar$)

$$\chi = \begin{pmatrix} \cos(\theta/2) \\ e^{i\phi}\sin(\theta/2) \end{pmatrix} \tag{19.30}$$

where θ and ϕ are arbitrary parameters. Let us calculate the expectation value of $\mathbf{s}$:

$$\langle s_x \rangle = \chi^\dagger s_x \chi = \sin(\theta/2)\cos(\theta/2)\cos\phi = \frac{1}{2}\sin\theta\ \cos\phi \tag{19.31}$$

$$\langle s_y \rangle = \chi^\dagger s_y \chi = \sin(\theta/2)\cos(\theta/2)\sin\phi = \frac{1}{2}\sin\theta\ \sin\phi \tag{19.32}$$

$$\langle s_z \rangle = \chi^\dagger s_z \chi = (\cos^2(\theta/2) - \sin^2(\theta/2)) = \frac{1}{2}\cos\theta. \tag{19.33}$$

Thus we see an exact mapping of the spinor ($SU(2)$ group) onto the Bloch sphere surface ($SO(3)$ group), in analogy to the Poincaré sphere for light polarization, as shown in Fig. 19.3. This shows that we can describe a two-level system by an equivalent vector, which we usually normalize to unit length.

An interesting consequence of this representation is that if θ is varied in the spinor, it does not return to its original state until $\theta = 4\pi$, as can be seen in Eq. (19.30).

19.3.1 Time evolution operator

Alternatively, we can describe the time evolution of a quantum state through the evolution operator

$$U(t) = e^{-iHt/\hbar} \tag{19.34}$$

if H is explicitly time-independent. Writing H in terms of the Pauli matrices,

$$H = \frac{-\hbar\gamma}{2}\hat{\sigma}\cdot\mathbf{B}, \tag{19.35}$$

where γ is the gyromagnetic ratio and can be either positive or negative, the time evolution operator can be expanded as

$$\begin{aligned} U(t) = 1 + i\frac{\gamma t}{2}(B_x\sigma_x + B_y\sigma_y + B_z\sigma_z) - \frac{1}{2!}\left(\frac{\gamma t}{2}(B_x\sigma_x + B_y\sigma_y + B_z\sigma_z)\right)^2 + \\ + \frac{i}{3!}\left(\frac{\gamma t}{2}(B_x\sigma_x + B_y\sigma_y + B_z\sigma_z)\right)^3 + \dots . \end{aligned} \tag{19.36}$$

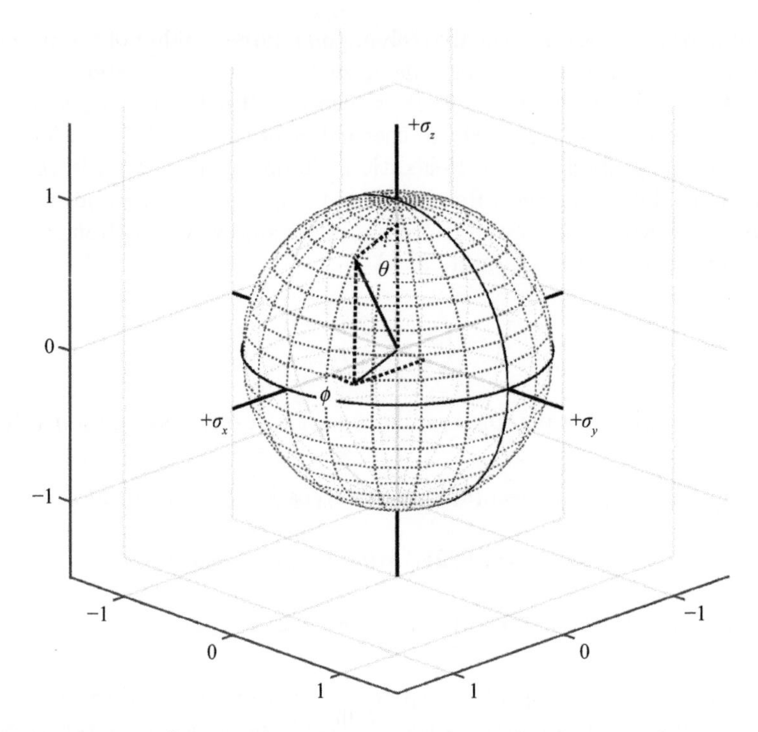

Fig. 19.3 The Poincaré/Bloch sphere shows the relationship between the state projections and the polarization vector, as described in the text.

Noting that $\sigma_x^2 = \sigma_y^2 = \sigma_z^2 = 1$, and terms such as $\sigma_x\sigma_y + \sigma_y\sigma_x = 0$ (the Pauli matrices anticommute), it is straightforward to show that

$$U(t) = \hat{I}\cos\left(\frac{\gamma Bt}{2}\right) + i\frac{\hat{\sigma}\cdot\mathbf{B}}{B}\sin\left(\frac{\gamma Bt}{2}\right) \tag{19.37}$$

where $\hat{I}$ is the 2×2 identity matrix and $B = \sqrt{B_x^2 + B_y^2 + B_z^2}$. We can further define $\omega = \gamma B$ and $\hat{n} = \mathbf{B}/B$ yielding

$$U(t) = \hat{I}\cos\left(\frac{\omega t}{2}\right) - i\,(\hat{n}\cdot\,\hat{\sigma})\sin\left(\frac{\omega t}{2}\right). \tag{19.38}$$

This is an important result that has broad applications.

19.3.2 Spinor rotation

If we think of applying the field **B** for a very short time so that the Larmor precession angle $\gamma Bt = \omega t = d\phi \ll 1$, we can interpret the above results as a rotation operator. In this limit, we have the generator of infinitesimal rotations,

$$R(d\phi_x, d\phi_y, d\phi_z) = \hat{I} - \frac{i}{2}\hat{n}\cdot\hat{\sigma}d\phi = \hat{I} - \frac{i}{2}(d\phi_x\sigma_x + d\phi_y\sigma_y + d\phi_z\sigma_z) \tag{19.39}$$

which is what we expect from an angular momentum operator, in analogy to linear momentum being the generator of finite displacement.

Let us now consider a rotation about the $\hat{z}$ axis. The rotation operator is

$$R_z(\theta) = \hat{I}\cos\frac{\theta}{2} - i\sigma_z \sin\frac{\theta}{2} \tag{19.40}$$

and we can immediately note that if $\theta = 2\pi$ a spin-1/2 wave function (spinor) will change sign. We had anticipated this result earlier when we introduced the vector spin representation.

The symmetry of a 4π rotation is closely related to the exchange symmetry of fermions. Consider the system in Fig. 19.4 where two particles with spin, A and B, which are affixed to the ends of a ribbon, have their positions swapped without any rotations. By observing the effects on a ribbon connecting the particles, it is easy to see that there is a 2π twist in the ribbon rotation which shows that a spatial exchange of two particles leads to a relative rotation by 2π between them, and thus an overall sign change of the wave function for spin-1/2 particles. Of course this is a crude model but it illustrates a principle that can be fully explained in the context of relativistic field theory. Nonetheless this simple picture is compelling. It is interesting to note that continuing the exchange process back to the original position in the clockwise sense shown, the ribbon will acquire a 4π twist; the system will return to its original state by untwisting the ribbon by rotating one end of the ribbon, hence one of the particles, by 4π, which will return a spin-1/2 particle to its original state.[8]

We can ask the meaning of this. In our usual three-dimensional Euclidean space, it is well known that in some situations the effect of a rotation by 2π can be undone by a second 2π rotation in the same direction. It is possible to construct an anti-twister device so that the twisting of a wire connecting into a rotating system can be continuously undone.

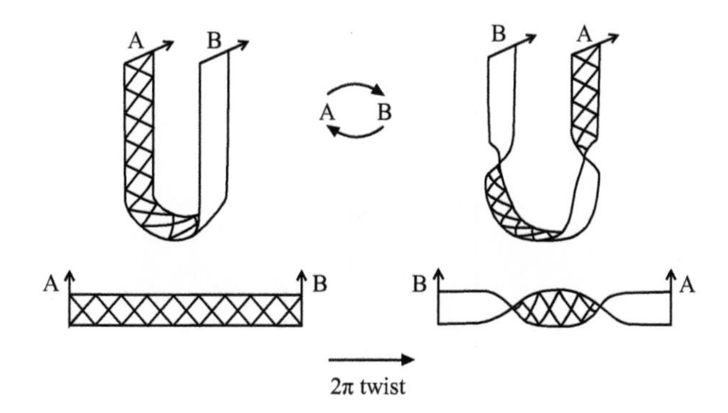

Fig. 19.4 Two particles are connected by a ribbon which is white on one side and marked on the other. It is possible to make an anti-twister device by rotating one end of the ribbon by 4π for every AB to BA to AB exchange that occurs in the same (clockwise) sense.

[8]According to R.P. Feynman in *Elementary Particles and the Laws of Physics* (Cambridge: Cambridge University Press, 1987), p. 57, this idea is due to David Finkelstein.

19.4 The effects of near-resonant oscillating magnetic fields

The effects of an oscillating magnetic field on an atomic beam were famously studied by I. I. Rabi and collaborators, with their first publication appearing in 1938. The first correct calculation of the effect of a linear oscillating magnetic field was done by the then 19-year-old J. Schwinger in 1937.[9] In the intervening years, it was recognized by Rabi's group, Felix Bloch, and others, that a simpler method is to transform to a rotating coordinate system by the use of Larmor's theorem. A review of this method was presented by Rabi, Ramsey, and Schwinger in 1954.[10] The basic idea is outlined in Fig. 19.5.

An oscillating magnetic field of the form (a rotating field)

$$\mathbf{B}(t) = B_1(\cos\omega t\ \hat{x} \pm \sin\omega t\ \hat{y}) \tag{19.41}$$

is applied in the $\hat{x} - \hat{y}$ plane, with $\pm$ is chosen so that the oscillating field rotates in the same direction as the spin precession, and a static field B_z is applied along the $\hat{z}$ axis. We can use the vector model, along with Larmor's theorem, to describe the subsequent motion of the spin. Assume a unit spin vector initially points along $\hat{z}$ when the oscillating field is applied. We can describe the subsequent motion most easily by transforming to a frame rotating about $\hat{z}$ at ω. This is a slight modification of Larmor's theorem in that we are not canceling the effects of B_z by transforming to the rotating frame but are making the Hamiltonian appear as time-independent.

In the rotating frame, the static magnetic field becomes

$$B_z' = B_z - \omega/|\gamma| \tag{19.42}$$

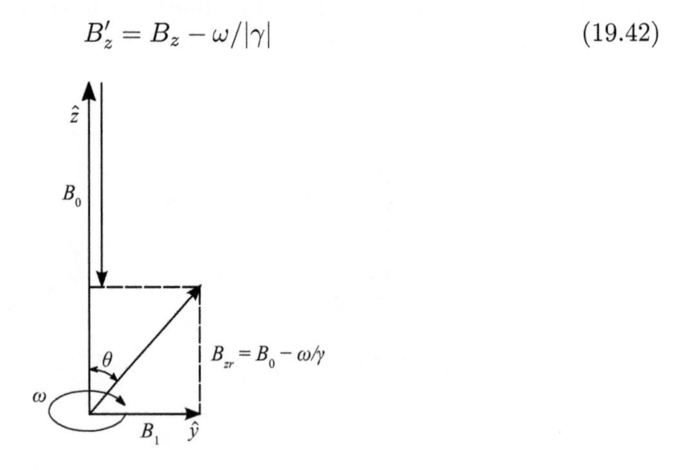

Fig. 19.5 The effect of the transformation to a rotating coordinate system as described in Ref. 10.

[9] Schwinger, J., *On Nonadiabatic Processes in Inhomogeneous Fields*, Phys. Rev. **51**, 648 (1937). However, he did not use the rotating coordinate system but the *interaction representation* that we will discuss later in this chapter.

[10] Rabi, I. I., Ramsey, N. F. and Schwinger, J., *Use of Rotating Coordinates in Magnetic Resonance Problems*, Rev. Mod. Phys. **26**, 167 (1954).

if we chose the oscillating field rotation direction correctly. Let us also choose the phase such that the oscillating field becomes $B'_y = B_1$ which is, of course, static in the rotating frame.

Let us first consider the *on-resonance* case where $\omega = \omega_o = \gamma B_z$, so $B'_z = 0$. The spin vector, originally along $\hat{z}$, will precess in the $\hat{x} - \hat{z}$ plane at a frequency

$$\omega_1 = \gamma B_1 \tag{19.43}$$

which is called the on-resonance *Rabi frequency*.

In the case where $\omega_1 \neq \omega$, the spin vector precesses about the vector

$$\mathbf{B} = (B_z - \omega/\gamma)\hat{z} + B_1\hat{y} \tag{19.44}$$

so the spin precesses in a cone with its apex at the origin, tilted in the $\hat{y} - \hat{z}$ plane by half the cone apex angle, with Larmor frequency given by the Rabi frequency

$$\Omega = \gamma\sqrt{B_1^2 + (B_z - \omega/\gamma)^2} \tag{19.45}$$

and tilt angle by

$$\theta = \arctan\left[\frac{B_1}{B_z - \omega/\gamma}\right] = \arcsin\left[\frac{\omega_1}{\Omega}\right]. \tag{19.46}$$

The transition probability as a function of time is given by the *Rabi formula* and can be easily worked out from the geometrical picture. In particular, the projection of the spin onto the $\hat{z}$ axis is not affected by the transformation to the rotating coordinate system. Also, the spin projection can be thought of as the probability that the spin points along $-\hat{z}$ after starting along $+\hat{z}$. For the on-resonance case, the probability that the spin (a unit vector) points down is simply determined as

$$P_{-z} = \frac{1 - S_z}{2} = \frac{1 - \cos\omega_1 t}{2} = \sin^2\left(\frac{\omega_1 t}{2}\right) \tag{19.47}$$

which has the maximum oscillation amplitude of 0 to 1. For the off-resonance case,

$$P_{-z} = \frac{\omega_1^2}{\Omega^2}\sin^2\left(\frac{\Omega t}{2}\right), \tag{19.48}$$

which is the Rabi formula that describes the transition between two levels due to an oscillating field.

Note that a linear oscillating field along $\hat{y}$ could also be used. Such a field can be decomposed into two counter-rotating fields as

$$\mathbf{B}_y = \frac{B_1}{2}(\hat{y}\cos\omega t + \hat{x}\sin\omega t) + \frac{B_1}{2}(\hat{y}\cos\omega t - \hat{x}\sin\omega t). \tag{19.49}$$

In the rotating frame, these rotating fields will appear to be stationary or twice as fast, depending on the precession direction,

$$\mathbf{B}_y = \frac{B_1}{2}(\hat{y}\cos(\omega \pm \omega)t + \hat{x}\sin(\omega \pm \omega)t) + \frac{B_1}{2}(\hat{y}\cos(\omega \mp \omega)t - \hat{x}\sin(\omega \mp \omega)t). \tag{19.50}$$

When we choose the appropriate rotation direction, one or the other of the rotating fields will become static, and we can ignore the fast oscillating 2ω counter-rotating field (in the case where $\gamma B_1 \ll \omega$). The effective amplitude of the oscillating field is reduced by a factor of two compared to the rotating field case.

The foregoing embodies all of the principles of Nuclear Magnetic Resonance (NMR), Magnetic Resonance Imaging (MRI), which represents a renaming of NMR to avoid the word nuclear, and is used ubiquitously, e.g., in atomic spectroscopy. The notion of using an oscillating field in this way, to drive transitions, developed from the atomic beam work of I. I. Rabi and collaborators, who used a modified version of the Stern-Gerlach apparatus. This modification involved a spatially alternating field which, in the frame of a moving particle, would appear as oscillatory. C. J. Gorter, during a visit to Rabi's laboratory in early 1937, asked why he was not using a time oscillating field. Gorter had been attempting to measure high frequency magnetic susceptibility due to absorption of radiofrequency energy by nuclear spin in solids, but was never able to attain high enough sensitivity (the first successful observation of a solid state nuclear magnetic resonance signal was obtained by Purcell and Pound, and shortly after by Bloch and collaborators, in 1946). Gorter is cited in Rabi et al.'s paper as,

> 3. C. J. Gorter, *Physica* **9**, 995 (1936). We are very much indebted to Dr. Gorter who, when visiting our laboratory in September 1937, drew our attention to his stimulating experiments in which he attempted to measure nuclear moments by observing the rise in temperature of solids placed in a constant magnetic field on which an oscillating field was superimposed. Dr. F. Bloch has independently worked out similar ideas but for another purpose (unpublished).

19.5 Effects of time-dependent, nonresonant variations of the potential

19.5.1 The interaction representation

Transforming to the interaction representation, which we will use several times in this chapter, is closely related to the transformation to the rotating coordinate system. The interaction representation is somewhere between the Schrödinger (time-dependent wave functions, static operators) and Heisenberg (static wave functions, time dependence is in the expectation values of operators) representations.

If we have eigenstates of a Hamiltonian H_0 that does not depend on time, we can remove the "trivial" time dependence of the states by use of the interaction representation. We can then study the effect of a perturbation H_1, which itself is possibly time-dependent, separately from the less interesting dynamics due to H_0.

Let us start with the Schrödinger picture and transform to the interaction representation,

$$|\psi\rangle_i = e^{iH_0 t/\hbar}|\psi\rangle_s = U_i(t)|\psi\rangle_s \tag{19.51}$$

where $U_i(t) = U^\dagger(t)$, the usual time evolution operator. We transform all operators, including H_1, with $U_i(t)$,

$$H_{1i} = U_i(t) H_1 U_i^\dagger(t). \tag{19.52}$$

Using the time-dependent Schrödinger equation for the combined $H = H_0 + H_1$

$$i\hbar\frac{\partial}{\partial t}|\psi(t)\rangle_s = (H_0 + H_1)|\psi(t)\rangle_s, \tag{19.53}$$

the time derivative in the interaction representation becomes

$$i\hbar\frac{\partial}{\partial t}|\psi(t)\rangle_i = i\hbar\left[\frac{\partial U_i(t)}{\partial t}|\psi\rangle_s + U_i(t)\frac{\partial}{\partial t}|\psi(t)\rangle_s\right] \tag{19.54}$$

$$= -U_i(t)H_0|\psi\rangle_s + U_i(t)\left[H_0 + H_1\right]|\psi\rangle_s \tag{19.55}$$

$$= U_i(t)H_1|\psi\rangle_s = U_i(t)H_1U_i^\dagger(t)U_i(t)|\psi\rangle_s \tag{19.56}$$

$$= H_{1i}|\psi\rangle_i. \tag{19.57}$$

Note that H_1 can be time-dependent. In relation to the rotating frame, if the oscillating field frequency is exactly at the Larmor frequency, then H_1 acquires a constant component in the interaction representation.

19.5.2 Effects of faster variations; the secular approximation

19.5.2.1 Bloch-Siegert shift

In the discussion of the effect of an oscillating field earlier in this chapter, we left the effects of the counter-rotating field for further study. The counter-rotating field will produce a shift in resonance frequency, which we can estimate as follows.

First, we transform to the counter-rotating frame,

$$B_z' = B_z + \frac{\omega}{\gamma} \tag{19.58}$$

When this is added to $B_1\hat{x}$ in this frame, the total field is

$$B' = \sqrt{(B_z + \omega/\gamma)^2 + B_1^2}, \tag{19.59}$$

which is at an angle very close to the $\hat{z}$ axis for B_1 relatively small. Transforming back to the nonrotating frame, the effective field is

$$B_z' = B' - \frac{\omega}{\gamma} = \sqrt{(B_z + \omega/\gamma)^2 + B_1^2} - \frac{\omega}{\gamma} \approx B_z + \frac{1}{2}\frac{B_1^2}{B_z + \omega/\gamma} \approx B_z + \frac{B_1^2}{4B_z} \tag{19.60}$$

which is the accepted lowest order correction.

19.5.2.2 Dressed spin and the secular approximation

The term *secular approximation* comes from classical mechanics, and it is applied to systems where rapid oscillatory motions, e.g., planets in their orbits around the sun, are replaced by their average positions (mass rings at the orbit locations) in calculations of very slow perturbations of one planet's orbit by the others. The fast motion is not relevant on the time scale over which significant orbital perturbations occur. It is not surprising that a similar approximation can be made in the context of the Schrödinger equation.

Usually, when we construct the time evolution operator, it is assumed that the Hamiltonian is time-independent. We cannot assume for an arbitrary time-dependent Hamiltonian that

$$U(t) = e^{-i\int^t dt' H(t')/\hbar} \tag{19.61}$$

because $H(t')$ is an operator with the integral representing a series of time steps, and the Hamiltonian at two different times might not commute, that is, $[H(t), H(t')] \neq 0$.

In this case, there is no simple form for the operator expansion of the exponential but it can be done as the Dyson sum, for example. However, in the case where $[H(t), H(t')] = 0$ the evolution operator can be written in terms of the time integral of the Hamiltonian in the exponent. The operator to transform into the interaction representation is the adjoint of $U(t)$. This transformation can be used to make the dominant term in the Hamiltonian time-independent but will lead to (non-trivially) time-dependent eigenstates.

Let us consider a case where there is a strong oscillating field at frequency ω, with maximum magnitude $\omega_1 = \gamma B_1$, along the $\hat{x}$ direction, and a weak static field, with magnitude $\omega_0 = \gamma B_0 \ll \omega_1$, along $\hat{z}$, where we will quantify strong and weak in a moment. The Hamiltonian is then

$$H = \frac{\hbar\omega_0}{2}\sigma_z + \frac{\hbar\omega_1 \cos\omega t}{2}\sigma_x \tag{19.62}$$

where $\omega_0 \ll \omega_1$ and $\omega_1 \sim \omega$. We can make the dominant part of the Hamiltonian time-independent by use of the interaction representation transformation,

$$U_i = e^{-i\omega_1\sigma_x \int^t dt' \cos\omega t'/2} = \hat{I}\cos\left(\frac{\omega_1}{2\omega}\sin\omega t\right) + i\sigma_x \sin\left(\frac{\omega_1}{2\omega}\sin\omega t\right) \tag{19.63}$$

which is allowed because σ_x commutes with itself. The static part of the Hamiltonian can be transformed to the interaction representation as, letting $\eta = \omega_1/2\omega$,

$$\frac{2}{\hbar}H_i = U_i(t)\omega_0\sigma_z U_i^\dagger(t) \tag{19.64}$$

$$= \omega_0 \begin{pmatrix} \cos(\eta\sin\omega t) & i\sin(\eta\sin\omega t) \\ i\sin(\eta\sin\omega t) & \cos(\eta\sin\omega t) \end{pmatrix} \sigma_z \begin{pmatrix} \cos(\eta\sin\omega t) & -i\sin(\eta\sin\omega t) \\ -i\sin(\eta\sin\omega t) & \cos(\eta\sin\omega t) \end{pmatrix} \tag{19.65}$$

$$= \omega_0 \begin{pmatrix} \cos^2(\eta\sin\omega t) - \sin^2(\eta\sin\omega t) & -2i\cos(\eta\sin\omega t)\sin(\eta\sin\omega t) \\ 2i\cos(\eta\sin\omega t)\sin(\eta\sin\omega t) & -(\cos^2(\eta\sin\omega t) - \sin^2(\eta\sin\omega t)) \end{pmatrix}$$

$$= \omega_0 \begin{pmatrix} \cos(2\eta\sin\omega t) & -i\sin(2\eta\sin\omega t) \\ i\sin(2\eta\sin\omega t) & -\cos(2\eta\sin\omega t) \end{pmatrix}. \tag{19.66}$$

It should be noted that there are no approximations to this point, and we might be tempted to say that the eigenvalues appear to be $\pm\hbar\omega_0/2$. However, that would be incorrect because the oscillating terms are so rapid compared to ω_0 that the spin does not precess any appreciable amount between subsequent oscillatory field reversals which occur with period $T = 2\pi/\omega$. Since $\omega_0 T/2\pi = \omega_0/\omega \ll 1$, we can think of B_z as oscillating in an arc in the $y - z$ plane, around and centered on $\hat{x}$ at frequency ω, with an average value of B_z along $\hat{z}$. It is this average to which the spin principally responds; this is where the secular approximation comes in (in the earlier case of magnetic resonance, $\omega \approx \omega_0$ and that approximation is not valid). We can determine the static and oscillating components of the Hamiltonian by noting, with $\alpha = 2\eta$,[11]

[11] Abramowitz, M. and Stegun, I. A., eds., *Handbook of Mathematical Functions* (US Department of Commerce, 10th Printing, 1970). Eqs. (9.1.42) and (9.1.43).

$$\cos(\alpha \sin \omega t) = J_0(\alpha) + 2 \sum_{k=1}^{\infty} J_{2k}(\alpha) \cos(2k\omega t) \tag{19.67}$$

$$\sin(\alpha \sin \omega t) = 2 \sum_{k=0}^{\infty} J_{2k+1}(\alpha) \sin((2k+1)\omega t) \tag{19.68}$$

where the $J_k(\alpha)$ are the Bessel functions of the first kind. Taking a time average, we are left with only the first term of the cosine expansion, so

$$\overline{H_i} = \frac{\hbar\omega_0}{2} J_0(\alpha)\sigma_z. \tag{19.69}$$

The reduction in the apparent magnetic moment is referred to as *spin dressing* and has been experimentally demonstrated many times. The corrections due to finite B_z and static fields in the $\hat{x}$, $\hat{y}$ directions have been calculated.[12]

19.5.3 The "sudden" approximation

We have already made use of the so-called sudden approximation in the foregoing discussion. Basically, if we change the Hamiltonian very rapidly compared to any frequency in the problem, the system remains in the same state but will evolve under the action of the new Hamiltonian. Except in trivial cases, the original Hamiltonian's eigenstates are not the same as those of the new Hamiltonian, so we need to rewrite the original state in the new basis. If we start in an eigenstate of the original Hamiltonian, in the new basis the state will be a superposition of the new Hamiltonian's eigenfunctions and the system will immediately begin to evolve under the new Hamiltonian's influence.

For example, if we initially have a system prepared in a spin state $|s_z+\rangle$ along a magnetic field $B_z\hat{z}$, if we suddenly turn off B_z and turn on a field $B_y\hat{y}$, the spin will precess in the $\hat{x} - \hat{z}$ plane at the Larmor frequency $\omega = \gamma B_y$. In fact, we could have slowly reduced B_z to zero, and in the absence of any other fields, s_z would remain constant. If B_y was slowly turned on, the system would again begin to precess in the $\hat{x} - \hat{z}$ plane, with the frequency increasing as the field increased. In practice, this does not work because if there was any residual magnetic field in a random direction, the spin would tend to be aligned with that field (depending on how slowly the field was reduced and the magnitude of the residual field). The sudden approximation simply requires that the fields be turned off very quickly compared to the Larmor frequency and avoids the problems due to stray fields that are encountered if the applied fields are ramped to zero and then increased.

The sudden approximation can be also applied to time-dependent fields, for example, a short pulse of a resonant oscillating field can be used to rotate a spin vector and form a superposition state from an initial eigenstate. Such oscillating field pulses are used in magnetic resonance and in quantum computing to form gates. We will see many examples as we progress. Of course, the length of the pulse must be sufficiently long to define an oscillating field (a few oscillation periods). The oscillating field can

[12]Golub, R. and Lamoreaux, S.K., *Neutron electric-dipole moment, ultracold neutrons and polarized* ^{3}*He*, Phys. Rep. **237**, 1 (1994) and references therein.

also be applied for a very long time with very small amplitude because the effects of random background fields are of much less concern than in the static case. We will show later the effects of random fluctuating fields in the context of the density matrix.

We can also picture the sudden approximation when applied to time-dependent fields by transforming into a rotating frame even before the time-dependent field is turned on. If we take the case of a spin-1/2 magnetic moment in a static field B_z along $\hat{z}$ and consider an oscillating field B_x along $\hat{x}$, even if we set $B_x = 0$ we can still transform to the rotating frame, the effect being $B_z \to B_z - \omega/\gamma$ as we derived before. Neglecting the counter-rotating field component, when we set $B_x \neq 0$ in the rotating frame we immediately see the field as $\mathbf{B} = (B_z - \omega/\gamma)\hat{z} + B_x/2\hat{x}$ and the spin precesses in a cone around the vector sum of the two static (in the rotating frame) fields as long as the oscillating field is turned on. When the oscillating field is turned off suddenly (B_x suddenly set to zero) whatever state the spin was in, it begins to precess around $B_z\hat{z}$; there can in general be a static component along $\hat{z}$ and a component in the $x-y$ plane, precessing aroung the $\hat{z}$ direction.

In the case of a pulsed resonant field, $B_z \to B_z - \omega/\gamma = 0$ in the rotating frame. Thus, when B_x is turned on, the spin precesses in the $y-z$ plane about $\hat{x}$ at a frequency $\gamma B_x/2$. If the B_x is on for a time such that $\gamma B_x \tau/2 = \pi/2$, the spin will lie along the $\hat{y}$ axis, after which $B_x = 0$ and the spin precesses in the $x-y$ plane around $\hat{x}$ at the Larmor frequency, γB_z. In this case, there is no static component along the $\hat{z}$ axis.

Therefore, the sudden approximation is essential to the understanding of $\pi/2$ etc. pulses that are used in magnetic resonance, quantum computing, etc.

For mass of charge e in a harmonic oscillator potential, if an electric field $\mathcal{E}$ is suddenly applied to the system (along the oscillation direction), the effect is to shift the equilibrium point of the oscillator from $x = 0$ to $x_0 = e\mathcal{E}/(m\omega^2)$. As was shown in Sec. 18.6.4, this leaves the oscillator in a coherent superposition.

As a final example,[13] let us take an atom in its ground state and give an impulse to the nucleus (e.g., by collision with a high-energy neutron that does not directly affect the electrons), causing it to suddenly start moving with velocity v. In the case where the impulse or "jolt" is of very short duration $\tau \ll a/v$ where a is the diameter of the atom, we can use the sudden approximation to determine the final state of the system. Instead of following the motion of the nucleus, by use of the Galilean invariance of the Schrödinger equation (see Appendix B), we can assign the motion to the atom's electrons, so the wavefuntion becomes, taking $\mathbf{q} = m_e v/\hbar$,

$$|0'\rangle = e^{-i\mathbf{q}\cdot\sum_{i=1}^{Z}\mathbf{r}_i}|0\rangle \tag{19.70}$$

where the sum i is over the Z electrons of the atom labeled by $\mathbf{r}_i$. The probability P_{k0} that the atom is excited to the k level (where k represents all possible eigenvalues to specify a state, including angular momentum $j = l + s$, principal quantum number n, etc.) is

$$P_{k0} = |\langle k|\, 0'\rangle|^2 = \left|\langle k|e^{-i\mathbf{q}\cdot\sum_{i=1}^{Z}\mathbf{r}_i}|0\rangle\right|^2. \tag{19.71}$$

[13]Landau, L. D. and Lifshitz, E. M., *Quantum Mechanics (Non-relativisitic Theory)* (Pergamon, 1977), pp. 149-150.

For the ground state of the hydrogen atom,

$$\psi_0(r) = (\pi a^2)^{-1/2} e^{-r/a} \tag{19.72}$$

the probability that the atom is left in an excited state or is ionized is

$$1 - P_{00} = 1 - \left| \int \psi_0^2(r) e^{-i\mathbf{q}\cdot\mathbf{r}} dV \right|^2 = 1 - \frac{1}{(1 + \frac{1}{4}q^2a^2)^4} \tag{19.73}$$

which, in the limit $qa \ll 1$, goes to zero as $(qa)^2$ (remains in ground state), and in the limit $qa \gg 1$ the probability to be in an excited or ionized state goes to unity as $1 - (2/qa)$.[8]

In the limit $qa \ll 1$,

$$P_{k0} \approx |\langle k| \left(1 - i\mathbf{q} \cdot \sum_{i=1}^{Z} \mathbf{r}_i\right) |0\rangle|^2 = |\langle k| \left(\mathbf{q} \cdot \sum_{i=1}^{Z} \mathbf{r}_i\right) |0\rangle|^2. \tag{19.74}$$

for $k \neq 0$ because $\langle k|0\rangle = 0$.

19.5.4 Berry's phase

Let us assume a polarized spin pointing along $\hat{z}$ in a static magnetic field $B_z\hat{z}$, so the system is in an eigenstate. At time $t = 0$ let us suddenly apply a field $B_1\hat{x}$ that rotates in the $x - y$ plane exactly once, in a time T much larger than the Larmor period. In the rotating frame, for rotation in the same direction as the Larmor precession, the total field is

$$\mathbf{B}' = \left[B_z - \frac{2\pi}{\gamma T}\right] \hat{z} + B_1\hat{x}. \tag{19.75}$$

The spin will precess around the net field during the rotation of B_1 and accumulates a phase in the lab frame

$$\phi = T\gamma\sqrt{\left[B_z - \frac{2\pi}{\gamma T}\right]^2 + B_1^2} \approx \gamma BT - \frac{2\pi B_z}{B} + 2\pi \tag{19.76}$$

with $B = \sqrt{B_z^2 + B_1^2}$ and neglecting $(1/T)^2$ (2π is added so that the phase is not altered when $B_1 = 0$). The extra phase that appears in addition to the dynamic phase γBT is related to the geometrical path that the magnetic field vector traces out in the laboratory reference frame. It is equal to the solid angle Ω that is traced out by $\mathbf{B}(t)$ in the laboratory frame:

$$2\pi \left(1 - \frac{B_z}{\sqrt{B_z^2 + B_1^2}}\right) = 2\pi(1 - \cos\theta) = \int_0^{2\pi} \int_0^{\theta} d\phi\, d\theta' \sin\theta' = \Omega \tag{19.77}$$

where θ is the constant angle between the $\hat{z}$ axis and the total field direction during the rotation period, in the laboratory frame.

This extra phase due to the motion of the B field direction is Berry's phase, or the geometrical phase, which for spin-1/2 is

$$\phi_b = \Omega. \tag{19.78}$$

Note that we could vary $\mathbf{B}$ in any manner, on a closed path, and we would get the same correction to the dynamic phase as long as the variation is slow enough to be in the adiabatic limit.

This result can be generalized to any quantum system that depends on a three or more dimensional parameter that can be continuously varied in parameter space. This was a surprisingly general result, obtained by M. V. Berry in 1984, although specific cases had been studied earlier.[14] Until Berry's work brought these issues to general attention, the phases that accumulated in the adiabatic limit of perturbation theory were generally discarded as uninteresting. In fact, this result has several important applications.[15]

As we have discussed already, the phase of a wave function cannot be thought of as an eigenvalue or a conserved quantity; phase is closely related to time, and time is a parameter, not an operator, in quantum mechanics. Thus Berry's phase depends on the specific representation that is used in its calculation, and in that sense cannot be considered fundamental.

Similar geometric effects occur in optics with polarized light, and even in mechanics; a rope that is wrapped in a spiral (helix) along a cylinder receives a twist that depends on the pitch angle of the helix.[16]

19.5.5 Landau-Zener effect: transitions under adiabatic variation of the potential energy

In classical mechanics, the effects of slowly varying potentials are relatively easily calculated, typically by using the principle of constant action. In quantum mechanics, treatment of adiabatic problems is subtly difficult and remains an area of research; the fairly recent discovery of Berry's phase is an example of such.

The prototypical problem is the Landau-Zener effect, whereby a particle transitions between two (fictitious or real) spin states when subjected to a time-dependent Hamiltonian of the form

$$H = \frac{\hbar\gamma}{2}\begin{pmatrix} \alpha' t & B_x \\ B_x & -\alpha' t \end{pmatrix} = \begin{pmatrix} \alpha t & E_1 \\ E_1 & -\alpha t \end{pmatrix} \tag{19.79}$$

where we have incorporated a linear time dependence to the z component of the magnetic (or fictitious) field, with $d(E_+ - E_-)/dt = 2\alpha$ in the large $|t|$ limit, and a

[14]Berry, M.V., *Quantal phase factors accompanying adiabatic changes*, Proc. R. Soc. Lond. A **392**, 45 (1984); Kato, T., *On the Adiabatic Theorem of Quantum Mechanics*, J. Phys. Soc. Jpn. **5**, 435-439 (1950); Pancharatnam, S.,*Generalized Theory of Interference, and Its Applications. Part I. Coherent Pencils*. Proc. Indian Acad. Sci. A. **44**(5), 247-262 (1956); Longuet-Higgins, H.C., Öpik, U., Pryce, M.H.L. and Sack, R. A., *Studies of the Jahn-Teller effect II. The dynamical problem*, Proc. R. Soc. Lond. A. **244**(1236), 1-16 (1958).

[15]A special case of the importance of this phase in the presence of a motional $v \times E$ field is discussed in Commins, E., *Berry's geometric phase and motional fields*, Am. J Phys. **59**, 1077 (1991).

[16]In optics, reference is made to the Pancharatnam phase.

constant magnetic (or fictitious) field along x. This Hamiltonian can be diagonalized, yielding two eigenvalues,

$$E_\pm = \pm\sqrt{(\alpha t)^2 + E_1^2}\,, \tag{19.80}$$

that are shown in Fig. 19.6 as a function of time.

The dashed lines show the asymptotic values, and the vertical dotted line indicates the minimum energy difference between the states, which is equal to $2E_1$.

Our goal is to calculate the probability that a system starting in state $|E_+\rangle$ at $t \to -\infty$ ends up in state $|E_-\rangle$ for $t \to +\infty$.

This is a deceptively simple system, the original solution of which came with great effort. However, Landau later described a general way to deal with this and similar problems.[17] His procedure is to consider the solutions of the Schrödinger equation as general functions of a complex variable, which is allowed because these solutions are meromorphic functions, meaning they are single-valued and analytic on all but discrete regions on the complex plane, and can thus have only a finite number of poles, zeros, and branch cuts.

When α is very small, if we start with the system in an eigenstate, it tends to stay in that state, with the phase accumulating as

$$\phi_\pm = \frac{1}{\hbar}\int E_\pm(t)dt + \phi_0; \quad |E_\pm(t)\rangle = e^{-i\phi_\pm}|E_\pm\rangle. \tag{19.81}$$

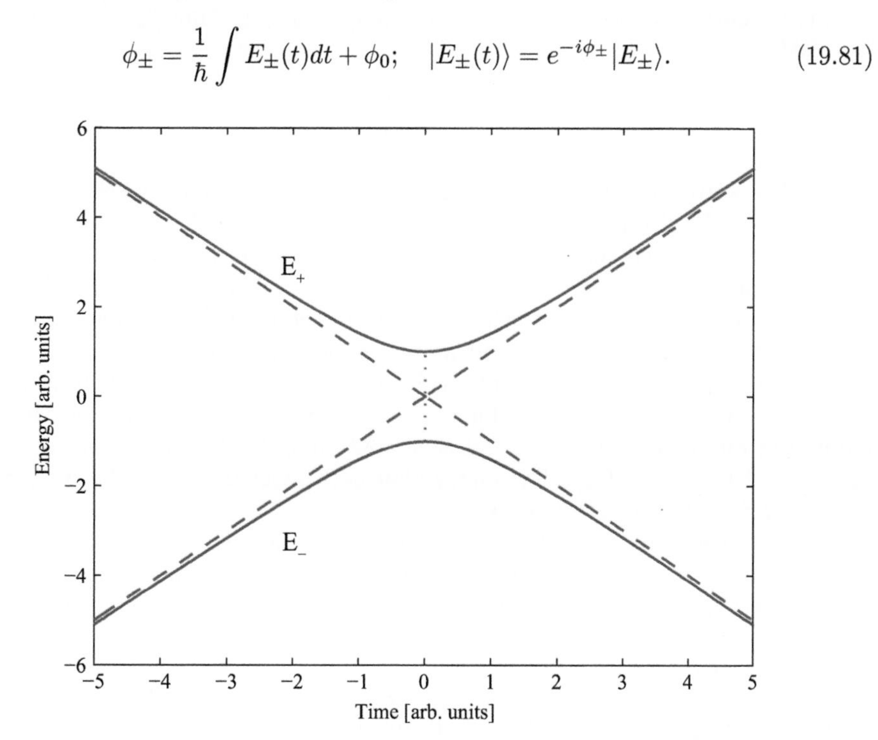

Fig. 19.6 The energy eigenvalues as a function of time when $\alpha = 1$ and $E_1 = 1$.

[17]Landau, L.D. and Lifshitz, E.M., *Quantum Mechanics*, 3rd ed. (English) (Oxford: Pergamon Press, 1977), §53.

This can be seen by considering a spin precessing at some small angle relative to a magnetic field vector. If the field direction is slowly varied, classically we would expect the spin projection to be an adiabatic invariant, resulting in the angle being constant relative to the field direction, in the limit of infinitesimally slow variation of the magnetic field direction.

Landau recognized that if t is allowed to take on complex values, the phase integral is independent of the path connecting two points t_1 and t_2 on the real axis, up to certain conditions that will be elaborated upon shortly. This, together with the realization that a transition from $|E_+\rangle$ to $|E_-\rangle$ at the energy difference minimum is analogous to tunneling of a particle through a classically forbidden region, provides the basis for a relatively simple calculation of the transition probability, and in this region ϕ is complex.

In order to determine the transition probability, let us start in state $|E_+\rangle$ for $t \ll 0$. We can then treat this as a transmission and reflection problem, with the probability for a transition between states given by the transmission, or tunneling, probability T, and the probability of remaining in the same state by the reflection probability R, with $R + T = 1$. We will assume that $T \ll 1$ so $R \approx 1$. We need to connect the incident state $|E_+\rangle$ to the transmitted and reflected states. This can be done by noting that $E_+ - E_- = 0$ at some imaginary time t_0, determined from Eq. (19.80)

$$t_0 = i\frac{E_1}{\alpha}. \tag{19.82}$$

The incident wave function is a spin eigenstate, and the boundary conditions for the transmitted and reflected states need to be matched to it. We can assume that the reflected $|E_+\rangle$ state and transmitted $|E_-\rangle$ state correspond to a two-level system when using the z axis as the quantization axis. Our goal is to find the probability for the spin to not follow the change in field direction. In the limit $T \ll 1$, a small amplitude as $t \to \infty$ for $|E_-\rangle$ is matched to unit amplitude for $|E_+\rangle$ as $t \to -\infty$.

Let us calculate the complex phase by integrating along the contour shown in Fig. 19.5. Moving from the right, for the $|E_-\rangle$ state, we bring it from its final state amplitude to its amplitude at t_0, and then continuing to the left, go on to match an amplitude of unity for $|E_+\rangle$. The integration path must not cross the branch cut of $\sqrt{(\alpha t)^2 + E_1^2}$ as shown in the figure.

For the integral shown in (b), the integral along the real axis gives a phase angle, while the integral along the imaginary axis is pure imaginary and leads to the variation in wave function amplitude. We thus need to evaluate the integral along the imaginary axis as

$$\hbar\phi_- = -\int_C \sqrt{(i\alpha t)^2 + E_1^2}\; idt = -\int_C \sqrt{(\alpha t)^2 - E_1^2}\; dt = -\frac{E_1^2}{\alpha}\int_C \sqrt{t'^2 - 1}\; dt' \tag{19.83}$$

where the substitution $\frac{\alpha t}{E_1} = t'$ was used, and $|t'| > 1$ to avoid the branch cut. This integral can be evaluated by substituting $t' = 1/w$, $dt' = -dw/w^2$, with $|w| < 1$, so we can collapse the contour to a small half-circle around the origin.

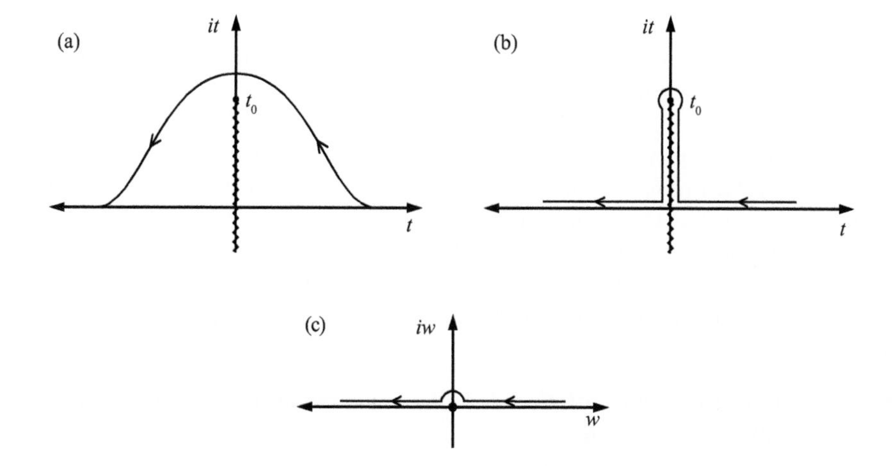

Fig. 19.7 (a) The integration path of t must go around the branch cut of $\sqrt{t'^2-1}$. The boundary values are matched at $(0, t_0)$ and the integration path does not cross the branch cut but passes around $(0, t_0)$. (b) The path can be collapsed to vertical lines along the branch cut and a semicircle around $(0, t_0)$. (c) The integral around (one half) of the branch cut can be evaluated by use of the residue theorem by transforming $t' = 1/w$ which leads to a pole at (0, 0).

$$-\int_C \sqrt{t'^2 - 1}dt' = \int_C \frac{\sqrt{w^{-2}-1}}{w^2}dw \tag{19.84}$$

$$\frac{\sqrt{w^{-2}-1}}{w^2} = \frac{\sqrt{1-w^2}}{w^3} \approx \frac{1}{w^3}(1 - \frac{1}{2}w^2 + ...) = \frac{1}{w^3} - \frac{1}{2}\frac{1}{w} + ... \tag{19.85}$$

and there is a pole at $w = 0$ with residue $-1/2$. (The term, $1/\omega^3$ does not contribute.) The full integral around the circle is therefore $2\pi i\left(-\frac{1}{2}\right) == -\pi i$, and we take half of this, so

$$\phi_- = -\frac{i\pi}{2}\frac{E_1^2}{\hbar\alpha} \tag{19.86}$$

and we find that transition probability from $|E_+\rangle$ to $|E_-\rangle$ is

$$T = |e^{-i\phi_-}|^2 = e^{-\pi E_1^2/\hbar\alpha} \tag{19.87}$$

which is the known result from more complicated methods (note that the definition of α used here is a factor of 2 larger than often used).

This calculation was presented in some detail as it is a useful and practical general technique. Similar techniques can be used to derive the WKB approximation.

19.6 The density matrix

The density matrix ρ was introduced in Chapter 14 as a direct means of addressing the statistical nature of quantum mechanics. In general, we could do almost any statistical problem by solving the Schrödinger equation directly and then performing a statistical averaging of the states subjected to, for example, random time-dependent fields or other perturbations. The linearity of quantum mechanics allows us to perform this averaging first, which defines the density matrix, and we will now derive its equation of motion. The introduction of the density matrix can result in much computational simplicity in situations where there are random fluctuations. One criticism of the density matrix is that different quantum states can produce an identical density matrix. It should be noted, however, that such states have the same expectation values, and in the end, it is those expectation values that define the physical system. It is further noted that, usually, solving for the wave function is easier than solving for the density matrix ρ, but the calculation of statistical averages of expectation values is much easier with ρ.

19.6.1 Time dependence of the density matrix: the Liouville-von Neumann equation

The density matrix is defined as

$$\rho = |\psi\rangle\langle\psi| \tag{19.88}$$

and its properties have been discussed in Chapter 14.

Let us now explore its time dependence. The time derivative is

$$\frac{\partial \rho}{\partial t} = \frac{\partial}{\partial t}|\psi\rangle\langle\psi| = \left[\frac{\partial}{\partial t}|\psi\rangle\right]\langle\psi| + |\psi\rangle\left[\frac{\partial}{\partial t}\langle\psi|\right]. \tag{19.89}$$

From the Schrödinger equation, we know that states evolve under the action of a Hamiltonian as

$$\frac{\partial}{\partial t}|\psi\rangle = -\frac{i}{\hbar}H|\psi\rangle; \qquad \frac{\partial}{\partial t}\langle\psi| = \frac{i}{\hbar}H\langle\psi| \tag{19.90}$$

which allows determination of the time derivative

$$\frac{\partial \rho}{\partial t} = -\frac{i}{\hbar}[H, \rho]. \tag{19.91}$$

This is the Liouville-von Neumann equation (often simply called the Liouville equation), and it looks a lot like the Heisenberg equation except that it has a minus sign. In some sense this is not surprising because ρ can be thought of as an operator.

Alternatively, we can obtain the time dependence of ρ by using the time translation operator on the wave functions, which yields

$$\rho(t) = U(t)|\psi\rangle\langle\psi|U^\dagger(t). \tag{19.92}$$

Alternatively, we can transform measurement operators and leave the density matrix static, as the trace which is used to calculate observables is cyclically invariant,

$$\langle\hat{A}(t)\rangle = \text{Tr}\ U(t)\rho(0)U^\dagger(t)\hat{A} = \text{Tr}\rho(0)U^\dagger\hat{A}U = \text{Tr}\rho(0)\hat{A}(t). \tag{19.93}$$

This is the Heisenberg representation, and it is also useful in the interaction representation.

For a time-independent Hamiltonian, the explicit time evolution of the density matrix can be determined by a straightforward application of $U(t)$,

$$\rho(t) = \sum_{m,n} c_n c_m^* \, e^{-i\omega_n t} |\psi_n\rangle\langle\psi_m| e^{i\omega_m t} \tag{19.94}$$

where $|\psi_{m,n}\rangle$ are the (static) energy eigenstates and $\omega_n = E_n/\hbar$. From this it can be seen that the diagonal elements, which are the state populations, are static, while the off-diagonal elements or *coherences* oscillate at frequencies determined by the energy splitting between the eigenstates.

19.6.2 The spin-1/2 density matrix

The simplest density matrix could be considered that for a single state; however, that is trivial. The advantage of introducing the density matrix is that it allows the dynamics of the ensemble average to be determined without having to solve the Schödinger equation for every system possibility and then taking the average. The linearity of quantum mechanics allows the reversal of the order and allows the system evolution to be determined from the time-varying statistical characteristics of perturbing fields, for example.

The spin-1/2 density matrix can be simply written as (normalized to one particle)

$$\rho = \frac{\hat{I}}{2} + a_x\sigma_x + a_y\sigma_y + a_z\sigma_z = \frac{\hat{I}}{2} + a_z\sigma_z + a_+\sigma_- + a_-\sigma_+ \tag{19.95}$$

where $a_+ = a_x + ia_y$ and $a_- = a_x - ia_y$. Using the polar angle vector notation introduced earlier in this chapter (19.30), we can also write, for a unit vector $\hat{n}$,

$$\rho_{\hat{n}}|\hat{n}\rangle\langle\hat{n}| = \begin{pmatrix} \cos^2(\theta/2) & \cos(\theta/2)\sin(\theta/2)e^{-i\phi} \\ \cos(\theta/2)\sin(\theta/2)e^{i\phi} & \sin^2(\theta/2) \end{pmatrix} \tag{19.96}$$

$$= \frac{\hat{I}}{2} + \frac{\cos\theta}{2}\sigma_z + \frac{\sin\theta}{2}(\sigma_x\cos\phi + \sigma_y\sin\phi) \tag{19.97}$$

$$= \frac{\hat{I} + \hat{n}\cdot\sigma}{2} \tag{19.98}$$

which is a nice compact form. Note that

$$\rho_{\hat{n}}^2 = \rho_{\hat{n}}\,, \tag{19.99}$$

which can be easily shown because the Pauli matrices anticommute, which means this is a pure state.

A density matrix for a state in thermal equlibrium in an energy basis is given by the Boltzmann distribution,

$$\rho = \sum_{n=1}^{d} p_n|n\rangle\langle n| = \frac{1}{Z}\sum_{n=1}^{d} e^{-\beta E_n}|n\rangle\langle n| \tag{19.100}$$

where the partition function is

$$Z = \text{Tr } e^{-\beta H} \tag{19.101}$$

and $\beta = 1/k_B T$. For spin-1/2, in the high temperature limit, in a magnetic field $B_0\hat{z}$,

$$\rho \approx \frac{\hat{I}}{2} - \frac{\beta\hbar\gamma B_0}{2}\sigma_z. \tag{19.102}$$

A general diagonal spin-1/2 density matrix can be written as

$$\rho = \frac{1}{2} + \frac{p}{2}\sigma_z = \frac{1}{2}\begin{pmatrix} (1+p) & 0 \\ 0 & (1-p) \end{pmatrix}, \tag{19.103}$$

where $p_\pm = (1 \pm p)/2$ are the probabilities to be in the ± 1 states, and $p_+ + p_- = 1$. The entropy can be calculated as (in units of k_B)

$$S(p) = -\text{Tr}\rho \ln \rho = -((p+1)/2)\ln((p+1)/2) - ((1-p)/2)\ln((1-p)/2). \tag{19.104}$$

This is very similar to the Shannon entropy for a binary bit stream,

$$S(p) = -p \log_2 p - (1-p)\log_2(1-p) \tag{19.105}$$

given a probability p of, for example, finding 1 vs. 0 in the stream, or the probability of an error for each bit in the stream. In the latter case, the Shannon entropy can be used to estimate the fraction of bits of information that survive after introducing the errors, e.g., as $p \to 0$, $S(p) \to 0$, and as $p \to 1$ (which is a simple NOT operation on the data stream), $S(p) \to 0$. However, $p = 1/2$ represents a random flip of each bit in the stream, and $S(1/2) = 1$ which means *all* information is lost. A principal difference between the quantum and Shannon entropies is that the latter replaces the natural logarithm with $\log_2$, which amounts to a multiplicative factor.[18]

19.6.3 Effects of random time-dependent perturbations: spin relaxation

The effects of a weak fluctuating potential on the evolution of the density matrix have been well addressed in the literature. Since the treatments are often opaque, we start from the beginning.[19]

Let us assume that the perturbing potential can be written as an effective field $B'_{x,y,z}(t)$, and we assume a field with magnitude B_0 along z. The Hamiltonian is thus, in frequency units ($\omega = \gamma B$),

$$H = -\frac{\omega_0 + \omega_z(t)}{2}\sigma_z - \frac{\omega_x(t)}{2}\sigma_x - \frac{\omega_y(t)}{2}\sigma_y = H_0 + H_1(t). \tag{19.106}$$

Let us further assume that the time average of $H_1(t)$ is zero.

[18]It is rumored Shannon asked von Neumann what he should call his measure of information uncertainty. Von Neumann suggested entropy because nobody knows what that is. The similarity to von Neumann's entropy, which was already formulated, suggests that there was a deeper reason.

[19]Abragam, A., *Principles of Nuclear Magnetism* (Oxford: Oxford University Press, 1961), p. 276; Slichter, C.P., *Principles of Magnetic Resonance* (New York: Harper, third edition, 1989).

Defining

$$-2b = \omega_x(t) + i\omega_y(t); \qquad -2b^* = \omega_x(t) - i\omega_y(t); \qquad -b_z = \omega_z(t) \tag{19.107}$$

the perturbing Hamiltonian can be rewritten as

$$H_1(t) = b^*\sigma_+ + b\sigma_- + b_z\sigma_z \tag{19.108}$$

where $\sigma_\pm$ are defined earlier in the chapter, and it is understood that b and b_z are intrinsically time-dependent, with zero mean. Furthermore, the density matrix can be expanded in the spherical Pauli basis as

$$\rho = \frac{1}{2} + \rho_{1,0}\sigma_z + \rho_{1,1}\sigma_+ + \rho_{1,-1}\sigma_- \tag{19.109}$$

where $\rho_{1,1} = \rho^*_{1,-1}$.

The time evolution of the density matrix is

$$\frac{d\rho}{dt} = -i[H_0 + H_1(t), \rho]. \tag{19.110}$$

The explicit dependence on the constant H_0 can be eliminated by transforming to the interaction representation (i.e., in this case, to the rotating frame), with

$$H_1(t) \to e^{iH_0t}H_1(t)e^{-iH_0t}; \qquad \rho \to e^{iH_0t}\rho e^{-iH_0t} \tag{19.111}$$

where

$$e^{iH_0t} = \begin{pmatrix} e^{-i\omega_0 t/2} & 0 \\ 0 & e^{i\omega_0 t/2} \end{pmatrix}. \tag{19.112}$$

Henceforth we will work in the interaction representation, with

$$H_1(t) = e^{-i\omega_0 t}b^*\sigma_+ + e^{i\omega_0 t}b\sigma_- + b_z\sigma_z. \tag{19.113}$$

The time evolution of the density matrix in the rotating frame (interaction representation for the static field) is

$$\frac{d\rho}{dt} = -i[H_1(t), \rho] \tag{19.114}$$

which can be integrated by successive approximations to

$$\rho(t) = \rho(0) - i\int_0^t [H_1(t'), \rho(0)]dt' \\ - \int_0^t dt' \int_0^{t'} dt''[H_1(t'), [H_1(t''), \rho(0)]]. \tag{19.115}$$

We are interested in the relaxation rates and frequency shifts due to the perturbing fields, which can be found through the time derivative of ρ. Introducing a new variable $\tau = t - t''$, the time derivative is

$$\frac{d\rho}{dt} = -i[H_1(t), \rho(0)] - \int_0^t d\tau [H_1(t), [H_1(t-\tau), \rho(0)]]. \tag{19.116}$$

We will assume that the ensemble average of $H_1(t) = 0$; a constant part of the perturbation can be added to H_0. In addition, if we assume the perturbation is weak, $\rho(0)$ can be replaced by $\rho(t)$, which introduces errors below second order. We then have

$$\frac{d\rho(t)}{dt} = -\int_0^t d\tau [H_1(t), [H_1(t-\tau), \rho(t)]] \equiv -\Gamma \rho(t) \tag{19.117}$$

where Γ is the "relaxation matrix," the real parts of which describe decay of coherence (off-diagonal) or relaxation to equilibrium (diagonal), and the imaginary parts of the off-diagonal elements describe frequency shifts.

Using the spherical basis Pauli matrices together with the expansion of the density matrix Eq. (19.109), and redefining (in the interaction representation)

$$\begin{aligned} 2b &= e^{-i\omega_0 t} \left(\omega_x(t) + i\omega_y(t)\right) \\ 2b' &= e^{-i\omega_0 (t-\tau)} \left(\omega_x(t-\tau) + i\omega_y(t-\tau)\right) \\ 2b_z &= \omega_z(t); \quad 2b'_z = \omega_z(t-\tau), \end{aligned} \tag{19.118}$$

the perturbing Hamiltonian can be written as

$$H_1(t) = b^* \sigma_+ + b\sigma_- + b_z \sigma_z; \quad H_1(t-\tau) = b'^* \sigma_+ + b'\sigma_- + b'_z \sigma_z. \tag{19.119}$$

The time derivative of ρ, correct to second order, is (neglecting terms with factors of $e^{\pm i\omega_0 t}$ and $e^{\pm 2i\omega_0 t}$, via the secular approximation; we assume also that the fluctuations in any direction are uncorrelated with those in the other two directions)

$$\dot{\rho}_{1,-1} = -\rho_{1,-1} \int_0^t 2(bb'^* + 2b_z b'_z) d\tau \tag{19.120}$$

$$\dot{\rho}_{1,1} = -\rho_{1,1} \int_0^t 2(b^* b' + 2b_z b'_z) d\tau \tag{19.121}$$

$$\dot{\rho}_{1,0} = -\rho_{1,0} \int_0^t 2(b^* b' + bb^{*\prime}) d\tau. \tag{19.122}$$

These equations describe both frequency shifts and relaxations of the density matrix. Let us also assume τ in the random process is effectively so small that the density matrix does not significantly evolve over that time; then we can use the secular approximation to replace the integrals by their averages over some time longer than τ but less than the time for significant evolution of the density matrix. Assume that the fluctuations are stationary random processes, and that they are uncorrelated. Writing the b's in terms of of $\omega_x(t)$, $\omega_y(t)$, and $\omega_z(t)$, then we can substitute

$$b_z b_z' \to \frac{\omega_z(t)\omega_z(t-\tau)}{4} = \frac{1}{4}R_{zz}(\tau) \tag{19.123}$$

$$bb'^* \to e^{-i\omega_0\tau}\frac{\omega_x(t)+i\omega_y(t)}{2}\,\frac{\omega_x(t-\tau)-i\omega_y(t-\tau)}{2} \tag{19.124}$$

$$= \frac{1}{4}e^{-i\omega_0\tau}(\omega_x(t)\omega_x(t-\tau)+\omega_y(t)\omega_y(t-\tau)) = \frac{1}{4}e^{-i\omega_0\tau}(R_{xx}(\tau)+R_{yy}(\tau))$$

$$b^*b' \to \frac{1}{4}e^{i\omega_0\tau}(R_{xx}(\tau)+R_{yy}(\tau)) \tag{19.125}$$

where

$$R_{ii}(\tau) = \langle\omega_i(t)\omega_i(t-\tau)\rangle; \quad R_{ij}(\tau) = 0 \text{ for } i \neq j \tag{19.126}$$

where $i = x, y,$ or z and $\langle ...\rangle$ represents a time average, and we assumed that ρ does not vary significantly over the field correlation time. It is straightforward to generalize to the case where the fluctuations in different directions are correlated, as described by their cross correlations, e.g., R_{xy}.

Presently, we are most interested in the relaxation and would like to make a theoretical connection to the phenomenological T_1 and T_2 introduced earlier. The imaginary part of Γ gives frequency shifts, while the real part describes relaxation. We can see immediately that the relaxation rates $\frac{1}{T_2}$ of the off-diagonal elements are equal, and faster than the diagonal element relaxation rate $\frac{1}{T_1}$. Let us further take $R_{xx} = R_{yy} = R_{zz} = R$. Then

$$\frac{1}{T_2} = \int_0^t (R(\tau)\cos\omega_0\tau + R(\tau))d\tau = J(\omega_0) + J(0) \tag{19.127}$$

$$\frac{1}{T_1} = \int_0^t R(\tau)\cos\omega_0\tau \, d\tau = J(\omega_0) \tag{19.128}$$

where $J(\omega)$ is the power spectral density at frequency ω (we have two frequencies of interset, 0 and ω_0). Fluctuations at ω_0 can drive transitions between states per the Rabi formula and tend to equalize the probability distribution in the density matrix or drive it toward the thermal equilibrium value. This effects both T_1 and T_2. Slow fluctuations of the $\hat{z}$ directed field that are different for each spin in the ensemble (if it is a truly random process) can cause the precessing spin vectors to dephase; dephasing only affects the off-diagonal density matrix elements. This increases the relaxation rate $1/T_2$ relative to $1/T_1$ by the additional contribution from $J(0)$, the power spectral density at $\omega \approx 0$.

Note that we have not included the effect of fixed spatial gradient or motion of the spin, nor have we included interactions between the spins or other atoms in the system. As such, we introduced an ad hoc random fluctuating field that is truly random in direction and magnitude at each spin, without specifying how it might be generated. Nonetheless, this is a reasonable starting point for discussing fluctuation-induced relaxation and captures most of the gross features. In particular, it is almost always true that

$$J(0) \geq J(\omega_0) \tag{19.129}$$

and we see immediately that

$$\frac{1}{T_2} \geq \frac{2}{T_1} \tag{19.130}$$

or $T_2 \leq T_1/2$ when relaxtion comes from random uncorrelated processes and is a usual assumption in magnetic resonance. Often, relaxation is due to random interactions of a spin with a very large atomic magnetic field from, for example, paramagnetic atom impurities in a gas or liquid. This interaction causes a spin precession rate ω_a that only lasts for a short duration τ_c, and these interactions occur at a relative slow rate rate $1/\tau_0 \ll 1/\tau_c$. We can parametrize this interaction through a correlation function, $R(\tau) = \omega_a^2 e^{-\tau/\tau_c}$, for which the power spectral density is

$$J(\omega) \sim \frac{\tau_c^2}{\tau_0} \frac{\omega_a^2}{(\omega\tau_c)^2 + 1}. \tag{19.131}$$

It is possible to measure the correlation time of the perturbing Hamiltonian by applying a large static magnetic field B, so that $\gamma B = \omega \sim 1/\tau_c$, in which case T_1 increases.

The foregoing shows that there is a sound theoretical basis for introducing the transverse and longitudinal relaxation rates to the semiclassical equation of motion for a spin-1/2 particle. This analysis can be also applied to higher spin systems, for which the density matrix is of higher dimension. The density matrix can be expanded in a spherical multipole basis, and the relaxation rates for the different multipoles are related through angular momentum algebra, e.g., the Wigner-Eckart theorem.[20]

19.7 General application to two-level systems: fictitious spin-1/2

Many problems in quantum mechanics can be closely approximated by considering only two states that are, for example, similar in energy or otherwise isolated, or are coupled by a resonant oscillating field. We can apply the spin-1/2 formalism to any quasi-two-level system, but when we do so, the concept of a vector representation of a state must be considered as fictitious since it is not a representation of a true spatial vector as in the case of, say, an electron spin for which such a representation (the expectation value of the spin vector) is physically meaningful; nonetheless the concept is generally valid and useful. We will explore the dynamics of spin-1/2 in general in this chapter, and apply the formalism to varied problems. The idea of such a fictitious vector, which serves as a geometrical representation of the Schrödinger equation, was first elucidated by Feynman, Vernon, and Hellworth[21] in connection with the solution of physical problems associated with MASERs. It should be noted that this representation is rigorous and not an approximation in any sense.

The simplest quantum system is a single uncharged point particle; even this simple system has an infinite number of possible quantum states via the particle momentum. The next simplest system is a particle with two *internal* states, and it is *the* prototypical system in quantum mechanics. As we have seen with the Einstein A and B coefficients, we often isolate two quantum states within a complex system and ignore the other states as best as possible. Doing so provides a powerful and useful simplification. Let us consider two cases that illustrate the general application of fictitious spin-1/2.

[20]Happer, W., *Optical pumping*, Rev. Mod. Phys. **44**, 169 (1972).

[21]Feynman, R.P., Vernon, F.L. and Hellworth, R.W., *Geometrical Representation of the Schrödinger Equation for Solving Maser Problems*, J Appl. Phys. **28**, 49 (1957).

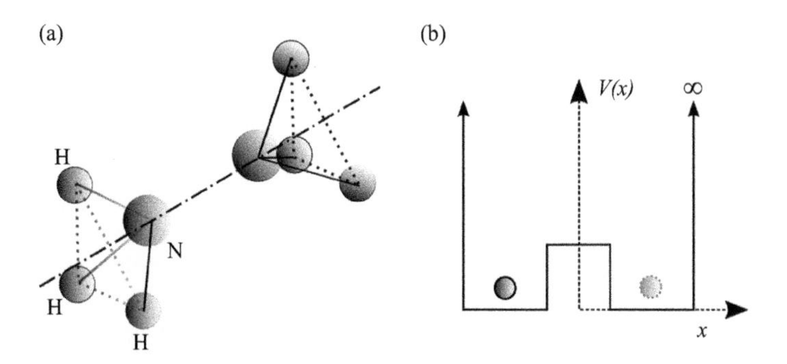

Fig. 19.8 Two examples of two-state systems. A: Inversion of the hydrogen atom plane relative to the nitrogen in ammonia, along the symmetry axis that intersects the Ns (a nonrotating system is shown here). B: A particle in a double potential well.

19.7.1 Ammonia molecule

The ammonia molecule can be considered as a two-level system when treated as a one-dimensional problem, as shown in Fig. 19.8A. In this treatment, rotations of the molecule are neglected but the analysis even applies if the molecule is rotating around the axis connecting the nitrogen atom to the hydrogen's plane (in this toy model, we are neglecting the fact that such transitions might not be allowed from particle symmetry considerations; however, this model serves to illustrate a principal qualitative feature of the ammonia molecule).

We can write two states for the orientation of the hydrogen plane relative to the nitrogen, say left and right, as

$$|L\rangle = \begin{pmatrix} 1 \\ 0 \end{pmatrix} \qquad |R\rangle = \begin{pmatrix} 0 \\ 1 \end{pmatrix} \tag{19.132}$$

where the choice was completely arbitrary. These states have the same energy; however, for the inversion to occur, the hydrogen atoms must tunnel through a potential barrier. We can represent the barrier by the Hamiltonian[22]

$$H = \begin{pmatrix} E_0 & -\Delta/2 \\ -\Delta/2 & E_0 \end{pmatrix} = E_0\mathbf{I} - \frac{\Delta}{2}\sigma_x \tag{19.133}$$

where $\Delta > 0$. It is easy to determine the eigenvalues of this matrix,

$$E_\pm = E_0 \pm \Delta/2 \tag{19.134}$$

which are the energies of the two eigenvectors, and it is straightforward to show that the higher energy corresponds to the antisymmetric state,

[22]While in the tunneling region, the total particle kinetic energy is less than zero, meaning the particle momentum is complex, leading to an exponential suppression of the transmission of the wave function through that region, i.e., tunneling through a classically forbidden region. The off-diagonal elements are therefore negative. Also, from the numerical solution to the double potential well problem, we know that the symmetric state has lower energy, and the negative off-diagonal elements will reproduce that.

$$|E_+\rangle = \frac{1}{\sqrt{2}}\begin{pmatrix} 1 \\ -1 \end{pmatrix} = \frac{|L\rangle - |R\rangle}{\sqrt{2}} \tag{19.135}$$

and that the lower energy corresponds to the symmetric state,

$$|E_-\rangle = \frac{1}{\sqrt{2}}\begin{pmatrix} 1 \\ 1 \end{pmatrix} = \frac{|L\rangle + |R\rangle}{\sqrt{2}}. \tag{19.136}$$

Therefore the ground state of the ammonia molecule is the symmetric state, with equal probability to find the hydrogen plane on either side of the nitrogen. The antisymmetric state, although the probabilities are again the same, has a higher energy. The important point is that the states $|L\rangle$ and $|R\rangle$ are not eigenstates, so if the molecule is prepared in, say, $|L\rangle$, it will be in a superposition of energy eigenstates and will evolve in time, oscillating between L and R at a frequency $\Delta/\hbar$.

When an electric field $\mathcal{E}$ pointing from left to right (as defined previously) is applied to the system, the Hamiltonian becomes

$$H = \begin{pmatrix} E_0 + d\mathcal{E} & -\Delta/2 \\ -\Delta/2 & E_0 - d\mathcal{E} \end{pmatrix} = E_0\mathbf{I} - \frac{\Delta}{2}\sigma_x + d\mathcal{E}\sigma_z \tag{19.137}$$

where d is the dipole moment of the molecule which is along the symmetry axis and points away from the net negatively charged nitrogen. In the following, we will set $E_0 = 0$ to simplify the calculation of the eigenvalues and eigenvectors. The eigenvalues are easily calculated as

$$E_\pm = \pm\sqrt{(d\mathcal{E})^2 + (\Delta/2)^2}. \tag{19.138}$$

The eigenvectors can be readily calculated by introducing the notation

$$\tan\theta = \frac{|\Delta|}{|2d\mathcal{E}|} \tag{19.139}$$

and a rescaled Hamiltonian,

$$\begin{pmatrix} 1 & -\tan\theta \\ -\tan\theta & -1 \end{pmatrix}\begin{pmatrix} a \\ b \end{pmatrix} = \lambda'\begin{pmatrix} a \\ b \end{pmatrix}. \tag{19.140}$$

The eigenvalues of the rescaled Hamiltonian are then

$$\lambda_\pm = \pm\frac{1}{\cos\theta}. \tag{19.141}$$

The components of the eigenvector therefore must satisfy

$$a\left(1 \mp \frac{1}{\cos\theta}\right) - b\tan\theta = 0. \tag{19.142}$$

For the λ_- case we have

$$\begin{aligned} a\left(1 - \frac{1}{\cos\theta}\right) - b\tan\theta = 0 &= a(\cos\theta - 1) - b\sin(\theta) \\ &= -2a\sin^2(\theta/2) - 2b\sin(\theta/2)\cos(\theta/2) \\ &= -a\sin(\theta/2) - b\cos(\theta/2). \end{aligned} \tag{19.143}$$

Noting that this equation is of the form $-aA - bB = 0$ which has the solution $a = B,\ b = -A$, the E_- eigenvector is

$$|E_-\rangle = \cos(\theta/2)|L\rangle - \sin(\theta/2)|R\rangle \tag{19.144}$$

and is obviously normalized. For λ_+ case, noting that $1 + \cos\theta = 2\cos^2(\theta/2)$, it is straightforward to work out that

$$|E_+\rangle = \sin(\theta/2)|L\rangle + \cos(\theta/2)|R\rangle\,. \tag{19.145}$$

This state is normalized, and it is easy to show that it is orthogonal to the $|E_-\rangle$ state. Note that when $\mathcal{E} \to 0$, $\theta \to \pi/2$ and we directly recover the previous case when there was no electric field applied. In the opposite extreme, when $\mathcal{E} \to \infty$, $\theta \to 0$, we see that the eigenstates are the L, R states, as expected because the energy depends exclusively on the left/right orientation of the molecule relative to the applied electric field.

We have illustrated that we can treat terms in a Hamiltonian as a fictitious magnetic field. The magnetic field magnitude is determined by diagonalizing the Hamiltonian, and the eigenstates are determined by the direction of the field in a fictitious spin space. It is tempting to think that the σ_x component of the Hamiltonian represents a real field and that we can think of the molecule oscillating as its axis rotates in two- or three-dimensional space. Such thinking would be incorrect, as a simple rotation would be equivalent to angular momentum, which we have set to zero in this problem; we have treated the system as a purely one-dimensional problem. Nonetheless, thinking of the system in a fictitious vector space allows the use of standard spin-1/2 precession and magnetic resonance techniques.

19.7.2 Neutrino oscillations

There are three known types or "flavors" of neutrinos: electron ν_e, muon ν_μ, and tau ν_τ. When neutrinos are produced in a nuclear or high-energy reaction, they are in a pure flavor eigenstate, but the pure flavor states are not the same as the free-space mass eigenstates.

The general problem of oscillation between three states is mathematically complicated. However, in many situations only two flavors are of importance, for example, ν_e and ν_μ. In such situations, we can use a fictitious spin-1/2 to describe the system.

Let us write the mass eigenstates of the two neutrinos of interest ν_1 and ν_2 as

$$|1\rangle = \begin{pmatrix} 1 \\ 0 \end{pmatrix}; \quad |2\rangle = \begin{pmatrix} 0 \\ 1 \end{pmatrix} \tag{19.146}$$

which are the eigenvectors of the mass Hamiltonian

$$H = \begin{pmatrix} m_1c^2 & 0 \\ 0 & m_2c^2 \end{pmatrix} = m_0\mathbf{I} + \frac{\Delta m}{2}\sigma_z \tag{19.147}$$

where $m_0 = (m_1 + m_2)/2$, c is the velocity of light, and $\Delta m = m_1 - m_2$, which we make positive by assigning the heavier mass to m_1. Note that the kinetic energies of the two mass states must also be added to this Hamiltonian, and the total energies for each state are equal, meaning that the velocities are different if the masses are different.

When neutrinos are produced by a weak interaction reaction, they are generated in a pure flavor state which is a mixture of mass eigenstates. The pure flavor state can be represented by a superposition of mass eigenstates as

$$\begin{pmatrix} c_1 \\ c_2 \end{pmatrix} = \begin{pmatrix} \cos(\theta/2) & \sin(\theta/2) \\ -\sin(\theta/2) & \cos(\theta/2) \end{pmatrix} \begin{pmatrix} 1 \\ 0 \end{pmatrix} = \begin{pmatrix} \cos(\theta/2) \\ -\sin(\theta/2) \end{pmatrix}. \tag{19.148}$$

Note that this is the *initial* flavor state in the mass basis,

$$|\psi_1\rangle = c_1|1\rangle + c_2|2\rangle \tag{19.149}$$

and we can construct an orthogonal flavor state,

$$|\psi_2\rangle = c_2|1\rangle - c_1|2\rangle. \tag{19.150}$$

Together these represent the two free space neutrino flavor states in the mass basis. The time or spatial dependence of the free-space states can be determined by introducing the dynamic phases, e.g.,

$$|\psi_1(x)\rangle = c_1 e^{-i\phi_1}|1\rangle + c_2 e^{-i\phi_2}|2\rangle \tag{19.151}$$

where $\phi_{1,2}$ are a function of time and/or space ($\phi_{1,2} = [p_{1,2}x - E_{1,2}t]/\hbar$), and similarly for $|\psi_2\rangle$. The probability that the system is in flavor state 1 is simply given by

$$P_1 = |\langle\psi_1|\psi_1(x)\rangle|^2 \tag{19.152}$$

$$= \left[\cos^2(\theta/2)e^{-i\phi_1} + \sin^2(\theta/2)e^{-i\phi_2}\right]\left[\cos^2(\theta/2)e^{i\phi_1} + \sin^2(\theta/2)e^{i\phi_2}\right] \tag{19.153}$$

$$= \cos^4(\theta/2) + \sin^4(\theta/2) + 2\sin^2(\theta/2)\cos^2(\theta/2)\cos(\phi_1 - \phi_2) \tag{19.154}$$

$$= 1 + \frac{\sin^2\theta}{2}(-1 + \cos(\phi_1 - \phi_2)) = 1 - \sin^2\theta\sin^2((\phi_1 - \phi_2)/2). \tag{19.155}$$

In the center of mass frame, the total mass-energy E is constant ($E_1 = E_2$).[23] Therefore

$$E^2 = p_1^2c^2 + m_1^2c^4 = p_2^2c^2 + m_2^2c^4. \tag{19.156}$$

Neutrinos are ultra-relativistic, so the mc^2 contribution to the energy is very small, which means the momentum can be calculated as

$$p_1 = \sqrt{(E/c)^2 - m_1^2c^2} \approx \frac{E}{c} - \frac{1}{2}\frac{m_1^2c^3}{E} \tag{19.157}$$

and similarly for p_2. We therefore obtain

$$\phi_2 - \phi_1 = \frac{(m_1^2 - m_2^2)c^3}{2\hbar E}x \tag{19.158}$$

[23]Because gravity couples the same way to all forms of mass energy, a gravitational field does not cause an additional term in the Hamiltonian due to the difference between m_1 and m_2.

which, defining $\Delta m^2 = m_1^2 - m_2^2$, yields the well-known neutrino flavor oscillation formula,

$$P_1(x) = 1 - \sin^2\theta \sin^2\left[\frac{\Delta m^2 c^3}{4\hbar E}x\right] = 1 - \sin^2\theta\sin^2\left[1{,}27\frac{\Delta m^2}{E}x\right], \tag{19.159}$$

where in the last step Δm^2 is given in $(\text{eV})^2$, E in GeV, and x in km. Note that the Hamiltonian was not directly used in this calculation.

This problem can be analyzed from another perspective, that is, the precession of a vector around the effective pseudo-magnetic field that is determined by the mass difference. Because the energy is constant, the velocity and hence the time dilation factor is different for the two mass states. The time dilation factor in the moving frame for time t measured in the lab frame is determined from the relativistic energy relationship, $E = \gamma mc^2$ and $t' = t/\gamma$. As observed from the rest frame, the precession in the moving frame accumulates phase as

$$\phi = \hbar^{-1}\frac{\Delta mc^2}{2}\frac{m_1 c^2}{E}t - \hbar^{-1}\frac{-\Delta mc^2}{2}\frac{m_2 c^2}{E}t = \frac{m_0 \Delta mc^4}{\hbar E}t = \frac{\Delta m^2 c^4}{2\hbar E}t = \omega_\ell t \tag{19.160}$$

and ω_ℓ is the precession frequency as observed in the laboratory frame. Note that the polarization vector is fictitious and not of any spatial extent, and therefore it is not subject to Lorentz contraction. The only low-order correction between the moving and lab frames is time dilation.

Let us take the $\hat{z}$ axis as along the effective magnetic field, so initially the polarization vector representing $|\psi_1\rangle$ at time $t = 0$ is

$$\vec{P}(0) = \cos\theta\ \hat{z} + \sin\theta\ \hat{y} \tag{19.161}$$

and the magnitude of P is 1. The $\hat{z}$ component is constant, while the perpendicular component precesses in the $x-y$ plane at a frequency ω_ℓ, so the tip of the polarization moves on a cone as

$$\vec{P}(t) = \cos\theta\ \hat{z} + \sin\theta(\cos\omega_\ell t\ \hat{y} + \sin\omega_\ell t\ \hat{x}). \tag{19.162}$$

The projection along the original polarization direction is related to the probability P_1 that the system is in state 1 as follows, noting that $P_1 + P_2 = 1$:

$$\vec{P}(t)\cdot\vec{P}(0) = \cos^2\theta + \sin^2\theta\ \cos\omega_\ell t = P_1 - P_2 = 2P_1 - 1 \tag{19.163}$$

and we find that

$$P_1(t) = (\cos^2\theta + \sin^2\theta\ \cos\omega_\ell t)/2 + 1/2 \tag{19.164}$$

$$= 1 + \sin^2\theta(-1 + \cos\omega_\ell t)/2 = 1 - \sin^2\theta\sin^2(\omega_\ell t/2), \tag{19.165}$$

which is essentially the Rabi formula that we introduced when discussing spin precession earlier in this chapter. Noting that, to good approximation in the lab frame, $t = x/c$,

$$\omega_\ell t/2 = \frac{\Delta m^2 c^4}{4\hbar E}x/c = \frac{\Delta m^2 c^3}{4\hbar E}x \tag{19.166}$$

and our result for $P_1(t)$ is the same as obtained previously. There was a slight added complication in this case because we had to use the relativistic transformation of time between the lab and moving frame. In the two-level systems that are normally encountered, for example the ammonia molecule, this complexity is unnecessary.

We could have approached the vector problem in yet other ways; for example, we could have set $|\psi_1\rangle$ as along $\hat{z}$ with the effective magnetic field in the $y-z$ plane. The result would be the same but slightly more complicated. One point of particular interest is that the probabilities to be in pure mass eigenstates $|1\rangle$ or $|2\rangle$ are constant and do not depend on time; only the flavor quantum number oscillates as the phases of the mass eigenstates in the superposition change. This oscillation can be detected through changes in neutrino reaction rates, which depend on the neutrino flavor.

20 Path integrals and scattering

20.1 Introduction

In this chapter we will see how Feynman was able to derive the Schrödinger equation using a wave optics approach that has classical mechanics as the geometric optics limit. This technique is now known as the "path integral" method and is useful in many fields of physics. We then turn to particle scattering and outline van Hove's demonstration that the scattering cross section is related to the Fourier transform of the density-density correlation function. The usual treatment of scattering uses the van Hove result, which considers the incident and scattered beams to be plane waves, for the scattering, and treats the beam propagation through the apparatus as that of classical particles.

This is followed by an alternate derivation applying a full wave optics approach, which yields the van Hove result modified by the instrument resolution function, as would be observed in an experiment.

The treatment of scattering in quantum mechanics is a vast field with many techniques available.[1] Here we have concentrated on the van Hove approach.[2]

20.2 Path integrals

20.2.1 Introduction

In 1948, as a prelude to his formulation of quantum electrodynamics (QED), Richard Feynman published a paper[3] outlining a derivation of the non-relativistic Schrödinger equation based on what might seem like a radically different viewpoint but which is really another expression of the idea that classical mechanics is the geometric optics limit of an underlying wave motion. This paper was based on his PhD thesis.[4] In doing this he introduced what would later be called the path integral method for calculating effects in quantum systems.

[1]A classic text is Goldberger, M.L. and Watson, K.H., *Collision Theory, John Wiley*, (1964, reissued Dover, 2004). See also Newton, R.G., *Scattering Theory of Waves and Particles*, (Springer, 1982).

[2]Van Hove, L., *Correlations in Space and Time and Born Approximation Scattering in Systems of Interacting Particles*, Phys. Rev. 95, 249, (1954) received 16 March 1954.

[3]R. P. Feynman, *Space-time approach to non-relativistic quantum mechanics*, Rev. Mod. Phys. **20**, 367-387 (1948); R. P. Feynman and A. R. Hibbs, *Quantum Mechanics and Path Integrals* (New York: McGraw Hill, 1965).

[4]Brown, L.M., ed. Feynman's Thesis, *A New Approach to Quantum Theory*, World Scientific, 2003, (contains Feynman's article of 1948 and Dirac's of 1933).

The Historical and Physical Foundations of Quantum Mechanics. Robert Golub and Steven K. Lamoreaux, Oxford University Press.
 DOI: 10.1093/oso/9780198822189.003.0020

20.2.2 Feynman's derivation of the Schrödinger equation

20.2.2.1 Summing over paths

In our discussion of the two-slit experiment in Chapter 2 (Fig. 20.1) we saw that the amplitude for a particle to arrive at a point D on the second screen is the sum of the amplitudes to arrive at D after passing through either slit 1 or slit 2. The amplitude to go from a source S to D is then given by

$$\langle D \,|S\rangle = \langle D \,|1\rangle \langle 1 \,|S\rangle + \langle D \,|2\rangle \langle 2 \,|S\rangle \tag{20.1}$$

where $\langle j \,|i\rangle$ is the amplitude to go from point i to point j.

The probability of arriving at D is given by $P(D) = |\langle D \,|S\rangle|^2$, which results in the interference pattern.

If we now consider the case with two screens, each with several slits, we will have to take the amplitude for each possible path going through every possible combination of slits in the two screens. So if the first screen has n_1 slits and the second has n_2 slits, we would have for the amplitude

$$\langle D|\, s\rangle = \sum_{i=1}^{n_1}\sum_{j=1}^{n_2} \langle D|y_j^{(2)}\rangle\langle y_j^{(2)}|y_i^{(1)}\rangle\langle y_i^{(1)}|S\rangle$$

where $y_i^{(1)}$ is the location of the ith slit in screen 1, etc.

Extending the argument to more and more screens with more and more slits (Fig. 20.2), we can see that we eventually arrive at a description of an empty space and we are led to consider the amplitude to travel between two points in space as the sum of the amplitudes for all possible paths between the two points.[5]

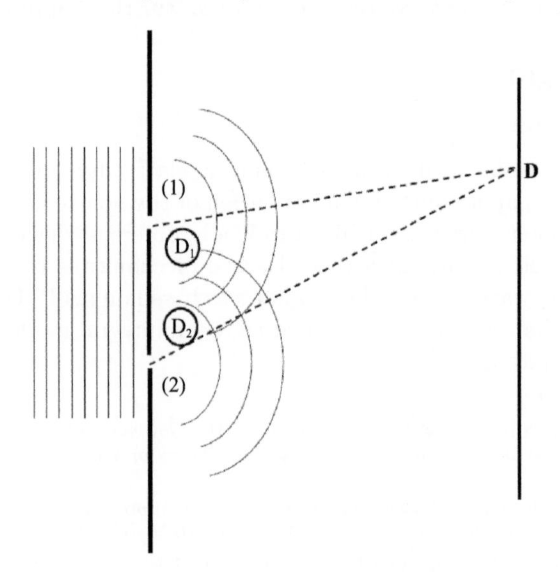

Fig. 20.1 Two-slit interference.

[5]Feynman, R.P. and Hibbs, A.R., *Quantum Mechanics and Path Integrals*, (Dover, 1965).

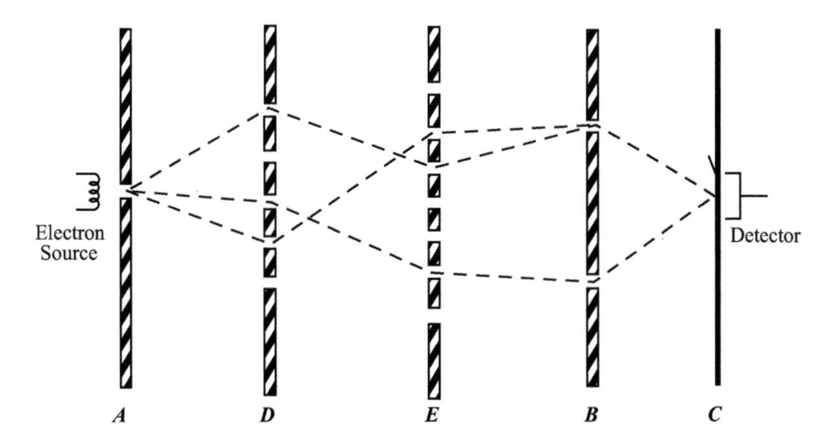

Fig. 20.2 Many screens with many holes.

Thus if the particle starts at x_o and arrives at x_1 at time t_1, x_2 at t_2, x_i at t_i, etc., ending at x_f at time t_f, where the difference between adjacent times is taken to be infinitesimal, the amplitude would be given by the sum of amplitudes over all possible paths leading between the initial and final points:

$$\langle x_f | \, x_o \rangle = \sum_{x_{f-1}} \cdots \sum_{x_i} \cdots \sum_{x_1} \langle x_f | \, x_{f-1} \rangle \cdots \langle x_{i+1} | \, x_i \rangle \cdots \langle x_1 | \, x_o \rangle$$

or

$$\langle x_f | \, x_o \rangle = B \int dx_{f-1} \cdots \int dx_i \cdots \int dx_1 \, \langle x_f | \, x_{f-1} \rangle \cdots \langle x_{i+1} | \, x_i \rangle \, \langle x_i | \, \dot{x}_{i-1} \rangle \cdots \langle x_1 | \, x_o \rangle \tag{20.2}$$

where B is a normalization factor which will be determined later.

20.2.2.2 The amplitude for a single path

We have seen in Chapter 7, in connection with Schrödinger's construction of the wave equation, how the formulation of classical mechanics in terms of the Hamilton-Jacobi equation leads to the expression of the solution in terms of the generating function $S(q, P) = \int L dt$ and that the surfaces of constant S propagate in space as the surfaces of constant phase for a wave (Section 7.2.3). In other words, classical mechanics could be regarded as the short wavelength (geometric optics) limit of a wave theory in which the phase of the wave is proportional to S. We previously found an expression for the phase velocity in terms of the energy and potential. Thus Feynman was led to propose that the amplitudes for all paths should have the same magnitude (=1) since all paths should be equally probable, but the amplitude for each path should have a phase proportional to S:

$$\langle x_j | \, x_i \rangle = e^{\frac{i}{\hbar} S} = e^{\frac{i}{\hbar} \int_{t_i}^{t_j} L dt} \tag{20.3}$$

Feynman was influenced by an earlier paper of Dirac[6] where Dirac stated that the lhs of (20.3) "corresponds to" the rhs. Feynman asked Dirac, when they met a conference, if "corresponds to" could mean "equals" and Dirac agreed that it could. Dirac extended the idea in a later work

It will be seen that the choice of $1/\hbar$ as the constant of proportionality, in addition to having the correct units to make the phase dimensionless, leads to results in agreement with other formulations of quantum mechanics and with experiment.

The constant $\hbar$ sets the scale at which quantum effects kick in. This can be seen from the fact that if S is much larger than $\hbar$, fractionally small changes in S will lead to rapid fluctuations in the phase for neighboring paths and the amplitudes for nearby paths will cancel. Thus the amplitude will only be significant for paths whose neighbors have the same S (within $\hbar$), i.e., in the neighborhood of the path whose S is stationary, $\delta S = 0$, and this, as we have seen, is the equation of the classical path. When $\hbar \ll S$ the classically allowed path will dominate the sum of amplitudes and the system will follow this path with very high probability. On the other hand, if $S \sim \hbar$ significantly different paths will have similar phases and all will contribute to the sum, leading to interference and other quantum effects.

20.2.2.3 Sum over paths (path integral)

Using the amplitude for a single path (20.3) we will calculate the amplitude for a particle to travel between two points, leaving x_o at time t_o and arriving at x_f at time t_f, using (20.2)

$$\begin{aligned}\langle x_f | \, x_o \rangle &= \sum_{\substack{\text{all}\\ \text{paths}}} e^{\frac{i}{\hbar}\int_{t_o}^{t_f} L dt} \\ &= B \int dx_{f-1} \cdots \int dx_i \cdots \int dx_1 e^{\frac{i}{\hbar}\int_{t_{f-1}}^{t_f} L dt} \cdots e^{\frac{i}{\hbar}\int_{t_i}^{t_{i+1}} L dt} \cdots \\ &\cdots e^{\frac{i}{\hbar}\int_{t_o}^{t_1} L dt}. \end{aligned} \tag{20.4}$$

In order to evaluate this we will take the time intervals to be of equal size, $(t_{i+1} - t_i) = \epsilon$, and will consider this as an infinitesimal quantity (the paths are determined by specifying the position at every time). In addition we will take the integrals in the exponents of (20.4) over the classical path between the two endpoints, that is, we take the minimum value of the integral over the infinitesimal path.

20.2.2.4 Particle moving in a potential V(x)—Feynman's derivation of the Schrödinger equation

To proceed further we turn our attention to a specific point on the path, say x_i. The sum over all paths which end at x_i is $\langle x_i | \, x_o \rangle$, the probability amplitude for the particle

[6]Dirac, P.A.M., *The Lagrangian in quantum mechanics.* Phys. Zeit. Sov. Union, 3, 64 (1933), received 9 November, 1932.

to be found at x_i at time t_i, given the fact that it started at x_o at time t_o. We write this amplitude as $\langle x_i | x_o \rangle = \psi(x_i)$. Then the amplitude to be at x_{i+1} is

$$\psi(x_{i+1}) = \frac{1}{A} \int dx_i \, e^{\frac{i}{\hbar} \int_{t_i}^{t_{i+1}} L dt} \psi(x_i). \tag{20.5}$$

Turning now to a particle moving under the influence of a potential $V(x)$, in one dimension for clarity, the Lagrangian is

$$L = \frac{m\dot{x}^2}{2} - V(x)$$

$$\int_{t_i}^{t_{i+1}=t_i+\epsilon} L \, dt = \frac{m}{2} \frac{(x_{i+1} - x_i)^2}{\epsilon} - V(x)\,\epsilon$$

so that we can write, using (20.5)

$$\begin{aligned} \psi(x_{i+1}, t_i + \epsilon) &= \frac{1}{A} \int dx_i \, e^{\frac{i}{\hbar} \int_{t_i}^{t_{i+1}} L(x_{i+1}, x_i) dt} \psi(x_i, t_i) \\ &= \frac{1}{A} e^{-\frac{i}{\hbar} V(x_{i+1})\epsilon} \int dx_i \, e^{\frac{i}{\hbar} \frac{m}{2} \frac{(x_{i+1}-x_i)^2}{\epsilon}} \psi(x_i, t_i) \end{aligned}$$

assuming V is not a function of the time. Now we substitute

$$x_{i+1} - x_i = \eta$$
$$dx_i = -d\eta$$

obtaining

$$\psi(x_{i+1}, t_i + \epsilon) = -\frac{e^{-\frac{i}{\hbar} V(x_{i+1})\epsilon}}{A} \int d\eta \, e^{\frac{i}{\hbar} \frac{m}{2} \frac{\eta^2}{\epsilon}} \psi(x_{i+1} - \eta, t_i). \tag{20.6}$$

Since ϵ is very small, only small values of η contribute to the integral since larger values of η correspond to very rapid oscillation of the exponent. Thus we can expand

$$\psi(x_{i+1} - \eta, t_i) = \psi(x_{i+1}, t_i) - \eta \left. \frac{\partial \psi}{\partial x} \right|_{x_{i+1}, t_i} + \frac{\eta^2}{2} \left. \frac{\partial^2 \psi}{\partial x^2} \right|_{x_{i+1}, t_i}$$

$$e^{-\frac{i}{\hbar} V(x_{i+1})\epsilon} = 1 - \frac{i}{\hbar} V(x_{i+1})\,\epsilon$$

using the Taylor series expansion. Substituting into (20.6), we obtain

$$\begin{aligned} \psi(x_{i+1}, t_i + \epsilon) &= -\frac{1}{A} \left[1 - \frac{i}{\hbar} V(x_{i+1})\,\epsilon \right] \int d\eta \, e^{\frac{i}{\hbar} \frac{m}{2} \frac{\eta^2}{\epsilon}} \times \ldots \\ &\cdots \times \left[\psi(x_{i+1}, t_i) - \eta \left. \frac{\partial \psi}{\partial x} \right|_{x_{i+1}, t_i} + \frac{\eta^2}{2} \left. \frac{\partial^2 \psi}{\partial x^2} \right|_{x_{i+1}, t_i} \right]. \end{aligned} \tag{20.7}$$

Now

$$\int_{-\infty}^{\infty} d\eta \, e^{\frac{i}{\hbar}\frac{m}{2}\frac{\eta^2}{\epsilon}} \eta = 0$$

by symmetry since the integrand is an odd function. We will use the fact that

$$\int_{-\infty}^{\infty} e^{-ax^2} dx = \frac{1}{\sqrt{a}}\sqrt{\pi} \qquad \int_{-\infty}^{\infty} x^2 e^{-ax^2} dx = \frac{1}{2\left(\sqrt{a}\right)^3}\sqrt{\pi}.$$

Setting

$$a = -\frac{i}{\hbar}\frac{m}{2\epsilon} = \frac{m}{i2\hbar\epsilon},$$

we obtain

$$\begin{aligned}\psi\left(x_{i+1}, t_i+\epsilon\right) = \psi\left(x_{i+1}, t_i\right) + \epsilon\frac{\partial\psi\left(x_{i+1}, t_i\right)}{\partial t} = \\ -\frac{1}{A}\left[1-\frac{i}{\hbar}V\left(x_{i+1}\right)\epsilon\right] \times \\ \times\left[\sqrt{\frac{2\pi i\hbar\epsilon}{m}}\psi\left(x_{i+1}, t_i\right) + \frac{\sqrt{\pi}}{4}\left(\frac{2i\hbar\epsilon}{im}\right)^{3/2}\left.\frac{\partial^2\psi}{\partial x^2}\right|_{x_{i+1}, t_i}\right].\end{aligned} \tag{20.8}$$

Considering the terms of lowest order in ϵ on both sides of the equation, we find

$$\psi\left(x_{i+1}, t_i\right) = -\frac{1}{A}\sqrt{\frac{2\pi i\hbar\epsilon}{m}}\psi\left(x_{i+1}, t_i\right)$$

which means that A must be equal to

$$A = -\sqrt{\frac{2\pi i\hbar\epsilon}{m}}$$

and

$$\epsilon\frac{\partial\psi\left(x_{i+1}, t_i\right)}{\partial t} = -\frac{i\epsilon}{\hbar}V\left(x_{i+1}\right)\psi\left(x_{i+1}, t_i\right) + \left(\frac{i\hbar\epsilon}{2m}\right)\left.\frac{\partial^2\psi}{\partial x^2}\right|_{x_{i+1}, t_i}$$

where we have neglected the term proportional to $\frac{i}{\hbar}V\left(x_{i+1}\right)\frac{\partial^2\psi}{\partial x^2}$ in (20.8) because it is of higher order in ϵ. Then

$$i\hbar\frac{\partial\psi\left(x_{i+1}, t_i\right)}{\partial t} = V\left(x_{i+1}\right)\psi\left(x_{i+1}, t_i\right) - \left(\frac{\hbar^2}{2m}\right)\left.\frac{\partial^2\psi}{\partial x^2}\right|_{x_{i+1}, t_i}$$

or

$$i\hbar\frac{\partial\psi\left(x, t\right)}{\partial t} = -\left(\frac{\hbar^2}{2m}\right)\frac{\partial^2\psi\left(x, t\right)}{\partial x^2} + V\left(x\right)\psi\left(x, t\right)$$

which is the Schrödinger equation for one dimension.

What we have seen is the surprising and interesting result that if the probability amplitude of finding a particle at a point is determined by summing the amplitude for every possible path arriving at that point, and the amplitude for the individual paths is proportional to $e^{\frac{i}{\hbar}\int L\,dt}$, the amplitude will develop in time according to the Schrödinger equation. In this way we again see that classical mechanics, which picks out those paths for which the action $S = \int L\,dt$ is a minimum, represents the limiting case of a wave theory. This is what we expected when we examined classical mechanics from the viewpoint of the Hamilton-Jacobi theory in Chapter 6 and Appendix A.

20.2.2.5 Path integral for a free particle: the Green's function or propagator

Now we will show that the path integral over finite path lengths leads to the Green's function for the Schrödinger equation.

For a free particle the Lagrangian is

$$L = \frac{mv^2}{2} = \frac{m}{2}\left(\frac{x_{i+1}-x_i}{\epsilon}\right)^2$$

so that

$$\int_{t_i}^{t_{i+1}} L\,dt = \int_{t_i}^{t_i+\epsilon} L\,dt = \frac{m}{2}\frac{(x_{i+1}-x_i)^2}{\epsilon}$$

and

$$\langle x_f|\, x_o\rangle = B\int dx_{f-1}\cdots\int dx_i\cdots\int dx_1 e^{\frac{i}{\hbar}\frac{m}{2}\frac{\left(x_f-x_{f-1}\right)^2}{\epsilon}}\cdots e^{\frac{i}{\hbar}\frac{m}{2}\frac{(x_{i+1}-x_i)^2}{\epsilon}}\times\cdots$$
$$\cdots\times e^{\frac{i}{\hbar}\frac{m}{2}\frac{(x_1-x_o)^2}{\epsilon}}.$$

Remembering that $\psi(x_i) = \langle x_i|\, x_o\rangle$, the amplitude $\langle x_{i+1}|\, x_o\rangle$ can be written as

$$\langle x_{i+1}|\, x_o\rangle = B\int dx_i\, e^{\frac{i}{\hbar}\frac{m}{2}\frac{(x_{i+1}-x_i)^2}{\epsilon}}\psi(x_i) \tag{20.9}$$

which can be rewritten as

$$\begin{aligned}\langle x_{i+1}|\, x_o\rangle &= B\int dx_i\, e^{\frac{i}{\hbar}\frac{m}{2}\frac{(x_{i+1}-x_i)^2}{\epsilon}}\int dx_{i-1}e^{\frac{i}{\hbar}\frac{m}{2}\frac{(x_i-x_{i-1})^2}{\epsilon}}\psi(x_{i-1})\\ &= B\int dx_i\int dx_{i-1}e^{\frac{i}{\hbar}\frac{m}{2}\frac{\left[(x_{i+1}-x_i)^2+(x_i-x_{i-1})^2\right]}{\epsilon}}\psi(x_{i-1}).\end{aligned}$$

The term in square brackets in the exponential can be expanded as

$$\begin{aligned}[\cdots] &= x_{i+1}^2 + 2x_i^2 + x_{i-1}^2 - 2x_i(x_{i+1}+x_{i-1})\\ &= x_{i+1}^2 + x_{i-1}^2 + 2(x_i - b/2)^2 - b^2/2\end{aligned}$$

where $b \equiv (x_{i+1} + x_{i-1})$. Then using

$$b^2/2 = \left(x_{i+1}^2 + x_{i-1}^2 + 2x_{i+1}x_{i-1}\right)/2$$
$$x_{i+1}^2 + x_{i-1}^2 - b^2/2 = \left(x_{i+1}^2 + x_{i-1}^2\right)/2 - x_{i+1}x_{i-1}$$
$$= \frac{1}{2}\left(x_{i+1} - x_{i-1}\right)^2$$

we have

$$\psi(x_{i+1}) = B \int dx_{i-1} e^{\frac{i}{\hbar}\frac{m}{2}\frac{(x_{i+1}-x_{i-1})^2}{2\epsilon}} \psi(x_{i-1}) \int_{-\infty}^{\infty} dx_i\, e^{\frac{i}{\hbar}m\frac{(x_i-b/2)^2}{\epsilon}}$$
$$= BK \int dx_{i-1} e^{\frac{i}{\hbar}\frac{m}{2}\frac{(x_{i+1}-x_{i-1})^2}{2\epsilon}} \psi(x_{i-1}) \tag{20.10}$$

where we have noted that $\int_{-\infty}^{\infty} dx_i e^{\frac{i}{\hbar}m\frac{(x_i-b/2)^2}{\epsilon}}$ is independent of b and hence is equal to some constant K. We see that the integral in (20.10) has the square of the distance taken by *two* steps (from x_{i-1} to x_{i+1}) in the exponent and the time in the denominator is the time for two steps, 2ϵ. By repeating this procedure, we find the amplitude at a given time in terms of the amplitude at a time earlier by a macroscopic amount:

$$\psi(x,t) = \frac{1}{A} \int_{-\infty}^{t} dt' \int dx' \left[e^{\frac{i}{\hbar}\frac{m}{2}\frac{(x-x')^2}{(t-t')}} \right] \psi(x',t') \tag{20.11}$$

where we have rewritten the normalization factor. This is the amplitude to find the particle at x at time t in terms of the amplitude $\psi(x',t')$ that the particle was at x' at an earlier time t'. The factor $\left[e^{\frac{i}{\hbar}\frac{m}{2}\frac{(x-x')^2}{(t-t')}} \right]$ is the Green's function $G(x-x', t-t')$ for the Schrödinger equation for a free particle. It is possible to verify by direct substitution that $\psi(x,t)$ as given by (20.11) is a solution of the Schrödinger equation (with $A = (2\pi i\hbar t/m)^{1/2}$ in one dimension). When normalized as

$$G(r,t) = \left(\frac{m}{2\pi i\hbar t}\right)^{3/2} e^{\frac{i}{\hbar}\frac{m}{2}\frac{r^2}{t}}, \tag{20.12}$$

appropriate for three dimensions, $G(r,t)$ satisfies the equation

$$i\hbar\frac{\partial G}{\partial t} + \frac{\hbar^2}{2m}\nabla^2 G = i\hbar\delta^{(3)}(\vec{r})\,\delta(t). \tag{20.13}$$

Now consider the variation of $G(r,t)$ in the neighborhood of a relatively large distance and time delay $\left(\vec{R}, T\right)$. Neglecting the small change in the prefactor, we have

$$G(r,t) \approx \exp\left[\frac{i}{\hbar}\frac{m}{2}\frac{\left|\vec{R}+\vec{\delta}\right|^2}{T+\tau}\right] \approx \exp\left[\frac{i}{\hbar}\frac{m}{2}\frac{\left|\vec{R}\right|^2}{T}\right] \exp\frac{i}{\hbar}m\left[\frac{\vec{\delta}\cdot\vec{R}}{T} - \frac{1}{2}\frac{\left|\vec{R}\right|^2}{T^2}\tau\right]$$
$$\sim \exp\left[i\left(\vec{k}\cdot\vec{\delta} - \omega\tau\right)\right] \tag{20.14}$$

with $\vec{k} = \left(\frac{m}{\hbar}\right)\frac{\vec{R}}{T}$ and $\omega = \frac{m}{\hbar}\frac{|\vec{R}|^2}{2T^2}$, so that at large distances and times from the source event, the Green's function behaves like a plane wave.

20.3 An introduction to scattering of nonrelativistic particles by a many-body system

20.3.1 A heuristic look at scattering

A nonrelativistic scattering event can be described by saying that an incident monoenergetic beam of particles, represented by a plane wave $e^{i\left(\vec{k}_1\cdot\vec{r}-\omega_{k_1}t\right)}$, will produce an outgoing wave $b\left(\theta\right)e^{i\left(k_2 r-\omega_{k_2}t\right)}/r$ as a result of scattering from a particle located at $r_s = 0$ at time $t_s = 0$. $b\left(\theta\right) = b$ is the scattering amplitude, which we take as constant for simplicity. If our scattering system consists of particles distributed with a density $\rho\left(r_s, t_s\right)$, so that there are many particles in a volume $d^3 r_s$ of dimensions small compared to the wavelength of the incident particles, the total scattering amplitude will be

$$\psi_{sc}\left(\vec{r},t\right) = \int d^3 r_s\, dt_s\, b\,\rho\left(\vec{r}_s,t_s\right)e^{i\left(\vec{k}_1\cdot\vec{r}_s-\omega_{k_1}t_s\right)}\frac{e^{i\left(k_2|\vec{r}-\vec{r}_s|-\omega_{k_2}(t-t_s)\right)}}{|\vec{r}-\vec{r}_s|} \tag{20.15}$$

In the usual case where $\vec{r} \gg \vec{r}_s$, we can expand

$$|\vec{r}-\vec{r}_s| \approx r - \frac{\vec{r}\cdot\vec{r}_s}{r}, \tag{20.16}$$

so that

$$\psi_{sc}\left(\vec{r},t\right) = b\frac{e^{i\left(k_2 r-\omega_{k_2}t\right)}}{r}\int d^3 r_s\, dt_s\, b\,\rho\left(\vec{r}_s,t_s\right)e^{i\left(\vec{q}\cdot\vec{r}_s-\omega t_s\right)} \tag{20.17}$$

where $\vec{k}_2 = k_2\vec{r}/r$, $\vec{q} = \vec{k}_1 - \vec{k}_2$ and $\omega = \omega_{k_1} - \omega_{k_2}$, that is, an outgoing spherical wave expanding with time. The probability of detecting a scattered particle, and hence the scattering cross section, is proportional to

$$\begin{aligned} r^2\left|\psi_{sc}\left(\vec{r}_d\right)\right|^2 &= b^2\int d^3 r_s\, dt_s \rho\left(\vec{r}_s,t_s\right)e^{i\left(\vec{q}\cdot\vec{r}_s-\omega t_s\right)}\int d^3 r'_s\, dt'_s \rho\left(\vec{r}'_s,t'_s\right)e^{-i\left(\vec{q}\cdot\vec{r}'_s-\omega t'_s\right)} \\ &= b^2\int d^3 r'_s\, dt'_s\int d^3 r_s\, dt_s\left\langle\rho\left(\vec{r}_s,t_s\right)\rho\left(\vec{r}'_s,t'_s\right)\right\rangle e^{i\left(\vec{q}\cdot\left(\vec{r}_s-\vec{r}'_s\right)-\omega\left(t_s-t'_s\right)\right)}. \end{aligned} \tag{20.18}$$

We make the substitutions $\vec{r}'_s = \vec{r}_s - \vec{\delta}$, $t'_s = t_s - \tau$. $\langle\ldots\rangle$ represents the ensemble average, which for stationary systems is a function only of $\vec{\delta},\tau$: $\left\langle\rho\left(\vec{r}_s,t_s\right)\rho\left(\vec{r}'_s,t'_s\right)\right\rangle = \left\langle\rho\left(\vec{r}_s,t_s\right)\rho(\vec{r}_s-\vec{\delta},t_s-\tau)\right\rangle = G(\vec{\delta},\tau)$. Thus

$$r^2\left|\psi_{sc}\left(\vec{r}_d\right)\right| = b^2\int d^3 r_s\, dt_s\int d^3\delta\, d\tau\, e^{i\left(\vec{q}\cdot\vec{\delta}-\omega\tau\right)}G(\vec{\delta},\tau), \tag{20.19}$$

that is, the Fourier transform of the density-density correlation function. We will see below that this is the van Hove result.

20.3.2 Van Hove's treatment of the scattering problem

A huge body of work on scattering is based on the work of van Hove,[7] who first showed that in the Born approximation the scattering cross section as a function of momentum transfer ($\hbar q$) and energy transfer ($\hbar\omega$) is given by the space and time Fourier transforms of the density-density correlation function. This correlation function is defined to be the probability of finding a particle within a volume d^3r around $\vec{r}$ at time t if there had been a particle in a similar volume located at the origin ($\vec{r}=0$) at time $t=0$.

We start with the result of first order time-dependent perturbation theory, given in Eq. (8.64):

$$i\hbar\dot{c}_k(t) = V_{kn}e^{i(E_k-E_n)t/\hbar}, \tag{20.20}$$

which is valid when the system is in the state $|n\rangle$ at $t=0$ ($c_k(0)=\delta_{kn}$). Here the states refer to the states of the whole system (scatterer plus particle),

$$|n\rangle = e^{i\vec{k}_n\cdot\vec{r}}\,|n\rangle_s\,, \tag{20.21}$$

where $|n\rangle_s$ are the states of the scatterer and $E_n = E_n^{(s)} + E_n^{(p)}$ with $E_n^{(s,p)}$ the energies of the scatterer and the scattering particle in the state $|n\rangle$. The matrix element is

$$V_{kn} = \sum_j {}_s\langle k|\,V(\vec{r}-\vec{r}_j)\,|n\rangle_s\,e^{i\vec{q}\cdot\vec{r}} = V(\vec{q})\;{}_s\langle k|\sum_j e^{i\vec{q}\cdot\vec{r}_j}\,|n\rangle_s \tag{20.22}$$

with $\vec{q}=\vec{k}_n-\vec{k}_k$. $\vec{r}_j$ is the position of the jth scatterer in the system, and $V(\vec{q}) = \int_s V(\vec{r})\,e^{i\vec{q}\cdot\vec{r}}d^3r$. The last step results from a change of variables ($\vec{r}-\vec{r}_j\to\vec{r}$) and the assumption that the scatterer is homogeneous so that V is a function only of the separation $\vec{r}$. Then (dropping the subscript s on the states of the scattering system) we find

$$i\hbar\dot{c}_k(t) = V(\vec{q})\,\langle k|\sum_j e^{i\frac{E_k^{(s)}}{\hbar}t}e^{i\vec{q}\cdot\vec{r}_j}e^{-i\frac{E_n^{(s)}}{\hbar}t}\,|n\rangle\,e^{-i\omega t} \tag{20.23}$$

$$= V(\vec{q})\,\langle k|\sum_j e^{i\vec{q}\cdot\vec{r}_j(t)}\,|n\rangle\,e^{-i\omega t} \tag{20.24}$$

where $\hbar\omega = E_n^{(p)} - E_k^{(p)}$ and $\vec{r}_j(t)$ is the Heisenberg operator, $\vec{r}_j(t) = e^{iHt/\hbar}\vec{r}_je^{-iHt/\hbar} = \vec{r}_j\exp\left(i\frac{E_k^{(s)}-E_n^{(s)}}{\hbar}t\right)$. Then

$$i\hbar c_k(t) = V(\vec{q})\int^t dt'\sum_j\langle k|\,e^{i\vec{q}\cdot\vec{r}_j(t')}\,|n\rangle\,e^{-i\omega t'} \tag{20.25}$$

and the scattering cross section is proportional to $W = d/dt\left(\sum_k|c_k|^2\right) = \sum_k c_k\dot{c}_k^* + \dot{c}_kc_k^*$. We will only work out the first term $c_k\dot{c}_k^*$; the second term can be obtained from this one by interchanging t and t'. The first term is

[7] Van Hove, L., *Correlations in space and time and Born approximation scattering in systems of interacting particles*, Phys. Rev. **95**, 249 (1954), received 16 March 1954.

$$
\begin{aligned}
W(\vec{q},\omega) &= \frac{V^*(\vec{q})}{-i\hbar}\frac{V(\vec{q})}{i\hbar}\int^t dt' \sum_k \sum_{i,j} \langle k| e^{i\vec{q}\cdot\vec{r}_j(t')} |n\rangle e^{-i\omega t'} \langle n| e^{-i\vec{q}\cdot\vec{r}_i(t)} |k\rangle e^{i\omega t} \\
&= \frac{|V(\vec{q})|^2}{\hbar^2}\int^t dt' \sum_{i,j}\sum_k \langle n| e^{-i\vec{q}\cdot\vec{r}_i(t)} |k\rangle \langle k| e^{i\vec{q}\cdot\vec{r}_j(t')} |n\rangle e^{i\omega(t-t')} \\
&= \frac{|V(\vec{q})|^2}{\hbar^2}\sum_{i,j}\int^t dt' \langle n| e^{i\vec{q}\cdot[\vec{r}_j(t')-\vec{r}_i(t)]} |n\rangle e^{i\omega(t-t')}.
\end{aligned} \tag{20.26}
$$

We can extend the integration over time from $t' = -\infty$ to t. In the case of a stationary system, the integrand will only be a function of $(t'-t) = \tau$, so we can set $t = 0$ and $t' = \tau$, giving

$$
W(\vec{q},\omega) = \frac{|V(\vec{q})|^2}{\hbar^2}\int_{-\infty}^{0} d\tau e^{-i\omega\tau} \sum_{i,j} \langle n| e^{i\vec{q}\cdot\vec{r}_j(\tau)} e^{-i\vec{q}\cdot\vec{r}_i(0)} |n\rangle \tag{20.27}
$$

for the first term. Now we replace each of the sums $\sum_{i,j}$ by $\int d^3r \rho(\vec{r},t)$ where $\rho(\vec{r},t) = \sum_i \delta(\vec{r} - \vec{r}_i(t))$. The second (complex conjugate) term is obtained by interchanging $t \Leftrightarrow t'$, the integral goes from $\tau = 0 \to \infty$ so the sum of the two terms is:

$$
W(\vec{q},\omega) = \frac{|V(\vec{q})|^2}{\hbar^2}\int_{-\infty}^{\infty} d\tau e^{-i\omega\tau} \int d^3r \int d^3r' \langle n| \rho(\vec{r},\tau)\rho(\vec{r}',0) |n\rangle e^{i\vec{q}\cdot(\vec{r}-\vec{r}')}. \tag{20.28}
$$

For a system in thermodynamic equilibrium, we have to take an ensemble average of the initial state expectation value in (20.28). For homogeneous systems this is a function of $\vec{R} = (\vec{r} - \vec{r}')$ only. Then

$$
W(\vec{q},\omega) = \frac{|V(\vec{q})|^2}{\hbar^2}\int_{-\infty}^{\infty} d\tau e^{-i\omega\tau} \int d^3R\, e^{i\vec{q}\cdot\vec{R}} \int d^3r' \langle n| \rho(\vec{r}'+\vec{R},\tau)\rho(\vec{r}',0) |n\rangle \tag{20.29}
$$

The result is seen to be $|V(\vec{q})|^2 S(\vec{q},\omega)$ where $S(\vec{q},\omega)$ is the real part of the space -time Fourier transform of

$$
G(\vec{r},\tau) = \int d^3r \langle n| \rho(\vec{r}+\vec{R},\tau)\rho(\vec{r},0) |n\rangle, \tag{20.30}
$$

that is, the density-density correlation function, or the probability of finding a particle at position $\vec{r}+\vec{R}$ at time τ, given that there was a particle at $\vec{r}$ at time $\tau = 0$.

20.3.3 Space-time approach to scattering[8]

So far we have described the usual treatment of scattering from many-particle systems, which assumes the initial beam to be represented by a plane wave and shows

[8]See Gähler, R., Felber, J., Mezei, F. and Golub, R. et al, *Space-time approach to scattering from many-body systems*, Phys. Rev. A 58, 280, (1998) received 19 September 1997, for a more detailed treatment.

that the scattering cross section is proportional to the Fourier transform of the correlation function of the density fluctuations, which is often called the van Hove time-dependent correlation function. The usual treatment then integrates the cross section for monochromatic plane waves over the spread of momentum in the incident and scattered beams, i.e., the resolution of the apparatus.

In this section we will show that another approach is both possible and interesting. As we do the entire calculation in real space-time, our approach sheds new light on the emergence of the Fourier transform in the scattering cross section. A crucial feature of our approach is that the influence of beam preparation on the resolution of the measurement is determined by the autocorrelation function of the beam as it propagates through the apparatus. This function is determined by the optical properties of the beam preparation elements, e.g., slits and choppers.

We begin by noting that the Schrödinger equation

$$i\hbar\frac{\partial\psi\left(\vec{r},t\right)}{\partial t}+\frac{\hbar^2}{2m}\nabla^2\psi\left(\vec{r},t\right)=V\psi\left(\vec{r},t\right) \tag{20.31}$$

can be written as an integral (Lippmann-Schwinge equation[9])

$$\psi\left(\vec{r},t\right)=\psi_o\left(\vec{r},t\right)+\frac{1}{i\hbar}\int d^3r'dt'G_o\left(\vec{r}-\vec{r}',t-t'\right)V\left(\vec{r}',t'\right)\psi\left(\vec{r}',t'\right), \tag{20.32}$$

as follows from Eq. (20.13) when $\psi_o\left(\vec{r},t\right)$ satisfies $i\hbar\frac{\partial\psi_o}{\partial t}+\frac{\hbar^2}{2m}\nabla^2\psi_o=0$, $G_o\left(\vec{r},t\right)$ is the Green's function for the unperturbed problem ($V=0$), and $V\left(\vec{r},t\right)$ is the interaction potential. The integration is carried out over the sample volume and the time interval during which the sample is exposed to the beam. Eq. (20.32) is exact. The Born approximation consists of replacing $\psi\left(\vec{r}',t'\right)$ in the integrand by the incident unscattered wave $\psi_{in}\left(\vec{r}',t'\right)$ at the position of the scatterer:

$$\begin{aligned}\psi\left(\vec{r},t\right)&=\psi_o\left(\vec{r},t\right)+\frac{1}{i\hbar}\int d^3r'dt'G_o\left(\vec{r}-\vec{r}',t-t'\right)V\left(\vec{r}',t'\right)\psi_{in}\left(\vec{r}',t'\right)\\&=\psi_o\left(\vec{r},t\right)+\psi_{sc}\left(\vec{r},t\right).\end{aligned} \tag{20.33}$$

Eq. (20.33) describes the scattered wave as a superposition of waves produced by many scattering events occurring at the different space-time points $(\vec{r}',t')$. This is in contrast to the van Hove treatment described above, which begins with time-dependent perturbation theory.

We write $\vec{r_s}-\vec{r_s}'=\vec{\delta}$, $t_s-t_s'=\tau$. Using (20.33), the intensity at the detector at $(\vec{r_d},t_d)$ is

$$\begin{aligned}\left|\psi_{sc}\left(\vec{r_d},t_d\right)\right|^2&=\int d^3r_s\,dt_s\int d^3r_s'\,dt_s'\,G_o\left(\vec{r_d}-\vec{r_s},t_d-t_s\right)G_o^*\left(\vec{r_d}-\vec{r_s}',t_d-t_s'\right)\times\\&\quad\times V\left(\vec{r_s},t_s\right)V^*\left(\vec{r_s'},t_s'\right)\psi_{in}\left(\vec{r_s},t_s\right)\psi_{in}^*\left(\vec{r_s'},t_s'\right)\\&=\int d^3r_s\,dt_s\int d^3\delta\,d\tau\,G_o\left(\vec{r_d}-\vec{r_s},t_d-t_s\right)G_o^*\left(\vec{r_d}-\vec{r_s}+\vec{\delta},t_d-t_s+\tau\right)\\&\quad\times V\left(\vec{r_s},t_s\right)V^*\left(\vec{r_s}-\vec{\delta},t_s-\tau\right)\psi_{in}\left(\vec{r_s},t_s\right)\psi_{in}^*\left(\vec{r_s}-\vec{\delta},t_s-\tau\right).\end{aligned} \tag{20.34}$$

[9]Lippmann, B. A. and Schwinger, J., *Variational Principles for Scattering Processes. I&II*, Phys. Rev. **79**, 469 & 481 (1950) received 10 April, 1950.

(We have dropped the factor of $1/\hbar^2$ and will continue to neglect constant factors in the following sections.[10]) In the case of a short-range interaction, the potential $V(\vec{r_s}, t_s)$ will be proportional to the density of scatterers at $(\vec{r_s}, t_s)$, $V(\vec{r_s}, t_s) = C\rho(\vec{r_s}, t_s)$. Thus we can write

$$|\psi_{sc}(\vec{r_d}, t_d)|^2 = C^2 \int d^3 r_s\, dt_s \int d^3\delta\, d\tau\, G_o(\vec{r_d} - \vec{r_s}, t_d - t_s) \\ G_o^*\left(\vec{r_d} - \vec{r_s} + \vec{\delta}, t_d - t_s + \tau\right) \times G_s(\vec{\delta}, \tau) R_{in}(\vec{\delta}, \tau). \tag{20.35}$$

Here

$$G_s(\vec{\delta}, \tau) = \left\langle \rho(\vec{r_s}, t_s)\, \rho(\vec{r_s} - \vec{\delta}, t_s - \tau) \right\rangle_s \tag{20.36}$$

is the (van Hove) density-density correlation function, and it is independent of $\vec{r_s}$, t_s for the usual case of a homogeneous, stationary system. The result depends on the pair correlation function (20.36) due to the fact that the intensity is proportional to the square of the amplitude, as can readily be seen in the above derivation. Because we restrict ourselves to the Born approximation, the result is independent of higher order correlation functions.

$$R_{in}(\vec{\delta}, \tau) = \left\langle \psi_{in}(\vec{r_s}, t_s)\, \psi_{in}^*(\vec{r_s} - \vec{\delta}, t_s - \tau) \right\rangle_{in} \tag{20.37}$$

is the autocorrelation function of the incident beam. The brackets $\langle\rangle_s$ and $\langle\rangle_{in}$ indicate statistical averages over the sample and incoming beam ensembles, respectively. We will see that $R_{in}(\vec{\delta}, \tau)$ is responsible for the contribution of the incident beam to the resolution of the measurement: according to (20.35), the intensity only depends on the correlation function G_s at values of $(\vec{\delta}, \tau)$ where R_{in} is significant.

To calculate the wave function incident on the scattering sample, we consider the beam incident on the aparatus as a completely incoherent wave $\psi_o(\vec{r}_o, t_o)$, i.e.,

$$\langle \psi_o(\vec{r_o}, t_o)\, \psi_o^*(\vec{r_o}', t_o') \rangle_{in} \propto \delta^{(3)}(\vec{r_o} - \vec{r_o}')\, \delta(t_o - t_o') \tag{20.38}$$

entering the apparatus through a slit and/or chopper. The solution for the wave problem with the specified boundary conditions is given by

$$\psi_{in}(\vec{r}, t) = \frac{-1}{4\pi} \int_0^{t_+} dt_o \int d^2 S_o\, \psi_o(\vec{r}_o, t_o)\, \vec{n} \cdot \vec{\nabla}_o \widetilde{G}(\vec{r} - \vec{r}_o, t - t_o) \tag{20.39}$$

where the integral over $d^2 S_o$ is taken over the plane of the entrance slit located at $z_o = 0$, $\vec{n}$ is the normal to this plane, $\vec{\nabla}_o$ is the gradient with respect to $\vec{r}_o$, and ψ_o is nonzero only in the entrance slit $(-a < y_o < a)$ and for a time interval determined by the input chopper (when applicable). (We make the Kirchhoff approximation as is usual in optics.) In some cases the integral over dt_o will be limited by the opening time

[10] As we will see, the important effects arise from the interference between waves that take different paths through the apparatus, so we will be focusing on the phase factors.

of a chopper. The function $\widetilde{G}$ satisfies the boundary condition $\widetilde{G}(z_o = 0) = 0$ and can be taken as

$$\widetilde{G}(\vec{r} - \vec{r}_o, t) = g(\vec{r}_+, t) - g(\vec{r}_-, t) \tag{20.40}$$

where

$$(\vec{r}_\pm)^2 = (x - x_o^2) + (y - y_o)^2 + (z \mp z_o)^2 \tag{20.41}$$

and $g(\vec{r}, t)$ satisfies

$$\nabla^2 g - \frac{2m}{i\hbar}\frac{\partial g}{\partial t} = -4\pi\delta^{(3)}(\vec{r})\,\delta(t). \tag{20.42}$$

$g(\vec{r}, t)$ is then normalized as[11]

$$g(\vec{r}, t) = \left(\frac{2\pi i\hbar}{m}\right)\left(\frac{m}{2\pi i\hbar t}\right)^{3/2} e^{i\frac{mr^2}{2\hbar t}}\,u(t) \tag{20.43}$$

where $u(t)$ is the unit step function at $t = 0$. In what follows we replace it with the condition $t > 0$. Eq. (20.39) becomes

$$\psi_{in}(\vec{r}, t) = z\left(\frac{m}{2\pi i\hbar}\right)^{3/2}\int_0^{t+}\frac{dt_o}{(t - t_o)^{5/2}}\int d^2S_o\,\underline{G}(\vec{r} - \vec{r}_o, t - t_o)\,\psi_o(\vec{r}_o, t_o) \tag{20.44}$$

where $\underline{G}(\vec{r}, t) = \exp\left(i\frac{mr^2}{2\hbar t}\right)$ is the Green's function without the prefactor. We then find

$$R_{in}(\vec{\delta}, \tau, \vec{r_1}, t_1) = \frac{z_1^2}{\underline{t_1^5}}\left(\frac{m}{2\pi\hbar}\right)^3\int d^2r_o\,dt_o\,\underline{G}(\vec{r_1}, t_1)\,\underline{G}^*(\vec{r_1} - \vec{\delta}, t_1 - \tau) \tag{20.45}$$

where $\vec{r_1} = \vec{r_s} - \vec{r_o}$, $t_1 = t_s - t_o$; $\underline{t_1}$ is the average flight time between the entrance and the scatterer. We have neglected the small variation in $(t_s - t_o)^{-5/2}$ over the interval of length $2T$ so that we can replace it by $\underline{t_1^{-5/2}}$ and take it out of the integral. We have made the assumption that the incident beam is completely uncorrelated, i.e.,

$$\langle\psi_o(\vec{r}_o, t_o)\,\psi_o^*(\vec{r}'_o, t'_o)\rangle_{in} = \delta^3(\vec{r}_o - \vec{r}'_o)\,\delta(t_o - t'_o). \tag{20.46}$$

By the Wiener-Khintchine theorem (which states that the Fourier transform of an autocorrelation function is the power spectrum of the function), we see this is equivalent to $\left|A(\vec{k})\right|^2 = \text{const}$, i.e., we are assuming that the incident spectrum is broad compared to that selected by the instrument.

[11] Morse, P. M. and Feshbach, H., *Methods of Theoretical Physics* (New York: McGraw-Hill, 1953), Vol. 1, Chapter 7.

In a scattering experiment we measure the integral of $|\psi_{sc}(\vec{r_d}, t_d)|^2$ over the entrance area of the detector and, in the case of a time-of-flight experiment, a definite time interval dt_d. From (20.35) the signal at the detector is given by

$$\int d^2 r_d \, dt_d \, |\psi_{sc}(\vec{r_d}, t_d)|^2 = \int d^3 r_s \, dt_s \int d^3\delta \, d\tau R_{out}(\vec{\delta}, \tau, \vec{r_2}, t_2) G_s(\vec{\delta}, \tau) R_{in}(\vec{\delta}, \tau, \vec{r_1}, t_1) \tag{20.47}$$

where

$$R_{out}(\vec{\delta}, \tau, \vec{r_2}, t_2) = \int d^2 r_d \, dt_d \, G_o(\vec{r_2}, t_2) \, G_o^*(\vec{r_2} + \vec{\delta}, t_2 + \tau) \tag{20.48}$$

and $\vec{r_2} = \vec{r_d} - \vec{r_s}$, $t_2 = t_d - t_s$. Although (20.47) contains an explicit dependence on only space and time variables, the dependence on the energy and momentum transfer will be seen to emerge from the variation of the phases of the $\underline{G}$ functions.

Now we see there is a symmetry between the integral over the detector parameters $(\vec{r_d}, t_d)$ of the product of Green's functions depending on $(\vec{r_2}, t_2)$, $R_{out}(\vec{\delta}, \tau)$ (Eq. 20.48), and the integral over the input slit $(\vec{r_o}, t_o)$ of the Green's function product depending on $(\vec{r_1}, t_1)$, $R_{in}(\vec{\delta}, \tau)$ (Eq. 20.45), so that we only have to calculate one of the pairs, say the latter. We then obtain the former integral by applying the symmetry rules

$$\vec{\delta} \to -\vec{\delta}, \; \tau \to -\tau, \; t_2 \to t_1, \; \vec{r_2} \to \vec{r_1}, \; \vec{r_s} \to -\vec{r_s}, \; t_s \to -t_s. \tag{20.49}$$

While functions such as $R_{in,out}(\vec{\delta}, \tau)$ are usually called "correlation functions," they really represent a loss of correlation as their arguments increase. In the present context it is instructive to keep this in mind by referring to them as "(loss of) correlation functions." Thus while $R_{in}\left(\vec{\delta}, \tau, \vec{r_1}, t_1\right)$ represents the loss of correlation between pairs of rays starting at the entrance slit $(\vec{r_o})$ and propagating to neighboring space-time points in the sample $(\vec{r_s}, t_s)$, $(\vec{r_s}', t_s')$, $R_{out}\left(\vec{\delta}, \tau, r_2, t_2\right)$ represents the loss of correlation due to the difference in path lengths in propagating from two neighboring points in the sample $(\vec{r_s}, t_s)$, $(\vec{r_s}', t_s')$ to the detector $(\vec{r_d}, t_d)$. $G_s\left(\vec{\delta}, \tau\right)$ is the loss of correlation between the waves scattered at the points $(\vec{r_s}, t_s)$, $(\vec{r_s}', t_s')$ due to the internal dynamics of the scattering system (see Fig. 20.3). Implicit in the above discussion is the fact that the detector responds to the value of $|\psi_{sc}|^2$ at a single point integrated over the detector area.

We see that the scattering process is intrinsically an interference phenomenon, with the cross section being a measure of the correlation between waves following paths separated by $(\vec{\delta}, \tau)$ at the sample.

In the interests of simplicity we have chosen to work with a real "hydrodynamic" density function $\rho(\vec{r}, t)$. Replacement of this by $\sum_i \delta(\vec{r} - \vec{r_i}(t))$ yields results consistent with the usual van Hove approach.

The result (20.47) represents a kind of path integral for intensity, as indicated in Fig. 20.3. We see that the intensity at the detector is the result of an interference pattern between every possible pair of paths through the apparatus.

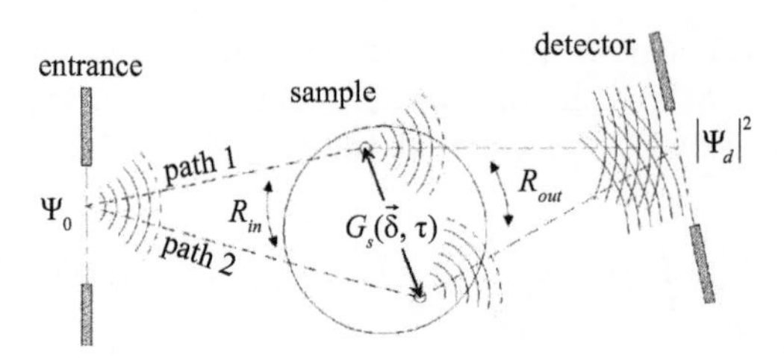

Fig. 20.3 Illustrating Eq. (20.47). The functions R_{in}, R_{out} represent the effects of the phase differences between the paths 1 and 2 on the correlation functions of the beam. G_s is the loss of correlation due to the density fluctuations at the different scattering points and $|\psi_d|^2$ is the sum of the contributions from paths going through all possible pairs of points in the sample. Only pairs whose separation lies within the correlation volumes contribute.

20.3.4 Scattering from a time-dependent system

We begin with a simple example to show how the ideas presented in the previous section can be applied to scattering from systems fluctuating in time. In scattering from such systems, the energy of the scattering radiation is changed (inelastic or quasi-elastic scattering). For simplicity, in this section we will neglect the position dependence of the scattering, considering the scattering system to be concentrated at a single point $\vec{r_s}$ and likewise for the source (detector) at $\vec{r_o} = 0$ $(\vec{r_d})$. We will restrict ourselves to massive particle scattering where the energy is defined by choppers.[12] This is equivalent to calculating the scattering integrated over all values of $\vec{q}$. We will treat the case of time- and position-dependent scattering in the next section.

Under these restrictions, Eq. (20.33) for the wave function at the detector located at $\vec{r_d}$ at time t_d can be written as

$$\psi_{sc}(\vec{r_d}, t_d) = \int dt_s G_o(\vec{r_d} - \vec{r_s}, t_d - t_s) V(\vec{r_s}, t_s) \psi_{in}(\vec{r_s}, t_s) \tag{20.50}$$

where

$$G_o(\vec{r_d} - \vec{r_s}, t_d - t_s) = \left(\frac{m}{2\pi i\hbar(t_d - t_s)}\right)^{3/2} e^{i\frac{m|\vec{r_d} - \vec{r_s}|^2}{2\hbar(t_d - t_s)}} \tag{20.51}$$

is the Green's function for the time-dependent unperturbed Schrödinger equation and satisfies

$$i\hbar\frac{\partial G_o}{\partial t} + \frac{\hbar^2}{2m}\nabla^2 G_o = i\hbar\,\delta^{(3)}(\vec{r})\,\delta(t). \tag{20.52}$$

The beam will be taken to be incident on the scatterer through a slit (chopper) located at the origin that opens during the time interval $-T \le t_o \le T$. (See Fig. 20.4 for a definition of the notation used here.)

We consider the wave function incident on the chopper to be completely uncorrelated for different values of t_o. That is, we assume that the correlation time of the

[12]The use of crystals as energy-defining elements can be discussed according to the present viewpoint. See Felber, J. Gähler, R., Golub, R. and. Prechtel, K., *Coherence volumes and neutron scattering*, Physica B **252**, 34 (1998).

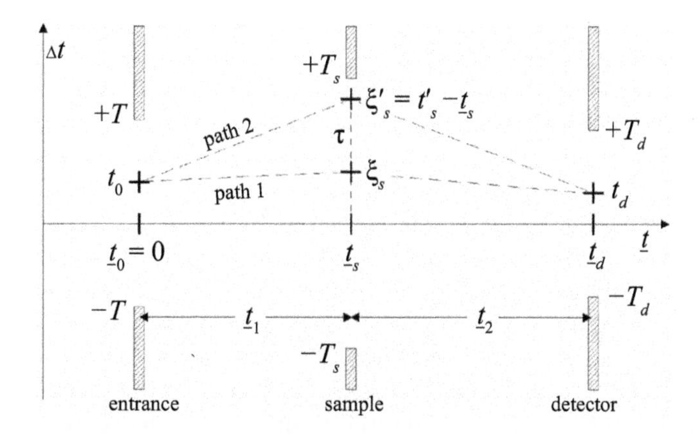

Fig. 20.4 Scattering from a time-dependent system. The scatterer is assumed to be concentrated at a single point ($\vec{r}_s$). The resolution is determined by the phase difference between paths (1) and (2). τ is the delay between the two scattering events at t_s, t'_s. ξ_s, ξ'_s are the delays relative to the nominal scattering time, $\underline{t_s}$.

beam incident on the chopper is much shorter than the correlation time that will be imposed on the beam by the action of the chopper, or, in other words, the beam incident on the chopper contains a much broader energy spectrum than will be selected by the chopper system.

The wave function leaving the chopper and arriving at a point $\vec{r_s}$ at time t_s is given by (20.51, 20.44)

$$\psi_{in}(\vec{r}_s, t_s) = A \int_0^{t_+} \frac{dt_o}{(t_s - t_o)^{5/2}} e^{i\frac{mr_s^2}{2\hbar(t_s - t_o)}} \psi_o(t_o) \tag{20.53}$$

with $A = z_1 \left(\frac{m}{2\pi i\hbar}\right)^{3/2}$. To calculate R_{in} we note that

$$G_o(\vec{r}_s, t_s - t_o) G_o^*(\vec{r}_s, t_s - t_o - \tau) = \exp\left(i\frac{mr_s^2}{2\hbar}\left[\frac{1}{(t_s - t_o)} - \frac{1}{(t_s - t_o - \tau)}\right]\right). \tag{20.54}$$

Writing $t_s = \underline{t_1} + \xi$, we expand the term in square brackets as

$$\left[\frac{1}{(t_s - t_o)} - \frac{1}{(t_s - t_o - \tau)}\right] = -\frac{1}{\underline{t_1}^2}\tau + \frac{2}{\underline{t_1}^3}\xi\tau - \frac{2}{\underline{t_1}^3}t_o\tau - \frac{1}{\underline{t_1}^3}\tau^2. \tag{20.55}$$

The last term is of higher order and can be neglected. Then

$$R_{in}(\tau, t_1) = \langle \psi_{in}(\vec{r}_s, t_s) \psi_{in}^*(\vec{r}_s, t_s - \tau) \rangle = |A|^2 e^{-i\underline{\omega_1}\tau} e^{i2\underline{\omega_1}\xi_s\frac{\tau}{\underline{t_1}}} \int_{-T}^{T} dt_o e^{-i2\underline{\omega_1}\tau t_o/\underline{t_1}} \tag{20.56}$$

$$= |A|^2 e^{-i\underline{\omega_1}\tau} e^{i2\underline{\omega_1}\xi_s\frac{\tau}{\underline{t_1}}} 2T\frac{\sin(\tau/\tau_{c1})}{(\tau/\tau_{c1})}$$

where $\underline{\omega_1} = mr_s^2/2\hbar\underline{t_1}^2$, and $\tau_{c1} = \underline{t_1}/(2\underline{\omega_1}T)$ is the correlation time (the inverse of the spectral width) that is imposed on the wave as a result of being limited in time by passing through a chopper. The pre-factor t_1^{-5} has been absorbed into $|A|^2$. Only values of $\tau \lesssim \tau_{c1}$ will contribute to the integral over $d\tau$ in (20.47). In deriving (20.56) we have taken

$$\langle \psi_o (t_o) \psi_o^* (t_o') \rangle = \delta (t_o - t_o').$$

Using the symmetry mentioned above, we can write

$$R_{out} (\tau, t_2) = \int_{-T_d}^{T_d} dt_d \, G_o (\vec{r_s}, t_2) \, G_o^* (\vec{r_s}, t_2 + \tau) \tag{20.57}$$

$$= |B|^2 e^{i\underline{\omega_2}\tau} e^{-i2\underline{\omega_2}\xi_s \frac{\tau}{\underline{t_2}}} 2T_d \frac{\sin (\tau/\tau_{c2})}{(\tau/\tau_{c2})}. \tag{20.58}$$

Substituting into (20.47) and neglecting the position dependence, we find for the intensity at the detector integrated over a detector time channel of width $2T_d$:

$$\int dt_d \, |\psi_{sc} (\vec{r_d}, t_d)|^2 = \int_s dt_s \int d\tau R_{out} (\tau, t_2) \, G_s (\tau) \, R_{in} (\tau, t_1) \tag{20.59}$$

$$= \int d\xi_s \int d\tau |B|^2 e^{-i\underline{\omega_1}\tau} e^{i\underline{\omega_2}\tau} e^{i2\underline{\omega_1}\xi_s \frac{\tau}{\underline{t_1}}} e^{-i2\underline{\omega_2}\xi_s \frac{\tau}{\underline{t_2}}} \times .. \tag{20.60}$$

$$.. \times 2T_d \frac{\sin (\tau/\tau_{c2})}{(\tau/\tau_{c2})} G_s (\tau) \, |A|^2 \, 2T \frac{\sin (\tau/\tau_{c1})}{(\tau/\tau_{c1})}$$

$$\int_{-T_d}^{T_d} dt_d \, |\psi_{sc} (\vec{r_d}, t_d)|^2 = N \int_{-T_s}^{T_s} d\xi_s \int d\tau e^{i\omega\tau} e^{i\mu\tau\xi_s} G_s (\tau) \, 2T \frac{\sin (\tau/\tau_{c1})}{(\tau/\tau_{c1})} 2T_d \frac{\sin (\tau/\tau_{c2})}{(\tau/\tau_{c2})} \tag{20.61}$$

where $\omega = \underline{\omega_2} - \underline{\omega_1}$ is the mean energy transfer of the scattered particles and

$$\mu = 2 \left(-\underline{\omega_2}/\underline{t_2} + \underline{\omega_1}/\underline{t_1} \right). \tag{20.62}$$

N is the normalization constant. To perform the measurement we require a third chopper at the scatterer which opens for $\underline{t_s} - T_s \le t_s \le \underline{t_s} + T_s$. The integral over $d\xi_s$ is seen to give $2T_s \frac{\sin \mu T_s \tau}{\mu T_s \tau}$. The term $e^{i\mu\xi_s}$ comes from the phase factor which is normally neglected in optics. In the present case this factor has the effect that moving the correlation interval τ (varying ξ_s) in (20.61) causes an additional difference in the two optical paths going from the entrance chopper to the detector chopper via the two points separated by τ, i.e., the paths *1* and *2* in Fig. 20.4. This defines the influence of the sample chopper width $(2T_s)$ on the resolution.

We then have

$$\int_{-T_d}^{T_d} dt_d \, |\psi_{sc} (\vec{r_d}, t_d)|^2 = NT_s T_d T \int d\tau e^{i\omega\tau} G_s (\tau) \, \mathbb{H} \tag{20.63}$$

$$\mathbb{H} = \frac{\sin (\tau/\tau_{c1})}{(\tau/\tau_{c1})} \frac{\sin (\tau/\tau_{c2})}{(\tau/\tau_{c2})} \frac{\sin (\tau/\tau_D)}{(\tau/\tau_D)} \tag{20.64}$$

where we have written $G_s (\tau)$ for $\langle \rho (\vec{r_s}, t_s) \, \rho^* (\vec{r_s}, t_s - \tau) \rangle_s$ the time-dependent density-density correlation function of the scattering system and $\tau_D = 1/(\mu T_s)$.

20.3.4.1 Discussion

Thus the scattering cross section is found to be proportional to the Fourier transform of the density-density correlation function of the scattering system multiplied by a function of τ which is the product of three functions representing the effects of the opening times of the three choppers on the resolution of the measurement. The argument of each term is (τ/τ_{ci}), where the τ_{ci} are different correlation times. Clearly the measurement will be optimized when all the τ_{ci} are approximately equal. If the τ_{ci} are all large compared to the values of τ for which $G_s(\tau)$ is significant (i.e., the opening times T_i are short), we reproduce the van Hove result. Since the Fourier transform of a product of functions is the convolution of the Fourier transforms of the functions, we see that the result (20.63) is the convolution of $S(\omega)$ (the Fourier transform of $G_s(\tau)$) with the Fourier transform of $\mathbb{H}$, which in turn is the convolution of the Fourier transforms of the $(\sin x/x)$ functions, each of which is a square pulse $[u(\omega + 1/\tau_{ci}) - u(\omega - 1/\tau_{ci})]$ with $u(x)$ the unit step. Our method then gives the result that would be obtained in a measurement, including the instrument resolution.

The scattering is seen as the interaction of the incoming wave with the time fluctuations of the scattering system. The wave function of the incoming state has an autocorrelation function determined by the (time-dependent) optical properties of the defining choppers. Only pairs of scattering events separated by times τ less than the correlation time of the incoming wave contribute to the scattered wave, and their contribution is weighted by the beam autocorrelation function evaluated at τ.

The opening times of the source and detector choppers, T and T_d, influence the resolution by an amount depending on $\underline{t}_{1,2}$ respectively, while T_s contributes to the uncertainty in the travel times in both arms and its influence depends on a weighted combination of the two travel times.

20.3.5 Scattering from a system fluctuating in space and time

We consider a scattering apparatus that consists of an entrance slit with a width $2a$ and a chopper which opens for a time between $-T_o$ and T_o, a detector with a slit of width $2d$ which is sensitive for a time between $\underline{t_d}-T_d$ and $\underline{t_d}+T_d$. To distinguish the velocities before and after the beam encounters the sample, we need a chopper at the sample which will open for a time between $\underline{t_s}-T_s$ and $\underline{t_s}+T_s$. See Fig. 20.5.

We take the sample to be a cylinder of radius R_s with its axis perpendicular to the scattering plane.

We now use Eq. (20.47) to calculate the detected intensity in such a system. To calculate R_{in} from (20.45),

$$R_{in}(\vec{\delta}, \tau) \sim \int d^2r_o\, dt_o\, G_o(\vec{r_1}, t_1)\, G_o^*(\vec{r_1} - \vec{\delta}, t_1 - \tau), \tag{20.65}$$

we need to calculate the phase factor Λ_{in} using

$$\vec{r'_s} = \vec{r_s} - \vec{\delta}, \quad t'_s = t_s - \tau \tag{20.66}$$

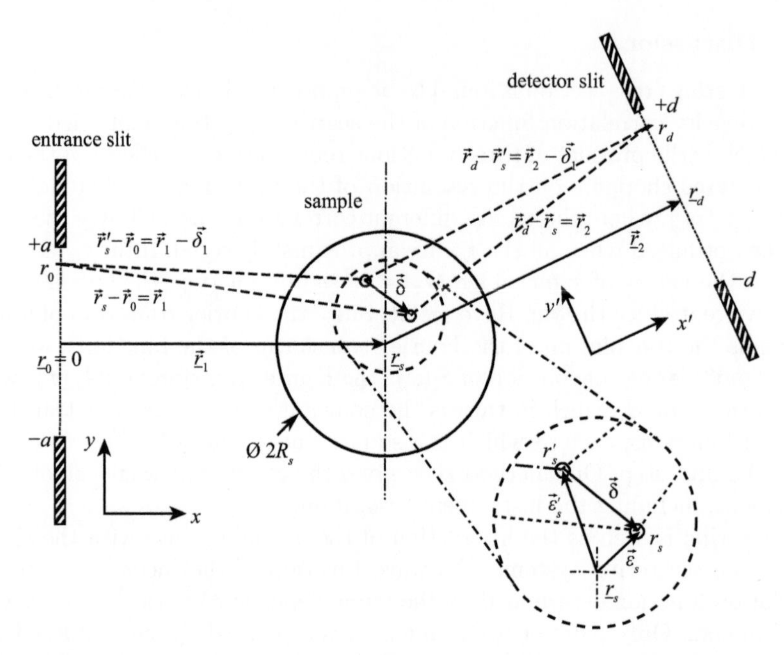

Fig. 20.5 Showing the meaning of the symbols used in the text to describe the spatial dependence. $\vec{\delta}$ is the distance between the two scattering events at $\vec{r}_s, \vec{r}'_s$. $\vec{\epsilon}_s, \vec{\epsilon}'_s$ are the positions relative to the center of the sample.

The phase factor is then

$$\Lambda_{in} = \frac{|\vec{r_s} - \vec{r_o}|^2}{(t_s - t_o)} - \frac{\left|\vec{r'_s} - \vec{r_o}\right|^2}{(t'_s - t_o)} \tag{20.67}$$

$$= -\frac{|\vec{r_s} - \vec{r_o}|^2 \, \tau}{(t_s - t_o)^2} + \frac{2\vec{\delta} \cdot (\vec{r_s} - \vec{r_o})}{(t_s - t_o)}, \tag{20.68}$$

where we have neglected higher order terms in $\vec{\delta}, \tau$. Introducing

$$t_s \Rightarrow \underline{t_s} + \xi_s = \underline{t_1} + \xi_s \tag{20.69}$$

$$\vec{r_s} \Rightarrow \underline{\vec{r_s}} + \vec{\epsilon_s} = \underline{\vec{r_1}} + \vec{\epsilon_s}\,, \tag{20.70}$$

we have

$$\Lambda_{in} = -\frac{|\vec{r_s} - \vec{r_o}|^2 \, \tau}{\underline{t_1}^2} \left(1 - 2\frac{\xi_s - t_o}{\underline{t_1}}\right) + \frac{2\vec{\delta} \cdot (\vec{r_s} - \vec{r_o})}{\underline{t_1}} \left(1 - \frac{\xi_s - t_o}{\underline{t_1}}\right). \tag{20.71}$$

Collecting terms, the phase of the integrand of R_{in} (Eq. 20.45) is

$$\frac{m}{2\hbar}\Lambda_{in} = -\underline{\omega}_1 \tau + \underline{\vec{k}}_1 \cdot \vec{\delta} + (t_o - \xi_s)\,\alpha_1 + (\vec{r_o} - \vec{\epsilon_s}) \cdot \vec{\beta_1} \tag{20.72}$$

where

$$\underline{\omega}_1 = \frac{m}{2\hbar}\frac{|\underline{\vec{r}}_1|^2}{\underline{t}_1^2}, \quad \underline{\vec{k}}_1 = \frac{m}{\hbar}\frac{\underline{\vec{r}}_1}{\underline{t}_1} \tag{20.73}$$

$$\vec{\beta}_1 = \left(\frac{\underline{\vec{k}}_1\tau}{\underline{t}_1} - \frac{m\vec{\delta}}{\hbar\underline{t}_1}\right), \quad \alpha_1 = \left(-\frac{2\underline{\omega}_1\tau}{\underline{t}_1} + \frac{\vec{\delta}\cdot\underline{\vec{k}}_1}{\underline{t}_1}\right) = -\frac{\hbar}{m}\vec{\beta}_1\cdot\underline{\vec{k}}_1. \tag{20.74}$$

Then

$$R_{in} = \int_{y=-a}^{y=a} d^2r_o \int_{-T_o}^{T_o} dt_o\, e^{i\frac{m}{2\hbar}\Lambda_{in}} = e^{i\left(-\underline{\omega}_1\tau + \underline{\vec{k}}_1\cdot\vec{\delta}\right)} e^{-i\left(\xi_s\alpha_1 + \vec{\epsilon}_s\cdot\vec{\beta}_1\right)} 2al\frac{\sin\beta_{1y}a}{\beta_{1y}a}2T_o\frac{\sin\alpha_1 T_o}{\alpha_1 T_o} \tag{20.75}$$

where we have specialized to the case of a one-dimensional input slit going from $y=-a$ to $y=a$ (i.e., $\vec{\delta}\,||\,\hat{y}$). l ($l \gg a$) is the height of the slit. Note that $\underline{\vec{k}}_1$ is defined to be parallel to the x axis so that $\left(\underline{\vec{k}}_1\right)_y = 0$. Comparison with the purely time-dependent case (20.56) shows that this is a special limit of the general case (20.75). Note that the functions ($\sin x/x$) determining the resolution are functions of the "correlation variables" τ and $\vec{\delta}$, and in that space the sensitive region (we call this the correlation volume) is moving with a velocity $v_1 = \underline{r}_1/\underline{t}_1$ in the direction of the incident beam. By our symmetry rules (20.49), R_{out} can be written as

$$R_{out} = \int_{Det} d^2r_d \int_{-T_d}^{T_d} dt_d\, e^{i\frac{m}{2\hbar}\Lambda_{out}} = e^{i\left(\underline{\omega}_2\tau - \underline{\vec{k}}_2\cdot\vec{\delta}\right)} e^{i\left(\xi_s\alpha_2 + \vec{\epsilon}_s\cdot\vec{\beta}_2\right)} 2dl\frac{\sin\beta_{2y'}d}{\beta_{2y'}d}2T_d\frac{\sin\alpha_2 T_d}{\alpha_2 T_d} \tag{20.76}$$

where y' indicates the direction perpendicular to the output arm ($\underline{\vec{r}}_2 = \underline{\vec{r}}_d - \underline{\vec{r}}_s$) and

$$\vec{\beta}_2 = \left(-\frac{\underline{\vec{k}}_2\tau}{\underline{t}_2} + \frac{m\vec{\delta}}{\hbar\underline{t}_2}\right), \quad \alpha_2 = -\frac{\hbar}{m}\vec{\beta}_2\cdot\underline{\vec{k}}_2. \tag{20.77}$$

The resolution volume for Eq. (20.76) is moving with a velocity $v_2 = \underline{r}_2/\underline{t}_2$ in the direction of the scattered beam.

The integration over the sample volume $\int d^3r_s = \int d^3\epsilon_s$ involves only the factor $e^{i\vec{\epsilon}_s\cdot(\vec{\beta}_2 - \vec{\beta}_1)}$ so we have to evaluate

$$\int d^3\epsilon_s\, e^{i\vec{\epsilon}_s\cdot(\vec{\beta}_2 - \vec{\beta}_1)}. \tag{20.78}$$

In this case we have a cylindrical sample of radius R and height L, with the cylinder axis perpendicular to the scattering plane. Introducing cylindrical coordinates, we have

$$\int d^3\epsilon_s\, e^{i\vec{\epsilon}_s\cdot(\vec{\beta}_2 - \vec{\beta}_1)} = L\int_0^{R_s} \epsilon_s d\epsilon_s \int_0^{2\pi} d\theta e^{i\epsilon_s\eta\cos\theta} = 2\pi L\int_0^{R_s} \epsilon_s\, d\epsilon_s\, J_o\left(\epsilon_s\eta\right) \tag{20.79}$$

$$= 2\pi L R_s^2 \frac{J_1\left(R_s\eta\right)}{R_s\eta} \tag{20.80}$$

with

$$\eta = \left|\vec{\beta}_2 - \vec{\beta}_1\right| = \left|\frac{m\vec{\delta}}{\hbar}\left(\frac{1}{\underline{t}_1} + \frac{1}{\underline{t}_2}\right) - \tau\left(\frac{\vec{k}_1}{\underline{t}_1} + \frac{\vec{k}_2}{\underline{t}_2}\right)\right| \tag{20.81}$$

The integral over the scattering time yields

$$\int_{-T_s}^{T_s} d\xi_s e^{i\mu\xi_s} = 2T_s \frac{\sin \mu T_s}{\mu T_s}. \tag{20.82}$$

In Eq. (20.82),

$$\mu = \alpha_2 - \alpha_1 = \left(-\vec{\delta}\cdot\left(\frac{\vec{k}_1}{\underline{t}_1} + \frac{\vec{k}_2}{\underline{t}_2}\right) + 2\tau\left(\frac{\underline{\omega}_1}{\underline{t}_1} + \frac{\underline{\omega}_2}{\underline{t}_2}\right)\right). \tag{20.83}$$

Putting it all together, the result of evaluating (20.47) is then

$$\int d^3 r_d \, dt_d \left|\psi_{sc}\left(\vec{r}_d, t_d\right)\right|^2 \approx \int d^3\delta \, d\tau e^{i\left(\omega\tau - \vec{q}\cdot\vec{\delta}\right)} G_s\left(\vec{\delta}, \tau\right) \mathbb{H} \tag{20.84}$$

where

$$\omega = \underline{\omega}_2 - \underline{\omega}_1, \quad \vec{q} = \underline{\vec{k}}_2 - \underline{\vec{k}}_1 \tag{20.85}$$

$$\begin{aligned}\mathbb{H} = 2^5 &\left(d\frac{\sin \beta_{2y'} d}{\beta_{2y'} d}\right)\left(T_d \frac{\sin \alpha_2 T_d}{\alpha_2 T_d}\right)\left(T_s \frac{\sin \mu T_s}{\mu T_s}\right)\left(2\pi R_s^2 \frac{J_1\left(\eta R_s\right)}{\eta R_s}\right) \\ &\left(T_o \frac{\sin \alpha_1 T_o}{\alpha_1 T_o}\right)\left(a\frac{\sin \beta_{1y} a}{\beta_{1y} a}\right).\end{aligned} \tag{20.86}$$

We have left out the normalization constants; the details are described in Gähler et al.[13] $\mathbb{H}$ represents the effect of the apparatus resolution on the detected signal.

20.3.6 Discussion

For convenience we rewrite below

$$\begin{aligned}\vec{\beta}_1 &= -\frac{m}{\hbar \underline{t}_1}\left(\vec{\delta} - \vec{v}_1\tau\right), \quad \alpha_1 = -\frac{\hbar}{m}\vec{\beta}_1 \cdot \underline{\vec{k}_1} \\ \vec{\beta}_2 &= \frac{m}{\hbar \underline{t}_2}\left(\vec{\delta} - \vec{v}_2\tau\right), \quad \alpha_2 = -\frac{\hbar}{m}\vec{\beta}_2 \cdot \underline{\vec{k}_2} \\ \mu &= (\alpha_2 - \alpha_1), \quad \eta = \left|\vec{\beta}_2 - \vec{\beta}_1\right|\end{aligned} \tag{20.87}$$

where α_1 and β_1 represent the influence of the input slit and chopper on the overall resolution, μ and η the influence of the sample size and chopper, while α_2 and β_2 represent the effects of the detector slit and integration time.

[13] Gähler, R., Felber, J., Mezei, F. and Golub, R., *Space-time approach to scattering from many-body systems*, Phys. Rev. A **58**, 280 (1998) received 19 September 1997.

Each of these variables has a term proportional to τ and a term proportional to δ. Keeping in mind the Fourier transform in Eq. (20.84), we see that the terms proportional to τ will give the energy resolution and those proportional to δ will give the q resolution. Thus, for example, the τ term in α_1 represents the effect of the first chopper opening time T_o on the energy resolution, just as in the purely time-dependent case (Eqs. (20.64), (20.63)), while the δ term represents the effect of the first chopper slit width on the q resolution, i.e., the chopper slit width results in an uncertainty in $\vec{k}_1$ that gives a contribution to the q resolution. Similar remarks hold for each of the terms in (20.87). Note that by definition $\underline{\vec{k_1}}_y, \underline{\vec{k_2}}_{y'} = 0$ but we display these terms in (20.87) to keep the symmetry.

This approach yields directly the mutual influence of the q and ω resolutions in a compact analytic form. Since each element of the apparatus is associated with a definite factor in $\mathbb{H}$, we can clearly see the influence of the individual elements on the overall resolution.

If we wish we can follow the more usual procedure of replacing each term by a Gaussian function with the appropriate width. This allows an easy way to estimate the overall width of the ω, q resolution (given by the convolution of the Fourier transform of each term in $\mathbb{H}$) as the square root of the sum of the squares of the widths of each term and shows that the optimum is when all the widths are equal.

Each factor in (20.86) is associated with a certain velocity in $(\vec{\delta}, \tau)$ space, so that the relevant correlation volumes can be considered as moving as a function of τ. For example, the volume associated with α_2 is moving with a velocity $\vec{v}_2$, the nominal velocity of the scattered wave. Thus we see that the coherence volume associated with the detector parameters can be considered as consisting of the region limited by $\pm\hbar\underline{t}_2/md$ in the δ'_y direction and $\pm\underline{t}_2/T_d k_2$ in the $\delta_{x'}$ direction, moving (as a function of τ) with velocity v_2 in the x' direction (direction of the scattered beam). Each of the other pairs of terms contributes a similar moving limit to the region of $(\vec{\delta}, \tau)$ space.

20.4 Conclusion

In this chapter we have shown how Feynman demonstrated that, if the probability amplitude for a particle to travel between two space-time points is taken to be a sum over all possible paths connecting these points, with each path assigned a phase proportional to the action integral over that path, $S = \int_{t_i}^{t_j} L dt$, then the amplitude is given by the Green's function of the Schrödinger equation. In this way we obtain the solution to the Schrödinger equation without ever introducing that equation.

We then saw how, starting with the result for first order time-dependent perturbation theory and assuming the incident and final wave functions of the scattered particles to be plane waves, van Hove showed that the scattering cross section is proportional to a function $S(q, \omega)$ of the momentum and energy transfer, which in turn is the Fourier transform of the time- and space-dependent autocorrelation function of the density fluctuations of the scattering system.

However, since no experiment is ever done with perfect resolution, measurements yield the convolution of $S(q, \omega)$ with an instrument resolution function. The determination of the resolution function and its use in the interpretation of the data are left to further work in this treatment.

In this conventional approach, a scattering experiment is divided up into two separate parts which are treated by two different, mutually exclusive, approximations. The scattering event in the sample is treated by assuming the incident and scattered beams to be infinitely extended plane waves, while the beam is treated as consisting of classical point particles in order to describe its motion through the apparatus.

We have shown that the entire scattering experiment can be described as a wave-optical phenomenon. This more general approach provides a rigorous justification for the usual dual approximation since, in most cases, the correlation volumes are much smaller than the geometrical dimensions (beam cross sections, chopper pulse lengths, etc.) involved. However, in some cases this is not true and the complete wave-optical description of the global system of instrument and scattering sample must be used. In addition to precision light scattering experiments, such cases include perfect crystal neutron optics (e.g., interferometry) where the definition of the wave vector is so precise that the correlation lengths can be comparable to the geometrical dimensions. These cases are sometimes referred to as "spherical wave effects." In fact, these effects are natural consequences of the full wave-optical analysis presented here.

Usual methods of determining the experiment resolution include measurement of a known calibration sample, Monte Carlo calculations, and estimates based on assuming a Gaussian form for the separate contributions to the overall resolution, so that the separate contributions can be combined using the sum of squares rule.

In this chapter we calculated the incoming wave function as it is shaped by transmission through the elements of the apparatus according to the laws of optics. For example, on propagating through a slit and/or a chopper, a wave acquires a correlation function with a finite width. After scattering, the individual waves scattered from a neighboring pair of points in the sample lose correlation on passing through the post-sample elements of the apparatus. We have shown that the probability of a scattered particle arriving at a detector at a given space-time location is proportional to the Fourier transform of the density fluctuation autocorrelation function (as shown by van Hove) but multiplied by a function $\mathbb{H}(\vec{\delta}, \tau)$ which represents the resolution of the measurement and is the Fourier transform of the instrument resolution as it is normally considered. This function is the product of a series of functions, each representing the effect of a single element of the apparatus on the measurement. Changes in one element can be easily accommodated by changes in the appropriate function. The regions of $(\vec{\delta}, \tau)$ space where each function is significant are the correlation volumes for the different elements of the apparatus. A measurement can be optimized by maximizing the region of overlap of all the correlation volumes. The fact that these volumes are seen to be moving as a function of the correlation time τ is a reflection of the fact that the ω and $\vec{q}$ resolutions depend on each other to some extent, and the wave-optical approach provides an analytic description of this dependence. The present analysis offers a graphical method of optimizing each measurement, with the influence of all parameters such as the scattering angle included. The model of moving volumes is a way to visualize the four dimensional space-time correlation functions.

In conclusion we note that particles and their trajectories are absent from the analysis.

21

Introduction to quantum computing (with the assistance of Edward D. Davis)

21.1 Overview

Currently (2022) quantum computing is one of the hottest topics in physics and is the object of an enormous research effort. Many academic institutions and internationally prominent corporations are devoting large amounts of resources to the attempt to realize a practical quantum computer. The science-supporting agencies of many governments around the world have made quantum computing one of their highest priorities and have instituted special funding programs to support it. On the scientific side, the interest in quantum computing has led to new ways of thinking about quantum mechanics and the development of new mathematical techniques.

In the last few years there have been hundreds, perhaps even thousands of publications in the most eminent refereed scientific journals, many prominent universities offer semester long courses in the subject and many textbooks have been published, one of which is 700 pages long in spite of leaving most of the important results to be worked out by the reader in the form of problems and exercises[1]

One reason for this interest is:

> In quantum computing, we witness an exciting and very promising merger of two of the deepest and most successful scientific and technological developments of this century: quantum physics and computer science. In spite of the fact that its experimental developments are in their infancy, there has already been a variety of concepts, models, methods and results obtained at the theoretical level that clearly have lasting value[2]

which is essentially still valid twenty years after its publication.

The "golden chalice" of quantum computing is the likelihood that a working quantum computer would be able to factor large numbers in a reasonable time. This is attractive because most currently used schemes for encrypting data owe their security to the difficulty of solving the factorization problem. In fact, interest in quantum computing got a tremendous boost from the publication of Shor's algorithm, which would

[1]Nielsen, M.A. and Chuang, I.L., *Quantum Computation and Quantum Information, 10th Annniversary Edition* (Cambridge University Press, 2010).

[2]Brown, J., *Minds, machines and the multiverse: The Quest for a Quantum Computer* (Simon & Schuster, 2002).

The Historical and Physical Foundations of Quantum Mechanics. Robert Golub and Steven K. Lamoreaux, Oxford University Press.
 DOI: 10.1093/oso/9780198822189.003.0021

be able to accomplish this in significantly shorter times than can be done with classical computing. Basically, this is because a quantum computer works with quantum superpositions of binary representations of numbers, so that it can calculate all values of a function $f(x_i)$ for a range of x_i $(i = 0, 1....N)$ simultaneously. The problem is that a measurement collapses the superposition state to a single (randomly chosen) value. In this simple case it would be necessary to repeat the calculation and measurement many times which will be no more efficient than a classical computer. However, it is possible to determine general properties of the function, such as its period, in a relatively straightforward manner. The genius of Schor's algorithm is that he showed how to apply this feature to the factoring of large numbers (hundreds of binary digits).

How did it all begin?

In the 1970s computer scientists began to ask about the limits, if any, to achievable computing power. One issue was the fact that many classical gates AND, OR, etc. with one output were irreversible. Looking at the truth tables for the AND and XOR (exclusive or) gates,

AND		*XOR*	
input	*output*	*input*	*output*
00	0	00	0
01	0	01	1
10	0	10	1
11	1	11	0

, (21.1)

we see it is impossible to construct the input given the output. Some people suggested this irreversibility was an essential feature of computing but in 1973 Bennett[3] showed this was not the case and computers could act reversibly. We can see this by considering the controlled NOT gate (CNOT) (two outputs) with a truth table:

input	*output*
00	00
01	01
10	11
11	10

. (21.2)

When the first bit is 0 the input bits are transmitted to the output but when the first bit is 1 the second bit is inverted. Here we see that given the output we can reconstruct the input. The CNOT gate is reversible. The first mention of quantum mechanics in connection with computers seems to have been by Paul Benioff.[4] He was interested in the question of process, which is connected to the reversibility question and was also motivated by the thought

> ... that if one is to make any progress giving a quantum mechanical description of intelligent beings—if it is possible at all—then one must first give such a description of computers and the computation process.

[3]Bennett, C.H., *Logical reversibility of computation*, IBM J. Res. Dev.17, 525 (1973)

[4]Benioff, P., *The Computer as a Physical System: A Microscopic Quantum Mechanical Hamiltonian Model of Computers as Represented by Turing Machines*, J. Stat. Phys. 22,563 (1980)

He then proceeded to give several Hamiltonian models that would run as Turing machines.[5,6]

Richard Feynman became interested in computing[7] and in 1982 published a lecture *Simulating Physics with Computers*[8] in which he investigated the possibilities of simulating physical systems with computers. He was interested in exact simulations in which the computer does exactly the same thing as nature—this requires that everything that happens in a given finite space-time volume be exactly analyzable with a finite number of operations. This disagrees with present theories of physics which involve infinitesimal distances. So if exact computer simulation is possible then current physics is wrong. In quantum mechanics, an N particle system represented by $\psi\,(x_1, x_2 ... x_N, t)$ cannot be simulated with a number of elements proportional to N. Here Feynman introduced the idea of a computer built on quantum elements obeying quantum laws.

He then turned to the "interesting" question of whether a quantum system can be "probabilistically simulated" by a classical (probabilistic) universal computer. In order to show this was impossible Feynman invoked the EPR result and gave a proof of Bell's theorem as follows:

Consider photons of horizontal (H) or vertical (V) polarization incident on a calcite crystal, schematically shown in Fig. 21.1. The photon exits the crystal in one of two different paths depending on the polarization. Now we imagine an atom emitting 2 photons in succession in opposite directions, e.g., a $3s \to 2p \to 1s$ transition.

Each photon goes through a calcite crystal set at angles ϕ_1 and ϕ_2 respectively. Quantum theory predicts (confirmed by experiment) that the probability to register H on both crystals (P_{HH}) or V on both (P_{VV}) is

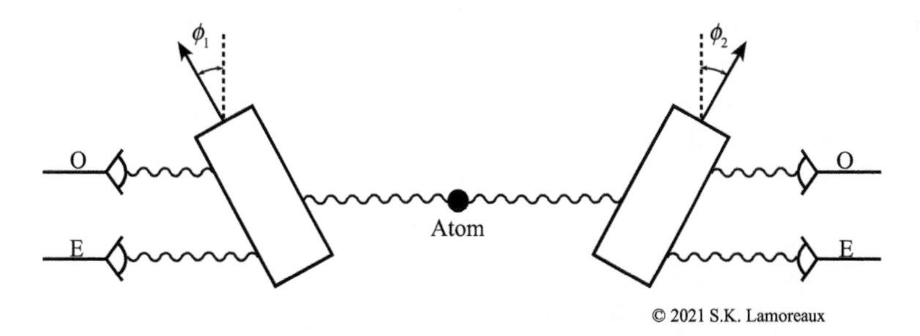

Fig. 21.1 Showing two photons traveling in opposite directions going through Calcite crystals set at the angles ϕ_1, ϕ_2. (As described in Feynman, op. cit., footnote p. 661.)

[5]Benioff, P. J. *Quantum Mechanical Models of Turing Machines That Dissipate No Energy.* Stat. Phys 29, 515, (1982) and PRL 48,1581, (1982)

[6]A Turing machine is a generic computer that works by reading and writing numbers on an infinite tape according to a given program and can be usd to compute any computable problem. Turing, A. M., 1936–7, "*On Computable Numbers, With an Application to the Entscheidungsproblem,*" Proceedings of the London Mathematical Society, s2-42: 230–265; correction ibid., s2-43: 544–546 (1937).

[7]Feynman, R. P., Hey, A. J. G. and Allen, R. W. eds., *The Feynman Lectures on Computation* (Addison-Wesley 1996)

[8]Feynman, R. P., *Simulating physics with computers*, Int. J. of Theo. Phys. **21**, 467 (1982)

$$P_{HH} = P_{VV} = \frac{1}{2}\cos^2(\phi_1 - \phi_2) \tag{21.3}$$

while the probability to measure H on one and V on the other ($P_{HV} = P_{VH}$) is given by

$$P_{HV} = P_{VH} = \frac{1}{2}\sin^2(\phi_1 - \phi_2). \tag{21.4}$$

If we choose $\phi_1 = \phi_2$ the results agree and an observer on one side can always predict, exactly, the result of the observer on the other side.

In a classical deterministic theory the photons would have to have some variable specifying how they will behave for each angle of the crystals. One condition is that the probabilities would have to be either zero or one in order that one observer can predict exactly the results of the other observer.

Consider the set of angles $0^o, 30^o, 60^o, 90^o, 120^o, 150^o$. A photon approaching one of the crystals will have its behavior specified exactly, i.e., what it will do for each angle of the crystal. Assume the photon is set up so that it would acts as H at 0° and 30° (dark circles in the Fig. 21.2), as V at 60° (open circles), and so on as shown in the Fig. 21.2. The same diagram must apply to both photons as they must always agree when the crystals are set to the same angle.

The second row represents another possible assignment. Note that points differing by 90° must be different as shifting by 90° converts H to V and vice-versa. Every set of photons can be represented by a different diagram as long there are three dark and three white dots. Now let's measure with a difference of $(\phi_1 - \phi_2) = 30^o$ between the crystals and ask how many times will the results agree. In the first row there are four out of six pairs of equal neighbors for a probability of match =2/3, while in the second case there are no matches. It is easy to see, by rearranging the dots that 2/3 is the maximum probability of matching for any arrangement. However, the quantum mechanical result is $\cos^2(\phi_1 - \phi_2) = \cos^2(30^o) = 3/4$. Feynman concludes that "That's why quantum mechanics can't seem to be imitable by a local classical computer."

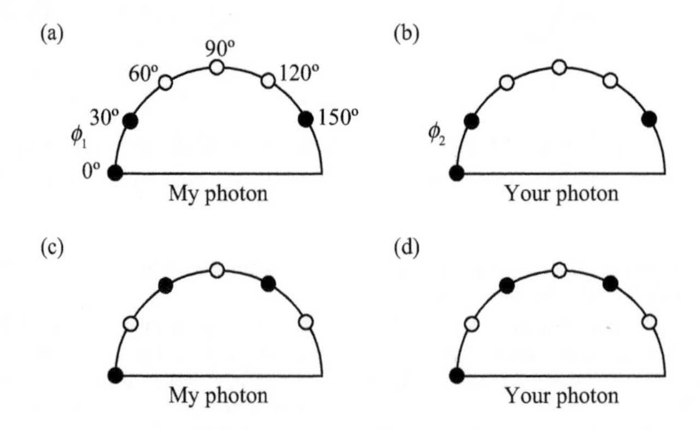

Fig. 21.2 Assignment of classical properties (a specifec polarization at each crystal angle) to each photon. Dark (light) circles represent $H(V)$ polarizations. (Feynman, op. cit., p. 661.)

It is interesting to note that there is a theorem called the Gottesman-Knill theorem[9] that states that any quantum computer circuit that consists of:

1. preparation of qubits in computational basis states;
2. hadamard, cNOT and phase gates, (see below for definition of these);
3. measurement of Pauli operators;

can be efficiently simulated (i.e., in polynomial time—a length of time proportional to some power of the number of bits involved in the calculation) on a classical computer. As we will see, entangled Bell type states (e.g., a two-bit state such as $[|00\rangle + |11\rangle]$) can be created by these operators. This would seem to contradict Feynman's assertion except for the fact that the observation of the correlations in an EPR-type state requires measurements in bases that are rotated through arbitrary angles and the theorem does not apply to arbitrary rotations. This sheds some light on the poorly understood reason for the speed-up of quantum computers. It is not the creation of entangled states *per se* that is decisive; rotations through arbitrary angles play a role.

In a second paper discussing the possibility of quantum computers[10] Feynman showed how a normal digital computer could be imitated using atoms or other quantum objects without using "the specific qualities of the differential equations of quantum mechanics." The motive was to show that quantum mechanics would not inhibit the construction of computers using elements on the atomic scale.

One of the first to realize that a quantum computer might offer significant advantages relative to classical computers was David Deutsch. Deutsch is a strong supporter of the Everett interpretation and he began his first paper[11] on the subject with the statement that a quantum computer using what Feynman called "The specific qualities of the differential equations of quantum mechanics," would be capable of what he called "quantum parallelism," which means calculations can be carried out in parallel, on a superposition of quantum states. An intuitive explanation of this puts an "intolerable strain on all interpretations of quantum mechanics except Everett's." He proposed a system of registers made up of two state systems representing bits and a series of operations on these bits. He defines an operation whereby a state $|\Pi(f), i, 0\rangle$, where $\Pi(f)$ is a program to calculate $f(i)$ evolves into the state $|\Pi(f), i, f(i)\rangle$ so that a superposition of such states will evolve as:

$$\frac{1}{\sqrt{N}}\sum_{i=1}^{N}|\Pi(f), i, 0\rangle \Rightarrow \frac{1}{\sqrt{N}}\sum_{i=1}^{N}|\Pi(f), i, f(i)\rangle \tag{21.5}$$

and will take the same amount of resources (time, memory space and other hardware) as the calculation of a single value of $f(i)$.[12] But only one of the results is accessible in

[9]Gottesman, D., *The Hesenberg Representation of Quantum Computers*, arXiv:quant-ph/9807006v.1, 1998

[10]Feynman, R.P., Optics News, Feb. 1985, 11-20, reprinted in *The Feynman Lecctures on Computation*, op. cit p.185.

[11]Deutsch, D., *Quantum theory the Church-Turing principle and universal quantum computing*, Proc. Roy. Soc., 400,97 (1985).

[12]The notation here is that each entry in the ket represents a register of a fixed number of qubits.

each Everettian universe. So in practice the calculation has to be repeated and nothing is gained. However, there is the possibility of calculating some general properties of the function which Deutsch first explored in a talk given in March 1984,[13] devoted to aspects of the Everett interpretation. He introduced a notional quantum computer programmed to calculate $f(i)$, defined to be a one-bit function (i.e., value either 0 or 1) of a one-bit input (i). The algorithm to calculate it is assumed to be very long and complex and takes a long time, T. The operation of the quantum computer is defined by the evolution

$$|\psi(t=0)\rangle = |i,0,t=0\rangle \Rightarrow |i,f(i),t=T\rangle = |\psi(T)\rangle. \tag{21.6}$$

We could calculate both values of the function in a single pass by starting with the system in a superposition state:

$$\frac{1}{\sqrt{2}}\sum_{i=0}^{1}|i,0,t=0\rangle \Rightarrow \frac{1}{\sqrt{2}}\sum_{i=0}^{1}|i,f(i),t=T\rangle. \tag{21.7}$$

Now assume we were interested in some function of the values of the function, say

$$G(f(0),f(1)) = f(0)\oplus f(1) \tag{21.8}$$

where $\oplus$ represents addition modulo 2 ($0\oplus 0 = 1\oplus 1 = 0$ and $0\oplus 1 = 1\oplus 0 = 1$).

Deutsch then showed that quantum parallelism can be used to calculate this in one round of calculation but with a probability less than 1. This is accomplished by measuring at time T an observable whose eigenstates are:

$$|zero\rangle = \frac{1}{2}\{|0,0\rangle - |0,1\rangle + |1,0\rangle - |1,1\rangle\} \tag{21.9}$$

$$|one\rangle = \frac{1}{2}\{|0,0\rangle - |0,1\rangle - |1,0\rangle + |1,1\rangle\} \tag{21.10}$$

$$|fail\rangle = \frac{1}{2}\{|0,0\rangle + |0,1\rangle + |1,0\rangle + |1,1\rangle\} \tag{21.11}$$

$$|error\rangle = \frac{1}{2}\{|0,0\rangle + |0,1\rangle - |1,0\rangle - |1,1\rangle\}. \tag{21.12}$$

Designating these states as $|R_i\rangle$, $i = 1..4$ the state of the system at time T can be written as a superposition of these eigenstates:

$$|\psi(T)\rangle = \sum_i c_i |R_i\rangle \tag{21.13}$$

with $c_i = \langle R_i|\psi(T)\rangle$. For all possible values $f(i)$ we can construct the following table

$f(0)$	$f(1)$	$f(0)\oplus f(1)$	$\langle zero\|\psi\rangle$	$\langle one\|\psi\rangle$	$\langle fail\|\psi\rangle$	$\langle error\|\psi\rangle$
0	0	0	1	0	1	0
0	1	1	0	1	1	0
1	0	1	0	−1	1	0
1	1	0	−1	0	1	0

(21.14)

[13] Deutsch, D., 'Three connections between Everett's interpretation and experiment,' in Penrose, R. and Isham, C.J., eds., *Quantum Concepts in Space and Time* (Clarendon, Oxford 1986), p.215.

So if the result of the measurement is the eigenvalue of $|zero\rangle$, $(|one\rangle)$ we know the result is definitely $G = 0, (G = 1)$ and this will occur with a probability $P = 1/2$. But, also with $P = 1/2$ we will obtain the result $|fail\rangle$. However, it is interesting to note that the expectation value of the time to get a definite result is

$$\langle t \rangle = \sum_{n=1}^{\infty} \frac{n}{2^n} T = 2T \tag{21.15}$$

so the question of whether this technique is advantageous over just calculating $f(0)$, $f(1)$ is a matter of luck. Deutsch contrasts the physical explanation of what is happening in the Everett interpretation with other interpretations. In the first case two different algorithms computing $f(0), f(1)$ are executed simultaneously in two different Everett branches. The measurement of one of the R_i by an observer transfers the information $f(0) \oplus f(1)$ into a single branch erasing all information about $f(i)$ in the other branch ($P_{fail} = 1/2$). According to Deutsch the standard interpretation tells us that the probability of the possible outcomes are given by the values of the different $|c_i|^2$. It is impossible to describe the internal processes, only the final result exists. "Describing the process internal to the quantum computer as an objective sequence of phenomena is impossible." We should point out that the standard interpretation does supply the time dependence of the state of the computer, e.g., the c_j above. On the occasions when the result is definitive after a single run Deutsch asks, "Where was the computation done?"

This example tells us a lot about quantum computing and quantum mechanics as well. It is similar to the two-slit problem in that it highlights the features of a quantum superposition.

However, as we will see later, its limitation is a property of the measurement Deutsch chose to make. In fact, there is an algorithm that can compute $f(0) \oplus f(1)$ with certainty in one interval T (see below).

In a further paper[14] Deutsch discussed the properties of a number of quantum gates and showed that there were a large number of two-bit gates that were "universal," that is that any computation capable of being performed in a finite time could be performed by a system of these gates and a number of one-bit gates. He demonstrated the possibilities of obtaining the results of irreversible classical gates with quantum reversible ones and emphasized the possibility that the output state of a gate could be a superposition of basis states even if the input is in a single such state. For example the gate M shown in Fig. 21.3 will set both outputs to the state b when a is in the state $a = |0\rangle$ thus producing a quantum fan-out.

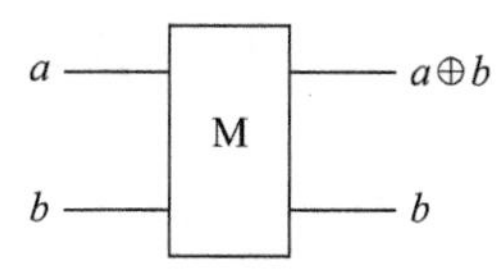

Fig. 21.3 A two-bit reversible gate called the measurement gate.

[14]Deutsch, D., *Quantum Computational Networks*, Proc. Roy. Soc. **A 425**, 73 (1989).

Deutsch distinguished between a quantum gate and a quantum network. An assembly of gates can form another gate or a network, distinguished by the fact that a network has a memory which is necessary for a universal computing machine, whether quantum or classical. He was thinking of the bits as being carried by some physical systems (atoms or spins perhaps) that move from one gate to another. Inside the gates they can interact with other bits that are also in the gate. Any two-bit quantum gate can be represented by a unitary matrix, S (4×4 for a two-bit gate), which in turn can be derived from a Hamiltonian, H,

$$S = e^{-iHT} \tag{21.16}$$

where T is the time of action of the gate. S should be local in the sense that it only has nonzero elements between neighboring bits. The Hamiltonian

$$H = \frac{i}{T} \ln S \tag{21.17}$$

will be necessarily non-local as the expansion of $\ln S$ contains all powers of $(1 - S)$. The conclusion is that any quantum computational network or universal computer must have a time-dependent Hamiltonian.

In a further study of the advantages of a quantum computer Deutsch and Jozsa[15] noted that up until that time all known computations that could be performed faster with quantum parallelism than by classical methods had the features:

1. The answer is not obtained with certainty, there is a finite probability that the calculation will fail and need to be repeated.
2. On average the quantum algorithm is no more efficient than a classical one—as we have seen with the example above.

They then proceeded to study a problem (that we will discuss in detail later) which can be solved with certainty in a single pass. They start by introducing a black box (called an oracle for $f(i)$) to calculate a two-valued function of a $2N$-valued input variable, U_f which acts on a basis state $|i, j\rangle$ where i goes from 0 to $N - 1$ so that $U_f |i, j\rangle = |i, j \oplus f(i)\rangle$, (see Fig. 21.12, p.680). Starting with a state

$$|\phi\rangle = \frac{1}{\sqrt{2N}} \sum_{i=0}^{2N-1} |i, 0\rangle \tag{21.18}$$

$$U_f |\phi\rangle = \frac{1}{\sqrt{2N}} \sum_{i=0}^{2N-1} |i, f(i)\rangle \tag{21.19}$$

and introducing an operation $S|i, j\rangle = (-1)^j |i, j\rangle$ we see that the final state, after applying $U_f S U_f$, will be

$$|\psi\rangle = U_f S U_f |\phi\rangle = \frac{1}{\sqrt{2N}} \sum_{i=0}^{2N-1} (-1)^{f(i)} |i, 0\rangle \tag{21.20}$$

[15]Deutsch, D. and Jozsa, R., *Rapid Solution of Problems by Quantum Computation*, Proc. Math. and Phys. Sciences (Royal Society) **439**, 553 (1992).

since $f(i) \oplus f(i) = 0$, and that

$$|\langle \phi | \psi \rangle| = \frac{1}{2N} \left| \sum_{i=0}^{2N-1} (-1)^{f(i)} \right| \tag{21.21}$$

$$= 0 \text{ when B is false} \tag{21.22}$$

$$= 1 \text{ when A is false} \tag{21.23}$$

where A, B are the statements:

A) $f(i)$ is not a constant.

B) The sequence $f(0)...f(2N-1)$ does not contain exactly N zeros.

Thus when B is false the sequence does contain exactly N zeros and there are equal numbers of positive and negative terms in the sum in (21.21). When A is false $f(i) =$ either 0 or 1 for all values of i so all the terms in the sum have the same sign.

Thus it is possible to solve this rather obtuse problem much faster than can be done with a classical computer.

These arcane discussions took a much more practical turn with Peter Shor's proposal of an algorithm for factoring large numbers. Schor[16] credits an idea of Simon's[17] for inspiration. Factoring of large numbers is of immense practical importance because the security of many currently used encryption methods is based on the mathematical difficulty of factoring large numbers. There is a classical method of factoring large numbers based on modular arithmetic. The computationally difficult step in this algorithm is modular exponentiation, i.e., given n, x, and r find $(x^r \mod n)$. As we will see in this chapter, Schor's method for calculating this is based on a quantum Fourier transform.

There is a large amount of work considering methods of error correction and fault-tolerant computing. In this chapter we give an introduction to quantum computing and an explanation of Schor's algorithm as well as the simplest example of quantum error correction.

A detailed timeline for the development of quantum computing and communication can be found at:

https://en.wikipedia.org/wiki/Timeline_of_quantum_computing_and_communication.

The basic idea of quantum computing is due to David Deutsch, who developed the circuit model of quantum computing that we will explain here. Deutsch has written that the existence of quantum computing is proof of the existence of an infinitude of parallel universes, a quantum superposition decomposing into a series of definite states, each in its own parallel universe. The extraordinary classical computing power invoked during the quantum factoring of a very large number comes from the calculation being carried out simultaneously in all the parallel universes. Steane[18] disputes this contention: The fact that a computation done in a certain way requires large

[16]Shor, P.W., "Algorithms for quantum computation: Discrete algorithms and factoring," Proc. 35th Annual Symp.on Foundations of Computer Science, IEEE, (1994), revised in "Polynomial Time Algorithms for Prime Factorization and Discrete Logarithms on a Quantum Computer," arXiv:quant-phy/9508027v2, (1996).

[17]Simon, D.R., *On the Power of Quantum Computation*, 35th Ann. Symp. Found. Comp. Science, loc.cit. p.118.

[18]Steane, A.M., *A quantum computer only needs one universe*, Studies in Hist. and Phil. of Mod. Phys. **34B** (3), pp. 469-478 (2003).

amounts of resources does not mean that an alternate way of doing the computation necessarily uses the same amount of resources.

Readers are referred to Nielsen and Chuang, op. cit., p. 659 for a more detailed discussion of the topics presented here.

21.2 The basic ideas

The two main concepts in quantum computing are the qubit and the unitary logic gate. Proponents of the field praise this mixing of computer science and quantum mechanics as giving new insights into both fields. It is necessary to emphasize that a qubit cannot be represented by an isolated spin 1/2 system and a quantum logic gate is a set of operations (rotations) applied to a qubit.

21.2.1 Qubits

The main insight into quantum computing is to represent numbers in a register consisting of a row of cells where each cell, instead of storing a 0 or 1 as in a classical computer, is represented by a two-level quantum system in a superposition state. We will see how in this way a quantum register can hold a superposition of all numbers from $0...2^n - 1$ where n is the number of cells in the register. Introductory lectures often start with the statement that qubits can be thought of as spin-1/2 systems. In addition it is necessary to have a means of coupling at least two of the qubit systems as we will see below.

The quantum state of any two-level system can be represented by the two eigenstates, $|\pm\rangle$ of the σ_z operator of an equivalent spin-1/2 system. (Section 19.7) The state $|+\rangle_z$ is taken to represent a logical 0, written as $|0\rangle$, while a logical 1 is represented by $|1\rangle = |-\rangle_z$. So a state

$$[a\,|0\rangle + b\,|1\rangle] = \begin{pmatrix} a \\ b \end{pmatrix}$$

represents a quantum superposition of the two possible qubit values.

A "register" consisting of n qubits can store numbers from 0 to $2^n - 1$. For example the binary number, 1011001....1 would be represented by the n qubit state

$$|1011001....1\rangle = |1\rangle\,|0\rangle\,|1\rangle\,|1\rangle\,|0\rangle\,|0\rangle\,|1\rangle\,....\,|1\rangle$$

However, the register can be put in a superposition state

$$\sum_{j=0}^{2^n-1} a_j\,|j\rangle\,.$$

To specify this state requires 2^n complex numbers or 2^{n+1} real numbers. To store them with a precision of one byte on a classical computer would require the same number of bytes. So the state of a 50-bit quantum computer would require $2^{51} = 2.25 \times 10^{15}$ bits (2,000 terabytes) of memory on a classical computer. Examples like this are used to demonstrate the alleged unbelievable potential power of quantum computers. However, it is not yet certain that we can produce an arbitrary state of the quantum register and even if we could we certainly could not extract all the information implied by this argument. The art of quantum computing is to invent ways to extract a small fraction of this information in a useful way.

21.2.2 Quantum logic gates

Since the evolution of quantum systems is determined by unitary transformations any set of operations carried out on a quantum computer will have to be reversible. That means we will have to redesign many classical logic gates to perform the logical operations reversibly. For example we have seen that a classical AND gate is irreversible (it is in general impossible to reconstruct the input from the output) so one needs to find a reversible system that is logically equivalent. What is called a logic gate in quantum computing is just a series of rotations of the spin-1/2 system representing a qubit. For example the operation NOT applied to the above state would give

$$NOT\begin{pmatrix} a \\ b \end{pmatrix} = \begin{pmatrix} b \\ a \end{pmatrix}.$$

Already in this example we see that quantum gates are really rotation operations which can be applied by subjecting the qubit system to external fields and other interactions.

21.2.3 Quantum circuit model

In this model single qubits are represented by horizontal lines and time moves from left to right. A spin rotation operation applied to perform some logical function is represented by a square box with the name of the function. Fig. 21.4 represents two qubits, one (A) on which a NOT operation is performed and the second (C) which is left alone.

21.2.4 Controlled (conditional) operations

Controlled operations are essential to the whole concept of quantum computing. They are what enable logical operations to be controlled reversibly and produce the entanglement between quantum states that are thought to lead to the expected increases of computing power. As discussed above the Gottesman-Knill theorem indicates that superposition, by itself, cannot account for the "quantum speedup."

A controlled operation is an operation on one qubit that is performed conditionally on the state of a second qubit. The control bit is left unchanged. The fundamental controlled operation is the controlled NOT (CNOT). In this operation the control qubit "C" is unchanged and the target qubit has the NOT operation applied if $|C\rangle=|1\rangle$ and is unchanged if $|C\rangle=|0\rangle$. It is represented as shown in Fig. 21.5.

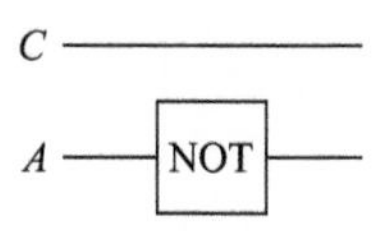

Fig. 21.4 Primitive quantum circuit.

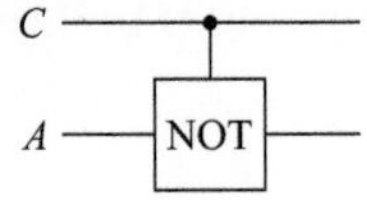

Fig. 21.5 CNOT gate as as quantum circuit element.

Thus using the notation of the joint state as $|CA\rangle$ we have

input	*output*
$\|00\rangle$	$\|00\rangle$
$\|01\rangle$	$\|01\rangle$
$\|10\rangle$	$\|11\rangle$
$\|11\rangle$	$\|10\rangle$

or written as a matrix in the basis (from left to right and top to bottom) $|00\rangle, |01\rangle, |10\rangle, |11\rangle$:

$$\begin{bmatrix} 1\,0\,0\,0 \\ 0\,1\,0\,0 \\ 0\,0\,0\,1 \\ 0\,0\,1\,0 \end{bmatrix}$$

We will see that it is possible to generate a controlled U operation (CU) where U is any unitary operation starting from the CNOT and single qubit rotations.

21.2.5 $\pi/2$ rotations and Hadamard gates

As is well known $\pi/2$ rotations take a spin-1/2 from pointing in say the z direction to the x, y plane where they precess about the z axis taken as the field direction. This is a way to produce a superpositions state starting with a σ_z eigenstate. A rotation of $\pi/2$ about the y axis is represented by

$$R_y(\pi/2) = \frac{1}{\sqrt{2}} \begin{bmatrix} 1 & -1 \\ 1 & 1 \end{bmatrix}.$$

Quantum computing people like to work with what they call the Hadamard gate which is given by

$$H = \frac{1}{\sqrt{2}} \begin{bmatrix} 1 & 1 \\ 1 & -1 \end{bmatrix}$$

which can be produced from rotations as we will see. The Hadamard transformation takes the state $|0\rangle, (\sigma_z = +1)$ to the state $[|0\rangle + |1\rangle]$ and the state $|1\rangle, (\sigma_z = -1)$ to $[|0\rangle - |1\rangle]$. Note that the Hadamard transform is its own inverse, $H^2 = 1$:

$$\frac{1}{\sqrt{2}} \begin{bmatrix} 1 & 1 \\ 1 & -1 \end{bmatrix} \times \frac{1}{\sqrt{2}} \begin{bmatrix} 1 & 1 \\ 1 & -1 \end{bmatrix} = \begin{bmatrix} 1\,0 \\ 0\,1 \end{bmatrix}.$$

21.2.6 Simple quantum "algorithms"

Assume we have a unitary operation U with eigenstates $|\pm\rangle$ so that $U\,|\pm\rangle = \pm 1$.

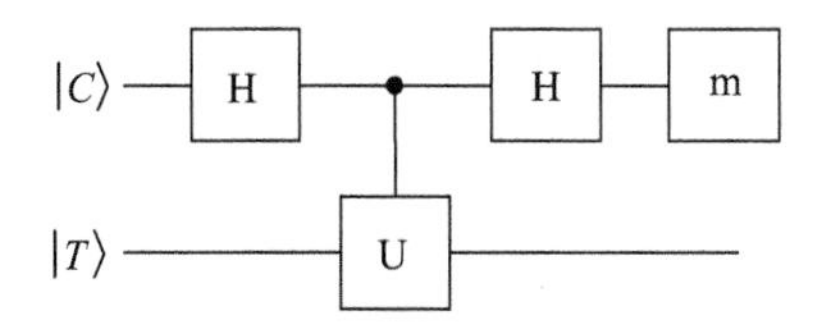

Fig. 21.6 Simple quantum algorithm showing the use of Hadamard gates to produce superposition and a controlled gate to produce entangelment, with m representing the masurement of state C.

Then the circuit shown in Fig. 21.6 operates as follows. C is the control qubit and T is the target. The system starts in the state $|\psi_0\rangle = |0\rangle_C [a|+\rangle + b|-\rangle]_T$. The H transformation on C produces the state

$$|\psi_1\rangle = \frac{1}{\sqrt{2}}[|0\rangle + |1\rangle]_C [a|+\rangle + b|-\rangle]_T. \tag{21.24}$$

The U transformation is then applied only if $C = 1$, so that the state of the system after the U transformation is

$$\begin{aligned}|\psi_2\rangle &= \frac{1}{\sqrt{2}}[|0\rangle_C [a|+\rangle + b|-\rangle]_T + |1\rangle_C U [a|+\rangle + b|-\rangle]_T] = \\ &= \frac{1}{\sqrt{2}}[|0\rangle_C [a|+\rangle + b|-\rangle]_T + |1\rangle_C [a|+\rangle - b|-\rangle]_T] = \\ &= \frac{1}{\sqrt{2}}[[|0\rangle + |1\rangle]_C a|+\rangle_T + [|0\rangle - |1\rangle]_C b|-\rangle_T]\end{aligned} \tag{21.25}$$

and the second H transformation on C $\left(|0\rangle \leftrightarrow \frac{1}{\sqrt{2}}[|0\rangle + |1\rangle], \quad |1\rangle \leftrightarrow \frac{1}{\sqrt{2}}[|0\rangle - |1\rangle]\right)$ produces the state

$$|\psi_3\rangle = |0\rangle_C a|+\rangle_T + |1\rangle_C b|-\rangle_T. \tag{21.26}$$

Measuring the state of C in the σ_z basis (called the computational basis because it represents the binary value of the qubit), represented in Fig. 21.6 by the box with "m," yields the result $|0\rangle$ with probability $|a|^2$ and $|1\rangle$ with probability $|b|^2$. In the former case the target qubit is known to be in the state $|+\rangle_T$ and in the latter case it is in the state $|-\rangle_T$.

This simple example shows all the essential features of quantum computation. The first H operation produces a superposition state of the C qubit (21.24), which is then entangled with the state of the target by the controlled U operation (21.25). The final H operation produces the desired state by means of quantum interference (21.26).

A more useful variant of this procedure occurs if we take the unitary operation U to have an eigenstate $|u\rangle$ so that $U|u\rangle = e^{i\phi}|u\rangle$.

Then the circuit shown in Fig. 21.6 is operated as follows. The system starts in the state $|\psi_0\rangle = |0\rangle_C |u\rangle_T$. The H transformation on C produces the state $|\psi_1\rangle = [|0\rangle + |1\rangle]_C |u\rangle_T$ (superposition). The U transformation is then applied only if $C = 1$, so that the state of the system after the U transformation is (entanglement)

$$|\psi_2\rangle = [|0\rangle_C |u\rangle_T + |1\rangle_C U|u\rangle_T] = \left[|0\rangle_C |u\rangle_T + |1\rangle_C e^{i\phi}|u\rangle_T\right].$$

Here we see how the phase shift applied to the second qubit is "kicked back" to the $|1\rangle_C$ component of the first qubit.[19]

The second H transformation produces the state (interference)

$$\begin{aligned}|\psi_2\rangle &= |0\rangle_C \left(1+e^{i\phi}\right)|u\rangle_T + |1\rangle_C \left(1-e^{i\phi}\right)|u\rangle_T \\ &= e^{i\phi/2}\left[|0\rangle_C\left(e^{-i\phi/2}+e^{i\phi/2}\right)+|1\rangle_C\left(e^{-i\phi/2}-e^{i\phi/2}\right)\right]|u\rangle_T \\ &= e^{i\phi/2}\left[|0\rangle_C \cos\phi/2 - i\,|1\rangle_C \sin\phi/2\right]|u\rangle_T\,.\end{aligned}$$

Thus in this case measurement of C produces the state $|0\rangle_C$ with probability $\cos^2\phi/2$ and the phase can be determined to arbitrary accuracy be repeating the procedure. Phase determination will be seen to play an important role in the factoring algorithm.

21.3 Unitary operations

In this section and the next we will show in more detail how unitary operations can be realized and will outline a plausible possibility for a physical realization.

21.3.1 Single qubit unitary transformations

These are operations carried out on single qubits independent of the other qubits in the system. An example is the H gate described in the previous section.

Since qubits are represented by systems with spin-1/2 or more correctly, by two selected states of a more complex system, whose behavior can be modeled by an effective spin-1/2 (as long as the matrix elements of the interactions with other states can be neglected, any two state system can be described by a spin-1/2, as we have seen in Section 19.7). Unitary operations can be represented by rotations of the equivalent spin and can be produced by the application of appropriate external magnetic or laser field pulses.

The building blocks for unitary operations are then the rotations around the three coordinate axes

$$\begin{aligned}R_x(\theta) &= e^{-i\sigma_x\theta/2} = \cos\frac{\theta}{2}I - i\sigma_x\sin\frac{\theta}{2} = \begin{bmatrix}\cos\frac{\theta}{2} & -i\sin\frac{\theta}{2}\\ -i\sin\frac{\theta}{2} & \cos\frac{\theta}{2}\end{bmatrix}_z \\ R_y(\theta) &= e^{-i\sigma_y\theta/2} = \cos\frac{\theta}{2}I - i\sigma_y\sin\frac{\theta}{2} = \begin{bmatrix}\cos\frac{\theta}{2} & -\sin\frac{\theta}{2}\\ \sin\frac{\theta}{2} & \cos\frac{\theta}{2}\end{bmatrix}_z \\ R_z(\theta) &= e^{-i\sigma_z\theta/2} = \cos\frac{\theta}{2}I - i\sigma_z\sin\frac{\theta}{2} = \begin{bmatrix}e^{-i\frac{\theta}{2}} & 0\\ 0 & e^{i\frac{\theta}{2}}\end{bmatrix}_z\end{aligned} \tag{21.27}$$

where $R_i(\theta)$ is the operator for a rotation by an angle θ about the i axis and the matrices are written in the basis where σ_z is diagonal (computational basis). These equations follow simply from the rule

$$e^{i\sigma_k\theta} = \cos\theta + i\sigma_k\sin\theta$$

[19] Cleve, R., Ekert, A., Macchiavello, C. and Mosca, M., *Quantum Algorithms Revisited*, Proc. Roy. Soc. **A459**, 339 (1998).

which in turn follows from expanding the exponential in a power series and the fact that $(\sigma_k)^2 = 1$. Since the σ matrices anticommute, e.g.,

$$\sigma_x\sigma_y + \sigma_y\sigma_x = 0$$
$$\sigma_x\sigma_z + \sigma_y\sigma_z = 0$$

we have

$$\sigma_x\sigma_y\sigma_x = -\sigma_y \tag{21.28}$$
$$\sigma_x\sigma_z\sigma_x = -\sigma_z. \tag{21.29}$$

From this we find

$$\sigma_x R_y(\theta)\,\sigma_x = R_y(-\theta) \tag{21.30}$$
$$\sigma_x R_z(\theta)\,\sigma_x = R_z(-\theta) \tag{21.31}$$

using (21.27). Note that

$$\sigma_x = \begin{bmatrix} 0 & 1 \\ 1 & 0 \end{bmatrix} = iR_x(\pi)$$

so that the operation σ_x can be realized as a simple rotation.

21.3.1.1 The phase gate

This is the unitary gate given by

$$\begin{bmatrix} 1 & 0 \\ 0 & e^{i\frac{\pi}{2}} \end{bmatrix} = \begin{bmatrix} 1 & 0 \\ 0 & i \end{bmatrix}. \tag{21.32}$$

21.3.1.2 The Hadamard gate

From equations (21.27) we have

$$R_z(\pi) = \begin{bmatrix} -i & 0 \\ 0 & i \end{bmatrix}_z$$
$$R_y(\pi/2) = \frac{1}{\sqrt{2}}\begin{bmatrix} 1 & -1 \\ 1 & 1 \end{bmatrix}_z$$

thus

$$R_y(\pi/2)\cdot R_z(\pi) = \frac{1}{\sqrt{2}}\begin{bmatrix} 1 & -1 \\ 1 & 1 \end{bmatrix}\begin{bmatrix} -i & 0 \\ 0 & i \end{bmatrix} = \frac{1}{\sqrt{2}}\begin{bmatrix} -i & -i \\ -i & i \end{bmatrix} = \tag{21.33}$$
$$= -i\frac{1}{\sqrt{2}}\begin{bmatrix} 1 & 1 \\ 1 & -1 \end{bmatrix} = -iH$$

so that the Hadamard transform can be produced by rotations up to an unimportant phase.

21.3.1.3 General single qubit unitary transform

Any single qubit operation can be written as

$$A = \begin{bmatrix} a & b \\ c & d \end{bmatrix}$$

and thus depends on four complex (eight real) quantities.

The unitary condition $AA^* = 1$ yields four relations.

$$\begin{bmatrix} a & b \\ c & d \end{bmatrix} \begin{bmatrix} a^* & b^* \\ c^* & d^* \end{bmatrix} = \begin{bmatrix} aa^* + bc^* & ab^* + bd^* \\ ca^* + dc^* & cb^* + dd^* \end{bmatrix} = \begin{bmatrix} 1 & 0 \\ 0 & 1 \end{bmatrix}$$

so that a unitary operations is specified by four real parameters. Thus we can write an arbitrary unitary transformation as

$$\begin{aligned} U &= e^{i\alpha} R_z(\beta) R_y(\gamma) R_z(\delta) = \\ &= e^{i\alpha} \begin{bmatrix} e^{-i\beta/2} & 0 \\ 0 & e^{i\beta/2} \end{bmatrix} \begin{bmatrix} \cos\frac{\gamma}{2} & -\sin\frac{\gamma}{2} \\ \sin\frac{\gamma}{2} & \cos\frac{\gamma}{2} \end{bmatrix} \begin{bmatrix} e^{-i\delta/2} & 0 \\ 0 & e^{i\delta/2} \end{bmatrix} \\ &= \begin{bmatrix} e^{i\alpha} e^{-\frac{1}{2}i\beta} \left(\cos\frac{1}{2}\gamma\right) e^{-\frac{1}{2}i\delta} & -e^{i\alpha} e^{-\frac{1}{2}i\beta} \left(\sin\frac{1}{2}\gamma\right) e^{\frac{1}{2}i\delta} \\ e^{i\alpha} e^{\frac{1}{2}i\beta} \left(\sin\frac{1}{2}\gamma\right) e^{-\frac{1}{2}i\delta} & e^{i\alpha} e^{\frac{1}{2}i\beta} \left(\cos\frac{1}{2}\gamma\right) e^{\frac{1}{2}i\delta} \end{bmatrix} \end{aligned} \tag{21.34}$$

which can represent any unitary matrix as it contains 4 independent parameters and is clearly unitary.

21.3.2 Controlled unitary operations

21.3.2.1 Controlled general unitary transform

We may rewrite (21.34) as

$$U = e^{i\alpha} R_z(\beta) R_y(\gamma/2) R_y(\gamma/2) R_z\left(\frac{\delta+\beta}{2}\right) R_z\left(\frac{\delta-\beta}{2}\right)$$

using

$$R_k(\theta + \phi) = R_k(\theta) R_k(\phi) \tag{21.35}$$

then

$$\begin{aligned} U &= e^{i\alpha} R_z(\beta) R_y(\gamma/2) \sigma_x R_y(-\gamma/2) R_z\left(-\frac{\delta+\beta}{2}\right) \sigma_x R_z\left(\frac{\delta-\beta}{2}\right) \\ &\equiv e^{i\alpha} A \sigma_x B \sigma_x C \end{aligned} \tag{21.36}$$

using equations (21.30). Now the form (21.34) was chosen so that

$$ABC = 1 \tag{21.37}$$

as follows from (21.35). The idea is that pre- and post-multiplying by σ_x changes the sign of the angle of any rotation (21.30) around the y or z axes. The reason that this result is important is that if we can produce a controlled σ_x operation i.e., a controlled

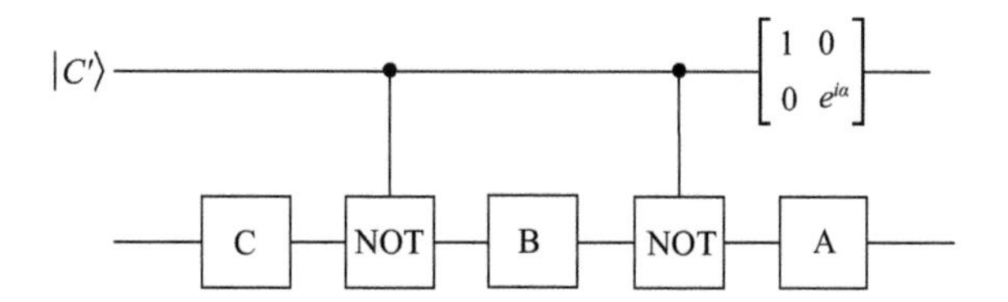

Fig. 21.7 Circuit to produce an arbitrary controlled unitary operation: $U = e^{i\alpha} A\sigma_x B\sigma_x C$.

not, CNOT, as discussed above, we can, using (21.36) and (21.37), produce a controlled U, (CU), from the CNOT and rotations.

$$\begin{aligned} CU &= e^{i\alpha} A(CNOT)B(CNOT)C \\ &= \left\{ \begin{array}{cc} e^{i\alpha} ABC = e^{i\alpha} & (C' = 0) \\ e^{i\alpha} U & (C' = 1) \end{array} \right\} \end{aligned}$$

where U is any single qubit unitary operation which is always realizable in the form (21.34) and C' is the control bit.

Fig. 21.7 shows the quantum circuit representation of this procedure.

21.3.2.2 Controlled Hadamard transform

Rewriting (21.33) we have

$$\begin{aligned} H &= iR_y(\pi/2) \cdot R_z(\pi) \\ &= e^{i\pi/2} R_z(0) \cdot R_y(\pi/2) \cdot R_z(\pi) \\ &= e^{i\pi/2} R_z(0) \cdot R_y(\pi/4) \cdot R_y(\pi/4) \cdot R_z(\pi/2) \cdot R_z(\pi/2) \\ &= e^{i\pi/2} R_z(0) \cdot R_y(\pi/4)\, \sigma_x R_y(-\pi/4) \cdot R_z(-\pi/2)\, \sigma_x \cdot R_z(\pi/2). \end{aligned}$$

Thus we can produce a controlled Hadamard transform with

$$\begin{aligned} \alpha &= \pi/2 \\ A &= R_z(0) \cdot R_y(\pi/4) = R_y(\pi/4) \\ B &= R_y(-\pi/4) \cdot R_z(-\pi/2) \\ C &= R_z(\pi/2) \end{aligned}$$

where $ABC = 1$ as required.

21.3.2.3 Controlled NOT from controlled Z

Assuming we have a controlled σ_z operation we can produce a controlled NOT by placing a Hadamard operation before and after the controlled σ_z (Fig. 21.8). When the control qubit is in the state $|0\rangle$ the σ_z does not operate and since $H^2 = 1$ the

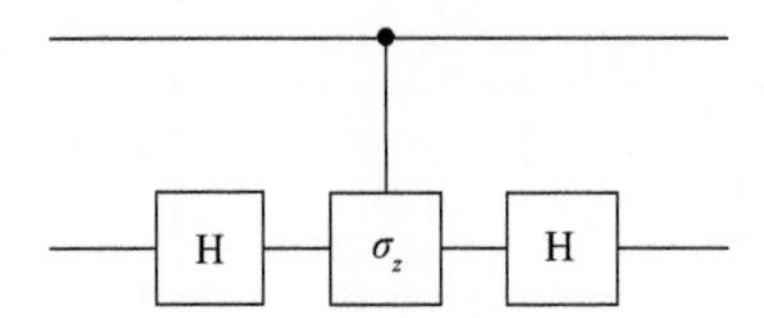

Fig. 21.8 Circuit producing a CNOT gate from a controlled Z gate.

target bit is unchanged. However, when the control bit is set (i.e., in the state $|1\rangle$ then we apply

$$H\sigma_z H = \frac{1}{\sqrt{2}}\begin{bmatrix}1 & 1\\ 1 & -1\end{bmatrix}\begin{bmatrix}1 & 0\\ 0 & -1\end{bmatrix}\frac{1}{\sqrt{2}}\begin{bmatrix}1 & 1\\ 1 & -1\end{bmatrix}$$
$$= \begin{bmatrix}0 & 1\\ 1 & 0\end{bmatrix} = \sigma_x = NOT.$$

21.4 A physical model of a quantum computer

Several physical systems have been proposed and are being intensively studied as possible realizations of quantum computers. Among these are single photons interacting by means of non-linear optical materials, photons in a cavity interacting by means of a single atom stored in the cavity, trapped ions, molecules in a liquid (NMR) as well as the nuclear spins of impurity atoms in a silicon matrix, electrons floating on the surface of ^{4}He and superconducting circuits. Each of these has serious drawbacks as the requirements for a quantum computer are somewhat contradictory and it obviously is not a simple task to find an effective compromise. Note that a system of isolated spins-1/2 cannot be the basis of qubit as we need to perform controlled operations, which require interactions between spins.

As an illustration we will show how an idealized system of trapped ions can, in principle, be used to perform computations. Ions can be contained in a linear trap where they can be held under certain circumstances at more or less fixed positions equally spaced along a straight line. This is reminiscent of a classical data register.

21.4.1 Interacting qubits

The positions of each atom can oscillate around their equilibrium position. Due to the mutual coulomb repulsion between the ions phonons can propagate along the chain. The lowest energy mode of this motion would be where the atoms keep a constant distance from each other and oscillate in the trapping potential. This mode is called the center of mass mode (it is the zero q mode) and is an essential component of the model of quantum computing we want to discuss. This mode can be excited with an arbitrary number, n, of phonons to an energy $n\hbar\omega_z$, but we will assume that only the state $n = 1$ is excited.

We select two internal states of the ion to represent a qubit and the states of the center of mass mode represent another qubit. Note that since all the ions in the trap partake of the motion of the c of m mode, this mode can serve as a bus ("quantum bus") passing information between the qubits represented by the ions in the chain. The states of the two qubits (one ion and the c of m mode) will be designated $|m, n\rangle$ where $m = 0, 1$ designates the state of the atom and $n = 0, 1$ that of the c of m mode. Then the energy levels are as shown in Fig. 21.9.

The Hamiltonian for the system in the absence of an external field is

$$H_o = \hbar\omega_o S_z + \hbar\omega_z \widehat{n} \tag{21.38}$$

where $\widehat{n}$ is the number operator of the c of m mode.

In the presence of an external magnetic field, $B_1\widehat{x}\cos(kz - \omega t + \phi)$, the interaction Hamiltonian will be

$$\begin{aligned} H_I &= \mu B_1 S_x \cos(kz - \omega t + \phi) \\ &= \mu B_1 \left(\frac{S_+ + S_-}{2}\right)\left(\frac{e^{i(kz-\omega t+\phi)} + e^{-i(kz-\omega t+\phi)}}{2}\right). \end{aligned}$$

Now the ions are trapped in the trap potential where the quantum states are those of a harmonic oscillator. The position z of an ion is then an operator and can be written

$$z = z_o\left(a + a^\dagger\right)$$

where $a, a^\dagger$ are the usual quantum oscillator ladder operators. For $kz_o \ll 1$ we can write

$$\begin{aligned} H_I &= \mu B_1 \left(\frac{S_+ + S_-}{2}\right)\left(\frac{(1 + ikz)\, e^{i(-\omega t+\phi)} + (1 - ikz)\, e^{-i(-\omega t+\phi)}}{2}\right) \\ &= H_1^A + H_1^B \end{aligned} \tag{21.39}$$

where

$$\begin{aligned} H_1^A &= \frac{\mu B_1}{2 \cdot 2}\left(S_+ e^{i(-\omega t+\phi)} + S_- e^{-i(-\omega t+\phi)}\right) + \quad \text{counter} - \text{rotating terms} \\ H_1^B &= i\frac{\mu B_1}{2}\left(\frac{S_+ + S_-}{2}\right) kz_o\left(a + a^\dagger\right)\left(e^{i(-\omega t+\phi)} - e^{-i(-\omega t+\phi)}\right). \end{aligned} \tag{21.40}$$

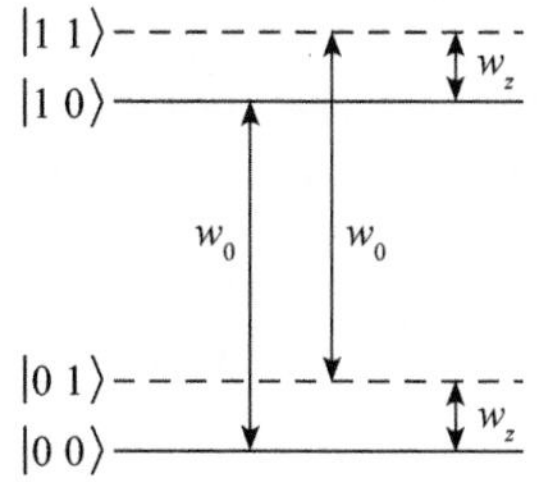

Fig. 21.9 Energy levels for a 2 qubit register based on trapped ions.

21.4.2 Single qubit operations

The Hamiltonian H_o causes the operators $S_\pm$ to evolve as

$$S_\pm(t) = S_\pm e^{\pm i(\omega_o)t}$$

and the operators $a^\dagger, a$ to evolve as

$$a(t) = ae^{-i\omega_z t}$$
$$a^\dagger(t) = a^\dagger e^{i\omega_z t}$$

so the terms $S_\pm(t)e^{\pm i\omega t}$ in H_1^A vary at a frequency of $(\omega + \omega_o)$ and so can be neglected (counter-rotating terms) compared to the terms we kept $\left(S_\pm(t)e^{\mp i\omega t}\right)$ for which the time variation is at a frequency $(\omega - \omega_o)$. This is the rotating wave approximation. Applying an external field at a frequency of $(\omega = \omega_o)$ effectively turns on a Hamiltonian

$$H_1^A = \frac{\mu B_1}{2 \cdot 2}\left(S_+ e^{i\phi} + S_- e^{-i\phi}\right)$$

so that by choosing ϕ and the strength and duration of the field we can apply any combinations of $R_x(\theta) = e^{-iS_x\theta}$ and $R_y(\theta) = e^{-iS_y\theta}$ and thus perform arbitrary transformations on the m qubit in the $|m, n\rangle$ notation defined above.

For example

$$R_x(\pi) \cdot R_y\left(\frac{\pi}{2}\right) = -i\begin{bmatrix} 0 & 1 \\ 1 & 0 \end{bmatrix}\begin{bmatrix} 1 & -1 \\ 1 & 1 \end{bmatrix}\frac{1}{\sqrt{2}}$$
$$= \frac{-i}{\sqrt{2}}\begin{bmatrix} 1 & 1 \\ 1 & -1 \end{bmatrix} = -iH,$$

i.e., another way of producing the Hadamard transform.

21.4.3 Two qubit operations

21.4.3.1 Swapping qubits between the ion and the c of m mode

The four terms in the interaction H_1^B $\left(\sim (S_+ + S_-)(a + a^\dagger)\right)$ produce transitions that couple the two qubits represented by the internal states and the c of m mode and resonate at $\omega_o \pm \omega_z$. By applying a pulse at $(\omega_o - \omega_z)$ we can induce transitions between the states $|01\rangle$ and $|10\rangle$, i.e., the terms in $H_1^B \sim S_+ a e^{i\phi} + S_- a^\dagger e^{-i\phi}$. By choosing ϕ and the strength of the pulse we can apply

$$R_y(\pi) = \begin{bmatrix} 0 & -1 \\ 1 & 0 \end{bmatrix}$$

to these two states effectively swapping the qubits so we can swap information between the c of m mode and any of the ions in the trap. Note individual ions can be addressed by focusing the laser pulse so that it only illuminates the desired ion.

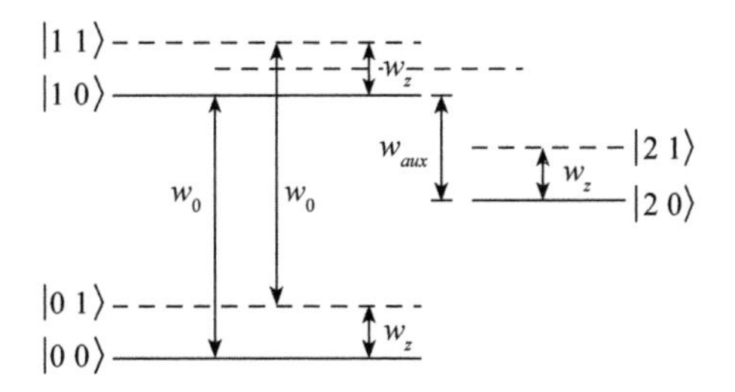

Fig. 21.10 Showing the use of a second internal state of the ion to produce controlled operations.

21.4.3.2 Controlled operations

In order to perform controlled operations it is convenient to introduce another internal energy state $m = 2$ as in Fig. 21.10.

Then applying a pulse at the frequency $\omega_{aux} + \omega_z$ can induce transitions between $|11\rangle$ and $|20\rangle$. Choosing the strength of the pulse to give a rotation $R_x(2\pi)$ or $R_y(2\pi)$ we change the state $|11\rangle$ into $-|11\rangle$ with all other states being unaffected. This is just the controlled σ_z operation:

$$\begin{bmatrix} 1 & 0 & 0 & 0 \\ 0 & 1 & 0 & 0 \\ 0 & 0 & 1 & 0 \\ 0 & 0 & 0 & -1 \end{bmatrix} \quad \begin{matrix} |00\rangle \\ |01\rangle \\ |10\rangle \\ |11\rangle \end{matrix}$$

Using this we can produce a controlled NOT operation by combining with 2 Hadamard transforms as in Fig. 21.8, Section (21.3.2.3).

21.5 Some additional algorithms

21.5.1 Deutsch's problem revisited

In the introduction we discussed a problem introduced by Deutsch to demonstrate some properties of quantum computing. This was given a quantum circuit that could calculate a two-valued function of a two-valued variable determine if the function is constant (i.e., $f(0) = f(1)$) or not. We have seen this can be accomplished by calculating $G = f(0) \oplus f(1)$. In the original publication Deutsch gave an algorithm that could calculate G using only a single evaluation of the function but gave the result with only a 50% probability.

Many years later Cleve et al.,[20] in a very interesting paper, gave a solution of the problem that gave the correct answer with probability equal to 1 while only requiring one application of the oracle.

The argument again makes use of a so-called oracle. This is a placeholder for an algorithm for calculating the desired function and is treated as a black box which transforms the state of two quantum registers $|x\rangle |y\rangle$ to

[20] Cleve, R., Ekert, A., Macchiavello, C. and Mosca, M., *Quantum Algorithms Revisited*, Proc. Roy. Soc. **A459**, 339 (1998).

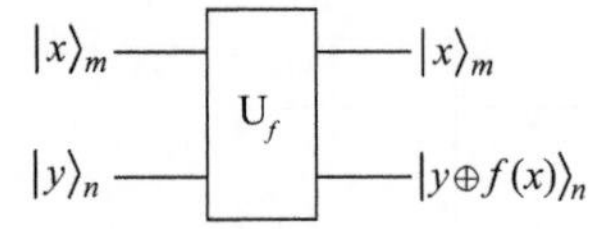

$|x\rangle_m$ — U_f — $|x\rangle_m$

$|y\rangle_n$ — U_f — $|y \oplus f(x)\rangle_n$

Fig. 21.11 A circuit for the oracle U_f. The subscripts m, n indicate the number of qubits in each register for the general case.

$$U_f |x\rangle |y\rangle = |x\rangle |y \oplus f(x)\rangle. \tag{21.41}$$

Deutsch considered the case where the operation U_f was very long and complicated so that the goal was to obtain results with as few applications of the oracle as possible. In the present case both registers consist of a single qubit ($m = n = 1$, see Fig. 21.11).

Applying U_f to the state $|x\rangle |-\rangle = |x\rangle [|0\rangle - |1\rangle]$ yields

$$U_f |x\rangle |-\rangle = |x\rangle [|0 \oplus f(x)\rangle - |1 \oplus f(x)\rangle] = |x\rangle [|f(x)\rangle - |NOT f(x)\rangle] \tag{21.42}$$

so with $f(x) = 0\,(1)$ the result is given by

$$U_f |x\rangle |-\rangle = (-1)^{f(x)} |x\rangle (|0\rangle - |1\rangle) = (-1)^{f(x)} |x\rangle |-\rangle. \tag{21.43}$$

Thus the states $|x\rangle |-\rangle$ are eigenstates of U_f with eigenvalues $(-1)^{f(x)}$.

The circuit for solving the Deutsch problem with 100% certainty and one use of the oracle is shown in Fig. 21.12.

At first glance this looks very strange. How can the output of the first bit depend on $f(x)$ when U_f just transmits this bit unchanged? The answer is in the "kick back" mechanism described above.

Using an obvious notation:

$$[A \otimes B] [|x\rangle \otimes |y\rangle] = A |x\rangle \otimes B |y\rangle. \tag{21.44}$$

(We will occasionally omit the $\otimes$ between the kets.) The action of this circuit can be described as follows.

The input to the oracle is given by

$$[H \otimes HX] [|0\rangle |0\rangle] = \frac{1}{2} [|0\rangle + |1\rangle] \otimes [|0\rangle - |1\rangle] = \frac{1}{2} [|0\rangle + |1\rangle] \otimes |-\rangle \tag{21.45}$$

and the output of the oracle is then seen to be

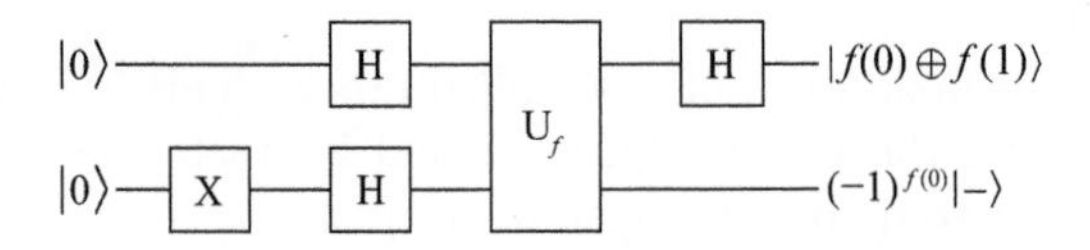

Fig. 21.12 Quantum circuit for solving the Deutsch problem with 100% probability using only one pass through the oracle.

$$
\begin{aligned}
2U_f\left[H\otimes HX\right]|0\rangle|0\rangle &= \sum_{x=0,1}|x\rangle\otimes\left[\left(|0\rangle\oplus|f(x)\rangle\right)-|1\rangle\oplus|f(x)\rangle\right] \\
&= \sum_{x=0,1}|x\rangle\otimes\left[\left(|f(x)\rangle\right)-|NOTf(x)\rangle\right] \\
&= \sum_{x=0,1}|x\rangle\otimes\left[\left[|0\rangle-|1\rangle\right]\right](-1)^{f(x)} \\
&= \left[(-1)^{f(0)}|0\rangle+(-1)^{f(1)}|1\rangle\right]\left[|0\rangle-|1\rangle\right].
\end{aligned} \tag{21.46}
$$

The final operation - $H\otimes I$ applied to this result yields for the first register

$$(-1)^{f(0)}\left[|0\rangle+|1\rangle\right]+(-1)^{f(1)}\left[|0\rangle-|1\rangle\right]\frac{1}{2}= \tag{21.47}$$

$$\frac{1}{2}\left((-1)^{f(0)}+(-1)^{f(1)}\right)|0\rangle+\left((-1)^{f(0)}-(-1)^{f(1)}\right)|1\rangle=|f(0)\oplus f(1)\rangle. \tag{21.48}$$

The final state of the system is then

$$|f(0)\oplus f(1)\rangle\otimes\frac{1}{\sqrt{2}}\left[|0\rangle-|1\rangle\right] \tag{21.49}$$

so that measurement of the first register yields the desired result.

21.5.2 Simple error correction

Physically existing realizations of qubits are notoriously prone to errors due to unavoidable interactions with the environment (e.g., decoherence—relaxation) so that any practical realization of a quantum computer will have to involve robust error correction and what is called fault-tolerant computing. These subjects have evolved appreciably and reached a high level of sophistication. Here we will only attempt to explain the simplest case in order to give a flavor of what is involved.

An obvious way to correct for errors (and this applies to classical computers as well) is to repeat the calculation many times and take the result given by the majority of the repetitions. In the simplest case of three repetitions a single error will be eliminated by taking the majority result as correct. Of course this is only useful when the probability of two errors is sufficiently small. Classically errors appear as bit flips.

We can apply this idea to a quantum computer by replacing each qubit by three equivalent qubits. Thus what is called the logical qubit in an arbitrary state is replaced by three physical bits

$$|\psi\rangle_L=\left[a|0\rangle+b|1\rangle\right]_L=a|000\rangle+b|111\rangle. \tag{21.50}$$

This (Bell-like) entangled state can be produced by the circuit in Fig. 21.13:

The first CNOT changes the original state $\left[a|000\rangle+b|100\rangle\right]$ to $\left[a|000\rangle+b|110\rangle\right]$ and the second CNOT produces the desired entangled state (21.50). These three qubits will then be sent through some quantum circuit resulting in the state $|x_2x_1x_0\rangle$ and a

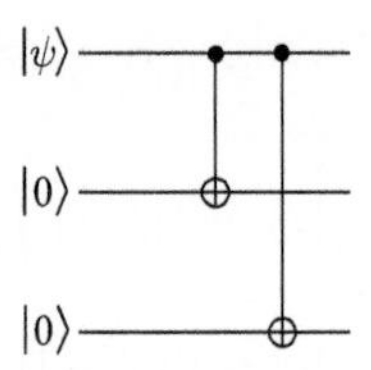

Fig. 21.13 Encoding a single qubit into three qbits for error detection.

bit flip of any bit will be apparent. Now the problem is that if we made a measurement on one of these qubits we would destroy any superposition and thus lose possible vital information. What we wish is to find out if the three qubits are in the same state or not; we need to determine $s_1 = x_2 \oplus x_1$ as well as $s_0 = x_2 \oplus x_0$. These are called error syndromes.

If either of these is nonzero then we have an error. In fact, for $s_1 s_0 = (01, 10, 11)$ the error is seen to be in (x_0, x_1, x_2). The circuit shown in Fig. 21.14 allows the determination of the values of the two syndromes without changing the state of the three "signal" qubits.

We can take the initial state of the syndrome bits to be $|a_1\rangle = |a_2\rangle = |0\rangle$ and we can understand the operation of the circuit by noting that if $x_1 = x_2$ the two CNOTs in the syndrome bit $|a_1\rangle$ do nothing so the output is $|a_1\rangle = |0\rangle = |x_2 \oplus x_1\rangle$ while, if $x_1 \neq x_2$, one of the CNOTs will operate and we will have $|a_1\rangle = |1\rangle = |x_2 \oplus x_1\rangle$. The second syndrome bit works in the same way. Thus we can detect which bit has flipped without altering the values of the bits. The faulty bit can then be corrected by applying the bit flip operator, X, to that bit.

In quantum computing bit flips are not the only error. We can also have random changes of phase, i.e., the state $|\psi\rangle = [a\,|0\rangle + b\,|1\rangle]$ can randomly change to $[a\,|0\rangle - b\,|1\rangle]$ which can be represented by the application of the operator $Z = \sigma_z$ to the qubit. Transforming to the basis $H\,|0\rangle\,, (|1\rangle) = |+\rangle\,, (|-\rangle)$ where $|\pm\rangle = |0\rangle \pm |1\rangle$ we see that we can write $|\psi\rangle = (a+b)\,|+\rangle + (a-b)\,|-\rangle$ and the phase flip changing the sign of b is equivalent to flipping the states $|+\rangle \Leftrightarrow |-\rangle$, i.e., a bit flip in the new basis. For error correction we use the three qubit states $|+++\rangle\,, |---\rangle$ obtained by applying Hadamard gates to each qubit in the output of Fig. 21.13, and apply the bit flip algorithm described above.

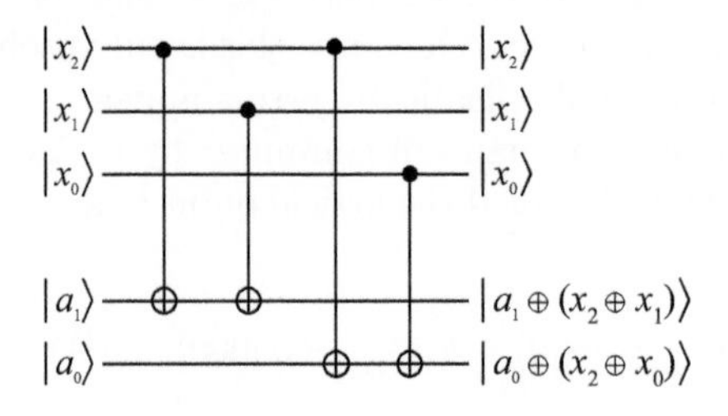

Fig. 21.14 Quantum algorithm for detecting a bit flip error

These are¡?TeX ?¿ only the most primitive methods available for error detection. Much more practical methods use coding into more qubits, for example Peter Shor has proposed an algorithm using nine qubits that can correct bit flip and phase errors at the same time.[21] In fact it can correct any arbitrary single-bit error The minimum number of qubits needed to protect a qubit from an arbitrary single bit error was later found to be five.[22] There are several books on quantum error correction.[23]

21.6 Factoring—the Holy Grail of quantum computing

The most important step to date in the development of quantum computing was the demonstration by Schor that a quantum computer would, in principle, be capable of factoring large numbers much more efficiently than classical computers. Factoring large numbers is a "difficult" problem for a classical computer; the amount of computation grows as the size of the number to be factored (exponentially in the number of bits used to represent the number) whereas by using the Schor algorithm a quantum computer should be able to solve the problem using resources that grow as the logarithm of the number (polynomial in the number of bits). The publication of this proposal[24] led to a rapid increase of interest in quantum computing because the difficulty of factoring large numbers is the basis of the very widely used public key-private key method of encrypting information. If large numbers could be factored these codes could be broken and the mere prospect of this is enough to release large amounts of research resources for this field. However, the opinion has been expressed that a research program whose sole goal is the destruction of another technology does not have a rosy future. So the real question is: Will there be other applications where quantum computing can offer significant advantages? There is an algorithm for searching databases due to Grover that offers some minor improvements over classical computation, but it does not seem terribly exciting.

However, the possibility of simulating quantum systems is very interesting. Our goal here is to understand the principle of the Schor algorithm.

21.6.1 Some number theory

Some years ago it was commonplace among physics students to consider number theory as the branch of mathematics that was most distant from any possible applications in the physical world. But now, with the widespread use of public key-private key codes and the heightened interest in quantum computing, number theory has taken center stage as a subject of applied research. In addition there are many other vital applications of number theory to real world problems.

21.6.1.1 Modular arithmetic

Modular arithmetic is the arithmetic of remainders after division. What we mean by this is that two numbers are said to be equal $\mod N$ if they both have the same remainder when divided by N:

[21] Shor, P.W., *Scheme for reducing decoherencee in quantum computer memory*, Phys. Rev. **A52**, R2493, (1995).

[22] Knill, E., Laflamme, R., Martinez, R. and Negrevergne, C., *Benchmarking Quantum Computers: The Five-Qubit Error Correcting Code*, Phys. Rev. Lett. **88**, 5811 (2001).

[23] For a recent book on quantum error correction see La Guardia, G.G. *Quantum Error Corrections: Symmetric, Asymmetric, Synchronizable and Convolutional Codes* (Springer, Nature, (2020)).

[24] Shor, P.W., (1994), op. cit., footnote 15.

$$a = b \mod N$$

means

$$a = b + kN$$

where k is an integer. One of the most amazing results in mathematics is the following

$$a^{p-1} = 1 \mod p \tag{21.51}$$

where p is a prime number. Rewriting this as

$$a^p = a \mod p \tag{21.52}$$

we can prove it by using the binomial theorem in a proof by induction. Assume (21.52) holds for a and show that this implies that it holds for $a+1$. It clearly holds for $a = 1$.

$$(a+b)^p = \sum_{k=0}^{p} \frac{p!}{k!\,(p-k)!} a^{p-k} b^k \tag{21.53}$$

$$(a+1)^p = a^p + \sum_{k=1}^{p-1} \frac{p!}{k!\,(p-k)!} a^{p-k} + 1 \tag{21.54}$$

Since $p! = 0 \mod p$ we have

$$\begin{aligned}(a+1)^p &= (a^p + 1) \mod p \\ &= (a+1) \mod p\end{aligned}$$

assuming the validity of (21.52). Thus we have established (21.52) and (21.51). The relation (21.52) also holds if a is a multiple of p since then both a^p and $a = 0 \mod p$.

Following this it can be shown that for any two integers a, N with no common factors there is a number r, such that

$$a^r = 1 \mod N. \tag{21.55}$$

The lowest integer r for which this holds is called the order of a modulo N.

21.6.1.2 From the order to the factors

Eq. (21.55) can be written

$$a^r - 1 = 0 \mod N = kN$$

or

$$\left(a^{r/2} - 1\right)\left(a^{r/2} + 1\right) = kN.$$

Thus if we knew r we could calculate (gcd = greatest common denominator), gcd $(a^{r/2} - 1, N)$, and gcd $(a^{r/2} + 1, N)$, one of which should be a factor of N. The gcd can be calculated efficiently using classical computation methods.

As a trivial example we take the infamous case of factoring 15.

For $a = 2$ we have

$$2^4 = 16 = 1 \mod 15$$

so that

$$\begin{aligned} 2^{r/2} &= 2^2 = 4 \\ 2^{r/2} - 1 &= 3 \\ 2^{r/2} + 1 &= 5. \end{aligned}$$

On the other hand if we took $a = 7$ we would have

$$\begin{aligned} &7^2 = 49 = 4 \mod 15 \\ &7^3 = 7 \cdot 7^2 = 7 \cdot 4 \mod 15 = 28 \mod 15 = 13 \\ &7^4 = 7 \cdot 13 \mod 15 = 91 = 1 \mod 15, \end{aligned}$$

so

$$\begin{aligned} 7^{r/2} &= 7^2 = 49 \\ 7^{r/2} - 1 &= 48 = 16 \cdot 3 \\ 7^{r/2} + 1 &= 50 = 5.5.2, \end{aligned}$$

so both $\left(7^{r/2} - 1\right)$ and $\left(7^{r/2} + 1\right)$ share a common factor with 15.

21.6.2 Quantum calculation of the order of a number

To see how a quantum computer can find the order of a number we consider that we have a series of n qubits forming a register that can store numbers up to $2^n - 1$. If the qubit register is storing the number y we have to produce a series of operations on the register that would yield the representation of $(xy \mod N)$. The argument is that this can be performed by a series of classical logical gates and any series of classical gates can be duplicated by a set of quantum gates consisting of the controlled NOT and single qubit gates. So we assume that we can produce the operation U_x on our register, defined so that

$$U_x \left|y\right\rangle = \left|xy \mod N\right\rangle. \tag{21.56}$$

Applying this to the register in the state $|1\rangle$ we obtain

$$U_x \left|1\right\rangle = \left|x \mod N\right\rangle \tag{21.57}$$

and by repeated applications we obtain

$$U_x^k \left|1\right\rangle = \left|x^k \mod N\right\rangle. \tag{21.58}$$

This sequence of operations is periodic with period $k = r$ since $x^r \mod N = 1$ when r is the order of x mod N. It is helpful to look for the eigenstates of U:

$$U_x \left|u_s\right\rangle = u_s \left|u_s\right\rangle. \tag{21.59}$$

We try to write these in the form

$$|u_s\rangle = \frac{1}{\sqrt{r}} \sum_{k=0}^{r-1} a_k \left|x^k \mod N\right\rangle \tag{21.60}$$

where $\sqrt{r}$ is a normalization factor and

$$\langle u_s | u_s \rangle = \frac{1}{r} \sum_k \sum_{k'} a_k a_k'^* \left\langle x^{k'} \mod N \middle| x^k \mod N \right\rangle \tag{21.61}$$

$$= \frac{1}{r} \sum_k |a_k|^2 \tag{21.62}$$

$$\sum_k |a_k|^2 = r \tag{21.63}$$

where we used the orthogonality of states representing different numbers in the qubit register.

Then

$$\begin{aligned} U_x |u_s\rangle &= \frac{1}{\sqrt{r}} \sum_{k=0}^{r-1} a_k \left|x^{k+1} \mod N\right\rangle \\ &= \frac{1}{\sqrt{r}} \sum_{k'=0}^{r-1} a_{k'-1} \left|x^{k'} \mod N\right\rangle \\ &\equiv u_s \frac{1}{\sqrt{r}} \sum_{k=0}^{r-1} a_k \left|x^k \mod N\right\rangle \end{aligned}$$

where in the first step we used the fact that $a_{k+r} = a_k$. Thus we have

$$\begin{aligned} a_{k-1} &= u_s a_k \\ a_{k-2} &= u_s a_{k-1} \\ a_{k-2} &= u_s^2 a_k \\ a_{k-m} &= u_s^m a_k \end{aligned} \tag{21.64}$$

as well as $(u_s)^r = 1$. Thus we can write

$$u_s = e^{i2\pi \frac{s}{r}} \tag{21.65}$$

with s an integer and if we take $a_o = 1$ we have

$$\begin{aligned} a_{-k} &= u_s^k = e^{i2\pi \frac{ks}{r}} \\ a_k &= u_s^{-k} = e^{-i2\pi \frac{ks}{r}}. \end{aligned} \tag{21.66}$$

Then

$$|u_s\rangle = \frac{1}{\sqrt{r}} \sum_{k=0}^{r-1} e^{-i2\pi \frac{ks}{r}} \left|x^k \mod N\right\rangle \tag{21.67}$$

and we can check this is an eigenstate of U_x

$$\begin{aligned} U_x |u_s\rangle &= \frac{1}{\sqrt{r}} \sum_{k=0}^{r-1} e^{-i2\pi \frac{ks}{r}} \left|x^{k+1} \mod N\right\rangle \\ &= \frac{1}{\sqrt{r}} \sum_{k'=0}^{r-1} e^{-i2\pi \frac{(k'-1)s}{r}} \left|x^{k'} \mod N\right\rangle \\ &= e^{i2\pi \frac{s}{r}} \frac{1}{\sqrt{r}} \sum_{k'=0}^{r-1} e^{-i2\pi \frac{k's}{r}} \left|x^{k'} \mod N\right\rangle \\ &= e^{i2\pi \frac{s}{r}} |u_s\rangle = u_s |u_s\rangle. \end{aligned} \tag{21.68}$$

Now if we could produce an eigenstate $|u_s\rangle$ of U_x we could find the phase, $2\pi\frac{s}{r}$, by applying the procedure discussed at the end of Section 21.2.6 after Fig. 21.6, and from this we can try to determine r. But we cannot produce an eigenstate $|u_s\rangle$ as we do not know r.

However, we see that

$$\begin{aligned} \sum_{s=0}^{r-1} |u_s\rangle &= \sum_{s=0}^{r-1} \frac{1}{\sqrt{r}} \sum_{k=0}^{r-1} e^{-i2\pi \frac{ks}{r}} \left|x^k \mod N\right\rangle \\ &= \frac{1}{\sqrt{r}} \sum_{k=0}^{r-1} r\delta_{k0} \left|x^k \mod N\right\rangle \\ &= \sqrt{r} \left|x^0 \mod N\right\rangle = \sqrt{r} \left|1 \mod N\right\rangle \end{aligned} \tag{21.69}$$

and the ability to produce states like $|1\rangle$ or $|0\rangle$ is one of the basic requirements for realizing a quantum computer.

Since we don't have an eigenstate we have to modify the phase finding procedure. If we apply U_x j times in succession to the state $|1\rangle$, i.e., apply U_x^j, we would obtain the state

$$U_x^j |1\rangle = U_x^j \frac{1}{\sqrt{r}} \sum_{s=0}^{r-1} |u_s\rangle = \frac{1}{\sqrt{r}} \sum_{s=0}^{r-1} e^{i2\pi \frac{s}{r} j} |u_s\rangle. \tag{21.70}$$

Now we introduce another register $|j\rangle$ and we have to set up an operation $\tilde{U}$ so that

$$\tilde{U} |j\rangle |1\rangle = |j\rangle U_x^j |1\rangle = |j\rangle \frac{1}{\sqrt{r}} \sum_{s=0}^{r-1} e^{i2\pi \frac{s}{r} j} |u_s\rangle \tag{21.71}$$

that is the value in the $|j\rangle$ register controls the application of U_x^j to the other register.

If instead we started with a state that was a superposition of $|j\rangle$ states, $\sum_j |j\rangle\,|1\rangle$ we would have

$$\tilde{U}\sum_{j=0}^{2^n-1}|j\rangle\,|1\rangle=\sum_{j=0}^{2^n-1}|j\rangle\,U_x^j\,|1\rangle=\frac{1}{\sqrt{r}}\sum_{s=0}^{r-1}\left[\sum_{j=0}^{2^n-1}e^{i2\pi\frac{s}{r}j}\,|j\rangle\right]|u_s\rangle. \tag{21.72}$$

The n bit register holding the values j can be put into a superposition state by starting all the qubits in the state $|0\rangle$ and then applying H to each qubit. The initial state of the register is then

$$|0\rangle_{2^n-1}\cdots\cdot|0\rangle_1\,|0\rangle_0 \;\underset{\rightarrow}{H}\; (|0\rangle+|1\rangle)_{2^n-1}\cdots(|0\rangle+|1\rangle)_1\,(|0\rangle+|1\rangle)_o$$

$$=\sum_{j=0}^{2^n-1}|j\rangle \tag{21.73}$$

because every combination of $|0\rangle$ and $|1\rangle$ occurs in the product. In Fig. 21.15 we see how we can produce the state (21.72), the Hadamard gate acting on the first register producing the state (21.73), while the controlled operations on the second register starting in the state $|1\rangle$ produce the state (21.72).

The expression in square brackets in (21.72) reminds us of a Fourier transform, so that we now turn to a discussion of the quantum Fourier transform which is interesting in its own right and will be seen to provide the method to extract the value of r, actually the value of s/r for some s from (21.72).

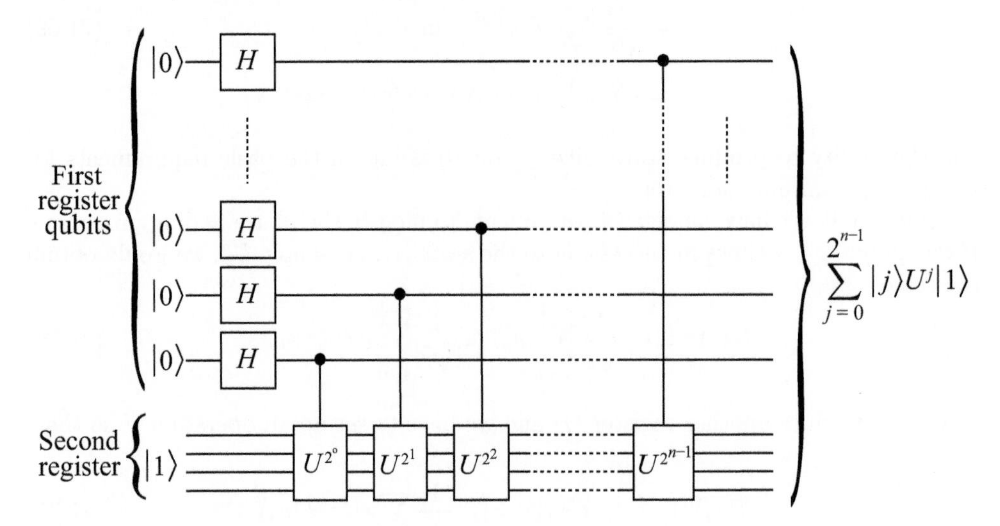

Fig. 21.15 Quantum circuit for producing the state $\sum_{s,j}|j\rangle\,U^j\,|1\rangle$. (As decribed in Nielsen and Chuang, op. cit., footnote p. 659.)

21.6.2.1 The quantum Fourier transform

Classically the discrete Fourier transform takes a set of numbers x_j ($j = 0..N-1$) and calculates the Fourier transform

$$y_k = \frac{1}{\sqrt{N}} \sum_{j=0}^{N-1} x_j e^{i2\pi jk/N}. \tag{21.74}$$

Quantum mechanically we want to construct an operator U_{FT} that acting on a computational state $|j\rangle$ produces the state

$$U_{FT} |j\rangle = \frac{1}{\sqrt{N}} \sum_{k=0}^{N-1} e^{i2\pi jk/N} |k\rangle. \tag{21.75}$$

Then if we had an arbitrary superposition of $|j\rangle$ states with amplitudes x_j we would obtain

$$\begin{aligned} U_{FT} \sum_{j=0}^{N-1} x_j |j\rangle &= \sum_{j=0}^{N-1} x_j \sum_{k=0}^{N-1} e^{i2\pi jk/N} |k\rangle \\ &= \sum_{k=0}^{N-1} y_k |k\rangle \end{aligned} \tag{21.76}$$

where y_k is given by the classical Fourier transform of x_j, Eq. (21.74). We can see that U_{FT} defined by (21.75) is unitary as follows:

$$\begin{aligned} \langle l| U_{FT} |j\rangle &= \frac{1}{\sqrt{N}} \sum_{k=0}^{N-1} e^{i2\pi jk/N} \langle l\, |k\rangle \\ &= \frac{1}{\sqrt{N}} e^{i2\pi jl/N}. \end{aligned} \tag{21.77}$$

Then

$$\langle l| U_{FT}^{+} |j\rangle = \langle j| U_{FT} |l\rangle^* = \frac{1}{\sqrt{N}} e^{-i2\pi jl/N} \tag{21.78}$$

and

$$\begin{aligned} \langle l| U_{FT}^{+} U_{FT} |j\rangle &= \sum_k \langle l| U_{FT}^{+} |k\rangle \langle k| U_{FT} |j\rangle = \frac{1}{N} \sum_k e^{-i2\pi lk/N} e^{i2\pi jk/N} \\ &= \frac{1}{N} \sum_{k=0}^{N-1} e^{i2\pi \frac{k(j-l)}{N}} = \delta_{jl} = \langle l| 1 |j\rangle \end{aligned} \tag{21.79}$$

so that we have $U_{FT}^{+} U_{FT} = 1$, i.e., U_{FT} is unitary, and the inverse of U_{FT} satisfies

$$U_{FT}^{-1} |j\rangle = U_{FT}^{+} |j\rangle = \frac{1}{\sqrt{N}} \sum_{k=0}^{N-1} e^{-i2\pi jk/N} |k\rangle. \tag{21.80}$$

21.6.2.2 The inverse Fourier transform leads to the order

Applying this to the first register of the state (21.72) we have

$$\begin{aligned}\frac{1}{\sqrt{r}}\sum_{s=0}^{r-1}\left[\sum_{j=0}^{2^n-1} e^{i2\pi\frac{s}{r}j}U_{FT}^{-1}|j\rangle\right]|u_s\rangle &= \frac{1}{\sqrt{r}}\sum_{s=0}^{r-1}\left[\sum_{j=0}^{2^n-1} e^{i2\pi\frac{s}{r}j}\frac{1}{\sqrt{2^n}}\sum_{k=0}^{N-1}e^{-i2\pi\frac{jk}{N}}|k\rangle\right]|u_s\rangle \\ &= \frac{1}{\sqrt{r}}\frac{1}{\sqrt{N}}\sum_{s=0}^{r-1}\sum_{k=0}^{N-1}\left[\sum_{j=0}^{2^n-1}e^{i2\pi j\left(\frac{s}{r}-\frac{k}{N}\right)}|k\rangle\right]|u_s\rangle\end{aligned} \tag{21.81}$$

where $N = 2^n$ and the sum over j is zero unless $k = \frac{s}{r}N$ so that we obtain the state

$$\frac{1}{\sqrt{r}}\frac{1}{\sqrt{N}}\sum_{s=0}^{r-1}\left|\frac{s}{r}N\right\rangle|u_s\rangle.$$

Measurement of the first register then yields the value of r, i.e., we get one of the values $\phi_s = \left(N\frac{s}{r}\right)$ with s chosen at random. Then we can find r by a continued fraction expansion of ϕ_s/N. (See e.g., Nielsen and Chuang.[25])

Note how this procedure makes use of the features of quantum computing. First we have the superposition of the states $|j\rangle$ in Eq. (21.73) and the superposition of eq. (21.69). The series of controlled U operations then produces the highly entangled state (21.72), while quantum interference in the state (21.81) selects out the state containing the desired information. We have calculated a function , $x^j \mod N$ simultaneously for a range of j values using superposition and were able to extract the desired result, which was not the values of the function but its period.

21.6.2.3 Implementation of the quantum Fourier transform

To complete our treatment of the factoring algorithm we show how it is possible to carry out a quantum Fourier transform.

An n-bit register represents a number j as

$$\sum_{i=1}^{n} j_i 2^{n-i} = j_1 2^{n-1} + j_2 2^{n-2} + ..j_n 2^0 \tag{21.82}$$

where $j_i = 0, 1$ is the value of the ith bit. We can represent the quantum Fourier transform as

$$\begin{aligned}U_{FT}|j\rangle &= \frac{1}{\sqrt{2^n}}\sum_{k=0}^{2^n-1} e^{i2\pi jk/2^n}|k\rangle \\ &= \frac{1}{\sqrt{2^n}}\sum_{k_1=0}^{1}\sum_{k_2=0}^{1}..\sum_{k_n=0}^{1} e^{i2\pi j\sum_{i=1}^{n}k_i 2^{-i}}|k_1k_2..k_n\rangle\end{aligned} \tag{21.83}$$

[25]Nielsen, M.A. and Chuang, I.L., op. cit., footnote p. 659.

where we used the representation (21.82) for k and wrote the state of the register by displaying the individual qubits. This state is a product state of the states of the individual qubits

$$\begin{aligned}|k_1k_2..k_n\rangle &= |k_1\rangle\,|k_2\rangle\,..\,|k_n\rangle \\ &\equiv \bigotimes_{i=1}^{n}|k_i\rangle\end{aligned}$$

so that (21.83) can be written

$$\begin{aligned}U_{FT}\,|j\rangle &= \frac{1}{\sqrt{2^n}}\sum_{k_1=0}^{1}\sum_{k_2=0}^{1}..\sum_{k_n=0}^{1}\bigotimes_{i=1}^{n}e^{i2\pi jk_i2^{-i}}\,|k_i\rangle \\ &= \frac{1}{\sqrt{2^n}}\bigotimes_{i=1}^{n}\sum_{k_i=0}^{1}e^{i2\pi jk_i2^{-i}}\,|k_i\rangle \\ &= \frac{1}{\sqrt{2^n}}\bigotimes_{i=1}^{n}\left(|0\rangle + e^{i2\pi j2^{-i}}\,|1\rangle\right) \\ &= \frac{1}{\sqrt{2^n}}\bigotimes_{i=1}^{n}\left(|0\rangle + e^{i2\pi\sum_{s=1}^{n}j_s2^{n-s-i}}\,|1\rangle\right) \qquad (21.84)\end{aligned}$$

using (21.82) for j. To see the meaning of this expression consider the term with $i = 1$. Then we see that the exponent is $2\pi i\,(j_n/2)$ for $s = n$ as the other s terms $(j_{n-1} + j_{n-2}2 + ..)\,2\pi$ are all integer multiples of 2π. For $i = 2$ we have $\left(j_n/2^2 + j_{n-1}/2)\right)$ as the only nontrivial terms. Thus if we introduce the notation

$$0.j_lj_{l+1}...j_m = j_l/2 + j_{l+1}/4.... + j_m/2^{m-l+1}$$

we can write (21.84) as

$$\begin{aligned}U_{FT}\,|j\rangle &= \frac{1}{\sqrt{2^n}}\left(|0\rangle + e^{i2\pi 0.j_n}\,|1\rangle\right)\left(|0\rangle + e^{i2\pi 0.j_{n-1}j_n}\,|1\rangle\right).... \\ &\quad ....\left(|0\rangle + e^{i2\pi 0.j_1j_2.......j_{n-1}j_n}\,|1\rangle\right). \qquad (21.85)\end{aligned}$$

Using this representation we can see how to produce a circuit implementing this transform. If we apply H to a qubit in the state $|j_1\rangle$ we obtain the state

$$H\,|j_1\rangle = \left(|0\rangle + e^{i2\pi 0.j_1}\,|1\rangle\right) \qquad (21.86)$$

since if $j_1 = 0$ we have $H\,|0\rangle = (|0\rangle + |1\rangle)$ whereas $j_1 = 1$ yields $H\,|1\rangle = (|0\rangle - |1\rangle)$. We introduce the operations

$$R_k = \begin{bmatrix}1 & 0 \\ 0 & e^{2\pi i/2^k}\end{bmatrix}. \qquad (21.87)$$

Then applying the operation R_2 to the state (21.86) controlled by the state of a second qubit $|j_2\rangle$, i.e., we apply R_2 only if $j_2 = 1$, we obtain

$$\left(|0\rangle + e^{i2\pi 0.j_1 j_2}\,|1\rangle\right).$$

We repeat the procedure applying the operation R_3 controlled by the state of the qubit $|j_3\rangle$, R_4 controlled by the state of the qubit $|j_4\rangle$, up to R_n controlled by the state $|j_n\rangle$ obtaining finally for the state of the first qubit

$$\left(|0\rangle + e^{i2\pi 0.j_1 j_2 \dots\dots j_{n-1} j_n}\,|1\rangle\right).$$

For the second qubit we apply the Hadamard gate obtaining $\left(|0\rangle + e^{i2\pi 0.j_2}\,|1\rangle\right)$ and then the operations R_2 to R_{n-1} controlled by the qubits $|j_3\rangle$ to $|j_n\rangle$ obtaining the state

$$\left(|0\rangle + e^{i2\pi 0.j_2 \dots\dots j_{n-1} j_n}\,|1\rangle\right)$$

for the second qubit. For the nth qubit we only apply H obtaining the state

$$\left(|0\rangle + e^{i2\pi 0.j_n}\,|1\rangle\right).$$

Thus we have obtained the state (21.85) with the order of the qubits reversed. The circuit for these operations is shown in Fig. 21.16.

Thus we have all the elements necessary for the Schor factoring algorithm. Counting up the number of operations necessary for this shows that it grows proportionally to the number of bits n necessary to represent the number to be factored N, whereas all known algorithms for solving this problem on a classical computer require the amount of resources to grow exponentially with n, i.e., proportional to N.

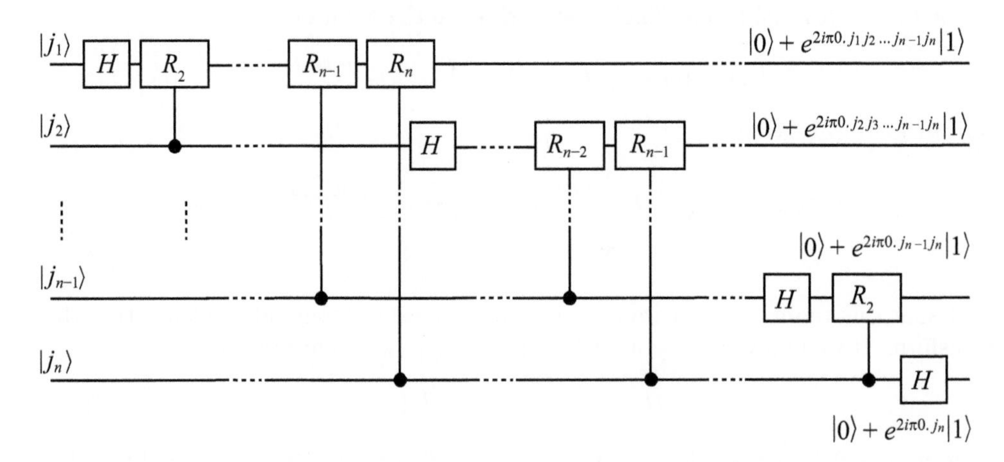

Fig. 21.16 Circuit to perform quantum Fourier transform. The input register represents a number $j = j_n.....j_1$. Each qubit is operated on by a Hadamard transform followed by a number of rotations R_k. The qubits of the result have to have their order reversed in order to complete the Fourier transform. (As decribed in Nielsen and Chuang, op. cit., footnote p. 659.)

21.7 Conclusion

We have tried to explain the basic ideas of quantum computing in order to obtain a general idea of the principles involved and help the reader develop some intuition in understanding how quantum computing circuits work. As previously mentioned the intellectual power behind many of these ideas is truly impressive. However, the question of putting these ideas into practice remains a question of highly active research.

Appendices

A
Classical mechanics

A.1 Introduction

Driven by, among other things, the desire to calculate planetary orbits, classical mechanics had been developed to a high degree of sophistication by the end of the nineteenth century. As this was the starting point for physicists' attempts to understand the physics of atoms and served as a source for many of the ideas leading up to quantum mechanics we give a review of Lagrangian and Hamiltonian mechanics. It will be seen that the subject has a certain beauty in its own right.[1]

A.2 Lagrangian mechanics

A.2.1 Definition of the Lagrangian

Lagrangian mechanics is derived from the following idea. Take a function $L(\dot{r}, r)$ to be the difference between the kinetic energy and the potential energy of a particle.

$$L(\dot{r}, r) = T - V \tag{A.1}$$

The amazing and exciting idea behind Lagrangian mechanics is that a particle or object moving between two fixed points in a field of force follows a path such that this integral:

$$\int_{t_1}^{t_2} L dt$$

is a minimum (least action principle). This means that of all possible paths starting at a given fixed point at time t_1 and finishing at another fixed point at time t_2 the particle follows that path which makes the integral a minimum.

This seems to be a contradiction of the approach based on Newton's law, $F = ma$ where the acceleration at a point is determined by the force at that point and the motion follows. The minimum principle says the motion at every point is determined by the requirement that a certain (global) property of the entire trajectory is a minimum. We will see that in fact there is no contradiction, the two approaches are equivalent.

The minimum principle makes it seem that the particle tries out alternate paths and decides to follow the one that makes the integral minimum. Classically this seems a bit spooky when we consider the viewpoint of the motion being determined by the force at

[1]There are many excellent books on these topics, e.g., Goldstein, H., Poole, C. and Safko, J., *Classical Mechanics* third edition, Pearson, (2001)

each point. However, we will see that according to quantum mechanics this is exactly what the particle does. All paths contribute to the motion in quantum mechanics, but the contribution of most paths cancels in cases where classical mechanics applies. We will see that in the classical case only those paths in the neighborhood of the minimum path contribute.

A.2.1.1 Derivation of Newton's law, $F = ma$

A particle is to move on a path between two points. There must be a true path determined by the laws of mechanics which we will call $x(t)$. This is the path a particle put in the given situation would follow in the physical world. The task is to predict this path without having to observe the particle in motion.

We will now calculate the path which will cause $\int Ldt$ to be a minimum. Since it is a minimum of a curve any path that is only slightly different should have a very similar value of $\int Ldt$. This is because a minimum is an inverted bell shape as the gradient near the minimum approaches zero. This small gradient means that for a small change in the x-axis value there will be a smaller change in the y-axis value. We take $x(t) + \epsilon(t)$ to be a path slightly different to the true path. We know that the kinetic energy $T = \frac{1}{2}mv^2$, which in this case becomes $T = \frac{1}{2}m(\dot{x} + \dot{\epsilon})^2$. Also the potential energy is $V = V(x+\epsilon)$ along the varied path. We then calculate the *variation* of the integral along the path, that is the difference between the Lagrangian integral for the true path and that for the varied path.

$$\begin{aligned}
\delta \int_{t_1}^{t_2} Ldt &= \int_{t_1}^{t_2} L(x+\epsilon)dt - \int_{t_1}^{t_2} L(x)dt \\
&= \int_{t_1}^{t_2} \left(T(x+\epsilon) - T(x)\right) dt - \cdots \\
&\quad \cdots - \int_{t_1}^{t_2} \left(V(x+\epsilon) - V(x)\right) dt
\end{aligned}$$

$$\begin{aligned}
\int_{t_1}^{t_2} \left(T(x+\epsilon) - T(x)\right) dt &= \int_{t_1}^{t_2} \left(\frac{1}{2}m\left(\dot{x}^2 + 2\dot{x}\dot{\epsilon} + \dot{\epsilon}^2\right) - \frac{1}{2}m\dot{x}^2\right) dt \\
&= \int_{t_1}^{t_2} m\dot{x}\dot{\epsilon}dt
\end{aligned} \tag{A.2}$$

$$\begin{aligned}
\int_{t_1}^{t_2} \left(V(x+\epsilon) - V(x)\right) dt &= \int_{t_1}^{t_2} \left(\left(V(x) + \frac{dV}{dx}\epsilon\right) - V(x)\right) dt \\
&= \int_{t_1}^{t_2} \left(\frac{dV}{dx}\epsilon\right) dt
\end{aligned} \tag{A.3}$$

so

$$\int_{t_1}^{t_2} L(x+\epsilon)dt - \int_{t_1}^{t_2} L(x)dt = \int_{t_1}^{t_2} m\dot{x}\dot{\epsilon}dt - \int_{t_1}^{t_2} \frac{dV}{dx}\epsilon dt. \tag{A.4}$$

Notice that in the second line of (A.2) the $\frac{1}{2}m\dot{x}^2$ cancel. The $\dot{\epsilon}^2$ term is ignored because ϵ is taken to be very small, making the second-order terms insignificant.

Next we evaluate $\int m\dot{x}\dot{\epsilon}dt$ integration by parts.

$$\int_{t_1}^{t_2} m\dot{x}\dot{\epsilon}dt = m\dot{x}\epsilon\Big|_{t_1}^{t_2} - \int_{t_1}^{t_2} \frac{d}{dt}(m\dot{x})\epsilon dt = -\int_{t_1}^{t_2} \frac{d}{dt}(m\dot{x})\epsilon dt. \tag{A.5}$$

The limits of this integral, t_1 and t_2 are the times at which the particle is at the end points of the motion. By definition every path between two points must start and end at the same points, so at these times t_1 and t_2, ϵ must be zero. This means the $m\dot{x}\epsilon$ term in (A.5) can be ignored, since it is zero when the integral is evaluated.

We have now shown that the variation due to the path difference is as follows:

$$\int_{t_1}^{t_2} L(x+\epsilon)dt - \int_{t_1}^{t_2} L(x)dt = -\int_{t_1}^{t_2} \left(\frac{d}{dt}(m\dot{x}) + \frac{dV}{dx}\right)\epsilon dt. \tag{A.6}$$

This should be equal (or very close) to zero for all possible choices of ϵ (where ϵ is small). This can only hold if $\frac{d}{dt}(m\dot{x}) + \frac{dV}{dx} = 0$. Since $\frac{d}{dt}(m\dot{x}) = ma$ and $F = -\frac{dV}{dx}$ this leads directly to Newton's equation.

A.2.2 Derivation of the general case—the Lagrangian equations of motion

Following the argument above but generalizing further leads to another important result. The difference of $\int_{t_1}^{t_2} Ldt$ for two paths differing by $\epsilon\,(t)$ is

$$\int_{t_1}^{t_2} L(x+\epsilon, \dot{x}+\dot{\epsilon})dt - \int_{t_1}^{t_2} L(x,\dot{x})dt = \int_{t_1}^{t_2} dt\left(\frac{\partial L}{\partial \dot{x}}\dot{\epsilon} + \frac{\partial L}{\partial x}\epsilon\right)$$

which on integrating the first term on the right hand side by parts becomes

$$\int \epsilon dt\left[\frac{\partial L}{\partial x} - \frac{d}{dt}\left(\frac{\partial L}{\partial \dot{x}}\right)\right],$$

which should equal zero for all values of ϵ for the path for which $\int_{t_1}^{t_2} Ldt$ is a minimum. This path will then be given by

$$\left[\frac{\partial L}{\partial x} - \frac{d}{dt}\left(\frac{\partial L}{\partial \dot{x}}\right)\right] = 0 \tag{A.7}$$

In the usual case involving several coordinates this represents a set of equations, one for each coordinate, called Lagrange's equations. These are fully equivalent to Newton's equation of motion as shown above for the case of one dimension.

The practical advantage of this formulation is that it offers a straightforward method for writing Newton's equations in any coordinate system.

For a generalized coordinate system, dq_i, possibly involving coordinates for many particles, defined by the differential distance element between two adjacent points:

$$ds^2 = g_{11}dq_1^2 + g_{22}dq_2^2 + g_{33}dq_3^2 + ...$$

we have the kinetic energy

$$T = \frac{m}{2}\left(\frac{ds}{dt}\right)^2$$

A.2.3 Example 1: 2D polar coordinates

As a first example we consider polar coordinates in two dimensions, r, θ

$$ds^2 = dr^2 + r^2 d\theta^2$$

so that

$$T = \frac{m}{2}\left(\dot{r}^2 + r^2\dot{\theta}^2\right)$$

and

$$L = \frac{m}{2}\left(\dot{r}^2 + r^2\dot{\theta}^2\right) - V(r,\theta).$$

Then the Lagrange equation (A.7) for the coordinate r is

$$\begin{aligned}\frac{d}{dt}\left(\frac{\partial L}{\partial \dot{r}}\right) - \frac{\partial L}{\partial r} &= 0\\ \frac{d}{dt}(m\dot{r}) - mr\dot{\theta}^2 + \frac{\partial V}{\partial r} &= 0\\ m\ddot{r} &= F_r + mr\dot{\theta}^2\end{aligned} \tag{A.8}$$

where $F_r = -\frac{\partial V}{\partial r}$ is the force exerted by the potential in the radial direction and $mr\dot{\theta}^2$ is the centrifugal force which is seen to follow automatically from the Lagrange equations and the nature of the coordinate system.

For the θ coordinate we have

$$\frac{d}{dt}\left(\frac{\partial L}{\partial \dot{\theta}}\right) = \frac{d}{dt}\left(mr^2\dot{\theta}\right) = \frac{\partial L}{\partial \theta} = -\frac{\partial V}{\partial \theta} = 0$$

where the last equation holds for the potential V independent of θ, i.e.

$$V(r,\theta) = V(r),$$

which is the important case of central (independent of angle) forces. Then

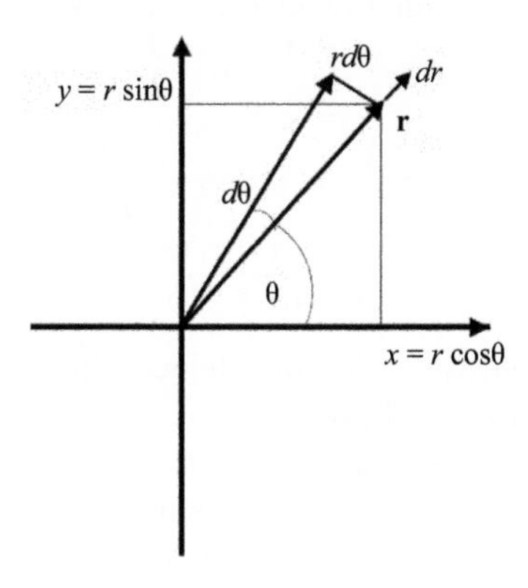

Fig. A.1 Polar plot, two dimensions.

$$mr^2\dot{\theta} = l = const$$

defines the angular momentum which is constant in the case of central forces and we have

$$\dot{\theta} = l/mr^2. \tag{A.9}$$

Substituting into (A.8) we have

$$m\ddot{r} = -\frac{\partial V}{\partial r} + l^2/mr^3 = -\frac{\partial}{\partial r}\left[V + l^2/2mr^2\right]. \tag{A.10}$$

The term in square brackets can be considered as an effective potential V with the "centrifugal potential" given by the second term in the bracket. We see that the Lagrangian formalism automatically yields the correct equations of motion including the centrifugal forces and conservation of angular momentum (if the potential only depends on r).

To calculate an orbit we need to find r as a function of θ, $r(\theta)$. We use

$$\frac{d}{dt} = \left(\frac{d\theta}{dt}\right)\frac{d}{d\theta}$$
$$\frac{d}{dt}\left(\frac{dr}{dt}\right) = \left(\frac{d\theta}{dt}\right)\frac{d}{d\theta}\left(\frac{d\theta}{dt}\frac{dr}{d\theta}\right) = \frac{l}{mr^2}\frac{d}{d\theta}\left(\frac{l}{mr^2}\frac{dr}{d\theta}\right)$$

using (A.9).

Substituting $u = 1/r$, $dr/d\theta = -\left(1/u^2\right)du/d\theta$ we find

$$m\ddot{r} = -\frac{l^2u^2}{m}\frac{d^2u}{d\theta^2} = u^2\frac{d}{du}\left[V + l^2u^2/2m\right]$$
$$\frac{d^2u}{d\theta^2} = -\left[\frac{m}{l^2}\frac{dV}{du} + u\right] = \left[\frac{m\kappa}{l^2} - u\right]. \tag{A.11}$$

The last step holding in the case when $V = -\kappa/r = -\kappa u$, the common case of a $1/r$ potential (force $\sim 1/r^2$, gravitational or electrostatic force). The solution of (A.11) is seen to be

$$u = \frac{1}{r} = A\cos\theta + \frac{m\kappa}{l^2} \tag{A.12}$$
$$r = \frac{1}{A\cos\theta + \frac{m\kappa}{l^2}}$$

which is the equation of an ellipse, reflecting the well-known fact that the closed orbits in the case of a $1/r$ potential are ellipses.

The case of motion in a $1/r$ potential is often called the Kepler problem because it applies to the motion of the planets around the sun under the influence of gravitational force. It also applies to the motion of a negatively charged electron around a nucleus in an atom under the influence of the electrostatic (Coulomb) potential, in particular to the hydrogen atom.

A.2.4 Example 2: generalized coordinates, normal modes

Another interesting case is that of two pendulums hanging side by side, connected by a spring. We assume the two pendulums are of equal length Λ, and equal masses, m.

The total kinetic energy of the system is the sum of the kinetic energies for each pendulum:

$$T = \frac{m}{2}(\Lambda^2\dot{\theta}_1^2 + \Lambda^2\dot{\theta}_2^2)$$

where $\theta_{1,2}$ are the angles of the two pendulums.

The potential energy for the two masses in the earth's gravitational field is

$$V_g = mg\Lambda\left[(1-\cos\theta_1) + (1-\cos\theta_2)\right] \approx mg\Lambda\left(\theta_1^2+\theta_2^2\right)/2$$

while the potential energy of the spring is

$$V_s = \frac{1}{2}k\Lambda^2\left(\theta_1-\theta_2\right)^2,$$

the latter two relations being valid for small θ. Thus the Lagrangian is

$$L = \frac{m\Lambda^2}{2}(\dot{\theta}_1^2+\dot{\theta}_2^2) - mg\Lambda\left(\theta_1^2+\theta_2^2\right)/2 - \frac{1}{2}k\Lambda^2\left(\theta_1-\theta_2\right)^2.$$

Lagrange's equations are:

$$\frac{d}{dt}\left(m\Lambda^2\dot{\theta}_1\right) + mg\Lambda\theta_1 + k\Lambda^2\left(\theta_1-\theta_2\right) = 0$$
$$\frac{d}{dt}\left(m\Lambda^2\dot{\theta}_2\right) + mg\Lambda\theta_2 - k\Lambda^2\left(\theta_1-\theta_2\right) = 0.$$

We see that each equation contains both variables. While these equations can be solved in their present form it is interesting to introduce new coordinates in which the equations of motion are independent. Let $q_1 = \theta_1 - \theta_2$, $q_2 = \theta_1 + \theta_2$, then

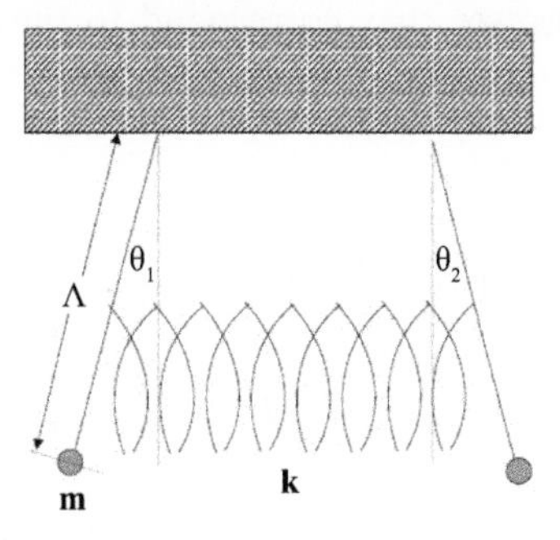

Fig. A.2 Two pendulums.

$$q_1^2 + q_2^2 = \theta_1^2 - 2\theta_1\theta_2 + \theta_2^2 + \theta_1^2 + 2\theta_1\theta_2 + \theta_2^2 = 2\left(\theta_1^2 + \theta_2^2\right)$$

so that we can write

$$L = \frac{m\Lambda^2}{4}\left(\dot{q}_1^2 + \dot{q}_2^2\right) - \frac{mg\Lambda}{4}\left(q_1^2 + q_2^2\right) - \frac{k\Lambda^2}{2}q_1^2.$$

Lagrange's equation for q_1 is

$$\frac{d}{dt}\left(\frac{m\Lambda^2}{2}\dot{q}_1\right) = \frac{m\Lambda^2}{2}\ddot{q}_1 = -\frac{mg\Lambda}{2}q_1 - k\Lambda^2 q_1.$$

Similarly for q_2 we find

$$\frac{m\Lambda^2}{2}\ddot{q}_2 = -\frac{mg\Lambda}{2}q_2$$

with the solutions (choosing the initial conditions, $\theta_1 = \theta_2 = 0$),

$$\begin{aligned}
q_1 &= \theta_1 - \theta_2 = A_1 \sin\omega_1 t \\
q_2 &= \theta_1 + \theta_2 = A_2 \sin\omega_2 t \\
\omega_1 &= \sqrt{\frac{g}{\Lambda} + \frac{2k}{m}} \\
\omega_2 &= \sqrt{\frac{g}{\Lambda}}.
\end{aligned}$$

The solutions to the original problem are

$$\begin{aligned}
2\theta_1 &= (q_1 + q_2) = (A_1 \sin\omega_1 t + A_2 \sin\omega_2 t) \\
2\theta_2 &= (q_2 - q_1) = (A_2 \sin\omega_2 t - A_1 \sin\omega_1 t)\,.
\end{aligned}$$

Thus in general, if only one of the coordinates is excited, e.g., $A_1 = 0$, or $A_2 = 0$, the system oscillates with a single frequency. Coordinates with this property are called normal modes. In this case we see they correspond to the masses swinging in such a way that they keep the same distance, $q_1 = 0$, or they move symmetrically in opposite directions, $q_2 = 0$. If, for simplicity, we set $A_1 = A_2$, the solution has the form

$$\begin{aligned}
\theta_1 &= B \sin\left(\frac{\omega_1 + \omega_2}{2}\right) t \bullet \cos\left(\frac{\omega_1 - \omega_2}{2}\right) t \\
\theta_2 &= B \sin\left(\frac{\omega_1 - \omega_2}{2}\right) t \bullet \cos\left(\frac{\omega_1 + \omega_2}{2}\right) t,
\end{aligned}$$

which, for a weak spring ($2k/m \ll g/\Lambda$), corresponds to θ_1 oscillating with a frequency $\left(\frac{\omega_1+\omega_2}{2}\right)$ and a large amplitude for times $(\omega_1 - \omega_2)\,t \ll 1$, while θ_2 begins oscillating at the same frequency but a small amplitude. At later times, $(\omega_1 - \omega_2)\,t \sim \pi$, the energy has been transferred from $\theta_1$to θ_2. As time goes on energy is continuously transferred between the motion of the two masses on a timescale given by $\pi/(\omega_1 - \omega_2)$. This behavior is seen to be quite different than the case where a single normal mode is excited.

A.3 Hamiltonian mechanics

A.3.1 Definition of the Hamiltonian

The Lagrangian $L(\dot{q}, q)$ is a function of the velocities, $\dot{q}$, and coordinates, q, describing the system. So if we calculate the rate of change of L, we find

$$\frac{dL}{dt} = \left(\frac{\partial L}{\partial q}\right)\dot{q} + \left(\frac{\partial L}{\partial \dot{q}}\right)\ddot{q}.$$

From Lagrange's equation of motion

$$\left(\frac{\partial L}{\partial q}\right) = \frac{d}{dt}\left(\frac{\partial L}{\partial \dot{q}}\right)$$

so that (we are calculating dL/dt along a trajectory satisfying the equations of motion)

$$\frac{dL}{dt} = \dot{q}\frac{d}{dt}\left(\frac{\partial L}{\partial \dot{q}}\right) + \left(\frac{\partial L}{\partial \dot{q}}\right)\ddot{q} = \frac{d}{dt}\left(\dot{q}\frac{\partial L}{\partial \dot{q}}\right)$$

and

$$\frac{d}{dt}\left(\dot{q}\frac{\partial L}{\partial \dot{q}} - L\right) = 0$$

$$\dot{q}\frac{\partial L}{\partial \dot{q}} - L = const = H$$

where $H =$ a constant which will be seen to be the energy of the motion. Defining

$$p = \frac{\partial L}{\partial \dot{q}} \tag{A.13}$$

the last equation can be written

$$p\dot{q} - L = H.$$

For example in Cartesian coordinates

$$L = \frac{m}{2}\left(\dot{x}^2 + \dot{y}^2 + \dot{z}^2\right) - V(x, y, z)$$

so

$$p_x = m\dot{x}, \quad p_y = m\dot{y}, \quad p_z = m\dot{z}$$

$$H = p\dot{q} - L = \frac{m}{2}\left(\dot{x}^2 + \dot{y}^2 + \dot{z}^2\right) + V = K.E. + P.E. = total\ energy$$

while in Polar coordinates: $ds^2 = dr^2 + r^2 d\theta^2$

$$
\begin{aligned}
L &= \frac{m}{2}\left(\dot{r}^2 + r^2\dot{\theta}^2\right) - V(r,\theta) \\
p_r &= m\dot{r}, \quad p_\theta = mr^2\dot{\theta} \\
H &= p\dot{q} - L = \frac{m}{2}\left(\dot{r}^2 + r^2\dot{\theta}^2\right) + V(r,\theta) \\
&= \frac{p_r^2}{2m} + \frac{p_\theta^2}{2mr^2} + V(r,\theta)
\end{aligned}
\tag{A.14}
$$

where $p_\theta = l = mr^2\dot{\theta}$ is the angular momentum.

A.3.2 The equations of motion in Hamiltonian form

From the definition

$$H = p\dot{q} - L$$

we find that

$$dH = \dot{q}dp + \left[pd\dot{q} - \frac{\partial L}{d\dot{q}}d\dot{q}\right] - \frac{\partial L}{dq}dq.$$

The terms in the square brackets cancel because of the definition of p (A.13) so that

$$dH = \dot{q}dp - \frac{\partial L}{dq}dq$$

which means that H is a function of p and q. The definition of H and p have resulted in a change of variables from $\dot{q}, q$ to p, q. Changes of variable of this type are called Legendre transformations. Further, since we have from Lagrange's equation (A.7)

$$\frac{\partial L}{dq} = \frac{d}{dt}\left(\frac{\partial L}{d\dot{q}}\right) = \dot{p}$$

we find

$$dH = \dot{q}dp - \dot{p}dq$$

so that

$$\frac{\partial H}{\partial p} = \dot{q}, \quad \frac{\partial H}{\partial q} = -\dot{p}.$$

These are called Hamilton's equations of motion.

As an example consider the Hamiltonian in polar coordinates (A.14). Then

$$
\begin{aligned}
\dot{r} &= \frac{\partial H}{\partial p_r} = \frac{p_r}{m}, & \dot{p}_r &= -\frac{\partial H}{\partial r} = -\frac{\partial}{\partial r}\left(V + \frac{p_\theta^2}{2mr^2}\right) \\
\dot{\theta} &= \frac{\partial H}{\partial p_\theta} = \frac{p_\theta}{mr^2}, & \dot{p}_\theta &= -\frac{\partial H}{\partial \theta} = -\frac{\partial}{\partial \theta}(V) = 0 \text{ for a central force,}
\end{aligned}
$$

which are identical to the equations of motion obtained from the Lagrangian (A.9, A.10).

A.4 Transformations of coordinates—canonical transformations and the Hamilton-Jacobi equation

It is sometimes useful to take a transformation of coordinates (p_i, q_i), mapping them to new coordinates (P_i, Q_i). Assuming that (p_i, q_i) satisfy Hamilton's equations, the transformation is said to be canonical if (P_i, Q_i) also satisfy Hamilton's equations.

Such a transformation can be generated by a function F if it is a function of one old variable and one new variable. There are therefore four possibilities for this function, $F(q_i, Q_i)$, $F(q_i, P_i)$, $F(p_i, Q_i)$, $F(p_i, P_i)$, each of which leads to a different set of equations. We will look at the first two of these.

A.4.1 Taking $F(q_i, Q_i)$ as a generating function

In order for Hamilton's equations to be valid in terms of the new coordinates, given that they are valid in the old ones, the condition $\delta \int_{t_1}^{t_2} L dt = 0$ must hold in both the old and new coordinate systems. Therefore

$$\delta \int_{t_1}^{t_2} L dt = \delta \int_{t_1}^{t_2} [p\dot{q} - H(p, q)] dt = \delta \int_{t_1}^{t_2} [P\dot{Q} - K(P, Q)] dt = 0$$

where $K(P, Q)$ is the Hamiltonian in the transformed coordinates system. In order for this to be true the integrands can differ only by the time derivative of a function, since $\delta \int_a^b (dF/dt)\, dt = \delta\, [F(b) - F(a)] = 0$, since the variation δ is defined to leave the endpoints fixed. Then

$$p\dot{q} - H = P\dot{Q} - K + \frac{dF(q, Q, t)}{dt} = P\dot{Q} - K + \frac{\partial F}{\partial q}\dot{q} + \frac{\partial F}{\partial Q}\dot{Q} + \frac{\partial F}{\partial t} \tag{A.15}$$

by comparing coefficients we see that

$$p = \frac{\partial F}{\partial q}, \qquad -P = \frac{\partial F}{\partial Q}, \qquad K = H + \frac{\partial F}{\partial t}.$$

This example is of limited use. However, it is an important stepping stone for calculation of the next case, which will prove to be of great importance.

A.4.2 Taking $S(q, P, t)$ as a generating function—the Hamilton-Jacobi equation

Using the method of Legendre transformation we take as a generating function

$$S(q, P, t) = F(q, Q, t) + PQ$$

and our condition that the transformation be canonical is (Eq. A.15)

$$\begin{aligned}
p\dot{q} - H &= P\dot{Q} - K + \frac{d}{dt}\left(S(q, P, t) - PQ\right) \\
&= P\dot{Q} - K + \frac{\partial S}{\partial q}\dot{q} + \frac{\partial S}{\partial P}\dot{P} - \dot{P}Q - P\dot{Q} + \frac{\partial S}{\partial t} \\
&= -K + \frac{\partial S}{\partial q}\dot{q} + \frac{\partial S}{\partial P}\dot{P} - \dot{P}Q + \frac{\partial S}{\partial t}.
\end{aligned}$$

So equating coefficients the transformation is given by

$$H = K - \frac{\partial S}{\partial t}, \qquad p = \frac{\partial S}{\partial q}, \qquad Q = \frac{\partial S}{\partial P}. \tag{A.16}$$

The idea now is to try to find a transformation to a set of new coordinates (P, Q) which are all constants. This can be accomplished if the new Hamiltonian $K(P, Q)$ is set to zero. In this case we will have, writing Hamilton's equations in the new coordinates,

$$\dot{Q} = \frac{\partial K}{\partial P} = 0, \qquad \dot{P} = -\frac{\partial K}{\partial Q} = 0.$$

So we use

$$H\left(p = \frac{\partial S}{\partial q}, q\right) + \frac{\partial S}{\partial t} = K(P, Q) = 0 \tag{A.17}$$

to find the function $S(q, P)$ generating the desired transformation. Equation (A.17) is called the Hamilton-Jacobi equation. Its solution, S, is seen to be related to the Lagrangian as follows: calculate

$$\begin{aligned} \frac{dS}{dt} &= \frac{\partial S}{\partial q}\dot{q} + \frac{\partial S}{\partial P}\dot{P} + \frac{\partial S}{\partial t} \\ &= p\dot{q} - H\left(p = \frac{\partial S}{\partial q}, q\right) = L \end{aligned}$$

where we have made use of the facts that $\dot{P} = 0$ and that S satisfies (A.17). Thus

$$S = \int L dt,$$

the integral being considered as a function of q.

Since the constants P, Q are determined by the initial conditions, the transformation (A.16) transforms the true coordinates (q, p) back to the initial conditions. Thus the transformation generates a backward motion that "cancels" the motion of the system and knowledge of the transformation is equivalent to knowledge of the motion.

In the common case where H is independent of time, we can separate variables in the H-J equation by writing

$$S(q, P, t) = W(q, P) + f(t)$$

so that substituting back into the Hamilton-Jacobi equation yields

$$H\left(\frac{\partial W}{\partial q}, q\right) + \frac{\partial f}{\partial t} = 0.$$

Since the first term is a function of q alone and the second a function of t alone, the equation can only hold if each term is a constant. Thus we can put

$$\frac{\partial f}{\partial t} = -E, \qquad f = -Et$$

$$H\left(\frac{\partial W}{\partial q}, q\right) = E \tag{A.18}$$

where E is a constant which is seen to be the energy.

A.4.3 Example: the simple harmonic oscillator

The Hamiltonian for a one dimensional simple harmonic oscillator can be written

$$H(p,q) = \frac{p^2}{2m} + \frac{1}{2}kq^2$$

where k is the force constant of the oscillator. Equation (A.18) then becomes

$$\frac{1}{2m}\left(\frac{\partial W}{\partial q}\right)^2 + \frac{1}{2}kq^2 = E$$

$$\frac{\partial W}{\partial q} = \sqrt{2mE - mkq^2}$$

$$W = \int \sqrt{2mE - mkq^2}dq.$$

Now since

$$S = W - Et, \tag{A.19}$$

we have, identifying the new constant momentum P with the constant E,

$$Q = \frac{\partial S}{\partial P} = \frac{\partial S}{\partial E} = \frac{\partial W}{\partial E} - t$$

and

$$\frac{\partial W}{\partial E} = m\int \frac{1}{\sqrt{2mE - mkq^2}}dq$$

$$Q = m\int \frac{1}{\sqrt{2mE - mkq^2}}dq - t = \beta$$

where β is a constant.

Thus

$$m\int \frac{1}{\sqrt{2mE - mkq^2}}dq = (\beta + t)$$

$$\sqrt{\frac{m}{k}}\sin^{-1}\left(q\sqrt{\frac{k}{2E}}\right) = (\beta + t)$$

$$q = \sqrt{\frac{2E}{k}}\sin\sqrt{\frac{k}{m}}(\beta + t)$$

and we have found the solution for the motion of the harmonic oscillator, i.e., oscillation at a frequency $\omega = \sqrt{k/m}$.

A.4.4 Example: 3-D polar coordinates and the Kepler (hydrogen atom) problem

In 3-D polar coordinates, r, θ, ϕ, also called spherical coordinates, defined by (see Fig. 4.1) the distance between two neighboring points separated by $dr, d\theta, d\phi$ is given by

$$ds^2 = dr^2 + r^2 (d\theta)^2 + r^2 \sin^2\theta d\phi$$

so that the kinetic energy is

$$T = \frac{m}{2}\left(\left(\frac{dr}{dt}\right)^2 + r^2\left(\frac{d\theta}{dt}\right)^2 + r^2\sin^2\theta\left(\frac{d\phi}{dt}\right)^2\right)$$

and the Lagrangian is

$$L = T - V.$$

Then the momentum is defined by $p_i = \partial L/\partial \dot{q}_i$:

$$p_r = m\dot{r}, \quad p_\theta = mr^2\dot{\theta}, \quad p_\phi = mr^2\sin^2\theta\dot{\phi} \tag{A.20}$$

and the Hamiltonian is given by

$$H = \sum_i p_i\dot{q}_i - L = \frac{p_r^2}{2m} + \frac{p_\theta^2}{2mr^2} + \frac{p_\phi^2}{2mr^2\sin^2\theta} + V.$$

We wish to consider the case where $V = -\kappa/r$ ($\kappa > 0$), that is the Kepler problem or a negative electron orbiting around a positive, massive nucleus.

Since the Hamiltonian is independent of the time we write $S = W - Et$ and the Hamilton-Jacobi equation is

$$H\left(p_i = \frac{\partial W}{\partial q_i}, q_i\right) = \frac{1}{2m}\left\{\left(\frac{\partial W}{\partial r}\right)^2 + \frac{1}{r^2}\left(\frac{\partial W}{\partial \theta}\right)^2 + \frac{1}{r^2\sin^2\theta}\left(\frac{\partial W}{\partial \phi}\right)^2\right\} - \frac{\kappa}{r}.$$
$$= E \tag{A.21}$$

In order to solve this we use the technique of separation of variables, that is we write

$$W(r, \vartheta, \phi, P_1, P_2, P_3) = W_r(r) + W_\theta(\theta) + W_\phi(\phi).$$

The first step then is to rewrite (A.21) as

$$r^2\sin^2\theta\left\{\left(\frac{\partial W_r}{\partial r}\right)^2 + \frac{1}{r^2}\left(\frac{\partial W_\theta}{\partial \theta}\right)^2 - 2m\left(E + \frac{\kappa}{r}\right)\right\} = -\left(\frac{\partial W_\phi}{\partial \phi}\right)^2 \tag{A.22}$$
$$= -m_\phi^2$$

where the last step follows from the fact that the left-hand side is a function of r, θ alone, and the right-hand side is a function of ϕ so that the only way the equation

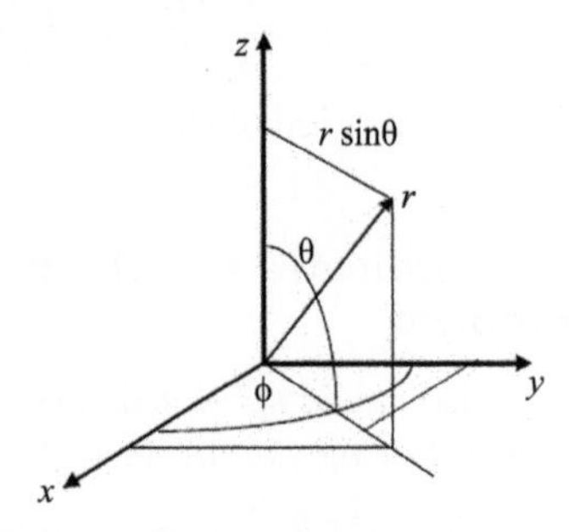

Fig. A.3 Three-dimensional polar coordinates.

can hold for all values of r, θ, ϕ is for both sides to be a constant which we set equal to $-m_\phi^2$. Then

$$\frac{\partial W_\phi}{\partial \phi} = m_\phi \tag{A.23}$$

and we can apply the same trick again to write (A.22)

$$r^2 \left\{ \left(\frac{\partial W_r}{\partial r} \right)^2 - 2m\left(E + \frac{\kappa}{r}\right) \right\} = -\left(\frac{\partial W_\theta}{\partial \theta} \right)^2 - \frac{m_\phi^2}{\sin^2 \theta} = -l^2$$

since the left-hand side is a function of r alone and the right-hand side a function of θ, so the equation can only hold if both sides are constant. Then

$$\left(\frac{\partial W_\theta}{\partial \theta} \right)^2 = l^2 - \frac{m_\phi^2}{\sin^2 \theta}$$

and

$$\begin{aligned} W_\theta &= \int d\theta \sqrt{l^2 - \frac{m_\phi^2}{\sin^2 \theta}} \\ W_r &= \int dr \sqrt{2m\left(E + \frac{\kappa}{r}\right) - \frac{l^2}{r^2}} \\ W_\phi &= m_\phi \phi. \end{aligned} \tag{A.24}$$

We identify the three constants m_ϕ, l, E with the three new canonical momenta, P_i, so that the (constant) values of the new coordinates, Q_i, are given by: (remember $S = W - Et$)

$$\begin{aligned} Q_1 = \beta_1 &= \frac{\partial S}{\partial E} = \frac{\partial W}{\partial E} - t = \frac{\partial W_r}{\partial E} - t \\ t + \beta_1 &= \frac{\partial W_r}{\partial E} = m \int \frac{r dr}{\sqrt{2m\left(Er^2 + \kappa r\right) - l^2}} \end{aligned}$$

and

$$\begin{aligned} Q_2 = \beta_2 &= \frac{\partial S}{\partial l} = \frac{\partial W}{\partial l} = \\ &= l \int \frac{d\theta}{\sqrt{l^2 + \frac{m_\phi^2}{\sin^2 \theta}}} - l \int \frac{dr}{r^2 \sqrt{2m\left(E + \frac{\kappa}{r}\right) - \frac{l^2}{r^2}}} \end{aligned} \tag{A.25}$$

with a similar relation for Q_3. We now take the special case $m_\phi = 0$. This means that $p_\phi = \partial W_\phi / \partial \phi = m_\phi = 0$ or that ϕ is a constant (see Eq. A.23), i.e., the plane of the orbit goes through the z axis (see Fig.A.4.4).

By this assumption we reduce the problem to that of 2D polar coordinates, Section A.2.3. We then see that (A.25) will furnish the equation of the orbit $r(\theta)$, if we evaluate the r integral. We do this by substituting $u = 1/r$ in (A.25) and thus obtain

$$\theta - \beta_2 = l \int \frac{du}{\sqrt{2m(E + \kappa u) - l^2 u^2}}.$$

Remembering that $E < 0$ we can evaluate the integral as

$$\theta - \beta_2 = \cos^{-1}\left[\frac{l^2 u - m\kappa}{D}\right]$$

where

$$D = \sqrt{m^2\kappa^2 + 2mEl^2},$$

which can be easily verified by differentiating. Then

$$l^2 u = D\cos(\theta - \beta_2) + m\kappa$$

$$u = \frac{m\kappa}{l^2}\left[1 + \cos(\theta - \beta_2)\sqrt{1 + \frac{2mEl^2}{m\kappa^2}}\right],$$

which is the equation of an ellipse, just as we obtained in Eq. (A.12), Section A.2.3.

A.5 Action-angle variables

We now introduce a technique for dealing with systems that have Hamiltonians independent of time, resulting in periodic motion, and that are *separable*. This last condition means that the function W can be written

$$W(q_1, q_2, q_3) = W_1(q_1) + W_2(q_2) + W_3(q_3)$$

as in the previous section. For such a system we can introduce a new set of canonical variables (Θ_i, J_i) where we define

$$J_i = \oint p_i dq_i = \oint \frac{\partial W_i}{\partial q_i} dq_i \tag{A.26}$$

where the symbol $\oint$ indicates the integral is to be taken over a complete period of the motion. As the motion is periodic, such integrals are constants of the motion.

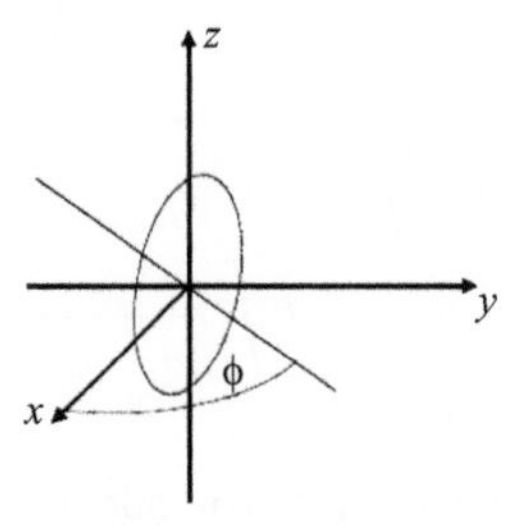

Fig. A.4 Orbits.

We consider J_i as new momentum and Θ_i as new coordinate variables. Then the Hamiltonian can be written as a function of the J_i and

$$\dot{\Theta}_i = \frac{\partial H(J)}{\partial J_i} = \alpha_i \tag{A.27}$$

where the α_i are constants because the J_i are constants of the motion. Thus we can write

$$\Theta_i = \alpha_i t + \beta_i$$

with β_i a set of constants of integration. From our transformation Eq. (A.16) we have

$$\Theta_i = \frac{\partial W}{\partial J_i}.$$

We now consider the change in Θ_i over one period of the motion, $\Delta\Theta_i$,

$$\begin{aligned}\Delta\Theta_i &= \sum_j \oint \frac{\partial \Theta_i}{\partial q_j} dq_j = \sum_j \oint \frac{\partial}{\partial q_j}\left(\frac{\partial W}{\partial J_i}\right) dq_j = \frac{\partial}{\partial J_i}\sum_j \oint \frac{\partial W}{\partial q_j} dq_j \\ &= \frac{\partial}{\partial J_i}\sum_j \oint p_j dq_j = \frac{\partial}{\partial J_i}\sum_j J_j = \sum_j \begin{Bmatrix} 0 & i \neq j \\ 1 & i = j \end{Bmatrix}. = 1\end{aligned}$$

Thus, if at time $t = 0$, we have $\Theta_i = \beta_i$ then after a period T we will have $\Theta_i = \alpha_i T + \beta_i = \Delta\Theta_i + \beta_i = 1 + \beta_i$ so that $\alpha_i T = 1$ and $\alpha_i = \dot{\Theta}_i = \frac{\partial H(J)}{\partial J_i} = 1/T$ is the frequency of the motion. So the action variables represent a useful technique to find the frequencies of a complex motion without solving for the motion exactly.

Starting with a Hamiltonian we can find the J_i and then we substitute them into the Hamiltonian to find the frequencies $\alpha_i = \frac{\partial H(J)}{\partial J_i}$. This technique has found many applications in astronomy and serves as the starting point for perturbation calculations of astronomical orbits.

Since the great problem in pre-quantum atomic physics was to determine the frequencies of the emitted light, and this frequency should be the same as the frequency of motion of the electrons according to classical electromagnetic theory, the action-angle variables were the starting point for many investigations. As we have seen, they form the basis for the Bohr-Sommerfeld quantization method and were the starting point for Heisenberg's matrix mechanics.

A.5.1 Example: the simple harmonic oscillator

As a simple example of the technique described above we take the one dimensional harmonic oscillator Hamiltonian

$$H = \frac{p^2}{2m} + \frac{1}{2}kx^2.$$

We then have the Hamilton-Jacobi equation, as above (Section A.4.3)

$$\frac{1}{2m}\left(\frac{\partial W}{\partial x}\right)^2 + \frac{1}{2}kx^2 = E$$

$$\frac{\partial W}{\partial x} = \sqrt{2mE - mkx^2}.$$

So that

$$J = \oint dx \frac{\partial W}{\partial x} = \oint dx \sqrt{2mE - mkx^2} = \sqrt{mk} \oint dx \sqrt{\frac{2E}{k} - x^2}.$$

Substituting $x = \sqrt{\frac{2E}{k}} \sin \eta$ we obtain

$$J = \sqrt{mk} \left(\frac{2E}{k} \right) \oint \cos^2 \eta d\eta = 2\pi E \sqrt{\frac{m}{k}}$$

and

$$E = \frac{J}{2\pi} \sqrt{\frac{k}{m}} = \frac{J}{2\pi} \omega \tag{A.28}$$

where $\omega = \sqrt{\frac{k}{m}}$ is the natural frequency of the oscillator. Now the frequency of motion is given by

$$\dot{\Theta} = \frac{\partial H}{\partial J} = \frac{\partial E}{\partial J} = \frac{\omega}{2\pi},$$

which of course is the well-known result.

A.5.2 Application of action-angle variables to the Kepler problem

Using the definition of the action variables Eq. (A.26), we can write the action variables for the Kepler problem (see Eq. (A.24)) as

$$J_r = \oint dr \frac{\partial W_r}{\partial r} = \oint dr \sqrt{2m \left(E + \frac{\kappa}{r} \right) - \frac{l^2}{r^2}} \tag{A.29}$$

$$J_\theta = \oint d\theta \frac{\partial W_\theta}{\partial \theta} = \oint d\theta \sqrt{l^2 - \frac{m^2}{\sin^2 \theta}}$$

$$J_\phi = \oint d\phi \frac{\partial W_\phi}{\partial \phi} = \oint d\phi m = 2\pi m.$$

The integral over $d\theta$ can be evaluated by substituting $u = \sin^2 \theta$ and comparing the resulting integral with that obtained from it by substituting $v = 1/u$. It will be seen that the integral over v differs from the integral over u by having l and m interchanged and the opposite sign. Since both integrals must be equal we see that J_θ must be a function of $(l - m)$. Since for $m = 0$, we have $J_\theta = 2\pi l$, we find

$$J_\theta = 2\pi (l - m). \tag{A.30}$$

To investigate the integral over dr we plot $V_{eff} = -\kappa/r + l^2/mr^2$ in Fig. A.4.4.

The integrand is the square root of the distance from E to the curve, which is shown by the arrow in the figure. The points where the curve intersects the line representing the constant E (remember $E < 0$ for a closed orbit), labeled (r_1, r_2), are called the turning points. At these points the momentum, p_r, represented by the integrand in

(A.29) is zero. So the integral around the closed orbit consists of going from r_1 to r_2 with the positive sign for the square root in (A.29) and in the reverse direction with the negative sign. So the integral is twice that obtained from r_1 to r_2. We can evaluate the integral by writing it as (remember $E < 0, \quad r_2 > r_1$)

$$J_r = \sqrt{2m\,|E|}\oint \frac{dr}{r}\sqrt{-r^2 + \frac{\kappa r}{|E|} - \frac{l^2}{2m\,|E|}} = 2\sqrt{2m\,|E|}\int_{r_1}^{r_2} \frac{dr}{r}\sqrt{(r-r_1)(r_2-r)}.$$

This can be evaluated (try to tease the answer out of Maple or Mathematica) as:

$$J_r = 2\sqrt{2m\,|E|}\left[\frac{\pi}{2}(r_1+r_2) - \pi\sqrt{r_1 r_2}\right]$$

and using the formula for the product and sum of the roots of a quadratic equation, or otherwise we find:

$$J_r = 2\pi\left[\frac{m\kappa}{\sqrt{2m\,|E|}} - l\right].$$

Now we solve for the energy:

$$E = \frac{-2\pi^2 m\kappa^2}{(J_r + 2\pi l)^2} = \frac{-2\pi^2 m\kappa^2}{(J_r + J_\theta + 2\pi m)^2} = \frac{-2\pi^2 m\kappa^2}{(J_r + J_\theta + J_\phi)^2} \tag{A.31}$$

where we used (A.29) and (A.30). Now we choose the new momenta, P_i, to be the J_i and find the rates of change of the corresponding angle variables

$$\nu_i = \dot{\Theta}_i = \frac{\partial H}{\partial P_i} = \frac{\partial E}{\partial J_i} = \frac{4\pi^2 m\kappa^2}{(J_r + J_\theta + J_\phi)^3} = \sqrt{\frac{2}{m}}\frac{(-E)^{3/2}}{\pi\kappa}$$

so that the frequencies are the same for all the coordinates, implying that the orbits are closed. This is referred to as degeneracy. The degeneracy with respect to ν_θ and ν_ϕ results from the fact that we have a central force. That with respect to ν_r is a special feature of the $1/r$ potential. This result is equivalent to Kepler's third law when we consider the relation between E and the semi-major axis of the elliptical orbit.

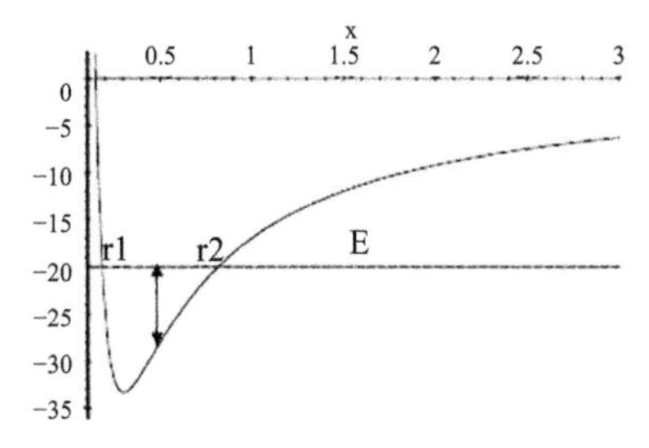

Fig. A.5 Plot of V_{eff}.

A.6 Conclusion

In this appendix we have introduced the important concept of normal modes, combinations of the usual coordinates that oscillate at a single frequency. We have already applied this concept in discussing Rayleigh's treatment of the radiation field where the energy in the field was considered to be the sum of energies in individual normal modes each oscillating at a single frequency and in the discussion of second quantization.

We have also introduced the technique of action-angle variables. This is an immensely powerful technique for analyzing periodic systems and has played a major role in the development of quantum mechanics.

We have also seen in an earlier chapter how the Hamilton-Jacobi theory shows that classical mechanics can be considered as the limit of a wave-motion, the analog to geometric optics. While the analogy was noticed by Hamilton it was only after the development of wave mechanics that its relevance became clear.

B

Galilean invariance of the Schrödinger equation

Let us consider a quantum system comprising two particles (the ensuing discussion can easily and obviously extended to any number of particles).[1] Let us also consider isotropic space (constant fields everywhere, for example) so that the system is translationally invariant. We can define a transformation $D(\vec{v})$ that adds a velocity $\vec{v}$ (note that $\vec{v}$ is a constant vector, not an operator) to both particles, and determine its effects. This operator does nothing to spin or angular momentum states, and does not alter the internal interactions between the particles, as we will show. Working in a momentum space representation,

$$D(\vec{v})|\vec{p}_1, \vec{p}_2\rangle = |\vec{p}_1 + m_1\vec{v}, \vec{p}_2 + m_2\vec{v}\rangle. \tag{B.1}$$

If we start in the spatial representation $|\vec{x}_1, \vec{x}_2\rangle$ and apply $D(\vec{v})$ to that, we know that the transformed state can be determined from the Fourier transform of $|\vec{p}_1 + m_1\vec{v}, \vec{p}_2 + m_2\vec{v}\rangle$. In the Fourier transform, the positions $x_1,\ x_2$ are operators. Setting $\hbar = 1$,

$$D(\vec{v})|\vec{x}_1, \vec{x}_2\rangle = \int \frac{d^3p_1}{(2\pi)^{3/2}} \frac{d^3p_2}{(2\pi)^{3/2}}\ e^{-i\vec{p}_1 \cdot \vec{\mathbf{x}}_1}\ e^{-i\vec{p}_2 \cdot \vec{\mathbf{x}}_2}\ |\vec{p}_1 + m_1\vec{v}, \vec{p}_2 + m_2\vec{v}\rangle. \tag{B.2}$$

Because $\vec{p}_1$ and $\vec{p}_2$ are dummy integration variables (numbers, not operators) over all momentum space, we can make a change of variables to $\vec{p}_1' = \vec{p}_1 + m_1\vec{v}_1$ and similarly for particle 2, and again note that $\vec{v}$ is a constant vector, so

$$D(\vec{v})|x_1, x_2\rangle = \int \frac{d^3p_1{}'}{(2\pi)^{3/2}} \frac{d^3p_2{}'}{(2\pi)^{3/2}} e^{-i(\vec{p_1}' - m_1\vec{v}) \cdot \vec{\mathbf{x}}_1} e^{-i(\vec{p_2}' - m_2\vec{v}) \cdot \vec{\mathbf{x}}_2} |\vec{p_1}'\rangle|\vec{p_2}'\rangle \tag{B.3}$$

$$= e^{im_1\vec{v} \cdot \vec{\mathbf{x}}_1} e^{im_2\vec{v} \cdot \vec{\mathbf{x}}_2} \int \frac{d^3p_1{}'}{(2\pi)^{3/2}} \frac{d^3p_2{}'}{(2\pi)^{3/2}} e^{-i\vec{p_1}' \cdot \vec{\mathbf{x}}_1} e^{-i\vec{p_2}' \cdot \vec{\mathbf{x}}_2} |\vec{p_1}'\rangle|\vec{p_2}'\rangle \tag{B.4}$$

$$= e^{im_1\vec{v} \cdot \vec{\mathbf{x}}_1} e^{im_2\vec{v} \cdot \vec{\mathbf{x}}_2} |\vec{x}_1, \vec{x}_2\rangle = e^{i\vec{v} \cdot (m_1\vec{\mathbf{x}}_1 + m_2\vec{\mathbf{x}}_2)} |\vec{x}_1, \vec{x}_2\rangle. \tag{B.5}$$

This can easily be generalized to any number N of particles, as

$$D(\vec{v})|\vec{x}_1, \vec{x}_2, ...\vec{x}_N\rangle = e^{i\vec{v} \cdot \sum_i m_i \vec{\mathbf{x}}_i} |\vec{x}_1, \vec{x}_2 ..., \vec{x}_N\rangle. \tag{B.6}$$

If we introduce the operator that determines the center of mass of the system,

[1] The discussion presented here has its basis from *Galilean Boost Symmetry* by Soper, D.E., 2011, unpublished

$$\vec{\mathbf{R}} = \frac{1}{M}\sum_i m_i \vec{\mathbf{x}}_i; \qquad M = \sum_i m_i \tag{B.7}$$

then the velocity "boost" operator takes on a simple form,

$$D(\vec{v}) = e^{-M\vec{v}\cdot\vec{\mathbf{R}}} \tag{B.8}$$

which we will call the center of mass velocity boost operator.

If a system of interacting particles is Galilean invariant, the Hamiltonian will commute with $D(\vec{v})$. To test this, we need to see how the operators $\vec{\mathbf{x}}_i$ and $\vec{\mathbf{p}}_i$ transform:

$$D(\vec{v})^{-1}\vec{\mathbf{x}}_i D(\vec{v}) = \vec{\mathbf{x}}_i \tag{B.9}$$

because the $\vec{\mathbf{x}}_i$ commute with themselves and each other, and therefore are invariant under a boost. In the case of $\vec{\mathbf{p}}_i$, we can use the relationship that

$$[\mathbf{p}_{i,k}, x_{j,l}^n] = -inx_{j,l}^{n-1}\delta_{i,j}\delta_{k,l} \tag{B.10}$$

where subscripts i, j label a particle and k, l label the coordinate ($\hat{x}, \hat{y}$, or $\hat{z}$ component of the vector $\mathbf{x}$), to show that

$$D(\vec{v})^{-1}\mathbf{p}_{i,k}D(\vec{v}) = e^{-im_i v_k \mathbf{x}_{i,k}}\mathbf{p}_{i,k}e^{im_i v_k \mathbf{x}_{i,k}} \tag{B.11}$$

$$= e^{-im_i v_k \mathbf{x}_{i,k}}\mathbf{p}_{i,k}(1 + \sum_{n=1}^{\infty}\frac{1}{n!}(im_i v_k \mathbf{x}_{i,k})^n) \tag{B.12}$$

$$= e^{-im_i v_k \mathbf{x}_{i,k}}\left[(1 + \sum_{n=1}^{\infty}\frac{1}{n!}(im_i v_k \mathbf{x}_{i,k})^n)\mathbf{p}_{i,k} + \sum_{n=1}^{\infty}\frac{1}{n!}\frac{n}{i}im_i v_k (im_i v_k \mathbf{x}_{i,k})^{n-1}\right] \tag{B.13}$$

$$= e^{-im_i v_k \mathbf{x}_{i,k}}e^{im_i v_k \mathbf{x}_{i,k}}(\mathbf{p}_{i,k} + m_i v_k) = (\mathbf{p}_{i,k} + m_i v_k). \tag{B.14}$$

Because this holds for each vector component and for each particle, we can write

$$D^{-1}(\vec{v})\vec{\mathbf{p}}_i D^(\vec{v}) = \vec{\mathbf{p}}_i + m_i\vec{v} \tag{B.15}$$

as expected.

The total momentum of the system is then

$$\vec{\mathbf{P}} = \sum_i \vec{\mathbf{p}}_i, \tag{B.16}$$

which transforms as

$$D^{-1}(\vec{v})\vec{\mathbf{P}}D^(\vec{v}) = \sum_i D^{-1}(\vec{v})\vec{\mathbf{p}}_i D^(\vec{v}) = \sum_i(\vec{\mathbf{p}}_i + m_i\vec{v}), \tag{B.17}$$

and we can further write

$$D^{-1}(\vec{v})\vec{\mathbf{P}}D(\vec{v}) = \vec{\mathbf{P}} + M\vec{v}. \tag{B.18}$$

We are now in position to determine the transformation properties of the Hamiltonian. We have assumed that the system is in a homogeneous region of space so that the system is translationally invariant. We have already shown that the boost operator commutes with the position operator. Let us now consider a spatial-position-invariant potential V that is a function of the pairwise differences in the center of mass coordinates, $\vec{x}_i - \vec{x}_j$. Therefore, the total momentum commutes with V,

$$\left[\vec{\mathbf{P}}, V\right] = 0, \tag{B.19}$$

and V will therefore commute with $D(\vec{v})$.

If we have a Hamiltonian

$$H = \sum_i \frac{\vec{\mathbf{p}}_i^2}{2m_i} + V(\vec{x}_1, \vec{x}_2, ...\vec{x}_N), \tag{B.20}$$

the velocity-displaced or boosted Hamiltonian is

$$D^{-1}(\vec{v})HD(\vec{v}) = \sum_i \frac{(\vec{\mathbf{p}}_i + m_i v)^2}{2m_i} + V(\vec{x}_1, \vec{x}_2, ...\vec{x}_N) \tag{B.21}$$

$$= H + \sum_i \vec{\mathbf{p}}_i \cdot \vec{v} + \frac{1}{2}\sum_i m_i \vec{v}^2 \tag{B.22}$$

$$= H + \vec{\mathbf{P}} \cdot \vec{v} + \frac{1}{2} M \vec{v}^2, \tag{B.23}$$

and it can be seen that the Hamiltonian is not boost invariant, which is not surprising because the total energy of the system has a contribution from the center of mass motion.

We can, however, separate the center of mass (or momentum) motion from the internal motion; it is the latter that determine the energy levels of the system. This means that the motions can be decoupled. Let us define

$$H' = H - \frac{1}{2M}\vec{\mathbf{P}}^2 \tag{B.24}$$

where the prime indicates it is the internal Hamiltonian, in the center of momentum frame. H' determines the total energy in that frame. Although $\vec{\mathbf{P}}$ is an operator and we can't set it to zero, we can perform a transformation that forces its eigenvalue to be zero. If the center of mass is moving at velocity $\vec{v}$ we can perform a boost by $-\vec{v}$ so expectation value of $\vec{\mathbf{P}}$ vanishes. Under such a boost,

$$D^{-1}(\vec{v})\frac{1}{2M}\vec{\mathbf{P}}^2 D(\vec{v}) = \frac{1}{2M}(\vec{\mathbf{P}} + M\vec{v})^2 = \frac{1}{2M}\vec{\mathbf{P}}^2 + \vec{\mathbf{P}} \cdot \vec{v} + \frac{M\vec{v}^2}{2} \tag{B.25}$$

so that

$$D^{-1}(\vec{v})\left(H - \frac{1}{2M}\vec{\mathbf{P}}^2\right) D(\vec{v}) = H + \vec{\mathbf{P}}\cdot\vec{v} + \frac{1}{2}M\vec{v}^2 - \left(\frac{1}{2M}\vec{\mathbf{P}}^2 + \vec{\mathbf{P}} \cdot \vec{v} + \frac{M\vec{v}^2}{2}\right) = H - \frac{1}{2M}\vec{\mathbf{P}}^2 \tag{B.26}$$

and therefore, from the definition of H',

$$D^{-1}(\vec{v})H'D(\vec{v}) = H' \tag{B.27}$$

and H' is boost invariant, i.e., is invariant under Galilean transformations.

B.1 Alternative test of Galilean invariance

We can also determine H' explicitly by determining the momentum and kinetic energy of the center of mass frame. The momenta are

$$\vec{\mathbf{p}}_i{}' = \vec{\mathbf{p}}_i - m_i\vec{v} = \vec{\mathbf{p}}_i - \frac{m_i}{M}\vec{\mathbf{P}} \tag{B.28}$$

which is boost invariant,

$$D^{-1}(\vec{v})\vec{\mathbf{p}}_i{}'D(\vec{v}) = (\vec{\mathbf{p}}_i + m_i\vec{v}) - \left(\frac{m_i}{M}\vec{\mathbf{P}} + m_i\vec{v}\right) = \vec{\mathbf{p}}_i{}'. \tag{B.29}$$

We further note that $\vec{\mathbf{p}}_i{}' = \vec{\mathbf{p}}_i$ when $\langle\vec{P}\rangle = 0$. We can therefore always boost the system to $\vec{P} = 0$ by transforming to a frame moving at $\vec{v} = -\vec{P}/M$. From the foregoing, we see immediately that the kinetic part of the center of mass system is

$$\sum_i \frac{\vec{\mathbf{p}}_i{}'^2}{2m_i} = \sum_i \frac{\vec{\mathbf{p}}_i{}^2}{2m_i} + \sum_i \frac{\vec{\mathbf{P}}^2}{2M}, \tag{B.30}$$

which follows directly from the definition of H'.

It is always possible to find a stationary frame with $\langle\vec{\mathbf{R}}\rangle = 0$ so that the coordinate system is coincident with the center of mass (or, equivalently, the center of momentum) of the system. Therefore, the center of mass or center of momentum variables are the same as in classical mechanics for a moving system,

$$H = H' + \frac{1}{2M}\vec{\mathbf{P}}^2. \tag{B.31}$$

The first term on the right-hand side gives the energy levels, and the second term gives the energy due to the center of mass motion, which can be trivially included as an exponential phase factor in the wave function.

B.2 Internal coordinates and momenta for a two- and multi-particle system

Let us choose to work in a frame where $\vec{P} = 0$ which we are always free to do for a translationally invariant potential V, i.e., that does not depend on the center of mass position $\langle\vec{R}\rangle$. In this frame, $\vec{p}_i' = \vec{p}_i$, allowing us to drop the primes in the center of mass frame and take $H = H'$.

In this frame, the expectation value of the center of mass position is constant (we can choose it to be 0) so that

$$\dot{\vec{R}} = 0 = \frac{1}{M}\sum_i m_i\dot{\vec{x}}_i \Rightarrow \sum_i \vec{p}_i = 0 \tag{B.32}$$

and therefore $\vec{p}_1 + \vec{p}_2 = 0$ for a two-particle system. This means we can use a single momentum operator

$$\vec{\mathbf{p}}_1 = -\vec{\mathbf{p}}_2 = \vec{\mathbf{p}}. \tag{B.33}$$

Similarly, $\vec{x}_1$ and $\vec{x}_2$ can be written in terms of a single parameter $\vec{x}_1 - \vec{x}_2$. For a Hamiltonian with a potential that is a function only of the separation of the two particles (with factors, for example, spin functions, that do not depend on position or momentum), the Hamiltonian can be taken as the internal one,

$$H = \left[\frac{1}{m_1} + \frac{1}{m_2}\right]\frac{\vec{\mathbf{p}}^2}{2} + V(|\vec{r}|) = \frac{1}{\mu}\frac{\vec{\mathbf{p}}^2}{2} + V(|\vec{r}|), \tag{B.34}$$

where $\vec{r} = \vec{x}_1 - \vec{x}_2$ and $\mu^{-1} = m_1^{-1} + m_2^{-1}$ is the reduced mass. Moving to the center of mass frame results in the loss of three momenta coordinates and three position coordinates, due to the totals of each set being zero.

Note that this simple separation of variables only works for two particles. This result can, however, be generalized for a greater number of particles. In cases where the internal potential V is only a function of pairwise separations (which is usually true), if we label the internal coordinates as $\vec{r}_1...\vec{r}_N$, from the foregoing it can be seen that the center of mass wavefunction can always be written as

$$\psi(\vec{x}_1...\vec{x}_N) = e^{i\vec{P}\cdot\vec{R}}\psi(\vec{r}_1....\vec{r}_N). \tag{B.35}$$

Therefore, a boost from the center of mass frame produces an overall and constant wavefunction phase factor that does not affect the expectation values of the state.

C
Universality of Planck's constant

One of the remarkable features of quantum mechanics is that Planck's constant $\hbar = h/2\pi$ appears to be a universal constant, for all interactions, particles, and at all energies. This universal feature provided motivation to *define* the value of $\hbar$ in the 2019 CODATA adjusted physical constants described in Chapter 2. However, we can ask what the consequences would be if different realms of physics have disparate quantization constants.[1] Because multiplicity of quantization constants would alter the agreement between theory and experimentally known results, we can place limits on this possibility.

For a multiparticle system with the dynamics of each particle governed by a different $\hbar$, there can be a violation of the usual space-time conservation laws. In particular, energy-momentum conservation is violated unless the different types of particles do not interact with each other, which can be seen by the following: Consider a simple non-relativistic one-dimensional system of two spinless particles with the same mass m but different Planck constants h_A and h_B, interacting through a potential $V(r)$. Assuming a Hamiltonian that includes an interaction between the particles, we can write

$$H = \frac{p_A^2}{2m} + \frac{p_B^2}{2m} + V(x_A - x_B) = \frac{P^2}{2M} + \frac{p^2}{m} + V(r) \tag{C.1}$$

where $r = x_A - x_B$, $p = (p_A - p_B)/2$, $M = 2m$, and $P = p_A + p_B$. The Planck constants for each particle are related to their momentum and position through the commutation relations

$$[x_A, p_A] = i\hbar_A, \quad [x_B, p_B] = i\hbar_B, \quad \text{with} \quad [x_A, x_B] = [p_A, p_B] = 0. \tag{C.2}$$

A quantity is conserved if it commutes with the Hamiltonian, but we find that

$$[H, P] = [V(r), P] = i(h_A - h_B)\frac{\partial V}{\partial r}$$

which is not zero unless either $h_A = h_B$ *or* if $V(r)$ is independent of r, i.e., there is no force between the particles. Therefore, for momentum to be conserved, the two particles must have the same Planck constant or they must not interact. As such, for an atom, the atomic nucleus and the electrons most certainly have the same $\hbar$ to high degree of precision. This has been implicitly tested to extremely high precision by use

[1]Fischbach, E., Greene, G.L. and Hughes, R.J., *New test of quantum mechanics: Is Planck's constant unique?*, Phys. Rev. Lett. **66** 256-259 (1991).

of atom interferometery to measure the recoil of, e.g., ^{133}Cs due to photon absorption. The latter has been used to determined the fine structure constant

$$\alpha = \frac{e^2}{\hbar c} \approx \frac{1}{137} \tag{C.3}$$

(which is a proxy for $\hbar$ in the present context) to an accuracy of 0.12 parts per billion.[2]

Direct measurements of the ratio of the velocity to wavelength of neutrons in a cold beam (time of flight for velocity, and Bragg reflection from Si for the wavelength) has determined $\hbar/m_n$ (m_n is the neutron mass) has produced an uncertainty in α of about 80 ppb.[3]

Experimental constraints on differences in the Planck constant are also set by how well quantum electrodynamics (QED) calculations of $g-2$ for the electron and for the muon agree with experiment, which are at the 0.3 parts per billion and 0.4 part per million level, respectively. However, the full QED calculation of varying $\hbar$ has not been done, so it is difficult to assess the overall sensitivity. We might expect effects to appear in the QED expansion only above the lowest order, and as such the net fractional effects will be reduced. Of course a variation between the QED calculation and the experimental result might arise due to new, unknown high-energy interactions that contribute to $g-2$ at higher orders. The QED calculation, to achieve the experimental accuracy, requires 12,672 tenth-order Feynman diagrams.

The extremely good agreement between measurements of α in systems involving various types of particles, and with different internal interactions and energy scales, means that any differences in $\hbar$ must be very small. Direct comparisons suggest that possible variations must be $< 10^{-8}$, but could be smaller.

[2] Richard H. Parker, Chenghui Yu, Weicheng Zhong, Brian Estey and Holger Müller, *Measurement of the fine-structure constant as a test of the Standard Model*, Science **6385**, pp. 191-195 (2018).

[3] Kruger, E., Nistler, W. and Weirauch, W., *Determination Of The Fine-structure Constant by measuring the quotient of the Planck constant and the neutron mass*, IEEE Transactions on Instrumentation and Measurement, **46**, 101 (1997).

D
Conservation laws

The notions of conservation of particles (mass), and of angular and linear momentum, can be seen through daily experiences, and were known to Newton. Conservation of energy came a century later, and still has an aura of mystery around it, primarily because energy cannot be defined in a simple manner based on everyday experience. However, all of these laws have been subject to experimental tests and are accepted as valid. In fact these laws have their basis in a fundamental mathematical principle.

Principal features of quantum mechanics are its probabilistic interpretation and uncertainty relationships of measured quantities. In the earliest days of the theory, it was not at all clear that the classical conservation laws, in particular the conservation of energy, carried over into the quantum regime. For example, as described in Section 13.3, the Bohr-Kramers-Slater (BKS) theory that was put forward in 1924 to explain the interaction of atoms with electromagnetic radiation conserved energy and momentum only on average. The goal of the BKS theory was to avoid the quantization of light (which at the time was not accepted by Bohr, Kramers, or even Planck). This theory was soon ruled out by measurements of the Compton effect.

Later, studies of $\beta-$decay of atomic nuclei showed a continuous energy spectrum, an effect that was studied by many, but especially by Lise Meitner who concluded that such a continuous spectrum from a quantum system of known energy appeared to violate energy conservation, something she regarded as unacceptable (Bohr apparently also brought up this point, and is sometimes credited with it. Meitner was much closer to the problem). On 4 December 1930, Wolfgang Pauli, in an open letter to researchers of radioactivity, proposed that the continuous spectrum was caused by the emission of a second particle during $\beta-$decay:[1]

Dear Radioactive Ladies and Gentlemen,

As the bearer of these lines, to whom I graciously ask you to listen, will explain to you in more detail, how because of the "wrong" statistics of the N and Li^6 nuclei and the continuous beta spectrum, I have hit upon a desperate remedy to save the "exchange theorem" of statistics and the law of conservation of energy. Namely, the possibility that there could exist in the nuclei electrically neutral particles, that I wish to call neutrons, which have spin 1/2 and obey the exclusion principle and which further differ from light quanta in that they do not travel with the velocity of light. The mass of the neutrons should be of the same order of magnitude as the electron mass and in any event not larger than 0.01 proton masses. The continuous beta spectrum would then become understandable by the assumption that in beta decay a neutron is emitted in addition to the electron such that the sum of the energies of the neutron and the electron is constant...

[1]Wolfgang Pauli, *Scientific Correspondence with Bohr, Einstein, Heisenberg a.o. Volume II: 1930-1939*, eds. by Meyenn, K. von, Hermann, A. and Weisskopf, V.F., (Springer, 1985), p.31.

I agree that my remedy could seem incredible because one should have seen these neutrons much earlier if they really exist. But only the one who dares can win and the difficult situation, due to the continuous structure of the beta spectrum, is highlighted by a remark of my honoured predecessor, Mr Debye, who told me recently in Brussels: "Oh, It's well better not to think about this at all, like new taxes". From now on, every solution to the issue must be discussed. Thus, dear radioactive people, look and judge.

Unfortunately, I cannot appear in Tübingen personally since I am indispensable here in Zurich because of a ball on the night of 6/7 December. With my best regards to you, and also to Mr Back.

Your humble servant,
W. Pauli

This second particle has no electric charge and no, or very small, rest mass, and was dubbed a "neutrino" by Fermi in his theory of β-decay, which proved to be successful.

What is surprising is that the mathematical tools to theoretically investigate energy and other conservation laws already existed at the time, due to several mathematical theorems by Emmy Noether.[2] [3] David Hilbert and Felix Klein brought Emmy Noether to Göttingen in 1915 to help them in understanding Einstein's general theory of relativity. Because gravitational energy itself can gravitate, Hilbert surmised that energy conservation might be violated in general relativity. The hope was that Noether could use her expertise in invariant theory to provide a resolution of this problem (which was only fully solved in the 1960s with an exact formulation of a non-linear gravitational wave that showed energy conservation). Noether's first theorem, that she proved in 1915, but did not publish until 1918, provided a means to do this for general relativity, but also determined the conserved quantities for every physical system that can be described by a Hamiltonian or Lagrangian formulation that possesses continuous symmetries. In light of Noether's theorem, the BKS theory was untenable because underlying symmetries in the system imply conservation laws, including energy conservation. The broad applicability of Noether's theorems in quantum mechanics and in field theory were not widely appreciated until the 1960s, but are now a mainstay of modern theoretical physics.

In non-relativistic quantum mechanics, we have the following conservation and invariance laws, and some are based on symmetries as indicated,

- Conservation of energy (time translation invariance).
- Conservation of linear momentum (spatial translation invariance).
- Conservation of angular momemtum (rotational invariance).
- Conservation of number of particles (electrons, nuclei, etc.) except photons.
- Invariance under parity inversion ($\vec{r} \to -\vec{r}$).
- Invariance under time reversal ($t \to -t$).

The conservation of particle number is a corollary of the conservation of mass–matter cannot be created or destroyed (in the nonrelativistic limit). The last two are

[2] Emmy Noether, *Invariante Variationsprobleme [Invariant Variation Problems]*, Nachrichten der Königlichen Gesellschaft der Wissenschaften zu Göttingen, Math.-phys. Klasse, 235-257 (1918).

[3] Chris Quigg, *Colloquium: A Century of Noether's Theorem*, arXiv:1902.01989v2 [physics.hist-ph]; Katherine Brading *A Note on General Relativity, Energy Conservation, and Noether's Theorems*, Einstein Stud., 11:125-135 (2005). doi:10.1007/0-8176-4454-7_8.

examples of discrete symmetries. Although we are considering non-relativistic quantum mechanics, it is possible to investigate low-energy relativistic corrections by, for example, taking the first order relativistic correction to the momentum of a particle. Photons are necessarily relativistic, with $E = pc$.

The time-translation invariance of the Schrödinger equation for a non-time varying potential immediately tells us that energy is conserved. Although the interaction that leads to $\beta-$decay was not known at the time Pauli proposed the neutrino, conservation of energy together with its quantization was taken as such fundamental necessities that the observed continuous spectrum motivated the proposal of a new particle with certain characteristics that was not directly observed until some 25 years later.

The symmetry relations can be very useful in quantum mechanics calculations, and allow the determination of the effects of perturbations. For example, Kramer's degeneracy theorem tells us that for every energy eigenstate of a time-reversal symmetric system with an odd number of half-integer spins, the eigenvalues are at least doubly degenerate, and the degeneracy is of even order.[4]

Symmetries can also be directly investigated by considering the commutator with the system Hamiltonian. For example, if H commutes with the parity operator, then states of H must be of definite parity. There are no known fundamental microscopic interactions that are time-reversal asymmetric, and the observation of such an interaction would be the discovery of a new physical interaction.

Sometimes there is a misconception that the uncertainty principle (UP) implies that conservation laws are broken. In some sense, the opposite is true in that the UP is a direct consequence of $[x, p] = i\hbar$, which ties the momentum and position of a particle together in a precise manner that has no classical analog. The time-energy UP comes from the Fourier spectrum of a time window. In an analogous manner, x and p are Fourier conjugate variables, i.e., a precise, sharp value for p implies a broad range of x, and this is reflected in the measurement process.

The finite Fourier transform widths of spectral lines, for example due to finite lifetime, implies that photons that are slightly off resonant with an atomic transition can still be absorbed. Such effects are of significant importance in atom and ion trapping and cooling experiments, for example. The recoil on photon absorption/emission of a moving atom can be large due to the Doppler effect, and has been used in laser cooling of the motion of atoms.

Recoil effects were essential in Einstein's development of the theory of light quanta as we have seen. Photon absorption/emission recoil are responsible for the Poynting-Robertson effect, which causes drag on dust grains in orbit around a star, and energy is conserved in the process.[5] With this effect, light is absorbed by a dust grain which causes its temperature to increase. This tendency to heat is balanced by black-body radiation from the dust grain. In the frame of the orbiting particle, the re-radiation is isotropic, but when viewed from a rest frame, the radiation in the forward direction appears as blue-shifted due to the Doppler effect of the moving source, and the backward direction is red-shifted. This results in a tangential force on the particle against its velocity, reducing its angular momentum at a rate determined by its veloc-

[4] A. Messiah, *Quantum Mechanics* (Wiley, New York, 1976), vol 2, p. 675.

[5] https://en.wikipedia.org/wiki/PoyntingRobertson_effect

ity (Doppler frequency shifts) and the photon emission rate. This is an instance where all of the fundamental interactions are easy to identify, and it is straightforward to verify that at every step in the process energy, momentum, and angular momentum are conserved. However, this process appears as time reversal asymmetric because the inverse process is nearly impossible, as it would require sending appropriately shifted light back onto the particle from two directons. Thus, the Robertson-Poynting effect is a frictional process that is irreversible in a practical sense, arising from a quite fundamental process of absorption and re-emission of light.

Time reversal or parity asymmetry in an atomic or other purely quantum system would be evidence of those asymmetries in some perturbing Hamiltonian, reflecting high-energy interactions that persist to lower energies. Low-velocity and low-energy measurements on atoms, neutrons, and molecules have shown parity asymmetry and have provided values to parameterize the Weinberg-Salam theory of electromagnetism and weak interactions. Such measurements have also provided a sensitive test for time reversal asymmetry and have helped constrain the parameters in supersymmetric models in high-energy particle theory.[6]

[6]Khriplovich, I.B. and Lamoreaux, S.K., *CP Violation without Strangeness* (Springer, Berlin, 1997).

E

Lagrangian and Hamiltonian formalism for classical fields

E.1 Lagrangian for a classical continuous field: example of a vibrating string

For a particle moving in one dimension (single degree of freedom), the Lagrangian is given by

$$L = \frac{m\dot{y}^2}{2} - V(y), \tag{E.1}$$

and for a system of many such degrees of freedom, the Lagrangian is

$$L = \sum_i \frac{m\dot{y}_i^2}{2} - V\left(y_1, y_2, ...y_i...\right). \tag{E.2}$$

The vibrating string is described by an infinite set of coordinates $y_i \longrightarrow y\left(x\right)$: in order to describe the state of the string we have to specify the value of the displacement y for every position x along the string. The equation of motion of the string is given by the wave equation

$$\rho\frac{\partial^2 y\left(x,t\right)}{\partial t^2} - \tau\frac{\partial^2 y\left(x,t\right)}{\partial x^2} = 0 \tag{E.3}$$

where ρ is the density of the string and τ is the tension. In the case of the continuous string, the sum in (E.2) has to be replaced by an integral. We introduce

$$L = \int dx \left[\rho\left(\dot{y}\left(x,t\right)\right)^2 - \tau\left(\frac{\partial y\left(x,t\right)}{\partial x}\right)^2\right] = \int \mathcal{L}\left(\dot{y}, \frac{\partial y}{\partial x}\right) dx \tag{E.4}$$

as a trial Lagrangian, which we hope will yield the wave equation (E.3). The action is $\int L\,dt$ and we want to find a function $y\left(x,t\right)$ so that this is a minimum. We introduce a variation $\eta\left(x,t\right)$ and write $y \to y + \eta$, $\dot{y} \to \dot{y} + \dot{\eta}$. Then the variation in the action will be

$$\delta\int L\,dt = 2\int dt \int dx \left[\rho\,\dot{y}\,\dot{\eta} - \tau\frac{\partial y}{\partial x}\frac{\partial \eta}{\partial x}\right]. \tag{E.5}$$

Integrating the first term by parts in the integral with respect to t and the second term with respect to x, we obtain

$$\delta\int L\,dt = -2\int dt \int dx \left(\rho\frac{\partial \dot{y}}{\partial t} - \tau\frac{\partial}{\partial x}\frac{\partial y}{\partial x}\right)\eta = 0 \tag{E.6}$$

where the integrated terms vanish because of the boundary conditions. In order for the variation to be zero for all possible variations η, the term in brackets must vanish, thus yielding the wave equation and demonstrating that our choice of Lagrangian (E.4) makes sense.

E.2 Lagrange's equations for a classical continuous field

As the Schrödinger equation is also a form of wave equation for a continuous field, we can try to derive it from a Lagrangian in a similar fashion. Writing the Lagrangian for a continuous field in the general form,

$$L = \int d^3x \mathcal{L}\left(\psi, \dot{\psi}, \vec{\nabla}\psi\right) \tag{E.7}$$

and

$$\int L\, dt = \int dt \int d^3x \mathcal{L}\left(\psi, \dot{\psi}, \vec{\nabla}\psi\right). \tag{E.8}$$

Then we can write the variation ($\psi \to \psi + \delta\psi$)

$$\delta \int L\, dt = \int dt \int d^3x \left[\frac{\partial \mathcal{L}}{\partial \psi}\delta\psi + \frac{\partial \mathcal{L}}{\partial \dot{\psi}}\frac{d}{dt}\delta\psi + \frac{\partial \mathcal{L}}{\partial \vec{\nabla}\psi}\vec{\nabla}\delta\psi\right]. \tag{E.9}$$

Integrating the second term by parts, we find

$$\int dt \int d^3x \frac{\partial \mathcal{L}}{\partial \dot{\psi}}\frac{d}{dt}\delta\psi = \int d^3x \left[\frac{\partial \mathcal{L}}{\partial \dot{\psi}}\delta\psi\bigg|_{t_1}^{t_2} - \int dt \frac{d}{dt}\frac{\partial \mathcal{L}}{\partial \dot{\psi}}\delta\psi\right]. \tag{E.10}$$

The integrated term vanishes because the variation $\delta\psi$ is defined to vanish at the endpoints just as in our previous applications of the variational method. For simplicity we specialize the last term to one spatial dimension, $\vec{\nabla}\psi \to \partial\psi/\partial x \equiv \psi'$, so that the last term is

$$\int dt \int dx \frac{\partial \mathcal{L}}{\partial \psi'}\delta\psi' = \int dt \left[\frac{\partial \mathcal{L}}{\partial \psi'}\delta\psi\bigg|_{t_1}^{t_2} - \int dx \frac{\partial}{\partial x}\left(\frac{\partial \mathcal{L}}{\partial \psi'}\right)\delta\psi\right]. \tag{E.11}$$

The integrated term vanishes as before. Returning to three dimensions, we have finally

$$\delta \int Ldt = \int dt \int d^3x \delta\psi \left[\frac{\partial \mathcal{L}}{\partial \psi} - \frac{d}{dt}\frac{\partial \mathcal{L}}{\partial \dot{\psi}} - \sum_{i=1}^{3}\frac{\partial}{\partial x_i}\left(\frac{\partial \mathcal{L}}{\partial \psi_i'}\right)\right] \tag{E.12}$$

where the summation is over $x_{1,2,3} = x, y, z$ and $\psi_i' = \partial\psi/\partial x_i$.

As before, if the variation is to be zero for every possible $\delta\psi$, the expression in square brackets must be identically zero. In this way we arrive at Lagrange's equations for a continuous field $\psi(\vec{r}, t)$:

$$\frac{\partial \mathcal{L}}{\partial \psi} - \frac{d}{dt}\frac{\partial \mathcal{L}}{\partial \dot{\psi}} - \sum_{i=1,2,3}\frac{\partial}{\partial x_i}\left(\frac{\partial \mathcal{L}}{\partial \psi_i'}\right) = 0\,. \tag{E.13}$$

E.2.1 A Lagrangian for the Schrödinger equation

We now show that an astute choice of Lagrangian can lead to Schrödinger's equation by applying (E.13). Note that we are going to consider the Schrödinger wave function $\psi(\vec{r}, t)$ as a continuous field. We try

$$\mathcal{L} = -\left[\frac{\hbar^2}{2m}\left(\vec{\nabla}\psi \cdot \vec{\nabla}\psi^*\right) + V\psi\psi^* + \frac{\hbar}{2i}\left(\psi^*\dot{\psi} - \psi\dot{\psi}^*\right)\right]. \tag{E.14}$$

Now in applying (E.13) to (E.14) we have to consider ψ and ψ^* as independent variables. Taking the variation with respect to ψ^* (i.e., replacing ψ in Eq. E.13 by ψ^*) we have

$$V\psi + \frac{\hbar}{2i}\dot{\psi} + \frac{\hbar}{2i}\frac{d\psi}{dt} - \frac{\hbar^2}{2m}\nabla^2\psi = 0\,, \tag{E.15}$$

which of course is the Schrödinger equation. Note that the overall minus sign in (E.14) plays no role here.

E.3 Hamiltonian formulation for classical continuous fields

Just as in ordinary classical mechanics, we can introduce a Hamiltonian formalism for continuous fields. In fact, the reason for us going into the Lagrangian formalism in the last section was to prepare the ground for the introduction of the Hamiltonian formalism. Proceeding as in the case of classical mechanics, we calculate

$$\begin{aligned}\frac{dL}{dt} &= \frac{d}{dt}\int d^3x\mathcal{L}\left(\psi, \dot{\psi}, \vec{\nabla}\psi, \psi^*, \dot{\psi}^*, \vec{\nabla}\psi^*\right) \\ &= \int d^3x\left(\frac{\partial\mathcal{L}}{\partial\psi}\dot{\psi} + \frac{\partial\mathcal{L}}{\partial\psi'}\frac{d}{dt}\psi' + \frac{\partial\mathcal{L}}{\partial\dot{\psi}}\ddot{\psi}\right).\end{aligned} \tag{E.16}$$

Substituting for $\frac{\partial\mathcal{L}}{\partial\psi}$ from Lagrange's field equations (E.13), we obtain

$$\frac{dL}{dt} = \int d^3x\left[\left(\frac{d}{dt}\frac{\partial\mathcal{L}}{\partial\dot{\psi}} + \sum_{i=1}^{3}\frac{\partial}{\partial x_i}\left(\frac{\partial\mathcal{L}}{\partial\psi_i'}\right)\right)\dot{\psi} + \sum_{i=1}^{3}\frac{\partial\mathcal{L}}{\partial\psi_i'}\frac{d}{dt}\psi_i' + \frac{\partial\mathcal{L}}{\partial\dot{\psi}}\ddot{\psi}\right]. \tag{E.17}$$

Now we can write

$$\frac{d}{dt}\psi_i' = \frac{d}{dt}\frac{\partial\psi}{\partial x_i} = \frac{\partial\dot{\psi}}{\partial x_i} \tag{E.18}$$

$$\int d^3x\frac{\partial\mathcal{L}}{\partial\psi_i'}\frac{d}{dt}\psi_i' = \int d^3x\frac{\partial\mathcal{L}}{\partial\psi_i'}\frac{\partial\dot{\psi}}{\partial x_i} = -\int d^3x\,\dot{\psi}\frac{\partial}{\partial x_i}\frac{\partial\mathcal{L}}{\partial\psi_i'} \tag{E.19}$$

where the last step follows by integration by parts (the integrated part vanishes as usual). Then

$$\begin{aligned}\frac{dL}{dt} &= \int d^3x\left[\left(\frac{d}{dt}\frac{\partial\mathcal{L}}{\partial\dot{\psi}} + \sum_{i=1}^{3}\frac{\partial}{\partial x_i}\left(\frac{\partial\mathcal{L}}{\partial\psi_i'}\right)\right)\dot{\psi} - \dot{\psi}\sum_{i=1}^{3}\frac{\partial}{\partial x_i}\frac{\partial\mathcal{L}}{\partial\psi_i'} + \frac{\partial\mathcal{L}}{\partial\dot{\psi}}\ddot{\psi}\right] \\ &= \int d^3x\left(\dot{\psi}\frac{d}{dt}\frac{\partial\mathcal{L}}{\partial\dot{\psi}} + \frac{\partial\mathcal{L}}{\partial\dot{\psi}}\ddot{\psi}\right) = \int d^3x\frac{d}{dt}\left(\frac{\partial\mathcal{L}}{\partial\dot{\psi}}\dot{\psi}\right)\end{aligned} \tag{E.20}$$

so that

$$\frac{d}{dt}\left(\int d^3x \left(\frac{\partial \mathcal{L}}{\partial \dot{\psi}}\dot{\psi}\right) - L\right) = 0$$

$$\int d^3x \left(\frac{\partial \mathcal{L}}{\partial \dot{\psi}}\dot{\psi} - \mathcal{L}\right) = \text{const} \equiv H, \tag{E.21}$$

and we can write

$$H = \int d^3x \left(\frac{\partial \mathcal{L}}{\partial \dot{\psi}}\dot{\psi} - \mathcal{L}\right) \equiv \int d^3x\, \mathcal{H} \tag{E.22}$$

$$\mathcal{H} = \frac{\partial \mathcal{L}}{\partial \dot{\psi}}\dot{\psi} - \mathcal{L}\left(\psi, \dot{\psi}, \psi'\right). \tag{E.23}$$

We now introduce the momentum conjugate to ψ,

$$\Pi = \frac{\partial \mathcal{L}}{\partial \dot{\psi}} \tag{E.24}$$

so that

$$\mathcal{H} = \Pi\dot{\psi} - \mathcal{L}\left(\psi, \dot{\psi}, \psi'\right). \tag{E.25}$$

Then, proceeding as in the case of the mechanics of point particles, we have

$$\begin{aligned} dH &= \int d^3x \left[\Pi d\dot{\psi} + \dot{\psi} d\Pi - \frac{\partial \mathcal{L}}{\partial \psi}d\psi - \frac{\partial \mathcal{L}}{\partial \dot{\psi}}d\dot{\psi} - \sum_{i=1}^{3}\frac{\partial \mathcal{L}}{\partial \psi'_i}d\psi'_i\right] \\ &= \int d^3x \left[\dot{\psi} d\Pi - \left(\dot{\Pi} + \sum_{i=1}^{3}\frac{\partial}{\partial x_i}\left(\frac{\partial \mathcal{L}}{\partial \psi'_i}\right)\right)d\psi - \sum_{i=1}^{3}\frac{\partial \mathcal{L}}{\partial \psi'_i}d\psi'_i\right] \qquad \text{(E.26)} \\ &= \int d^3x \left[\dot{\psi}\, d\Pi - \dot{\Pi}\, d\psi\right], \qquad \text{(E.27)} \end{aligned}$$

where we used Eqs. (E.24) and (E.13) and integrated the last term in (E.26) by parts. Thus we see that $\mathcal{H}$ is not a function of $\dot{\psi}$ $\left(\partial\mathcal{H}/\partial\dot{\psi} = 0\right)$, so we can write

$$H = \int d^3x\, \mathcal{H}\left(\psi, \Pi, \psi', \Pi'\right). \tag{E.28}$$

Then

$$\begin{aligned} dH &= \int d^3x \left[\frac{\partial \mathcal{H}}{\partial \psi}d\psi + \sum_{i=1}^{3}\frac{\partial \mathcal{H}}{\partial \psi'_i}d\psi'_i + \frac{\partial \mathcal{H}}{\partial \Pi}d\Pi + \sum_{i=1}^{3}\frac{\partial \mathcal{H}}{\partial \Pi'_i}d\Pi'_i\right] \qquad \text{(E.29)} \\ &= \int d^3x \left[\left(\frac{\partial \mathcal{H}}{\partial \psi} - \sum_{i=1}^{3}\frac{\partial}{\partial x_i}\frac{\partial \mathcal{H}}{\partial \psi'_i}\right)d\psi + \left(\frac{\partial \mathcal{H}}{\partial \Pi} - \sum_{i=1}^{3}\frac{\partial}{\partial x_i}\frac{\partial \mathcal{H}}{\partial \Pi'_i}\right)d\Pi\right] \qquad \text{(E.30)} \end{aligned}$$

where we once again integrated by parts. Comparing this with equation (E.27) we see that

$$\dot{\Pi} = -\left(\frac{\partial \mathcal{H}}{\partial \psi} - \sum_{i=1}^{3} \frac{\partial}{\partial x_i}\frac{\partial \mathcal{H}}{\partial \psi_i'}\right) \tag{E.31}$$

$$\dot{\psi} = \left(\frac{\partial \mathcal{H}}{\partial \Pi} - \sum_{i=1}^{3} \frac{\partial}{\partial x_i}\frac{\partial \mathcal{H}}{\partial \Pi_i'}\right), \tag{E.32}$$

which are Hamilton's equations of motion for the field.

Index